Canadian Edition

Ecology
The Economy of Nature

Seventh Edition

Robert Ricklefs
University of Missouri—St. Louis

Rick Relyea
University of Pittsburgh

Christoph Richter
University of Toronto—Mississauga

W. H. Freeman and Company

A Macmillan Higher Education Company

Publisher: Kate Parker
Senior Acquisitions Editor: Bill Minick
Associate Marketing Director: John Britch
Developmental Editor: Rebecca Kohn
Art Development Editor: Lee Wilcox
Market Development: Claire Hunter
Supplements Editor: Rebecca Kohn
Manager of Digital Development: Amanda Dunning
Assistant Editor: Tue Tran
Photo Editor: Christine Buese
Photo Researcher: Deborah Goodsite
Art Director: Diana Blume
Text Design: MGMT. design, Diana Blume
Cover Design: Diana Blume
Project Editor: Elizabeth Geller
Illustrations: MGMT. design and Nicolle R. Fuller
Art Manager: Matthew McAdams
Production Manager: Susan Wein
Composition: codeMantra
Printing and Binding: RR Donnelley

Front cover photo: © Glen and Rebecca Grambo/Getty Images
Back cover photos (clockwise from top left): Design Pics/Deddeda/
Deddeda/Newscom; Michelle Valberg-www.agefotostock.com;
Barrett & MacKay-www.agefotostock.com; John E Marriott-www.
agefotostock.com; Gary Fiegehen/All Canada Photos/SuperStock

Library of Congress Control Number: 2014939220

ISBN-13: 978-1-4641-5424-9
ISBN-10: 1-4641-5424-4

W. H. Freeman and Company
41 Madison Avenue
New York, NY 10010
Houndmills, Basingstoke RG21 6XS, England
www.whfreeman.com

Appendices

DETAILED CONTENTS

PART IV SPECIES INTERACTIONS

Photo by Maria W. Pil

ROBERT RICKLEFS is Curators' Professor of Biology at the University of Missouri–St. Louis, where he has been a member of the faculty since 1995. His teaching at Missouri, and previously at the University of Pennsylvania, has included courses in introductory and advanced ecology, biogeography, evolution, and biological statistics. Bob's research has addressed a broad range of topics in ecology and evolutionary biology, from the adaptive significance of life-history traits of birds, to island biogeography and the community relationships of birds, herbivorous insects, and forest trees. In particular, he has championed the importance of recognizing the impact of large-scale processes on local ecological assemblages of species. Bob has published in numerous journals including *Science*, *Nature, PNAS, Evolution, Ecology, Ecology Letters,* and the *American Naturalist.* His contributions have been recognized by honorary doctorates from the Université Catholique de Louvain (Belgium), Aarhus University (Denmark), and the University of Burgundy (France). He is a member of the American Academy of Arts and Sciences and the National Academy of Sciences of the United States. Bob published the first edition of *The Economy of Nature* in 1976 and is joined by a coauthor for the first time with this seventh edition.

Photo by Christine Relyea

RICK RELYEA is Professor of Biology at the University of Pittsburgh, where he has been on the faculty since 1999. In 2005 he was named the Chancellor's Distinguished Researcher at the University of Pittsburgh, and he was the 2014 recipient of the David and Tina Bellet Teaching Excellence Award. Since 2007, he has also served as the director of the University of Pittsburgh's field station, the Pymatuning Laboratory of Ecology, where he oversees a diverse set of ecological field courses and facilitates researchers from around the world. Rick has taught thousands of undergraduate students in introductory ecology, behavioural ecology, and evolution. His research is recognized throughout the world and has been published in the leading ecological journals including *Ecology, Ecology Letters, American Naturalist,* and *PNAS*. The research spans a wide range of ecological and evolutionary topics including animal behaviour, sexual selection, ecotoxicology, disease ecology, phenotypic plasticity, community ecology, ecosystem ecology, and landscape ecology. Rick's research focuses on aquatic habitats and the diversity of species that live in these ecosystems. He strives to integrate different areas of ecology in ways that provide new discoveries and applications.

Photo by Silvia Caicedo

CHRISTOPH RICHTER is Lecturer in the Department of Biology at the University of Toronto–Mississauga, where he has been on the faculty since 2011. He began his teaching career in 2002 at Queen's University, where he was recognized for his excellence in teaching. He has taught undergraduate courses on many topics including introductory ecology, animal behaviour, vertebrate biology, marine mammalogy, biological diversity, and statistics. He strives to engage students by making course content relevant for them and by demonstrating the dynamic nature of scientific research. His teaching has taken him from the Arctic to the Gulf of Maine and to Ecuador. Christoph earned a MSc from Memorial University of Newfoundland and a PhD from Otago University in New Zealand. His research focused on the impacts of human activities, such as fishing, whale watching, and oil exploration, on the behaviour of cetaceans. Christoph's current research focuses on students' opinions and knowledge of evolutionary concepts and on factors that limit academic success. He is also interested in involving undergraduate students in collecting and using long-term ecological data sets.

From the Authors...

Since *The Economy of Nature* debuted in 1976, it has enjoyed a strong and loyal following. This edition, the seventh, begins a new collaboration. When we first talked about the book in 2009, it was immediately clear that we had a similar vision to retain everything instructors loved while updating it for today's instructors and students, and to make it as relevant as possible for students and instructors across Canada. We asked hundreds of ecology instructors what they want to see in a textbook and we listened carefully to the challenges they face. We thought about ways to make Canadian students connect more deeply with the material and how we might bring the Canadian context to bear on many of the ecological issues important to Canadians today. You hold in your hands the result of a five-year effort to bring an exciting and completely new textbook to you and your students.

A new vision.

Chapters are organized around four to six key concepts that are repeated as major headings and repeated again in streamlined summaries. This new structure allows the instructor and student to focus on the essential messages of each chapter.

The chapters contain substantially *more figures showing real data with a better balance between terrestrial and aquatic systems.* These changes will give students more experience in seeing and interpreting scientific data from a wide range of systems. We have also increased our coverage of species interactions, community ecology, and ecosystem ecology to provide a broader scope for an introductory course.

You told us that you want ecological applications integrated throughout the entire book rather than relegated to a final chapter. We agree, *so every chapter now contains multiple studies with real-world applications that underscore the value of understanding ecology and that help students understand why ecology is relevant to their lives.*

Like you, we strive to include global change throughout our ecology course. To make this possible, we now *discuss global change issues in nearly every chapter.* These changes help instructors and students make the connections between basic ecology and real ecological issues that affect their lives.

We know that students appreciate clear writing and strong visuals. To make the text an appealing study tool that students will want to read, we worked with superb editors who helped us put *the science into plain language without diminishing the complexity of the concepts.* We have also replaced nearly every photo, graph, and illustration in the book. We now give you and your students *an abundance of spectacular photos, beautiful color illustrations, and easy-to-interpret graphs.*

Presents the process of scientific discovery through a global lens.

As active researchers ourselves, we want students to know that science is an ongoing process. In every chapter *we show students how scientists begin with hypotheses and test them with data that the students can view and interpret.* We demonstrate how early hypotheses are continually revised as new observations and new data are produced. To help convey the dynamic nature of ecology to your students, *we have retained all of your favorite classic studies while adding a large number of new, captivating studies from around the world,* including many from 2010 to 2013.

Takes a learning-by-doing approach to basic quantitative tools and the use of data.

You have told us that many of today's students lack the background to apply basic quantitative tools. This edition contains a new feature, "Analyzing Ecology," *to help students learn basic statistical and mathematical techniques that real ecologists use every day.* We show students how to do the math and then challenge them to apply it in "Your Turn." We make a point of integrating these tools with research studies that are discussed in the chapters.

To help students gain more practice with quantitative skills, we created a feature at the end of each chapter, "Practicing Ecology," in which *students are challenged to understand how to create and interpret different types of graphs.* To provide immediate feedback to students, answers to both "Analyzing Ecology" and "Practicing Ecology" exercises are provided at the back of the book.

Considers many scales and diverse habitats.

To make the seventh edition reflect the broad range of modern ecology, *we have included examples from a wider diversity of organisms—from microbes to vertebrates.* The section on species interactions includes new chapters on mutualism and parasitism to reflect the importance of these interactions in ecological communities. The research examples also come from a diversity of terrestrial and aquatic habitats.

Fosters ecological literacy through environmental applications.

As ecology instructors, we want to make the material interesting and relevant to our students—most of whom will not go on to become professional ecologists. *Our addition of hundreds of applied studies demonstrates the relevance of ecology in students' lives.* Every chapter now opens with an attention-grabbing case study that highlights important and relevant research to pique student interest. Chapters end with an application called *Ecology Today: Connecting the Concepts*—examples of applied ecology that bring together the major concepts of the chapter and demonstrate their practical importance in a variety of arenas including human health, conservation, and managing our environment.

While making all of these improvements, we were committed to *streamlining the book from 27 to 23 chapters, so it will be more manageable for a one-semester course.* Our goal is to give you a text that presents the material most relevant to class lectures, discussions, and activities. Based on feedback from hundreds of ecology instructors whom we asked to read the chapters, there is broad agreement that this new edition is a beautiful book that will intrigue your students and capture their interest. We look forward to hearing from you as you review the book and we encourage you to share your thoughts with us.

Robert Ricklefs   Rick Relyea   Christoph Richter

Acknowledgments

We were incredibly fortunate to work with a tremendous group of people who made this new edition possible. As with every textbook, each chapter began with a text manuscript and an art manuscript that moved through many revisions to make the final version clear and interesting to undergraduate students. We had the privilege of working daily with developmental editor Rebecca Kohn and art editor Lee Wilcox. These two individuals played a major role in dramatically revising this new edition. Aaron Stoler was instrumental in bringing our vision for the "Practicing Ecology" feature to the page. Claire Hunter handled all the reviewing and surveys and conducted focus groups on the text's design. Other individuals who played key roles in the book's success include senior acquisitions editor Bill Minick, project editor Elizabeth Geller, copy editor Fred Burns, art director Diana Blume, and production manager Susan Wein. Our replacement of nearly every photograph, graph, and illustration required a tremendous effort by a spectacular team; MGMT produced beautiful graphs, Nicolle Fuller of Sayo Art produced stunning illustrations, and the team of Deborah Goodsite and Christine Buese were tireless in hunting down just the right photo to meet each of our requests.

During the past six editions, countless colleagues and instructors have helped shape *Ecology: The Economy of Nature* into a book that has introduced tens of thousands of students to the wonders of ecology. We are extremely grateful to these colleagues. As we undertook a major revision with this seventh edition, we once again received extensive help at every stage of development from many of our colleagues and fellow instructors. We extend our sincere thanks to the following people who graciously offered their time:

Loreen Allphin, *Brigham Young University*

Marty Anderies, *Arizona State University*

Tom Arsuffi, *Texas Tech University*

Betsy Bancroft, *Southern Utah University*

Paul Bartell, *Pennsylvania State University*

David Baumgardner, *Texas A&M University*

Christopher Beatty, *Santa Clara University*

Marc Bechard, *Boise State University*

Mark Belk, *Brigham Young University*

Michael F. Benard, *Case Western Reserve University*

Ritin Bhaduri, *Spelman College*

Andrew Blaustein, *Oregon State University*

Steve Blumenshine, *California State University–Fresno*

Michelle D. Boone, *Miami University of Ohio*

Jennifer Borgo, *Coker College*

Victoria Borowicz, *Illinois State University*

Alison Boyer, *University of Tennessee*

Judith Bramble, *DePaul University*

Shannon Bros-Seemann, *San Jose State University*

Ken Brown, *Louisiana State University*

Romi Burks, *Southwestern University*

Willodean Burton, *Austin Peay State University*

David Byres, *Florida State College–Jacksonville*

Walter P. Carson, *University of Pittsburgh*

Michael F. Chislock, *Auburn University*

George Cline, *Jacksonville State University*

Clay Corbin, *Bloomsburg University of Pennsylvania*

Douglas Crawford-Brown, *University of North Carolina at Chapel Hill*

William Currie, *University of Michigan*

Richard Deslippe, *Texas Tech University*

Hudson DeYoe, *University of Texas–Pan American*

James Dunn, *Grand Valley State University*

Kenneth Ede, *Oklahoma State University–Tulsa*

James Elser, *Arizona State University*

Rebecca Ferrell, *Metropolitan State University of Denver*

Kerri Finlay, *University of Regina*

Ben Fitzpatrick, *University of Tennessee*

Lloyd Fitzpatrick, *University of North Texas*

Matt Forister, *University of Nevada*

Norma Fowler, *University of Texas at Austin*

Rachel Elizabeth Gallery, *University of Arizona Tucson*

Pamela Geddes, *Northeastern Illinois University*

Linda Green, *Georgia Institute of Technology*

Danny Gustafson, *The Citadel*

Monika Havelka, *University of Toronto–Mississauga*

Stephen Hecnar, *Lakehead University*

Colleen Hitchcock, *Boston College*

Gerlinde Hoebel, *University of Wisconsin–Milwaukee*

Claus Holzapfel, *Rutgers University–Newark*

Robert Howard, *Middle Tennessee State University*

Jon Hubbard, *Gavilan College*

Anthony Ippolito, *DePaul University*

John Jaenike, *University of Rochester*

Steven Juliano, *Illinois State University*

Thomas Jurik, *Iowa State University*

Doug Keran, *Central Lakes College*

Tigga Kingston, *Texas Tech University*

Christopher Kitting, *California State University–East Bay*

Jamie Kneitel, *California State University–Sacramento*

Ned Knight, *Linfield College*

William Kroll, *Loyola University of Chicago*

Hugh Lefcort, *Gonzaga University*

Mary Lehman, *Longwood University*

Dale Lockwood, *Colorado State University*

Eric Long, *Seattle Pacific University*

Genaro Lopez, *University of Texas–Brownsville*

C. J. Lortie, *York University*

Roy Mason, *Mt. San Jacinto College*

Robert McGregor, *Douglas College*

L. Maynard Moe, *California State University–Bakersfield*

Don Moll, *Missouri State University*

Peter Morin, *Rutgers University*

Patrick Osborne, *University of Missouri–St. Louis*

Peggy Ostrom, *Michigan State University*

Michael Palmer, *Oklahoma State University*

Mitchell Pavao-Zuckerman, *University of Arizona*

William Pearson, *University of Louisville*

Bill Perry, *Illinois State University*

Kenneth Petren, *University of Cincinnati*

Raymond Pierotti, *University of Kansas*

David Pindel, *Corning Community College*

Craig Plante, *College of Charleston*

Thomas Pliske, *Florida International University*

Diane Post, *University of Texas–Permian Basin*

Mark Pyron, *Ball State University*

Laurel Roberts, *University of Pittsburgh*

Robert Rosenfield, *University of Wisconsin–Stevens Point*

Tatiana Roth, *Coppin State University*

Arthur N. Samel, *Bowling Green State University*

Nate Sanders, *University of Tennessee*

Mark Sandheinrich, *University of Wisconsin–La Crosse*

Thomas Sasek, *University of Louisiana at Monroe*

Kenneth Schmidt, *Texas Tech University*

Robert Schoch, *Boston University*

Erik Scully, *Towson University*

Kathleen Sealey, *University of Miami*

Kari A. Segraves, *Syracuse University*

Chrissy Spencer, *Georgia Institute of Technology*

Janette Steets, *Oklahoma State University*

Juliet Stromberg, *Arizona State University*

Keith Summerville, *Drake University*

Carol Thornber, *University of Rhode Island*

David Tonkyn, *Clemson University*

William Tonn, *University of Alberta*

James Traniello, *Boston University*

Stephen Vail, *William Paterson University*

Michael Vanni, *Miami University*

Eric Vetter, *Hawaii Pacific University*

Joe von Fischer, *Colorado State University*

Stephen Wagener, *Western Connecticut State University*

Diane Wagner, *University of Alaska–Fairbanks*

Sean Walker, *California State University–Fullerton*

Xianzhong Wang, *Indiana University–Purdue University Indianapolis*

John Weishampel, *University of Central Florida*

Gene Weller, *Brigham Young University of Idaho*

Marcia Wendeln, *Wright State University*

Tom Wentworth, *North Carolina State University*

Yolanda Wiersma, *Memorial University*

Frank Williams, *Langara College*

Alan E. Wilson, *Auburn University*

Kelly Wolfe-Bellin, *College of the Holy Cross*

Lan Xu, *South Dakota State University*

Todd Yetter, *University of the Cumberlands*

Teaching in a Canadian Context

Uniquely Canadian Examples

ECOLOGY The Economy of Nature, Canadian Edition has been created to meet the needs of instructors teaching the basic ecology course to biology students in a Canadian context. It is filled with relevant Canadian case studies, photos, and graphics.

The following is a list of some of the uniquely Canadian examples and case studies covered in the text.

Chapter 2 Adaptations to Aquatic Environments

- Example of limestone sediments, Figure 2.8: *Percé Rock, Gaspe Peninsula*, Quebec.
- Example of thermophiles, Figure 2.19: *Rabbitkettle Hotsprings, Yukon*.
- Migration of salmon on *Fraser River* affected by water temperatures.

Chapter 3 Adaptations to Terrestrial Environments

- Examples of structural adaptations of plants against heat and drought, Figure 3.13: Native Canadian species.
- Example of basking, Figure 3.19: Five-lined skink.
- Example of counter-current circulation, Figure 3.22: Caribou.

Chapter 4 Adaptations to Variable Environments

- Examples of spatial variation, Figure 4.2: Various locations across Canada.
- Adaptations to fluctuating salt environments, Figure 4.12: *Vancouver Island, British Columbia*.
- Adaptability of food choices in sea otters living off the coast of *Newfoundland*.

Chapter 5 Climates and Soils

- Hardiness zones for plants across Canada, chapter opener and map.
- Example of regressions, using Canada temperature information, "Analyzing Ecology."
- Example of rainshadow, the prairie in *southern Alberta*.
- Inclusion of Ae soil horizon, Figures 5.16 and 5.17.
- Polar bears in *Labrador*.

Chapter 6 Terrestrial and Aquatic Biomes

- Climate diagram examples:
  - Figure 6.5, Tundra biome: *Iqaluit, Nunavut*
  - Figure 6.6, Boreal forest biome: *Whitehorse, Yukon*
  - Figure 6.7, Temperate rainforest biome: *Carmanah Point, British Columbia*
  - Figure 6.10, Temperate grassland/cold desert biome: *Brandon, Manitoba*
- Example of a river, Figure 6.14: *Athabasca River, Alberta*

- Example of a lake, Figure 6.14: *Lake Louise, Alberta*
- Examples of freshwater wetlands, Figure 6.17: Swamp in *Campbellville, Ontario*.
- Marsh in *Point Pelee, National Park in Leamington, Ontario*; bog in the *Mingan Archipelago National Park Reserve in Quebec*.
- Example of salt marsh, Figure 6.18: *Prince Edward Island*.
- Examples of intertidal zones, Figure 6.20: *Gros Mornee National Park, Newfoundland and Labrador, and Bay of Fundy, New Brunswick*.

Chapter 7 Evolution and Adaptation

- Example of evolution through selection, Figures 7.4 and 7.5: Spirit bears on coast of *British Columbia*.
- Example of bottleneck effect, Figure 7.7: Whooping crane
- Case study on founder effect: Founder effect in the *Saguenay–Lac-Saint-Jean region of northeastern Quebec*.
- Case study on sympatric speciation, Figure 7.19: Stickleback species living in some of *British Columbia's coastal lakes*.
- Case study on drug resistant tuberculosis in Inuit communities.

Chapter 8 Life Histories

- Case study on offspring number versus offspring size: Salmon in *Nova Scotia*.
- Case study on comparisons within species: Pumpkinseed sunfish in *Ontario lakes*.
- Canadian data on senescence in humans, Figure 8.11.
- Case study on Canadian fisheries with regard to life history and commercial fishing.

Chapter 9 Reproductive Strategies

- Example of a Canadian plant that uses vegetative reproduction, Figure 9.1: Arctic Ross's sandwort.
- Case study on adaptive sex determination, Figure 9.10: Atlantic silverside fish.
- Example of sexual dimorphism among caribou, Figure 9.17.

- Case study of sexual conflict among water striders, Figure 9.21

Chapter 10 Social Behaviours

Case study on competition among killer whales off the coast of *British Columbia*, Figure 10.5

Chapter 11 Population Distributions

- Case study on distribution of horse gentian showing how small-scale variation in the environment can create geographic ranges that are composed of small patches of suitable habitat, Figure 11.2
- Case study on suitability of habitat using four species of moss in *Alberta peatland*, Figure 11.3
- Case study of the geographic range of the red-winged blackbird throughout Canada, Figure 11.6
- Discussion of *Yukon Territory to Yellowstone corridor*, Figure 11.15.

Chapter 12 Population Growth and Regulation

- Case study of growth in sea otter population in *British Columbia* waters.
- Case study of negative density dependence in deer mouse population in *Algonquin Park, Ontario*, Figure 12.8.

Chapter 13 Population Dynamics over Space and Time

- Case study of wolverine hunt in Canada as a measure of population size, Figure 13.1
- Dating the ages of existing trees in an *old-growth forest in Manitoba* to determine how age structures of populations changed over time, Figure 13.4
- Case study of duck populations on three *archipelagos off the Labrador coast* to examine extinction in small populations, Figure 13.13.
- Case study on a metapopulation of the Rocky Mountain Parnassian butterfly in meadows on the eastern slopes of the Canadian Rocky Mountains to discuss the importance of patch size and patch isolation, Figure 13.19.

Chapter 14 Predation and Herbivory

- Case study on the effects of herbivores, fencing out deer in *Gwaii Haanas National Park Preserve and Heritage Site, British Columbia*, Figure 14.6.
- Case study on plant response to herbivory, the lesser snow goose in the *wetlands of Hudson Bay Lowlands*.

Chapter 15 Parasitism and Infectious Disease

- Case study on Dutch Elm disease in *Toronto*.
- Case study on SARS in *Toronto*.
- Discussion of Lyme disease in Canada.

Chapter 16 Competition

- Competition for space on *Whaler Islets off the coast of Vancouver Island, British Columbia*, Figure 16.1.

- Case study on competition among mosses in *northeastern Alberta*.
- Case study on disturbances and competition in *fescue prairie in Alberta and Saskatchewan*, Figure 16.11.
- Case study on herbivory and plants with tent caterpillars and trembling aspens in *Alberta*.
- Case study on apparent competition and variation over time with woodland caribou, wolves, and moose, west of the *Athabasca River in northeastern Alberta*, Figure 16.18.

Chapter 17 Mutualism

- Case study on the mutualistic relationship between grasses and their arbuscular mycorrhizal fungi in *Grassland National Park, Saskatchewan*.
- Case study on how a change in fungal abundance might affect plant and insect communities in an *Alberta grassland*, Figure 17.21.
- Case study on how communities are also influenced by the responses of different plant species to arbuscular mycorrhizal fungi, discussion of an experiment carried out at Guelph University.

Chapter 18 Community Structure

- Changes in vegetation as one walks from the base to the top of a *mountain on the eastern side of the Rocky Mountains in Alberta*, Figure 18.1
- Case study on recolonization after fire of areas in the *boreal mixed-wood forest in Alberta*.
- Case study on distribution of fish abundance based on survey from Canada's Department of Fisheries and Oceans, Figure 18.8.
- Case study of how a community can move to an alternative stable state in the lichen woodlands in *Quebec's Parc des Grands-Jardins*.

Chapter 19 Community Succession and Development

- Succession at *Hudson Bay*, Figure 19.2
- Case study on observed changes in the animal community that coincide with plant succession in the black spruce forests in *western Quebec*.
- Discussion of succession in the *Saint Lawrence estuary*.
- Persistence of fire-tolerant species in the *Yukon Territory*.
- Case study of how intense grazing by greater snow geese reduces dominant grass species, Figure 19.17.

Chapter 20 Movement of Energy in Ecosystems

- Case study on invasive species and food web alteration with worms introduced into the forests of the *Kananaskis Valley in Alberta*.
- General patterns in the distribution of primary productivity across Canada, Figure 20.6.

Chapter 21 Movement of Elements in Ecosystems

- Dead zone in the *lower St. Lawrence estuary in Quebec*.

- Natural Resources Canada study of ground water and aquifer use in Canada.
- Canada limits the use of phosphorus in detergents.
- Nutrient travel in *streams in British Columbia*.

Chapter 22 Landscape Ecology, Biogeography, and Global Biodiversity

- Case study on causes of habitat heterogeneity in the *boreal forest of the Lake Nipigon region in northwestern Ontario*, Figure 22.2.
- An example of legacy effects, *an esker near Whitefish Lake, Northwest Territories*, Figure 22.1.
- Case study on relationship between habitat heterogeneity and species diversity among bird species in the *boreal forest of Alberta*.
- Case study of marbled murrelet threatened by increasing edge habitat in old growth forests on the *British Columbia coast*, Figure 22.10.

- Case study on matrix of habitats between fragments with Alpine Parnassius butterfly on the *eastern slopes of the Rocky Mountains in Alberta*.

Chapter 23 Global Conservation

- Economic benefit of boreal forests in Canada.
- Variety in Canadian apple farming as an example of declining agricultural diversity.
- Reduction in the amount of old-growth forest in Canada.
- Loss of *tallgrass prairie in Manitoba*.
- Restriction on DDT use in Canada.
- Case study on conservation in *Wood Buffalo National Park, in Alberta and the Northwest Territories*, Figure 23.22.

Focus on Canadian Research: Practicing Ecology

The "Practicing Ecology" feature highlights the work of researchers in Canada. These extended problems give students the opportunity to work with real data, build graphs, and think critically to interpret experimental results.

Chapter 2 Adaptations to Aquatic Environments

Seeing Underwater Shai Sabbah and his colleagues from Queen's University investigate how the visual system of fish is adapted to light transmission underwater.

Chapter 3 Adaptations to Terrestrial Environments

Habitat Choices of Lizards Stephen Lougheed and his lab at Queen's University have been studying how skinks cope with extreme temperatures.

Chapter 4 Adaptations to Variable Environments

Plastic Mouths in Fish Dolph Schluter and his colleagues at the University of British Columbia have studied the morphological plasticity of fish mouths in response to changing prey abundances.

Chapter 5 Climates and Soils

Why Are the Polar Bears Disappearing? Ian Stirling and his colleagues from Environment Canada have examined possible reasons why polar bear populations have declined in recent years.

Chapter 6 Terrestrial and Aquatic Biomes

Methane Production in Peatlands Tim Moore, from McGill University, and his colleagues have studied how changes in precipitation influence the decomposition of peat.

Chapter 7 Evolution and Adaptation

Predator Preferences and Prey Phenotypes Elizabeth Boulding and her lab at the University of Guelph study the response of phenotypes to selection in intertidal and marine environments.

Chapter 8 Life Histories

Trophy Hunting for Bighorn Sheep David Coltman from the University of Alberta and his colleagues have studied how sport hunting has altered the evolutionary trajectory of bighorn rams.

Chapter 9 Reproductive Strategies

How to Handle Sneaky Males Bryan Neff and his lab from the University of Western Ontario study behavioural ecology and the evolution of mating systems.

Chapter 10 Social Behaviours

Plants That Know Their Kin As a researcher at the University of Toronto, Jay Biernaskie studied the fitness consequences of kin recognition in plants.

Chapter 11 Population Distributions

Blowing Across the Geographic Range Chris Eckert and his lab at Queen's University in Ontario study evolutionary adaptations that enable plants to disperse across a landscape.

Chapter 12 Population Growth and Regulation

Acidification and Zooplankton Shelley Arnott from Queen's University and her colleagues study the factors that determine the distribution and abundance of freshwater invertebrates.

Chapter 13 Population Dynamics over Space and Time

Voles and Berries Charles Krebs at the University of British Columbia and his colleagues study the causes and consequences of rodent population fluctuations in North American forests.

Chapter 14 Predation and Herbivory

Indirect Effects of Deer Herbivory Peter Arcese from the University of British Columbia and his colleagues study the effects of animal management and conservation on ecosystems.

Chapter 15 Parasitism and Infectious Disease

The Cost of Having a Bodyguard Jacques Brodeur from Université de Montréal and his colleagues study interactions between herbivores and their enemies.

Chapter 16 Competition

The Importance of Diversity Marc Cadotte at the University of Toronto studies how plant diversity alters competition and primary productivity.

Chapter 17 Mutualism

The Benefits of a Backup Plan Megan Frederickson at the University of Toronto studies the factors that maintain mutualistic relationships between plants and animals.

Chapter 18 Community Structure

Invasive Crayfish and Trophic Cascades Jonathan Moore at Simon Fraser University studies how the invasion or extinction of species alters the functioning of aquatic ecosystems.

Chapter 19 Community Succession and Development

Determining Conservation Baselines Mark Vellend and his lab at the Université de Sherbrooke study the historical and contemporary processes that alter plant communities.

Chapter 20 Movement of Energy in Ecosystems

Viruses and Marine Nutrient Cycling Chris Suttle and his lab at the University of British Columbia study the ecological importance of viruses in the open ocean.

Chapter 21 Movement of Elements in Ecosystems

Leaf after Death of Salmon John Richardson and his lab at the University of British Columbia study the factors that alter stream community composition and ecosystem function.

Chapter 22 Landscape Ecology, Biogeography, and Global Biodiversity

Maintaining Nightshade Nash Turley, a graduate student at the University of Toronto Mississauga, studies the ecology and conservation of plants and herbivorous insects.

Chapter 23 Global Conservation

A Critical Threshold for Cod Boris Worm at Dalhousie University in Nova Scotia studies global trends in marine conservation.

The seventh edition welcomes Rick Relyea and Christoph Richter.

Every chapter of *ECOLOGY The Economy of Nature*, Canadian edition has been restructured and rewritten to create an accessible, successful learning experience for a wide range of students.

Chapters are organized around four to six key concepts.

An organized learning experience.

---

344 | CHAPTER 15

An infected carpenter ant. In Thailand, carpenter ants that become infected by a fungus crawl down from the canopy and attach themselves to the underside of a leaf by biting the leaf vein and then dying. After death, a spore-producing stalk grows out of the ant's head and releases its spores into the environment. Photo by David Hughes/Penn State University.

canopy, its nest mates would remove the corpse before the fungus could grow the spore-producing structure that is critical to its reproduction.

The ability of parasites to act as puppet masters over the behaviour of their victims is just one way that parasites have evolved to improve their fitness. As we will discover in this chapter, parasites come in a wide variety of forms and their effects on hosts can range from mild to lethal. Adaptations that enable parasites to infect hosts and adaptions that help hosts to resist infections by parasites offer intriguing insights into the strategies of parasite–host interactions.

SOURCES: S. B. Andersen et al., The life of a dead ant: The expression of an adaptive extended phenotype, *American Naturalist* 174 (2009): 424–433.

D. P. Hughes et al., Behavioural mechanisms and morphological symptoms of zombie ants dying from fungal infection, *BMC Ecology* 11 (2011): 13.

T. Lefevre, and F. Thomas, Behind the scene, something else is pulling the strings: Emphasizing parasitic manipulation in vector-borne diseases, *Infection, Genetics and Evolution* 8 (2008): 504–519.

CHAPTER CONCEPTS

- Many different types of parasites affect the abundance of host species.
- Parasite and host dynamics are determined by the parasite's ability to infect the host.
- Parasite and host populations commonly fluctuate in regular cycles.
- Parasites have evolved offensive strategies while hosts have evolved defensive strategies.

The struggle between parasites and hosts has produced many fascinating examples of ecological interactions and evolutionary adaptations. In Chapter 1, we defined a parasite as an organism that lives in or on another organism, called the host, and causes harmful effects as it consumes resources from the host. Some hosts have **infection resistance**, which is the ability of a host to prevent the infection from occurring, while other hosts have **infection tolerance**, which is the ability of a host to minimize the harm that an infection can cause. The number of parasites of a given species that an individual host can harbour is known as the host's **parasite load**. A parasite typically has only one or a few species of hosts, although a given host could contain dozens of species of parasites. In fact, it is estimated that approximately half of all species on Earth are parasites.

When parasites cause an infectious disease, we call them *pathogens*. However, infection by a pathogen does not always result in a disease. For example, humans can be infected with human immunodeficiency virus (HIV) but they may never experience the disease symptoms known as Acquired Immune Deficiency Syndrome (AIDS). In many cases, it is not known what causes a host to transition from being infected by a pathogen to experiencing the disease.

Infectious diseases take a large toll on people; the World Health Organization estimates that more than 25 percent of all human deaths are caused by infectious diseases. Note that only infectious diseases are caused by pathogens; there are many noninfectious diseases in which pathogens do not play a role.

In this chapter, we will focus on the interaction between parasites and their hosts. In later chapters, we will discuss the larger role that parasites can play in communities and ecosystems. We will begin by looking at the many different types of parasites that exist, including those that have large effects on ecological communities, crops, domesticated animals, and human health. We will then examine the factors that determine whether parasites can infect hosts, spread rapidly through a population, and cause widespread harmful effects. Because mathematical

Infection resistance The ability of a host to prevent an infection from occurring.

Infection tolerance The ability of a host to minimize the harm that an infection can cause.

Parasite load The number of parasites of a given species that an individual host can harbour.

models can help us understand the population dynamics of interacting species, we will also discuss parasite–host models. Finally, we will consider how parasites have evolved to increase their chances of infecting hosts, and how hosts have evolved to combat the risk of infection.

Many different types of parasites affect the abundance of host species

Parasites typically have specific habitat needs and, as a result, often live in particular places on a host organism. In humans, for instance, parasites find certain parts of the body to be highly suitable habitats. As illustrated in Figure 15.1, head lice live in the hair, liver flukes reside in the liver, the fungus that causes athlete's foot resides in the feet, and so on.

We can categorize the wide variety of parasites as either *ectoparasites* or *endoparasites*. Both are common in plants and animals. **Ectoparasites** are

Ectoparasite A parasite that lives on the outside of an organism.

Brain
Prion
Creutzfeldt-Jakob disease

Scalp
Head lice
(*Pediculus humanus*)

Lungs
Bacterium (*Mycobacterium* spp.)
Tuberculosis

Liver
Liver fluke (*Fasciola* spp.)
Fasciolosis

Heart and blood
Protist (*Plasmodium* spp.)
Malaria

Multiple body systems
Virus (EBOV)
Ebola virus disease

Virus (HIV)
AIDS

Digestive tract
Bacterium (*Helicobacter pylori*)
Some stomach ulcers

Tapeworms (several genera)

Genitalia
Fungus (*Candida albicans*)
Candidal vulvovaginitis

Many body areas
Black-legged tick (*Ixodes scapularis*)
Lyme disease

Skin
Itch mite (*Sarcoptes scabiei*)
Scabies

Chigger mite (*Trombicula alfreddugesi*)
Chiggers

Feet
Fungus (*Trichophyton rubrum*)
Athlete's foot

Figure 15.1 Preferred habitats. Parasites have preferred habitats on their host. The human body, for example, offers a wide range of habitats for parasites.

chipmunks, and mice. In high-acorn years, tick-carrying deer gather under the oak trees, which causes an aggregation of reproducing ticks that drop their eggs under the trees. Chipmunks and mice are also attracted to the acorns since the abundant food permits higher survival and increased reproduction. More rodents and an increased density of tick eggs cause an increase in the number of infected rodents the following summer. In low-acorn years when both deer and rodents spend more time in maple forests, the prevalence of infected ticks shifts from the oak forests.

To examine what steps can be taken to reduce the risk of Lyme disease to humans, the researchers also used mathematical models that included all of the key species. These models suggested that, while reductions in deer densities would have little effect on the population of infected ticks unless the deer population were completely eradicated, reductions in the rodent population could cause a large reduction in the population of infected ticks.

The story of Lyme disease is an excellent example of how understanding the ecology of parasites allows us to predict when and where they pose the greatest risk to wild organisms and to humans.

SOURCES: Ostfeld, R. 1997. The ecology of Lyme-disease risk. *American Scientist* 85: 338–346.

Ostfeld, R., et al. 2006. Climate, deer, rodents, and acorns as determinants of variation in Lyme-disease risk. *PLOS Biology* 4: 1058–1068.

Chapter concepts are repeated in both major headings and streamlined summaries.

SUMMARY OF CHAPTER CONCEPTS

■ **Many different types of parasites affect the abundance of host species.** Ectoparasites live on host organisms whereas endoparasites live in host organisms. As a group, parasites include a wide range of species that include plants, fungi, protozoa, helminths, bacteria, viruses, and prions. Among parasites that cause diseases—known as pathogens—those that have recently become abundant are called emerging infectious diseases.

■ **Parasite and host dynamics are determined by the parasite's ability to infect the host.** The transmission of parasites can be horizontal— either through direct transmission or transmission by a vector—or vertical from parent to offspring. The ability to infect a host also depends on the parasite's mode of entering the host, its ability to infect reservoir species, its ability to jump to new host species, and its ability to avoid the host's immune system.

■ **Parasite and host populations commonly fluctuate in regular cycles.** These fluctuations occur because transmission increases with host density but decreases as an increased proportion of the host population develops immunity. These fluctuations can be modelled using the S-I-R model.

■ **Parasites have evolved offensive strategies while hosts have evolved defensive strategies.** Natural selection has favoured parasites that can improve their probability of transmission, including manipulations of host behaviour. Hosts have evolved both specific and general immune responses to combat host infection. Hosts also can employ mechanical and biochemical defences against parasites. Coevolution occurs when the parasite and host continually evolve in response to each other.

REVIEW QUESTIONS

1. Compare and contrast the advantages and disadvantages of life as an ectoparasite versus an endoparasite.

2. Why are parasites often not very harmful to hosts in their native range but highly detrimental to hosts in an introduced range?

3. What allows a parasite to jump to a new species of host?

4. If prions are not normally transmittable among cattle, what procedure caused the spread of mad cow disease?

5. What constitutes an emerging infectious disease?

6. Contrast horizontal versus vertical transmission of a parasite.

7. Using the basic S-I-R model of parasite and host dynamics, explain why the proportion of infected individuals in the population declines over time.

8. In the S-I-R model of parasite and host dynamics, how does the outcome change if we allow new susceptible individuals to be born into the population?

9. When using a *t*-test, what is an important assumption regarding the distribution of the data?

10. Explain why pathogens that are highly lethal to their hosts might become less lethal over time.

xxvi

Hundreds of aquatic examples

survival rate during the first year, followed by high survival for the next 7 years. After 7 years of age, the annual survival rate began to drop, with the exception of those at 12 years of age. However, only four animals made it to 12 years, which gives us a poor estimate of the typical survival rate for 12-year-old sheep. Because these data were taken from the skeletons of sheep that presumably died over a relatively short period, we do not see the large variation in survival rate that we saw in the cohort life table of the cactus finch.

Throughout this chapter we have examined how ecologists use their understanding of populations to develop mathematical models that mimic natural patterns of population growth. Although populations commonly grow rapidly when at low densities, they become limited as the populations grow larger. The simplest population models are helpful starting points, but they are not sufficient for most species that have survival and fecundity rates that vary with age, size, or life history stage. Using life tables helps to incorporate this complexity. As we will see below in "Ecology Today: Connecting the Concepts," the analysis of life tables can be very helpful in setting priorities for management strategies to save species from extinction.

ECOLOGY TODAY CONNECTING THE CONCEPTS

SAVING THE SEA TURTLES

Sea turtles. Population models have identified the stages in life that need the most protection to allow turtle populations to recover. These models are relevant to all species of sea turtles, including this green sea turtle (*Chelonia mydas*) off the coast of Hawaii. Photo by Masa Ushioda/Image Quest Marine.

Under the cover of darkness, the swimmers come to shore. After spending a year out in the ocean, the female sea turtles crawl up the shore and dig a nest to lay their eggs. After about 2 months, these eggs will hatch and the hatchlings will scramble toward the ocean where they will face a gauntlet of predators. If they survive, they will need 2 or 3 decades to become reproductively mature. They may live for 80 years.

Not long ago, many more females came to shore. Of the six species of sea turtles, all have declined over the past several centuries. For example, scientists estimate that there were once 91 million green turtles (*Chelonia mydas*) in the Caribbean; today there are about 300 000 which is a decline of more than 99 percent. Sea turtles, which live in temperate and tropical waters, are declining worldwide. Historically, people hunted them for food

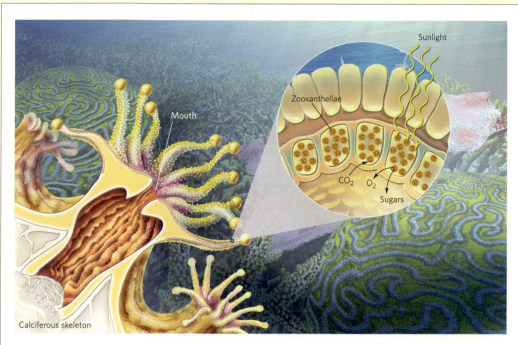

Figure 17.1 **A mutualism between coral and zooxanthellae.** A coral grabs food with its tentacles, which contain stinging cells, and pulls the food into its mouth. Zooxanthellae algae live along the surface of a coral's tentacles where they obtain sunlight for photosynthesis. Corals provide a place for algae to live and emit CO_2 that algae use in photosynthesis. As algae photosynthesize, they provide sugars and O_2 that corals consume.

Whale ancestor. The ancestor of modern whales, in the genus *Indohyus*, was a terrestrial animal that spent part of its time in the water. Over time, the descendants of this animal developed numerous adaptations for living in water that are found in modern-day whales.

Figure 17.2 **The green sea slug.** When the slug first hatches from an egg, it is brown. However, as it begins to eat algae, it stores the chloroplasts from the algae in its own tissues. As the slug accumulates a large number of chloroplasts, its body turns green and it is able to acquire most of its energy through photosynthesis rather than through herbivory. Photo by Nicholas E. Curtis and Ray Martinez, University of South Florida.

Each chapter contains a wealth of contemporary and classic examples.

A modern approach to community and ecosystem ecology

CHAPTER 18

Community Structure

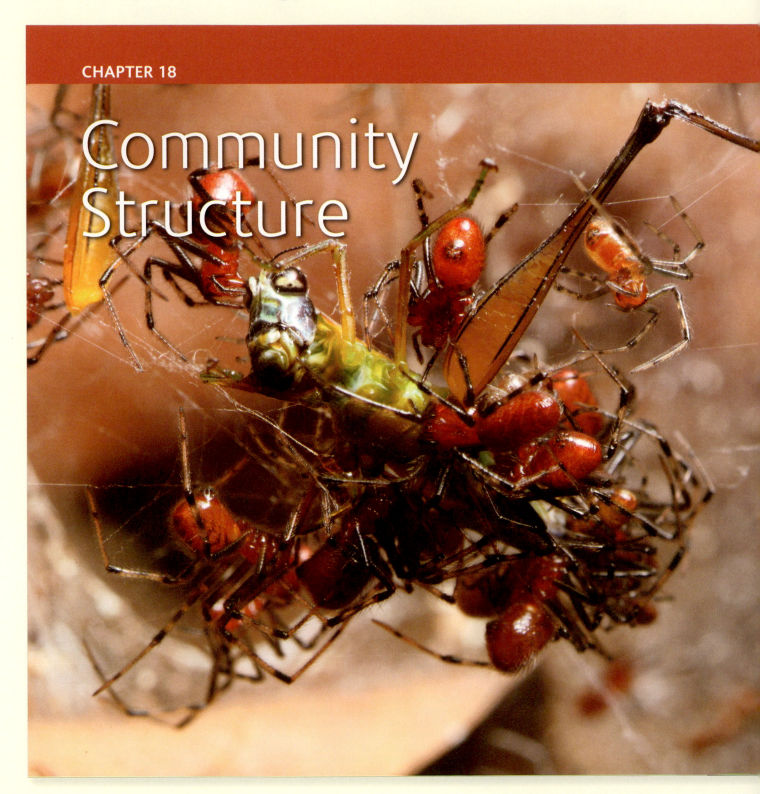

A Web of Interactions in Social Spiders

If you have ever walked through a forest or a field, you have undoubtedly seen a spider perched on a web, waiting for an unsuspecting insect to come by. Most spiders live solitary lives except when they mate. However, some species of social spiders live in large groups composed of thousands of spiders. These spiders build webs that are typically 1 to 4 m wide, and they occasionally make massive webs up to 100 m wide. These webs are so large that a single web can serve as a habitat for more than 100 species of invertebrates that include many other species of spiders.

The interactions that occur in such a web are diverse. Some of the species that join the community of invertebrates are spiders that catch prey. The scraps they leave behind can be consumed by the spider species that builds the web. Other species that join the community are predators of the web-building spider species. In addition, the individual spiders that build and maintain the web have specific behavioural traits. For example, the webs of the tangle web spider (*Anelosimus studiosus*), which lives in riparian areas from the northeastern United States to southern Argentina, are inhabited almost entirely by females. These female spiders have different personalities; some are quite aggressive whereas others are docile. Aggressive individuals are better at capturing prey and defending against predators whereas docile individuals spend more of their time building the massive web.

> "Some species of social spiders live in large groups that make massive webs up to 100 m wide."

The composition of the spider community and the specific behavioural traits of the spiders that build the web both influence how long the community can persist. Although docile individuals grow and reproduce better in the presence of other species of spiders, colonies containing only docile individuals ultimately go extinct sooner. In 2012, researchers discovered that the higher extinction rate in webs containing docile spiders is related to the diversity of other spider genera that live in the web. When there are more spider genera, it is more likely the web will contain spiders that are predatory on the docile females; these predatory genera will ultimately drive the docile females to extinction.

In contrast, webs that contain a higher proportion of aggressive females successfully maintain a lower diversity of spider genera because they attack new spiders that attempt to intrude. As a result, these webs are less likely to contain predatory spider genera. Ultimately, a mixture of docile spiders and aggressive spiders strikes the best balance between the need to catch and subdue prey and the need to avoid too many aggressive interactions.

The tangle web spiders. As social spiders, docile and aggressive individuals work together to build massive webs that catch a wide variety of prey including grasshoppers. Photo by Alexander Wild.

Enhanced coverage of species interactions, community ecology, and ecosystem ecology.

Figure 14.18 Chemical defences. The bombardier beetle (*Stenaptinus insignis*) mixes two chemicals inside its abdomen that react to release a boiling hot fluid to deter predators.

Highlights contemporary approaches and research throughout.

Integrated applications to medicine and public health

ECOLOGY TODAY CONNECTING THE CONCEPTS

DRUG-RESISTANT TUBERCULOSIS

Tuberculosis in Inuit communities. In Kangiqsualujjuaq, Quebec, more than 30 active cases of TB were reported in less than a year. Photo by Bryan and Cherry Alexander Photography/ArcticPhoto/All Canada Photos.

Tuberculosis, or TB, is a highly infectious disease caused by a bacterium (*Mycobacterium tuberculosis*). TB has killed people for thousands of years. In 2009, for example, researchers discovered that the preserved tissues of a woman who died 2600 years ago and was mummified have genetic markers of the TB bacterium. Today, experts estimate that nearly one-third of the world's human population is infected by the extensive tissue damage, weakness, night sweats, and bloody sputum. It is highly contagious; when an infected individual coughs or talks, bacteria are expelled and can survive in the air for several hours and infect other people.

Around the world each year, 2 million people die of TB. Nationally, Canada has a rate of infection comparable to other industrialized nations. Canada-born, non-native groups show an incidence of 1 TB infection per 100 000 people. However, in Inuit communities, the rate is 155.8 per 100 000. In Nunavut, the rate of infection is 62 times higher than the national average. Northern communities are isolated and less likely to receive medical treatment and supplies to treat TB infections. Incomes in most Inuit communities are low, resulting in houses that are frequently in poor condition and lack insulation. The inability to properly heat these homes can make inhabitants more vulnerable to diseases. Many homes also lack proper ventilation, which increases the exposure to airborne bacteria, and most people live in cramped conditions, which can facilitate the spread of the bacterium. Despite efforts to improve these conditions and reduce the rate of infection, researchers estimate that the goal of reducing the rate to 1 per 100 000 Inuit may still be a long way off.

Drug-resistant tuberculosis is a rapidly growing problem around the world, particularly in Africa, Russia,

individuals who possess mutations. Occasionally, a mutation makes a bacterium more resistant. Antibiotics represent a strong selective force that can quickly kill the vast majority of sensitive bacteria, thereby leaving resistant bacteria to flourish.

One of the biggest contributors to the evolution of TB resistance is thought to be the behaviour of TB patients. The typical drug treatment of TB requires taking pills daily for 1 year. Although many bacteria are killed early in the treatment, continued treatment helps eliminate every pathogen. Sometimes patients stop taking their pills because they feel better after a few months or they simply lack the money to pay for an entire year of treatment. In either case the most resistant bacteria will still survive in their bodies.

Drug-resistant TB is becoming a major problem. Researchers have developed new types of TB drugs to try to select different TB traits with the hope that evolving resistance to one drug will still make the pathogen susceptible to other drugs. However, there is now an increase in "Multiple Drug Resistant TB" or MDR-TB, a strain of the bacterium that has evolved resistance to several different drugs. In Russia, for example, nearly 20 percent of all people infected with TB are carrying the MDR-TB strain. These strains are much harder for doctors to kill and the medicines required to kill them are 100 times more expensive than the traditional drugs. Even more sobering is the discovery of what is being called "Extensively Drug Resistant TB." This version of TB has been detected in 45 countries including Russia and currently there are no drugs available to kill it.

The evolution of TB resistance is an excellent example of why we need to understand the process of evolution. Knowing the sources of genetic variation and how selection operates on this variation helps us develop drug treatment programs that are better able to control pathogens without producing strains that have multiple drug resistance.

SOURCES: Altman, L. K. Drug-resistant TB rates soar in former Soviet regions. 2008. *New York Times*, February 27. http://www.nytimes.com/2008/02/27/health/27tb.html

Goozner, M. 2008. A report from the Russian front in the global fight against drug-resistant tuberculosis. *Scientific American*, August 25. http://www.scientificamerican.com/article.cfm?id=siberia-drug-resistant-tuberculosis

Anonymous. Nunavut TB infections break record. December 17, 2010. http://www.cbc.ca/news/health/story/2010/12/17/nunavut-tuberculosis.html

Anonymous. Tuberculosis (TB). http://www.hc-sc.gc.ca/fniah-spnia/diseases-maladies/tuberculos/index-eng.php

Clark, M., P. Riben, and E. Nowgesic. 2002. The association of housing density, isolation and tuberculosis in Canadian First Nations communities. *International Journal of Epidemiology* 31: 940–945.

Numerous examples show the relevance of ecological concepts to contemporary issues in public health.

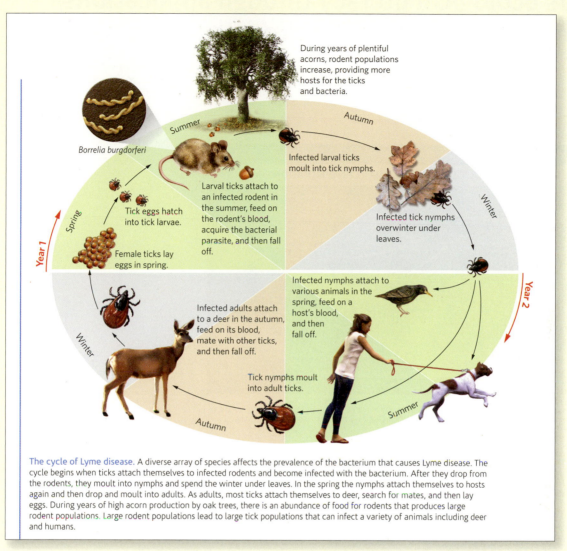

During years of plentiful acorns, rodent populations increase, providing more hosts for the ticks and bacteria.

Borrelia burgdorferi

Year 1

Summer

Spring

Tick eggs hatch into tick larvae.

Female ticks lay eggs in spring.

Larval ticks attach to an infected rodent in the summer, feed on the rodent's blood, acquire the bacterial parasite, and then fall off.

Autumn

Infected larval ticks moult into tick nymphs.

Winter

Infected tick nymphs overwinter under leaves.

Year 2

Infected nymphs attach to various animals in the spring, feed on a host's blood, and then fall off.

Winter

Infected adults attach to a deer in the autumn, feed on its blood, mate with other ticks, and then fall off.

Autumn

Tick nymphs moult into adult ticks.

Summer

The cycle of Lyme disease. A diverse array of species affects the prevalence of the bacterium that causes Lyme disease. The cycle begins when ticks attach themselves to infected rodents and become infected with the bacterium. After they drop from the rodents, they moult into nymphs and spend the winter under leaves. In the spring the nymphs attach themselves to hosts again and then drop and moult into adults. As adults, most ticks attach themselves to deer, search for mates, and then lay eggs. During years of high acorn production by oak trees, there is an abundance of food for rodents that produces large rodent populations. Large rodent populations lead to large tick populations that can infect a variety of animals including deer and humans.

Students see how ecological thinking is essential to our understanding of infectious disease.

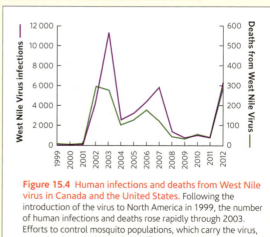

Figure 15.4 Human infections and deaths from West Nile virus in Canada and the United States. Following the introduction of the virus to North America in 1999, the number of human infections and deaths rose rapidly through 2003. Efforts to control mosquito populations, which carry the virus, caused a sharp decline in these effects on humans through 2011. In 2012, however, unusually warm temperatures with high precipitation in some regions resulted in large mosquito populations and a new surge in West Nile infections and deaths. Note that the two *y* axes have different scales. Data from http://www.cdc.gov/ncidod/dvbid/westnile/surv&control.htm#maps.

Fosters ecological literacy through applications to conservation and the environment

ECOLOGY TODAY CONNECTING THE CONCEPTS

LETHAL EFFECTS OF PESTICIDES AT NONLETHAL CONCENTRATIONS

Leopard frog. Low concentrations of an insecticide that cannot kill the tadpoles of the leopard frog can kill other species in the community. This sets off a chain of events that alters the food web and indirectly prevents the leopard frogs from achieving metamorphosis. Photo by Lee Wilcox.

Appreciating the connectedness of species in a food web can help us understand some of the effects our activities have on ecological communities. Pesticides, for example, provide important benefits in protecting crops and improving human health, and their widespread use has made them common in ecological communities. Maintaining pesticide concentrations below levels lethal to nonpest organisms is the key to ensuring that pesticides harm only pests and not other species. Pesticides are tested on a range of species in the laboratory to determine concentrations that cause death. However, laboratory tests do not consider the fact that species are part of a food web. Therefore, researchers have asked if a pesticide that is not directly lethal to a species can nonetheless have indirect impacts that cause mortality.

Some of the best-studied pesticides are those used to control terrestrial insects that damage crops as well as insects such as mosquitoes that carry infectious diseases. The most commonly applied insecticide is malathion, which impairs the nervous system of animals and is highly lethal to insects and other invertebrates. Based on single-species laboratory tests, scientists thought that concentrations found in nature were not lethal to vertebrate animals.

In a series of experiments, researchers examined the effects of malathion in large outdoor tanks that contained many of the components of natural wetlands including algae, zooplankton, and tadpoles (see Figure 1.18b). It had been long known that insecticides such as malathion were highly toxic to zooplankton, which are tiny crustaceans, but little was known about how large reductions in zooplankton might affect other species in the food web.

In one such experiment, researchers added tadpoles of wood frogs and leopard frogs (*R. pipiens*) to tanks and then manipulated malathion in one of four ways: no malathion to serve as a control, 50 parts per billion (ppb) at the beginning of the experiment, 250 ppb at the beginning of the experiment, or 10 ppb once per week. These manipulations allowed the researchers to ask whether single, large applications and multiple, small applications could alter the food web in ways that could affect the amphibians.

Soon after malathion was applied, the zooplankton populations declined to very low numbers in all of the tanks treated with the insecticide. This decline was not surprising since zooplankton were known to be quite sensitive to insecticides, but it set off a chain of events that ultimately affected the amphibians. Zooplankton feed on phytoplankton. When malathion was added and the zooplankton were killed, the phytoplankton populations rapidly grew over the next few weeks. This rapid growth, known as an algal bloom, caused the water to turn pea green.

Each chapter closes with ECOLOGY TODAY: CONNECTING THE CONCEPTS, a case study that applies multiple chapter concepts to a contemporary ecological issue.

COMMUNITY STRUCTURE | 439

Insecticide

Zooplankton   Tadpoles

Phytoplankton   Periphyton

Resources
(light and nutrients)

(b)

...unity. **(a)** In the absence of an insecticide, zooplankton are
...keeps the water relatively clear and allows sunlight to reac...
... food for tadpoles. **(b)** When an insecticide is added, the z...
...ws an increase in phytoplankton. Abundant phytoplankton...
...aching the periphyton at the bottom of the wetland. With l...
...its tadpole growth. The number of individuals pictured in e...

...ankton,   when the insecticide was present. In fact,
...the   frogs had such stunted growth in the tan...
...known   weekly addition of 10 ppb of malathion t...
...h as soil   half of them failed to metamorphose bef...
...ry well   aquatic environment dried in late summe...
...eeks   This research demonstrated for the first
...   low concentration of an insecticide that i...
...in those   lethal to amphibians can be indirectly let...
...n every   insecticide initiated a chain of events that
...ankton   through the food web and caused nearly
...ause the   leopard frogs to die. Moreover, the smalle...
...duced the   of the insecticide, which was applied each
...   much larger effect on the leopard frogs th...
...   applications containing 25 times more ins...
...develop   These results underscore the importan...
...orphosed   understanding ecology at the communi...
...re able   Quantifying the insecticide's direct effects on the
...plankton   amphibians suggested that the animals would not be
...ce. As a   harmed, but tests that incorporated a food web
...ot   approach with indirect effects found that the
                   insecticide could kill nearly half of the animals.

...rtunate.
...growth   **SOURCES:** Relyea, R. A. 2005. The impact of insecticides and herbicides on the
...hey   biodiversity and productivity of aquatic communities. *Ecological Applications* 15:
...riphyton   618–627.

Relyea, R. A., and N. Diecks. 2008. An unexpected chain of events: Lethal effects of
pesticides on frogs at sublethal concentrations. *Ecological Applications* 18: 1728–1742.

The effects and challenges of global climate change are considered throughout the text.

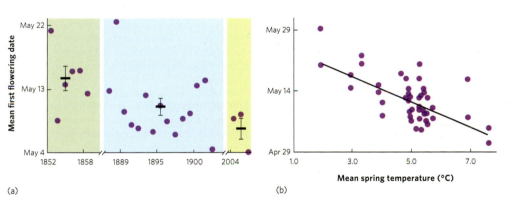

(a)   (b)

Figure 8.15 First flowering dates for plants in Concord, Massachusetts. **(a)** The mean flowering time today is 7 days earlier than in the 1850s. Error bars are standard errors. **(b)** The variation in first flowering time is associated with the mean temperature of the 1 or 2 months preceding each species' flowering time. Data from A. J. Miller-Rushing and R. B. Primack, Global warming and flowering times in Thoreau's Concord: A community perspective, *Ecology* 89 (2008): 332–341.

A learning-by-doing approach to basic quantitative tools and the use of data

ANALYZING ECOLOGY

Regressions

As we have discussed, latitudes closer to the equator receive more solar radiation than latitudes closer to the poles. Given this observation, lower latitudes should also have warmer temperatures than higher latitudes. In fact, understanding the nature of this relationship would help us determine exactly how much temperature changes with latitude. When we want to know how one variable changes in relation to another, we use a statistical tool called **regression.** In Chapter 4 we saw that a correlation determines if there is a relationship between two variables. A regression determines whether there is a relationship and also describes the nature of that relationship.

To help illustrate this idea, we can use data on the average January temperature from 59 locations spanning Canada from north to south. If we plot the relationship between location latitude and average location temperature in January, we obtain the following graph:

In this case, the relationship between the two variables follows a straight line; we have drawn a line of best fit through the distribution of the data points. This is a regression line because it represents the relationship between the two variables. It informs us about the nature of the relationship through the slope and intercept of the line. For these data, the regression can be described using the equation of a straight line, where Y is the dependent variable, X is the independent variable, m is the slope of the line, and b is the Y-intercept of the line at the point where $X = 0$. In this example, the slope is -0.9 and the intercept is 38:

$$Y = mX + b$$

$$\text{Temperature} = -0.9 \times \text{Latitude} + 39$$

This regression equation tells us that for every 1 degree increase in latitude, the average temperature in January decreases by 0.9°C. Note that while the simplest form of a regression is a straight line, regression lines can also be curvilinear.

> **YOUR TURN** Based on the relationship between latitude and temperature, use the regression equation to estimate the average January temperature at 40, 60, and 80 degrees of latitude.

Regression A statistical tool that determines whether there is a relationship between two variables and that also describes the nature of that relationship.

ANALYZING ECOLOGY introduces mathematical and statistical techniques in relevant research contexts. YOUR TURN self-guided study problems provide opportunities for practice.

PRACTICING ECOLOGY problems help students to gain experience graphing and interpreting real data.

Over 500 graphs throughout the text present and describe actual research data.

PRACTICING ECOLOGY: THE COST OF HAVING A BODYGUARD

Jacques Brodeur from Université de Montréal and his colleagues study interactions between herbivores and their enemies. Photo by Anne Audy.

As we have seen in this chapter, many parasites complete part of their life cycle by manipulating the body of a host. However, in order to control its host, a parasite must expend energy. Often, a host ceases to feed once infected, so the energy that the parasite uses must come from stored energy that the parasite could otherwise use for its own development. In addition, the body manipulation necessarily requires that the parasite transfer chemical compounds to the host's brain, and these chemicals are likely expensive to produce. Do these costs detract from the fitness of the parasite?

Dinocampus coccinellae is a parasitic wasp that lays a single egg inside the spotted lady beetle (*Coleomegilla maculata*). After feeding on the beetle's eggs, gonads, and fat reserves, the larva paralyzes the beetle and emerges from the anal gland as a pupa. The pupa then spins a cocoon around the legs of the beetle and develops for an average of 7 days until it emerges as an adult wasp. Only 25 percent of beetles survive during the time that the pupa is in its cocoon. A beetle that does survive exhibits bodyguard behaviour over the cocoon; the pupa makes the beetle twitch when provoked by an external stimulus and this behaviour reduces predation on the pupa. However, the pupa must use energy to control the beetle's behaviour. We might hypothesize that this energy expense comes at a cost to the future fitness of the parasitic wasp.

Jacques Brodeur from the Université de Montréal and his colleagues tested this hypothesis by observing parasitized beetles in the laboratory. They collected wasps and beetles from the field and allowed the wasps to parasitize the beetles in the lab. After pupae emerged from the beetles and spun cocoons, the researchers recorded the number of days each beetle exhibited twitching behaviour, which corresponds to the amount of time the beetle survived and the pupa was able to control its movement. Once the wasps emerged from their cocoons, the researchers counted the number of eggs in female wasps as a measure of wasp fitness. You can view their data in the table.

DURATION OF BEETLE TWITCHING BEHAVIOUR (DAYS) AND NUMBER OF EGGS IN EMERGENT FEMALE WASPS

DURATION OF TWITCHING BEHAVIOUR	NUMBER OF EGGS
2.9	103
4.9	102
4	102
7.9	101
8.9	99
9	98
7	100
5.9	100
6.1	101
7	103
4.9	101
8	99

GRAPH THE DATA

1. Create a scatterplot graph with duration of twitching behaviour on the *x* axis and number of eggs in emergent wasps on the *y* axis.

2. Visually estimate and draw a regression line of best fit through the data.

INTERPRET THE DATA

1. Explain how the data support or refute the hypothesis that the energy expense of controlling a lady beetle host comes at a cost to future fitness of the developing wasp.

2. Unlike animals that are born with a fixed number of eggs, female *D. coccinellae* wasps are able to produce eggs throughout their lifetime. How might the later production of eggs compensate for the fitness cost of controlling the behaviour of a host beetle?

3. In the same study, the researchers found no relationship between duration of beetle twitching behaviour and the emergent wasp's lifespan. However, the amount of food after emergence positively correlated with wasp lifespan. What does this suggest about the allocation of energy for fecundity versus energy allocated to survival in *D. coccinellae*?

SOURCE: Maure, F., J. Brodeur, N. Ponlet, J. Doyon, A. Firlej, E. Elguero, and F. Thomas. 2011. The cost of a bodyguard. *Biology Letters*, 7: 843–846.

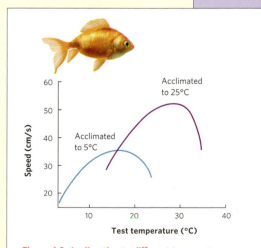

Figure 4.9 Acclimation to different temperatures. Goldfish raised at low temperatures swim faster in cold water and more slowly in warmer water. Individuals raised at high temperatures swim faster in warmer water than in cold water.
After F. E. J. Fry and J. S. Hart, Cruising speed of goldfish in relation to water temperature, *J. Fish. Res. Bd. Can.* 7 (1948): 169–174.

Focus on scientific method and the process of discovery

CHAPTER 19

Community Succession

The rate of glacier retreat. Historic observations by Captain George Vancouver, John Muir, William Cooper, and current researchers indicate that the glacier at Glacier Bay has retreated rapidly during the past 200 years.

responded to the massive disturbance of an advancing and retreating glacier.

The observations of Vancouver, Muir, and Cooper paved the way for more than a century of subsequent ecological research at Glacier Bay. In fact, the original terrestrial quadrats set up by Cooper are still monitored today, as are local streams that have been created by the melting glacier. As we will see in this chapter, long-term changes in ecological communities follow predictable patterns, which are important for us to understand the ways in which terrestrial and aquatic communities change over time and the processes that underlie these changes.

SOURCES: W. S. Cooper, The recent ecological history of Glacier Bay, Alaska II: The present vegetation cycle, *Ecology* 4 (1923): 223–246.

C. L. Fastie, Causes and ecosystem consequences of multiple pathways of primary succession at Glacier Bay, Alaska, *Ecology* 76 (1995): 1899–1916.

Students learn how researchers formulate and test their hypotheses. The text shows how researchers challenge and extend the work of those who came before them.

Retreating Glaciers in Alaska

In the panhandle of Alaska near Juneau, a stretch of land has undergone incredible changes during the past 200 years. In 1794, Captain George Vancouver found an inlet that headed toward modern-day Juneau. To the north of the inlet was a massive glacier that measured more than 1 km thick and 32 km wide. When the naturalist John Muir visited the glacier site in 1879, 85 years after Captain Vancouver's expedition, he was shocked to find that the glacier had receded nearly 80 km and left behind a large bay. Moreover, while most bays in Alaska were heavily forested, the shores of this new bay were relatively barren. As Muir explored, he found several locations with large stumps that were the remnants of hemlock, black cottonwood (*Populus trichocarpa*), and Sitka spruce trees that had been sheared off by the glacier as it advanced centuries earlier. Muir had no way of knowing that he was looking at one of the fastest melting glaciers in the world.

> "When the naturalist John Muir visited the glacier site in 1879, 85 years after Captain Vancouver's expedition, he was shocked to find that the glacier had receded nearly 80 km and left behind a large bay."

John Muir's writings about the natural world were widely read. In 1916 William Cooper, an ecologist who was particularly intrigued by Muir's stories of Alaska, made the first of several expeditions to the site that came to be known as Glacier Bay. On these expeditions, he surveyed the plants growing at various sites along the shore of the bay, including sites where the glacier existed during Vancouver's visit, sites where the glacier existed during Muir's visit, and sites where the glacier had receded only recently. He also set up quadrats around the bay to record the changes in vegetation at various locations that he planned to check when he returned on subsequent expeditions.

Cooper reasoned that sites exposed for the longest duration would have the most time to grow back into the kind of forest that had existed before the glaciers had advanced centuries earlier. In contrast, he expected recently exposed sites to be bare rock and gravel, representing the first stages of a developing forest. On these recently exposed sites, Cooper found mosses, lichens, herbs, and low shrubs. Sites that had been exposed for 35 to 45 years had tall species of willow and alder shrubs (*Alnus sinuata*) and black cottonwood trees. Sites older than 100 years contained Sitka spruce trees, and sites older than 160 years had hemlock in the understorey. By examining changes in the plant community at sites that had been exposed for different lengths of time, he was able to hypothesize how the forests of Alaska

Glacier Bay, Alaska. In 1794, a glacier that was more than 1200 m thick covered all but the inlet of the bay. Since then, the glaciers have melted and retreated, leaving behind a large body of water. Photo by Ron Niebrugge/ wildnatureimages.com.

Chapters present research from across the globe.

Figure 17.11 Oxpeckers. On the African savannah oxpeckers, such as this red-billed oxpecker in Kruger National Park in South Africa, remove ticks from a variety of grazing mammals, such as this impala (*Aepyceros melampus*). Photo by Robert Harding Picture Library/SuperStock.

large source of food and the larger fish benefit by having fewer parasites.

A similar situation exists for large terrestrial animals in Africa. Two species of birds, the red-billed oxpecker (*Buphagus erythrorhynchus*) and the yellow-billed oxpecker (*B. africanus*), perch on the backs of grazing animals such as rhinos and antelopes (Figure 17.11). While perched on the backs of the grazing animals, the birds consume ticks that are attached to the mammals. However, because the birds also peck at the wounds caused by the ticks, scientists wondered whether the birds were mutualists or parasites. If oxpeckers act primarily as mutualists, their preferences for certain species of grazing mammals should be related to the number of ticks carried by each species. Alternatively, if the birds act primarily as parasites that seek to peck at mammal flesh, they should prefer mammals with thinner hides, which their beaks can penetrate more easily. In a 2011 study, researchers examined these relationships in both species of oxpeckers using up to 15 species of grazing mammals in Africa. The researchers quantified the oxpecker preferences for different species of mammals by observing grazing mammals of several species and dividing the number of oxpeckers on a given species by the total number of species grazing. Then they quantified the abundance of ticks on an individual of each mammal species. Although they found no relationship between the species preferences of oxpeckers and the thickness of the animals' hides, they found positive correlations between the species preferences of oxpeckers and tick abundance, as you can see in Figure 17.12. These results suggest that the birds are acting primarily as mutualists whose preferences are geared toward feeding on ticks.

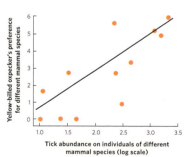

Figure 17.12 Oxpecker preferences for different species of mammals. The preferences of red-billed oxpeckers and yellow-billed oxpeckers are positively correlated to the abundance of ticks on the different mammals. Neither bird showed a preference for mammals with thin hides. These data suggest that oxpeckers seek out particular species of mammals primarily to consume ticks as a mutualist rather than to consume bits of mammal flesh as a parasite. Data from C. L. Nunn et al., Mutualism or parasitism? Using a phylogenetic approach to characterize the oxpecker-ungulate relationship, *Evolution* 65 (2011): 1297–1304.

Resources for Instructors and Students Create an Enhanced Teaching and Learning Experience

Test bank More than 1400 questions test student understanding and integration of the concepts at six different levels from knowledge and comprehension checks to application, analysis, synthesis, and evaluation.

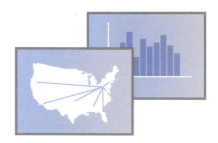

e-Book A complete online version of the textbook that allows the reader to highlight, bookmark, and add notes.

Activities Classroom activities use proven active learning techniques to engage students with the material and to encourage critical thinking.

Pre-lecture quizzes Quizzes for each concept heading help ensure that students have read the material before attending class.

Enhanced lectures for PowerPoint presentations Lecture slides using dynamic, layered art taken directly from the text

Optimized art Photos and illustrations from the text in both JPEG and PowerPoint formats.

LaunchPad An intuitive on-line homework, e-Book, and class-management system that provides curated resources compiled into ready-to-use teaching units, complete with problem sets, activities, e-book sections, and state-of-the-art online quizzing and testing.

Flash cards Key terms and definitions in interactive flashcards.

LearningCurve A set of assessment activities that uses a game-like interface to guide students through a series of questions tailored to individual understanding. Personalized study plans are generated based on quiz results.

Canadian Edition

Ecology
The Economy
of Nature

CHAPTER 1

Introduction: Ecology, Evolution, and the Scientific Method

Searching for Life at the Bottom of the Ocean

In the early 1800s, scientists hypothesized that deep ocean waters were devoid of life. Although unable to explore the deepest regions of the ocean at that time, they knew that sunlight could not penetrate depths greater than 275 m. Without sunlight there can be no photosynthesis, and without photosynthesis there can be no plants or algae to serve as food for other organisms. The cold temperatures and extreme pressures of deep ocean waters were also thought to contribute to the absence of deep-sea life. Given that ocean depths can exceed 10 000 m, it was reasonable to hypothesize that the deepest areas of the ocean could not support life.

As exploration continued throughout the nineteenth century, scientists began to collect organisms from ever greater ocean depths and their ideas about the limits of life began to change. In an 1873 expedition, scientists aboard the British research ship HMS *Challenger* dragged a large, open-sided, heavy box that was suspended from long ropes behind the ship across the floor of the Atlantic Ocean. This box—known as a dredge—sampled the sea floor in different parts of the ocean at depths of up to 4572 m. The

> ## "How could so much life exist at the bottom of the ocean?"

scientists were astonished to discover nearly 5000 previously unknown species. When it became clear that life flourished at depths beyond the penetration of light, scientists were forced to reject their earlier hypothesis that no life existed in the deep ocean waters.

After discovering this rich abundance of deep-sea life, scientists were faced with the need to understand how it could exist. The lack of light suggested that deep-sea organisms were somehow sustained by energy that did not come from photosynthesis on the ocean floor. Scientists had observed that the surface waters of the ocean produced a steady descent of tiny particles that were produced by the death and decomposition of organisms living in the surface waters of the ocean. These particles are known as "marine snow." In addition to marine snow, large organisms, such as whales, occasionally died and fell to the ocean floor. Scientists hypothesized that marine snow and the remains of large organisms such as whales provided the energy needed to sustain organisms in the depths of the ocean.

In the 1970s, scientists were finally able to send submersibles—small manned submarines—to take a first-hand look at the deepest ocean areas. Their discoveries were shocking. They not only confirmed that much of the ocean floor supported living organisms, but that areas near openings in the floor of the ocean, which later came to be known as *hydrothermal vents*, contained a great diversity of deep-sea species. Hydrothermal vents release plumes of hot water with high concentrations of sulphur compounds and other mineral nutrients. A tremendous diversity of species surrounded these hydrothermal vents, including tubeworms, clams, crabs, and fish. Indeed, the total amount of life at these depths rivaled that seen in some of the most diverse places on Earth. It became clear that the amount of energy contained

A deep-sea vent. In some regions of the ocean floor, hot water containing sulphur compounds is released from the ground. The sulphur compounds provide energy for chemosynthetic bacteria, which then serve as food for many other species that live near the vents, including these rust-coloured tube worms (*Tevnia jerichonana*) that have been stained orange by iron compounds emitted from the vents. Photo by Peter Batson/Image Quest Marine.

in the descending organic matter—the marine snow—was not sufficient to support such a diverse and abundant set of life forms. That earlier hypothesis now had to be rejected.

How could so much life exist at the bottom of the ocean? That this life existed near the hydrothermal vents suggested that the vents were somehow responsible. Scientists had known for a long time that some species of bacteria could obtain their energy from chemicals rather than from the Sun. The bacteria use the energy in chemical bonds, combined with carbon dioxide (CO_2), to produce organic compounds—a process known as *chemosynthesis*—similar to the way that plants and algae use the energy of the Sun and CO_2 to produce organic compounds through photosynthesis. Based on this knowledge, scientists hypothesized that the hot vents, which release water with dissolved hydrogen sulphide gas and other chemicals, provided a source of energy for bacteria and that these bacteria could be consumed by the other organisms living around the vents. After several years of investigations, scientists found that the immediate area around the hot vents contained a group of organisms known as tubeworms, which can grow to more than 2 m long. These animals have no digestive system, but possess specialized organs that house vast numbers of chemosynthetic bacteria that live in a symbiotic relationship with the tubeworms. The tubeworms capture the sulphide gases and CO_2 from the surrounding water and pass these compounds to the bacteria, which then use the sulphide gases and CO_2 to produce organic compounds. Some of these organic compounds are passed to the tubeworms, which use them as food. These bacteria also represent a food source for many of the other animals that live near the vents. In turn, these bacteria-consuming animals can be consumed by larger animals, such as fish.

The story of the deep-sea vents reveals how scientists work: They make observations, devise hypotheses, test the hypotheses to confirm or reject them, and, if a hypothesis is rejected, devise a new hypothesis. As you will see throughout this chapter and subsequent chapters, science is an ongoing process that often leads to fascinating discoveries about how nature works.

SOURCES: Dubilier, et al., Symbiotic diversity in marine animals: The art of harnessing chemosynthesis, *Nature Reviews Microbiology* 6 (2008): 725–740.

R. R. Dunn. *Every Living Thing* (Harper Collins, 2002).

CHAPTER CONCEPTS

- Ecological systems exist in a hierarchical organization.

- Ecological systems are governed by physical and biological principles.

- Different organisms play diverse roles in ecological systems.

- Scientists use several approaches to studying ecology.

- Humans influence ecological systems.

The story of deep-sea vents offers an excellent introduction to the science of ecology. **Ecology** is the scientific study of the abundance and distribution of organisms in relation to other organisms and environmental conditions. The word ecology is taken from the Greek oikos, meaning "house," and thus refers to our immediate surroundings, or environment.

Although Charles Darwin never used the word *ecology* in his writings, he appreciated the importance of beneficial and harmful interactions among species. In his book, *On the Origin of Species*, published in 1859, Darwin compared the large number of interactions among species in nature to the large number of interactions among consumers and businesses in human economic systems. He

Ecology The scientific study of the abundance and distribution of organisms in relation to other organisms and environmental conditions.

described species interactions as "the economy of nature."

In 1870, the German zoologist Ernst Haeckel gave the word a broader meaning:

> By ecology, we mean the body of knowledge concerning the economy of nature—the investigation of the total relations of the animal both to its organic and to its inorganic environment; including above all, its friendly and inimical relation with those animals and plants with which it comes directly or indirectly into contact—in a word, ecology is the study of all the complex interrelationships referred to by Darwin as the conditions of the struggle for existence.

The word *ecology* came into general use in the late 1800s. Since that time, the science of ecology has grown and diversified. Professional ecologists now number in the tens of thousands and have produced an immense body of knowledge about the world around us. Ecology is an active, modern science that continues to yield fascinating new insights about the environment and our impact on it. As we saw in the chapter opening story about life in the ocean depths, science is an ongoing process through which our understanding of nature constantly changes. Scientific investigation uses a variety of tools to understand how nature works. This understanding is never complete or absolute, but changes constantly as scientists make new discoveries. At the same time, rapid growth of the human population and the increasing sophistication of technological advances have caused major changes in our environment, frequently with dramatic consequences. With the knowledge that ecologists provide through their study of the natural world, we are in a better position to develop effective policies to manage environmental concerns related to land use, water, natural catastrophes, and public health.

This chapter will start you on the road to thinking like an ecologist. Throughout this book, we will consider the full range of **ecological systems**—biological entities that have both their own internal processes and yet interact with their external surroundings. Ecological systems exist at many different levels, ranging from an individual organism to the entire globe. Despite tremendous variations in size, all ecological systems obey the same principles with regard to their physical and chemical attributes and the regulation of their structure and function.

We begin this journey by examining the many different levels of organization for ecological systems, the physical and biological principles that govern ecological systems, and the different roles species play in ecological systems. Once we understand these basics of ecological systems, we will consider the many approaches to studying ecology and then consider the importance of understanding ecology when faced with the wide variety of ways that humans affect ecological systems.

Ecological systems exist in a hierarchy of organization

An ecological system may be an *individual,* a *population,* a *community,* an *ecosystem,* or the entire *biosphere.* As you can see in Figure 1.1, each ecological system is a subset of the next larger one, so the different types of ecological systems form a hierarchy. In this section we will examine the individual components of ecological systems, and how we study ecology at different levels in the ecological hierarchy.

INDIVIDUALS

An **individual** is a living being—the most fundamental unit of ecology. Although smaller units in biology exist—for example an organ, a cell, or a macromolecule—none of them has a separate life in the environment. Every individual has a membrane or other covering across which it exchanges energy and materials with its environment. This boundary separates the internal processes and structures of the ecological system from the external resources and conditions of the environment. In the course of its life, an individual transforms energy and processes materials. To accomplish this, it must acquire energy and nutrients from its surroundings and rid itself of unwanted waste products. This process alters the conditions of the environment and affects the resources available for other organisms. It also contributes to the movement of energy and chemical elements.

POPULATIONS AND SPECIES

Scientists assign organisms to particular *species.* Historically, the term **species** was defined as a group of organisms that naturally interbred with each other and produced fertile offspring. Over time, scientists have realized that this definition does not fit all species because no single definition can apply to all organisms. For example, some species of salamanders

---

Ecological systems Biological entities that have their own internal processes and interact with their external surroundings.

Individual A living being; the most fundamental unit of ecology.

Species Historically defined as a group of organisms that naturally interbreed with each other and produce fertile offspring. Current research demonstrates that no single definition can be applied to all organisms.

Figure 1.1 The hierarchy of organization in ecological systems. At each level of complexity, ecologists study different processes.

are all female and only produce daughters, which are clones of their mothers. In this case, individuals do not interbreed, but we consider these individuals to be of the same species because they are genetically very similar to each other. In addition, some organisms that we consider distinct species can interbreed. In cases where this interbreeding occurs, we cannot use reproductive isolation to draw the line between species.

Defining a species becomes even more complicated when we consider organisms such as bacteria, which are prokaryotes—that is to say, individuals consisting of a simple, single cell without a nucleus or any other membrane-bound organelles. Scientists increasingly appreciate that prokaryotic organisms can transfer bits of their DNA to other bacteria that are not closely related, a process known as *horizontal gene transfer.* This can happen in a number of ways: when a bacterium engulfs genetic material from the environment, when two bacteria come into contact and exchange genetic material, or when a virus transfers genetic material between two bacteria. Such cases make it difficult to group prokaryotic organisms into distinct species. Despite these difficulties, the term *species* has still proven useful to ecologists.

A **population** consists of individuals of the same species living in a particular area. For example, we might talk about a population of catfish living in a pond, a population of wolves living along the British Columbia coast, or a population of tubeworms living near a hydrothermal vent on the ocean floor. The boundaries that determine a population can be natural, for example where a continent meets the ocean. Alternatively, a population might be defined by another criterion such as a political boundary. For example, a scientist might want to study the coastal population of wolves in British Columbia, whereas biologists with the Canadian Wildlife Service might want to study the wolf population across all of Canada.

Populations have five distinct properties that are not exhibited by individuals: geographic range, abundance, density, change in size, and composition. The *geographic range* of a population—also known as its distribution—is the extent of land or water within which a population lives. For example, the geographic range of the North American grizzly bear (*Ursus*

Population The individuals of the same species living in a particular area.

arctos) includes western Canada, Alaska, Montana, and Wyoming. The *abundance* of a population refers to the total number of individuals. The *density* of a population refers to the number of individuals per unit of area. For instance, we might count the grizzly bears in an area and determine that there is 1 bear/100 km². The *change in size* of a population refers to increases and decreases in the number of individuals in an area. Finally, *composition* of a population describes the makeup of the population in terms of gender, age, or genetics. For example, we can ask what proportion of the grizzly population is male versus female, or juvenile versus adult.

COMMUNITIES

At the next level of the ecological hierarchy, we identify an ecological **community**, which is composed of all populations of species living together in a particular area. The populations in a community interact with each other in various ways. Some species eat other species while others, for example bees and the plants they pollinate, have cooperative relationships that benefit both parties. These types of interactions influence the number of individuals in each population. A community may cover large areas, such as a large forest, or may be enclosed within a very small area, such as the community of tiny organisms that live in the digestive systems of animals, or in the tiny amount of water found in tree holes. In practical terms, ecologists who study communities do not study every organism in the community. Instead, they generally study a subset of the organisms in the community, such as the trees, the insects, or the birds, as well as the interactions between certain groups of organisms.

The boundaries that define a community are not always rigid. For example, if you were to examine the species of plants and animals that live at the base of a mountain in Banff National Park in Alberta, you would find that most differ from the species of plants and animals that live at the top of this mountain. That is, the base and the peak appear to have distinctly different communities. However, if you were to walk up the mountain, you would notice that some species of trees, such as Douglas firs (*Pseudotsuga menziesii*), are abundant at the beginning of your hike and then gradually dwindle as you move higher. Other species, such as subalpine firs (*Abies lasiocarpa*), begin to take their place as the number of Douglas firs declines. In other words, the boundaries of the upper and lower forest communities are not distinct. Because of this, scientists must often decide on the boundaries of a community they want to study. For instance, an ecologist might decide to study the community of plants and animals in a section of boreal forest in northern Saskatchewan, or the community of aquatic organisms that lives along a designated stretch of coastline in Labrador. In these cases, no distinct boundary separates the studied community from the area that surrounds it.

ECOSYSTEMS

From communities we move on to *ecosystems*. An **ecosystem** is composed of one or more communities of living organisms interacting with their nonliving physical and chemical environments, which include water, air, temperature, sunlight, and nutrients. Ecosystems are complex ecological systems that can include many thousands of different species living under a great variety of conditions. For example, we may speak of the Great Lakes ecosystem or the boreal forest ecosystem.

At the ecosystem level, we typically focus on the movement of energy and matter between physical and biological components of the ecosystem. Most energy that flows through ecosystems originates with sunlight and eventually escapes Earth as radiated heat. In contrast, matter cycles within and between ecosystems. With the exception of places such as deep-sea vents where energy is acquired through chemosynthesis, the energy for most ecosystems comes from the Sun and is converted to organic compounds by photosynthetic plants and algae. These organisms can then be eaten by *herbivores*—animals that eat plants—which are in turn eaten by *carnivores*—animals that eat other animals. In addition, dead organisms and their waste products can be consumed by *detritivores,* which themselves can be consumed by other animals. In all of these cases, each step results in some of the energy originally assimilated from sunlight being converted into growth or reproduction of consumers; the remainder of the energy is lost to the surroundings as heat, and is eventually radiated back into space.

In contrast to the movement of energy, the movement of matter largely cycles within an ecosystem. When considering matter in an ecosystem, we often look at the most common elements that organisms use, such as carbon, oxygen, hydrogen, nitrogen, and phosphorus. These elements comprise a major portion of the most important compounds for living organisms and include water, carbohydrates, proteins, and DNA. These elements can be held in many different places, or *pools,* on Earth, including in living organisms, and in the atmosphere, water, and rocks. The movement of these elements among these

---

Community All populations of species living together in a particular area.

Ecosystem One or more communities of living organisms interacting with their nonliving physical and chemical environments.

Figure 1.2 **The biosphere.** The biosphere consists of all ecosystems on Earth, which are linked together by movements of air, water, and organisms.

pools is known as the *flow* of matter. For instance, many nutrients that are in the soil are taken up by plants, and these plants are consumed by animals. The nutrients exist in an animal's tissues, and many leave as excreted waste. When the animal dies, the nutrients in its tissues are returned to the soil, thereby completing the cycling of nutrients.

The boundaries of ecosystems, like those of populations and communities, are often not distinct. Scientists generally distinguish ecosystems by their relative isolation with respect to flows of energy and materials, but, in reality, few ecosystems are completely isolated. Even aquatic and terrestrial ecosystems exchange materials and energy by runoff from the land and the harvesting of aquatic organisms by terrestrial consumers, as when bears capture salmon on their upstream spawning runs.

THE BIOSPHERE

At the highest level of the ecological hierarchy is the **biosphere**, which includes all of the ecosystems on Earth. As Figure 1.2 illustrates, distant ecosystems are linked together by exchanges of energy and nutrients carried by currents of wind and water and by the movements of organisms, such as migrating whales, birds, and fish. Such movement connects terrestrial, freshwater, and marine ecosystems by carrying soil, nutrients, and organisms.

The biosphere is the ultimate ecological system. All transformations of the biosphere are internal,

with two exceptions: the energy that enters from the Sun, and the energy that is lost to space. The biosphere holds practically all materials that it has ever had, and retains whatever waste materials we generate.

STUDYING ECOLOGY AT DIFFERENT LEVELS OF ORGANIZATION

Each level in the hierarchy of ecological systems is distinguished by unique structures and processes. As a result, ecologists have developed different approaches for exploring these levels and for answering the questions that arise. The five approaches to studying ecology match the different levels of hierarchy: the *individual approach*, the *population approach*, the *community approach*, the *ecosystem approach*, and the *biosphere approach*.

The **individual approach** to ecology emphasizes the way in which an individual's morphology (the size and shape of its body), physiology, and behaviour enable it to survive in its environment. This approach also seeks to understand why an organism lives in some environments but not in others. For example, an ecologist studying plants at the organism level might

Biosphere All of the ecosystems on Earth.

Individual approach An approach to ecology that emphasizes the way in which an individual's morphology, physiology, and behaviour enable it to survive in its environment.

ask why trees are dominant in warm, moist environments, while shrubs with small, tough leaves are dominant in environments with cool, wet winters and hot, dry summers.

Ecologists who use the individual approach are often interested in **adaptations**—the characteristics of an organism that make it well-suited to its environment. For example, desert animals have enhanced kidney function, which helps them to conserve water. The cryptic colouration of many animals helps them avoid detection by predators. Flowers are shaped and scented to attract certain kinds of pollinators. Adaptations are the result of evolutionary change through the process of *natural selection*, which we will consider later in this chapter.

The **population approach** to ecology examines variation over time and space in the number of individuals, the density of individuals, and the composition of individuals, which includes the sex ratio, the distribution of individuals among different age classes, and the genetic makeup of a population. Changes in the number or density of individuals can reflect the balance of births and deaths within a population, as well as immigration and emigration of individuals from a local population. This can be influenced by a number of factors, including interactions with other species and the physical conditions of the environment, such as temperature or the availability of water. In the process of evolution, genetic mutations may alter birth and death rates, genetically distinct types of individuals may become common within a population, and the overall genetic makeup of the population may change. Because other species might serve as food, pathogens, or predators, interactions among species can also influence the births and deaths of individuals within a population.

The **community approach** to ecology is concerned with understanding the diversity and relative abundances of different kinds of organisms living together in the same place. The community approach focuses on interactions between populations, which can either promote or limit the coexistence of species (Figure 1.3). For example, in studying the Serengeti Plains of Africa, an ecologist taking the community approach might ask how the presence of zebras, which consume grasses, might affect the abundance of other species, such as gazelles, that also consume the grasses.

The **ecosystem approach** to ecology describes the storage and transfer of energy and matter, including the various chemical elements essential to life, such as oxygen, carbon, nitrogen, phosphorus, and sulphur. These movements of energy and matter occur through the activities of organisms and

Figure 1.3 The community approach to ecology. Ecologists using the community approach study interactions among plants and animals that live together. For example, on an African plain, ecologists might ask how cheetahs (*Acinonyx jubatus*) affect the abundance of gazelles and how the gazelles affect the abundance of the plants that they consume. Photo by Michel & Christine Denis-Huot/Photo Researchers, Inc.

through the physical and chemical transformations that occur in the soil, atmosphere, and water.

The **biosphere approach** to ecology is concerned with the largest scale in the hierarchy of ecological systems. This approach tackles the movements of air and water—and the energy and chemical elements they contain—over Earth's surface. Ocean currents and winds carry the heat and moisture that define the climates at each location on Earth, which in turn govern the distributions of organisms, the dynamics of populations, the composition of communities, and the productivity of ecosystems.

We have described these five approaches as distinct. However, most ecologists use multiple approaches to study the natural world. A scientist who wants to understand how an ecosystem will respond to a drought, for example, will likely want to know how individual plants and animals respond to a lack of

Adaptation A characteristic of an organism that makes it well-suited to its environment.

Population approach An approach to ecology that emphasizes variation over time and space in the number of individuals, the density of individuals, and the composition of individuals.

Community approach An approach to ecology that emphasizes the diversity and relative abundances of different kinds of organisms living together in the same place.

Ecosystem approach An approach to ecology that emphasizes the storage and transfer of energy and matter, including the various chemical elements essential to life.

Biosphere approach An approach to ecology concerned with the largest scale in the hierarchy of ecological systems, including movements of air and water—and the energy and chemical elements they contain—over Earth's surface.

water, how these individual responses affect the populations of plants and animals, how a change in the populations might affect interactions among species, and how a change in species interactions might affect a change in the flow of energy and matter.

Ecological systems are governed by physical and biological principles

Although ecological systems are complex, they are governed by a few basic principles. Life builds on the physical properties and chemical reactions of matter. The diffusion of oxygen across body surfaces, the rates of chemical reactions, the resistance of vessels to the flow of fluids, and the transmission of nerve impulses all obey the laws of thermodynamics. Within these constraints, life can pursue many options and has done so with astounding innovation. In this section we briefly review the three major biological principles that you may recall from your introductory biology course: *conservation of matter and energy, dynamic steady states,* and *evolution.*

CONSERVATION OF MATTER AND ENERGY

The **law of conservation of matter** states that matter cannot be created or destroyed, but can only change form. For example, as you drive a car, gasoline is burned in the engine; the amount of fuel in the tank declines, but you have not destroyed matter. The molecules that comprise gasoline are converted into new forms including carbon monoxide (CO), carbon dioxide (CO_2), and water (H_2O).

Another important biological principle, the **first law of thermodynamics**—also known as the **law of conservation of energy**—states that energy cannot be created or destroyed. Like matter, energy can only be converted into different forms. Living organisms must constantly obtain energy to grow, maintain their bodies, and replace energy lost as heat.

The law of conservation of matter and the first law of thermodynamics imply that ecologists can track the movement of matter and energy as it is converted into new forms through organisms, populations, communities, ecosystems, and the biosphere. At every level of organization, we should be able to determine how much matter enters the system and account for its movement. For example, consider a field full of cattle (*Bos taurus*) eating grass. At the organismal level, we can determine how much energy an individual animal consumes and then calculate the proportion of this energy that is converted into growth of its body, the maintenance of its physiology, and waste. At the population level, we can calculate how much energy the entire herd of cattle consumes by eating grass. At the community

level, we can evaluate how much energy each species of grass creates via photosynthesis and how much of this energy is passed on to cattle and other plant-eating species, such as rabbits that might coexist with the cattle. At the ecosystem level, we can estimate how elements such as carbon flow from the grasses to herbivores (cattle and rabbits) and then on to predators. We can then track how dead grass, the waste products of herbivores and predators, and the dead bodies of herbivores and predators decompose and return to the soil.

DYNAMIC STEADY STATES

Although matter and energy cannot be created or destroyed, ecological systems continuously exchange matter and energy with their surroundings. When gains and losses are in balance, ecological systems are unchanged and the system is said to be in a **dynamic steady state**. For example, birds and mammals continuously lose heat in a cold environment. However, this loss is balanced by heat gained from the metabolism of foods, so the animal's body temperature remains constant. Similarly, the proteins of our bodies are continuously broken down and replaced by newly synthesized proteins, so our appearance remains relatively unchanged.

The principle of the dynamic steady state applies to all levels of ecological organization, as illustrated in Figure 1.4. For individual organisms, assimilated food and energy must balance energy expenditure and metabolic breakdown of tissues. A population increases with births and immigration, and it decreases with death and emigration. At the community level, the number of species living in a community decreases when a species becomes extinct, and increases when a new species colonizes the area. Ecosystems and the biosphere receive energy from the Sun, and this gain of energy is balanced by heat energy radiated by Earth back out into space. One of the most important questions ecologists ask is how the steady states of ecological systems are maintained and regulated. We will return to this question frequently throughout this book.

An understanding of dynamic steady states helps provide insights regarding the inputs and outputs of

---

Law of conservation of matter Matter cannot be created or destroyed; it can only change form.

First law of thermodynamics Energy cannot be created or destroyed; it can only change form. *Also known as* Law of conservation of energy.

Dynamic steady state When the gains and losses of ecological systems are in balance.

Inputs	Level	Outputs
Food	Individual	Energy expended; Waste
Births; Immigration	Population	Deaths; Emigration
Colonization by new species	Community	Extinction of species
Solar energy and matter entering ecosystem	Ecosystem	Energy and matter leaving ecosystem
Solar energy entering system	Biosphere	Energy re-radiated to space

Figure 1.4 Dynamic steady states. At all levels of organization, the inputs to the systems must equal the outputs.

ecological systems. Of course, ecological systems also change. Organisms grow, populations vary in abundance, and abandoned fields revert to forest. Yet all ecological systems have mechanisms that tend to maintain a dynamic steady state.

EVOLUTION

Although matter and energy cannot be created or destroyed, what living systems do with matter and energy is as variable as all the forms of organisms that have ever existed on Earth. To understand the variation among organisms—the diversity of life—we turn to the concept of *evolution*.

An attribute of an organism, such as its behaviour, morphology, or physiology, is the organism's **phenotype**. A phenotype is determined by the interaction of the organism's **genotype**, or the set of genes it carries, with the environment in which it lives. For example, your height is a phenotype that is determined by your genes and the nutrition you received in the environment where you were raised.

Over the history of life on Earth, the phenotypes of organisms have changed and diversified dramatically. This is the process of **evolution**, which is a change in the genetic composition of a population over time. Evolution can happen through a number of different processes that we will discuss in detail in later chapters. Perhaps the best known process is evolution by **natural selection**, which is a change in the frequency of genes in a population through differential survival and reproduction of individuals that possess certain phenotypes. As outlined by Charles Darwin in his book *On the Origin of Species*, evolution by natural selection depends on three conditions:

1. Individual organisms vary in their traits.

2. Parental traits are inherited by their offspring.

3. The variation in traits causes some individuals to experience higher **fitness**, which we define as the survival and reproduction of an individual.

When these three conditions exist, an individual with higher survival and reproductive success will pass more copies of its genes to the next generation. Over time, the genetic composition of a population changes as the most successful phenotypes come to predominate. As a result, the population becomes better suited to the surrounding environmental conditions. Phenotypes that are well suited to their environment and, in turn, confer higher fitness are known as adaptations. Consider the example in

Phenotype An attribute of an organism, such as its behaviour, morphology, or physiology.

Genotype The set of genes an organism carries.

Evolution Change in the genetic composition of a population over time.

Natural selection Change in the frequency of genes in a population through differential survival and reproduction of individuals that possess certain phenotypes.

Fitness The survival and reproduction of an individual.

Figure 1.5 in which some individuals in a population of caterpillars are coloured in such a way that they blend in with their surroundings and escape the notice of predators while other individuals are not. If colour is inherited, over time the population will consist of a progressively larger proportion of caterpillars that blend in with their environment.

Species do not evolve in isolation. Evolution in one species opens up new possibilities for evolutionary change, both in the evolving species and in other species with which the evolving species interacts. For example, caterpillars of the monarch butterfly (*Danaus plexippus*) have evolved the ability to eat the leaves of milkweed plants, which are toxic to caterpillars of other species. Monarch caterpillars not only tolerate these toxic compounds, but also sequester these compounds in their bodies and use the toxins to defend themselves, as larvae as well as

adults, against bird predators. In addition, both larvae and adults have evolved conspicuous "warning" colouration to advertise their toxicity. After the caterpillar species evolved these defensive abilities, predatory birds evolved a new ability to discriminate between caterpillars and butterflies that were toxic and those that were edible.

The complexity of ecological communities and ecosystems builds on, and is fostered by, existing complexity. Ecologists seek to understand how these complex ecological systems came into being and how they function in their environmental settings.

Different organisms play diverse roles in ecological systems

Transformations of matter and energy in ecological systems are performed by both large and small forms of life. These different organisms can play unique roles within ecological systems. In this section, we will examine how organisms interact with each other and the environment. We will see how ecologists categorize species based on how they obtain their energy and how they interact with other species, describe the major groups of organisms that have evolved and the diverse ecological roles found within each group, and examine the concepts of *habitat* and *niche*.

BROAD EVOLUTIONARY PATTERNS

Early in the history of Earth, ecosystems were dominated by bacteria. The ancient evolution of these groups is still debated by scientists, but a leading hypothesis, illustrated in Figure 1.6, is that the bacteria are the oldest group of organisms. Over time, the bacteria gave rise to the archaea. Bacteria and archaea are both *prokaryotic* organisms, which are single-cell organisms that do not contain distinct cell organelles, such as a nucleus. These evolutionary events likely happened in the ocean and may have occurred near the deep-sea vents that we discussed in the chapter opener. If so, it may also be the case that the first bacteria used chemosynthesis and this later gave rise to the evolution of photosynthesis.

Over time, *eukaryotic* organisms, which have distinct cellular organelles, evolved. The key event in the evolution of eukaryotes occurred when one bacterium engulfed another. The engulfed bacterium became what we now know as the *mitochondrion*, an important organelle for cellular respiration in eukaryotic organisms. This ancestor subsequently gave rise to all modern organisms that contain mitochondria including red algae, green algae, plants, fungi, and animals. As the evolution of the eukaryotes progressed, there was a second key event. A eukaryotic cell engulfed a bacterium that was capable

(a)

(b)

(c)

Figure 1.5 Evolution by natural selection. In this example, the caterpillar population is initially quite variable in colour (**a**). Individuals that better match the twig are less obvious to the bird hunting for food and therefore more likely to survive. If colour is genetically inherited, the next generation (**b**) of the caterpillar population will be better matched to resemble twigs. As this natural selection continues over many generations, the colour of the caterpillar population will closely match the twigs (**c**). At this stage, the colour of the caterpillar represents an adaptation against predation.

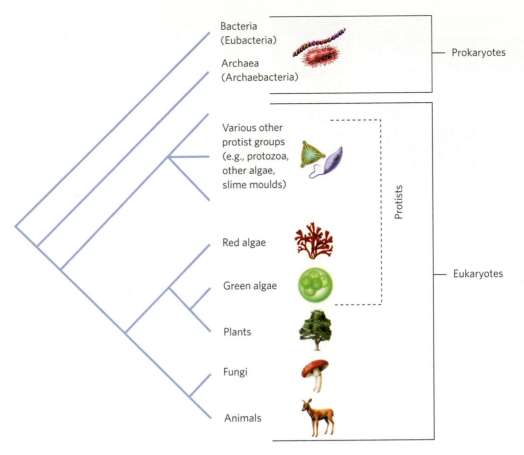

Figure 1.6 The evolution of life on Earth. Bacteria are the oldest forms of life on Earth. The engulfing of one bacterial species by another led to the rise of eukaryotes containing cell organelles such as mitochondria and chloroplasts.

of photosynthesis and the engulfed bacterium evolved into what we now know as the chloroplast. The group of eukaryotic organisms that contained chloroplasts subsequently gave rise to modern-day red algae, green algae, and plants. Those species that didn't contain chloroplasts gave rise to modern-day fungi and animals.

Bacteria also modified the biosphere, making it possible for other forms of life to exist. For example, photosynthetic bacteria that were present more than 3 billion years ago produced oxygen as a byproduct of photosynthesis. The higher concentrations of oxygen in the atmosphere and oceans favoured the evolution of additional life forms that required oxygen, such as plants and animals. Despite all these changes, bacteria have persisted to the present day. As we shall see, their unique biochemical capabilities allow them to consume resources that their more complex descendants cannot use, and to tolerate ecological conditions that are beyond the capacities of other organisms.

Ecosystems depend on the activities of many forms of life. Each major group fills a unique and necessary role in the biosphere. We will now briefly review the major groups of organisms including bacteria, protists, plants, fungi, and animals.

Bacteria

Although bacteria are tiny, their enormous range of metabolic capabilities enables them to accomplish many unique biochemical transformations and to occupy parts of the biosphere where plants, animals, fungi, and most protists cannot survive. Some bacteria can assimilate molecular nitrogen (N_2) from the atmosphere, which they use to synthesize proteins and nucleic acids. Other species of bacteria, such as those living in deep-sea hydrothermal vents, can use inorganic compounds such as hydrogen sulphide (H_2S) as sources of energy in chemosynthesis. Furthermore, many bacteria can live under anaerobic conditions—in which free oxygen is lacking—such as in mucky soils and sediments, where their metabolic activities release nutrients that can be taken up by plants. Finally, some bacteria, including cyanobacteria (colloquially known as blue-green algae), can conduct photosynthesis. Cyanobacteria account for a large fraction of the photosynthesis that occurs in aquatic ecosystems. When bodies of water contain high

Figure 1.7 Cyanobacteria. Also known as blue-green algae, cyanobacteria are bacteria capable of conducting photosynthesis. These organisms can grow rapidly under highly fertile conditions, producing large floating mats in the water that can be toxic to animals, as illustrated in this photo of Simon Lake, Naughton, Ontario. Photos by Don Johnston-www.agefotostock.com (top) and M I (Spike) Walker/Alamy (inset).

amounts of nutrients, cyanobacteria can form dense populations that turn the water green, an event known as an **algal bloom** (Figure 1.7). Later in the book we will have much more to say about the special place of bacteria in ecosystems.

Protists

Protists are a highly diverse group of mostly single-celled eukaryotic organisms that includes the algae, slime moulds, and protozoans. This bewildering variety of protists fills almost every ecological role. For instance, algae are the primary photosynthetic organisms in most aquatic ecosystems. Some algae can form large plant-like structures, such as the seaweed known as kelps, which can grow up to 100 m in length. Because of this large size, regions of the ocean containing large amounts of kelps, such as that shown in Figure 1.8, are called kelp forests. Although

kelps may resemble large plants, the actual organization of kelp tissues is less structurally complex than that of trees and other plants.

Other protists are not photosynthetic. Foraminifera and radiolarians are protists that feed on tiny particles of organic matter or absorb small dissolved organic molecules. Some of the ciliate protozoans are effective predators of other microorganisms. Many protists live in the guts or other tissues of a host organism where they might be helpful or cause harm. For example, termites are a type of insect that consumes large amounts of cellulose. Cellulose is a very difficult substance for animals to digest, but the termite has a community of protists—as well as bacteria—in its gut that are very effective at breaking down the cellulose. Some of the best known harmful protists include *Plasmodium,* which causes malaria in humans, and *Trypanosoma brucei,* a protist that produces sleeping sickness.

Plants

Plants are well known for their role in using the energy of sunlight to synthesize organic molecules from CO_2 and water. On land, most plants have structures with large exposed surfaces—their leaves—to capture the energy of sunlight (Figure 1.9a). Leaves are thin because surface area is more important than leaf thickness for capturing light energy. To obtain carbon, terrestrial plants take up gaseous CO_2 from the atmosphere. At the same time, they lose large amounts of water by evaporation from their leaf tissues to the atmosphere. Thus, plants need a steady supply of water to replace the water lost during photosynthesis. Most plants are firmly rooted in the

Figure 1.8 Kelp forests. Although most protists are very small organisms, some protists such as seaweeds can grow very large and look like large plants. This kelp forest is off the coast of southern California. Photo by Mark Conlin/Image Quest Marine.

Algal bloom A rapid increase in the growth of algae in aquatic habitats, typically due to an influx of nutrients.

Figure 1.9 Plants. Plants can play many roles in an ecosystem. **(a)** Most plants, such as this garlic mustard (*Alliaria petiolata*), are rooted in the soil and make organic compounds by photosynthesis. **(b)** Epiphytes, such as this Blood spattered beard lichen (*Usnea wirthii*) on Vancouver Island, also conduct photosynthesis but grow above the ground and are attached to other plants. **(c)** Carnivorous plants, such as this Purple pitcher plant in Newfoundland, can both photosynthesize and obtain nutrients by trapping and digesting invertebrates. **(d)** Some plants, such as dodder, act as parasites by taking nutrients from other plants. Photos by (a) Zoonar/Lothar Hinz/age fotostock; (b) Matthew Maran/naturepl.com; (c) Michelle Gilders/Alamy; and (d) A Jagel/age fotostock.

ground and in constant contact with water in the soil. Others, including orchids and a variety of tropical "air plants" (known as *epiphytes*), typically grow by attaching themselves to other plants—often the trunks of trees—and can photosynthesize only in humid environments that are bathed in clouds (Figure 1.9b).

Although we typically envision plants as organisms that obtain their energy from sunlight, plants can also obtain energy in other ways. For example, several groups of plants have evolved to be simultaneously photosynthetic and carnivorous. These plants include the Venus fly trap (*Dionaea muscipula*), several species of sundews, and several species of pitcher plants, such as the northern pitcher plant (*Sarracenia purpurea*) (Figure 1.9c). They often live in locations that are low in nutrients, so the invertebrates they trap and consume provide an additional source of nutrients and energy. In addition, more than 400 species of plants—including more than 200 species of orchids—lack chlorophyll and therefore cannot photosynthesize to obtain energy. Scientists once thought that these plants were acting as

decomposers and obtained their organic carbon from dead organic matter, but we now know that many of these plants are in fact acting as parasites on fungi, which are the real decomposers in the ecosystem. These parasitic plants obtain the vast majority of their organic carbon from the fungi. Other plants, such as dodder, have little chlorophyll and no roots (Figure 1.9d). Instead, the stringy plant—also known as strangle weed or devil's guts—winds around other plants, penetrates their tissues, and sucks up water, nutrients, and products of photosynthesis. The dodder is a serious pest for many farm crops.

Fungi

Fungi assume unique roles in the biosphere because of their distinctive growth form. With the exception of unicellular yeasts and their relatives, fungi—like plants and animals—are multicellular. Most fungal organisms consist of threadlike structures called *hyphae* that are only a single cell in diameter. These hyphae may either form a loose network, which can invade plant or animal tissues or dead leaves and

Figure 1.10 Fungi. The mushrooms produced by this sulphur tuft fungus *(Hypholoma fasciculare)* in Belgium are fruiting bodies of much larger, unseen masses of threadlike hyphae that penetrate decaying wood and leaf litter. Fungi are effective decomposers. Photos by Philippe Clement/naturepl.com (top) and Steve Gschmeissner/Photo Researchers, Inc. (inset).

wood on the soil surface, or grow together into reproductive structures such as mushrooms (Figure 1.10). Because fungal hyphae are able to penetrate deeply into tissue, they readily decompose dead plant material, eventually making nutrients available to other organisms. Fungi digest their foods externally and secrete acids and enzymes into their immediate surroundings. Such digestion allows fungi to decompose dead organisms and dissolve nutrients from soil minerals.

Although most fungi function as decomposers, they can interact with other species in both positive and negative roles. Many fungi live in mutualistic relationships with plants, living either around or within the roots of plants. Using their extensive network of hyphae, they obtain scarce nutrients from the surrounding soil and provide them to the plant.

In exchange, the plant provides the products of photosynthesis. Other fungi, though, can act as pathogens. Several closely related species of fungus cause Dutch elm disease, which has caused widespread death in several species of elm trees throughout North America and Europe during the past 100 years. Fungal pathogens are also a major problem for many crops including potatoes, wheat, and rice.

Animals

Animals play a wide range of roles as consumers in ecological systems. Some animals, for example, elephants, gazelles, and voles, eat plants. Other animals, such as mountain lions, rattlesnakes, and frogs, eat other animals. Ticks, lice, and tapeworms are animals that live on or in other organisms. Finally, animals such as flies, bees, butterflies,

Producers (Autotrophs)		Mixed Nutrition (Mixotrophs)	Consumers (Heterotrophs)	
Chemosynthetic archaea and bacteria	Cyanobacteria	Many algae	Fungi	Bacteria
Most algae	Most plants	Other protists Some plants Some animals	Herbivore	Carnivore

Figure 1.11 Categories of species based on their energy sources. Species that obtain their energy from photosynthesis or chemosynthesis are known as producers or autotrophs. Species that obtain their energy from consuming other species are heterotrophs. Species that can take a mixed strategy of being producers and heterotrophs are mixotrophs.

Parasite
(winter tick–*Demacentor albipictus*)

Predator
(mountain lion–*Puma concolor*)

Herbivore
(bison–*Bison bison*)

Parasitoid
(braconid wasp–*Cotesia congregatus*)

Figure 1.12 **The four types of consumers.** Predators can be broken down into predators such as mountain lions, parasitoids such as braconid wasps shown here on a tomato hornworm, parasites such as winter ticks, and herbivores such as bison.

moths, and bats can serve as important plant pollinators and seed dispersers.

CATEGORIZING SPECIES BASED ON SOURCES OF ENERGY

Ecologists often categorize organisms according to how they obtain energy, as illustrated in Figure 1.11. Organisms that use photosynthesis to convert solar energy into organic compounds or use chemosynthesis to convert chemical energy into organic compounds are known as **producers** or **autotrophs.** Organisms that obtain their energy from other organisms are known as **consumers** or **heterotrophs.** There are many different kinds of heterotrophs. Some consume plants, some consume animals, and some consume dead organic matter. In the next section, we will discuss these various interactions in more detail.

Not all species fit neatly into categories of autotrophs or heterotrophs. Some species can obtain their sources of carbon through a variety of ways. Because these species take a mixed approach to obtaining their energy, they are called **mixotrophs.** Mixotrophs are quite common in nature. For example, some bacteria can switch back and forth between photosynthesis and chemosynthesis. In addition,

many species of algae can photosynthesize and obtain organic carbon by engulfing bacteria, protists, and bits of organic carbon that exist in the water. Other mixotrophs include the carnivorous plants that we discussed earlier in this chapter. These plants obtain their energy both from photosynthesis and from consuming invertebrates.

TYPES OF SPECIES INTERACTIONS

In considering the diversity of species that exist on Earth, we are often interested in the roles that they play. Ecologists categorize species by the types of interactions they have with other species, as you can see from the examples in Figure 1.12. Below is a brief introduction to these interactions, beginning with the

Producer An organism that uses photosynthesis to convert solar energy into organic compounds or uses chemosynthesis to convert chemical energy into organic compounds. *Also known as* **Autotroph.**

Consumer An organism that obtains its energy from other organisms. *Also known as* **Heterotroph.**

Mixotroph An organism that obtains its energy from more than one source.

Figure 1.13 Mutualism. A lichen is a symbiotic association of a fungus and algal cells. Photo by DUNCAN MCEWAN/ naturepl.com.

Algal cells

Fungal hyphae

various types of consumers, each of which will be covered in much greater detail in later chapters.

Predation

Predators are organisms that kill and partially or entirely consume another individual. The mountain lion, for example, is a predator that kills white-tailed deer (*Odocoileus virginianus*) and many other species of prey.

Parasitoids represent a special kind of predator. Parasitoids lay their eggs on or inside other animals, particularly insects, and the eggs hatch into larvae that consume the host individual from the inside, eventually killing it. Most parasitoid species are wasps and flies.

Parasitism

Parasites are organisms that live in or on another organism, called the *host*. An individual parasite rarely kills its host, although some hosts die when they are infected by a large number of parasites. Common parasites include tapeworms and ticks. When a parasite causes a disease, it is called a **pathogen.** Pathogens include several species of bacteria, viruses, protists, fungi, and a group of worms called *helminths*.

Herbivory

Herbivores are organisms that consume producers, such as plants and algae. When an herbivore consumes a plant, it typically consumes only a small portion of the plant, and does not kill the plant. For example, caterpillars consume a few leaves or parts of leaves, which the plant can regenerate. Cattle consume the tops of grass leaves, but do not destroy the growing region, located at the base of the plant.

Competition

Competition can be defined as an interaction with negative effects between two species that depend on the same limiting resource to survive, grow, and reproduce. For example, two species of grass might compete for nitrogen in the soil. As a result, the survival, growth, and reproduction of each is reduced when living with the other species of grass in the same area compared to when they are living in the area alone. Similarly, coyotes (*Canis latrans*) and wolves (*Canis lupus*) might compete for the same prey animals in the forest, such that they survive, grow, and reproduce better when they are living alone than when the other species is present. Competition for limited resources is a very common interaction in nature.

Mutualism

When two species interact in a way such that each species receives benefits from the other, their interaction is a **mutualism.** The lichens in Figure 1.13, for example, are composed of a fungus living together with green algal cells or cyanobacteria as a single

---

Predator An organism that kills and partially or entirely consumes another individual.

Parasitoid An organism that lives within and consumes the tissues of a living host, eventually killing the host.

Parasite An organism that lives in or on another organism, but rarely kills it.

Pathogen A parasite that causes disease in its host.

Herbivore An organism that consumes producers such as plants and algae.

Competition An interaction with negative effects between two species that depend on the same limiting resource to survive, grow, and reproduce.

Mutualism An interaction between two species in which each species receives benefits from the other.

organism. The fungus provides nutrients to the algae and the algae provide carbohydrates from photosynthesis to the fungus. Other examples of mutualisms include the bacteria that help digest plant material in the guts of cattle, fungi that help plants extract mineral nutrients from the soil in return for carbohydrate energy from the plant, and honeybees that pollinate flowers as they obtain nectar.

Commensalism

Commensalisms are interactions in which two species live in close association and one species receives a benefit while the other experiences neither a benefit nor a cost. For example, plants such as burdock (*Arctium lappa*) produce fruits containing tiny barbs that stick to the hair of mammals that brush up against it. The burdock receives a benefit of having its seeds dispersed while the mammal is neither helped nor harmed by carrying these fruits on its body.

Because organisms are specialized for particular ways of life, many different types of organisms are able to live together in close association. A close physical relationship between two different types of organisms is referred to as a **symbiotic relationship**. Many parasites, parasitoids, mutualists, and commensal organisms live in symbiotic relationships.

When considering the different types of interactions among species, it can be helpful to categorize interactions between the two participants as positive (+), negative (−), or neutral (0). Table 1.1 provides a summary of species interactions using this approach.

Consumers of Dead Organic Matter

Consumers of dead organic matter—including *scavengers, detritivores,* and *decomposers*—also play important roles in nature. **Scavengers,** such as vultures, consume dead animals. **Detritivores,** such as dung beetles and many species of millipedes, break down dead organic matter and waste products—known as *detritus*—into smaller particles. **Decomposers,** such as many species of mushrooms, break down dead organic material into simpler elements and compounds that can be recycled through the ecosystem.

HABITAT VERSUS NICHE

If you were to walk through a meadow in the southern regions of Ontario, Quebec, or Manitoba you would likely come across the eastern cottontail rabbit (*Sylvilagus floridanus*). This species thrives in abandoned farm fields that are full of grasses and other tall wildflowers interspersed with shrubs. These plants provide food for the rabbit and protection from its many predators, including coyotes and several species of hawk. In considering species in nature, ecologists find it useful to distinguish between where an organism lives and what it does.

The **habitat** of an organism is the place, or physical setting, in which it lives. In the case of the rabbit, the habitat consists of old fields that contain grasses, wildflowers, and shrubs. Habitats are distinguished by physical features, often including the predominant form of plant or animal life. Thus, we speak of forest habitats, desert habitats, stream habitats, and lake habitats (Figure 1.14).

During the early years of ecological research, scientists developed a complex system of classifying habitats. For example, they began by distinguishing between terrestrial and aquatic habitats. Among aquatic habitats, they identified freshwater and marine habitats. Among the marine habitats, they found coastal habitats, the open ocean, and the ocean floor. As such classifications became more finely divided, the distinctions began to break down since scientists found that habitat types overlap and that absolute distinctions rarely exist. However, the idea

TABLE 1.1
THE OUTCOME OF INTERACTIONS BETWEEN TWO SPECIES

TYPE OF INTERACTION	SPECIES 1	SPECIES 2
PREDATION/PARASITOIDISM	+	−
PARASITISM	+	−
HERBIVORY	+	−
COMPETITION	−	−
MUTUALISM	+	+
COMMENSALISM	+	0

Interactions that provide a benefit to a species are indicated by a "+" symbol, interactions that cause harm to a species are indicated by a "−" symbol, and interactions that have no effect on a species are indicated by a "0" symbol.

Commensalism An interaction in which two species live in close association and one species receives a benefit while the other experiences neither a benefit nor a cost.

Symbiotic relationship When two different types of organisms live in a close physical relationship.

Scavenger An organism that consumes dead animals.

Detritivore An organism that feeds on dead organic matter and waste products that are collectively known as detritus.

Decomposer Organisms that break down dead organic material into simpler elements and compounds that can be recycled through the ecosystem.

Habitat The place, or physical setting, in which an organism lives.

(a) Algonquin Park, Ontario

(c) Parc National des Volcans, Rwanda

(b) Oxbow bend in Snake River, Grand Teton National Park, Wyoming

(d) Amboseli National Park, Kenya

Figure 1.14 Habitats. Terrestrial habitats are distinguished by their dominant vegetation whereas aquatic habitats are distinguished by their depth and the presence or absence of flowing water. **(a)** Lakes are typically large bodies of water that have very little flow. **(b)** Freshwater streams contain flowing water. **(c)** In the tropical rain forest, warm temperatures and abundant rainfall support the highest productivity and biodiversity on Earth. **(d)** Tropical grasslands, which develop where rainfall is sparse, nonetheless support vast herds of grazing herbivores during the productive rainy season. Photos by (a) R. Linke-www.agefotostock.com; (b) George Sanker/naturepl.com; (c) Josef Friedhuber/Getty Images; and (d) Staffan Widstrand/naturepl.com.

of habitat is nonetheless useful because it emphasizes the variety of conditions to which organisms are exposed. For example, inhabitants of extreme ocean depths and tropical rain forest canopies experience altogether different conditions of light, pressure, temperature, oxygen concentration, moisture, and salt concentrations, as well as differences in food resources and predators.

An organism's **niche** includes the range of abiotic and biotic conditions it can tolerate. In the case of the cottontail rabbit, its niche includes the ranges of temperature and humidity that it can tolerate, the plants that it eats, and the predatory hawks and coyotes with which it coexists. An important ecological principle states that each species has a distinct niche. No two species have exactly the same niche because each has distinctive attributes of form and function that determine the conditions it can tolerate, how it feeds, and how it escapes enemies. Consider the hundreds of insect species that might live in a garden; each has a unique niche in terms of the food that it eats (Figure 1.15). For example, the

---

Niche The range of abiotic and biotic conditions that an organism can tolerate.

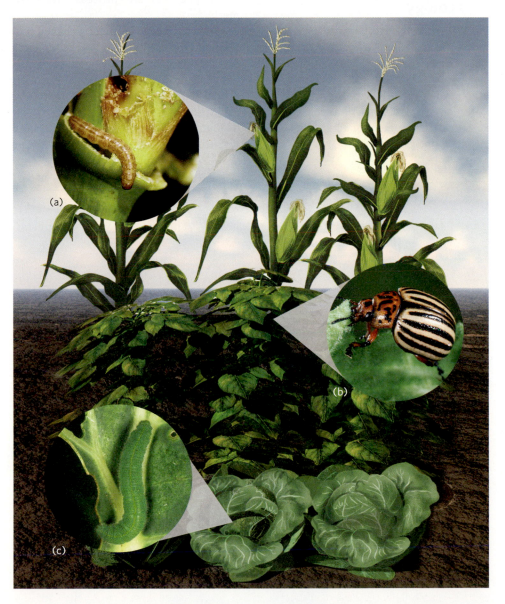

Figure 1.15 Niche. Even within a group of similar organisms, such as the insects, each species has a distinct niche. In the case of insects, the food they consume is only one aspect of their niche. **(a)** The European corn borer is specialized to feed on corn plants. **(b)** The Colorado potato beetle is specialized to feed on the leaves of potato plants. **(c)** The caterpillar of the cabbage white butterfly is specialized to feed on the leaves of cabbage, broccoli, and cauliflower. Photos by (a) Scott Sinklier/age fotostock; (b) blickwinkel/Alamy; and (c) Nigel Cattlin/Alamy.

caterpillar of the cabbage white butterfly (*Pieris rapae*) feeds on the group of plants that have been cultivated from the wild mustard plant (*Brassica oleracea*) including cabbage, broccoli, and cauliflower. However, the Colorado potato beetle (*Leptinotarsa decemlineata*) feeds almost exclusively on the leaves of potato plants (*Solanum tuberosum*). Similarly the European corn borer (*Ostrinia nubilalis*) feeds primarily on corn plants (*Zea mays*). The variety of habitats and niches holds the key to much of the diversity of living organisms.

Scientists use several approaches to studying ecology

Scientists have investigated the diverse roles that organisms play in the environment for more than a century. Ecologists investigate their subject matter using a systematic procedure, often referred to as the scientific method. The three steps of this process, shown in Figure 1.16, are (1) observations regarding a pattern in nature, (2) development of a hypothesis, and (3) testing the hypothesis.

OBSERVATIONS, HYPOTHESES, AND PREDICTIONS

Most research begins with a set of observations about nature that invite explanation. Usually these observations identify and describe a consistent pattern. As we learned in the history of research on the deep-sea vents, once it was discovered that the diversity and abundance of species living around the vents could not be sustained by the relatively small amount of organic matter drifting down from the sunlit surface, new hypotheses regarding chemosynthetic bacteria had to be developed and tested. In such cases, some hypotheses will be supported, while others will be rejected and require new hypotheses. This process is the scientific method.

To help you better understand the scientific method, imagine that you are walking around a pond on a warm, spring night following a rainstorm. You would likely hear male frogs making frog calls. If you returned to the same pond on cooler nights following a dry period, you would be less likely to hear frogs calling. If you travelled to many different ponds, you would observe this pattern over and over again. That

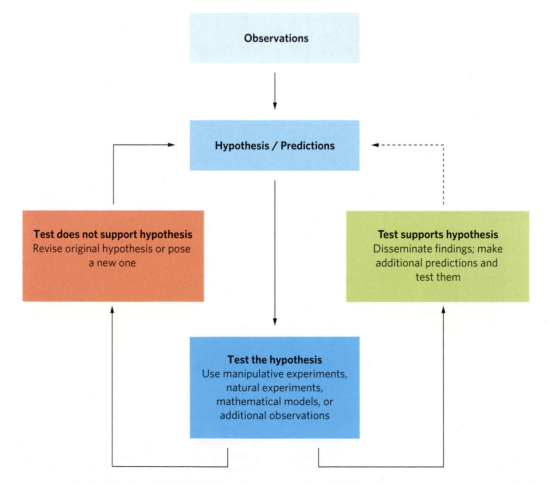

Figure 1.16 The scientific method. The scientific method begins with observing patterns in nature and developing a hypothesis that explains how or why the pattern exists. Predictions from a hypothesis are tested with manipulative experiments, natural experiments, mathematical models, or additional observations.

is, you would observe and describe a consistent pattern in nature. Repeated natural patterns lead scientists to hypothesize about the causes of these patterns.

Hypotheses are ideas that potentially explain a repeated observation. In the case of the frogs, we consistently observed that they called only on warm nights after a rainstorm. Having established the existence of this pattern, we want to understand it better. We might want to explain *how* frogs sense changes in temperature and rainfall, and how sensing these environmental changes stimulates frogs to call. We also might want to explain *why* frogs call on warm nights after it rains. How do the frogs benefit from calling—perhaps by attracting mates—and what, if any, are the costs of calling?

Hypotheses about how and why organisms respond to the environment represent different types of explanations. The "how" explanation addresses the details of the animal's sensory perception and changes in its hormone concentrations, nervous system, and muscular system. In the case of the frogs, we might hypothesize that the frog's nervous system detects warm temperatures and rain. This initiates changes in a male frog's hormones and physiology that cause him to contract muscles that make him call.

Hypotheses that address the immediate changes in an organism's hormones, physiology, nervous system, or muscular system are known as **proximate hypotheses**. If these hypotheses are correct, then we can make *predictions*. **Predictions** are statements that arise logically from hypotheses. For example, if our hypothesis about how a rainy night causes male frogs to call is correct, then we can predict that any frogs exposed to warm rain will respond by changing the concentration of specific hormones that stimulate the brain to send a signal to the muscles of the vocal apparatus to contract.

Ultimate hypotheses address why an organism responds in a certain way to its environment—that is, the fitness costs and benefits of a particular response. For example, we might hypothesize that male frogs call to attract females. Furthermore, if we suspect that male frogs call to attract females, then perhaps males sing on warm nights after a rainstorm because such nights produce the best conditions for laying eggs, which is when females are most interested in mating. The males benefit from calling on a warm, wet night because they will be more likely to attract females and therefore father more offspring. If the males call at other times, they will attract fewer female mates and receive a much smaller benefit. But what about costs? We might hypothesize that when male frogs call to attract females, they risk attracting the attention of predators. The increased risk of death represents a high fitness cost to male frogs.

We have now generated a number of predictions that logically follow from our ultimate hypotheses about male frog calling: (1) males that call will attract females; (2) females actively search for males only on warm, wet nights; (3) if singing imposes a cost, then males should save their singing for times when it will provide the greatest benefit.

TESTING HYPOTHESES WITH MANIPULATIVE EXPERIMENTS

A particular hypothesis can rarely be confirmed beyond a doubt. However, our confidence increases as we continue testing a hypothesis and repeatedly find that our observations support it. Although the methods of acquiring scientific knowledge appear to be straightforward, many pitfalls exist. For example, an observed relationship between two factors does not necessarily mean that one factor causes the other to change. The cause must be determined independently. To accomplish this, we can design **manipulative experiments** in which a hypothesis is tested by altering a factor that is hypothesized to be an underlying cause of the phenomenon.

To understand the process of a manipulative experiment, consider the observation that herbivorous insects often consume less than 10 percent of a plant's tissues. Ecologists have proposed several hypotheses to explain this. One is that predators consume insect herbivores at such a high rate that insect populations remain low. This low insect population cannot eat very much of the plant tissue. This seems like a reasonable hypothesis, but how do we test it with a manipulative experiment?

Researchers working on this question decided to explore whether the predation hypothesis applies to insects feeding on oak trees in the U.S. state of Missouri. They observed that birds consume many insects on oak leaves and then hypothesized that birds reduce the populations of insect herbivores. If this hypothesis were correct, when birds are absent, insect populations would increase and consume more leaf biomass. Confirmation of this prediction would support their hypothesis; a lack of confirmation would lead them to reject the hypothesis and propose a new one.

Hypothesis An idea that potentially explains a repeated observation.

Proximate hypothesis A hypothesis that addresses the immediate changes in an organism's hormones, physiology, nervous system, or muscular system.

Prediction A logical consequence of a hypothesis.

Ultimate hypothesis A hypothesis that addresses why an organism responds in a certain way to its environment in terms of the fitness costs and benefits of the response.

Manipulative experiment A process by which a hypothesis is tested by altering a factor that is hypothesized to be an underlying cause of the phenomenon.

To test the hypothesis that predation by birds lowers the abundance of insects on oak trees, the researchers decided to conduct a manipulative experiment in which they used cages that excluded birds from the trees (Figure 1.17a). The **manipulation**, also known as the **treatment**, is the factor that we want to vary in an experiment. Often one of the manipulations used is a *control*. A **control** is a manipulation that includes all aspects of an experiment except the factor of interest. In the oak tree experiment, the caged trees served as the treatment whereas the uncaged trees served as the control.

Once we decide on which manipulations we wish to do, we have to assign each manipulation to a specific **experimental unit**. An experimental unit is the object to which we apply the manipulation. In the case of the oak tree experiment, the researchers decided that they would use groups of three white oak saplings as their experimental units. After making this decision, each experimental unit was either caged—by surrounding the group of three saplings with bird-proof netting—or left uncaged to allow bird access.

A manipulation to a single experimental unit might provide exciting results, but the results may not be reliable unless the experiment is repeated and demonstrates a similar outcome. Being able to produce a similar outcome multiple times is known as **replication**, which is an integral feature of most experimental studies. In the oak tree study, the investigators decided to add cages to 10 groups of trees and leave 10 groups of trees uncaged. In doing so, they replicated their experiment 10 times.

When we assign different manipulations to our experimental units, the assignments must be made using **randomization**, which means that every experimental unit has an equal chance of being assigned a particular manipulation. In the oak tree experiment, the researchers randomly assigned groups of trees to be caged or left uncaged as controls. In this way, they could be assured that the caged trees were initially no different from the controls.

Once the researchers set up the experiment, they collected data on the number of herbivorous insects and the percentage of leaf tissue that had been consumed. They found that caged trees had about twice as many insect herbivores as the control trees. Moreover, the percentage of leaf tissue consumed at the end of the growing season was nearly twice as high on caged trees as on the control trees (Figure 1.17b). These findings led the researchers to conclude that the experiment supported their hypothesis.

Although many experiments are conducted in natural settings such as oak forests or lakes, other experiments are conducted at smaller scales (Figure 1.18). Many experiments make use of

(a)

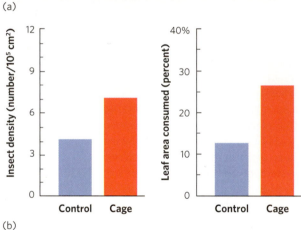

(b)

Figure 1.17 A manipulative experiment. Manipulative experiments provide the strongest tests of hypotheses. **(a)** In a study that tested if birds are an important factor in determining the number of insects on oak trees in Missouri, ecologists placed cages around some white oak saplings to exclude birds, and left other oak saplings uncaged to serve as a control. **(b)** From this experiment, the researchers measured the number of insect herbivores per leaf and the amount of leaf tissue that was consumed in each of the two treatments. After R. J. Marquis and C. J. Whelan, Insectivorous birds increase growth of white oak through consumption of leaf-chewing insects, *Ecology* 75 (1994): 2007–2014. Photo by Chris Whelan, University of Illinois.

Manipulation The factor that we want to vary in an experiment. *Also known as* **Treatment**.

Control A manipulation that includes all aspects of an experiment except the factor of interest.

Experimental unit The object to which we apply an experimental manipulation.

Replication Being able to produce a similar outcome multiple times.

Randomization An aspect of experiment design in which every experimental unit has an equal chance of being assigned to a particular manipulation.

(a)

(b)

(c)

Figure 1.18 Experimental venues. The choice of experimental venue often represents a trade-off between the complexity of natural conditions and the more highly controlled conditions of a laboratory experiment. **(a)** Manipulative experiments of entire lakes, such as the Experimental Lakes Area in western Ontario, include natural conditions, but such experiments are difficult to replicate. **(b)** Microcosm experiments, such as this experiment at the University of Pittsburgh's Pymatuning Lab of Ecology, include many features of a lake by containing communities of aquatic organisms. Using microcosms allows the manipulations to be replicated many times. **(c)** Laboratory experiments, such as this pesticide experiment conducted in Petri dishes, allow researchers to conduct highly controlled experiments, but they are conducted under very unnatural conditions. Photos (a) Fisheries and Oceans Canada; (b) and (c) courtesy of Rick Relyea.

microcosms, which are simplified ecological systems that attempt to replicate the essential features of an ecological system in a laboratory or field setting. In the case of experiments that study aquatic systems, for example, microcosms might consist of large outdoor tanks of water. These tanks would include many of the features of natural water bodies, including soil, vegetation, and a diversity of organisms (Figure 1.18b). The use of microcosms assumes that a response to manipulations in a microcosm is representative of responses that would occur in a natural habitat. For example, we might wish to understand how species of fish compete for food. Observing competition among the fish species in a murky lake might not be feasible, but a large tank of water that included many features of the lake might work well, providing that the fish behave similarly under both conditions. If so, the results of the microcosm experiment may yield results that can be generalized to the larger, natural system.

ALTERNATIVE APPROACHES TO MANIPULATIVE EXPERIMENTS

Many hypotheses cannot be tested by experiments, either because the amount of area or length of time needed to test the hypotheses is simply too large, or

because it is not possible to isolate particular variables and devise a suitable control. These limitations are common when we are trying to understand patterns that have occurred over long periods, or in systems such as entire populations or ecosystems that are too large to be manipulated.

Several different hypotheses might explain a particular observation equally well, so investigators must make predictions that distinguish among the alternatives. For example, many ecologists have observed a decrease in the number of species as one moves north or south, away from the equator. This repeated pattern has many potential explanations. As one travels north from the equator, average temperature and precipitation decrease, sunlight decreases, and seasonality increases. Each of these factors alone or together could affect the number of species that can coexist in a specific locality. Indeed, dozens of hypotheses have been proposed to explain the observed decrease in the number of species as one moves away from the equator. Isolating the effect of

---

Microcosm A simplified ecological system that attempts to replicate the essential features of an ecological system in a laboratory or field setting.

Why Do We Calculate Means and Variances?

As we saw in the oak tree experiment, when testing hypotheses ecologists make **observations**, which can be thought of as information, including measurements that are collected from organisms or the environment. These observations, also known as **data**, are then used to test hypotheses. In the case of the oak trees, the researchers collected data on the density of herbivorous insects and the amount of leaf tissue they consumed.

In asking questions in ecology, we often want to know the average value, or *mean,* of the data collected from different treatments or obtained under different conditions. In the case of the oak tree experiment, the researchers wanted to compare the mean density of insects in caged trees versus uncaged trees to determine whether birds depressed the numbers of insects consuming the tree leaves.

While a comparison of the different means tells us about the central tendencies of the data, ecologists often want to know if the data used to generate the mean have high or low variability. For example, if the mean density of insects on leaves was 10 insects per square metre of leaf surface in both of the following sets of data, which group is more variable?

Group A: 10, 9, 11, 10, 8, 12, 9, 11, 8, 12

Group B: 10, 5, 15, 10, 6, 14, 5, 15, 7, 13

Although both groups of data have the same mean, observations in Group A range from 8 to 12 whereas in Group B they range from 5 to 15. Therefore, the data in Group B are more variable.

Why do we care about the variability of the data that we collect? Given that each mean is calculated from a set of data that has either a narrow or a wide range, the variability gives us an idea of how much the distributions of data overlap with each other. If two groups of data have different means but the distributions of data overlap a great deal, then we cannot be confident that the two groups are actually different from each other. In contrast, if two groups of data have different means but the distributions of data do not overlap, then we might be very confident that the two groups are different.

One way to measure how widely the data points are spread around the mean is to calculate the *variance of the mean.* The **variance of the mean** is a measurement that indicates the spread of data around the mean of a population when every member of the population has been measured. Data points that are more widely spread around the mean will have a higher variance.

The easiest way to calculate the variance in a set of data (denoted as σ^2) is to do it in two steps:

1. Square each observed value (denoted as χ) and calculate the mean of these squared values (where E indicates that we are taking the mean of several values):

$$E[\chi^2]$$

2. From this mean, subtract the square of the mean observed value:

$$\sigma^2 = E[\chi^2] - [E(\chi)]^2$$

Observations Information, including measurements, that is collected from organisms or the environment. *Also known as* Data.

Variance of the mean A measurement that indicates the spread of the data around the mean of a population when every member of the population has been measured.

each factor has proved difficult because all of the other factors for which we might construct a hypothesis change together.

Ecologists have several alternative approaches that address these difficulties. One option, the **natural experiment**, relies on natural variation in the environment to test a hypothesis. For example, consider the hypothesis that the number of species on an island is influenced by the size of the island because larger islands have more available niches, support larger populations that resist extinction, and are easier for organisms to find and colonize. A manipulative experiment to test this hypothesis would be impossible since it would require both a massive manipulation of many islands as well as the ability to observe a difference in the number of colonizing species over hundreds or even thousands of years. Instead, we can "test" the hypothesis by comparing the number of species living on islands of different sizes that have been created over shorter periods by changes in sea or lake levels. Although a manipulative experiment is not possible in such cases, a natural experiment like this still allows researchers to determine if patterns in nature are consistent or inconsistent with hypotheses about the underlying causes.

Natural experiment An approach to hypothesis testing that relies on natural variation in the environment.

In words, $E[\chi^2]$ is the mean of the squared observed values and $[E(\chi)]^2$ is the square of the mean observed value.

When we calculate the variance of the mean, the calculation is based on the assumption that we have measured every member of a population. In reality, we often cannot measure every member, but instead measure a sample of the population. In the oak tree study, for example, the researchers did not measure the insects on all of the oak trees, but instead used a sample of 10 groups of oak trees. When we measure a sample of the population, the variation in the data is called the **sample variance**. The sample variance is very similar to the variance of the mean, except that we now account for how many samples of the population we measured (denoted as n). The sample variance (denoted as s^2) is calculated as:

$$s^2 = \frac{n}{n-1} \times \sigma^2$$

or

$$s^2 = \frac{n}{n-1} \times (E[\chi^2] - [E(\chi)]^2)$$

As you might notice from this equation, as the number of samples becomes very large, the value of the sample variance approaches the value of the variance of the mean for the entire population.

To help you understand how to calculate the sample variance, consider the following set of observations on the abundance of insects per leaf on caged and uncaged trees:

CAGED TREES	UNCAGED TREES
8	4
6	3
7	2
9	4
5	2

For the caged trees, we can calculate the mean of the values as:

$$(8 + 6 + 7 + 9 + 5) \div 5 = 7$$

and the mean of the squared values as:

$$(8^2 + 6^2 + 7^2 + 9^2 + 5^2) \div 5 = 51$$

We can then calculate the sample variance for the caged-tree data as:

$$s^2 = \frac{n}{n-1} \times (E[\chi^2] - [E(\chi)]^2)$$

$$s^2 = \frac{5}{5-1} \times (51 - (7)^2)$$

$$s^2 = (1.25) \times (51 - 49) = 2.5$$

YOUR TURN Using the data from the five replicates of uncaged trees, calculate the mean and sample variance of insect abundance.

Sample variance A measurement that indicates the spread of the data around the mean of a population when only a sample of the population has been measured.

Ecologists also use *mathematical models* to explore the behaviour of ecological systems. In a **mathematical model**, an investigator designs a representation of a system with a set of equations that corresponds to the hypothesized relationships among each of the system's components. For example, we might use a mathematical model to represent how births and immigration add to the growth rate of a population, and how deaths and emigration subtract from that growth rate. In this sense, a mathematical model is a hypothesis; it provides an explanation of the observed structure and functioning of the system.

We can test the accuracy of a mathematical model by comparing the predictions it yields with observations in nature. For example, epidemiologists have developed models to describe the spread of communicable diseases. These models include such factors as the proportions of a population that are susceptible, exposed, infected, and recovered from infections. The models also include the rates of transmission and the probability that the organism will cause a disease in an infected organism. By including all of these factors, such models can make predictions about the frequency and severity of disease outbreaks.

Mathematical model A representation of a system with a set of equations that correspond to hypothesized relationships among the system's components.

These predictions can then be tested by comparing them with real-world observations of disease outbreaks. This approach is being used for a number of important wildlife diseases, including the transmission of rabies in such animals as bats, raccoons, skunks, and foxes, and the transmission of Lyme disease in wildlife and human populations.

Mathematical models can be used on any scale. For example, at a larger scale, ecologists have created mathematical models to investigate how burning fossil fuels affects the CO_2 content of the atmosphere. To manage human impacts on our environment, it is critically important to understand this relationship. Models of global carbon content include, among other factors, equations that describe the uptake of CO_2 by plants and the dissolution of CO_2 in the oceans. The earlier versions of these models failed to match observations and overestimated the annual increase in atmospheric CO_2 concentrations. The real world evidently contains CO_2 "sinks" such as regenerating forests that remove CO_2 from the atmosphere. By including the effects of these CO_2 sinks, the refined carbon models more accurately describe observed atmospheric data and are more likely to predict future changes accurately.

For any model, we can support or reject the hypothesis by comparing the model's predictions against our observations. Rejected models can be further refined to incorporate additional complexities and better fit our observations.

Humans influence ecological systems

For more than a century, ecologists have laboured passionately to understand how nature works, from the level of the organism to the level of the biosphere. The wonders of the natural world summon our curiosity about life and our environment. For many ecologists, a curiosity about how nature works is reason enough to study ecology. Increasingly, however, ecologists find themselves struggling to understand how the rapidly growing human population, now more than 7 billion people, is affecting the planet. Our need to understand nature is becoming more and more urgent as the growing human population stresses the functioning of ecological systems. Environments dominated by humans or created by them—including urban and suburban living places, agricultural fields, tree farms, and recreational areas—are also ecological systems. The welfare of humanity depends on maintaining the proper functioning of these systems.

The human population currently consumes massive amounts of energy and resources, and produces large amounts of waste. As a result, virtually the entire planet is strongly influenced by human activities, as shown in Figure 1.19. These influences include the degradation of the natural environment and the disruption of many important functions that natural environments provide to humans. Growing human consumption of natural resources has caused a number of ecological problems. For example, removing plants from their natural environment to use as house plants and exploiting animals for human consumption and the pet trade have caused the decline of many species in their native habitats. The species affected are diverse, including the cacti of the American Southwest that are collected for sale as house plants, several species of reptiles and amphibians that are sold in the pet trade, and many species of fish that are overexploited by commercial fishing. As commerce has become more global, species have been introduced unintentionally to new locations at an increasing rate. Some of these species such as rats, snakes, and pathogens can have devastating effects on local species. To feed 7 billion people, we have converted a large amount of land for agricultural use. This conversion has brought with it a number of challenges including loss of natural habitats, pollution from fertilizers and pesticides, and questions about growing genetically modified crops. Some crops, such as corn, are now increasingly being used as sources of fuel, also known as *biofuels,* causing even more land to be converted to agricultural use. Humans also need land for housing, business, and industry. This has further reduced the amount of natural habitat available for other species, and has been a major contributor to the decline and extinction of many species. We will deal with these issues in greater detail throughout the book.

Another suite of ecological challenges has been caused by wastes produced by human activity. For example, untreated sewage and industrial processes can damage the air, water, and soil. In addition, the use of nuclear power plants to generate electricity produces substantial amounts of nuclear waste. Of all human wastes, perhaps none has a higher public awareness than the *greenhouse gases* responsible for global warming. **Greenhouse gases** are compounds in the atmosphere that absorb the infrared heat energy emitted by Earth and then emit some of the energy back toward Earth. In doing so, the gases prevent much of this energy radiated from the surface of Earth from escaping into space. Greenhouse gases include many different compounds, but an important player is CO_2, which is produced by burning fossil fuels in the cars we drive and in the coal-powered electricity-generating plants that provide electricity to so many of our homes and businesses. As the human population continues to grow and demands for energy increase,

Greenhouse gases Compounds in the atmosphere that absorb the infrared heat energy emitted by Earth and then emit some of the energy back toward Earth.

Figure 1.19 Human impacts on ecological systems. The growth of human population, particularly over the past two centuries, has altered much of the planet. Humans have destroyed habitats, converted land to agriculture, created air and water pollution, burned large amounts of fossil fuel, and overharvested plants and animals.

we burn more fossil fuels and produce more greenhouse gases. The more greenhouse gases we put into the atmosphere, the warmer our Earth becomes.

Because ecological systems are inherently complex, it is difficult to predict and manage the effects of a growing human population on ecological systems at every level. At the level of the organism, we might want to know how a pesticide sprayed in the environment could affect each of the many tissues and organ systems of an animal's body, leading to changes in behaviour, growth, and reproduction. At the community level, we might ask how a decrease in abundance of one species caused by commercial harvesting could affect the populations of other species in that community. At the biosphere level, we would like to quantify the large number of sources that emit CO_2 into the atmosphere and understand the processes that take CO_2 out of the atmosphere. Each of these cases presents a set of complex questions that are not easy to answer. Yet we need a

solid understanding of how the ecological system operates before we can predict the outcome of anthropogenic impacts on the system and recommend ways to minimize damages to the system.

THE ROLE OF ECOLOGISTS

The plight of individual species headed toward extinction arouses us emotionally. However, ecologists increasingly realize that the only effective means of preserving the species of the world is through the conservation of ecosystems and the management of large-scale ecological processes. Individual species, including those that humans rely on for food and other products, are themselves dependent on the maintenance of environmental support systems. Local effects of human activities on ecological systems can often be managed once we understand the underlying mechanisms responsible for change. Increasingly, however, our activities have led to multiple, widespread effects that are more difficult for scientists to characterize and for legislative and regulatory bodies to control. For this reason, a sound scientific comprehension of environmental problems is a necessary prerequisite to action.

The media is filled with reports of environmental problems: disappearing tropical forests, depleted fish stocks, emerging diseases, global warming, and wars that cause environmental tragedies and human suffering. But it is important to know that there are success stories as well. Many countries have made great strides in cleaning up their rivers, lakes, and air. Fish are once again migrating up major rivers in North America and Europe to spawn. Acid rain has decreased, thanks to changes in the combustion of fossil fuels. The release of chlorofluorocarbons, which damage the ozone layer that shields the surface of Earth from ultraviolet radiation, has decreased dramatically. The inevitability of global warming caused by increasing atmospheric CO_2 concentrations has provoked global concern and set off an international research effort. Conservation programs, including breeding endangered species in captivity, have saved some animals and plants from certain extinction. They have also heightened public awareness of environmental issues, and sometimes sparked public controversy.

These successes would not have been possible without a general consensus founded on evidence produced by scientific study of the natural world. Understanding ecology will not by itself solve our environmental problems, because these problems also have political, economic, and social dimensions. However, as we contemplate the need for global management of natural systems, our effectiveness in this enterprise critically depends on our understanding of their structure and functioning—an understanding that depends on knowing the principles of ecology.

This book introduces you to the study of ecology by building an understanding of all aspects of the discipline. We begin by looking at the individual level, including how species have adapted to the challenges of the aquatic and terrestrial environments. We will then explore the topic of evolution, including how species have evolved various strategies for mating, reproducing, and living in social groups. Next we move to the population level with a discussion of population distributions, population growth, and population dynamics over space and time. With a firm understanding of populations, we move on to examine species interactions, communities, and ecosystems. Finally, we consider ecology at the global level and investigate global patterns of biodiversity and global conservation.

ECOLOGY TODAY CONNECTING THE CONCEPTS

THE CALIFORNIA SEA OTTER

The California sea otter. This once abundant marine mammal has experienced large fluctuations in numbers as a result of human activities during the past three centuries. Photo by Neil A. Fisher.

At the end of each chapter, we want to reflect on the topics covered and explore how these topics are interconnected. In this first chapter, we have examined a wide range of topics including the hierarchy of perspectives in ecology, the biological and physical principles that govern natural systems, the variety of roles that different species play, the multiple approaches to studying ecology, and the influence of humans on ecological systems. To help you see how these topics interconnect, let's examine a case study of the sea otter (*Enhydra lutris*) off the Pacific coast. Humans have affected sea otter populations for hundreds of years. Several scientific approaches have been taken to understand these impacts and to help reverse them.

The sea otter was once abundant, with a geographic range that extended around the northern Pacific Rim, from Japan up to Alaska and down to Baja California. While First Nations people along the western coast of Canada have hunted sea otters for centuries, it was not until the second half of the eighteenth century that sea otter populations were depleted in an unsustainable way. By the time otters became protected in 1911, less than 2000 were left from an earlier population of approximately 300 000 individuals. Between 1969 and 1972, 89 otters were reintroduced off the coast of British Columbia and this reintroduction was very successful. There are now more than 2500 otters in British Columbia waters and the population continues to grow. These changes in the size of otter populations presented an opportunity for scientists to examine a natural experiment in action.

Ecologists quickly realized that to understand the causes and consequences of the sea otter's fluctuations in abundance, they needed to use a range of ecological approaches, from the individual to the ecosystem. Taking an individual approach, ecologists established that the sea otter was a predator on a wide range of prey species including abalone, spiny lobsters, small fish, crabs, sea urchins, and small snails. Among these prey items, observations of otter feeding behaviour revealed that otters prefer certain prey such as abalone, a large species of sea snail. They will only eat other small species of snails when their preferred prey becomes rare.

Once scientists understood the sea otter's niche, they were better able to protect it. However, not everyone was happy about the resurgence of sea otters through the 1990s. For instance, California anglers became upset; they argued that the growing otter population would cause a dramatic change in the marine community, including a drastic reduction in the populations of commercially valuable fish, abalone, and spiny lobsters—all harvested for human consumption. However, scientists who took a community approach to ecology found that an increasing otter population was also having other dramatic effects on the marine community. For example, the otter's consumption of sea urchins—marine invertebrates that eat kelps—was causing an

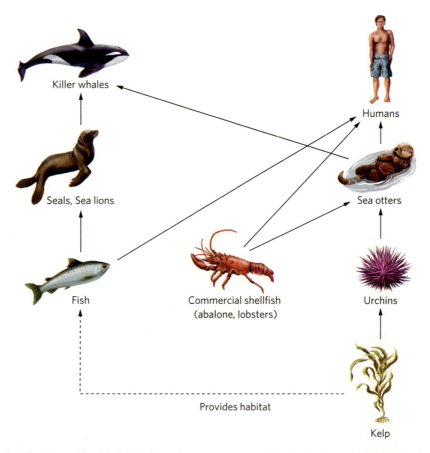

Sea otters and the species with which they interact. Once scientists determined the major species in the ocean that affected the abundance of otter populations, they could better protect the otter from extinction.

increase in kelps (Figure 1.8). Kelps can be harvested for making fertilizer, food, and pharmaceuticals. Thus, the growing otter population caused sea urchins to decrease, kelps to increase, and the commercial harvesting of kelps to increase. It turns out that the increase in kelps also provided young fish with a refuge from predators and a place to feed. Thus, the sea otter plays a key role in determining the community composition of coastal marine ecosystems.

In the 1990s, the sea otter population mysteriously began to decline. To understand these declines, scientists have used individual, community, and ecosystem approaches. In 1998, researchers showed that populations of otters in the vicinity of the Aleutian Islands, Alaska, had declined precipitously during the 1990s. The reason was that killer whales, or orcas (*Orcinus orca*), which previously had not preyed on otters, had begun to come close to shore where they consumed large numbers of otters. Why did killer whales adopt this new behaviour? The researchers pointed out that populations of the principal prey of killer whales—seals and sea lions—collapsed during the same period, perhaps inducing the whales to hunt the otters as an alternative food source. Why did the seals and sea lions decline? One can only speculate at this point, but intense human fisheries have reduced fish stocks exploited by the seals to levels low enough to seriously threaten seal populations.

There also were declines in otter populations along the California coast. Initially, declines in sea otters were attributed to the use of gill nets along the coast to exploit a new fishery that inadvertently killed otters in substantial numbers. Subsequent legislation moved the fishery farther offshore to help protect the otters. In this same region, the otters also were dying from infections by two protist parasites, *Toxoplasma gondii* and *Sarcocystis neurona*. These parasites cause a lethal inflammation of the brain. In 2010, for example, 40 dead and dying sea otters were found near Morro Bay, California, and 94 percent were infected with *S. neurona*. This was a surprising observation because the only known hosts of these parasites are opossums (*Didelphis virginiana*) and several species of cats. Given that these mammals live on land, how did sea otters become infected?

Scientists hypothesized links between the terrestrial and marine ecosystems that are allowing the parasites to infect sea otters. To date, two potential links have been suggested. First, cats that spend time outside defecate on land and their feces contain the parasites; when it rains, the parasites get washed into local streams and rivers, and eventually make their way to the ocean. Second, when humans flush cat feces and Kitty Litter down the toilet and into the sewer system, the waste water eventually enters the ocean. Although manipulative experiments found that the protists do not infect marine invertebrates and cause illness, the invertebrates can take the parasites into their bodies inadvertently while feeding. When invertebrates infested with parasites are consumed by otters, the otters get infected. New research indicates that abalone do not carry the parasites whereas small marine snails do. Thus, when otters have an abundance of their preferred food, such as abalone, they have a low risk of being infected by the deadly parasite. When abalone is scarce, however, the otters are forced to feed on small snails that carry the parasite, which dramatically increases the risk of infection and death.

The story of the sea otter highlights the importance of understanding ecology from multiple approaches using both manipulative and natural experiments. It also underscores the multiple roles that species can play in communities and ecosystems and how humans can dramatically influence the outcome. This understanding can then be used to take action to reverse harmful impacts on the environment. In the case of the sea otter, education campaigns now encourage the public to keep their cats inside more and to put used cat litter into the trash rather than flushing it down the toilet.

SOURCES: Johnson, C. K., et al. 2009. Prey choice and habitat use drive sea otter pathogen exposure in a resource-limited coastal system. *Proceedings of the National Academy of Sciences* 106: 2242–2247.

Miller, M. A. 2010. A protozoal-associated epizootic impacting marine wildlife: Mass mortality of southern sea otters (*Enhydra lutris nereis*) due to *Sarcocystis neurona* infection. *Veterinary Parasitology* 172: 183–194.

SUMMARY OF CHAPTER CONCEPTS

■ **Ecological systems exist in a hierarchical organization.** The hierarchy begins with individual organisms and moves up through higher levels of complexity including populations, communities, ecosystems, and the biosphere. At each of these levels, ecologists study different types of processes.

■ **Ecological systems are governed by physical and biological principles.** These principles include the conservation of matter and energy, dynamic steady states, a requirement to expend energy, and the evolution of new phenotypes and new species.

■ **Different organisms play diverse roles in ecological systems.** The major groups of organisms are plants, animals, fungi, protists, and bacteria. These organisms are involved in numerous species interactions, including competition, predation, mutualism, and commensalism. Each organism lives in specific habitats and has a particular niche.

■ **Scientists use several approaches to studying ecology.** Like all scientists, ecologists use the scientific method of developing and testing hypotheses. The testing of proximate and ultimate hypotheses can be accomplished using manipulative experiments, natural experiments, or mathematical models.

■ **Humans influence ecological systems.** The rapid growth of the human population during the past two centuries has increased human influence on ecological systems, particularly as the result of resources they consume and waste products they release.

REVIEW QUESTIONS

1. What is ecology?

2. Why do ecologists consider both individuals and ecosystems to be ecological systems?

3. What are the unique processes that are examined when taking the individual, population, community, and ecosystem approaches to studying ecology?

4. Describe how ecological systems are governed by general physical and biological principles.

5. How do the sources of energy acquired by plants, animals, and fungi differ?

6. Compare and contrast an organism's habitat and an organism's niche.

7. In the Northern Hemisphere, many species of birds fly south during the autumn months. Propose a proximate and an ultimate cause for this behaviour.

8. When experimental manipulations are conducted to test a hypothesis, what is the purpose of including a control?

9. In what ways do manipulative experiments differ from natural experiments?

10. How can our knowledge of ecological systems help humans manage these systems?

PRACTICING ECOLOGY

In this chapter we have seen that ecologists make observations, develop hypotheses, and then test their hypotheses in a variety of ways. When testing hypotheses, it is often helpful to use a graph to visualize the data collected. Graphs used in ecological studies can take many different forms, which you can read about in the appendix "Reading Graphs" at the end of the book.

To help you develop your graphing skills and learn to interpret real data, each subsequent chapter will present a "Practicing Ecology" problem. These problems are based on the work of ecologists at Canadian universities. You will be introduced to a specific study relevant to the chapter's key concepts. Using data from the study, you will construct a graph. Finally you will interpret the graph in light of the hypotheses the ecologists are testing. As you move through the chapters, these exercises will provide you with a better sense of how ecological studies are conducted and how ecologists graph and interpret their data.

CHAPTER 2

Adaptations to Aquatic Environments

The Evolution of Whales

Life on Earth began in the water. Of the many species that live in water, some of the most charismatic are the whales—a group that is particularly well suited to aquatic life. Surprisingly, the ancestor of modern whales can be traced back to a terrestrial mammal related to cattle, pigs, and hippos. Scientists first proposed an evolutionary relationship between whales and this group of land mammals in 1883, based on observations of similarities in their skeletons. In the 1990s, DNA technology revealed that the groups are related genetically. In 2007, scientists discovered a critical link between hippos and whales: the fossilized bones of a previously unknown large terrestrial mammal that may have spent at least some of its time in the water. Scientists speculate that over the subsequent 50 million to 60 million years, selection imposed by the aquatic environment led to the evolution of the whale as we know it today.

Modern whales have evolved a wide range of adaptations for aquatic life. One of the most obvious challenges is the ability to swim efficiently. For example, the killer whale is capable of swimming up to 48 km per hour. Such speeds are possible only with a highly streamlined body. Over evolutionary time, natural selection would have favoured any individuals that had a more streamlined body, including bodies with reduced hindlimbs. In modern whales, tiny remnants of the hindlimb bones can be found entirely within the whale's body. During the course of the whale's evolutionary history, whale ears became internalized. Whether this was due to selection for a streamlined body or for some other reason is not known, but the outcome made for a more streamlined body.

> "Surprisingly, the ancestor of modern whales can be traced back to a terrestrial mammal related to cattle, pigs, and hippos."

Obtaining oxygen is another challenge for whales because they need to dive for long periods in search of food. Over time, there was an evolutionary change in the location of the nostrils from the front of the head to the top. While scientists cannot be certain of the selective forces that caused this change, one hypothesis is that changes in nostril position may have been favoured over time to the point that the modern whale can more easily grab a breath of air when surfacing. In addition, whales can hold their breath for long periods. Sperm whales, for example, dive to depths of 500 m and may stay below the surface for more than an hour as they search for fish, squid, and other food. During a dive, the sperm whale relies on oxygen stored in its body. It might surprise you to learn

Whale ancestor. The ancestor of modern whales, in the genus *Indohyus*, was a terrestrial animal that spent part of its time in the water. Over time, the descendants of this animal developed numerous adaptations for living in water that are found in modern-day whales.

Killer whale. Modern whales, such as this killer whale (*Orcinus orca*) swimming off the coast of British Columbia, have a number of adaptations that enable them to live in their aquatic environment. Photo by Thomas Kitchin & Vict/All Canada Photos/SuperStock.

that very little of this oxygen resides in the lungs; most is bound to hemoglobin in the blood or to a similar oxygen storage molecule, myoglobin, in the muscles. Under water, whales slow their metabolism by reducing blood flow to nonvital organs such as the skin, viscera, lungs, and kidneys, while blood flow to the brain and heart continues. Consequently, during a dive the temperature of all but a few key organs drops, the heart rate slows, and demand for oxygen is reduced.

Regulating body temperature is yet another challenge. Because heat loss occurs much faster in water than in air, a thick layer of fat under the skin insulates most oceanic mammals living in cold waters. Like a warm coat, this insulation slows the loss of heat generated by its internal organs. Whales also maintain a higher metabolic rate than land mammals of similar size, and this helps generate extra heat. The whale's vascular structure also helps maintain heat; the veins and arteries of a whale's flippers and tail are next to each other. This allows the warm arterial blood that travels away from the heart to transfer heat to cooler blood in adjacent veins as it returns from a whale's extremities to the heart.

The evolution of the whale took place over a period of 50 million years. The selective forces occurring over this time cannot be known with certainty, but the fossils found to date suggest that an ancient terrestrial mammal slowly evolved a number of adaptations that gave rise to modern-day whales. In this chapter, we examine the challenges posed by living in an aquatic environment: the properties of water; the need to balance salt and water concentrations; the movement of gases, including carbon dioxide and oxygen; and the wide range of temperatures that can exist. We will see how organisms have evolved adaptations that meet these challenges and allow species to persist in the aquatic world.

SOURCES: Valley of the Whales, *National Geographic Magazine* (2010), http://ngm.nationalgeographic.com/2010/08/whale-evolution/mueller-text/8.

Whales descended from tiny deer-like ancestors, *ScienceDaily* (2007), http://www.sciencedaily.com/releases/2007/12/071220220241.htm).

CHAPTER CONCEPTS

- Water has many properties favourable to life.

- Aquatic environments challenge the balance of water and salt in animals.

- The uptake of gases from water is limited by diffusion.

- Temperature limits the occurrence of aquatic life.

Scientists generally agree that life began in the ocean and that the first forms of life were simple bacteria. Over millions of years, these bacteria gave rise to an incredible diversity of organisms, many of which still live in the water. Other species, as we saw in the case of the whale, evolved terrestrial forms that later evolved to return to life in the ocean. In this chapter, we begin an exploration of ecology at the individual level by examining the ways in which the properties of water support and constrain aquatic organisms and direct the evolution of adaptations.

Water has many properties favourable to life

Water is abundant over most of Earth's surface. Within the temperature range organisms usually encounter, it is liquid. Because water also has an immense capacity to dissolve inorganic compounds, it is an excellent medium for the chemical processes of living systems. In fact, it is hard to imagine a form of life that could exist without water. In this section we will look at how water makes life possible, including water's thermal properties, its density and viscosity, and its function as a solvent for inorganic nutrients.

THERMAL PROPERTIES OF WATER

On Earth, water can be found as a solid (ice), as a liquid, and as a gas (water vapour). No other common substance is liquid under most conditions at Earth's surface. Pure water—water not containing any dissolved minerals or other compounds—becomes a solid below 0°C and becomes a gas above 100°C at sea level. At higher elevations, the freezing point of water changes very little, but the boiling point can be several degrees lower.

When water contains dissolved compounds, such as salts, its freezing temperature drops below 0°C. This is why road salts are applied to ice- and snow-covered roads; the salts allow the ice and snow to melt at a lower temperature than they otherwise would. Dissolved compounds also raise the boiling point of water above 100°C.

Water stays liquid over a broad range of temperatures, from 0°C to 100°C. The temperature of water remains relatively steady even when heat is removed or added rapidly, as can happen at the air–water interface or at an organism's surface, such as on the surface of a whale in the ocean. This is because water has a high specific heat, which is the amount of heat required to increase its temperature by 1°C.

Water also resists changing from one state to another. For example, raising the temperature of 1 kg of liquid water by 1°C requires the addition of 1 Calorie of heat. However, converting 1 kg of liquid water into water vapour requires the addition of 540 Calories of heat. Similarly, lowering the temperature of 1 kg of liquid water by 1°C requires the removal of 1 Calorie of heat, but converting that amount of liquid water to ice requires the removal of 80 Calories of heat. In short, liquid water is very resistant to changing states. This resistance helps prevent large bodies of water from freezing solid during winter. In addition, because water transfers heat rapidly, heat tends to spread evenly throughout a body of water, which also slows local changes in temperature.

Another curious, yet fortunate, thermal property of water is the way it changes density with changes in temperature. Water achieves its highest density—that is, its molecules are the most densely packed together—at 4°C. Above and below 4°C, the water molecules are less tightly packed and the water becomes less dense. Below 0°C, pure water is converted into ice, which is less dense than liquid water as you can see in Figure 2.1. As a result of its lower density, ice floats on the surface of liquid water. This means that lakes experiencing cold winters will generally have a layer of 4°C water at their bottom. Above this layer will be water that is less than 4°C and on top of that will be a layer of ice.

Water's unusual thermal properties are especially important to aquatic plants and animals. In large water bodies containing fresh water, such as lakes, the bottom of the lake does not freeze, in part because of the insulation that ice provides from very cold air temperatures. The salt in ocean water lowers the freezing point of the water well below 0°C, which prevents oceans from freezing. In both cases, the

Figure 2.1 The density of water. As water cools, the molecules contract and become more dense. Below 4°C, they begin to expand and becomes less dense. Below 0°C, pure water is converted into ice that is even less dense. As a result of its lower density, ice floats on the surface of liquid water. Photo by Zoonar/Christa Kurtz/age fotostock.

available water provides a refuge for organisms during periods of cold temperatures.

DENSITY AND VISCOSITY OF WATER

The adaptations of aquatic organisms often exploit the density and viscosity of water. For example, animals and plants are comprised of bone, proteins, and other materials that are somewhat denser than salt water, and much denser than fresh water. However, organisms can also contain fats and oils that are less dense than water. In some cases, they also possess pockets of air, such as the lungs of the whales that were described at the beginning of this chapter. The combination of the materials that compose an animal's body and the presence of air pockets determine whether an organism will float or sink in water.

For those organisms that are denser than their surrounding water, a variety of adaptations can either reduce an organism's density or retard its rate of sinking. For example, many fish species have a gas-filled swim bladder that can adjust in size to make the density of the fish's body equal to that of the surrounding water. Human divers use this same concept when they wear air-filled inflatable vests to help match the density of the water. Divers can add air to the vest, which allows them to float on the surface of the water. Alternatively, divers can release some of the air to match the density of the water so that they neither float nor sink, or they can let even more air out so that they sink in the water. Some large kelps, like those we saw in Figure 1.8, have gas-filled bulbs that cause their leaf-like blades to float up into the sunlit surface waters. The whales discussed earlier become buoyant when they take a breath of air but a slow release of air bubbles will help them sink to a particular depth. At the other end of the size spectrum, many of the microscopic unicellular algae that float in great numbers in the surface waters of lakes and oceans use droplets of oil as flotation devices (Figure 2.2). Because oils are less dense than water, the algae can use the oil droplets to help offset their natural tendency to sink.

Aquatic organisms also possess adaptations to deal with the high *viscosity* of water. Viscosity is technically defined as the resistance of a fluid to being deformed by a stress, but we can think of viscosity as the thickness of a fluid that causes objects to encounter

Figure 2.3 Streamlined shape. Large, fast-moving aquatic organisms like the northern pike (*Esox lucius*) have evolved highly streamlined shapes to help them move through the highly viscous water. Photo by Christian GUY/imagebr–www. agefotostock.com.

resistance as they move through it. In response to living in water, fast-moving aquatic animals such as fish, penguins, and whales have evolved highly streamlined shapes that reduce the drag caused by the high viscosity of water and other factors (Figure 2.3). The viscosity of water is higher in cold water than in warm water, which can make swimming in cold water more difficult. Movement in water is even more difficult for smaller animals. However, the same high viscosity that impedes the progress of tiny organisms as they swim in the water also impedes them from sinking. Because these organisms are slightly denser than water, they are prone to sinking due to the force of gravity. To take advantage of water's viscosity, many tiny marine animals have evolved long, filamentous appendages that cause greater drag in the water. The appendages function like a parachute slowing the fall of a body through air (Figure 2.4).

DISSOLVED INORGANIC NUTRIENTS

Both aquatic and terrestrial organisms require a variety of nutrients to build necessary biological structures and maintain life processes. Large amounts of hydrogen, carbon, and oxygen are necessary for building most compounds found in organisms. Nitrogen, phosphorus, and sulphur are used in building proteins, nucleic acids, phospholipids, and bones. Other major nutrients—including potassium, calcium, magnesium, and iron—play important roles as solutes and as structural components of bones, woody plant cells, enzymes, and chlorophyll. Certain organisms need additional minor nutrients. For

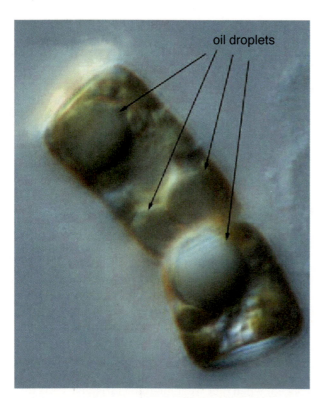

oil droplets

Figure 2.2 Adaptation to water density. These algae (*Cyclotella cryptica*) are able to float near the water's surface by using oil droplets that have a lower density than water. Photo by Bigelow Laboratory National Center for Marine Algae and Microbiota.

Figure 2.4 Adaptation to water viscosity. Some small aquatic organisms exploit the high viscosity of water by evolving large appendages, such as the antenna and feathery projections of this tiny marine crustacean. These appendages help slow down movement through the viscous water and thereby retard sinking. Photo by Solvin Zankl/naturepl.com.

example, diatoms are a group of algae that need silica to construct their glassy shells (Figure 2.5). Similarly, some species of bacteria require the element molybdenum, which makes up part of the enzyme they use to convert nitrogen from the atmosphere (N_2) into ammonia (NH_3).

The Solubility of Minerals

Water is a powerful solvent with an impressive capacity to dissolve substances, which makes them

Figure 2.5 Use of inorganic nutrients. Diatoms, such as this species of *Arachnoidiscus*, are a type of algae that require minor nutrients such as silica to build a hard, glassy shell. Image is magnified 175 times. Photo by Steve Gschmeissner/Science Source.

accessible to living systems. Because of this property, water also provides a medium in which substances can react chemically to form new compounds.

Water acts as a solvent because of its molecular structure. As you can see in Figure 2.6, water molecules consist of an oxygen atom in the middle and two hydrogen atoms connected in a V-shaped arrangement. This arrangement results in an unequal sharing of electrons: The oxygen end of the water molecule has a slightly negative charge and the hydrogen end has a slightly positive charge. When the two ends of the molecule possess opposite charges, we say that the molecule is *polar*. Water is a polar molecule: The negative oxygen end of one water molecule is strongly attracted to the positive hydrogen end of another. These forces of attraction are known as *hydrogen bonds*.

The polar nature of water molecules also allows them to be attracted to other polar compounds. Some solid compounds consist of electrically charged atoms, or groups of atoms, called **ions**. For example, common table salt—sodium chloride (NaCl)—contains positively charged sodium ions (Na^+) and negatively charged chloride ions (Cl^-). In their solid form, these ions are arranged in a crystal lattice. In water, however, the charged sodium and chloride ions are attracted by the charges of the water molecules. As

Ions Atoms or groups of atoms that are electrically charged.

Figure 2.6 Water molecules. Because of the configuration of water molecules, they are negatively charged on the oxygen end and positively charged on the hydrogen end. The attractive forces of these opposite charges, known as hydrogen bonds, allow water molecules to be attracted to each other and to the charged ions of other compounds such as salts and sugars.

shown in Figure 2.7, the attraction of these ions to water molecules is stronger than the attraction that holds the crystal together. As a result, when sodium chloride is added to water, its crystal lattice breaks apart, and water molecules surround the salt ions. In other words, when you put salt in water, it dissolves. This solvent ability of water is not restricted to ionic compounds such as salts; it occurs with any polar compound, including the various types of sugars that organisms commonly use. In contrast, water is not a good solvent for oils and fats because they are nonpolar compounds.

The solvent properties of water explain the presence of minerals in streams, rivers, lakes, and oceans. When water vapour in the atmosphere condenses to form clouds, the condensed water in the atmosphere is nearly pure. However, as it falls back to Earth as rain or snow, water acquires some minerals from dust particles in the atmosphere. Precipitation that hits land comes into contact with rocks and soils, and it dissolves some of the minerals in them. These additional minerals are carried toward the ocean.

Water in most lakes and rivers contains a dissolved mineral concentration of 0.01 to 0.02 percent whereas ocean water contains a dissolved mineral concentration of 3.4 percent. Oceans have much higher concentrations of dissolved minerals because mineral-laden water continuously flows in from streams and rivers, and the constant evaporation from the ocean's surface removes pure water, leaving the minerals behind. Over billions of years, this process has caused increased concentrations of minerals in the oceans.

Every mineral has an upper limit of solubility in water, known as **saturation**. This limit generally increases with higher temperatures. After a mineral achieves saturation, water cannot hold any more and the mineral precipitates out of solution. For some minerals, such as sodium, ocean concentrations are far below saturation. Most of the sodium that is washed into ocean basins remains dissolved and its concentration in seawater continues to increase. In contrast, the concentrations of other minerals in the oceans commonly exceed their saturation concentrations. For example, calcium ions (Ca^{+2}) readily combine with dissolved CO_2 to form calcium carbonate ($CaCO_3$), and calcium carbonate has a low solubility in ocean water. Over millions of years, the excess calcium carbonate that has washed into the oceans from streams and rivers has subsequently precipitated out of the water. This precipitated calcium carbonate, combined with the calcium carbonate from the bodies of countless tiny marine organisms, has resulted in massive limestone sediments (Figure 2.8). Today, these sediments are important sources of limestone for construction applications such as stone blocks and concrete, for agricultural uses such as fertilizer, and for numerous industrial processes.

Hydrogen Ions

Among dissolved substances in water, hydrogen ions (H^+) deserve special mention because they are extremely reactive with other compounds. In pure water, a small fraction of the water molecules (H_2O) break apart into their hydrogen (H^+) and hydroxide (OH^- ions. The concentration of hydrogen ions in a solution is referred to as its **acidity**. Acidity is commonly measured as **pH**, which is defined as the negative logarithm of hydrogen ion concentration

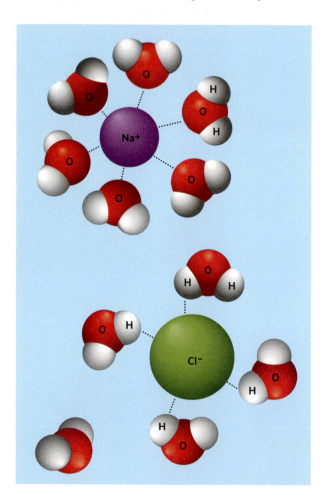

Figure 2.7 Dissolving ions in water. Because water molecules have negative and positive ends, they attract the negatively and positively charged ions, such as the sodium and chlorine ions found in sodium chloride. The forces of attraction to water molecules are stronger than the forces of attraction within the crystal, so the ions separate and become surrounded by water molecules.

Saturation The upper limit of solubility in water.

Acidity The concentration of hydrogen ions in a solution.

pH A measure of acidity or alkalinity; defined as pH = −log (H^+ concentration).

Figure 2.8 The formation of limestone sediments. The continuous addition of calcium minerals into oceans from streams and rivers causes calcium to combine with CO_2 to become calcium carbonate. Because calcium carbonate is not very soluble in water, it precipitates out of the water to form massive sediments over millions of years. This site of limestone sediments located at Percé Rock, off the Gaspé Peninsula in Quebec, formed in the ocean during the Devonian period between 419–358 million years ago. Photo by Michael Snell/Robert Harding/ Newscom.

(as measured in moles/L, where 1 mole $= 6.02 \times 10^{23}$ molecules):

$$pH = -\log(H^+ \text{ concentration})$$

As you can see in the pH scale shown in Figure 2.9, water containing a high concentration of hydrogen ions has a low pH value, whereas water containing a low concentration of hydrogen ions has a high pH value. Therefore, we categorize water with low pH values as *acidic,* water with a mid-range value of 7 as *neutral,* and water with a high pH value as *basic* or *alkaline.* As we will see, natural rain or snow can vary a great deal in pH, depending on the presence of different chemical compounds in the atmosphere that affect the pH value.

Hydrogen ions, because of their high reactivity, dissolve minerals from rocks and soils, enhancing the natural solvent properties of water. For example, in the presence of hydrogen ions, the calcium carbonate found in limestone dissolves readily, according to the following chemical equation:

$$H^+ + CaCO_3 \rightarrow Ca^{2+} HCO_3^-$$

Calcium ions are important to life processes, and their presence at high concentrations is vital to organisms such as snails that form shells made of calcium carbonate. Indeed, mollusks are less abundant in streams and lakes that are low in calcium. Therefore, hydrogen ions are essential for making certain nutrients available for life processes. At high concentrations, however, hydrogen ions negatively affect the activities of most enzymes. In addition, high concentrations of hydrogen ions cause many heavy metals to begin dissolving in water. These heavy metals, including arsenic, cadmium, and mercury, are highly toxic to most aquatic organisms.

The normal range of pH of lakes, streams, and wetlands is between 5 and 9. However, some bodies of water can have even lower pH values. Sometimes lower pH conditions have a natural cause. Bogs, for example, are aquatic habitats with vegetation such as sphagnum mosses that release H^+ ions into the water and thereby make the water more acidic and unsuitable for many other species of plants.

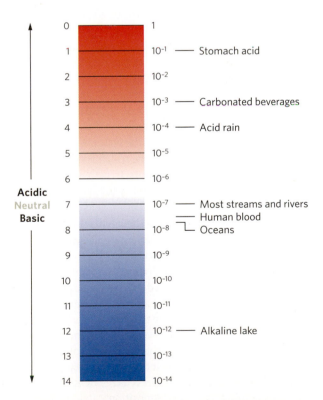

Figure 2.9 The relationship between pH and hydrogen ion concentration in water. The pH scale of hydrogen ion concentration extends from 0 (highly acidic) to 14 (highly alkaline). The pH of rainfall can vary a great deal around the world.

Other bodies of water have low pH as a result of human influences. For example the release of sulphur dioxide (SO_2) and nitrogen dioxide (NO_2) from coal-powered industrial plants became a major environmental issue in the 1960s. At that time, ecologists in Russia, China, northern Europe, the United States, and Canada began to notice that many bodies of water were becoming more acidic and less hospitable to numerous species of fish and other aquatic organisms. They also noticed that trees were dying, particularly in the spruce-fir forests that existed at high elevations in these regions of the world.

It turned out that the areas with more acidic bodies of water and dying trees were all far downwind of industrial areas with coal-powered factories that had tall smokestacks. Years of data collection revealed that the sulphur dioxide and nitrogen dioxide emitted into the atmosphere by these smokestacks were converted to sulphuric acid and nitric acid in the atmosphere. These acids were transported downwind and then came back down to Earth as **acid deposition**, also known as **acid rain**. The acid deposition occurred in two forms: as gases and particles that stuck to plants and soil, a form called *dry acid deposition,* and as rain and snow, a form called *wet acid deposition.* Acid deposition lowered the pH of precipitation and, as a result, water of unusually low pH was entering streams, lakes, and forests. Most aquatic species cannot tolerate water with a pH lower than 4, so these bodies of water became toxic to many aquatic organisms, which included insects and fish.

In forests, acid deposition has several effects. First, it leaches the calcium out of the needles of conifer trees such as spruce. It also causes increased leaching of soil nutrients that trees require, including calcium, magnesium, and potassium. Finally, acid deposition causes aluminum to dissolve in the water. Although aluminum naturally occurs in the soil, it is typically not in a form that is available to plants. Dissolved aluminum can negatively affect a plant's ability to take up nutrients. Collectively, these effects of acid deposition make trees more susceptible to the harmful effects of natural stressors that include drought, diseases, and extreme temperatures. In short, while the trees do not die directly from acid deposition they become more susceptible to other causes of death. Because acid deposition interacts with so many other natural stressors, scientists recognize that acid deposition has contributed to the death of trees in North America, Europe, and Asia. However, the complexity of the interactive effects has made it difficult to accurately estimate how much of the observed tree death is directly attributable to acid deposition.

Once scientists understood the causes and consequences of acid deposition, they began to explore solutions. In Canada, legislation required the provinces to reduce their sulphate depositions over specified time frames. Techniques such as smokestack scrubbers that force industrial gases through a slurry of limestone and water, helped reduce the amount of acidic compounds going into the atmosphere. Data from the National Pollutant Release Inventory at Environment Canada shows that from 1985 to 2010 Canada's sulphur oxide emissions declined by 63 percent. At the end of this chapter, in "Ecology Today," we discuss another example of how understanding environmental problems related to pH can help us develop effective solutions.

Animals and plants face the challenge of water and salt balance

SALT BALANCE IN AQUATIC ANIMALS

It might surprise you to learn that animals in an aquatic environment need specialized mechanisms to maintain the proper amount of water in their bodies. To understand why, we need to recognize that the water surrounding an aquatic animal and the water inside its body contain dissolved substances, known as **solutes**. These solutes affect the movement of water in and out of an organism. The movement of water occurs at the cellular level, where water passes across cell membranes from regions of low solute concentration to regions of high solute concentration. At the same time, solutes attempt to move across membranes to equalize the concentration of solutes. Cell membranes typically do not allow the free movement of large solute molecules such as carbohydrates and most proteins. Membranes that allow only particular molecules to pass through—for example, water and small solute ions and molecules—are known as **semipermeable membranes**. Cells are able to transport ions and small molecules through a membrane against a concentration gradient to maintain their concentrations, a process known as **active transport**. Active transport expends energy because it requires the cell to work against the concentration gradient of solutes.

The difference between the solutes inside and outside the body of an aquatic animal makes it difficult to maintain the right amount of water. The osmotic potential generated by an aqueous solution depends on its solute concentration, which is measured as the number of solute molecules in a

---

Acid deposition Acids deposited as rain and snow or as gases and particles that attach to the surfaces of plants, soil, and water. *Also known as* **Acid rain**.

Solute A dissolved substance.

Semipermeable membrane A membrane that allows only particular molecules to pass through.

Active transport Movement of molecules or ions through a membrane against a concentration gradient.

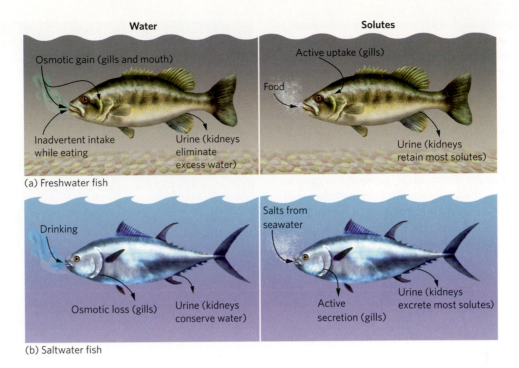

Figure 2.10 Osmoregulation in fish. Exchanges of water and solutes differ between freshwater and marine fish. **(a)** Freshwater fish are hyperosmotic: They have a higher salt concentration in their body than exists in the surrounding water. To maintain salt balance they must excrete large amounts of water and use their gills and kidneys to actively retain solutes. **(b)** Saltwater fish are hyposmotic: They have a lower salt concentration in their body than exists in the surrounding water. To maintain salt balance, they must excrete large amounts of solutes, so their gills and kidneys actively exclude solutes.

given volume of water. When the water inside a cell has a higher concentration of solutes than the water outside, water tries to move into the cell from the surrounding environment. In contrast, when the water inside a cell has a lower concentration of solutes than the water outside, water tries to move out of the cell. The movement of water across a semipermeable membrane is called **osmosis**. The force with which an aqueous solution attracts water by osmosis is known as its **osmotic potential**, which is expressed in megapascals (MPa), a unit of pressure.

The challenge for most aquatic animals is that they live in water with a solute concentration that differs from the solute concentration of their bodies. If they take in too many solutes, they must rid themselves of the excess. If they take in too few solutes, they must make up the deficit. Maintaining a particular solute concentration in the body is important because solute concentrations affect the way proteins interact with other molecules. In short, altered solute concentrations can disrupt cell functions. Because solutes determine the osmotic potential of body fluids, the mechanisms that organisms use to maintain a proper solute balance are referred to as **osmoregulation**.

ADAPTATIONS IN FRESHWATER ANIMALS
We have seen that the water balance of aquatic animals is closely tied to the concentrations of salts and other solutes in their body tissues and in the environment.

The body fluids of vertebrate animals have an osmotic potential of about −0.4 MPa. In comparison, fresh water has an osmotic potential close to 0 MPa and seawater has an osmotic potential of −1.2 MPa. Freshwater animals, such as bass and trout, that have higher solute concentrations in their tissues than the surrounding water are said to be **hyperosmotic** compared to their freshwater environment.

Because water and solutes move in the direction that will equalize the concentrations of solutes on both sides of a membrane, a hyperosmotic organism faces a constant challenge: water attempts to enter its body while solutes attempt to leave. Figure 2.10a shows how hyperosmotic fish regulate the balance of solutes in their tissues. Freshwater fish continuously gain water both when they consume food and when osmosis occurs across the mouth and gills, which are the most permeable tissues exposed to the water. Fish respond to this influx of water by eliminating the

Osmosis The movement of water across a semipermeable membrane.

Osmotic potential The force with which an aqueous solution attracts water by osmosis.

Osmoregulation The mechanisms that organisms use to maintain proper solute balance.

Hyperosmotic When an organism has a higher solute concentration in its tissues than the surrounding water.

excess water through their urine. They add solutes to their bloodstream by using their gill cells to actively transport solutes into the body from the water. In addition, their kidneys remove ions from their urine.

ADAPTATIONS IN SALTWATER ANIMALS

Saltwater animals—such as whales, sardines, and plankton—have lower solute concentrations in their tissues than exists in the surrounding water. Such organisms are said to be **hyposmotic** compared to their saltwater environment. Hyposmotic organisms face a constant challenge to maintain the balance of water and solutes in their tissues. As shown in Figure 2.10b, water tries to leave their bodies and solutes try to enter. To replace the loss of water, saltwater animals drink large amounts of saltwater and release only small amounts of urine. To counteract the accompanying influx of solutes, the excess solutes are actively excreted out of the body using the kidneys and, in the case of fish, the gills.

Some sharks and rays have evolved a unique adaptation to the challenge of water balance in a saltwater environment. Like all vertebrate animals, when sharks and rays digest proteins, they produce ammonia as a by-product, as shown in Figure 2.11. Aquatic vertebrates excrete ammonia in their urine, whereas terrestrial vertebrates generally convert this ammonia to urea, which they then excrete at high concentrations in their urine. Interestingly, sharks and rays convert ammonia into urea too, but they do not excrete all of it. Unlike terrestrial vertebrates, which maintain concentrations of urea below 0.03 percent in their bloodstream, sharks and rays actively retain urea and allow its concentration to increase to 2.5 percent in their bloodstream. The retained urea raises the osmotic potential of their blood to that of seawater without any increase in the concentrations of sodium and chloride ions. Consequently, the movement of water across the animal's body surface becomes balanced in relation to the surrounding saltwater, with neither gain nor loss. This adaptation frees sharks and rays from having to drink salt-laden water to replace water lost by osmosis.

Freshwater species of rays do not accumulate urea in their blood, though they are similar to their marine cousins in other respects. This confirms the importance of urea for osmoregulation in saltwater species. Most vertebrates do not retain much urea in their bloodstream because urea impairs protein function. However, sharks, rays, and many other marine organisms that use urea to maintain their water balance have an additional adaptation: They accumulate high concentrations of a compound called trimethylamine oxide to protect proteins from the harmful effects of urea.

Certain environments pose unusual osmotic challenges. For instance, the salt concentrations in

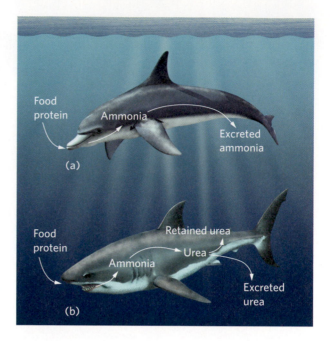

Figure 2.11 Unique adaptations to saltwater. When aquatic vertebrates digest proteins, they produce ammonia as a by-product. **(a)** Most aquatic organisms, such as dolphins, excrete this ammonia in their urine. **(b)** Sharks and rays convert this ammonia into urea and then retain some of the urea in their bloodstream. The result is a higher solute concentration that helps offset the challenge of being hyposmotic in saltwater.

some landlocked bodies of water greatly exceed those of the ocean. This is particularly common in arid regions where, because evaporation outpaces precipitation, very high concentrations of solutes accumulate in the water. The Great Salt Lake in the U.S. state of Utah, for example, contains 5 to 27 percent salt, depending on the water level. This is up to eight times more salt than in seawater. The osmotic potential of the water in the Great Salt Lake—in excess of −10 MPa—would cause most organisms to shrivel. However, a few aquatic creatures, such as brine shrimp, have adaptations that allow them to thrive in these conditions by excreting salt at a very high rate. This level of excretion comes at a high energetic cost, which they meet by feeding on the abundant photosynthetic bacteria that live in the lake.

The ability of an organism to cope with the osmotic potential of its environment reflects the outcome of evolutionary processes. As we will discuss in Chapter 4, some environments naturally experience large fluctuations in osmotic potential and the organisms living in these environments have evolved ways to adjust to these fluctuations. When changes in osmotic concentrations are not within the natural range that organisms have experienced over

Hyposmotic When an organism has a lower solute concentration in its tissues than the surrounding water.

(a)

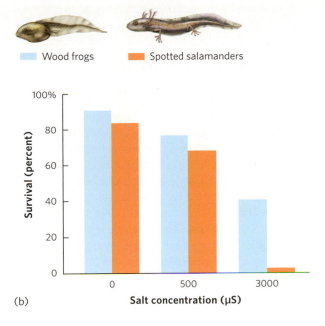

(b)

Figure 2.12 Effect of salt concentration on amphibians. (a) Based on a sample of ponds in the Adirondack Mountains of New York state, ponds closer to roads had higher conductivity, a measure of salt concentration. (b) At this range of salt concentrations, spotted salamanders and wood frogs experienced a decline in survival. The data displayed are means ± standard deviation. Data from: N. E. Karraker et al., Impacts of road deicing salt on the demography of vernal pool-breeding amphibians, *Ecological Applications* 18 (2008): 724–734.

evolutionary time, however, individuals typically lack appropriate adaptations and can be harmed. For example, in many places various mixtures of salts are spread on roads to melt the ice and snow to provide safer driving conditions during winter weather. With the arrival of warmer spring weather, however, all of this salt has to go somewhere. In 2008, researchers reported on the concentrations of road salt in ponds inhabited by amphibians at different distances from salt-treated roads. Because salt ions allow electricity to be conducted through water, the concentration of salt can be measured in units of microsiemens (μS). As you can see in Figure 2.12a, ponds close to roads had salt concentrations up to 3000 μS, which corresponds to 0.12 percent salt. Ponds at least 200 m from a road had essentially 0 μS. The researchers then conducted

experiments in which they exposed larval wood frogs (*Rana sylvatica*) and spotted salamanders (*Ambystoma maculatum*) to a range of relevant salt concentrations, from 0 μS to 3000 μS. The data, shown in Figure 2.12b, revealed that increases in salt concentration caused the larvae of both species to die in high numbers. Not having been exposed to high salt concentrations during their evolutionary history, these freshwater organisms are not adapted to these stressful conditions and are unable to survive in them.

SALT BALANCE IN AQUATIC PLANTS

Some aquatic plants also face major challenges of salt balance. Mangrove trees, for example, grow on coastal mudflats that are inundated by saltwater during high tides (Figure 2.13a). Not only does this habitat impose

(a)

(b)

Figure 2.13 Salt balance in mangrove trees. (a) The roots of mangrove trees are frequently submerged in saltwater at high tide. These trees are from Palau, an island in the South Pacific. (b) Specialized salt glands in the leaves of the button mangrove (*Conocarpus recta*) excrete a salty solution. As the solution evaporates, it leaves behind salt crystals on the outer surface of the leaves. Photos by (a) Reinhard Dirscherl/Alamy, and (b) Ulf Mehlig.

ANALYZING ECOLOGY

Standard Deviation and Standard Error

When the researchers tested the effects of road salt on larval amphibians, they exposed groups of larvae to three salt concentrations and replicated the experiment five times. In the previous chapter, we discussed how ecologists use data from such manipulative experiments to determine how different factors affect the means and variance of the variables that are measured. While the variance is a useful measure of how consistent measurements are among replicates, ecologists also use several other related measures of variation, including the *sample standard deviation* and the *standard error of the mean*. Each of these can be calculated from a measure of sample variance (s^2) as we discussed in Chapter 1.

When data are collected from limited samples of a much larger distribution of data, we can gain additional information about the data by calculating a *sample standard deviation*. The **sample standard deviation** gives us a standardized way of measuring how widely our data are spread from the mean. Large sample standard deviations indicate that many of our data are spread far from the mean. Small sample standard deviations indicate that most of our data points are close to the mean value.

If data are normally distributed—that is, if they follow a bell-shaped curve like the one in the figure at right—then about 68 percent of our data will fall within one standard deviation of the mean. Moreover, about 95 percent of our data will fall within two standard deviations of the mean and 99.7 percent of our data will fall within three standard deviations of the mean. For data that have a wide distribution of frequencies, as in the upper figure, the standard deviation value will be large. For data that have a narrow distribution of frequencies, as in the lower figure, the standard deviation value will be small.

The sample standard deviation, denoted as s, is defined as the square root of the sample variance:

$$s = \sqrt{s^2}$$

The **standard error of the mean** is a useful measurement of variation in our data because it

(a)

(b)

The normal distribution. In a normal distribution, the most frequent observations fall near the mean and the least frequent observations occur farther away from the mean. The dark blue areas represent one standard deviation from the mean and account for 68.3 percent of all data. The medium and dark blue areas combined represent two standard deviations from the mean and account for 95.5 percent of all data. The light, medium, and dark blue areas combined represent three standard deviations from the mean and account for 99.7 percent of all data. The remaining small amount of data fall outside of three standard deviations. **(a)** Data that have a wide distribution have a large standard deviation. **(b)** Data that have a narrow distribution have a small standard deviation.

Sample standard deviation A statistic that provides a standardized way of measuring how widely data are spread from the mean.

Standard error of the mean A measurement of variation in data that takes into account the number of replicates that were used to measure the standard deviation.

a high salt load on the trees, but the high osmotic potential of the saltwater environment also makes it difficult for the roots to take up water. To counter

these osmotic problems, many mangroves maintain high concentrations of organic solutes—various amino acids and small sugar molecules—in their roots and takes into account the number of replicates that were used to measure the standard deviation. The higher the number of replicates, the more precise the estimate we should have of the mean. As a result, an increase in the number of replicates in a given experiment produces a decrease in the standard error of the mean. As we will see in later chapters, the standard error of the mean is commonly used to determine whether two means are significantly different from each other.

The standard error of the mean (denoted as SE) is defined as the sample standard deviation divided by the square root of the number of observations or replicates (denoted as n):

$$SE = s \div \sqrt{n}$$

Consider the following set of observations on the percentage of surviving wood frogs that were exposed to salt concentrations of 0 μS or 3000 μS:

REPLICATE	0 μS	3000 μS
1	88	32
2	90	37
3	91	41
4	92	45
5	94	50
MEAN	91	41
VARIANCE	5.0	48.5

Using the data from the five replicates that exposed wood frog tadpoles to 0 μS, we see that the mean survival is 91 percent and the sample variance is 5 percent. Using this sample variance, we can calculate the sample standard deviation as:

$$s = \sqrt{s^2} = \sqrt{5} = 2.2$$

and we can calculate the standard error of the mean as:

$$SE = s \div \sqrt{n} = 2.2 \div \sqrt{5} = 1.0$$

> **YOUR TURN** Use the data collected from the five replicates that exposed wood frog tadpoles to 3000 μS and calculate the standard deviation of the mean and the standard error of the mean.

leaves to increase their osmotic potential so that water will diffuse into the plant's tissues. In addition, mangroves possess salt glands in their leaves that can secrete salt by active transport to the exterior surface of the leaf (Figure 2.13b). Many mangrove species also exclude salts from their roots by active transport. Because relatively few species of terrestrial plants have evolved this adaptation, mangrove forests do not contain numerous species of plants.

The uptake of gases from water is limited by diffusion

Almost 21 percent of Earth's atmosphere is oxygen, but because we live on land we rarely think about the task of obtaining this required element. Aquatic organisms also require oxygen to support their metabolism, but obtaining a sufficient supply can be a problem for them because of the limited solubility of oxygen in water. The same is true of the CO_2 required by aquatic plants for photosynthesis. Organisms obtain sufficient quantities of these necessary gases in the aquatic environment through several adaptations that we will examine in this section.

CARBON DIOXIDE

Getting enough CO_2 for photosynthesis is a particular challenge for aquatic plants and algae. The solubility of CO_2 in fresh water is about 0.0003 litres of gas per litre of water, which is 0.03 percent by volume, or about the same as its concentration in the atmosphere. The problem for aquatic plants is that CO_2 diffuses slowly through water, and plants can use the gas close to the surface of their leaves faster than it arrives by diffusion. Fortunately, however, when CO_2 dissolves in water, most of the molecules combine with water and are quickly converted to a compound called carbonic acid (H_2CO_3):

$$CO_2 + H_2O \rightarrow H_2CO_3$$

As shown in Figure 2.14, carbonic acid can increase to high concentrations and provide a reservoir of carbon required for photosynthesis. Depending on the acidity of the water, carbonic acid molecules can release hydrogen ions (H^+) to form either **bicarbonate ions** (HCO_3^-) or **carbonate ions** (CO_3^{2-}):

$$H_2CO_3 \rightarrow H^+ + HCO_3^-$$

Although some of the carbonate ions can combine with calcium ions to form calcium carbonate, as we saw in Figure 2.8, the bicarbonate ion is the most common form of inorganic carbon within the range of acidity that is typical of most aquatic habitats

---

Bicarbonate ion (HCO_3^-) An anion formed by the dissociation of carbonic acid.

Carbonate ion (CO_3^{2-}) An anion formed by the dissociation of carbonic acid.

Figure 2.14 An equilibrium reaction for carbon in water. The reaction converting CO_2 into bicarbonate ions (HCO_3^-) is an equilibrium reaction. When photosynthetic organisms preferentially use CO_2, because it is used more efficiently, the amount of dissolved CO_2 in the water declines. As the amount of CO_2 declines, some of the bicarbonate ions are converted to CO_2 to replenish the supply. The size of each circle represents the relative size of the carbon pool.

(pH values between 5 and 9). In addition, bicarbonate ions dissolve readily in water. The result, as shown in Figure 2.14, is a concentration of bicarbonate ions equivalent to 0.03 to 0.06 L of CO_2 gas per litre of water (3 to 6 percent)—more than 100 times the concentration of CO_2 in air. In short, this means that most of the CO_2 that is dissolved into water is rapidly converted to bicarbonate ions and these bicarbonate ions provide an enormous reservoir of inorganic carbon in aquatic systems. Dissolved CO_2 and bicarbonate ions are in a chemical equilibrium, which represents the balance achieved between H^+ and HCO_3^- on one hand, and CO_2 and H_2O on the other. When CO_2 is removed from the water during photosynthesis some of the abundant bicarbonate ions combine with hydrogen ions to produce more CO_2 and H_2O:

$$H^+ + HCO_3^- \rightarrow CO_2 + H_2O$$

Aquatic plants and algae use CO_2 and bicarbonate ions for photosynthesis. Although the bicarbonate ion is the most common form of inorganic carbon under moderate pH conditions, CO_2 is the most common form under more acidic conditions, such as in bogs. Some species of plants and algae can only use CO_2 for photosynthesis; these species live in aquatic habitats containing low pH. Species of plants and algae that live in moderate pH conditions, ranging from 5 to 9, can either directly uptake the bicarbonate ion and use it for photosynthesis or they can use adaptations that provide them with additional CO_2. One way to do this is by secreting an enzyme into the water that is highly effective at converting bicarbonate ions into CO_2, which can then be taken up by the organism for photosynthesis. Plants and algae can also obtain CO_2 by secreting hydrogen ions into the surrounding water. This helps drive the chemical equilibrium in a direction that converts bicarbonate ions into CO_2 that can then be taken up by the organism and used in photosynthesis.

Even when CO_2 and bicarbonate ions are abundant in water, the slow rate at which these carbon sources diffuse through the water prevents organisms from getting access to them. Indeed, carbon dioxide diffuses through unstirred water about 10 000 times more slowly than through air, and diffusion of HCO_3^- is even slower because larger molecules diffuse at a slower rate. Compounding this slow rate of diffusion is the fact that every surface of an aquatic plant, alga, or microbe is surrounded by a *boundary layer*. A **boundary layer** is a region of unstirred air or water that surrounds the surface of an object. In the water, the boundary layer ranges from as little as 10 micrometres (10 μm, or 0.01 mm) for single-celled algae in turbulent waters to 500 μm (0.5 mm) for a large aquatic plant in stagnant water. As Figure 2.15 illustrates, because this boundary layer is composed of unstirred water, CO_2 and HCO_3^- can be depleted within the boundary layer by uptake—especially in the region closest to the photosynthesizing organism—but the removed gases are slow to be replaced from the surrounding water. Without a boundary layer, the moving water in the surrounding environment would continually provide the plant with a supply of CO_2 and HCO_3^-. So, despite the generally high concentration of bicarbonate ions in the water, photosynthesis may still be limited by carbon availability within the boundary layer.

Boundary layer A region of unstirred air or water that surrounds the surface of an object.

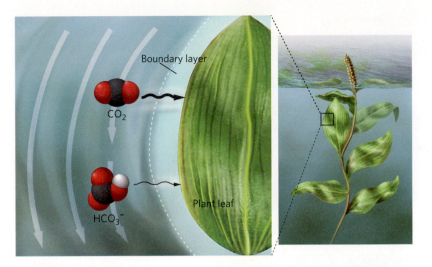

Figure 2.15 Boundary layer. A thin boundary of unstirred water along the surface of photosynthesizing organisms slows the rate of gas diffusion through the water. Smaller molecules, such as CO_2, diffuse faster than larger molecules, such as bicarbonate ions.

OXYGEN

Oxygen in the atmosphere has a concentration of 0.21 L per litre of air (21 percent by volume). In water, however, the maximum solubility of oxygen is 0.01 L per litre of water (1 percent), under the conditions of fresh water at 0°C. Oxygen's low solubility in water can limit the metabolism of organisms in aquatic habitats. For marine mammals such as whales, this is not a problem because they obtain O_2 from the air and they store copious amounts in their hemoglobin and myoglobin. For organisms that obtain O_2 from the water, however, the problem of low O_2 concentration is compounded by its slow diffusion in water, similar to that of CO_2. Oxygen is in even shorter supply in waters that cannot support photosynthesis—and therefore do not receive the O_2 it produces as a by-product—including in deep water that does not receive sunlight and in waterlogged sediments and soils. These habitats can become severely depleted of dissolved O_2, making them challenging environments for animals and microbes that use aerobic respiration.

One important adaptation that allows aquatic animals to deal with a limited amount of oxygen involves the direction of blood flow in the gills. Many aquatic animals have gills to extract oxygen from the water. When water passes over the gills, oxygen diffuses across the membranes of the gill cells and enters the capillaries, which are part of the bloodstream. The key to extracting the most oxygen from the water lies in the use of *countercurrent circulation*. In **countercurrent circulation**, two fluids move in opposite directions on either side of a barrier and heat or materials are exchanged. In contrast, **concurrent circulation** involves two fluids moving

in the same direction on either side of a barrier and heat or materials are exchanged. As illustrated in Figure 2.16, if blood and water were to flow in the same direction the concentration of oxygen would quickly come to an intermediate equilibrium. After this region of contact, there is no net movement of oxygen across the membrane. In contrast, when blood and water flow in opposite directions the concentration of oxygen in the water exceeds the concentration in the blood throughout most of the region of contact. This happens because even as the capillaries begin to develop high concentrations of oxygen, the adjacent water still has a higher concentration of oxygen. As a result, the oxygen continues to diffuse into the gill capillaries. Thus, countercurrent blood flow in animal gills allows much more oxygen to move from the water to the gills.

Species of animals that live in habitats with low amounts of oxygen have evolved a number of additional adaptations. In the deep oceans, many organisms have very low rates of activity, thereby reducing their need for oxygen. Many species of zooplankton, a group of tiny crustaceans, can increase the amount of hemoglobin in their bodies to the point that their normally transparent bodies turn red. Other animals, such as tadpoles and fish that live in oxygen-depleted swamps, swim to the surface and

Countercurrent circulation Movement of two fluids in opposite directions on either side of a barrier through which heat or dissolved substances are exchanged.

Concurrent circulation Movement of two fluids in the same direction on either side of a barrier through which heat or dissolved substances are exchanged.

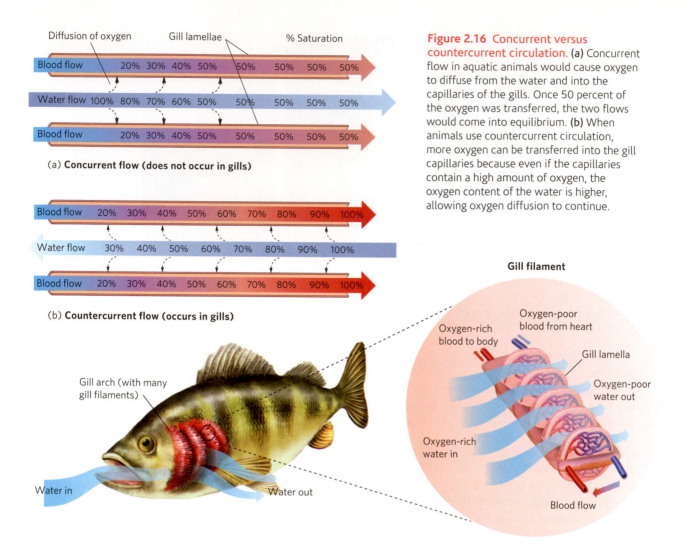

Figure 2.16 Concurrent versus countercurrent circulation. (a) Concurrent flow in aquatic animals would cause oxygen to diffuse from the water and into the capillaries of the gills. Once 50 percent of the oxygen was transferred, the two flows would come into equilibrium. **(b)** When animals use countercurrent circulation, more oxygen can be transferred into the gill capillaries because even if the capillaries contain a high amount of oxygen, the oxygen content of the water is higher, allowing oxygen diffusion to continue.

take gulps of air. Many tadpoles can use this air because they possess primitive lungs in addition to gills. Fish store this air in a swim bladder from which they extract the oxygen into their bloodstream.

One of the most surprising animal adaptations for obtaining oxygen was recently discovered in a species of North American salamander. For more than a century it was known that eggs of the spotted salamander, which are usually attached to sticks that are submerged in water, have a mutualistic relationship with a species of algae (*Oophila amblystomatis*). The algae obtain a place to live and photosynthesize while the developing embryo obtains oxygen from the photosynthesizing algae (**Figure 2.17**). This oxygen benefit is important because it allows the salamander embryos to have a higher survival rate and to hatch earlier and larger. In 2011, however, scientists reported that this relationship was much closer than they had appreciated. The algae not only live in the fluid of the egg surrounding the embryo, but also move into the embryo, positioning themselves in between the developing embryo's cells. This was the first discovery of an algae living within the tissues of a vertebrate animal.

Figure 2.17 Salamander embryos and algae. The embryos of the spotted salamander contain algae that live in the egg sac and within the cells of the embryo. The algae provide extra oxygen to the embryos, which improves embryo survival and growth. Photo by Ted Levin/Animals Animals-Earth Scenes.

When an environment becomes completely devoid of oxygen, it is referred to as **anaerobic** or **anoxic**. Anaerobic conditions pose problems for terrestrial

Anaerobic Without oxygen. *Also known as* **Anoxic**.

plants rooted in waterlogged soils, such as the many species of mangrove trees that live along coastal mudflats. The roots of these trees need oxygen for respiration, so the plants have evolved special air-filled tissues extending from the roots that rise above the waterlogged soils and exchange gases directly with the atmosphere (see Figure 2.13).

Many microbes are able to live in environments without oxygen because they use anaerobic respiration. A common product of anaerobic respiration by bacteria living in anoxic soils is hydrogen sulphide gas (H_2S). This gas is the cause of the rotten-egg smell that occurs when soils that are saturated with water become anaerobic.

Temperature limits the occurrence of aquatic life

At the beginning of this chapter, we saw that whales possess a number of adaptations that allow them to deal with the cold ocean temperatures, including thick layers of fat, high metabolism, and countercurrent circulation that exchanges heat between the warm arteries and cold veins. For aquatic organisms, most physiological processes occur only within the range of temperatures at which water is liquid. Relatively few organisms can survive temperatures above 45°C, which is the upper limit of the physiological range for most eukaryotic organisms. In this section we will look at how organisms have adapted to hot and cold temperatures.

HEAT AND BIOLOGICAL MOLECULES
Temperature influences physiological processes because of the way in which heat affects organic molecules. Heat imparts kinetic energy to living systems, causing biological molecules to change shape. Heat also accelerates chemical reactions by increasing the rate of molecule movement. In fact, the rate of most biological processes increases two to four times for each 10°C rise in temperature. We can see this when we examine the data for a fish known as the miiuy croaker (*Miichthys miiuy*), shown in Figure 2.18. The fish's rate of oxygen consumption approximately doubles when the environmental temperature increases from 15°C to 25°C. To understand the relationship between temperature and physiological process, we calculate the ratio of the physiological rate at one temperature to the physiological rate when the temperature is 10°C cooler, a ratio that is referred to as the Q_{10} of the physiological process. By knowing the Q_{10} values of different physiological processes, we can better understand which processes are more sensitive to a change in temperature.

Higher temperatures allow organisms to do many things more rapidly. They can swim, run, and fly

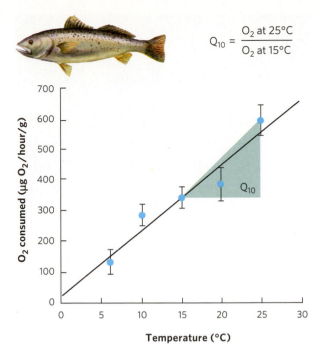

$$Q_{10} = \frac{O_2 \text{ at } 25°C}{O_2 \text{ at } 15°C}$$

Figure 2.18 Oxygen consumption as a function of temperature. For the miiuy croaker, the amount of oxygen it consumes increases as the temperature increases. By dividing the rate of oxygen consumption at 25°C by the rate of oxygen consumption at 15°C, one can arrive at the Q_{10} value for this physiological function. Error bars are standard deviations. Data from: Z. Zheng et al., Effects of temperature and salinity on oxygen consumption and ammonia excretion of juvenile miiuy croaker, Miichthys miiuy (*Basilewsky*), Aquaculture International 16 (2008): 581–589.

faster. They can also digest and assimilate more food and, as a result, grow and develop faster. Beyond a certain point, however, high temperatures can depress life processes. In particular, proteins and other biological molecules become less stable at higher temperatures, and may not function properly or may lose their structure. The molecular motion caused by heat tends to open up, or *denature,* the structure of these molecules.

Given that proteins denature at high temperatures, scientists have been intrigued by the fact that some organisms, such as **thermophilic** (heat-loving) bacteria, can live at very high temperatures. For example, some photosynthetic bacteria can tolerate temperatures as high as 75°C and some archaebacteria can live in hot springs at temperatures up to 110°C (Figure 2.19). The chemosynthetic bacteria that live near deep-sea vents are a group of thermophiles that we discussed in Chapter 1. Researchers have discovered that the proteins of thermophilic bacteria

Q_{10} The ratio of the rate of a physiological process at one temperature to its rate at a temperature 10°C cooler.

Thermophilic Heat-loving.

Figure 2.19 Thermophiles. Some species of bacteria and archaebacteria can live under very hot conditions, such as the Rabbitkettle Hotsprings in Yukon. Thermophiles often come in a variety of bright colours. Photo by Stephen J. Krasemann/Science Source.

have higher proportions of particular amino acids that form stronger bonds than the proteins of other, heat-intolerant species. These strong forces of attraction within and between molecules prevent them from being shaken apart under high temperatures, so the proteins do not denature.

Temperature also affects other biological compounds. For instance, the physical properties of fats and oils, which are major components of cell membranes and constitute the energy reserves of animals, depend on temperature. When fats are cold, they become stiff; when they are warm, they become fluid.

Organisms are negatively affected when exposed to temperatures that range above or below those to which they are adapted. For example, the Fraser River in British Columbia is home to one of the largest salmon runs in the world, and every year millions of sockeye salmon (*Oncorhynchus nerka*) migrate up the river to spawn. Many factors affect the success of migration and spawning, including temperature. Salmon exposed to temperatures outside the range of their natal stream exhibit decreased swimming performance, metabolism, and survival. Global climate change has caused mean summer water temperatures in the Fraser River to increase by 1.5°C over the last 6 decades. If temperatures continue to rise, sockeye may have an increasingly difficult time migrating and spawning.

Water used to keep nuclear power plants cool is another cause of exposure to unusually high temperatures. Many nuclear power plants extract water from nearby lakes, rivers, or oceans for cooling the power plants and then return the water—which has become much warmer in the process—to its source. But the organisms in these bodies of water have no evolutionary history of being exposed to such high water temperatures and many of them die. Discharging water that is too hot to sustain aquatic species is known as **thermal pollution**. Understanding the thermal optima for aquatic species has led to regulations that restrict how much a discharging power plant can raise the temperature of a lake or river.

COLD TEMPERATURES AND FREEZING

Temperatures on Earth's surface rarely exceed 50°C, except in hot springs and at the soil surface in hot deserts. However, temperatures below the freezing point of water are common, particularly on land and in small ponds, which may freeze solid during winter. When living cells freeze, the crystal structure of ice disrupts most life processes and may damage delicate cell structures, eventually causing death. Many organisms successfully cope with freezing temperatures either by maintaining their body temperatures above the freezing point of water, or by activating chemical pathways that enable them to resist freezing or to tolerate its effects.

It might surprise you to learn that marine vertebrates are susceptible to freezing in cold seawater. How can blood and body tissues freeze solid while immersed in liquid? The answer is that dissolved substances lower the temperature at which water freezes. While pure water freezes at 0°C, seawater, which contains about 3.5 percent dissolved salts, freezes at −1.9°C. Because the blood and body tissues of most vertebrates contain about half the salt content of seawater, the animals can freeze solid before the surrounding seawater does.

Marine animals have evolved a number of adaptations to combat the problem of freezing in water. We know that high salt concentrations interfere with many biochemical processes, so raising the concentration of solutes in blood and tissues is not a viable option. Instead, some polar fish prevent freezing by raising their blood and tissue concentrations of nonsalt compounds such as *glycerol*. **Glycerol** is a chemical that prevents the hydrogen bonds of water from coming together to form ice unless the temperatures are well below freezing. A 10 percent glycerol solution in the body lowers the freezing point of water to about −2.3°C without severely disrupting biochemical processes. This is enough to lower the freezing point of the body below the freezing point of seawater. **Glycoproteins** are

---

Thermal pollution Discharging water that is too hot to sustain aquatic species.

Glycerol A chemical that prevents the hydrogen bonds of water from coming together to form ice unless the temperatures are well below freezing.

Glycoproteins A group of compounds that can be used to lower the freezing temperature of water.

Figure 2.20 Adaptation to different water. The Arctic cod, which is commonly harvested in Russia for human consumption, can live in water that is less than 4°C. Antifreeze compounds in its tissues prevent it from freezing solid. Photo by Elizabeth Calvert Siddon/UAF/NOAA.

another group of compounds that can be used to lower the freezing temperature of water. Glycerol and glycoproteins act as antifreeze compounds, similar to the antifreeze used in automobiles, and allow fish such as the Arctic cod (*Boreogadus saida*) to remain active in seawater that is cold enough to cause most fish to freeze solid (Figure 2.20). Some terrestrial invertebrates also use the antifreeze approach; their body fluids may contain up to 30 percent glycerol as winter approaches.

Supercooling provides a second physical solution to the problem of freezing. Under certain circumstances, liquids can cool below the freezing point without developing ice crystals. Ice generally forms around an object, called a *seed,* which can be a small ice crystal or other particle. In **supercooling**, however, glycoproteins in the blood impede ice formation by coating any ice crystals that begin to form. In the absence of ice seeds, pure water may cool to more than −20°C without freezing. Such supercooling has been recorded down to −8°C in reptiles and −18°C in invertebrates.

THERMAL OPTIMA

Every organism is best suited to a narrow range of environmental conditions; this range defines the **optimum** environmental conditions. In terms of temperature, most organisms have a **thermal optimum**, meaning the range of temperatures in which they perform best. The thermal optimum is determined by the properties of enzymes and lipids, the structures of cells and tissues, body form, and other characteristics that influence the ability of an organism to function well under the particular conditions of its environment. Returning to the example of fish in polar ocean waters, many species swim actively and consume oxygen at a rate comparable to fish living in much warmer regions near the equator. If you put a tropical fish in cold water, it becomes sluggish and soon dies; conversely, Antarctic fish cannot tolerate temperatures warmer than 5°C to 10°C.

Certain adaptations allow fish in cold oceans to swim as actively as fish in warm oceans. Swimming involves a series of biochemical reactions, most of which depend on enzymes. These reactions generally proceed more rapidly at high temperatures, so cold-adapted organisms must either have more of the substrate for a biochemical reaction, more of the enzyme that catalyzes the reaction, or a different version of the enzyme that operates better under colder temperatures. Different forms of an enzyme that catalyze a given reaction are called **isozymes**.

Consider the case of the rainbow trout (*Oncorhynchus mykiss*), a fish that lives in cold streams throughout much of North America. Seasonal changes in temperature are predictable for the trout: During winter, water temperatures may drop close to the freezing point, while summer temperatures can become very warm. In response to these seasonal temperature changes, the trout has evolved the ability to produce different isozymes in winter and in summer. One of these enzymes is acetylcholinesterase, which plays an important role in ensuring proper functioning of the nervous system by binding with the neurotransmitter acetylcholine. To understand how well different isozymes function at different temperatures, we can examine the rate of the chemical reaction between acetylcholine and acetylcholinesterase, a measure known as *enzyme-substrate affinity.* The winter isozyme, shown as the

---

Supercooling A process in which glycoproteins in the blood impede ice formation by coating any ice crystals that begin to form.

Optimum The narrow range of environmental conditions to which an organism is best suited.

Thermal optimum The range of temperatures within which organisms perform best.

Isozymes Different forms of an enzyme that catalyze a given reaction.

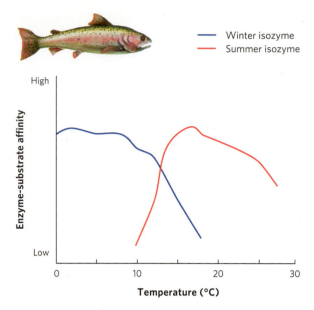

High

Enzyme-substrate affinity

Low

Temperature (°C)

0   10   20   30

— Winter isozyme
— Summer isozyme

Figure 2.21 Using isozymes as adaptations to changing water temperatures. In winter, the rainbow trout produces a form of the enzyme acetylcholinesterase that has a high substrate affinity at 0°C to 10°C, but a lower substrate affinity at warmer temperatures. In summer, the trout produces a different form of the enzyme, which has a high substrate affinity at 10°C to 20°C, but a lower substrate affinity at colder temperatures.

blue line in Figure 2.21, best catalyzes the reaction at 0°C to 10°C. This affinity drops rapidly at higher temperatures. In contrast, the summer isozyme, shown as the orange line, has a weak affinity for acetylcholine at 10°C and catalyzes reactions the best at 10°C to 20°C, though the affinity drops slowly at higher temperatures. As you might predict, the particular isozyme that a trout produces depends on the temperature of the water it lives in. When trout live at 2°C, they produce the winter isozyme, whereas when they live at 17°C, they produce the summer isozyme.

ECOLOGY TODAY CONNECTING THE CONCEPTS

THE DECLINE OF CORAL REEFS

Coral reef diversity. Coral reefs, such as this one off Beqa Island, Fiji, in the South Pacific, are among the most species-diverse places on Earth. Photo by Michael Patrick O'Neill/Photo Researchers, Inc.

Coral reefs are some of the most beautiful places on Earth and are home to an incredible diversity of species. Coral reefs are also one of the most diverse places on Earth, rivaling even the highly diverse tropical rainforests. For instance, the Great Barrier Reef off the eastern coast of Australia contains more than 400 species of coral, 200 species of birds, and 1500 species of fish. For these reasons, many people become concerned when the biodiversity of coral reefs is threatened by human activity. We have long understood that overfishing and pollution have affected the species that inhabit coral reefs. In the past 15 years, however, scientists have discovered that changes in the abiotic aquatic environment—including changes in temperature, pH, and salinity—are damaging coral reef ecosystems.

Corals are a group of tiny animals that secrete hard exoskeletons made of limestone (calcium carbonate). They can be found worldwide in relatively shallow ocean waters that are low in nutrients and food. Although each individual coral is small—only a few millimetres in size—the limestone bodies of dead corals accumulate over hundreds or thousands of years to form massive coral reefs that can exceed 300 000 km². The corals survive in these low nutrient waters by living in symbiotic relationships with several species of photosynthetic algae known as *zooxanthallae*. Corals have tubular bodies with tentacles that stick out and catch bits of food and detritus that go by. Their digestion produces CO_2, which can be used by symbiotic algae during photosynthesis. As we learned earlier in this chapter, CO_2 can often be difficult for aquatic producers to obtain. In exchange, the algae produce O_2 and sugars, some of which can be passed to the coral. In short, the algae get a safe place to live and a steady supply of CO_2 for photosynthesis, while the corals get a source of energy in the form of sugars and a steady supply of O_2 for respiration.

During the past two decades, scientists have learned that the symbiotic relationship between the corals and algae is very sensitive to environmental changes. When corals are experiencing stress in their environment, they expel the symbiotic algae from their bodies. Because corals obtain their bright colours from the symbiotic algae, corals that expel their algae often look white and are said to experience **coral bleaching**.

Coral bleaching is associated with unusually high ocean temperatures. As we discussed in this chapter, while increases in water temperature can increase the rate of chemical reactions, temperatures that exceed the thermal optima can be detrimental. Bleaching can begin if summer ocean temperatures are even just 1°C higher than the average maximum. If the temperature rise is brief—a few days or weeks—the algae can recolonize the corals. However, the corals will experience slower growth and reduced reproduction. If the temperatures are 2°C to 3°C higher than the average maximum, corals can die. During the past two decades, scientists have witnessed major bleaching events worldwide in 1998, 2003, 2005, and 2010. During the 1998 event, 42 percent of corals in the Great Barrier Reef experienced bleaching. With continuing increases in global temperatures, a topic we will cover in detail in later chapters, temperature-induced coral bleaching is expected to continue.

Changes in salt concentration are also an issue for corals. High ocean temperatures increase the evaporation of water from the ocean, which increases the ocean's salt concentration. In the case of corals, the stress of increased salt concentrations combined with the stress of high temperatures makes them increasingly vulnerable to coral bleaching and coral death. The rise in ocean temperatures also appears to stress corals in a way

Coral reef decline. Numerous changes in the ocean environment are causing declines in the numbers of live corals. In the Caribbean, for example, there has been a sharp drop in the number of live corals over a period of three decades. [After L. Alvarez-Filip, et al., Region-wide temporal and spatial variation in Caribbean reef architecture: Is coral cover the whole story? *Global Change Biology* 17 (2011): 2470–2477.]

that makes them more susceptible to pathogens that can kill them.

Another source of decline in corals is a decrease in the pH of seawater. Because atmospheric CO_2 is in equilibrium with dissolved CO_2 in the ocean, the recent increase in atmospheric CO_2 has caused an increase in dissolved CO_2. This, in turn, has caused an increase in carbonic acid, and a decline in the pH of the oceans. As we have seen, carbonic acid (H_2CO_3) disassociates to bicarbonate ions (HCO_3^-) and hydrogen ions (H^+). The hydrogen ions can then combine with other carbonate ions to make bicarbonate, making carbonate ions less available for the corals to produce their exoskeleton of calcium carbonate.

The decline in coral reefs has become a serious problem. In the Caribbean, for example, the percent of live corals has declined from more than 40 percent in 1977 to less than 10 percent in 2007. Only 20 years ago, scientists debated whether abiotic changes were playing a role in the decline of corals. As more data have been collected, there is now general agreement. Given that the changes in temperature, pH, and salinity are expected to rise even more in future decades, scientists currently predict that many species of coral will continue to decline. However, some species of coral may have sufficient genetic variation to allow evolutionary changes that permit them to adapt to the changing environmental conditions and persist.

SOURCES: Carpenter, K. E., et al. 2008. One-third of reef-building corals face elevated extinction risk from climate change and local impacts. *Science* 321: 560-563.

Hoegh-Guldberg, O., et al. 2007. Coral reefs under rapid climate change and ocean acidification. *Science* 318: 1737-1742.

Coral bleaching Loss of colour in corals as a result of the corals expelling their symbiotic algae.

SUMMARY OF CHAPTER CONCEPTS

■ **Water has many properties favourable to life.** These include thermal resistance to changes in temperature, a density and viscosity that select for adaptations for movement, and the ability to dissolve many elements and compounds that are essential to life.

■ **Aquatic environments challenge the balance of water and salt in animals.** This challenge occurs because animal tissues are generally hyperosmotic or hyposmotic compared to the solute concentration of the surrounding aquatic environment.

■ **The uptake of gases from the water is limited by diffusion.** This limitation can make it difficult for organisms to exchange gases such as CO_2 and O_2. Diffusion is slowed because organisms are surrounded by a thin boundary layer of unstirred water. Although CO_2 can be abundant in water—either as a dissolved gas or in the form of bicarbonate ions—oxygen's low solubility in water makes it less abundant. As a result, many animals have respiratory structures with large surface areas and countercurrent circulation to obtain oxygen from the water.

■ **Temperature limits the occurrence of aquatic life.** Although warmer temperatures increase the rate of chemical reactions, excessively high temperatures cause proteins and other important molecules to become unstable and denature. Cold temperatures also pose a challenge and many organisms living under near-freezing conditions have evolved adaptations that include the use of glycoproteins and supercooling to prevent the harmful effects of ice crystals growing inside their cells. Organisms that live under a wide range of temperature often use isozymes to permit proper physiological function at each temperature.

REVIEW QUESTIONS

1. As water cools below 4°C, it expands and becomes less dense. Why is this beneficial to many organisms?

2. How can the viscosity of water both hinder and facilitate movement in aquatic animals?

3. Describe the changes in mineral content of water as it moves from rainwater to lake water and, eventually, to ocean water.

4. Explain the difference between a sample standard deviation and the standard error of the mean.

5. Both saltwater and freshwater fish have adaptations to control the movement of water and salts across their external surfaces. Describe what would happen without these adaptations.

6. Why does the boundary layer surrounding a photosynthetic organism make it more difficult to exchange CO_2 and O_2?

7. Why are deep ocean waters typically low in oxygen?

8. Explain the adaptation that allows thermophiles to live under very high temperatures.

9. Describe the adaptations that enable fish to survive in very cold seawater.

10. How do isozymes help organisms function in habitats with a wide range of temperatures?

PRACTICING ECOLOGY: SEEING UNDERWATER

Shai Sabbah and his colleagues from Queen's University investigate how the visual system of fish is adapted to light transmission underwater. Photo by Omer Polak.

The way light travels in water is very different from how it travels on land, and this presents visual challenges to aquatic life. Much of the light that strikes the surface of water is scattered or absorbed near the surface and relatively little light penetrates to greater depths. Additionally, the light spectrum changes with increasing water depth. Long wavelengths of light, such as red, are absorbed near the surface. Shorter wavelengths of light, such as blue and ultraviolet, are able to penetrate to deeper waters. Given that different light wavelengths penetrate to different depths, does natural selection favour visual abilities adapted to the depth at which an animal spends most of its time?

Nile tilapia (*Oreochromis niloticus*) is a species of freshwater fish native to Africa. Tilapia relies on vision to detect animal prey and other resources. The newly hatched tilapia, known as fry, inhabit the surface waters and feed primarily on tiny zooplankton. These zooplankton absorb ultraviolet light, which should make them more noticeable against the surrounding water. As the tilapia grow into juveniles and then adults, they move into deeper waters where they feed primarily on plants. Hence, we might hypothesize that the differences in the depth and diet of tilapia at different stages of life are accompanied by a change in sensitivity to ultraviolet light.

Shai Sabbah and his colleagues from Queen's University in Ontario tested this hypothesis by examining the various types of light-recepting pigments in the eyes of tilapia. One pigment, SWS1, is sensitive to ultraviolet light, and a higher percentage of SWS1 in the retina makes individuals more sensitive to ultraviolet light. The researchers measured the percentage of SWS1 in fry, juveniles, and adults. You can view their data in the table.

PERCENTAGE OF SWS1 PIGMENT

FRY	JUVENILES	ADULTS
29.3	0.2	16.5
40.1	10.1	0.0
30.6	28.8	0.0
28.1	0.0	0.0
8.7	0.0	0.0
29.0	0.0	0.0
0.0	0.0	14.7
32.7	22.0	0.0
47.3	30.0	24.0
0.0	1.1	37.5
5.8	0.0	0.0
25.6	0.0	0.0
24.1	25.9	0.0
0.0	0.0	0.0
0.0	10.5	0.0
	0.0	
	0.0	
	0.0	
	0.0	

GRAPH THE DATA

1. Calculate the means and standard errors for fry, juvenile, and adult fish.

2. Plot the means and standard errors on a bar graph.

INTERPRET THE DATA

1. Based on the mean percentage of SWS1 pigment in each group, which group is most sensitive to UV light?

2. Does this result support the researchers' hypothesis?

3. How might the percentage of SWS1 vision pigment serve as an adaptation for the fry?

SOURCE: Sabbah, S., J. Hui, F. E. Hauser, W. A. Nelson, and C. W. Hawryshyn. 2012. Ontogeny in the visual system of Nile tilapia. *The Journal of Experimental Biology* 215: 2684–2695.

CHAPTER 3

Adaptations to Terrestrial Environments

The Evolution of Camels

When you think of camels you might envision the iconic animals of the African and Asian deserts. The ancestor of all camel species actually originated in North America about 30 million years ago and camels roamed many parts of North America until about 8000 years ago. Current evidence suggests that some of these ancestors crossed the Bering land bridge from Alaska about 3 million years ago and made their way into Asia and Africa. These individuals evolved into two modern-day species of camels: the endangered Bactrian camel (*Camelus bactrianus*) and the much more common dromedary camel (*C. dromedarius*). At the same time, other individuals moved down to South America and evolved into a second group: guanacos (*Lama guanicoe*), llamas (*L. glama*), vicuñas (*Vicugna vicugna*), and alpacas (*V. pacos*).

All of these animals live in dry environments and have evolved a number of adaptations that help them cope with the harsh conditions. For example, the dromedary camel—the most-studied of the group—risks overheating in the very hot deserts where it lives. During the day, the rays of the Sun strike its body and warm its surface. The camel can respond behaviourally by facing into the Sun and presenting a smaller profile for the Sun's rays to warm. The sandy desert soil is also hot and radiates the heat it absorbs, which makes the camel even hotter. Fortunately, the camel has a very large body relative to its surface area, so these heat inputs raise its body temperature slowly. Although camels, like all mammals, try to maintain a constant body temperature, they can tolerate a rise in body temperature of 6°C before experiencing any harmful effects. In contrast, most mammals can only tolerate an increase of about 3°C. At night, as the air and sand cool rapidly, the camel radiates its excess heat into the air or lies down and transfers its excess body heat to the sand.

> "One of the many amazing adaptations of camels is their ability to cool their brains."

One of the many amazing adaptations of camels and other mammals in hot environments is their ability to cool their brains. Whereas much of the body can tolerate short-term increases in temperature, the brain cannot. The camel has evolved an arrangement of veins and arteries that helps solve this problem. As the animal breathes, veins positioned beside the long nasal cavity are cooled by means of evaporating water vapour. The veins that carry cool blood then travel toward the back of the camel's head where they come into close contact with the arteries that supply blood to the brain. Although the blood does not mix between the veins and arteries, heat is exchanged between the two types of vessels. This cools the arterial blood before it travels to the brain and keeps the brain several degrees cooler than the rest of the body.

Lack of water is also a challenge in the desert environment. As an adaptation, the dromedary camel can store large amounts of water in its body, most of which resides in its tissues. As water is lost from the bloodstream, the water from the tissues enters the blood. As much as 30 to 40 percent of the camel's body mass is water that can be used over several days when it is unable to drink water. In other mammals, a mere 15 percent loss of water can be lethal. Camels also conserve the water that they take in

Adaptations of the dromedary camel. Dromedary camels, such as this individual in the Hajar Mountains of the United Arab Emirates, have a wide range of adaptations that allow them to live in hot, dry environments. Photo by Allan Hartley/age fotostock.

Guanacos of the Patagonia. Guanacos, from the Patagonian region of Chile, share a common ancestor with Asian camels. As a result of their common ancestry, guanacos and Asian camels also share a number of adaptations for coping with a dry environment. Photo by Morty Ortega.

by producing relatively dry feces and urine that is high in waste products and low in water. In addition, while sweating can be an effective way to cool the body through evaporation, camels that are low on water can reduce the amount that they sweat. Collectively, these adaptations allow the camels to survive when water is scarce.

Camels represent just one case in which terrestrial organisms have evolved numerous adaptations to the challenges posed by terrestrial environments. In this chapter, we will explore the challenges of living on land and the adaptations of terrestrial plants and animals that make their lives possible.

SOURCES: J. W. Cain et al., Mechanisms of thermoregulation and water balance in desert ungulates, *Wildlife Society Bulletin* 34 (2006): 570–581.

S. Ouajd and B. Kamel, Physiological particularities of dromedary (*Camelus dromedarius*) and experimental implications, *Scandinavian Journal of Laboratory Animal Science* 36 (2009): 19–29.

CHAPTER CONCEPTS

- Most terrestrial plants obtain nutrients and water from the soil.

- Sunlight provides the energy for photosynthesis.

- Terrestrial environments pose a challenge for animals to balance water, salt, and nitrogen.

- Adaptations to different temperatures allow terrestrial life to exist around the planet.

As we discussed in Chapter 2, life on Earth probably arose in the water. Following these early origins, life forms evolved adaptations that allowed them to live on land. The transition from water to land posed a number of new challenges. Plants evolved to obtain water and nutrients from the soil, and to conduct photosynthesis under hot and dry conditions. Animals evolved to balance water, salts, and wastes, and to adjust to extreme temperatures in terrestrial environments. In this chapter, we will examine the diversity of adaptations that allow plants and animals to live in terrestrial environments.

Most terrestrial plants obtain nutrients and water from the soil

A few unusual plants, for example the epiphytes discussed in Chapter 1 (Figure 1.9), obtain necessary water and nutrients without being rooted in the soil. However, the vast majority of plants obtain nutrients and water from the soil through their root systems. As a result, plants have a number of adaptations that help them perform this task.

SOIL NUTRIENTS

In addition to the oxygen, carbon, and hydrogen that plants incorporate into carbohydrates to fuel their survival and growth, plants require many other inorganic nutrients including nitrogen, phosphorus, calcium, and potassium to make proteins, nucleic acids, and other essential organic compounds. Whereas oxygen and carbon are available in the air, other nutrients are obtained as ions dissolved in the water held by the soil around plant roots. Nitrogen exists in soil as ammonium (NH_4^+) and nitrate (NO_3^-) ions, phosphorus exists as phosphate ions (PO_4^{3-}), and calcium and potassium exist as the elemental ions Ca^{2+} and K^+, respectively. The availability of these and other inorganic nutrients varies with their chemical form in the soil, and with temperature, pH, and the presence of other ions. A scarcity of inorganic nutrients, such as nitrogen, often limits plant production in terrestrial environments. We shall have much more to say about nutrient uptake by plants in later chapters.

SOIL STRUCTURE AND WATER-HOLDING CAPACITY

To understand how plants obtain water and nutrients, it is first necessary to understand how water behaves in the soil. The movement of water in the soil can be described in terms of its **water potential**, which is a measure of the water's potential energy. Water potential affects the movement of water in the soil from one location to another and depends on several

---

Water potential A measure of the water's potential energy.

(a) Saturation (0 MPa)   (b) Field capacity (−0.01 MPa)   (c) Wilting point (−1.5 MPa)

Figure 3.1 Soil water. (a) Immediately after a rain event, soils can become saturated with water and all spaces between soil particles are filled. **(b)** The field capacity of soil represents the amount of water remaining after it has been drained by gravity. **(c)** The wilting point occurs when the opposing attractive forces of the soil particles prevent plants from extracting any more water.

factors that include gravity, pressure, osmotic potential (discussed in Chapter 2), and *matric potential,* so named because the collection of all soil particles is known as the *soil matrix.* The **matric (or matrix) potential** is the potential energy generated by the attractive forces between water molecules and soil particles. It exists because water molecules, which have electrical charges, are attracted to the surfaces of soil particles, which also have electrical charges. This attraction explains why soil is able to retain water against the downward pull of gravity.

Because electrical charges are responsible for the attraction between water molecules and soil particles, those water molecules closest to the surfaces of soil particles adhere the most strongly. When water is plentiful, most of the water molecules are not close to the surfaces of soil particles. As a result, these water molecules are not held tightly and plant roots can easily take up the water. As more of the water is used up, however, the water molecules that remain are positioned close to the soil particles and adhere tightly.

As we learned when discussing osmotic pressure in Chapter 2, scientists quantify water potential in units of pressure, called megapascals (MPa). In a soil that is completely saturated with water, as illustrated in Figure 3.1a, the matric potential is 0 MPa. When a saturated soil drains under the force of gravity, the resulting matric potential is about −0.01 MPa. At this point, the force of gravity on the water molecules is equally opposed by the attractive force of soil particles on the water molecules. The maximum amount of

water held by soil particles against the force of gravity is called the **field capacity** of the soil. The field capacity, illustrated in Figure 3.1b, represents the maximum amount of water available to plants.

As water becomes less abundant, such as when plants take up some of the water from the soil, the matric potential values become more negative. Water always moves from areas of higher potential (less negative values) to areas of lower potential (more negative values). So, for plants to extract water from the soil, they must produce a water potential that is lower than that of the soil. As soils dry out, they hold the remaining water ever more tightly because a greater proportion of that water lies close to the surfaces of soil particles. Most crop plants can extract water from soils with water potentials down to about −1.5 MPa. At lower soil water potentials, these plants wilt, even though some water still remains in the soil, as shown in Figure 3.1c. Scientists refer to a water potential of −1.5 MPa as the **wilting point** of soil because this is the lowest water potential at which most plants can obtain water from the soil. The

Matric potential The potential energy generated by the attractive forces between water molecules and soil particles. *Also known as* Matrix potential.

Field capacity The maximum amount of water held by soil particles against the force of gravity.

Wilting point The water potential at which most plants can no longer retrieve water from the soil, which is about −1.5 MPa.

wilting point of −1.5 MPa, however, is only a general rule of thumb. Many species of drought-adapted plants can extract water when the water potential is less than −1.5 MPa.

The amount of water in soil and its availability to plants depends on the physical structure of the soil. It also explains why the amount of water the soil can hold depends on the soil's surface area; for a given volume of soil, the more surface area that soil has, the more water it can hold. The surface area of soil depends on the sizes of the particles that comprise the soil. Soil particles include sand, silt, and clay, in addition to organic material from decomposing organisms. As shown in Figure 3.2, sand particles are the largest, with diameters exceeding 0.05 mm. Silt particles have diameters of 0.002 to 0.05 mm and clay particles are the smallest, with a diameter of less than 0.002 mm. Rarely is a soil composed of a single particle size. Instead, as illustrated in Figure 3.3, soils are typically composed of mixtures of different ratios of each

(a)

(b)

Figure 3.2 Soil particle size. **(a)** Soil particles are separated into three sizes: clay, silt, and sand. **(b)** Each soil particle attracts a surface film of water around it. The greater surface area of small clay particles holds a greater total amount of water than the much larger sand particles, which have a much smaller surface area relative to their volume.

particle size. For example, a soil composed of 40 percent sand, 40 percent silt, and 20 percent clay is classified as a *loam* soil. In contrast, a soil containing a higher proportion of silt and a lower proportion of sand is classified as a *silt loam* soil.

Smaller particles have a larger surface area relative to their volume compared to larger particles. As a result, the total surface area of particles in a given volume of soil increases as particle size decreases. Therefore, soils with a high proportion of clay particles hold more water than soils with a high proportion of silt particles, which hold more water than soils with a high proportion of sand particles. Soils with a high proportion of sand particles tend to dry out because water quickly drains away, leaving many tiny pockets of air between the large sand particles. Clay soils represent the opposite extreme; each tiny particle of clay can attract a thin film of water on its surface, leaving little space for air pockets. Although clay soils retain a lot of water, clay particles can hold water molecules so tightly that it can be difficult for plants to extract the water from the soil.

In Figure 3.4, we can see how soil particle size affects the amount of water in the soil, measured in terms of the percent of the soil volume occupied by water. As we move from sand to silt to clay, there is an increase in field capacity. However, there is also an increase in the wilting point. The difference between the field capacity and the wilting point is the amount of water available to plants. Thus, even when precipitation is frequent, sandy soils cannot retain much of the water that enters the soil. At the other extreme, clay soils can retain a lot of water, but if precipitation is not frequent enough for the soil to reach its field capacity, most of the water in the soil will be unavailable. This means that soils high in sand or high in clay are both poor soils for growing many plants including crops that humans rely on for food. Instead, soils that contain a mixture of clay, silt, and sand particles—such as loam—are some of the best soils for growing plants.

OSMOTIC PRESSURE AND WATER UPTAKE

In Chapter 2 we noted that osmotic forces cause water molecules to move from areas of low solute concentration to areas of high solute concentration. At the same time, ions and other solutes diffuse through water from regions of high solute concentration to regions of low solute concentration. In the case of a plant, if a root cell has a higher solute concentration than the soil water, osmotic forces can draw the water into the root. It is this osmotic potential in plant roots that causes water to enter the roots from the soil against the attractive forces of soil particles and the downward pull of gravity.

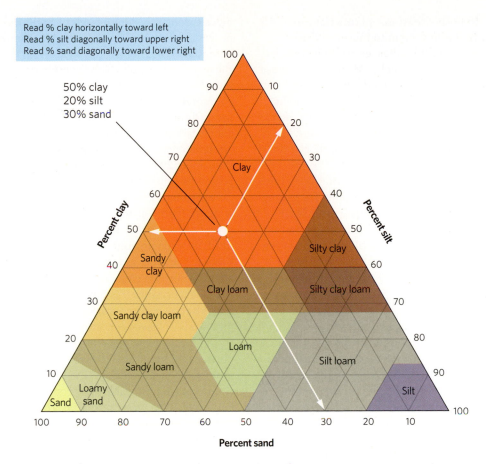

Read % clay horizontally toward left
Read % silt diagonally toward upper right
Read % sand diagonally toward lower right

50% clay
20% silt
30% sand

Figure 3.3 Combinations of soil particle sizes used to categorize soils. Most soils are composed of different percentages of sand, silt, and clay. Each name represents a category with a specific composition of the three particle sizes.

Without any other adaptations, we would expect the solute concentrations within the root cell and in

Figure 3.4 Water-holding capacity of different soils. Soils composed of different combinations of sand, silt, and clay differ in their water-holding capacity. Soils containing large amounts of sand have a low field capacity and a low wilting point. In contrast, soils containing large amounts of clay have a high field capacity and a high wilting point.

the soil water to eventually come into equilibrium. At this point, the osmotic potentials of the root cell and its surroundings would be equal, and there would be no net movement of water into the plant. However, root cells possess two adaptations that prevent this equalization. First, semipermeable cell membranes prevent larger solute molecules from leaving the plant's root. Second, cell membranes can actively transport ions and small molecules against a concentration gradient into the root cells. These two adaptations maintain high solute concentrations inside the roots and allow strong osmotic forces to continue.

As noted earlier, plants growing in places with very negative water potentials typically have adaptations to help them extract water beyond −1.5 MPa. Plants living in deserts, for example, can lower the water potential of their roots to as much as −6 MPa, thereby overcoming soil water potentials down to −6 MPa. Plants living in very salty environments can also meet the challenge of extracting water from an environment that contains unusually high concentrations of solutes in the form of salt ions. In both situations, plants have evolved adaptations that allow them to increase the concentrations of amino acids, carbohydrates, or organic acids in their root cells. Maintaining these high concentrations of dissolved substances, however, comes at a high metabolic price for the plants because they must divert some of the energy that would normally be used for growth and use it instead to manufacture these additional organic compounds.

Plants lacking the appropriate adaptations grow poorly when exposed to salty conditions. For example, in the deserts of the American southwest, a large amount of land is irrigated to grow crops, including cotton and orchard trees. Most well water, however, contains small amounts of salt. As this well water is sprayed over the fields, it moves down into the soil and dissolves any salts. If the irrigation system uses large amounts of water, the water can move deeply into the soil and the salts can be flushed away. If irrigation uses smaller amounts of water—just enough to feed the plant roots—the water stays near the soil's surface. As much of this water is then taken up by plants or evaporates, the salt is left behind on the soil's surface. With repeated irrigation events, the salt concentration of the soil continually increases. After many years, the soil can have such a high solute concentration that many crop plants cannot create a lower water potential than the soil, and therefore cannot obtain sufficient water. The process of repeated irrigation that causes increased soil salinity is known as **salinization**. High-salt soils occur across 831 million hectares (ha) of land and in 100 countries of the world. It is a particular problem throughout arid areas of the world where irrigation typically is limited to small amounts of water, which concentrates the salts at the soil surface.

TRANSPIRATION AND THE COHESION–TENSION THEORY

We have seen how osmotic potential draws water from the soil into the cells of plant roots. How does that water get from the roots to the leaves? As you might recall from a previous biology course, plants conduct water to their leaves through tubular xylem elements, which are the empty remains of xylem cells in the cores of roots and stems, connected to form the equivalent of water pipes. The movement of water through these xylem cells depends on the *cohesion* of water molecules and differences in water potential between the leaves and roots.

The **cohesion** of water is the result of mutual attraction among water molecules. The attraction of hydrogen bonds causes a water molecule moving up the xylem of a plant to pull other water molecules with it. Cohesion of water helps the entire column of water

move through the long vessels of a tall tree. The process, shown in Figure 3.5, begins when osmotic potential in the roots draws water from the soil into the plant and creates a **root pressure** that forces water into the xylem elements. However, this pressure is counteracted by gravity and the osmotic potential inside living root cells. Because of these two important counteracting forces, root pressure under the best circumstances can raise water to a height of no more than about 20 m, even though the tallest trees can achieve heights of more than 100 m.

Fortunately for plants, leaves can also generate water potential as water evaporates from the surfaces of leaf cells into the tiny air spaces inside the leaf, ultimately moving out of the leaf and into the air. This process is known as **transpiration**. The column of water in a xylem element is continuous from the roots to the leaves, since it is held together by hydrogen bonds between the water molecules. Thus, low water potentials in leaves can literally draw water upward through the xylem elements against the osmotic potential of both the living root cells and the pull of gravity. The water potential is low enough under most conditions to pull water up through the roots, xylem, and leaves. The water potential from transpiration creates a continuous gradient of water potential from leaf surfaces in contact with the atmosphere down to the surfaces of root hairs in contact with soil water. The movement of water is due to both water cohesion and water *tension* (which is another name for differences in water potential). This mechanism of water movement from roots to leaves due to water cohesion and water tension is known as the **cohesion–tension theory**.

Based on the cohesion–tension theory of water transport in plants, very tall plants should have a more difficult time moving water up their stems because the movement of a tall column of water in the plant is being opposed by the force of gravity. Recent research estimates that this system limits plants to a maximum height of 130 m. In support of this prediction, the

---

Salinization The process of repeated irrigation, which causes increased soil salinity.

Cohesion The mutual attraction among water molecules.

Root pressure When osmotic potential in the roots of a plant draws in water from the soil and forces it into the xylem elements.

Transpiration The process by which leaves can generate water potential as water evaporates from the surfaces of leaf cells into the air spaces within the leaves.

Cohesion–tension theory The mechanism of water movement from roots to leaves due to water cohesion and water tension.

Leaf
Tension pulls the water column up and out in the xylem of the leaf veins.

Water evaporates from the mesophyll cells and diffuses out of the stoma.

Mesophyll cells

Xylem

Stoma

Stem
Water molecules form a cohesive column in the xylem.

Tension pulls the water upward in the xylem of the stem.

Xylem

Root Hair
Water moves into the root by osmosis, and then into the xylem.

Tension pulls the water upward in the xylem of the root.

Xylem

Figure 3.5 Water movement in plants by cohesion and tension. Differences in water potential, also known as tension, cause water to move from the soil into the roots, from the roots into the stem, and from the stem into the leaves. The cohesion of water causes the water molecules to adhere to each other and to move as a single column of water up the xylem cells.

tallest tree that has been reliably measured was a 126-m Douglas fir tree.

Although transpiration generates a powerful force that moves water through a plant, when the soil reaches the wilting point, water lost from the leaves of a plant can no longer be replaced by new water moving into the roots. To prevent further water loss from the leaves, plants have various adaptations for controlling transpiration. Most of the cells on the exterior of a leaf are coated with a waxy cuticle that retards water loss. As a result, gas exchange between the atmosphere and the interior of the leaf primarily

Figure 3.6 Stomata. Stomata are pores in the surfaces of leaves, each bordered by two guard cells. Under conditions of low water availability, the guard cells close the opening and prevent the loss of water from the leaves. Photo by Callista Images Cultura/Newscom.

occurs through small openings on the surface of the leaves, called **stomata** (Figure 3.6). The stomata (stoma, singular) are the points of entry for CO_2 and the points of exit for water vapour escaping to the atmosphere by transpiration. When plants experience a scarcity of water, they can reduce water loss to the atmosphere by closing their stomata. As leaf water potential weakens, the *guard cells* that border a stoma collapse slightly, causing the guard cells to press together and close the stoma. Although closing the stomata provides the important benefit of reducing water loss, it comes at the cost of preventing CO_2 needed for photosynthesis from entering the leaf. As we will soon see, plants living in hot, dry

environments have evolved additional adaptations for dealing with this undesirable side effect.

Sunlight provides the energy for photosynthesis

Whether in water or on land, energy from the Sun is essential for the existence of most life on Earth. To understand how this energy is captured, we need to examine the energy that is available, the energy that is absorbed, and how this energy is converted into a usable form by photosynthesis. Plants have evolved several adaptive approaches for conducting photosynthesis in terrestrial environments; these mechanisms coincide with the environmental conditions of different regions of the world.

AVAILABLE AND ABSORBED SOLAR ENERGY
The energy from the Sun, known as **electromagnetic radiation**, is packaged in small particle-like units called photons. The energy of photons is positively related to their frequency and inversely to their wavelength; the highest energy photons have the highest frequency and the shortest wavelengths. Wavelengths are expressed in units of nanometres (nm); a nanometre is one-billionth of a metre. The different wavelengths of light can be separated using a prism. As you can see in Figure 3.7, infrared radiation

Stomata Small openings on the surface of leaves, which serve as the points of entry for CO_2 and exit points for water vapour.

Electromagnetic radiation Energy from the Sun, packaged in small particle-like units called photons.

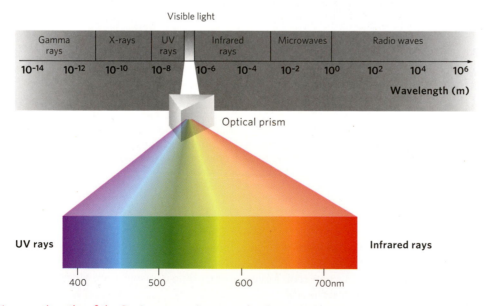

Figure 3.7 The wavelengths of the Sun's energy. The Sun emits electromagnetic radiation that spans a wide range of energy and wavelengths.

Cell walls

Chloroplasts

Thylakoid membranes

Stroma

Figure 3.8 Chloroplasts. Chloroplasts are the site of photosynthesis. The chloroplasts contain stacks of membranes called thylakoids surrounded by a fluid-filled space known as the stroma.

has long wavelengths, which we know contain lower energy. Short wavelengths, such as ultraviolet radiation, contain higher energy. Between the two extremes of infrared and ultraviolet radiation are wavelengths collectively known as **visible light**, which are visible to the human eye. Visible light represents only a small part of the spectrum of electromagnetic radiation.

The visible portion of the spectrum includes the **photosynthetically active region**, which consists of wavelengths of light that are suitable for photosynthesis. This range of wavelengths falls between about 400 nm (violet) and 700 nm (red). Plants, algae, and some bacteria absorb these wavelengths and assimilate their energy by photosynthesis. These also happen to be the wavelengths of greatest intensity at the Earth's surface.

Eukaryotic photosynthetic organisms contain specialized cell organelles known as **chloroplasts**. As you can see in Figure 3.8, chloroplasts contain stacks of membranes known as *thylakoids* and a fluid-filled space surrounding the thylakoids called the *stroma*. Embedded within the thylakoid membranes are several kinds of pigments that absorb solar radiation, including *chlorophylls* and *carotenoids*. The patterns of

absorption for several of these pigments are shown in Figure 3.9.

Chlorophylls, which are primarily responsible for capturing light energy for photosynthesis, absorb red and violet light. Chlorophylls reflect green and blue light, which is why leaves on most plants are predominantly green in colour. Over the past 60 years, scientists identified four types of chlorophyll that differ in the wavelengths that they absorb: chlorophyll *a, b, c,* and *d.* Chlorophyll *a* is found in all photosynthesizing organisms and is responsible for the actual steps in photosynthesis. The other types of chlorophyll act as *accessory pigments,* meaning that they capture light energy and then pass it on to chlorophyll *a.* In 2010, however, scientists reported the discovery of a fifth type of chlorophyll that they named

Visible light Wavelengths in between infrared and ultraviolet radiation that are visible to the human eye.

Photosynthetically active region Wavelengths of light that are suitable for photosynthesis.

Chloroplasts Specialized cell organelles found in photosynthetic organisms.

Figure 3.9 Light-absorbing pigments. Photosynthetic organisms contain a variety of photosynthetic pigments including chlorophyll *a* and *b* and carotenoids that serve as accessory pigments that capture the Sun's energy and pass it to the chlorophylls.

chlorophyll f. This pigment, discovered in algae that live in shallow rocky pools on the coast of Australia, absorbs light at longer wavelengths than the others.

Carotenoids are also accessory pigments and they include two major classes: carotenes and xanthophylls. Carotenoids absorb primarily blue and green light, thereby complementing the absorption spectrum of chlorophyll. Carotenoids, the pigments that give carrots their orange colour, reflect yellow and orange light. By containing carotenoids and several types of chlorophyll, producers can absorb a wider range of solar energy and use it to power photosynthesis.

PHOTOSYNTHESIS

In this section, we review the process of photosynthesis and the different pathways that have evolved under different ecological conditions. During photosynthesis, photosynthetic pigments absorb energy from photons of light, which is then converted into chemical energy stored in the high-energy bonds of organic compounds. In its simplest form, photosynthesis is the process of combining CO_2, H_2O,

and solar energy to produce glucose ($C_6H_{12}O_6$) and oxygen:

$$6\ CO_2 + 6\ H_2O + \text{photons} \rightarrow C_6H_{12}O_6 + 6\ O_2$$

This simple equation summarizes a long chain of complex chemical reactions. As shown in Figure 3.10, the process takes place in two parts: *light reactions* and the *Calvin cycle*.

Light reactions depend on light energy from the Sun and include a series of events from the absorption of light to the production of high-energy compounds and oxygen (O_2). First, photosynthetic pigments inside the thylakoids capture light energy. When chlorophyll molecules absorb photons, they release electrons, which are then passed along a chain of reactions. In the process, H_2O molecules are split into hydrogen (H^+) and oxygen (O_2^-) atoms. The individual oxygen atoms immediately combine with each other to become molecular oxygen (O_2). The energy absorbed from the released electrons, as well as the H^+ ions that are split from H_2O, are used to produce two high-energy compounds. ADP (adenosine diphosphate) and P (phosphate) combine to form *ATP* (adenosine triphosphate), whereas $NADP^+$ (nicotinamide adenine dinucleotide phosphate) and H^+ combine to form *NADPH* (the reduced form of $NADP^+$).

Once the cell has created the high-energy compounds ATP and NADPH, it uses the energy in these compounds to convert CO_2 into glucose in a process known as the Calvin cycle. This takes place in the stroma of the chloroplast. Over evolutionary time, three distinct biochemical pathways have evolved for the Calvin cycle: C_3, C_4, and *CAM photosynthesis*. As we will soon see, each of these pathways is suited to particular ecological conditions.

C_3 Photosynthesis

For most plants, photosynthesis begins with a reaction between CO_2 and a five-carbon sugar known as *RuBP* (ribulose bisphosphate) to produce a six-carbon compound. This reaction is catalyzed by the enzyme **RuBP carboxylase-oxidase** (also known as **Rubisco**). Once the six-carbon compound is created, it immediately splits into two molecules of a three-carbon sugar called *G3P* (glyceraldehyde 3-phosphate). We can represent this process as:

$$CO_2 + RuBP \rightarrow 2\ G3P$$

This photosynthetic pathway, in which CO_2 is initially assimilated into the three-carbon compound (G3P), is

---

RuBP carboxylase-oxidase An enzyme involved in photosynthesis that catalyzes the reaction of RuBP and CO_2 to form two molecules of glyceraldehyde 3-phosphate (G3P). *Also known as* Rubisco.

Inputs: Light H_2O CO_2

Outputs: O_2 G3P
Glucose and other sugars

Figure 3.10 Photosynthesis. The two-part process of photosynthesis begins with the light reactions that capture solar energy. When chlorophyll molecules absorb photons, they release electrons, which are passed along a chain of reactions that produces the high-energy molecules ATP and NADPH. In the process, H_2O molecules split and produce O_2 as a waste product. The second part of this process, known as the Calvin cycle, combines CO_2 with RuBP, using the enzyme Rubisco, to form G3P, a three-carbon sugar that is later converted to glucose and other sugars.

known as **C_3 photosynthesis**. The vast majority of plants on Earth use C_3 photosynthesis. In most plants, this process occurs in mesophyll cells in leaves.

One of the challenges for plants using C_3 photosynthesis is that Rubisco, the enzyme responsible for joining CO_2 and RuBP, has a low affinity for CO_2 compared to another enzyme that we will discuss shortly. Consequently, carbon assimilation using Rubisco is quite inefficient at the low concentrations of CO_2 found in the mesophyll cells of plant leaves. To achieve high rates of carbon assimilation, plants must therefore pack their mesophyll cells with large amounts of the enzyme Rubisco. In some plant species, Rubisco can constitute up to 30 percent of the dry weight of leaf tissue.

Rubisco's low affinity for CO_2 is not the only problem plants face. Under certain conditions—such as when temperatures are high, concentrations of O_2 are high, and CO_2 concentrations are low— Rubisco preferentially binds to O_2 rather than CO_2. This occurs when hot, dry conditions cause the leaf stomata to close, preventing new CO_2 from entering to replenish the CO_2 that has been consumed by the Calvin cycle. Closed stomata also prevent O_2 produced by the light reaction from leaving the leaf. Consequently, hot and dry conditions lead to changes in CO_2 and O_2 concentrations that cause Rubisco to preferentially bind to O_2 rather than CO_2. When the Rubisco enzyme binds to O_2, it initiates a series of reactions that reverses the outcome of photosynthesis in a process known as **photorespiration**:

$$2 \text{ G3P} \rightarrow \text{RuBP} + CO_2$$

The reverse reaction consumes energy, consumes O_2, and produces CO_2. This reverse reaction in plants is called photorespiration because it resembles

the process of respiration. In short, what is accomplished by photosynthesis when Rubisco binds to CO_2 is undone by photorespiration when Rubisco binds to O_2.

The problem of photorespiration is caused in part by closed stomata. This leads to high O_2 concentrations and low CO_2 concentrations in the leaves. One potential solution is to keep the leaf stomata open. This would permit free gas exchange, allowing CO_2 to enter the leaves and O_2 to exit the leaves. This strategy works as long as plants can replace the water they also lose by transpiration when the stomata are open. However, this solution may be too costly in hot, dry environments where water is scarce. When such costs are too high, natural selection will favour traits that can either reduce the demand for water or reduce the loss of water.

C_4 Photosynthesis

We have seen that hot and dry conditions cause stomata to close, which results in a decrease of CO_2, an increase of O_2, and an increase in photorespiration. To address this problem, many herbaceous plants, particularly grasses growing in hot climates, have evolved a modification to the process of photosynthesis. This modification has evolved at least 45 times during the past 30 million years in at least 19 different families of angiosperms.

The C_4 pathway is an adaptation that adds a step to the initial assimilation of CO_2 to make it more efficient when CO_2 is present at low concentrations.

C_3 photosynthesis The most common photosynthetic pathway, in which CO_2 is initially assimilated into a three-carbon compound, glyceraldehyde 3-phosphate (G3P).

Photorespiration The oxidation of carbohydrates to CO_2 and H_2O by Rubisco, which reverses the light reactions of photosynthesis.

(a) C₃ plant

(b) C₄ plant

Figure 3.11 Leaf cell arrangements for C₃ versus C₄ plants. (a) C₃ plants conduct all steps of photosynthesis in the chloroplasts of the mesophyll cells. (b) In C₄ plants, the initial step of carbon assimilation and the light reactions take place in the chloroplasts in mesophyll cells. However, the CO₂ that has been assimilated is then transported to the bundle sheath cells where the Calvin cycle takes place.

You can see an overview of this process in Figure 3.11. Biologists call this modification **C₄ photosynthesis** because the first step joins CO_2 with a three-carbon molecule called *PEP* (phosphoenol pyruvate) to produce a four-carbon molecule called *OAA* (oxaloacetic acid):

$$CO_2 + PEP \rightarrow OAA$$

This reaction is the key difference between C₃ and C₄ photosynthesis. It is catalyzed by the enzyme PEP carboxylase, which has a higher affinity for CO_2 than Rubisco. This additional assimilation step occurs in the mesophyll cells of the leaf, which is also the site of the light reaction. In most C₄ plants, however, the Calvin cycle takes place in the bundle sheath cells that surround the leaf veins. This means that the plant must move the CO_2 that has been assimilated in the mesophyll cells over to the bundle sheath cells. To do this, the plant converts OAA into malic acid, which then diffuses into the bundle sheath cells, where another enzyme breaks it down to produce CO_2 and *pyruvate,* a three-carbon compound. In the bundle sheath cells, the chloroplasts use the CO_2 that is brought over from the mesophyll cells for the Calvin cycle. To complete the cycle, pyruvate moves back into the mesophyll cells where it is converted back to PEP to be used once again.

This strategy solves the problem of photorespiration by creating concentrations of CO_2 in the bundle sheath cells that are three to eight times higher than is available to C₃ plants. It helps that there are multiple mesophyll cells for each bundle sheath cell, thereby providing a large number of sites for CO_2 assimilation that can provide CO_2 to each bundle sheath cell. At this higher CO_2 concentration, the Calvin cycle operates more efficiently. In addition, because the enzyme PEP carboxylase has a high affinity for CO_2, it can bind CO_2 at a lower concentration in the cell. This pathway allows the stomata to remain partially or completely closed for longer periods of time, which reduces water loss. However, C₄ photosynthesis has two disadvantages that reduce its efficiency: less leaf tissue is devoted to photosynthesis, and some of the energy produced by the light reactions is used in the initial C₄ carbon assimilation step.

Whereas C₃ plants are favoured in cool, moist climates, C₄ plants are favoured in climates that are hot or have less abundant water. When water is abundant, the C₄ pathway does not present a distinct advantage because the costs of the C₃ pathway are relatively low. When water is less abundant, however, the C₄ pathway has an advantage. Only about 4 percent of all plant species on Earth are C₄ plants and they are primarily found in two types of nonwoody plants: *grasses* and *sedges.* C₄ plants dominate tropical and subtropical grasslands and are important components of the plant communities found in arid regions of the world, including the Great Plains of North America. Plants using the C₄ pathway also include many of our most

C₄ photosynthesis A photosynthetic pathway in which CO_2 is initially assimilated into a four-carbon compound, oxaloacetic acid (OAA).

(a) C$_3$ photosynthesis   (b) C$_4$ photosynthesis   (c) CAM

Figure 3.12 Alternative photosynthetic pathways. (a) C$_3$ plants conduct photosynthesis in mesophyll cells during the day. **(b)** C$_4$ plants separate the steps of photosynthesis in space. The initial CO$_2$ assimilation step occurs in the mesophyll cells and the remaining steps occur in the bundle sheath cells. **(c)** CAM plants separate the steps of photosynthesis in time. Assimilation of CO$_2$ happens at night and the remaining steps occur during the day.

important crop plants such as corn (maize), sorghum, and sugarcane. These plants are highly productive during hot growing seasons. In fact, while only 4 percent of all plant species use the C$_4$ pathway, these plants account for 20 to 30 percent of all CO$_2$ fixation and 30 percent of all grain production. As a result, C$_4$ plants can play substantial roles in the ecosystems in which they live.

CAM Photosynthesis

Certain succulent plants that inhabit water-stressed environments, for example cacti and pineapple plants, use the same biochemical pathways as C$_4$ plants. However, rather than separating the steps of CO$_2$ assimilation and the Calvin cycle in space with mesophyll and bundle sheath cells, these succulent plants separate the steps in time. Plants that follow this pathway, known as **crassulacean acid metabolism**, or **CAM**, open their stomata for gas exchange during the cool night, when transpiration is minimal, and then conduct photosynthesis during the hot day. The discovery of this arrangement was first made in plants of the family Crassulaceae (the stonecrop family)—which includes the jade plant (*Crassula ovata*).

Like the C$_4$ plants, CAM plants use an initial step of assimilating CO$_2$ into OAA, which is then converted to malic acid and stored at high concentrations in vacuoles within the mesophyll cells of the leaf. The enzyme responsible for the assimilation of CO$_2$ works best at the cool temperatures that occur at night, when the stomata are open. During the day, the stomata close, and the stored organic acids are gradually broken down to release CO$_2$ to the Calvin cycle. A different enzyme with a higher optimum temperature, geared to promote daytime photosynthesis, regulates the regeneration of PEP from pyruvate following the release of CO$_2$. Because CAM plants can conduct gas exchange during the night when the air is cooler and more humid, a plant using CAM photosynthesis reduces its water loss. Thus, CAM photosynthesis is an adaptation that results in extremely high water use efficiencies and enables plants that use this pathway to live in very hot and dry regions of the world. While CAM allows photosynthesis to occur in water-limited conditions, it happens at a relatively slow rate. As a result, CAM plants typically grow much more slowly than C$_3$ or C$_4$ plants. Figure 3.12 compares the three alternative photosynthetic pathways.

Crassulacean acid metabolism (CAM) A photosynthetic pathway in which the initial assimilation of carbon into a four-carbon compound occurs at night.

(a) Woolly pussytoe
(Antennaria lanata)

(b) Arctic Bell-heather
(Cassiope tetragona)

(c) Purple saxifrage
(Saxifraga oppositifolia)

Figure 3.13 Adaptations of plants against temperature and water stress. **(a)** Spines and hairs on leaf surfaces, such as on the leaves of the woolly pussytoe plant, reduce evaporation and trap warm air close to the leaf surface. **(b)** Arctic Bell-heather retains its leaves, which reduces its need for water and nutrients. **(c)** Purple saxifrage grows low to the ground and has dark purple flowers, both adaptations that help with retention and gain of heat.

Plants possessing the C_3 pathway are better adapted to cool, wet conditions, whereas plants possessing the C_4 and CAM pathways are better adapted to warm and arid conditions. However, there is not a clear distinction in where these different types of plants grow. For example, regions that are hot and dry during the summer can be cool and moist during the winter and spring. As a result, such an area can be dominated by short-lived C_4 plants in the summer and short-lived C_3 plants in the winter and spring. Moreover, different photosynthetic pathways represent just one of several adaptations that plants have evolved to handle hot temperatures and a scarcity of water. As we will discuss in the next section, many plants also have evolved structural adaptations.

STRUCTURAL ADAPTATIONS TO WATER AND TEMPERATURE STRESS

Heat- and drought-adapted plants have anatomical and physiological modifications that improve water uptake and retention, reduce transpiration, and reduce the buildup of heat in their tissues. These adaptations include roots that can take advantage of different water sources, resistance to heat buildup, vein configurations that protect against air blockages, and morphological adaptations in leaves.

Plants living in arid regions often have very shallow or very deep roots, which represent two different adaptive strategies. Plants with very shallow roots, such as many species of cacti, are able to rapidly uptake water from brief rain events in which the rain does not penetrate very far into the soil. Cacti often pair this adaptation with thick, succulent tissues that can hold a great deal of water whenever it becomes available. In contrast to the cacti, some perennial shrubs, such as mesquite, have roots that can extend several metres down into the soil, allowing the shrubs to access water that is very deep below the surface.

Drought conditions can occur in both hot and cold environments. Some plants in such different habitats as deserts and the Arctic use similar adaptations to cope with extreme temperatures and lack of water. Figure 3.13 displays some of these adaptations. For example, spines and hair, like those on the woolly pussytoe, produce a boundary layer of still air that traps warm air and moisture and reduces evaporation. Many northern plants, such as the Arctic Bell-heather, are evergreen, which means they retain their leaves—an adaptation that reduces their need for nutrients and water. Arctic Bell-heather also grows low and in bunches, which reduces the effect of cold winds and traps warm air below the plant. Other species, such as the purple saxifrage, have dark-coloured flowers that gain heat quicker than do bright flowers, which allows it to bloom earlier than many other plants.

For a long time scientists observed that plants in habitats with high temperatures generally had smaller leaves than plants in habitats with abundant water. Scientists hypothesized that the smaller leaves represented an adaptation that permitted the dissipation of heat. However, small leaves are not only found in plants from hot, dry places, but also in plants from cool, dry places. Smaller leaves also contain a higher density of large veins that transport and distribute water to the many small veins. In 2011, an international team of scientists discovered that having small leaves with a high density of large veins is actually an adaptation to overcome the problem of air bubbles, known as *embolisms*, which can form in large veins. Under severe drought stress, air can move into the stomata and up into the large veins, causing a bubble to form in the vein and blocking water movement. A large density of veins allows the plant to get around this problem by sending water through adjacent veins. This suggests that the small leaf size is actually an adaptation to scarce water in both hot and cool environments, and the fact that small leaves can better dissipate heat in the hot environments may be a valuable secondary benefit. Collectively, these structural adaptations help make it possible for plants to live in regions of the world that have high temperatures or scarce water.

Terrestrial environments pose a challenge for animals to balance water, salt, and nitrogen

While plants have a variety of adaptations for conducting photosynthesis on land, terrestrial life presents many additional challenges, including the need to maintain a balance of water, salt, and nitrogen. This balance is known as *homeostasis*. **Homeostasis** is an organism's ability to maintain constant internal conditions in the face of a varying external environment. All organisms exhibit homeostasis to some degree—for example, the balance of water and salt, or the regulation of body temperature.

Although the occurrence and effectiveness of homeostatic mechanisms vary, all homeostatic systems exhibit **negative feedbacks**, meaning that when the system deviates from its desired state, or set point, internal response mechanisms act to restore that desired state. An example that might be familiar to you is the presence of negative feedbacks in the regulation of body temperature. Figure 3.14 shows how this works in mammals. The hypothalamus—a gland in the brain—determines whether the body temperature is above or below the desired set point, which differs among species of mammals. If body temperature drops below this set point, the hypothalamus uses neural and hormonal signals to trigger the body to generate more heat. When the body temperature reaches the set point, the hypothalamus triggers the body to stop producing heat. If the body temperature greatly exceeds the set point, the hypothalamus triggers the body to start

Homeostasis An organism's ability to maintain constant internal conditions in the face of a varying external environment.

Negative feedbacks The action of internal response mechanisms that restores a system to a desired state, or set point, when the system deviates from that state.

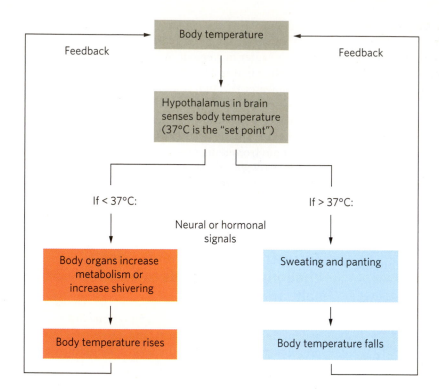

Figure 3.14 Negative feedback for regulating body temperature. In mammals, the hypothalamus acts like a thermostat. When the desired set point differs from the current body temperature, the hypothalamus signals the body to return its temperature to the desired set point.

using cooling mechanisms, including sweating and panting. This is much like how a thermostat regulates the temperature of a room; when the house is cold, a temperature-sensitive switch turns on a heater. Once the house warms up to the desired set point, the thermostat turns the heater off.

In Chapter 2 we saw that to maintain the proper amounts of water and dissolved substances in their bodies, aquatic organisms must balance their gains and losses. This is equally true for organisms on land. Organisms often take in water with a solute concentration that differs from that of their bodies, so they must either acquire additional solutes to make up the deficit or rid themselves of excess solutes. If they do not balance the concentration of solutes, many of their physiological functions will not operate correctly. When water evaporates from the surfaces of terrestrial organisms into the atmosphere, solutes are left behind and their concentration in the body tends to increase. Under such circumstances, organisms must excrete excess salts to maintain the proper concentrations in their bodies.

WATER AND SALT BALANCE IN ANIMALS

Water is as important to terrestrial animals as it is to terrestrial plants. Terrestrial animals, with their internalized gas exchange surfaces, are less vulnerable to respiratory water loss than plants. Moreover, because terrestrial animals are not immersed continuously in water, they have little trouble retaining

ions. They acquire the mineral ions they need in the water they drink and the food they eat, and they use urine to eliminate excess salts in their bodies. Where fresh water abounds, animals can drink large quantities of water to flush out salts that would otherwise accumulate in the body. Where water is scarce, however, animals make use of adaptations to conserve water.

As one would expect, desert animals have evolved a number of adaptations in response to water scarcity. Kangaroo rats, for example, are a group of small rodents that live in the dry regions of North America (Figure 3.15). Behavioural and physiological adaptations

Figure 3.15 Animal adaptations to conserve water. The Ord's kangaroo rat (*Dipodomys ordii*) lives in dry, desert environments in North America. These environments favour adaptations that allow the kangaroo rat to conserve water, including feeding at night and having large, efficient kidneys. Photo by Jim Zipp/Science Source.

allow them to live in these places. The kangaroo rat conserves water by hunting for food during the nights and staying in a cool and humid underground burrow during the hot days—a valuable behavioural adaptation. For both the kangaroo rat and the camel, the kidneys provide an additional physiological adaptation to extreme heat and water scarcity. In all mammals, the kidneys are responsible for removing salts and nitrogenous wastes from the blood. These solutes are dissolved in water, but because water is valuable, a structure known as the loop of Henle helps recover some of it before the mixture is excreted. Kangaroo rats and camels possess an unusually long loop of Henle, which provides a greater length over which a kidney can recover water from urine prior to excretion. Whereas human kidneys concentrate most solutes in urine to about 4 times the blood concentration level, the kidneys of kangaroo rats produce urine with solute concentrations as much as 14 times greater.

Although the loop of Henle plays a large role in concentrating the urine of mammals, additional adaptations help conserve water. The efficiency of water use is best determined by the overall size of the kidney relative to the size of a mammal's body. One way to assess the potential importance of relative kidney size as an adaptation for conserving water is to examine how relative kidney size differs among closely related species of rodents that live in habitats with differing amounts of precipitation. Figure 3.16 shows this relationship in a group of South American rodents. Rodents living in habitats with the lowest amount of precipitation have relatively large kidneys whereas rodents living in habitats with the greatest amount of precipitation have relatively small kidneys.

Because sodium and chloride ions participate in the water conservation mechanism of kidneys, the kidneys do not excrete these ions efficiently. Many

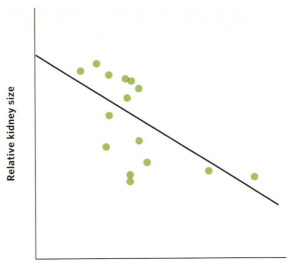

Figure 3.16 Kidney size in South American rodents. Species of South American rodents that live in relatively drier habitats have relatively larger, more efficient kidneys. [After G. B. Diaz et al., Renal morphology, phylogenetic history and desert adaptation of South American hystricognath rodents, *Functional Ecology* 20 (2006): 609–620.]

animals that lack ready access to fresh water have specialized salt-secreting organs that work on a different principle from the kidneys. In fact, these secretory organs are more like the salt glands of mangrove plants. For example, salt glands of birds and reptiles, which are particularly well developed in marine species, are actually modified tear glands located in the orbit of the eye that are capable of secreting a concentrated salt solution. These adaptations help animals to balance their salt and water budget on land (Figure 3.17).

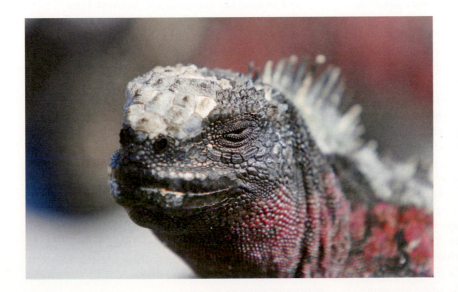

Figure 3.17 Animal adaptations to expel salt. Many animals that live on land and forage in salt water have evolved specialized glands to expel excess salt. The white crust on top of the head of this marine iguana (*Amblyrhynchus cristatus*) from the Galápagos Islands is salt that was produced in these glands and expelled through the nostrils. Photo by Michael S. Nolan/age fotostock.

ANALYZING ECOLOGY

Different Types of Variables

When we think about testing hypotheses in ecology, we often collect and analyze data to determine whether the hypothesis is supported or refuted. However, before we collect and analyze the data, we have to consider what types of data we are collecting.

The first distinction is between *independent variables* and *dependent variables*. **Independent variables** are factors that are presumed to cause other variables to change. **Dependent variables** are factors that are being changed. For example, in the examination of the relative kidney size of South American rodents (Figure 3.16), the hypothesis was that differences in precipitation have led to the evolution of different kidney sizes. In this case, precipitation is the independent variable and kidney size is the dependent variable.

A second distinction is whether a variable is continuous or categorical. **Continuous variables** can take on any numeric value, including values that are not whole numbers. In the case of the rodents, for example, precipitation represents a continuous variable because it can take on any value (Figure 3.16). Additional continuous variables include temperature, salinity, and light. In contrast, **categorical variables**, also known as **nominal variables**, fall into distinct groupings or categories. For example, if we wanted to know how the solute concentration of a kangaroo rat's urine was affected by its diet, we might provide kangaroo rats with one of three different species of seeds and then measure their urine concentrations. In this case, the diet is a categorical variable because the different diets fall into three distinct categories. Additional categorical variables include sex (e.g., males and females) and species (e.g., dromedary camels, Bactrian camels, and guanacos).

As we will see in later chapters, these distinctions between dependent versus independent variables and continuous versus categorical variables are important in the statistical analysis of ecological data.

> **YOUR TURN** If you were to conduct an experiment that examined how the water holding capacity differed among soil types, as in Figure 3.4, which would be the independent variable and which would be the dependent variable?
>
> If you were only to compare the water holding capacity of soils that contained 100 percent sand, 100 percent silt, or 100 percent clay, would this soil variable be considered continuous or categorical?

Independent variable A factor that causes other variables to change.

Dependent variable A factor that is being changed.

Continuous variable A variable that can take on any numeric value, including values that are not whole numbers.

Categorical variable A variable that falls into a distinct category or grouping. *Also known as* Nominal variable.

WATER AND NITROGEN BALANCE IN ANIMALS

Most carnivores, whether they eat crustaceans, fish, insects, or mammals, consume excess nitrogen from the proteins and nucleic acids in their diet. When these compounds are metabolized, the excess nitrogen must be eliminated from the body. Most aquatic animals produce ammonia (NH_3) because it is a simple metabolic by-product of nitrogen metabolism. Although ammonia is mildly poisonous to tissues, aquatic animals eliminate it rapidly either in copious dilute urine or directly across the body surface before it reaches a dangerous concentration within the body.

Terrestrial animals, however, rarely have access to large quantities of water to excrete excess nitrogen. Instead, they produce metabolic by-products that are less toxic than ammonia. This allows them to accumulate higher concentrations of metabolic by-products in their blood and urine without any harmful side effects. In mammals, this metabolic by-product is urea ($CO(NH_2)_2$), the same substance that sharks produce and retain to achieve osmotic balance in marine environments. Because urea dissolves in water, excreting urea still causes the loss of some water, although the amount depends on the concentrating power of the kidneys. Birds and reptiles require even less water; they excrete nitrogen in the form of uric acid ($C_5H_4N_4O_3$), which crystallizes out of solution and can then be excreted as a highly concentrated paste in the urine.

Although excreting urea and uric acid conserves water, it is costly in terms of the energy needed to form these compounds. One way that scientists use to quantify energy costs is to determine the amount of organic carbon that is consumed to produce the energy necessary for excretion. For example, for each atom of nitrogen excreted in the form of ammonia, no atoms of organic carbon are used. In contrast, to excrete nitrogen in the form of urea requires 0.5 atoms of organic carbon, and uric acid uses 1.25 atoms of organic carbon.

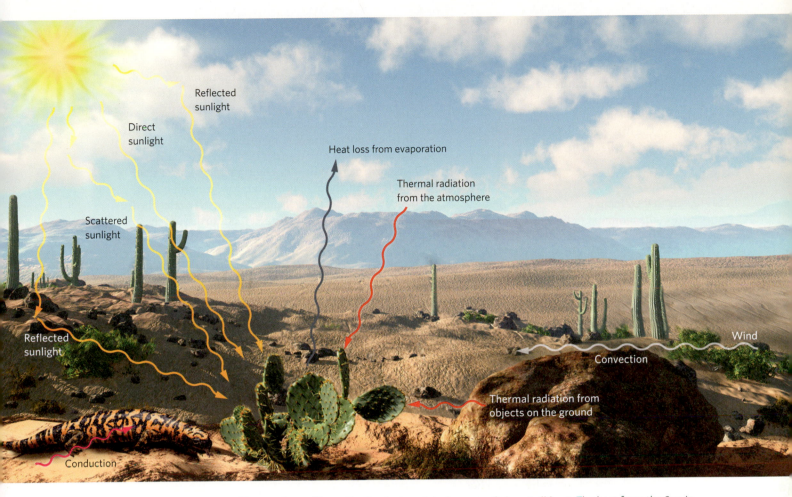

Figure 3.18 Sources of heat gain and loss. The Sun is the original source of almost all heat. The heat from the Sun is exchanged among objects across the landscape. In the case of the cactus, heat is gained by direct sunlight, scattered sunlight, and reflected sunlight. Heat can be lost by evaporation of water vapour to the atmosphere. Heat can also be gained or lost by radiation from surrounding objects such as rocks, by conduction where the cactus comes into contact with the soil, and by convection as winds move hot or cold air over the surface of the cactus and disrupt its boundary layer.

Adaptations to different temperatures allow terrestrial life to exist around the planet

On Earth, land temperatures can reach as high as 58°C in northern Africa and as low as −89°C in Antarctica. These extremes can limit the occurrence of life. To understand how organisms are affected by temperature and the adaptations that have evolved to deal with different temperatures, we first need to examine how they gain and lose heat.

SOURCES OF HEAT GAIN AND LOSS

Because body temperature impacts physiological functions, organisms must manage heat gain and heat loss carefully. The ultimate source of heat at the surface of Earth is sunlight, most of which is absorbed by water, soil, plants, and animals and converted to heat. Objects and organisms continuously exchange heat with their surroundings. When the temperature

of the environment exceeds the temperature of an organism, the organism gains heat and becomes warmer. When the environment is cooler than the organism, the organism loses heat to the environment and cools. As illustrated in Figure 3.18, this exchange of heat can occur through four processes: *radiation, conduction, convection,* and *evaporation.*

Radiation

Radiation is the emission of electromagnetic energy by a surface. The primary source of radiation in the environment is the Sun. As objects in the landscape are warmed by solar radiation, they emit more lower-energy radiation in the form of infrared light. The temperature of the radiating surface determines how rapidly an object loses energy by radiation to colder parts of the environment. We measure in units of

Radiation The emission of electromagnetic energy by a surface.

Kelvins (K), also known as absolute temperature, where 0°C = 273°K. The amount of heat radiation increases with the *fourth* power of absolute temperature. So, for example, we can consider the heat radiation of two small animals such as a mouse and a lizard. If the mammal has a skin temperature of 37°C (310°K) and the lizard has a skin temperature of 17°C (290°K), the difference in heat radiation between the mammal and the lizard is:

$$310^4 \div 290^4 = 130\%$$

This means that by having a 20°C higher body temperature, the mammal radiates 30 percent more heat than the lizard.

The relatively high amount of heat radiation produced by animals with a higher body temperature than their external environment has been used by ecologists in a variety of research endeavours, including estimates of population sizes. When biologists need to count the number of moose living in remote regions of Alaska, for example, planes equipped with infrared cameras fly over these regions in the winter and the warm bodies of the moose stand out as a bright signal of infrared radiation against their cold, snowy background.

Conduction

Conduction is the transfer of the kinetic energy of heat between substances that are in contact with one another. For example, lizards often lie flat on hot rocks to warm their bodies by conduction. Water, because it is so much denser than air, conducts heat more than 20 times faster. As a result, you would lose body heat much faster if you stood in 10°C water than if you stood in 10°C air.

The rate at which heat moves by conduction between an organism and its surroundings depends on three factors: its surface area, its resistance to heat transfer, and the temperature difference between the organism and its surroundings. An organism's surface area helps determine its rate of heat conduction because a greater amount of exposed surface allows a greater surface for the energy transfer to take place. This is why many animals curl up in a ball to lessen the amount of exposed surface when they are trying to stay warm on a cold night. An organism's resistance to heat transfer is just another way of saying how much insulation the organism has. Thick layers of fat, fur, or feathers have a high resistance to heat transfer and therefore slow the rate of heat loss due to conductance. Indeed, this is why you choose to wear insulated boots rather than walking barefoot in snow. Finally, the rate of heat loss is higher when there are large differences between the temperature of the organism and that of the environment. As we will see in Chapter 4, this feature of conductance is why some hibernating animals lower their body temperatures

during the winter. A lower body temperature results in less heat loss to the cold external environment.

Convection

Convection is the transfer of heat by the movement of liquids and gases: molecules of air or water next to a warm surface gain energy and move away from the surface. In still air, a boundary layer of air forms over a surface of organisms. Having a thicker boundary layer will tend to slow heat transfer between an organism and its environment. When the environment is colder than the organism, the organism tends to warm this boundary layer, effectively insulating itself against heat loss. If there is a current of air, it tends to disrupt the boundary layer so that heat can be carried away from the body by convection. This convection of heat away from the body surface is the basis of the wind chill factor we hear about in winter on the evening weather report. Wind on a cold day makes you feel as cold as you would on an even colder day without wind. For example, wind blowing at 32 km per hour in an air temperature of −7°C has the cooling power of still air at −23°C.

In the same way, air movement can add heat to an organism if the boundary layer is cooler than the surrounding air. If you were to stand in a hot desert, and your boundary layer were cooler than the air, for example, a hot wind would disrupt the boundary layer between your skin and the air and make your body even hotter.

Evaporation

Evaporation is the transformation of water from a liquid to a gaseous state with the input of heat energy. Evaporation removes heat from a surface. The evaporation of 1 g of water from a body's surface removes 2.43 kilojoules (kJ) of heat when the temperature of the surface is 30°C. As plants transpire and animals breathe, water evaporates from their exposed gas exchange surfaces, especially at higher temperatures. In dry air, the rate of evaporation nearly doubles with each 10°C increase in temperature.

As summarized in Figure 3.18, all of these sources of heat gain and loss can occur simultaneously. Radiation from the Sun can occur as direct sunlight, as well as sunlight that has been scattered as it interacts with gas molecules in the atmosphere or is reflected from clouds and the ground. Plants and animals in contact with rocks, soil, and each other can conduct heat to or from these objects, depending on whether their body temperatures are warmer or colder than the

---

Conduction The transfer of the kinetic energy of heat between substances that are in contact with one another.

Convection The transfer of heat by the movement of liquids and gases.

Evaporation The transformation of water from a liquid to a gaseous state with the input of heat energy.

surrounding objects. As winds move the air past the organisms, there can be an additional exchange of heat, depending again on the temperature of the air compared to the temperature of the organism. Finally, organisms that experience evaporation can lose heat because evaporation requires heat energy.

BODY SIZE AND THERMAL INERTIA

Most exchanges of energy and materials between an organism and its environment occur across body surfaces. Therefore, the volume and surface of an organism affect the rate of these exchanges. As an example, let's consider the differences between the body sizes of a mouse and an elephant. The elephant obviously has a much larger volume and it takes much more energy to meet its metabolic needs each day. However, relative to its volume, the elephant has a smaller surface area then the mouse. This relationship becomes more apparent if we make the simplifying assumption that all organisms are shaped like a box with sides of equal length. In this case, the surface area (*SA*) of an organism increases as the square of its length (*L*), but the volume (*V*) of an organism increases as the cube of its length:

$$SA = L^2$$

$$V = L^3$$

In short, as an organism grows larger, its volume grows faster than its surface area.

Of course, organisms are not shaped like boxes, but the same principles apply to the shapes that organisms do have. Because an organism's metabolic needs are related to its volume and volume increases faster than surface area, an organism's metabolic needs increase faster than the surface area that exchanges energy and materials between the organism and its environment.

The relationship between surface area and volume is particularly relevant when considering heat exchange. Because large organisms have a low surface-to-volume ratio, larger individuals lose and gain heat across their surfaces less rapidly than smaller individuals. In general, larger sizes and lower surface-to-volume ratios make it easier for organisms to maintain constant internal temperatures in the face of varying external temperatures. The resistance to a change in temperature due to a large body volume is known as **thermal inertia**. Although thermal inertia can be an important advantage in cold environments, in hot environments it causes moderately large individuals to have a harder time ridding themselves of excess heat. For this reason, large individuals run a greater risk of overheating. However, very large animals can benefit from thermal inertia under hot environmental conditions because their bodies heat up more slowly. We saw an example of this in the case of the dromedary camels, whose very large bodies slowly

added heat during the day but then released this heat during the night.

THERMOREGULATION

The ability of an organism to control the temperature of its body is known as **thermoregulation**. Some organisms, known as **homeotherms**, maintain constant temperature conditions within the cells. Maintaining a constant internal body temperature allows an organism to adjust its biochemical reactions to work most efficiently. In contrast, **poikilotherms** do not have constant body temperatures. These terms tell us whether an organism's temperature is constant or variable, not whether the temperature changes of the body are controlled internally or externally.

ECTOTHERMS

Ectotherms have body temperatures that are largely determined by their external environment. Ectotherms tend to be organisms with low metabolic rates—such as reptiles, amphibians, and plants—or small body sizes—such as insects—that cannot generate or retain sufficient heat to offset heat losses across their surfaces.

Although ectotherms have body temperatures that match the temperature of their environment, they are not powerless to alter their body temperature. Indeed, many species of ectotherms adjust their heat balance behaviourally by moving into or out of shade, by changing their orientation with respect to the Sun, or by adjusting their contact with warm substrates. When horned lizards are hot, for example, they decrease their exposure to the ground surface by standing erect on their legs. When they are cold, they lie flat against the ground and gain heat both by conduction from the ground and from direct solar radiation. This behaviour, known as basking, is widespread among reptiles and insects (Figure 3.19). Animals that bask in the radiation of the Sun can effectively regulate their body temperatures. Indeed, their temperatures may rise considerably above that of the surrounding air, well into the range of birds and mammals. Some larger species of ectotherms, such as tuna, can generate a substantial amount of heat by exercising their massive muscles. This flexing of large muscles allows such fish

---

Thermal inertia The resistance to a change in temperature due to a large body volume.

Thermoregulation The ability of an organism to control the temperature of its body.

Homeotherm An organism that maintains constant temperature conditions within its cells.

Poikilotherm An organism that does not have constant body temperatures.

Ectotherm An organism with a body temperature that is largely determined by its external environment.

Figure 3.19 Basking. Ectotherms such as this five-lined skink (*Eumeces fasciatus*) commonly lie in the sun to increase their body temperature. Photo by Ryan M. Bolton/Alamy.

to stay warmer than their external environment, making it possible for them to swim and feed in relatively cold waters.

Some plants can occasionally generate enough heat to make their tissues substantially warmer than the external environment. The skunk cabbage (*Symplocarpus foetidus*), for example, is a foul smelling plant that lives in wet soils in eastern North America (Figure 3.20). The odor attracts insect pollinators such as flies that typically feed on dead, rotting organisms. The skunk cabbage sprouts new leaves in early spring, even when snow still covers the ground. The plant's mitochondria generate enough metabolic heat in its tissues to raise its temperature more than 10°C above the external environment. This incredible achievement requires a great deal of energy, but it provides a number of substantial benefits, including earlier flowering in the spring, more rapid development of flowers, and protection from freezing temperatures. In one species of skunk cabbage, scientists have discovered that

Figure 3.20 Skunk cabbage. Using mitochondria to generate heat, the skunk cabbage can elevate its temperature by more than 10°C above the environmental temperature. The elevated temperature melts a path through the snow in early spring, making the skunk cabbage one of the first plants to sprout and attract pollinators to its flowers. Photo by JAPACK/age fotostock.

generating heat also improves the rate of pollen germination and pollen tube growth in the flowers. The heat also benefits the pollinators, which can absorb some of the heat produced by the plant. Collectively, heat generation in plants can be very beneficial to both the plants and their pollinators.

In considering the various adaptations of ectotherms, we should note that they are not necessarily poikilotherms; their internal temperature may not vary a great deal. While it is true that their temperature is determined by the environment, the environmental temperature may not be highly variable. For example, fish living in the polar oceans experience very cold waters with little temperature variation. These fish are certainly ectotherms, but their body temperatures are nearly homeothermic.

ENDOTHERMS

Endotherms are organisms that can generate sufficient metabolic heat to raise body temperature higher than the external environment. Most mammals and birds maintain their body temperatures between 36°C and 41°C, even though the temperature of their surroundings may vary from −50°C to +50°C. These organisms gain the benefit of accelerated biological activity in colder climates, where they may be better able to forage, escape predators, and compete with individuals other than ectotherms.

Sustaining internal conditions in organisms that differ significantly from conditions in the external environment requires a lot of work and energy. Consider the costs to birds and mammals of maintaining constant high body temperatures in cold environments. As air temperature decreases, the difference between the internal and external environments increases. Recall that heat is lost across body surfaces in direct proportion to this temperature difference. Consider, for example, an animal that maintains its body temperature at 40°C. At an outside temperature of 20°C, it loses heat much faster than it does at an outside temperature of 30°C. This is because at 20°C the outside temperature is 20°C lower than its body temperature, whereas an outside temperature of 30°C is only 10°C lower than its body temperature. To maintain a constant body temperature, endothermic organisms must replace heat lost to their environment either by generating metabolic heat or by gaining heat through other means such as solar radiation, conduction, or convection. The rate of metabolism required to maintain a particular body temperature increases in direct proportion to the difference between the temperature of the body and the temperature of the environment.

Endotherm An organism that can generate sufficient metabolic heat to raise body temperature higher than the external environment.

Figure 3.21 Blood shunting. In cold environments some animals can close certain blood vessels at their precapillary sphincters. This reduces the circular flow of blood from the arteries to the extremities and back to the veins, which limits the amount of chilled blood that returns to the heart.

ADAPTATIONS OF THE CIRCULATORY SYSTEM

You have probably noticed that when walking on a cold day your hands and feet are the first body parts to become cold. Similarly, because the legs and feet of most birds do not have feathers, these extremities would be major potential sources of heat loss in cold regions if they were held at the same temperature as the rest of the body. Exposed extremities lose heat rapidly, due to their high ratio of surface area to volume. The conduction of heat, particularly from exposed extremities, works against the maintenance of a constant warm body temperature. Ectotherms and endotherms have evolved a number of adaptations to minimize the impact of chilled extremities and thereby to help maintain a warm temperature in the core of the body where many vital organs are located.

One prominent adaptation is *blood shunting*. **Blood shunting** occurs when specific blood vessels can be shut off—at locations called *precapillary* sphincters—so that less of the animals' warm blood flows out to the cold extremities such as the forelimbs and hindlimbs. Instead, much of this blood is redirected into the veins before it ever reaches the extremities. From the veins, it returns to the heart, as shown in Figure 3.21. By sending less blood to nonvital areas such as the limbs, the blood experiences less cooling and allows the core of the animal's body to maintain a constant internal temperature while expending less energy.

Another adaptation to cold extremities is countercurrent circulation. In Chapter 2 we saw that fish maximize oxygen uptake by this mechanism; the blood in their gills flows in the opposite direction from water. A similar arrangement occurs with the position of veins and arteries in the extremities of many animals; arteries that carry warm blood away from the heart and toward the extremities are positioned alongside veins that carry chilled blood from the extremities back to the heart. Figure 3.22 shows an example. The large amount of surface area on the caribou's long, slender legs could cause it to lose a lot of body heat. Countercurrent circulation reduces this heat loss. Warm blood in arteries leading to the legs cools as it passes close to veins that return cold blood to the body. Rather than being lost to the environment, heat is transferred from the blood in the arteries to the blood in the veins. The legs themselves are kept only slightly above freezing, which minimizes heat transfer to the environment. Cells in caribou legs can withstand such low temperatures because they use lipids that remain liquid in lower temperatures.

The diverse range of adaptations for life on land is a fascinating testament to the ability of natural selection to favour those traits that improve the fitness of organisms. Whether we consider the adaptation of plants to obtain water and nutrients, the variety of

---

Blood shunting An adaptation that allows specific blood vessels to shut off so less of an animal's warm blood flows to the cold extremities.

40°C

30°C

20°C

10°C

Artery

Vein

Heat

Figure 3.22 Countercurrent blood circulation. The arteries in a caribou's leg that carry warm blood from the heart to the feet are positioned next to the veins that carry chilled blood away from the feet and back to the heart. This positioning of the arteries and veins allows the caribou to transfer heat from arteries to the veins.

scenarios for conducting photosynthesis under different environmental conditions, or the ability of animals to balance water, salt, nitrogen, and heat, these adaptations have evolved to help make the transition from life in water to life on land. Throughout this book, we will return to many of these adaptations as we seek to understand the ecology of communities and ecosystems.

ECOLOGY TODAY CONNECTING THE CONCEPTS

THE CHALLENGE OF GROWING COTTON

Cotton plants. Cotton has numerous adaptations that help it cope with hot and dry conditions. Photo by David Nance/ARS/USDA.

Throughout this chapter we have examined the challenges of the terrestrial environment and the adaptations that organisms have evolved to deal with these challenges. For plants, there are the challenges of extracting water and nutrients from the soil and then photosynthesizing at the cost of water lost due to transpiration. At the same time, plants need to balance their water and salts while also frequently coping with heat gains and losses that can place organisms at the limits of their temperature tolerance. To understand how an organism handles all of these challenges, we turn to the common cotton plant.

Cotton is a major crop throughout the world because of its many uses that range from clothing to cottonseed oil. Cotton is typically grown in regions of the world that have warm temperatures with either high or low amounts of precipitation. In the United States, nearly half of the cotton is grown in the panhandle of Texas. Like many arid regions, the Texas Panhandle experiences droughts every few years. In 2011, Texas experienced one of the worst droughts ever recorded, and this made it a tough place to grow cotton.

Although C_3 plants like cotton are typically associated with moister habitats, cotton can also grow well in relatively dry places. However, extended droughts often cause farmers to irrigate cotton fields because young cotton plants have relatively small and shallow root systems that have a hard time acquiring water. As we saw with cacti and other succulent plants, shallow root systems can be effective at taking up water from brief rain events, but only if a plant can store excess water in its tissues. However, cotton has a limited ability to store water.

The soils in the Texas Panhandle range from loam to clay. Recall from our earlier discussion and Figure 3.4 that such soils have a moderate to high field capacity. However, they also have a moderate to high wilting point, which makes it hard for cotton to extract water from these soils. In addition, the shallow watering can lead to soil salinization.

In Texas, cotton is traditionally planted in early May and harvested in July or August. Because July and August can be very dry months, farmers have been experimenting with earlier planting dates. When cotton is planted in April, the cotton plants bloom in late June and produce a larger crop, in part because the plants avoid the drier months of July and August. Plants that are stressed by a lack of water can abort the development of their flowers, and flowers are the source of the cotton fibers. In addition, cotton planted earlier matures in June, when days are longer and the plants have more daylight hours to conduct photosynthesis. In 2010, researchers reported that while an earlier planting date is effective, it only works if the seedling plants are irrigated to supplement scarce natural rainwater. Without irrigation, these seedling plants that have a small root system cannot survive.

To help cotton farmers achieve larger crop yields, scientists have conducted a great deal of research on how to make cotton more drought resistant. Researchers have recently found that when cytokinin, a natural plant hormone, is sprayed onto young cotton plants, it induces the plants to grow larger root systems. By growing more roots that can penetrate deeper into the soil, the cotton is less prone to experience the effects of a water shortage. Cytokinin also stimulates the cotton plants to build a waxy outer coating, which we know makes plants less susceptible to water loss. These two responses provide a 5 to 10 percent increase in crop yields under drought conditions.

Cotton plants also have to deal with hot temperatures. Researchers have found that the ideal temperature for growing cotton is a daytime high of 28°C. However, temperatures where cotton is grown can easily exceed 38°C. At these higher temperatures, the enzyme Rubisco does not work as well. As we discussed, Rubisco is a key enzyme for C_3 photosynthesis; poorly performing Rubisco results in a lower rate of photosynthesis and, in turn, lower cotton yields. Plant breeding efforts have developed varieties of heat-tolerant cotton with variations of the Rubisco enzyme that continue to perform well under higher temperatures.

Other varieties of cotton have been bred to transpire higher amounts of water vapour out of their stomata, which improves the plant's ability to cool itself through the process of evaporation. These varieties of cotton can only transpire more water vapour out of their stomata if they have an abundant supply of water coming in through their roots and must therefore be grown in soils that are either naturally moist or irrigated.

The story of cotton illustrates how adaptations allow a species to live under a range of challenging environmental conditions on land. It also illustrates how a knowledge of these adaptations can help farmers adjust their practices and help scientists cultivate varieties of plants that can better perform in the face of these challenges and produce higher crop yields. However, not all challenges faced by agricultural crops can be solved through plant breeding. When this happens, we are left to plant the crops in regions of the world where their adaptations are better suited to the environmental conditions.

SOURCES: Pettigrew, W. T. 2010. Impact of varying planting dates and irrigation regimes on cotton growth and lint yield production. *Agronomy Journal* 102: 1379–1387.

Salvucci, M. E., and S. J. Crafts-Brander. 2004. Inhibition of photosynthesis by heat stress: The activation state of Rubisco as a limiting factor in photosynthesis. *Physiologia Plantarum* 120: 179–186.

SUMMARY OF CHAPTER CONCEPTS

■ **Most terrestrial plants obtain nutrients and water from the soil.** Soil nutrients include nitrogen, phosphorus, calcium, and potassium. The water-holding capacity of soil, known as its field capacity, depends on the particle sizes in the soil. The ability of plants to absorb this water requires that the osmotic potential of the roots be stronger than the matrix potential of the soil. This water moves up the plant stem to its leaves through a combination of osmotic pressure, the cohesion of water molecules, and the force of transpiration.

■ **Sunlight provides the energy for photosynthesis.** Within the full range of electromagnetic radiation produced by the Sun, only a narrow range of wavelengths is used by the photosynthetic pigments of plants. This solar energy is used to power the process of photosynthesis by splitting water molecules and producing molecular oxygen and sugar. There are three pathways of photosynthesis: C_3, C_4, and CAM. Each differs in how it captures CO_2 and each works best in particular environmental conditions. These different pathways are often associated with structural adaptations that help plants from arid regions conserve water.

■ **Terrestrial environments pose a challenge for animals to balance water, salt, and nitrogen.** Organisms attempt to achieve homeostasis in all of these compounds, typically through the use of negative feedbacks. Plants and animals both possess a number of adaptations to balance their concentrations of salt, water, and nitrogenous wastes.

■ **Adaptations to different temperatures allow terrestrial life to exist around the planet.** Organisms can gain and lose heat through radiation, conduction, convection, and evaporation. These processes combine to form an individual's heat budget. Temperature can be regulated to different degrees by animals through the process of thermoregulation. Poikilotherms have variable body temperatures whereas homeotherms have relatively constant body temperatures. The body temperatures of ectotherms are largely determined by their external environment whereas endotherms can raise their body temperatures to be higher than the external environment. Additional adaptations to assist in thermoregulation include the shunting of blood and countercurrent circulation.

REVIEW QUESTIONS

1. Explain the relationship between soil particle size and the field capacity of soil.

2. Explain why the availability of water to plants is highest in soils with particle sizes intermediate between sand and clay.

3. How can we be sure that root pressure is not sufficient to explain the movement of water in trees?

4. Explain how light serves as the ultimate source of energy for a meat-eating animal.

5. Why is C_3 photosynthesis inefficient when the concentration of CO_2 in a leaf is low?

6. How do CAM plants solve the problem of obtaining CO_2 for photosynthesis while minimizing water loss?

7. Explain how plants use structural adaptations to reduce water loss.

8. Describe the costs and benefits associated with the different nitrogen products excreted by fish, mammals, and birds.

9. Explain the concept of a negative feedback.

10. If a snake is lying on a rock in the desert sun, how is the snake's body temperature affected by radiation, conduction, convection, and evaporation?

PRACTICING ECOLOGY: HABITAT CHOICES OF LIZARDS

Stephen Lougheed and his lab at Queen's University in Ontario have been studying how skinks cope with extreme temperatures. Photo by Lori Parker.

For animals in temperate ecosystems, failure to adapt to either hot or cold temperatures can be lethal. Overheating during the summer can lead to muscle spasms and death, while freezing during the winter can fatally slow all biological processes. This is a particular problem for ectotherms since their internal body temperatures are largely regulated by the temperature of the external environment. How do animals in these variable environments cope with the extreme temperatures?

Stephen Lougheed and his lab at Queen's University in Kingston, Ontario, have been studying how one particular ectotherm, the five-lined skink (see Figure 3.19), controls its body temperature. At the northernmost limit of their range in southern Ontario, skinks can experience summer temperatures that exceed their optimal temperature. Lougheed and his colleagues hypothesized that one strategy skinks use to cope with this extreme heat may be seeking habitats where air temperatures are cooler.

To test this hypothesis, the researchers determined the small-scale habitat features where five-lined skinks are most often found. They began their search by flipping over any loose material, such as rocks, logs, and leaves, until they had found 20 sites occupied by at least one skink. At each of those sites, the researchers then moved 3 m away in a random direction to identify a nearby unoccupied site. At both the occupied and unoccupied sites, they dropped a 1-m² frame on the ground and determined the percent of the marked area that contained different habitat features. You can view their data in the table.

GRAPH THE DATA

1. Create a bar graph showing the percent of each habitat feature for the occupied sites.

2. Create a second bar graph showing the percent of each habitat feature for the unoccupied sites.

INTERPRET THE DATA

1. What are the continuous and categorical variables in this study?

2. Based on your graph, what can you conclude about the habitat preferences of the skinks?

3. Suggest reasons why skinks might prefer or avoid certain habitat features.

4. Lougheed and colleagues also measured temperatures on occupied and unoccupied sites. Temperatures on all of the occupied sites reached the skinks' preferred temperature range of 28°C to 36°C, whereas temperatures on the unoccupied sites were either above or below this range. Does this new information provide evidence in support of the hypothesis that skinks seek habitats with cooler air temperatures?

SOURCES: Howes, B.J., and S.C. Lougheed. 2004. The importance of cover rock in northern populations of the five-lined skink (*Eumeces faciatus*). *Herpetologica* 60: 287–294.

Quirt, K.C., G. Blouin-Demers, B.J. Howes, and S.C. Lougheed. 2006. Microhabitat selection of five-lined skinks in northern peripheral populations. *Journal of Herpetology* 40: 335–342.

PERCENT OF AREA WITH HABITAT FEATURES

HABITAT FEATURE	OCCUPIED SITE	UNOCCUPIED SITE
COVER ROCK	23.0	3.0
LEAF LITTER	9.0	15.0
JUNIPER	0.2	4.5
OPEN ROCK	29.0	35.0
RAISED LICHEN	14.0	14.0
HERBACEOUS	16.0	19.0
OTHER	8.8	9.5

CHAPTER 4

Adaptations to Variable Environments

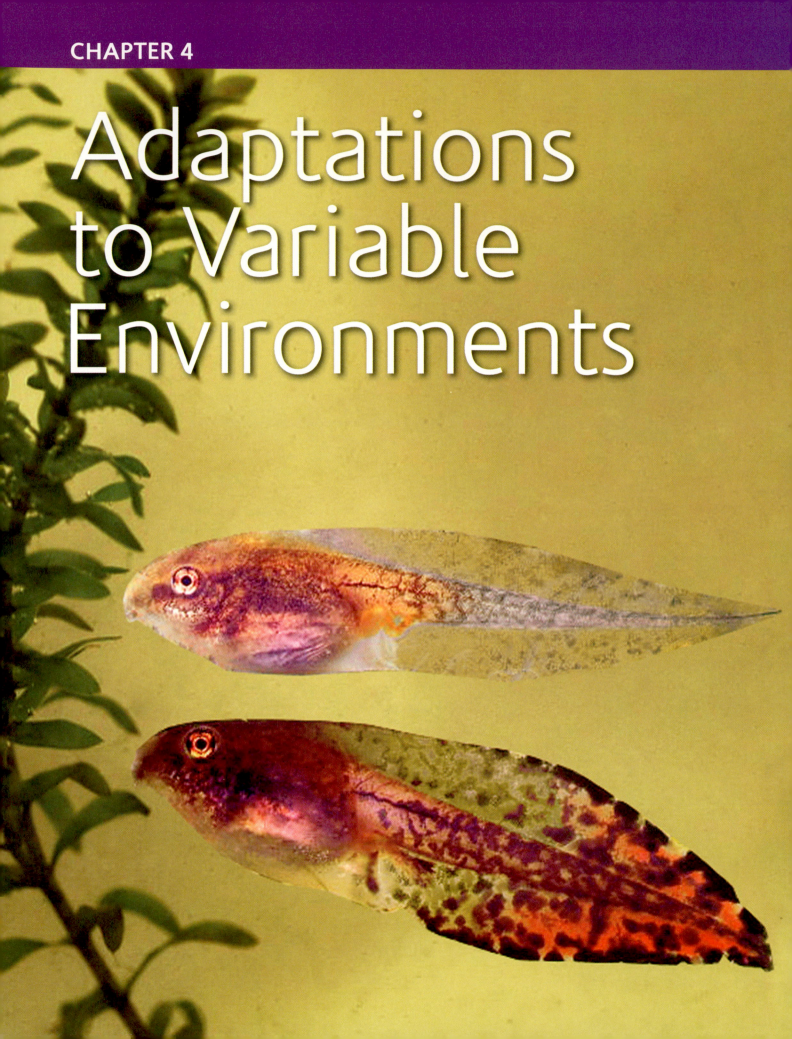

The Fine-Tuned Phenotypes of Frogs

Every spring the female grey treefrog (*Hyla versicolor*) must choose where she will lay her eggs. The treefrog is a medium-sized frog that lives in Canada from Manitoba to New Brunswick, and in the eastern United States south to the Gulf Coast of Texas. As adults, they spend most of their time in forests, feeding on insects in trees. However, in the spring the male and female frogs move to the water to breed. Under ideal conditions, females lay their eggs in ponds that remain free of predators throughout the two months it takes for them to hatch into tadpoles and then metamorphose into frogs. Unfortunately, the females have no way to predict whether a pond will contain predators in the weeks ahead. Their offspring, however, have evolved an amazing ability to adjust to a wide range of different predator environments.

After a female frog lays her eggs, the embryos experience rapid growth and development; in just a few days they are ready to hatch. The timing, however, can change depending on the presence of predators, such as crayfish, that commonly consume frog eggs. Embryos of grey treefrogs, like many species of frogs, can detect the presence of predators by sensing chemical cues that the predators produce. When the embryo detects a nearby predator, development accelerates and it hatches into a tadpole sooner than it would normally, thereby reducing the risk of predation as an embryo. Although it survives the egg predator, it emerges smaller than it would have been if it had remained an embryo for a longer period.

> "The presence or absence of predators influences the tadpole's phenotype."

The grey treefrog has also evolved an ability to respond to changing environmental conditions after eggs hatch into tadpoles. Like the embryos, grey treefrog tadpoles can sense predators in the water through chemical cues. When they detect the presence of a predator, tadpoles hide at the bottom of the pond, become less active, and start to change shape. Within a few days of first sensing a predator, the tadpoles develop big red tails. While the reason for the red colour remains a mystery, the large tails improve a tadpole's ability to escape from predators since they serve as a large sacrificial target that can be lost to a predator and regrown. However, the energy required for rapid development of a large tail is so great that other body parts cannot grow as fast. Consequently, tadpoles with large tails have smaller mouths and shorter digestive tracts, which limits their ability to eat and grow.

In short, the presence or absence of predators influences the tadpole's phenotype. In an environment without predators, it becomes highly active, small-tailed, and fast growing. In the presence of predators, it becomes inactive, large-tailed, and slow growing. But the flexibility does not end there. Tadpoles not only detect the presence of predators but also distinguish different species of predators. This allows them to distribute the use of their energy according to the level of risk; they adjust their defences most strongly

Grey treefrog tadpoles. Tadpoles of the grey treefrog that live without predators exhibit high activity and develop relatively small tails that are drab. In contrast, tadpoles raised with predators exhibit low activity and develop large, red tails. Photo by John I. Hammond.

to the most dangerous predators, and produce more modest defences against less dangerous predators. This strategy has the advantage of allowing tadpoles to use their energy where it will make the most difference to their survival.

Additional studies have also shown that tadpoles can even detect what a predator had for lunch. When a predator habitually feeds on tadpoles, the treefrog tadpoles spend more time hiding and undergo changes in shape as we have noted. However, if the predator is feeding on something else, such as snails, the treefrog tadpoles spend less time hiding and only undergo small shape changes. In essence, the tadpoles detect that they are in more danger from predators feeding on tadpoles than predators feeding on other prey, and they defend themselves accordingly.

Grey treefrog tadpoles also respond to other environmental conditions, including the presence of intraspecific and interspecific competitors. In ponds without predators, many tadpoles survive and compete for algae, the food on which they depend. In response to the relative scarcity of food, tadpoles develop larger mouths and longer intestines. The larger mouths contain wider rows of tooth-like projections, and these improve the ability to scrape algae from rocks and leaves. Longer intestines enable a more efficient extraction of energy from the limited amount of available algae. To grow a large body, however, the tadpole must divert energy away from the formation of its tail. As a result, tadpoles that live in an environment with high competition have smaller tails. There is a cost associated with the phenotypic adaptation best suited to survive high competition; if a predator does show up, the tadpoles with smaller tails will be more vulnerable because they do not have the proper phenotype.

The story of the grey treefrog represents a situation in which a species can experience a tremendous amount of environmental variation within and across generations. In response to this variation, the grey treefrog has evolved a wide range of strategies that help to improve its fitness. The responsiveness of the grey treefrog tadpole represents just one example of how organisms have evolved to respond to future variation in their environment. In this chapter, we will explore the wide range of environmental variation and look at how species have evolved the ability to alter phenotype in response to changing environments.

SOURCE: N. M. Schoeppner, and R. A. Relyea, Damage, digestion, and defense: The roles of alarm cues and kairomones for inducing prey defenses, *Ecology Letters* 8 (2005): 505–512.

CHAPTER CONCEPTS

- Ecological systems and processes vary in time and space.

- Variable environments favour the evolution of variable phenotypes.

- Many organisms have evolved adaptations to variation in enemies, competitors, and mates.

- Organisms can evolve adaptations to variable abiotic conditions.

- Migration, storage, and dormancy are strategies to survive extreme environmental variation.

- Variation in food quality and quantity is the basis of optimal foraging theory.

In Chapters 2 and 3 we discussed the range of environmental conditions in aquatic and terrestrial environments and the many adaptations that organisms have evolved to handle these conditions. However, the environmental conditions an organism faces can also vary considerably over time and among different places. In this chapter we will look at environmental variation and the adaptations that

organisms have evolved to respond to these changing environments.

Ecological systems and processes vary in time and space

Most properties of the environment change over time and across space, and they can change at different rates. For example, air temperature can plunge dramatically in a matter of hours as a cold front passes through a region. On the other hand, ocean water may require weeks or months to cool the same amount. In this section we will look at temporal and spatial variation in the environment, and then bring them together to see how they are related.

TEMPORAL VARIATION

Temporal variation in the environment describes how environmental conditions change over time. Some temporal variation is predictable, including alternation of day and night and seasonal changes in temperature and precipitation. Superimposed on these predictable cycles are irregular and unpredictable variations due to *weather*. **Weather** refers to the variation in temperature and precipitation over periods of hours or days. The weather on any particular day cannot be predicted very far in advance; it can vary over intervals of a few hours or days with the movement of cold and warm air masses. **Climate**, on the other hand, refers to the typical atmospheric conditions that occur throughout the year, measured over many years. For example, the climate of Saskatchewan is typically cold and snowy in the winter but hot and dry in the summer. However, the weather on any particular day cannot be predicted very far in advance; it can vary over intervals of a few hours or days with the movement of cold and warm air masses. Although climate describes the average conditions of a given location throughout the year, climate also can vary over long periods of time. For example, a location might experience a string of years that are much wetter or drier than the average year.

Some types of temporal variation can cause large impacts on ecosystems but occur infrequently in a particular place. For example, droughts, fires, tornadoes, and tsunamis can cause major changes in the landscape, yet their frequency at a particular location is rare. The massive tsunami that hit Japan in 2011, for example, was of a magnitude that has not been seen in more than a hundred years (Figure 4.1). Other sources of temporal variation, such as the current warming of Earth, occur very slowly over decades and centuries.

Some ecosystems can experience temporal variation at fairly regular intervals. In pine woodlands, for example, pine needles and twigs accumulate over time, increasing the probability of a destructive fire as the time since the last fire increases, and contributing to the occurrence of forest fires at fairly regular intervals. Similarly, communicable diseases can spread through a population at regular intervals. After an initial exposure to a disease, many individuals develop

Figure 4.1 Temporal variation in the environment. Some major environmental events are rare but have large effects on humans and other organisms. For example, the earthquake and subsequent tsunami that hit Japan in 2011 was an unlikely event, but it caused catastrophic losses. Photo by Reuters/Mainchi Shimbun/Landov.

immunity. However, as new individuals who lack immunity are produced or join the community, the disease can appear in the population once again. As a result, disease prevalence fluctuates over time every few years. In such cases, the temporal variation in the environment has a fairly regular cycle.

How organisms and populations respond to temporal variation in their environment depends on the severity of the change and how often it occurs. In general, the more extreme the event, the less frequently it occurs. However, both the severity and the frequency of events are relative measures; they depend on the organism that experiences them. Forest fires that occur every 20 years can affect a long-lived tree numerous times, but may only affect one out of 20 generations of an insect that has a one-year lifespan.

SPATIAL VARIATION

Environmental variation also occurs from place to place. Variations in climate, topography, and soil type cause large-scale heterogeneity (Figure 4.2). If you were to fly across Canada from British Columbia to Newfoundland, for instance, you would observe a series of major, large-scale environmental changes along the trip: the coastal forests on the British Columbia coast, the montane forests of the Rockies, Prairie planes, Arctic tundra, and boreal forests. At smaller scales, environmental variation is generated by the structures of plants, the activities of animals, the composition of soil, and the activities of humans.

Weather The variation in temperature and precipitation over periods of hours or days.

Climate The typical atmospheric conditions that occur throughout the year, measured over many years.

Figure 4.2 Spatial variation in the environment. As one travels from the west coast to the east coast of Canada, there is tremendous variation in the natural and human-altered habitats. Photos by (clockwise from top top left): Design Pics/Deddeda/Deddeda/Newscom; Michelle Valberg - www.agefotostock.com; Barrett & MacKay - www.agefotostock.com; John E Marriott - www.agefotostock.com; Gary Fiegehen/All Canada Photos/SuperStock.

As with temporal variation, a particular scale of spatial variation may be important to one organism but not to another. The difference between the top and the underside of a leaf, for example, is important to an aphid, but not to a moose, which happily eats the whole leaf, aphid and all.

Moving through environments that vary in space, an individual experiences environmental variation as a sequence in time. In other words, a moving individual perceives spatial variation as temporal variation. The faster an individual moves and the smaller the scale of spatial variation, the more quickly the individual encounters new environmental conditions and the shorter the temporal scale of the variation. This principle applies to plants as well as to animals. For example, as plant roots grow, they push their way through the soil, and soils commonly contain small-scale variation in moisture and nutrients. If a plant's roots grow quickly in soil with a fine scale of variation, the plant roots will frequently encounter new soil environments. Similarly, wind and animals disperse the seeds of plants. The variety of habitats in which the seeds might land depends on the distance the seeds travel and the scale of spatial variation in the habitat.

CORRELATION OF SPATIAL AND TEMPORAL DIMENSIONS

Scientists have observed that the extent of the space affected by an event is usually positively related to an event's duration in time. Looking at Figure 4.3, we see that thunderstorms typically affect relatively small areas and last only a few minutes, whereas hurricanes inflict devastation across hundreds of kilometres and last for days or weeks. Similar patterns can be

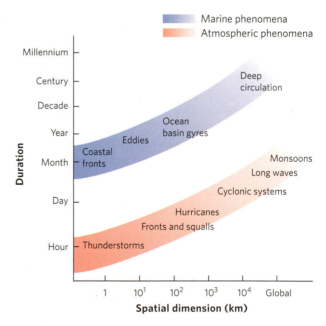

Figure 4.3 Spatial and temporal variation in different physical processes. There is a relationship between the length of time and the spatial dimension of physical phenomena in both marine and atmospheric environments. Processes that affect a large area also last for a long time. Atmospheric phenomena vary on shorter time scales than marine phenomena and, for a particular spatial dimension, marine phenomena have a much larger temporal duration than atmospheric phenomena.

observed in the oceans. Eddy currents, which are circular ocean currents that span 100 to 200 km, may last only a few days. Ocean gyres, which are large circular currents, can last several years. Deep circulation, which involves currents that mix all of the world's oceans, lasts for hundreds of years.

Topography and geology are transformed at a very slow pace by such processes as mountain building, volcanic eruptions, erosion, and even continental drift. Variations in landforms have very long temporal scales. In contrast, spatial heterogeneity in the open ocean is the result of physical processes in water that change more quickly than those in rock and soil. Because air is even more fluid than water, atmospheric processes have very short time scales at any given spatial scale.

Environmental variation favours the evolution of variable phenotypes

As we saw at the beginning of this chapter, survival and growth of treefrog tadpoles depend on the environmental conditions that they experience and on how they alter their phenotype in response. As we have noted in earlier chapters, all phenotypes are the product of genes interacting with environments. As a result, environmentally induced traits have a genetic basis, but reflect the ability of the environment to turn certain genes on or off, which causes different phenotypes to develop. The environments that induce these changes may change so rapidly that they occur within a generation or they may be somewhat slower and vary across generations. The experience of the tree frog is a good example; predators and competitors can differ substantially from pond to pond in a particular year, from year to year in a particular pond, and even from week to week during the time it takes for a generation of tadpoles to metamorphose and leave the pond as frogs. In this section we will examine how producing different phenotypes can be an effective strategy in changing environments. We will also examine the importance of reliable environmental cues, the speed at which different traits can change, and whether environmentally induced phenotypes can be reversed.

PHENOTYPIC TRADEOFFS

In Chapter 2 we saw that rainbow trout that express cold-water isozymes in their tissues perform well in cold water and poorly in warm water. In contrast, fish expressing warm-water isozymes perform well in warm water and poorly in cold water (Figure 2.21). Throughout the natural world we see that a phenotype that is well-suited to one environment may be poorly suited to other environments.

Figure 4.4 illustrates phenotypic fitness in relation to environment. In Figure 4.4a, an individual possessing phenotype X is well suited to environment X and therefore experiences high fitness. In environment Y, however, the phenotype is no longer well suited to the environment and it therefore experiences reduced fitness. In contrast, an individual that possesses phenotype Y is well suited to environment Y and experiences high fitness in environment Y. However, it is poorly suited to environment X, so it experiences reduced fitness in environment X. When a given phenotype experiences higher fitness in one environment whereas other phenotypes experience higher fitness in other environments, we say that there is a **phenotypic trade-off,** meaning that neither phenotype does well in both environments.

But what if an individual could produce a range of phenotypes and each phenotype could perform well in a specific environment? Individuals with mutations that allow them to produce unique phenotypes suited to both environments would experience high fitness in different environments and therefore be favoured by natural selection. The ability of a single genotype to produce multiple phenotypes is called **phenotypic plasticity.** Phenotypic plasticity is a widespread phenomenon in nature; nearly every organism—bacteria, protists, plants, fungi, and animals—possesses phenotypically plastic traits. Different traits can change at different rates and these environmentally induced traits can be either reversible or irreversible. By changing its traits, an individual often maintains a high level of performance when the environment changes. This means that phenotypically plastic traits often are a mechanism of achieving homeostasis, a concept we discussed in Chapter 3.

Figure 4.4b shows the advantage of being phenotypically plastic. In contrast to the two nonplastic genotypes, labelled as genotype X and genotype Y, the plastic genotype, labelled as genotype Z, has a relatively high fitness in both environments because it can produce a phenotype that is nearly as fit as genotype X in environment X and a phenotype that is nearly as fit as genotype Y in environment Y. If we examine the average fitness of the three different genotypes in Figure 4.4c, we see that it is higher for

Phenotypic trade-off A situation in which a given phenotype experiences higher fitness in one environment whereas other phenotypes experience higher fitness in other environments.

Phenotypic plasticity The ability of a single genotype to produce multiple phenotypes.

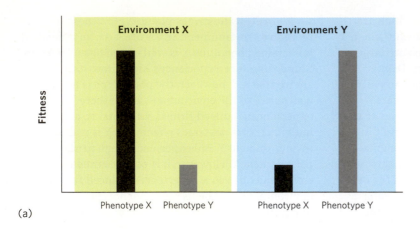

(a)

Figure 4.4 Environments, phenotypes, and fitness. Different environments cause phenotypes to experience different amounts of fitness. **(a)** Phenotypic plasticity evolves because a phenotype has high fitness in one environment and low fitness in another environment. **(b)** Given this trade-off, nonplastic genotypes have high fitness in one environment but low fitness in other environments. In contrast, a plastic genotype can have high fitness in both environments. **(c)** If we consider the average fitness across both environments, we see that the plastic genotype experiences higher average fitness than either of the two nonplastic genotypes.

(b)

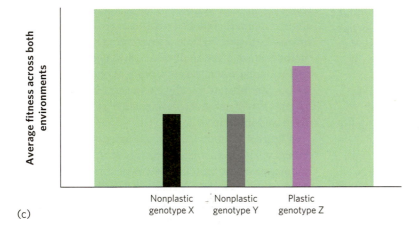

(c)

the plastic genotype. Whenever environmental variation coincides with phenotypic trade-offs across different environments, natural selection will favour the evolution of phenotypic plasticity.

For a long time, scientists applied the concept of phenotypic plasticity to certain types of traits, such as changes in morphology or physiology. However, scientists now recognize that many other types of traits—for example, behaviour, growth, development, and reproduction—frequently represent alternative

phenotypes that can change under different environmental conditions. As a result, the conceptual framework of phenotypic plasticity has expanded in recent years to consider all of these types of traits.

We can see the advantage of phenotypic plasticity in the example of the grey treefrog tadpoles discussed at the beginning of this chapter. In environments with predators, the tadpoles produce a phenotype that is well suited to escape. However, the cost of this phenotype is slower growth. In predator-free environments, the tadpoles produce a different phenotype that is well suited for faster growth. However, the cost of this phenotype is increased vulnerability to predators. With only one possible phenotype, the tadpoles would perform poorly whenever the environment changed. In contrast, a tadpole that can change its behaviour and body shape performs relatively well when the environment changes.

The fitness advantage of phenotypic plasticity occurs whenever environmental variation in space or time occurs frequently. If environmental conditions frequently change, then the phenotype favoured by natural selection also changes frequently and this gives the plastic genotype a higher average fitness than the nonplastic genotype. If spatial or temporal variation is not common, a single phenotype will be favoured; if a nonplastic genotype has higher fitness than any of the plastic genotypes, then the nonplastic genotype will come to dominate the population over time.

ENVIRONMENTAL CUES

For an organism to alter its phenotype in an adaptive way, it must first be able to sense its environmental conditions. For example, the grey treefrog tadpole first senses whether the pond contains predators or competitors and then alters its phenotype to improve fitness. As we will see throughout the rest of this chapter, environmental cues can take many forms, including smells, sights, sounds, and changes in abiotic conditions. Of the numerous potential cues an organism might use, the best are those that offer the most reliable information about the environment. For example, an organism that requires a reliable cue about competition for food could use the presence of a large number of conspecifics—members of its own species—that will be eating the same thing. But if ample food is available, even a large number of conspecifics would not result in competition for food and would be a poor indicator of the level of competition. A better environmental cue of competition for food might be the amount of food that an individual can acquire each day. In an environment with high competition, an individual will always experience a decline in the amount of food available per day. Therefore, the daily intake of food is a more reliable indicator of high competition than the number of conspecifics. When organisms have very reliable cues, they can more accurately produce a phenotype that is well suited to the environment.

RESPONSE SPEED AND REVERSIBILITY

Phenotypically plastic traits respond to changes in the environment at different rates. Some of the trait changes are irreversible. The most rapid responses are typically behavioural traits, which can be altered in seconds. For example, most prey rapidly respond to a predator's pursuit; often it takes less than one second for the prey to flee. Physiological plasticity, which is an environmentally induced change in an individual's physiology—sometimes referred to as **acclimation**—can also be relatively rapid. Consider the time it takes humans to acclimate to the low-oxygen conditions that are caused by lower air pressures at high altitudes, or the time required for human skin to tan. Both of these physiological changes can be accomplished in just a few days. In contrast to these behavioural and physiological changes, changes in morphology—including changes in body shape and the size of internal organs—and changes in life history—including time to sexual maturity and number of progeny produced—can take considerably more time, often on the scale of weeks, months, or years.

Differences in response speed have implications for the reversibility of the induced traits. Behavioural traits that are induced by a change in the environment typically can be rapidly reversed if the environment reverts to its original condition. For example, an animal can quickly and easily adjust its food intake as food conditions change over time. Induced changes in morphology and life history are more difficult to reverse. For many organisms such as plants, changes in morphology are difficult or impossible to undo. For example, plants commonly respond to low light conditions by growing taller in an attempt to rise above neighbouring plants that are casting shade. If the environment suddenly becomes sunny, a plant cannot make itself shorter. Even less reversible are life history decisions such as those related to the timing of reproductive maturity and the amount of reproduction. Once sexual maturity has been achieved, an organism cannot become sexually immature, although it can refrain from reproducing.

The differences in the speed of phenotypic changes, and the ability to reverse phenotypic changes, influence which traits are favoured by natural selection. When environments fluctuate rapidly relative to the length of an individual's lifetime, selection should favour plastic behavioural and physiological traits because these traits can often respond rapidly and reverse rapidly. When environments fluctuate more slowly, selection can favour many more types of traits, including morphological and life history traits that are slow to respond and are often much less reversible.

Acclimation An environmentally induced change in an individual's physiology.

Many organisms have evolved adaptations to variation in enemies, competitors, and mates

Many types of environmental variation can induce phenotypic plasticity. Among biotic environments, three of the best-studied types of environmental variation involve the occurrence of enemies, competitors, and mates.

ENEMIES

Because enemies—including predators, herbivores, parasites, and pathogens—pose a major risk to those organisms that are partially or entirely consumed, we would expect that many organisms have evolved defences against their enemies. Like the treefrog tadpoles, many aquatic animals including fish, salamanders, insects, zooplankton, and protists alter their growth and change shape in response to predators in their environment. These changes might improve a prey's ability to escape, make the prey difficult to fit in the predator's mouth, or deter consumption by producing sharp spines.

Ciliates in the genus *Euplotes* are tiny protists that live in lakes and streams. As you can see in Figure 4.5, these tiny organisms can sense chemical cues emitted by predators and, within hours, respond to the smell of predators by growing "wings" and numerous other projections that make them up to 60 percent larger. The larger size makes it difficult for a predator to fit *Euplotes* into its mouth, so the winged phenotype suffers less predation. However, because of the considerable amount of energy required to grow these projections, this phenotype takes 20 percent more time to develop. Thus, in a predator-free environment, the superior strategy for *Euplotes* is to have the nonwinged phenotype.

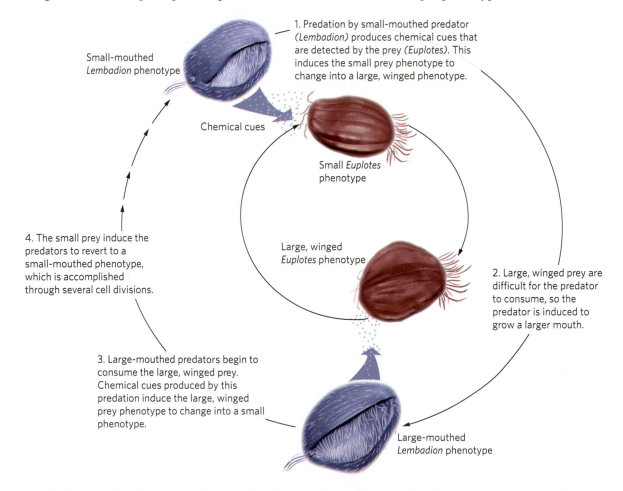

Small-mouthed *Lembadion* phenotype

1. Predation by small-mouthed predator (*Lembadion*) produces chemical cues that are detected by the prey (*Euplotes*). This induces the small prey phenotype to change into a large, winged phenotype.

Chemical cues

Small *Euplotes* phenotype

4. The small prey induce the predators to revert to a small-mouthed phenotype, which is accomplished through several cell divisions.

Large, winged *Euplotes* phenotype

2. Large, winged prey are difficult for the predator to consume, so the predator is induced to grow a larger mouth.

3. Large-mouthed predators begin to consume the large, winged prey. Chemical cues produced by this predation induce the large, winged prey phenotype to change into a small phenotype.

Large-mouthed *Lembadion* phenotype

Figure 4.5 Inducible defences and offences. The ciliate *Euplotes* serves as prey for a larger ciliate, *Lembadion*. When *Euplotes* detects the predatory *Lembadion* in the water via chemical cues, it changes shape from an uninduced small phenotype to a predator-induced phenotype that has large "wings" and other projections that make it too large to fit in the predator's mouth. When the predator begins to encounter the large, winged prey, mechanical cues induce it to increase the size of its mouth. This larger mouth can capture the large, winged prey but it is poor at capturing the small prey, which develop in response to chemical cues emitted by large-mouthed predators. As small prey become more abundant, the predatory *Lembadion* rapidly divides several times to once again have a small mouth. After H. W. Kuhlmann and K. Heckmann, *Hydrobiologica* 284 (1994): 219–227; M. Kopp and R. Tollrian, *Ecology* 84 (2003): 641–651; M. Kopp and R. Tollrian, *Ecology Letters* 6 (2003): 742–748.

How do predators respond when prey make themselves more difficult to catch or consume? It turns out that the predators have plastic abilities as well. For example, when the predatory ciliate *Lembadion* is surrounded by the predator-induced, winged *Euplotes*, *Lembadion* grows a larger body with a bigger mouth that can engulf the winged prey. In doing so, *Lembadion* benefits by consuming the prey. However, once most of the large prey have been consumed and only small prey remain, the large *Lembadion* is poorly suited to eat the small prey. As a result, the large *Lembadion* experiences lower fitness than the small *Lembadion*. When this occurs, a large *Lembadion* can go through several cell divisions and revert back to having a smaller phenotype with the smaller mouth; in doing so, it is better suited to eat the small prey.

Prey also use behavioural defences against predators. They may move away from areas containing predators or become less active to avoid detection. Some animals also congregate in refuges that are safe from predation. Although these behaviours usually reduce the risk of predation, increased safety comes at a price. When prey become less active or congregate in refuges, they spend less time feeding. Also, the supply of food in and around crowded refuges can be depleted quickly. As a result, behavioural defences commonly come at the cost of slower growth, development, or reproduction. In the absence of predators, prey become more active and leave the refuges to find more food; this change in behaviour allows more rapid growth.

Plants also have the ability to respond to the presence of organisms that consume them. The Virginia pepperweed, a member of the mustard family, is eaten by several species of herbivores, including caterpillars and aphids. As Figure 4.6a shows, when an herbivore chews on the leaves of pepperweed, the plant rapidly develops extra leaf hairs, called *trichomes,* that make it harder for the leaves to be consumed. It also increases its production of *glucosinolates,* a group of chemicals that gives mustard its strong flavour and functions as a natural insecticide. If previously attacked plants and nonattacked plants are subsequently placed in a garden, the attacked plants, which have more trichomes and more glucosinolates, attract fewer herbivores and exhibit better survival, as shown in Figure 4.6b.

COMPETITION FOR SCARCE RESOURCES

Most organisms face the challenge of scarce resources, which leads to competition. However, the intensity of competition varies across and within habitats. As a result, organisms have evolved a variety of phenotypically plastic strategies for high and low competition. As we would expect, responses to high competition often exhibit phenotypic trade-offs that

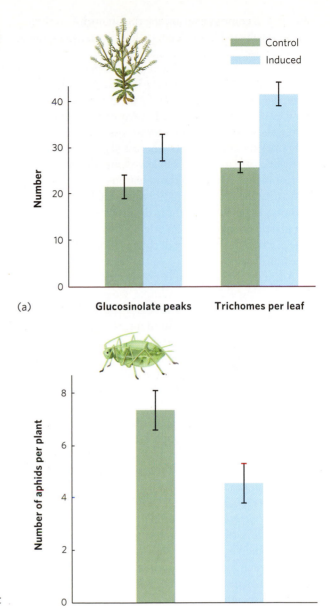

(a)

(b)

Figure 4.6 Herbivore-induced responses by plants.
(a) The Virginia pepperweed (*Lepidium virginicum*) responds to herbivore attacks by growing more leaf hairs, known as trichomes, and more types of defensive chemicals, called glucosinolates. **(b)** Plants attacked by herbivores in the past developed more trichomes and glucosinolates and have fewer aphids on them than plants that were not previously attacked by herbivores. Error bars are standard errors. After A. Agrawal, Benefits and costs of induced plant defense for *Lepidium virginicum* (Brassicaceae), *Ecology* 81 (2000): 1804–1813.

favour the evolution of phenotypically plastic responses.

Jewelweed (*Impatiens capensis*), a flowering plant with beautiful orange flowers, is found in moist habitats throughout much of North America. In nature, it can grow in clumps that are either very sparse or very dense. The range of densities has an effect on the intensity of competition among the plants for sunlight. When jewelweed is shaded by other plants, it responds by elongating its stems, which allows the plant to become taller and to rise above the competing plants. If competition for light causes the jewelweed to grow taller, we might wonder why the plant doesn't always grow tall. Researchers have discovered that elongated jewelweeds experience greater fitness in high-competition environments whereas short jewelweeds experience greater fitness in low-competition environments. Although the researchers were not able to identify the reasons why, it is clear that different phenotypes perform better in different environments, so a plastic phenotype is an effective way to gain high fitness when the intensity of competition varies over time and space.

Animals also respond to competition in a number of fascinating ways. For example, we might expect animals to spend more time searching for food when it is rare than when it is abundant. However, a second option is to find ways to extract more nutrients from the food that is available. One way to do this is to alter the size of the digestive tract, as we saw in the example of grey treefrogs, which respond to competition by increasing the length of their intestine. When the intestine is longer, food spends more time travelling through it and the organism can extract more nutrients from it.

A striking example of a plastic response to variation in food availability can be found in the digestive tracts of snakes. The Burmese python (*Python bivittatus*), for example, can consume a large rodent that is 25 percent of its own body weight, but it may find such a prey item only once a month (Figure 4.7). Between meals, the snake's stomach and intestine shrink a great deal, which reduces its weight and the associated carrying costs. Once a prey is consumed, however, the python can enlarge its cells and double the length of its digestive tract in just 24 hours. This dramatically increases the surface area of the intestine, which allows the snake to absorb more energy from the digested prey. The snake also sends 10 times more blood to the intestines to assist in the absorption of nutrients into the bloodstream. To handle this increase in blood flow, the snake increases the size of its heart by 40 percent, a remarkable achievement given that it occurs in 2 days.

MATES

When mates are rare, reproduction can be a challenge. Consider the situation of flowering plants that depend on pollinators to deliver the pollen that contain a male's sperm. In this situation, the probability of being pollinated can be highly variable, which means that the probability of finding a mate is also highly variable. There is a solution for flowering plants that are **hermaphrodites,** which are individuals that produce

(a)

Fasted
(b)

2 DPF
(c)

10 DPF
(d)

Figure 4.7 The plasticity of a python. (a) A Burmese python can consume large prey, but may do so only once a month. **(b)** A fasting animal has a shrunken intestine. **(c)** Within two days after eating, the intestine doubles in length and swells in diameter. (Note that DPF means days post feeding.) **(d)** Ten days later, digestion is complete and the intestine shrinks again. Photos by (a) Bryan Rourke, California State University; (b through d) Reproduced/adapted with permission Secor S M J Exp Biol 2008; 211: 3767–3774 © 2008 by The Company of Biologists Ltd.

both male and female gametes. Such individuals are self-compatible, that is, they are capable of fertilizing their eggs with their own sperm. This process, known as self-fertilization, often carries the potential cost of **inbreeding depression,** in which offspring can experience reduced fitness when deleterious alleles are inherited from both the egg and the sperm. Because

Hermaphrodite An individual that produces both male and female gametes.

Inbreeding depression The decrease in fitness caused by matings between close relatives due to offspring inheriting deleterious alleles from both the egg and the sperm.

mate availability can be quite variable and self-fertilization can have a substantial cost, the organism must develop one of two life history choices: wait to find a mate and enjoy higher fitness, or self-fertilize. In some cases, this choice is made at the level of the entire organism. In other cases, as in many hermaphroditic species of flowering plants, the choice is made by each flower. If a flower is not pollinated within a given period of time, the flower will self-fertilize. Although self-fertilizing may result in lower fitness than breeding with a mate, low fitness is better than the zero fitness the flower would obtain by avoiding self-fertilization altogether.

Reproductive adaptations to variation in mate availability have been studied in a variety of hermaphroditic plants and animals. For example, the common pond snail (*Physa acuta*) is a hermaphrodite that experiences large variation in population density, which means that it also experiences large variation in the availability of mates. When potential mates are abundant, a snail will normally mate with another individual; when potential mates are rare, it can fertilize its eggs with its own sperm. In an experiment designed to measure the fitness effects of the two alternative strategies, researchers assigned snails to one of two groups: snails living with mates or snails living without mates. The researchers then observed the time it took for each group to begin reproduction and the total number of eggs that each group of snails laid. As you can see in Figure 4.8a, snails living without mates delayed their reproduction for 2 weeks before they used the alternative strategy of self-fertilization. As predicted, this choice came at a cost. Looking at Figure 4.8b, we see that self-fertilizing snails laid many fewer eggs than snails that had mates available. However, the self-fertilizing snails did gain some fitness, which is better than deciding to forego reproduction. The good news for the pond snail, and many other hermaphroditic organisms, is that by evolving multiple strategies they can experience higher fitness over time than would be possible without a plastic strategy.

Many organisms have evolved adaptations to variable abiotic conditions

We have seen that variation in biotic conditions, including enemies, competitors, and mates, can be quite high. Abiotic conditions, including temperature, water availability, salinity, and oxygen, also vary. Faced with this abiotic variation, many species have evolved phenotypically plastic traits that allow them to improve their fitness.

TEMPERATURE
Organisms have evolved a number of plastic responses to temperature variation. As we saw in

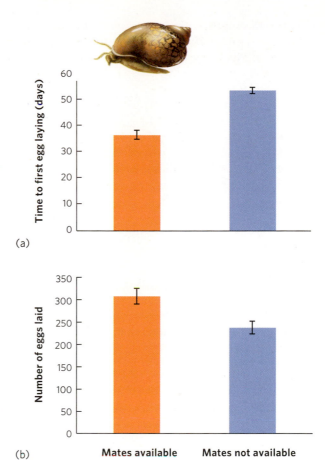

(a)

(b)

Mates available   Mates not available

Figure 4.8 Snail responses to variation in mates. In the common pond snail, mate availability can vary a great deal. **(a)** Snails raised without mates wait nearly 2 weeks before self-fertilizing their eggs. **(b)** Snails without mates produce fewer progeny compared to snails with mates, but this strategy is more fit than foregoing all reproduction. Error bars are standard errors. *After A. Tsitrone et al., Delayed selfing and resource reallocations in relation to mate availability in the freshwater snail Physa acuta, American Naturalist 162 (2003): 474–488.*

Chapter 2, isozymes in rainbow trout allow the fish to have proper nerve transmission in cold winter water and warm summer water. Isozymes are actually a form of phenotypic plasticity with a rapid response time, in some cases a matter of hours or days. For example, goldfish (*Carassius auratus*) can be held at either 5°C or 25°C for a few days and then tested to determine how fast they can swim in a variety of temperatures. As shown in Figure 4.9, fish acclimated to 5°C swim fastest at temperatures around 15°C but swim poorly at temperatures higher than that. In contrast, fish acclimated to 25°C swim fastest at 30°C but swim poorly at temperatures below 25°C. This demonstrates that goldfish can adjust their physiology to maintain relatively high swimming speeds across different environmental temperatures.

Many animals respond to changing temperatures by moving to habitats with more favourable

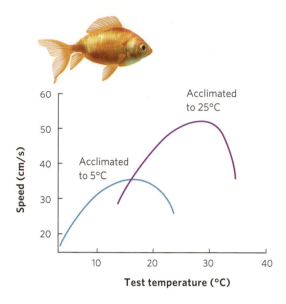

Figure 4.9 Acclimation to different temperatures.
Goldfish raised at low temperatures swim faster in cold water and more slowly in warmer water. Individuals raised at high temperatures swim faster in warmer water than in cold water.
After F. E. J. Fry and J. S. Hart, Cruising speed of goldfish in relation to water temperature, *J. Fish. Res. Bd. Can.* 7 (1948): 169–174.

temperatures. Migrating birds present an extreme example; they fly every autumn to warmer latitudes. Not all animals travel a long distance. Some move to a specific location within a habitat, called a **microhabitat,** that contains more favourable abiotic conditions.

The use of microhabitats can be illustrated by the daily behaviour cycle of the desert iguana (*Dipsosaurus dorsalis*), shown in Figure 4.10. Although the desert iguana cannot regulate its temperature by generating heat metabolically, it is able to take advantage of sunny and shady microhabitats to alter its temperature. The iguana has a preferred temperature range of 39°C to 43°C. In the southwestern United States, the air temperature can reach 45°C. As temperatures rise during the day, the lizard first moves to the shade of plants or rocks and then to a cooler underground burrow. If temperatures begin to cool, the iguana can move out of its burrow and bask in the sun to raise its temperature to its

Microhabitat A specific location within a habitat that typically differs in environmental conditions from other parts of the habitat.

Figure 4.10 Selecting microhabitats. The desert iguana regulates its body temperature by selecting microhabitats that contain favourable abiotic conditions. When the lizard is cool, it can bask in the sun to increase its internal temperature. As the temperature becomes hot during the day, the lizard can seek shade or move into a burrow to lower its temperature.

preferred range. Behavioural plasticity allows the iguana to remain within its preferred temperature range for a greater proportion of the day. In effect, the plasticity of the iguana's behaviour permits homeostasis in its body temperature.

WATER AVAILABILITY

When faced with changes in water availability, most animals can move among different microhabitats. However, plants are typically rooted in a single location and therefore face a tremendous challenge in locating water. As a result, plants have a number of phenotypically plastic adaptations for coping with water variability. Closing the stomata is one of the most common adaptations. As we saw in Chapter 3, when water is plentiful, the guard cells in a plant's leaves open and transpiration occurs through the stomata. However, this process causes water loss. When water is scarce, these cells change shape and the stomata close to conserve water. In this way, the

plant can transpire when water is abundant but stop transpiring when water is in short supply.

Some growth strategies of plants in response to a lack of water are even more dramatic. For example, plants living on coastal dunes in Europe commonly experience drought because water drains quickly through the sandy substrate. Three common plants of these coastal dunes—grey hair-grass (*Corynephorus canescens*), mouse-ear hawkweed (*Hieracium pilosella*), and sand sedge (*Carex arenaria*)—demonstrate how plants adjust their relative allocations of energy and materials to grow either roots or shoots. To demonstrate this phenomenon, researchers grew each of the three plants under conditions of abundant versus scarce water. After 5 months of growth, the researchers measured the ratio of root growth to shoot growth. Figure 4.11 shows the result. When water was abundant, the plants devoted more energy to the growth of shoots, which function primarily to photosynthesize. When water was

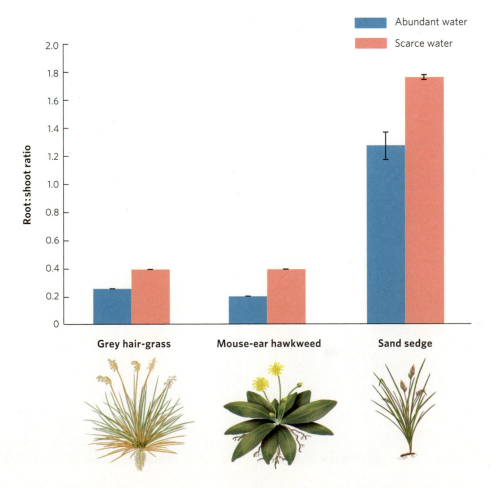

Figure 4.11 Morphological plasticity in response to water. Plants living under scarce water conditions have a higher root to shoot ratio. By devoting more growth into the roots than into the shoots, these plants can better obtain water when it is scarce. Error bars are standard errors. After A. Weigelt et al., Competition among three dune species: The impact of water availability on below-ground processes, *Plant Ecology* 176 (2005): 57–68.

scarce, the plants devoted more energy to the growth of roots, which expanded their ability to capture what little water was available. Given that these plants experience variation in water availability, it is clear that no single allocation strategy would be as beneficial as the phenotypically plastic strategy that they exhibit.

SALINITY

In Chapter 2 we learned that freshwater and saltwater organisms have evolved numerous adaptations to handle their aquatic environments. However, some organisms live in aquatic environments characterized by solute concentrations that fluctuate widely over short periods of time. To survive, these organisms must have the ability to make rapid physiological adjustments. For example, the copepod *Tigriopus* lives along rocky Pacific coasts in pools that receive seawater infrequently from the splash of high waves (Figure 4.12). As water in the pools evaporates, the salt concentration rises to high levels, but a heavy rainfall can lower the salt concentration—a rapid reversal of environmental conditions for the copepod.

Like sharks and rays, *Tigriopus* manages its water balance by changing the osmotic potential of its body fluids. When the salt concentration in a pool is high, individuals synthesize large quantities of certain amino acids such as alanine and proline. These small molecules increase the osmotic potential of the body fluids to match that of the environment without the deleterious physiological consequences that come with high levels of salts or urea. However, this response to excess salts in the environment is costly in terms of the energy it requires. When individual *Tigriopus* are switched from 50 percent seawater to 100 percent seawater, the respiration rate of the copepods initially declines, owing to salt stress, and then increases as they synthesize alanine and proline to restore their water balance. When switched from 100 percent seawater back to 50 percent seawater, the copepods' respiration rate immediately increases as they rapidly degrade and metabolize excess free amino acids to reduce their osmotic difference to be more in line with their new environment.

OXYGEN

If you have ever been to a location at a high elevation, you probably experienced the challenge of low oxygen pressure. As you continue to rise above sea level, the air pressure becomes lower, which reduces the amount of available oxygen. At thousands of metres above sea level, breathing becomes laboured and physical activity is very difficult. Animals that travel up and down mountains as part of their daily or seasonal movements,

(a)

(b)

Figure 4.12 Adaptations to fluctuating salt environments. **(a)** Pools of water on the rocky coasts of Vancouver Island, British Columbia, are filled with a mixture of rainwater and seawater from waves. Evaporation can cause high concentrations of salts and other solutes. **(b)** The tiny copepod *Tigriopus*, shown here carrying eggs, is able to handle the widely fluctuating solute concentrations in its environment by adjusting its solute concentrations through the production of amino acids. Photos by (a) All Canada Photos/Alamy; (b) Morgan Kelly.

for example the llamas of South America, are able to adjust their physiology to this variation in oxygen concentration. Similarly, human mountaineers who tackle Mount Everest experience an extreme oxygen challenge for which their bodies can only partially adjust. At its peak, the mountain is 8848 m above sea level and the oxygen pressure is only one-third of that found at sea level. To acclimate to the low oxygen conditions, climbers stop periodically for several days along the way. Initial changes include more rapid breathing and an increased heart rate. After a week or two, additional changes improve the body's ability to carry oxygen, including an increase in the number of red blood cells and an increased concentration of hemoglobin in the red blood cells. At the highest altitudes, where humans rarely spend much time, the body is not able to fully adjust to such low oxygen concentrations; this has contributed to the deaths of many climbers at high altitudes. When climbers later return to low altitudes, their physiological changes slowly revert to their original state.

Migration, storage, and dormancy are strategies to survive extreme environmental variation

In many parts of the world, extremes of temperature, drought, darkness, and other adverse conditions are so severe that individuals either cannot change enough to maintain their normal activities, or the change required would not be worth the cost. Under such conditions, organisms resort to a number of extreme phenotypically plastic responses. These include migration, storage, and dormancy.

MIGRATION

Migration is the seasonal movement of animals from one region to another. In this case, the phenotype is the behaviour of living in a particular location and the plasticity is displayed in the act of migrating, which allows the animal to express the alternative phenotype of living in multiple locations. Each fall, hundreds of species of land birds leave temperate and Arctic North America, Europe, and Asia for the south, in anticipation of cold winter weather and dwindling food supplies. In east Africa, many large herbivores, such as wildebeests (*Connochaetes taurinus*), migrate long distances, following the geographic pattern of seasonal rainfall and fresh vegetation. Some insects also migrate. Monarch butterflies (*Danaus plexippus*) offer a fascinating example of insect migration, as shown in Figure 4.13. The adult butterflies living in southern Canada and the northern United States migrate to wintering areas in the southern United States and Mexico. Here they hibernate through the winter and then begin heading back north. On their way north, they breed and produce a second generation of butterflies. This second generation completes the migration back to the summer breeding areas. In all cases, the decision to migrate is a plastic behaviour in response to environmental changes. For

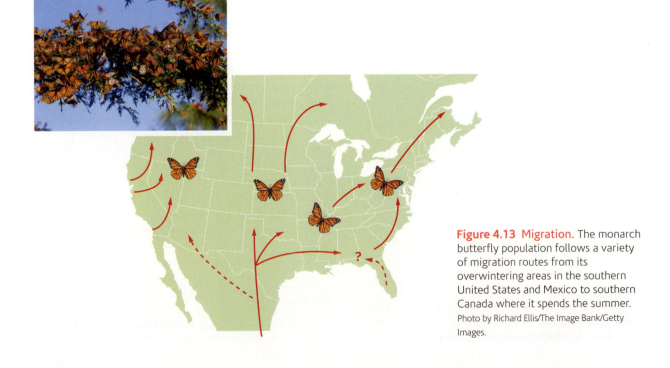

Figure 4.13 Migration. The monarch butterfly population follows a variety of migration routes from its overwintering areas in the southern United States and Mexico to southern Canada where it spends the summer. Photo by Richard Ellis/The Image Bank/Getty Images.

these animals, migration is a better strategy than staying put in either location year-round.

Some migratory movements occur in response to reductions in local food supplies. For example, locust migrations occur when the insects leave an area where they have a large population and a depleted food supply. The migrations can reach immense proportions and cause extensive crop damage over wide areas. This migratory behaviour in locusts is the result of a number of behavioural changes. In sparse populations, adult locusts are solitary and sedentary. In dense populations, however, frequent contact with other locusts stimulates young individuals to develop gregarious, highly mobile behaviour, which can grow into a mass migration.

STORAGE

Where environmental variation shifts the food supply from feast into famine and migration is not a possibility, storage of resources can be an adaptive strategy. For example, during infrequent rainy periods, desert cacti swell with water stored in their succulent stems, as discussed in Chapter 3. In habitats that frequently burn—such as the chaparral of southern California—perennial plants store food reserves in fire-resistant root crowns, shown in Figure 4.14. The surviving root crowns send up new shoots shortly after a fire has passed.

Many temperate and Arctic animals accumulate fat during mild weather as a reserve of energy for periods of harsher weather when snow and ice make food inaccessible. However, fat reserves can make an animal slower and less agile, and therefore more likely to be caught by predators. One way to avoid this problem is to store the food before consuming it. Some mammals and birds that are active during the winter—for example beavers (*Castor canadensis*), squirrels, arctic foxes (*Alopex lagopus*), black-capped chickadees (*Poecile atricapillus*), and jays—cache food supplies underground or under the bark of trees for later retrieval. These caches can be large enough to sustain individuals for long periods.

DORMANCY

Environments sometimes become so cold, dry, or depleted of nutrients that organisms can no longer function normally. Some species that do not migrate have evolved a strategy of **dormancy**, a condition in which they dramatically reduce their metabolic processes. One of the most obvious forms of dormancy occurs when many temperate and Arctic trees shed their leaves in the autumn before the onset of winter frost and long nights. Similarly, many tropical and subtropical trees shed their leaves during seasonal periods of drought. Plant seeds and spores of bacteria and fungi also exhibit dormancy. Indeed, there are many cases in which researchers have sprouted seeds

Root crown

(a)

(b)

Figure 4.14 Energy storage. (a) Some species of plants, including one known as death camas due to the toxic chemicals it produces, possess root crowns that are resistant to fire and store large amounts of energy. **(b)** This stored energy can be used by the plant to quickly resprout after a fire. When fire burns an area, such as this site in Angeles National Forest, California, the death camas plant can quickly resprout. Photo by Rob Sheppard/DanitaDelimont.com.

recovered from archeological excavations where the seeds have been dormant for hundreds of years. Whereas many plants experience dormancy, for animals ecologists distinguish among four specific types of dormancy: *diapause, hibernation, torpor,* and *aestivation.*

In most species, worsening environmental conditions are anticipated and individuals proceed through a series of physiological changes that prepare them for a partial or complete physiological shutdown. **Diapause** is a type of dormancy that is common in insects in response to unfavourable environmental conditions. For example, as winter approaches, insects shut down their metabolism to barely detectable levels. In doing so, the insects need to either reduce the quantity of water in their bodies or chemically bind the water in their bodies to prevent freezing. Similarly, insects facing drought conditions can enter a summer diapause by dehydrating themselves. They either tolerate the desiccated condition of their bodies or

---

Dormancy A condition in which organisms dramatically reduce their metabolic processes.

Diapause A type of dormancy in insects that is associated with a period of unfavourable environmental conditions.

secrete an impermeable outer covering to prevent further drying.

During **hibernation**, a less extreme type of dormancy that occurs in mammals, animals reduce the energetic costs of being active by lowering their heart rate and decreasing their body temperature. Many mammals, including ground squirrels and bats, hibernate during seasons when they are unable to find food (Figure 4.15). Prior to hibernation, the animal consumes enough food to produce a thick layer of fat that provides the energy necessary to survive the hibernation period without eating.

Some types of dormancy occur over short periods of time to deal with cold temperatures. At low air temperatures, some birds and mammals are not able to maintain a high body temperature. To maintain a high body temperature would require that the animal burn up its stored energy faster than it can consume and digest the food needed to replace the energy lost in creating body heat. In this situation the animal might go into a brief period of dormancy, known as **torpor**, during which the animal reduces its activity and its body temperature decreases. During torpor, reduced activity and reduced body temperature help conserve energy. Torpor may last as little as a few hours or extend for a few days and is a voluntary, reversible condition.

Many small birds and mammals use torpor. Hummingbirds, a group of tiny birds with body

Figure 4.16 Torpor. The tiny West Indian hummingbird has a high surface-to-volume ratio, causing it to lose much of the heat it generates during periods of cold air temperatures. To save energy, measured as the amount of energy consumed to fuel the metabolism of its food, the bird enters torpor when it is resting at night. Data from F. R. Hainsworth and L. L. Wolf. Regulation of oxygen consumption and body temperature during torpor in a hummingbird, *Eulampis jugularis. Science* 168 (1970): 368–369.

lengths of 7.5 to 13 cm, provide a good example. These small birds have a high surface to volume ratio. This causes a rapid loss of heat across the body surface relative to the volume of body that can produce heat. As the air temperature declines, hummingbirds must metabolize increasing amounts of energy to maintain a resting body temperature near 40°C. Figure 4.16 shows the relationship between air temperature and energy needs for the West Indian hummingbird (*Eulampis jugularis*), with its metabolic rate measured as the amount of oxygen consumed as it converts its stored energy into body heat. When the bird enters torpor, it reduces its resting body temperature by 18°C to 20°C. If the air temperature drops to 20°C, torpor allows the bird to stop burning extra energy to generate body heat and thereby conserve its energy reserves. Torpor does not mean that the animal ceases to regulate its body temperature; it merely changes the set point on its thermostat to reduce the

Figure 4.15 Hibernation. Some mammals, such as this chipmunk, spend the winter in a deep sleep. During this time, the chipmunk slows its breathing and heart rate and reduces its temperature close to 0°C to help conserve energy through the winter. Photo by Leonard Lee Rue III/Getty Images.

Hibernation A type of dormancy that occurs in mammals in which individuals reduce the energetic costs of being active by lowering their heart rate and decreasing their body temperature.

Torpor A brief period of dormancy that occurs in birds and mammals in which individuals reduce their activity and their body temperature.

Correlations

In the hummingbird example, we saw that as air temperature decreased, the bird's consumption of oxygen increased (Figure 4.16). This is an example of a statistical *correlation*. A **correlation** is a statistical description of how one variable changes in relation to another variable. For example, at the beginning of this chapter, we observed that when tadpoles faced more dangerous predators they exhibited larger phenotypic changes. Similarly, in our discussion of environmental variation, we noted that the size of the area affected by an event such as tornadoes or hurricanes correlated with the duration of the event (Figure 4.3).

Two variables can be related to each other in a variety of ways, shown in the graphs below. A *positive correlation* (a) indicates that as one variable increases in value, the second variable also increases. A *negative correlation* (b) indicates that as one variable increases in value, the second variable decreases. Such increases or decreases may be linear, meaning that the data fall along a straight line, as shown in the example. They may also be curvilinear, as shown in (c) and (d), meaning that the data follow a curved line.

Correlations do not tell us anything about causation. For example, the positive correlation between variable A and variable B in panel (a) might occur because a change in variable A causes the change in variable B. But it could also be that a change in variable B causes the change in variable A. Alternatively, a third, unmeasured variable might cause both A and B to change.

Consider the case of humans climbing Mount Everest. As a person climbs up the mountain over several weeks, the temperature continually decreases and the climber's efficiency in obtaining oxygen from the air increases. This is a correlation, but it is not causation. Lower temperatures do not cause a person to acquire oxygen more efficiently. We know that a third variable, the declining oxygen pressure at high altitudes, is the real cause of the increased efficiency in obtaining oxygen from the air.

YOUR TURN For the following set of data, construct a graph that demonstrates the relationship between the number of predators in a pond and the activity level of grey treefrog tadpoles (i.e., the proportion of their time spent moving). After constructing the graph, determine (a) whether it is a positive or negative correlation and (b) whether it is linear or curvilinear.

NUMBER OF PREDATOR	ACTIVITY LEVEL
0	40
1	20
2	10
3	5
4	3

Correlation A statistical description of how one variable changes in relation to another variable.

(a) **Positive correlation-linear**

Variable B / Variable A

(b) **Negative correlation-linear**

Variable B / Variable A

(c) **Positive correlation-curvilinear**

Variable B / Variable A

(d) **Negative correlation-curvilinear**

Variable B / Variable A

Correlations. (a) When an increase in one variable is associated with a linear increase in another variable, it is a positive linear correlation. (b) When an increase in one variable is associated with a linear decrease in another variable, it is a negative linear correlation. Correlations also can be (c) positive and curvilinear or (d) negative and curvilinear.

difference between ambient and body temperature, and thereby reduces the energy expenditure needed to maintain its temperature at the set point.

A fourth type of dormancy is **aestivation,** which is the shutting down of metabolic processes during the summer in response to hot or dry conditions. Well-known aestivating animals include snails, desert tortoises, and crocodiles.

By whatever mechanism it occurs, dormancy reduces exchange between organisms and their environments, enabling animals and plants to survive unfavourable conditions.

ADAPTATIONS TO PREVENT FREEZING

In Chapter 2 we discussed the adaptations that some aquatic animals have to avoid the damaging effects that freezing can cause to their tissues. In a similar fashion, some terrestrial animals survive very cold weather on land by using special adaptations, including the production of antifreeze chemicals that prevent or control the formation of ice crystals. For example, in cold northern climates many insects spend the winter living just beneath the bark of trees. This bark helps insulate insects to some degree, but they still experience temperatures below the freezing point of water. Similarly, many species of amphibians spend the winter buried a short distance below the soil surface. This location exposes the amphibians to below freezing temperatures. These animals can freeze solid underground in a state that requires very little metabolic activity (Figure 4.17). Two strategies—the use of antifreeze chemicals and forming ice crystals between rather than within cells—help insects and amphibians avoid tissue damage and are an important element in the dormancy of many species during cold weather. As

temperatures warm in the spring, the animals slowly thaw out and resume their normal activities.

Migration, storage, and dormancy represent phenotypically plastic strategies that allow organisms to cope with extreme changes in their environment. Such behavioural and physiological flexibility provides a substantial selective advantage.

Variation in food quality and quantity is the basis of optimal foraging theory

As we discussed earlier in this chapter, animal behaviour is a form of phenotypic plasticity and therefore can be considered under the same conceptual framework. Foraging is one of many important behaviours for animals and a great deal of research has been conducted on how animals search for food and must select from a diversity of choices. Because abundance of food items varies over space and time, no single feeding strategy can maximize an animal's fitness. Hence, feeding decisions represent phenotypically plastic behaviour because different feeding strategies represent different behavioural phenotypes. These phenotypes are induced by unique environmental cues and each alternative feeding behaviour is well suited to a particular environment but not well suited to other environments. Therefore, the alternative behavioural phenotypes experience fitness trade-offs.

Animals must determine where to forage, how long to feed in a certain patch of habitat, and which types of food to eat. Ecologists evaluate foraging decisions by estimating the costs and benefits of feeding in

Aestivation The shutting down of metabolic processes during the summer in response to hot or dry conditions.

Figure 4.17 Freezing solid. Many species of frogs, such as this grey treefrog, can become dormant during cold winter months and then thaw out when spring arrives. By producing antifreeze and forming ice crystals between their cells rather than within their cells, the frogs can dramatically reduce their metabolic activity through the winter. Photo by K. B. Storey, Carleton University.

particular environmental situations. They then compare these estimates to observations of foraging animals to see which strategy provides the highest fitness. Although it would be ideal to measure costs and benefits in terms of survival and reproduction, these components of evolutionary fitness can be difficult to measure. Consequently, ecologists usually look at factors correlated with fitness, such as foraging efficiency. This is based on an assumption that animals able to gather more food in less time should be more successful at survival and reproduction. The idea that animals should strive for the best balance between the costs and benefits of different foraging strategies is known as **optimal foraging theory**.

Animals have four responses to food variation in space and time: *central place foraging, risk-sensitive foraging, optimal diet composition,* and *diet mixing*.

CENTRAL PLACE FORAGING

When birds feed their offspring in a nest, the chicks are tied to a single location while the parents are free to search for food at a distance. This situation is known as **central place foraging** because acquired food is brought to a central place, such as a nest with young. As the parent ranges farther from the nest, it finds a greater amount of potential food sources. However, travelling a longer distance increases time, energy costs, and exposure to risk. The animal must choose the amount of time spent gathering food before returning to the nest as well as how much food to bring back with each trip.

Researchers have used these choices to investigate the feeding behaviour of European starlings (*Sturnus vulgaris*). During the summer, starlings typically forage on lawns and pastures for the larvae of craneflies, called leatherjackets. Starlings feed by thrusting their bills into the soft ground and spreading their beak to expose the prey. When they are gathering food for their young, they hold captured leatherjackets in the base of their bill. Researchers predicted that as a starling continued to capture more leatherjackets and to hold them in its bill, it would become more difficult to grab the next one. This is analogous to shopping for items in a grocery store without a cart or a basket; the more items you hold, the harder it is to add another. As a result, the number of prey caught by the starlings should slow down over time. The prediction, shown in Figure 4.18, was supported by the data. The shape of the curve shows that the rate of food gathering rises rapidly at first and then, as the starling fills its beak, begins to slow down. We say that the starling experiences diminishing benefits over time.

The rate at which the parent bird brings food back to its offspring is a function of how much food it obtains and how much time it takes to obtain it. The total time it takes to obtain the food depends on the

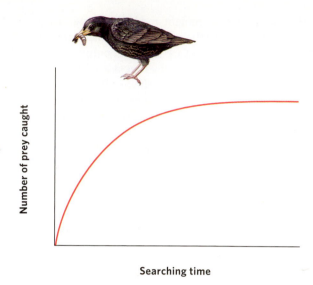

Figure 4.18 Diminishing benefits over time. The rate of food gathering for the European starling is rapid at first but, as time passes, it experiences diminishing benefits because the amount of prey gathered per unit of time decreases.

time needed to fly round-trip to the site that contains the food, known as the *travelling time,* plus the time spent obtaining the food once the bird arrives at the site, known as the *searching time*. Figure 4.19 shows a graphical model of how an animal should make decisions as an optimal forager. The diminishing benefits line is shown in orange. To this, we can add a fixed travelling time, which is the amount of time the bird needs to get to the feeding area. We can then draw a red line from the origin of the trip to intersect the benefits curve. If we draw the red line at the steepest slope that intersects the orange benefits curve, the two lines cross at the red dot on the figure. This intersection point—which is drawn tangent to the orange benefits curve—represents the highest rate of food capture the bird can obtain, including travelling time. If the starling expressed any alternative behavioural phenotypes—for example, if it stayed at the feeding site for longer or shorter periods indicated by the black dots in the figure—it would have a lower rate of food acquisition.

Given our understanding of how the starling should forage when the feeding location is at a fixed distance from its nest, how should the bird's behaviour change when the food source is closer or farther away? At sites that are farther away, the bird should spend more time searching for food and bring back more food to help offset the extra travel time. In contrast, as travel time decreases for sites that are closer to the

Central place foraging Foraging behaviour in which acquired food is brought to a central place, such as a nest with young.

Optimal foraging theory A model describing foraging behaviour that provides the best balance between the costs and benefits of different foraging strategies.

Figure 4.19 Optimal foraging. The optimal rate of foraging for an animal that leaves its nest to find food depends on the time needed to travel to a location that contains food and the time spent feeding once it has arrived. For a given benefits curve (the orange line), the optimal rate of prey capture is found by drawing a straight line from the origin of the trip tangent to the benefits curve. The point of tangency indicates the optimal time that the animal should spend searching and the optimal amount of food it should bring back. Spending more or less time feeding in the location, as indicated by black dots, results in suboptimal amounts of food obtained per unit of time.

nest, the bird should spend less time searching the site for food and bring back less food. Recall the example of the grocery store. If the store happened to be across the street from your house, you would probably make frequent trips, spend a short amount of time searching for food, and bring back a few items on each trip. If the store were an hour's drive away, you would likely make fewer trips, spend a longer time searching for food, and bring back an armload of items on each trip. Like the starling, these decisions improve your efficiency in bringing back food.

To what extent do organisms actually forage optimally? Researchers addressed this question in a clever experiment. They trained starlings to visit feeding tables that offered mealworms through a plastic tube at precisely timed intervals. A starling would arrive at the table, pick up the first mealworm, and then wait for the next mealworm to be delivered. Each successive mealworm was presented at progressively longer intervals, mimicking the longer intervals at which a starling would catch leatherjackets as its beak became increasingly full. The researchers set up feeding tables at different distances from the starling nests and observed how many mealworms a starling picked up before it departed back to its nest. The predicted number of prey brought back to the bird's nest, shown as a blue line in Figure 4.20, agrees with the actual number observed in the experiment, represented by the red data points.

RISK-SENSITIVE FORAGING

Our predictions of how animals should forage have assumed that animals are simply maximizing their rate of energy gain. However, most animals have other considerations, including predators. Because the act of feeding puts most animals at risk for predation, they must consider this danger when making their foraging decisions. Animals that incorporate the risk of predation into their foraging decisions are said to practice **risk-sensitive foraging**.

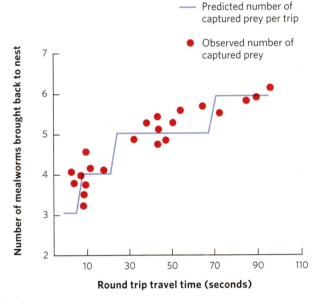

Figure 4.20 Predicted versus observed prey capture for a central place forager. Based on a central place foraging model, researchers predicted that longer travel times would cause starlings to return to their nests with a larger number of mealworms. The researchers offered mealworms to starlings on tables that were located at different distances from their nests. The observed number of mealworms brought back to the nests show agreement with the predictions. After A. Kacelnik, Central place foraging in starlings (*Sturnus vulgaris*), I. Patch residence time. *Journal of Animal Ecology* 53 (1984): 283–299.

The creek chub (*Semotilus atromaculatus*) is a fish that faces the common challenge of finding its lunch rather than becoming lunch for a predator. Small creek chubs feed on tubifex worms and they prefer to feed in locations that have more worms. But what if locations containing more worms also contain more predators including cannibalistic larger creek chubs? How much

Risk-sensitive foraging Foraging behaviour that is influenced by the presence of predators.

Figure 4.21 Risk-sensitive foraging. The sensitivity of young creek chubs to food density and predators was tested in artificial streams. All streams had one side that contained one predatory adult creek chub and a low density of food (0.17 worms/cm²). **(a)** When the right end of the stream contained two predators and the same low density of food, the young chubs moved to the left side. **(b)** When the right end contained two predators and a medium food density (0.17 worms/cm²), the young chubs still moved to the left side. **(c)** Only when the side with two predators contained either a high food density (0.50 worms/cm²) or **(d)** very high food density (1.0 worms/cm²) did the young chubs move to the right side of the stream. After J. F. Gilliam and D. F. Fraser, Habitat selection under predation hazard: Test of a model with foraging minnows, *Ecology* 68 (1987): 1856–1862.

food would it take to entice the small creek chubs to feed in the riskier location? To address this question, researchers placed small creek chubs into an artificial stream that contained a refuge from predators in the middle section of the stream. On one end of the stream, the researchers placed a large creek chub and a low density of worms. On the other end of the stream, they placed two large creek chubs and manipulated different densities of worms. As you can see in Figure 4.21, when the opposite end of the stream had two large creek chubs, the small creek chubs would not move to the side containing two predators until that side contained three times as much food.

(a) **Crop plants**   (b) **Natural plants**

Figure 4.22 Mixed diets. Young grasshoppers grow faster on mixed diets than on any single diet, regardless of whether comparisons were made using (a) crop plants or (b) natural plants. In general, mixed diets supply a more complete range of nutrients needed by animals than single diets. Error bars are standard errors. After E. A. Bernays et al., Dietary mixing in a generalist herbivore: Tests of two hypotheses, *Ecology* 75 (1994): 1997–2006.

OPTIMAL DIET COMPOSITION

Most animals do not consume a single food item but choose from a range of food items. For example, consider the food choices of sea otters living off Newfoundland. They can prey on a range of fish species, including small fish such as sculpins and rock gunnel (*Pholis gunnellus*), and larger species such as cod and hake (*Urophycis tenuis*). Fish species also differ in their swimming speed. Larger prey species provide larger energy benefits to the otter, but require more time and energy to subdue and consume. Given these options, the otter has to decide which prey species it should pursue and which prey species it should ignore.

To determine the optimal food decision we must balance the energy obtained from the prey and the **handling time**—the time required to subdue and consume the prey. Assuming the same amount of handling time for each choice, the optimal decision for the predator will depend on the energy obtained from each prey and the abundance of the prey. The optimal decision can change if handling time is not equal. In this situation, we need to consider the amount of energy gained per unit time for each prey species. We can do this by dividing the energy benefit of a prey item by its handling time. When we do this, we sometimes find that the smallest prey should be consumed because, although they provide a smaller energy benefit than larger prey, their low handling time can provide the predator with the highest energy gain per unit time. In the case of the sea otter prey, species also differ in their swimming speed. Sculpin and gunnel are slow swimmers, whereas cod and hake are fast. So otters should rank species not only by their size but also by their swimming speed, which would influence handling time.

Once we know how different food items compare in terms of energy gained per unit of handling time, we can make a number of predictions. For example, the predator should always eat the prey species that provides the highest energy benefit; if this prey is abundant, it is the only prey that the predator should consume. This strategy maximizes the animal's energy gain. However, if this highest energy prey is rare and the predator's energy needs are not met, the animal should include less profitable items in its diet. Prey species of very low energy value should never be included in the diet unless all higher energy prey are scarce. In the case of the sea otters, researchers found that otters preferred slow-swimming species over fast species, and chose larger individuals over smaller ones.

DIET MIXING

Some foragers consume a varied diet because one type of food might not provide all of the necessary nutrients. Humans, for example, can synthesize many amino acids, but other amino acids—known as essential amino acids—can only be obtained from one's diet. A diet of only rice or only beans does not possess the complete set of essential amino acids needed by humans. However, a diet that combines rice and beans contains all of the required essential amino acids because each contains the essential amino acids that are missing in the other.

The benefits of diet mixing have been demonstrated using nymphs (immature stages) of the American grasshopper (*Schistocerca americana*). As you can see in Figure 4.22a, grasshopper nymphs grew faster when fed a mixture of kale, cotton, and basil than when they were offered any one of these food plants alone. The effect was even more pronounced on lower-quality, natural food plants, such as mesquite and mulberry: Nymphs with mixed diets grew almost twice as fast as those feeding on either one of these plant species alone, shown in Figure 4.22b. Similar results were obtained on artificial diets that were low

Handling time The amount of time that a predator takes to consume a captured prey.

in either protein or carbohydrates, both of which are required for proper growth: Grasshoppers with mixed diets grew more rapidly. Based on these data, we might predict that when given a choice, these grasshoppers would decide to forage on a mixed diet to improve their fitness.

Throughout this chapter, we have seen that organisms commonly experience spatial and temporal variation in their environment. In response to this variation, many have evolved the ability to produce multiple phenotypes from a single genotype. The strategy of using multiple phenotypes—including changes in morphology, physiology, or behaviour—is effective when there are trade-offs such that no single phenotype performs well in all environments. The evolution of phenotypic plasticity is common among all groups of organisms on Earth, wherever there are reliable environmental cues.

ECOLOGY TODAY CONNECTING THE CONCEPTS

RESPONDING TO NOVEL ENVIRONMENTAL VARIATION

Elevated CO_2 experiment. Tall towers at the Duke University Forest have been emitting CO_2 into the atmosphere for several years and researchers have been tracking the effects on the plants. Photo by Jeffery S. Pippen, http:people.duke.edu/~jspippen/nature.htm.

Ecologists have a good understanding of phenotypically plastic adaptations to environmental variation that has been present for hundreds of thousands of generations, long enough to evolve an appropriate phenotypic response mechanism. But how do organisms respond to more recent environmental variation?

One of the most profound changes in our environment has been the global increase in atmospheric CO_2. In 1958, Charles Keeling began recording atmospheric CO_2 concentrations atop 3400-m-high Mauna Loa on the island of Hawaii. Keeling wanted to determine whether anthropogenic emissions were increasing the concentration of CO_2 in the atmosphere. At the time he began his study, scientists had no accurate long-term measurements of atmospheric CO_2 concentrations. In 1958, the CO_2 concentration was about 316 parts per million (ppm; 316 CO_2 molecules per million molecules of air, mostly nitrogen and oxygen). During the subsequent decades, the concentration of CO_2 in the atmosphere has increased dramatically, rising to 352 ppm by 1990 and 395 ppm by 2012, with no sign of levelling off. Other research indicates that this concentration of CO_2 has not been present on Earth for at least the past 10 000 years. As demand for energy and agricultural land increases, the concentration of CO_2 is expected to reach 500 to 1000 ppm by the year 2100.

How will organisms respond to such a change in their environment given that they have not experienced such high concentrations of CO_2 in the past 10 000 years? To address this question, scientists have been conducting large outdoor experiments in which tall towers emit CO_2

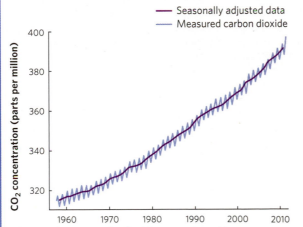

—— Seasonally adjusted data
—— Measured carbon dioxide

Changes in atmospheric CO_2 over time. Measurements on the island of Hawaii have shown that CO_2 concentrations have continuously increased during the past 50 years. After http://www.esrl.noaa.gov/gmd/ccgg/trends/

gas over forests and croplands to cause an elevation in CO_2 ranging from 475 to 600 ppm. Averaged across several such experiments that were initiated in the mid-1990s, researchers have found that elevated CO_2 causes an increase in the rate of photosynthesis by 40 percent. In addition, because plants open their stomata to obtain CO_2, the elevated CO_2 concentrations allow plants to keep their stomata closed more often, resulting in a 22 percent reduction in transpiration of water. All of this translates to improved plant growth. Plants experiencing elevated CO_2 experienced a 17 percent increase in the growth of their shoots and a 30 percent increase in the growth of their roots.

These growth responses represent averages across a variety of species, but not all species responded in the same manner. For example, growth improved in C_3 plants but not in C_4 plants. Researchers hypothesize that because the C_4 pathway of photosynthesis already pumps high concentrations of CO_2 into the bundle sheath cells of the leaf, higher atmospheric concentration of CO_2 has little additional effect. Given that most plant species use the C_3 pathway, most plants will experience higher growth unless other nutrients become limiting or if herbivory of the plants also increases and causes an increased loss of plant tissues. On the other hand, C_4 plants, which include corn, sugar cane, and many other important crops, are not expected to grow any faster as humans continue to elevate the concentration of CO_2 in the atmosphere.

The change in CO_2 concentrations is just one example of the many anthropogenic changes occurring on Earth today. Most organisms have flexible phenotypes that have been shaped by natural selection in response to past environmental variation. Organisms facing novel environmental variation from anthropogenic causes may be able to make use of these existing adaptations and they may also experience continued evolution for new types of flexible phenotypes. However, many other types of anthropogenic impacts—such as air and water pollution—may be far outside the range of historic environmental variation for a population. As a result, populations may not possess phenotypically plastic strategies that will allow them to perform well when facing these types of anthropogenic impacts.

SOURCES: Jaub, D. 2010. Effects of rising atmospheric concentrations of carbon dioxide on plants. *Nature Education Knowledge* 1: 21.

Ainsworth, E. A., and S. P. Long. 2005. What have we learned from 15 years of free-air CO_2 enrichment (FACE)? A meta-analytic review of the responses of photosynthesis, canopy properties and plant production to rising CO_2. *New Phytologist* 165: 351–372.

SUMMARY OF CHAPTER CONCEPTS

■ **Ecological systems and processes vary in time and space.** Temporal variation occurs across a range of hours to years with the most extreme variation being the least frequent. Spatial variation also exists due to differences in climate, topography, and soils. The extent of the space affected by an event is usually positively related to an event's duration in time.

■ **Variable environments favour the evolution of variable phenotypes.** Phenotypic plasticity, the ability to produce alternative phenotypes, is favoured when organisms experience environmental variation, when reliable cues indicate the current state of the environment, and when no single phenotype is superior in all environments. Phenotypically plastic traits include behaviour, physiology, morphology, and life history. Each type of trait differs in how fast it can respond to environmental change and whether the responses are reversible.

■ **Many organisms have evolved adaptations to variation in enemies, competitors, and mates.** Responses to enemies include changes in behaviour that make individuals harder to detect, morphological defence that makes prey harder to capture, and chemical defence that makes prey less palatable. Responses to competitors include morphological changes in plants to make them better able to obtain resources, morphological changes in animals that make them better able to consume and digest scarce food, and behavioural strategies in animals that improve their ability to find scarce food. Organisms generally favour breeding with another individual, but a scarcity of mates can make self-fertilization a viable alternative for some species.

■ **Organisms can evolve adaptations to variable abiotic conditions.** Variation in temperature has favoured the evolution of isozymes and switching between microhabitats. Variation in water availability has favoured plants that can open and close their stomata and alter their root to shoot ratios. Variation in salinity has favoured the evolution of novel ways of adjusting solute concentrations to minimize the cost of osmoregulation. Variation in oxygen can cause adaptive increases in red blood cells and hemoglobin to improve the uptake of oxygen at high altitudes.

■ **Migration, storage, and dormancy are strategies used to survive extreme environmental variation.** Migration allows organisms to leave areas with degrading environments, storage allows organisms to have an extra supply of energy to make it through periods of degrading environments, and dormancy allows organisms to shut down their metabolism until the harmful environmental conditions have passed.

■ **Variation in food quality and quantity is the basis of optimal foraging theory.** Optimal foraging predicts that the amount of time spent foraging at a site and the amount of food brought back to a central nest will depend on the benefits gained over time at the site and the round-trip travel time to the site. Risk-sensitive foragers consider not only the energy to be gained, but also the predation risk posed. Many animals must also consider a range of alternative food items, the energy and abundance of each food item, and whether they should consume a mixture of food items to meet all of their nutritional needs.

REVIEW QUESTIONS

1. Explain the difference between weather and climate in terms of temporal variation.

2. How could spatial heterogeneity be perceived by an organism as temporal heterogeneity?

3. What does it mean when an organism faces a phenotypic trade-off?

4. Why does the evolution of phenotypically plastic traits require reliable environmental cues?

5. Under what conditions might phenotypic plasticity not evolve and why?

6. If self-fertilization results in low fitness due to many nonviable offspring, why would an organism choose self-fertilization as a mating strategy?

7. If a plant can improve its ability to obtain water by growing more roots, why shouldn't the plant always grow more roots?

8. Explain how migration is a phenotypically plastic behaviour.

9. Distinguish between diapause, hibernation, torpor, and aestivation.

10. Explain why an optimal forager should alter its foraging behaviour as the location of food becomes farther away.

PRACTICING ECOLOGY: PLASTIC MOUTHS IN FISH

Dolph Schluter and his colleagues at the University of British Columbia have studied the morphological plasticity of fish mouths in response to changing prey abundances. Photo by Martin Dee.

Most species have mouthparts that are adapted for the type of food they eat. For example, fish that consume small prey near the top of the water column—known as pelagic prey—have small mouths that create a powerful suction that precisely targets their prey. In contrast, fish that consume the larger prey items on the bottom of a lake—known as benthic prey—have large mouths that can more easily engulf prey. Such morphological specialization makes it easier to consume one prey type and more difficult to consume the other. For species that have access to both benthic and pelagic prey, we might hypothesize that natural selection would favour plastic mouthparts that can change to feed on whichever prey type is more abundant. In contrast, selection would favour nonplastic mouthparts for species that have access to only one type of prey.

Dolph Schluter and his colleagues at the University of British Columbia tested this hypothesis using a small, freshwater fish species known as the threespine stickleback, which inhabits lakes throughout North America. Although taxonomists have traditionally considered all threespine sticklebacks as a single species (*Gasterosteus aculeatus*), Schluter has identified three distinct species. Two are specialists that generally live in the same lakes: a species that feeds only on benthic prey and a species that feeds only on pelagic prey. Distinct morphological features in the two specialist species make each well-adapted to the prey it consumes and both stay in their separate regions of the lake. The third species, which is only found in lakes without these specialists, is a generalist that feeds on both benthic and pelagic prey. The generalist species often has intermediate and variable morphological features relative to the specialists. Because the abundance of benthic and pelagic prey varies from year to year, and because the generalists have access to a more diverse array of prey items, Schluter predicted that the generalist sticklebacks would exhibit greater plasticity in their mouthparts than the two specialists.

To test this prediction, the researchers collected fertilized eggs from each of the three stickleback species. Three weeks after the eggs hatched, they placed individuals from each of the three species into separate aquaria—two tanks for each species. They fed each group either bloodworms, which are benthic prey, or brine shrimp, which are pelagic prey. After feeding the fish for 4 months, they measured the mouth sizes. You can view these data in the table below.

GRAPH THE DATA

1. Calculate the mean mouth size and standard error of the mean for each group.

2. Create a bar graph showing the means and standard errors for each group.

INTERPRET THE DATA

1. Which species exhibits the greatest plasticity in mouth size when fed different diets?

2. Docs this support the hypothesis of Schluter and his colleagues?

3. Suppose that the benthic specialist sticklebacks went extinct and, subsequently, more benthic prey became available in the lake. Assuming that the abundance of benthic and pelagic prey continues to fluctuate each year, what might you predict regarding the evolution of plasticity in the surviving specialist species?

SOURCE: Svanbäck, R., and D. Schluter. 2012. Niche specialization influences adaptive phenotypic plasticity in the Threespine Stickleback. *American Naturalist*, 180: 50–59.

STICKLEBACK MOUTH SIZE (mm)

GENERALIST		BENTHIC SPECIALIST		PELAGIC SPECIALIST	
BLOODWORMS	SHRIMP	BLOODWORMS	SHRIMP	BLOODWORMS	SHRIMP
2.70	2.41	2.98	3.31	2.24	3.47
3.40	2.86	3.06	2.58	2.62	2.85
4.09	2.77	2.93	2.90	3.02	2.85
3.41	2.97	3.32	2.76	2.62	2.94
3.62	2.96	3.18	3.15	3.29	2.91
3.75	2.54	3.47	3.73	2.72	3.40
3.04	2.70	2.75	2.85	3.02	2.84
2.64	3.18	3.72	2.79	2.72	2.91
3.57	2.67	3.85	2.69	3.35	2.85
3.31	3.23	3.22	2.48	3.29	2.61
3.30	2.84	2.96	2.73	2.79	2.72

Climates and Soils

Where Does Your Garden Grow?

If you have planted a garden, you know that you have a lot of decisions to make. You can choose from a vast selection of fruits and vegetables, not to mention a dizzying array of flowers, shrubs, and trees. Although you have a large number of choices, not all plants grow well in all locations. To help gardeners determine what plants can survive and flourish in their location, Agriculture Canada and the Canadian Forest Service have developed a map of plant hardiness zones.

Plant hardiness zones take into account a range of factors such as minimum winter temperature, maximum temperature and rainfall in summer, and length of ice-free period. The hardiest plants tolerate very cold temperatures whereas others are too sensitive to survive harsh winters. Zone 9, for example, is found in parts of Vancouver Island, where the average minimum temperature in winter is above −7°C. In contrast, zone 0, with an average minimum winter temperature below −54°C, is found across northern Canada.

> "To help gardeners determine what plants can survive and flourish in their location, Agriculture Canada and the Canadian Forest Service have developed a map of plant hardiness zones."

In looking at the map of plant hardiness zones, several patterns emerge. First, there appears to be an ordering of zones that is related to latitude. Higher latitudes receive less sunlight in the winter and have lower winter temperatures. However, a second pattern of plant hardiness zones curves up along the western and eastern coastlines. For instance, the highest hardiness zones in the prairies and western Ontario is 4, whereas both coasts have zones that are 6 or higher. These patterns occur because the two coasts are adjacent to oceans, which contain warm tropical waters that circulate up from the equator. These warm waters heat the air during winter and make the land along the coasts warmer than the interior of the continent at the same latitude. A third pattern can be seen in elevation; mountain tops have colder temperatures than lower elevations.

The hardiness zone map also shows that the west coast has warmer winter temperatures than the east coast. This allows farmers in British Columbia's valleys to grow fruits and vegetables during the winter. The difference in temperatures between the east and west coasts is caused by prevailing winds that blow from west to east. In the winter along the west coast, the winds carry the warmer ocean air toward the coast, making it warm. Along the east coast, however, winds carry cold air from the middle of the continent toward the

A beautiful park. This is the Azalea Garden path in Stanley Park, Vancouver, British Columbia. Photo by George Ostertag/www.agefotostock.com.

Average Annual Extreme Minimum Temperature 1961 – 1990

Zone	Temp (C)
0	< -51°C
1	-51 to -46
2	-46 to -40
3	-40 to -34
4	-34 to -29
5	-29 to -23
6	-23 to -18
7	-18 to -12
8	-12 to -9

Plant hardiness zones for Canada. The warmest zones occur in southern Ontario and along the western and eastern coasts.

coast and push the warm ocean air away from the coast. As a result, the east coast remains colder than the west coast during the winter.

The map of plant hardiness zones illustrates how climates around the world are the result of a complex combination of sunlight, latitude, elevation, air currents, and ocean currents. In this chapter, we will explore how global processes affect the distribution of climates on Earth and how climates affect the types of soils that form.

SOURCE: www.pacificbulbsociety.org/pbswiki/files/Maps/CanadaHZMap.pdf

CHAPTER CONCEPTS

- Earth is warmed by the greenhouse effect.

- There is an unequal heating of Earth by the Sun.

- The unequal heating of Earth drives air currents in the atmosphere.

- Ocean currents also affect the distribution of climates.

- Smaller-scale geographic features can affect regional and local climates.

- Climate and the underlying bedrock interact to create a diversity of soils.

As we saw in Chapter 3, the climate of a region on Earth refers to the average atmospheric conditions measured over many years. Climates can range widely, from very cold areas near the North and South Poles, to hot and dry deserts at approximately 30° N and 30° S latitudes, to hot and wet areas near the equator. In this chapter, we will examine factors that determine the location of climates around the world. With an understanding of climates, we will move on to look at how soils are formed. As we will see in subsequent chapters, these differences in climates and soils help determine the distribution of organisms around the globe.

A number of factors contribute to the different climate patterns. Some of the most important are the greenhouse effect, the unequal heating of Earth by solar energy, atmospheric convection currents, the rotation of Earth, ocean currents, and a variety of smaller-scale topographic features, including rain shadows.

Earth is warmed by the greenhouse effect

Solar radiation provides the vast majority of the energy that warms Earth and that organisms use. However, solar radiation alone is not sufficient to warm the planet; gases in the atmosphere also play a critical role.

THE GREENHOUSE EFFECT

Solar radiation warms Earth through a series of steps, illustrated in Figure 5.1. About one-third of the solar radiation emitted toward Earth is reflected by the atmosphere—the 600-km thick layer of air that surrounds the planet—and heads back into space. The remaining solar radiation penetrates the atmosphere. Much of the high-energy radiation—including ultraviolet radiation—is absorbed in the atmosphere. The rest passes with most of the visible light through the atmosphere. When this radiation subsequently strikes clouds and the surface of Earth, a portion is reflected back into space and the rest is absorbed. As

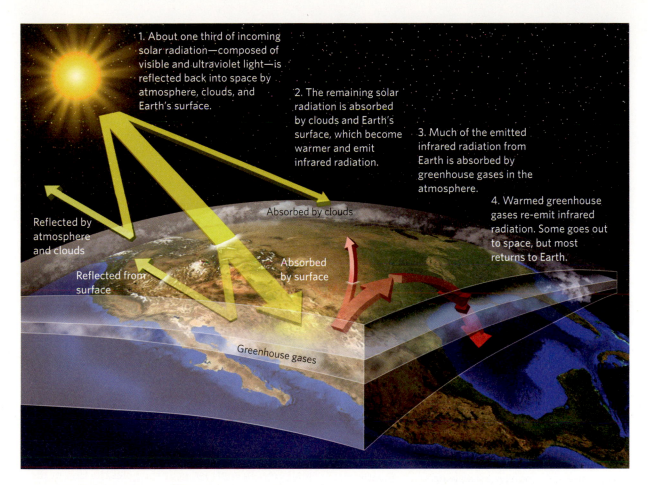

1. About one third of incoming solar radiation—composed of visible and ultraviolet light—is reflected back into space by atmosphere, clouds, and Earth's surface.

2. The remaining solar radiation is absorbed by clouds and Earth's surface, which become warmer and emit infrared radiation.

3. Much of the emitted infrared radiation from Earth is absorbed by greenhouse gases in the atmosphere.

4. Warmed greenhouse gases re-emit infrared radiation. Some goes out to space, but most returns to Earth.

Absorbed by clouds

Reflected by atmosphere and clouds

Reflected from surface

Absorbed by surface

Greenhouse gases

Figure 5.1 The greenhouse effect. Of the solar radiation that strikes Earth, some is reflected back into space and the rest penetrates the atmosphere where much of it warms clouds and the planet's surface. These warmed objects emit infrared radiation back toward the atmosphere where it is absorbed by greenhouse gases. The warmed greenhouse gases re-emit infrared radiation back toward Earth, which causes the surface to warm further.

clouds and Earth's surface absorb this radiation, they begin to warm and emit lower energy infrared radiation. The heat you feel radiating into the air if you stand on a hot asphalt road is an example of this infrared radiation.

The solar radiation that we call visible light easily passes through the gases of the atmosphere. Infrared radiation, however, is readily absorbed by gases in the atmosphere. The gases are warmed by the infrared radiation and then re-emit infrared radiation in all directions. Some of this energy goes out into space and some goes back toward the planet's surface. This process of solar radiation striking Earth, being converted to infrared radiation, and then being absorbed and reradiated by atmospheric gases is known as the **greenhouse effect.** The name comes from the fact that the effect resembles a gardener's greenhouse with windows that hold in the heat from solar radiation.

GREENHOUSE GASES

There are many different gases in the atmosphere, but only those that absorb and re-emit infrared radiation

and contribute to the greenhouse effect are known as greenhouse gases. In fact, if we exclude water vapour, 99 percent of the gases in the atmosphere are oxygen (O_2) and nitrogen (N_2) and neither gas functions as a greenhouse gas. This means that greenhouse gases, which play such a major role in keeping our planet warm, compose only a small fraction of the atmosphere. The two most prevalent greenhouse gases are water vapour (H_2O) and carbon dioxide (CO_2). Other naturally occurring greenhouse gases include methane (CH_4), nitrous oxide (N_2O), and ozone (O_3). These gases have a variety of natural sources: water vapour comes from large bodies of water and the transpiration of plants; CO_2 comes from decomposition, respiration of organisms, and volcanic eruptions; N_2O comes from wet soils and low-oxygen regions of water bodies; and CH_4 comes from

Greenhouse effect The process of solar radiation striking Earth, being converted to infrared radiation and being absorbed and re-emitted by atmospheric gases.

anaerobic decomposition. The naturally occurring greenhouse effect is quite beneficial to organisms on Earth. Without this phenomenon, the average temperature on Earth, which is currently 14°C, would be a much colder –18°C.

The concentrations of greenhouse gases in the atmosphere are increasing. As we noted in Chapter 4, the concentration of CO_2 in the atmosphere has substantially increased over the past 2 centuries due to increased combustion of fossil fuels by automobiles, electric generating plants, and other industrial processes. At the same time, there have also been increases in methane and nitrous oxide from a variety of anthropogenic sources that include agriculture, landfills, and the combustion of fossil fuels. Finally, there are gases that are not naturally produced, such as chlorofluorocarbons that have been manufactured to serve as propellants in aerosol cans and as refrigerants in freezers, refrigerators, and air conditioners. Although these human-created compounds exist at much lower concentrations than water vapour or CO_2, each molecule can absorb much more infrared radiation than H_2O and CO_2, and persists in the atmosphere for hundreds of years. A steady increase of these manufactured gases over the past 2 centuries has raised concerns among scientists, environmentalists, and policy makers.

Because greenhouse gases absorb and re-emit infrared radiation to Earth and its atmosphere, it makes sense that an increase in the concentration of greenhouse gases in the atmosphere would cause an increase in the average temperature of Earth. This expectation has been borne out. Based on thousands of measurements made throughout the world, the average air temperature of the planet's surface increased by approximately 1°C from 1880 to 2011. While some regions, such as Antarctica, have become 1°C to 2°C cooler, other regions, such as northern Canada, have become up to 4°C warmer. In fact, over the 131-year period of monitoring temperatures around the world, 9 of the 10 warmest years occurred from 2000 to 2011. As we will see at the end of this chapter, these temperature changes are altering global climates.

There is an unequal heating of Earth by the Sun

The differences in temperature around the globe are the result of how much solar radiation strikes the surface of Earth at a given location. Differences in solar radiation are determined by the angle of the Sun striking different regions of the globe, the depth of the atmosphere that the energy passes through, and seasonal changes in the position of Earth relative to the Sun.

THE PATH AND ANGLE OF THE SUN

Consider the position of the Sun during the March and September equinoxes when the Sun is positioned directly over the equator. At these times of the year, the equator receives the greatest amount of solar radiation and the poles receive the least. Three factors dictate this pattern: the distance that sunlight must pass through Earth's atmosphere, the angle at which the Sun's rays hit Earth, and the reflectivity of Earth's surface.

As illustrated in Figure 5.2, before the rays of the Sun reach Earth, they must travel through Earth's atmosphere. When they pass through the atmosphere, gases absorb some of the solar energy. Following the path of the rays, you can see that the distance travelled through the atmosphere is shorter at the equator than at the poles. This means that less solar energy is removed by the atmosphere before it strikes Earth at the equator.

The intensity of solar radiation that strikes an area also depends on the angle of the Sun's rays. Looking again at Figure 5.2, you can see that when the Sun is positioned directly above the equator, the rays of the Sun strike Earth at a right angle. This causes a large quantity of solar energy to strike a small area. In contrast, near the poles the rays of the Sun strike Earth at an oblique angle, which causes the solar energy to spread over a larger area. As a result, Earth's surface receives more solar energy per

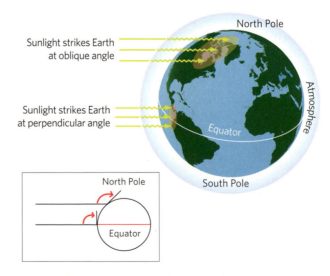

Figure 5.2 Unequal heating of Earth. When the Sun is directly over the equator, its rays travel through less atmosphere and are spread over a smaller area. Near the poles, however, the rays of the Sun must travel through more atmosphere and are spread over a larger area.

Figure 5.3 Albedo effect. Light-coloured objects such as fresh snow reflect a high percentage of incoming solar energy, and dark-coloured objects reflect very little. The average albedo of Earth is 30 percent.

square meter near the equator than near the poles. You can replicate this phenomenon by shining a flashlight onto a large ball. If you point the flashlight at the middle of the ball, you will see a small circle of intense light. However, if you then rotate the flashlight so that it points near the top or bottom of the ball, the beam of light spreads over a larger area and this light is less intense per unit area.

Finally, some surfaces of the globe reflect solar energy more than others. Light-coloured objects reflect a higher percentage of solar energy than dark-coloured objects, which absorb most incoming solar energy. For example, asphalt absorbs 90 to 95 percent of the total solar energy that strikes its surface, which explains why asphalt pavement becomes so hot on a sunny summer afternoon. On the other hand, cropland reflects 10 to 25 percent of the total solar energy that strikes its surface and fresh snow reflects 80 to 95 percent. The fraction of solar energy reflected by an object is its **albedo.** As you can see in Figure 5.3, the more solar energy reflected, the higher the albedo.

The unequal heating of Earth explains the general pattern of declining temperatures as we move from the equator to the poles. At the equator, the Sun's rays lose less energy to the atmosphere, solar energy is spread over a smaller area, and the low albedo of dark-coloured forests causes much of this energy to be absorbed. Near the poles, however, the Sun's rays

lose much more of their energy to the atmosphere, solar energy is spread over a larger area, and the high albedo of the snow-covered land causes much of this solar energy to be reflected. This helps explain why the plant hardiness zone numbers we discussed at the beginning of this chapter generally decrease as you move to higher latitudes.

SEASONAL HEATING OF EARTH
The relationship between the Sun and Earth also causes seasonal differences in temperatures on Earth. The axis of Earth is tilted 23.5° with respect to the path Earth follows in its orbit around the Sun. Figure 5.4 illustrates how this tilt affects the seasonal heating of Earth. During the March equinox, the Sun is directly over the equator. As we approach the June solstice, the orbit and tilt of Earth causes the Sun to be directly over 23.5° N latitude, which is also known as the Tropic of Cancer. In September, the Sun is back to being directly over the equator, and in December the Sun is directly over 23.5° S latitude, which is also known as the Tropic of Capricorn.

The tilt of Earth as it orbits around the Sun causes the Northern Hemisphere to receive more solar energy between March and September than the Southern Hemisphere. During this time the daylight

Albedo The fraction of solar energy reflected by an object.

Regressions

As we have discussed, latitudes closer to the equator receive more solar radiation than latitudes closer to the poles. Given this observation, lower latitudes should also have warmer temperatures than higher latitudes. In fact, understanding the nature of this relationship would help us determine exactly how much temperature changes with latitude. When we want to know how one variable changes in relation to another, we use a statistical tool called **regression.** In Chapter 4 we saw that a correlation determines if there is a relationship between two variables. A regression determines whether there is a relationship and also describes the nature of that relationship.

To help illustrate this idea, we can use data on the average January temperature from 59 locations spanning Canada from north to south. If we plot the relationship between location latitude and average location temperature in January, we obtain the following graph:

In this case, the relationship between the two variables follows a straight line; we have drawn a line of best fit through the distribution of the data points. This is a regression line because it represents the relationship between the two variables. It informs us about the nature of the relationship through the slope and intercept of the line. For these data, the regression can be described using the equation of a straight line, where Y is the dependent variable, X is the independent variable, m is the slope of the line, and b is the Y-intercept of the line at the point where $X = 0$. In this example, the slope is -0.9 and the intercept is 38:

$$Y = mX + b$$

$$\text{Temperature} = -0.9 \times \text{Latitude} + 39$$

This regression equation tells us that for every 1 degree increase in latitude, the average temperature in January decreases by 0.9°C. Note that while the simplest form of a regression is a straight line, regression lines can also be curvilinear.

> **YOUR TURN** Based on the relationship between latitude and temperature, use the regression equation to estimate the average January temperature at 40, 60, and 80 degrees of latitude.

Regression A statistical tool that determines whether there is a relationship between two variables and that also describes the nature of that relationship.

period in the Northern Hemisphere is greater than the nighttime period, and the Sun's angle is 90° somewhere over the Northern Hemisphere. This means that more intense solar radiation is produced per unit area and for a longer period of time. Between the fall equinox in September and the spring equinox in March, the situation reverses and the Southern Hemisphere has longer days and receives more direct solar energy than the Northern Hemisphere. The latitude that receives the most direct rays of the Sun, known as the **solar equator,** shifts throughout the year—from 23.5° N latitude in June to 23.5° S latitude

in December. These are the warmest latitudes on Earth and are known as the tropical latitudes.

Seasonal changes in temperature vary as Earth traces its annual path around the Sun. While the average temperatures of the warmest and coldest months in the tropics differ by as little as 2°C to 3°C, at higher latitudes in the Northern Hemisphere average monthly temperatures vary by an average of 30°C over the year and extreme temperatures vary by more than 50°C annually.

Solar equator The latitude receiving the most direct rays of the Sun.

March
equinox

23.5° Tilted axis

North Pole

Equator

South Pole

Tropic of Cancer

June
solstice

Sun

December
solstice

Tropic of Capricorn

September
equinox

Figure 5.4 Seasonal heating of Earth. The central axis of Earth is tilted 23.5°. Because of the tilt, the Northern Hemisphere receives the most direct sunlight during the June solstice and the Southern Hemisphere receives the most direct sunlight during the December solstice. Locations near the equator receive the most direct sunlight during the March and September equinoxes.

The unequal heating of Earth drives air currents in the atmosphere

Earth's uneven heating helps determine **atmospheric convection currents,** which are the circulations of air between the surface of Earth and the atmosphere. The patterns of air circulation play a major role in the location of tropical rainforests, deserts, and grasslands throughout the world. In this section, we will explore how the interaction of Earth's unequal heating and the properties of air create atmospheric convection currents.

PROPERTIES OF AIR

Four properties of air influence atmospheric convection currents: density, water vapour saturation point, adiabatic heating or cooling, and latent heat release.

Like all gases, when air warms it expands and becomes less dense. When air next to the surface of Earth warms, it becomes less dense than the air above it. This causes the warm air next to the surface of Earth to rise.

The graph in Figure 5.5 shows the relationship between air temperature and the maximum amount of water vapour that the air is able to contain. As air temperature increases, its capacity to contain water vapour—the gaseous form of water—increases. Although the capacity to contain water increases at higher temperatures, there is always a limit, known as the **saturation point.** When the water vapour content of air exceeds the saturation point, the excess condenses and changes phase to either liquid water or

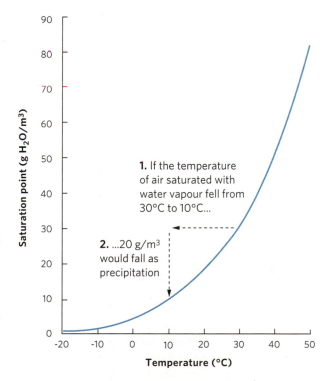

1. If the temperature of air saturated with water vapour fell from 30°C to 10°C...

2. ...20 g/m³ would fall as precipitation

Figure 5.5 Saturation point of water vapour in air. As the temperature of the air increases, the air is able to contain greater amounts of water vapour.

Atmospheric convection currents The circulations of air between the surface of Earth and the atmosphere.

Saturation point The limit of the amount of water vapour the air can contain.

ice. When the water vapour content is below the saturation point, liquid water or ice can be converted to water vapour. For example, at 30°C, air can contain up to 30 g of water vapour per m³. Air that contains the maximum amount of water vapour has reached its saturation point. If the air at 30°C cools to 10°C, the saturation point of the air will decrease to 10 g of water vapour per m³. As a result, the excess vapour changes phase to liquid water or ice and produces clouds and precipitation.

If you live in a region with abundant water, you probably expect warm summer days to be humid. Evaporation from the ground and nearby bodies of water, as well as transpiration from plants, add water vapour to the air. However, if you live in a drier region, such as a desert, you know that the hottest days are not humid; a scarcity of water keeps the atmosphere from becoming saturated, and plants and animals face the threat of losing too much precious body water to the atmosphere. Under these conditions, plants close their stomata, which reduces the amount of water that can be moved into the atmosphere via transpiration.

The relationship between temperature and water vapour saturation affects patterns of evaporation and precipitation around the world. This, in combination with air currents, determines the distribution of wet and dry environments.

The movement of air in response to changes in pressure also influences convection currents. Near the surface of Earth, the gravitational pull on all molecules in the atmosphere brings many of the molecules close to Earth's surface. An increase in the number of molecules causes an increase in air pressure near Earth's surface. As one moves higher into the atmosphere, the air contains fewer total molecules, which lowers the air pressure. As a result, when air moves between the surface of Earth and the atmosphere, it experiences changes in pressure.

Air pressure is related to the frequency of collisions among air molecules, which also influences temperature. Lower rates of collision cause lower temperatures. As a result, when air moves higher up into the atmosphere and experiences lower pressure, the air expands and the temperature decreases—a process known as **adiabatic cooling.** Conversely, when air moves down to Earth's surface and experiences higher pressure, the air compresses and the temperature increases in a process known as **adiabatic heating.**

The final property of air to consider is the release of heat. As you might recall from our discussion of the thermal properties of water in Chapter 2, converting liquid water to water vapour requires a great deal of energy. In the reverse process, known as **latent heat release,** water vapour converted back to liquid water releases energy in the form of heat. Latent heat release is significant because whenever water vapour exceeds its saturation point, condensation will cause a release of heat that warms the surrounding air.

FORMATION OF ATMOSPHERIC CONVECTION CURRENTS

Differences in climate are determined in part by patterns of worldwide air circulation. Let's look at the equator during the March or September equinoxes. As you can see in Figure 5.6, solar energy warms the air at the surface of Earth. This warming causes the air to expand and rise. As air rises into regions of decreased atmospheric pressure, it expands. The temperature of the air then cools through the mechanism of adiabatic cooling. This cooled air has a reduced capacity to contain water vapour, so the excess water vapour condenses and falls back to Earth as rain. This process, in which the surface air heats, rises, and releases excess water vapour in the form of rain, is the primary reason that latitudes near the equator often experience high amounts of rainfall.

Returning to Figure 5.6, we can see that when the water vapour condenses, it causes latent heat release, which further warms the rising air. This enhances the upward motion of the air, condensation, and rainfall. As the air pressure continues to fall with rising altitude, the air temperature continues to drop. At this altitude, the cool, dry air is pushed from below by more rising air and begins to move horizontally toward the poles.

The upward movement of air is the driving force behind atmospheric convection currents, but it is just the first of a series of steps in the process. Once the cool, dry air is displaced horizontally toward the poles, it begins to sink back toward Earth at approximately 30° N and 30° S latitudes. As Figure 5.6 illustrates, this air sinks toward Earth where increased pressure causes it to compress. As the air compresses, it experiences adiabatic heating and decreased humidity. By the time the air falls back to the surface of Earth, it is hot and dry. This explains why many of the major deserts of the world—which are characterized by hot, dry air—are located at approximately 30° N and 30° S latitudes.

Adiabatic cooling The cooling effect of reduced pressure on air as it rises higher in the atmosphere and expands.

Adiabatic heating The heating effect of increased pressure on air as it sinks toward the surface of Earth and decreases in volume.

Latent heat release When water vapour is converted back to liquid, water releases energy in the form of heat.

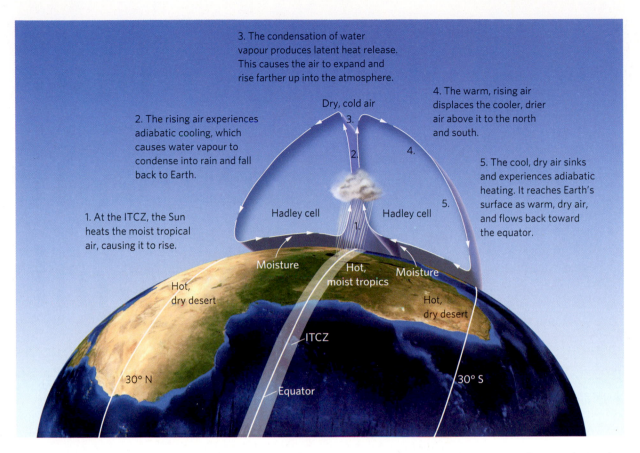

1. At the ITCZ, the Sun heats the moist tropical air, causing it to rise.

2. The rising air experiences adiabatic cooling, which causes water vapour to condense into rain and fall back to Earth.

3. The condensation of water vapour produces latent heat release. This causes the air to expand and rise farther up into the atmosphere.

4. The warm, rising air displaces the cooler, drier air above it to the north and south.

5. The cool, dry air sinks and experiences adiabatic heating. It reaches Earth's surface as warm, dry air, and flows back toward the equator.

Dry, cold air

Hadley cell   Hadley cell

Moisture   Hot, moist tropics   Moisture

Hot, dry desert   Hot, dry desert

ITCZ

30° N   30° S

Equator

Figure 5.6 The circulation of air in Hadley cells. In this example, the Sun is directly over the equator, as happens during the March and September equinoxes. At the latitude receiving the most direct sun, a column of warm air rises and the intertropical convergence zone (ITCZ) drops its precipitation. After rising more than 10 km into the atmosphere, the now cool, dry air circulates back to Earth at approximately 30° N and 30° S latitudes.

Once this hot, dry air reaches the ground, it flows back toward the equator, completing the air circulation cycle. The two circulation cells of air between the equator and 30° N and 30° S latitudes are known as **Hadley cells.** The area where the two Hadley cells converge and cause large amounts of precipitation is known as the **intertropical convergence zone (ITCZ).**

The intense sunlight at the solar equator drives the Hadley cells and the ITCZ, causing the warmed air to rise and precipitation to be released in the form of rain. As we saw in our earlier discussion about unequal seasonal heating of Earth, we know that the solar equator shifts throughout the year, from 23.5° N latitude in June to 23.5° S latitude in December. Because the latitude of the solar equator determines the latitude of the ITCZ, the latitude of the ITCZ also moves throughout the year. This means that the seasonal movement of the solar equator influences seasonal patterns of rainfall. The seasonal movement of the ITCZ between the Tropic of Cancer and the Tropic of Capricorn leads to one pronounced rainy season over most of the tropics and subtropics. In

contrast, latitudes near the equator experience two pronounced rainy seasons because the ITCZ passes over the equator in March and September.

You can see the effect of the ITCZ movement in **Figure 5.7** by examining the patterns of rainfall across three locations in the Western Hemisphere. The city of Mérida, Mexico, lies about 20° N of the equator. The intertropical convergence reaches Mérida only during June, which is why June is the rainy season for Mérida. In comparison, Rio de Janeiro, Brazil, lies about 20° S latitude. The intertropical convergence reaches Rio de Janeiro in December, which is the middle of the rainy season for that city. Close to the equator, in Bogotá, Colombia, the intertropical convergence zone passes overhead twice each year, during the March and September equinoxes. As a result, Bogotá experiences two rainy seasons.

Hadley cells The two circulation cells of air between the equator and 30° N and 30° S latitudes.

Intertropical convergence zone (ITCZ) The area where the two Hadley cells converge and cause large amounts of precipitation.

Figure 5.7 Rainy seasons and the ITCZ. As the solar equator moves throughout the year, so does the ITCZ. As a result, latitudes north and south of the equator have a single distinct rainy season whereas latitudes close to the equator have two rainy seasons.

The pattern of air circulation near the equator also exists near the two poles. At approximately 60° N and 60° S latitudes, air rises up into the atmosphere and drops moisture. The cold, dry air then moves toward the poles and sinks back to Earth at approximately 90° N and 90° S latitudes. This air then moves along the ground back to 60° N and 60° S latitudes where it rises again. The atmospheric convection currents that move air between 60° and 90° latitudes are called **polar cells.**

Between the Hadley cells and polar cells, from latitudes of approximately 30° to 60°, is an area of air circulation that lacks distinct convection currents. In this range of latitudes in the Northern Hemisphere—which includes much of Canada, the United States, Europe, and Central Asia—some of the warm air from the Hadley cells that descends at 30° latitude moves toward the North Pole while some of the cold air from the polar cells that is travelling toward 60° moves toward the equator. The movement of air in this region also helps redistribute the warm air of the tropics and the cold air of the polar regions toward the middle latitudes.

The region between Hadley cells and polar cells can have dramatic changes in wind direction and, therefore, can experience large fluctuations in temperature and precipitation. However, winds generally move from west to east. This wind direction contributes to warmer conditions on the west coast of North America than on the east coast, as you can see from the plant hardiness zones.

EARTH'S ROTATION AND THE CORIOLIS EFFECT

Hadley cells and polar cells are important drivers of wind direction on Earth. However, wind direction is also affected by the speed of Earth's rotation, which changes with latitude. Earth completes a single rotation in 24 hours. Because the circumference of the planet at the equator is much larger than its circumference near the poles, the speed of rotation is faster at the equator. As Figure 5.8 shows, an object at the equator rotates at 1670 km/hr, an object at 30° N rotates at 1445 km/hr, and an object at 80° N rotates at 291 km/hr.

The different rotation speeds deflect the direction of surface air circulation in the Hadley and polar cells. Imagine standing at the North Pole and throwing a baseball straight south to the equator, as shown in Figure 5.9. While the ball is flying through the air, the planet continues to rotate. As a result, the ball does not land straight south at the equator.

Polar cells The atmospheric convection currents that move air between 60° and 90° latitudes in the Northern and Southern Hemispheres.

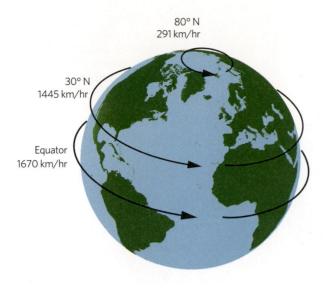

Figure 5.8 Earth's speed of rotation. An object sitting at the equator travels at a much higher speed to complete a rotation in 24 hours than do objects at higher latitudes.

Instead, it travels along a path that appears to deflect to the west. In fact, the ball is travelling straight, but because the planet rotates while the ball is in motion, the ball lands west of its intended target. With respect to the planet, the ball's path appears to be deflected. The deflection of an object's path due to the rotation of Earth is known as the **Coriolis effect.**

The deflected path of the ball in our example mimics the Coriolis effect on air moving to the north or

south. For example, Hadley cells north of the equator move air along the surface from north to south. As we can see in Figure 5.10, the Coriolis effect causes this path to deflect so that it moves from the northeast to the southwest. These winds are known as the *northeast trade winds*. Below the equator, the Hadley cells are moving air along the ground from the south to the equator. The Coriolis effect causes this path to deflect so that it moves from the southeast to the northwest. These winds are known as the *southeast trade winds*. A similar phenomenon occurs in the polar cells. In the latitudes between Hadley cells and polar cells, wind direction can be quite variable. However, these winds often move away from the equator and toward the poles, only to be deflected by the Coriolis effect. This causes winds known as *westerlies*. Thus, weather in the middle latitudes tends to move from west to east. When considering the Coriolis effect, the general rule is that surface winds are deflected to the right, relative to wind direction, in the Northern Hemisphere and to the left in the Southern Hemisphere.

Ocean currents also affect the distribution of climates

Like air currents, ocean currents distribute unequal heating of Earth's water and therefore influence the

Coriolis effect The deflection of an object's path due to rotation of Earth.

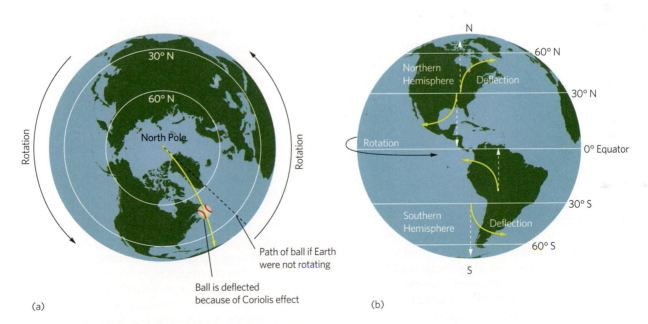

(a)

(b)

Figure 5.9 The Coriolis effect. (a) Because Earth rotates, the path of any object that travels north or south is deflected. **(b)** This deflection causes the predominant air circulation currents along Earth's surface to be deflected.

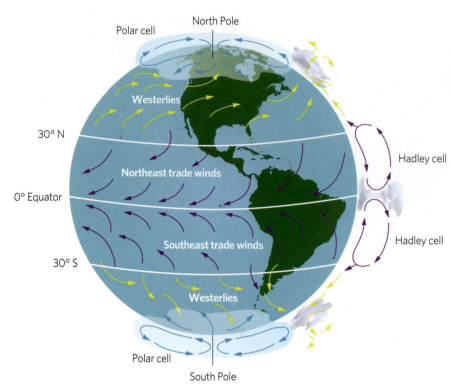

Figure 5.10 Air circulation patterns on Earth. Through a combination of atmospheric convection currents and the Coriolis effect, different latitudes have predominant winds that travel in different directions.

location of different climates. Figure 5.11 shows some general patterns of circulation: warm tropical water circulates up through the western reaches of ocean basins toward the poles and the cold polar water circulates down the eastern edges of ocean basins toward the tropics. Many factors create these currents, including unequal heating, Coriolis effects, predominant wind directions, the topography of the ocean basins, and differences in salinity. In this section, we will examine the drivers of major ocean currents including *gyres* and *upwelling.* We will then investigate how natural changes in ocean currents can have large effects on global climates through a process known as the *El Niño–Southern Oscillation.* Finally, we will explore the thermohaline ocean circulation, a deep ocean current that can take hundreds of years to complete a single path around the globe.

GYRES

We have noted how tropical regions of Earth receive more direct sunlight than regions at higher latitudes. This causes ocean waters near the equator generally to be warmer than ocean waters at higher latitudes. Because of the unequal heating, the tropical waters expand as they warm. This expansion causes the water near the equator to be approximately 8 cm higher in elevation than the water in mid-latitudes. Although the difference may seem small, it is enough for the force of gravity to cause movement of water away from the equator.

Around the globe, ocean circulation patterns are also affected by the dominant wind directions and Coriolis effects. North of the equator, for example, the northeast trade winds push surface water from the northeast to the southwest. At the same time, the Coriolis forces deflect ocean currents to the right. The combination of the two forces causes tropical water above the equator to move from east to west. The topography of ocean basins, particularly the locations of continents, causes these currents to change their direction. At mid-latitudes, the westerlies push surface waters to the northeast. As this occurs, Coriolis forces deflect the ocean currents to the right, which causes the ocean currents to move from west to east at mid-latitudes in the Northern Hemisphere. These large-scale water circulation patterns between continents are called **gyres.** The direction of the deflections caused by Coriolis forces depends on latitude; gyres move in a clockwise direction in the Northern Hemisphere and in a counterclockwise direction in the Southern Hemisphere.

Gyres redistribute energy by transporting both warm and cold ocean water around the globe. The proximity of these ocean waters to continents can make the continents considerably warmer or colder and therefore influence terrestrial climates. For example, England and Newfoundland are at similar latitudes. However, England is adjacent to a

---

Gyre A large-scale water circulation pattern between continents.

Figure 5.11 Ocean currents. Ocean currents circulate as a result of unequal heating, Coriolis effects, predominant wind directions, and the topography of the ocean basins. Each of the five major ocean basins contains a gyre. These gyres are driven by the trade winds in the tropics and the westerlies at mid-latitudes. This produces a clockwise circulation pattern in the Northern Hemisphere and a counterclockwise circulation pattern in the Southern Hemisphere. Along the west coasts of many continents, currents diverge and cause the upwelling of deeper and more fertile water.

warm-water current, the Gulf Stream, that comes out of the Gulf of Mexico, whereas Newfoundland is adjacent to a cold-water current, the Labrador Current, that comes down from the west side of Greenland and pushes the warmer Gulf Stream water offshore. As a result, England experiences winter temperatures that are on average 20°C warmer than Newfoundland.

UPWELLING

Any upward movement of ocean water is referred to as **upwelling.** Illustrated as dark blue areas in Figure 5.11, upwelling occurs in locations along continents where surface currents move away from the coastline. As surface water moves away from land, cold water from beneath is drawn upward. Strong upwelling zones occur on the western coasts of continents where gyres move surface currents toward the equator and then veer from the continents. As

surface water moves away from the continents, it is replaced by water rising from greater depths. Because deep water tends to be rich in nutrients, upwelling zones are often regions of high biological productivity. Major commercial fisheries are often located in these zones.

THE EL NIÑO–SOUTHERN OSCILLATION

Sometimes ocean currents are substantially altered and this can affect climatic conditions. One of the best-known examples is the **El Niño–Southern Oscillation (ENSO),** which is illustrated in

---

Upwelling An upward movement of ocean water.

El Niño–Southern Oscillation (ENSO) The periodic changes in winds and ocean currents in the South Pacific, causing weather changes throughout much of the world.

(a) **Normal year**

(b) **El Niño year**

Figure 5.12 The El-Niño–Southern Oscillation (ENSO). Changes in the strength of trade winds near the equator can have major impacts on the climates of the world. **(a)** In most years, strong trade winds push warm surface waters away from the west coast of South America. This causes cold, deep waters to upwell along the coast. **(b)** During an ENSO year, the trade winds weaken or reverse and the warm surface water moves from west to east. As a result, warm water builds up along the west coast of South America and prevents the upwelling of the cold, deep water. This change in ocean circulation alters climates around the world.

Figure 5.12. During most years in the South Pacific Ocean, southeast trade winds and Coriolis forces push the surface waters of the Peru Current that cause them to flow northwest along the west coast of South America with upwelling of cold water along the coast. Equatorial winds—powered by high air pressures in the eastern Pacific and low air pressures in the western Pacific—then push these surface waters offshore at Ecuador and west across the Pacific Ocean. As this water moves west, it warms. This warm water drives thunderstorm activity in the western Pacific, which results in high amounts of precipitation.

Every 3 to 7 years, however, this series of events changes. In the atmosphere, the normal difference in air pressures reverses and the equatorial winds weaken. In some years, these winds can even reverse direction. This change in air pressures in the Southern Hemisphere is the Southern Oscillation element of the ENSO. With either weakened or reversed equatorial winds, the warm surface waters of the western Pacific

move east toward South America. As a result, the upwelling of nutrients shuts down and the normally productive fisheries in the area become much less productive. The accumulating warm water also serves as a source of increased precipitation for this region. The unusually warm water is the El Niño ("the baby boy") element of the ENSO, so named because it typically occurs around Christmas time.

Because air and water currents are responsible for distributing energy throughout the world, the effects of an ENSO event extend over much of the world. For example, a strong ENSO event in 1982 to 1983 disrupted fisheries and destroyed kelp beds in California, caused reproductive failure of seabirds in the central Pacific Ocean, and killed off large areas of coral in Panama. Precipitation was also dramatically affected in many terrestrial ecosystems. Another ENSO event from 1991 to 1992—one of the strongest on record—was accompanied by the worst drought of the twentieth century in Africa, which caused poor crop production and widespread starvation. The event also brought extreme dryness to many areas of tropical South America and Australasia. Heat and drought in Australia reduced populations of red kangaroos (*Macropus rufus*) to less than half their pre-ENSO levels. The El Niño event of 1997 to 1998 was blamed for 23 000 human deaths—mostly from famine—and $33 billion in damages to crops and property worldwide.

Outside the tropics and subtropics, ENSO events tend to increase precipitation, which boosts the production of natural and agricultural systems but also causes flooding. In North America, ENSO events bring warm, dry conditions to southern Canada and the northern United States, and cooler, wetter, and often stormy weather to the southern United States.

THERMOHALINE CIRCULATION

Ocean currents are also driven by the **thermohaline circulation,** a global pattern of surface- and deep-water currents that flow as a result of variations in temperature and salinity that change the density of water. The thermohaline circulation, shown in Figure 5.13, is responsible for the global movement of great masses of water between the major ocean basins. As wind-driven surface currents—for example, the Gulf Stream—move toward higher latitudes, the water cools and becomes denser. In the far north, toward Iceland and Greenland, the surface

Thermohaline circulation A global pattern of surface- and deep-water currents that flow as a result of variations in temperature and salinity that change the density of water.

① Warm water flows from the Gulf of Mexico to the North Atlantic, where some of it freezes and evaporates.

② The remaining water, now saltier and denser, sinks to the ocean bottom.

③ The cold water travels along the ocean floor, connecting the world's oceans.

④ The cold, deep water eventually rises to the surface and circulates back to the North Atlantic.

Figure 5.13 Thermohaline circulation. This slow circulation of deep water and surface waters is driven by the sinking of cold, high salinity water near Greenland and Iceland.

of the ocean freezes in winter. Because ice does not contain salts, the salt concentration of the underlying water rises, which causes the cold water to become even denser. This high-density water begins to sink and acts as the driving force behind a deep-water current in the Atlantic Ocean known as the North Atlantic Deep Water. Similar sinking currents are formed around the edges of Antarctica in the Southern Ocean. These cold, dense waters then flow through the deep ocean basins and back into equatorial regions, where they eventually surface as upwelling currents. These upwelling currents become warm and begin to make their way back to the North Atlantic. Like a giant conveyor belt, the thermohaline circulation slowly redistributes energy and nutrients among the oceans of the world in a trip that can take hundreds of years.

Smaller-scale geographic features can affect regional and local climates

As we have seen, the primary global patterns in Earth's climate are the result of the unequal solar heating of Earth's surface. However, a number of other factors have secondary effects on local temperature and precipitation. These include continental land area, proximity to coasts, and rain shadows.

CONTINENTAL LAND AREA

The positions of continents exert important secondary effects on temperature and precipitation. For example, oceans and lakes, the sources of most atmospheric water vapour, cover 81 percent of the Southern Hemisphere but only 61 percent of the Northern Hemisphere. Because of this difference, more rain falls at any given latitude in the Southern Hemisphere than in the Northern Hemisphere. The presence of water has a moderating influence on land temperatures, so temperatures in the Northern Hemisphere vary more than they do in the Southern Hemisphere.

PROXIMITY TO COASTS

The interior of a continent usually experiences less precipitation than its coasts, simply because the

Figure 5.14 Rain shadows. When winds carry warm moist air up over a mountain, the air cools and releases much of its moisture as precipitation. After crossing the mountain, the now dry air descends down the mountain, which causes the environment on this side of the mountain to be very dry.

interior lies farther from the oceans, which are the major sources of atmospheric water. Furthermore, as we saw in our discussion of plant hardiness zones, coastal climates vary less than interior climates because the heat storage capacity of ocean water reduces temperature fluctuations near the coast. The ocean warms the air near the coasts during the winter and cools the air near the coasts during the summer. For example, the hottest and coldest average monthly temperatures at the Pacific coast of Canada at Vancouver, B.C., differ by only 19°C. Farther inland, this range increases to 33°C at Kelowna, B.C.; 38°C at Calgary, Alberta; and 47°C at Regina, Saskatchewan.

RAIN SHADOWS
Mountains also play a secondary role in determining climates, as we can see in Figure 5.14. When winds blowing inland from the ocean encounter coastal mountains, the mountains force the air upward, which causes adiabatic cooling, condensation, and precipitation. The air, which is now dry and warmed by latent heat release, descends the other side of the mountain, warms adiabatically, and travels across the lowlands beyond, where it creates relatively warm, arid environments called **rain shadows.** The prairie in southern Alberta, for example, lies in the rain shadow of the Rocky Mountains. The processes involved in creating rain shadows have much in common with the processes we saw occurring in Hadley cells including adiabatic cooling, adiabatic heating, and release of latent heat.

We can now use what we have learned to draw a complete picture of worldwide climate distribution. Looking at Figure 5.15, we see repeated patterns that

show where different climates exist. **Tropical climates,** characterized by warm temperatures and high precipitation, occur in regions near the equator. At approximately 30° N and 30° S latitudes, we commonly find the **dry climates** that experience a wide range of temperatures. Dry climates are not only affected by latitude, however. Many dry climates are caused by rain shadows, for example the extensive regions that lie just east of the Andes Mountains in western South America. **Moist subtropical mid-latitude climates** are characterized by warm, dry summers and cold, wet winters. **Moist continental mid-latitude climates** exist at the interior of continents and typically have warm summers, cold winters, and moderate amounts of precipitation. Finally, closest to the poles we find **polar climates**

Rain shadow A region with dry conditions found on the leeward side of a mountain range as a result of humid winds from the ocean causing precipitation on the windward side.

Tropical climate A climate characterized by warm temperatures and high precipitation, occurring in regions near the equator.

Dry climate A climate characterized by low precipitation and a wide range of temperatures, commonly found at approximately 30° N and 30° S latitudes.

Moist subtropical mid-latitude climate A climate characterized by warm, dry summers and cold, wet winters.

Moist continental mid-latitude climate A climate that exists at the interior of continents and is typically characterized by warm summers, cold winters, and moderate amounts of precipitation.

Polar climate A climate that experiences very cold temperatures and relatively little precipitation.

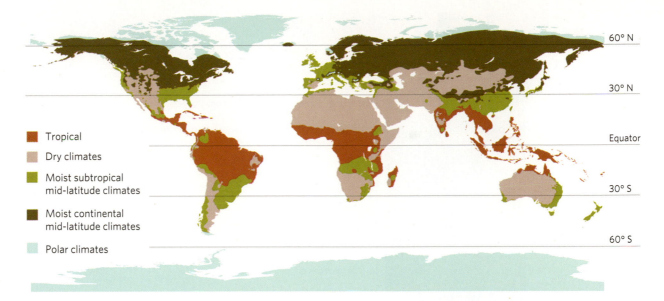

Figure 5.15 Broad climate patterns around the world. Near the tropics, the climate is warm with high amounts of precipitation. The world's great deserts are near 30° N and 30° S latitudes. Cold and snowy polar regions are located at even higher latitudes. In addition, we can see that regions of high precipitation sometimes occur on the western side of mountains, as in western Canada, and that deserts occur in the rain shadows of mountains, such as on the eastern sides of the Cascade Mountains and Sierra Nevadas in North America and the Andes Mountains in South America.

Legend:
- Tropical
- Dry climates
- Moist subtropical mid-latitude climates
- Moist continental mid-latitude climates
- Polar climates

that experience very cold temperatures and relatively little precipitation.

We have seen the processes that account for different climates around the world. Before we look more closely at the individual climate types and the plants that they support, we need to know something about the formation of soil, which supports all life.

Climate and the underlying bedrock interact to create a diversity of soils

Climate indirectly affects the distributions of plants and animals through its influence on the development of *soil,* which provides the substratum for plant roots to grow and in which many animals burrow. **Soil** defies a simple definition, but we may describe it as the layer of chemically and biologically altered material that overlies bedrock or other unaltered material at Earth's surface. Because the layer of bedrock that underlies soils plays a major role in determining the type of soil that will form above it, soil scientists call it the **parent material.**

SOIL FORMATION
Soil includes minerals derived from the parent material, modified minerals formed within the soil, organic material contributed by plants, air, and water within the pores of the soil, living roots of plants, microorganisms, and the larger worms and

arthropods that make the soil their home. For example, if you have ever seen a recently excavated road cutting through a hillside, you may have observed that soils have distinct layers, called **horizons,** as shown in Figure 5.16. Soil horizons are categorized by the components and processes that occur at each level.

Soils exist in a dynamic state and their characteristics are determined by climate, parent material, vegetation, local topography, and, to some extent, age. Groundwater removes some substances by dissolving them and moving them down through the soil to lower layers, a process known as **leaching.** Other materials enter the soil from vegetation, in precipitation, as dust from above, and from the parent rock below. Where little rain falls, the parent material breaks down slowly, and sparse plant production means that little organic material is added to the soil.

Soil The layer of chemically and biologically altered material that overlies bedrock or other unaltered material at Earth's surface.

Parent material The layer of bedrock that underlies soil and plays a major role in determining the type of soil that will form above it.

Horizon A distinct layer of soil.

Leaching A process in which groundwater removes some substances by dissolving them and moving them down through the soil to lower layers.

Horizon

O A layer composed primarily of dead organic litter

A A layer rich in humus, consisting of partly decomposed organic material mixed with mineral soil

Ae A layer of soil that has been leached of its minerals

B A layer of soil with a chemical composition that resembles the underlying parent material. Clay, minerals, and oxides of aluminum and iron leached out from the E horizon are sometimes deposited here.

C A layer composed primarily of large rocks of broken parent material

R Unaltered parent material, typically in the form of bedrock

Figure 5.16 Soil horizons. Soils develop distinct horizons that differ in thickness depending on climates and parent material.

Thus, dry climates typically have shallow soils, with bedrock lying close to the surface. In places where decomposed bedrock and organic material erode as rapidly as they form, soils may not form at all. Soil development also stops short on alluvial deposits, where fresh layers of silt deposited each year by floodwaters bury older material. At the other extreme, soil formation proceeds rapidly in tropical climates, where chemical alteration of parent material may extend to depths of 100 m. Most soils of mid-latitude climates are intermediate in depth, extending to an average of about 1 m.

WEATHERING

Weathering is the physical and chemical alteration of rock material near Earth's surface. Weathering occurs whenever surface water penetrates the parent material. In cold climates, for example, the repeated freezing and thawing of water in crevices causes the rock to break into smaller pieces and exposes a greater surface area of the rock to chemical reactions. The initial chemical alteration of the rock occurs when water dissolves some of the more soluble

minerals, such as sodium chloride (NaCl) and calcium sulphate ($CaSO_4$). Further chemical reactions continue the soil building process.

The weathering of granite illustrates some basic processes of soil formation. The minerals responsible for the grainy texture of granite—feldspar, mica, and quartz—consist of various combinations of oxides of aluminum, iron, silicon, magnesium, calcium, and potassium. The key aspect of the process of weathering is the displacement of many of these elements by hydrogen ions, followed by the reorganization of the remaining minerals into new types of minerals. The hydrogen ions involved in weathering are derived from two sources. One source is the carbonic acid that forms when carbon dioxide dissolves in rainwater, as discussed in Chapter 2. The other source of hydrogen ions is the decomposition of organic material in the soil itself. The metabolism of

Weathering The physical and chemical alteration of rock material near Earth's surface.

carbohydrates, for example, produces carbon dioxide. This carbon dioxide is converted into carbonic acid in water, which generates additional hydrogen ions.

As granite weathers, many of the positively charged elements in feldspar and mica—such as Fe^{3+} and Ca^{2+}—are replaced by hydrogen ions to form new, insoluble materials, such as the clay particles we discussed in Chapter 3. Clay particles are important to the water-holding capacity of soils. They accumulate negative charges on their surfaces that attract positively charged ions called *cations*. Cations—including calcium (Ca^{2+}), magnesium (Mg^{2+}), potassium (K^+), and sodium (Na^+)—are important nutrients for plants. The ability of a soil to retain these cations, called its **cation exchange capacity,** provides an index to the fertility of that soil. Young soils have relatively few clay particles and little added organic material; this low cation exchange capacity leads to relatively low fertility. Older soils generally have a higher cation exchange capacity and therefore relatively high fertility. Soil fertility improves with time, up to a point. Eventually weathering breaks down clay particles, cation exchange capacity decreases, and soil fertility drops.

Podsolization

Under mild temperatures and moderate precipitation, sand grains and clay particles resist weathering and become stable components of the soil. This allows soils to retain relatively high fertility. However, in acidic soils typical of cool, moist regions, clay particles break down in the Ae horizon, and their soluble ions are transported down to the lower B horizon. This process, known as **podsolization,** reduces the fertility of the soil's upper layers.

Acidic soils occur primarily in cool regions where needle-leaved trees such as spruces and firs dominate the forests. The slow decomposition of the spruce and fir needles produces organic acids that promote high concentrations of hydrogen ions. In the moist regions where podsolization occurs, rainfall usually exceeds evaporation. Because water continuously moves downward through the soil profile, little clay-forming material is transported upward from the weathered bedrock.

In North America, podsolization is most advanced under spruce and fir forests across a wide belt of southern and western Canada, in the Great Lakes region, and in New England. A typical profile of a highly podsolized soil, shown in Figure 5.17, reveals striking bands corresponding to the regions of leaching and redeposition. The A horizon is dark and rich in organic matter. It is underlain by a light-coloured Ae horizon that has been leached of most of its clay, leaving behind sandy material that holds neither

Figure 5.17 Podsolization. In cool, moist conditions with highly acidic soils, clay particles normally found in the E horizon are weathered and leached down, leaving a very sandy layer with little capacity to retain nutrients for plants.

water nor nutrients well. One usually finds a dark band immediately below the Ae horizon. This is the uppermost layer of the B horizon, where iron and aluminum oxides are redeposited. Other, more mobile minerals may accumulate to some extent in lower parts of the B horizon, which then grades almost

---

Cation exchange capacity The ability of a soil to retain cations.

Podsolization A process occurring in acidic soils typical of cool, moist regions, where clay particles break down in the E or Ae horizon, and their soluble ions are transported down to the lower B horizon.

imperceptibly into a C horizon and the parent material.

Laterization

In the warm, humid climates of many tropical and subtropical regions, soils weather to great depths. One of the most conspicuous features of weathering under these conditions is the breakdown of clay particles, which causes silicon to leach from the soil and leaves oxides of iron and aluminum to predominate throughout the soil profile—a process that is called **laterization.** The iron and aluminum oxides give such soils a characteristic reddish colouration as illustrated in Figure 5.18. Even though the rapid decomposition of organic material in tropical soils contributes an abundance of hydrogen ions, bases formed by the breakdown of clay particles neutralize them. Consequently, lateritic soils are not usually acidic, even though they may be deeply weathered. Regardless of the parent material, weathering reaches deepest, and laterization proceeds farthest, on low-lying soils, such as those of the Amazon basin, where highly weathered surface layers are not eroded away and the soil profiles are very old.

Laterization causes many tropical soils to have a low cation exchange capacity. In the absence of clay and organic matter, mineral nutrients are readily leached from the soil. Where soils are deeply weathered, new minerals formed by the decomposition of the parent material are simply too far from the surface to contribute to soil fertility. In addition, heavy rainfall in the tropics keeps water moving down through the soil profile, preventing the upward movement of nutrients. In general, the deeper the ultimate sources of nutrients in the unaltered bedrock, the lower the fertility of the surface layers. The high productivity of tropical rainforests depends more on rapid cycling of nutrients close to the surface of the soil than on the nutrient content of the soil itself. Rich soils do, however, develop in many tropical regions, particularly in mountainous areas where erosion continually removes nutrient-depleted surface layers of soil, and in volcanic areas where the parent material of ash and lava is often rich in nutrients such as potassium.

From our discussion of soils, you can see that the composition of the soils present in various parts of the world depends on differences in climate, underlying parent material, and vegetation. In the next chapter,

Figure 5.18 Laterization. Under conditions of warm temperatures and high precipitation, clay particles are broken down and leave behind a soil that has a low cation exchange capacity and low fertility.

we will discuss how these regional differences in climate and the associated effects on soils affect the types of plants and animals that can live in each region.

Laterization The breakdown of clay particles, which results in the leaching of silicon from the soil, leaving oxides of iron and aluminum to predominate throughout the soil profile.

GLOBAL CLIMATE CHANGE

A polar bear mother with her cubs in Ungava Bay, Labrador. Warming trends over the last few decades have caused the Arctic ice to melt earlier in the year; this means that polar bears have less time to hunt for seals, which make up a large part of their diet. Photo by Christoph Richter.

As we have seen in this chapter, a substantial number of interacting factors determine the different climates of the world. For example, the differential heating of Earth drives the movements of air and water, which are further modified by the Coriolis effect and the position of continents. Because these interactions are complex, any changes in these factors can have far-reaching effects on the entire system. Such is the case for global warming and *global climate change*. Global warming is the increase in the average temperature of the planet due to an increased concentration of greenhouse gases in the atmosphere. **Global climate change** is a much broader phenomenon that refers to changes in Earth's climates and includes global warming, changes in the global distribution of precipitation and temperature, changes in the intensity of storms, and altered ocean circulation. Throughout the history of Earth, long periods of gradual global warming and cooling have been associated with substantial global climate change. During the past 2 centuries, however, human activities have caused a rapid change in conditions that have led to global warming and global climate change.

Global warming is a major driver of current changes in global climates. One direct impact is the increase in temperatures in many parts of the world, particularly at high latitudes in the Northern Hemisphere. The rise in temperature in these regions has had a wide range of effects. For example, in high latitudes and at high altitudes, the lower layers of soil may be permanently frozen, a phenomenon that is known as **permafrost.** Warmer temperatures cause these highly organic soils to thaw and begin decomposing. Because these soils are waterlogged and anaerobic, the decomposition produces methane, a greenhouse gas that can further contribute to global warming.

Increased temperatures also affect the amount of ice melting around the world. From 2006 to 2012, the polar ice cap that exists between Canada, the United States, Europe, and Russia became 45 percent smaller than its average size during the 1980s and 1990s. The remaining ice has also become thinner. The ice of Greenland and Antarctica is also melting. NASA scientists found, from 1992 to 2009, that the two regions lost an average of 475 gigatonnes (Gt) of ice per year and that the annual rate of ice loss is accelerating. Similarly, glaciers are melting in many regions of the world. In Montana's Glacier National Park, for example, there were 150 glaciers in 1850. Today only 25 glaciers remain. All of this ice melting has caused sea levels to rise 220 mm since 1870 and scientists predict that continued melting could raise sea levels an additional 280 to 430 mm by the year 2100.

Because the complex nature of the global climate system can make it difficult to predict how climate will change in the coming decades, scientists have developed computer models that incorporate our best understanding of the processes that govern climate along with the changes that are being caused by increased atmospheric greenhouse gas concentrations. Although different models make somewhat different predictions, there is general agreement regarding several aspects of predicted change. For example, the increase in temperatures is expected to cause longer periods of hot weather and fewer days of extremely cold weather. Because heat is the driver of evaporation and air circulation that determine precipitation, precipitation patterns are also predicted to change across the globe, with some regions of the world receiving increased amounts of rain and snow while other regions receive less. The intensity of storms, such as hurricanes, is also predicted to increase due to an increased warming of the world's oceans.

Ocean currents may also be affected by global warming. As we discussed earlier in this chapter, ocean currents are driven by the differential heating of Earth

---

Global climate change A phenomenon that refers to changes in Earth's climates including global warming, changes in the global distribution of precipitation and temperature, changes in the intensity of storms, and altered ocean circulation.

Permafrost A phenomenon where layers of soil are permanently frozen.

The melting polar ice cap. Compared to the average amount of sea ice present from 1979 through the 1990s, there has been a 45 percent reduction in sea ice during the period of 2006 to 2012.

and, in turn, play a major role in determining the temperature of nearby continents. Of particular concern is the potential impact on thermohaline circulation. As

you may recall, thermohaline circulation is the slow, deep circulation of ocean water around the globe that is driven by the dense, salty water that sinks near Greenland. With the increased melting of the polar ice cap and the ice sheets of Greenland, climate scientists are concerned that the water in the North Atlantic may not be dense enough to sink, and therefore may cause the thermohaline circulation to shut down. The disappearance of this deep-water current would effectively stop the circulation of warm water from the Gulf of Mexico to Europe and cause a substantial cooling of Europe—with potentially devastating consequences for the people and environment of that region.

SOURCES: Climate Change 2007: Synthesis Report. *Fourth Assessment Report of the Intergovernmental Panel on Climate Change.* http://www.ipcc.ch/pdf/assessment-report/ar4/syr/ar4_syr.pdf
Global Climate Change Impacts in the United States. 2009. *U.S. Global Change Research Program.* http://downloads.globalchange.gov/usimpacts/pdfs/climate-impacts-report.pdf

SUMMARY OF CHAPTER CONCEPTS

■ **Earth is warmed by the greenhouse effect.** Much of the ultraviolet and visible light emitted by the Sun passes through the atmosphere and strikes clouds and the surface of Earth. Clouds and the planet begin to warm and emit infrared radiation back toward the atmosphere. The gases in the atmosphere absorb the infrared radiation, become warmer, and re-emit infrared radiation back toward Earth. These greenhouse gases allow the planet to become warmer than would otherwise be possible. The increase in production of greenhouse gases due to human activities increases the greenhouse effect and leads to global warming.

■ **There is an unequal heating of Earth by the Sun.** Each year, high latitude regions of the world receive solar radiation that is weaker in intensity due to a longer path through the atmosphere with a less direct angle, which causes the energy of the Sun to be spread over a larger area. In addition, the axis of Earth is tilted 23.5° and this causes seasonal changes in temperature.

■ **The unequal heating of Earth drives air currents in the atmosphere.** Due to the properties of air, the warmer

temperatures near the equator drive atmospheric convection currents known as Hadley cells between approximately 0° to 30° latitude in the Northern and Southern Hemispheres. Polar cells are at higher latitudes, between approximately 60° to 90°. These air convection currents cause the distribution of heat and precipitation around the globe. Their path is also influenced by Coriolis forces that are created by the rotation of Earth.

■ **Ocean currents also affect the distribution of climates.** Ocean currents are driven by the unequal warming of Earth combined with Coriolis effects, atmospheric convection currents, and differences in salinity. Gyres exist on both sides of the equator and help distribute heat and nutrients to higher latitudes. El Niño–Southern Oscillation (ENSO) events represent a disruption in normal ocean currents in the South Pacific and the impacts on climates can be felt around the world. Thermohaline circulation is a deep and slow circulation of the world's oceans driven by changes in salt concentration in the waters of the North Atlantic.

■ **Smaller-scale geographic features can affect regional and local climates.**

Increased land area of continents reduces the amount of evaporation possible, which causes the Northern Hemisphere to experience less precipitation than the Southern Hemisphere. Proximity to the coast can also affect climates; regions that are more distant from coastlines typically have lower precipitation and higher variation in temperature. Mountain ranges force air to rise over them, causing higher precipitation on one side of the mountain range and rain shadows on the opposite side.

■ **Climate and the underlying bedrock interact to create a diversity of soils.** Soils are comprised of horizons that contain different amounts of organic matter, nutrients, and minerals. Soils can be weathered by processes including freezing, thawing, and leaching. In acidic soils of cool, moist regions, soils can experience podsolization, a process that breaks down clay particles and reduces fertility. In warm, humid climates, soils can experience laterization, a process that breaks down clay particles and leaches nutrients from the soil.

REVIEW QUESTIONS

1. Explain the steps involved in the greenhouse effect.

2. How does the human production of greenhouse gases lead to global warming?

3. Why is solar energy input greater near the equator than near the poles?

4. Explain the factors that drive the movement of air in Hadley cells.

5. The position of the solar equator moves throughout the year. What does its changing position suggest about the location of the intertropical convergence zone throughout the year?

6. How do the unequal heating of Earth, the Coriolis effect, and atmospheric convection currents work together to drive the movement of ocean gyres?

7. Explain the process that causes El Niño–Southern Oscillation events.

8. Based on your knowledge of the ocean conveyor belt, how might melting of the ice in the Arctic Ocean affect the climate of Europe?

9. Why do many mountain ranges have high precipitation on one side and low precipitation on the other side?

10. Compare and contrast podsolization and laterization.

PRACTICING ECOLOGY: WHY ARE THE POLAR BEARS DISAPPEARING?

Ian Stirling examines reasons for the decline in polar bear populations. Photo by Dan Guravich/Corbis.

Small changes in Earth's climate can have large and lasting consequences for species and their communities. For example, the timing of interactions between species often relies on the timing of the seasons, which can be strongly affected by global climates. This is particularly true for many types of predator-prey interactions, since predators often rely on seasonal pulses in the abundance of prey.

Ian Stirling and his colleagues at Environment Canada hypothesized that a disruption in the timing of predator-prey interactions due to global warming has led to the recent decline in the population size of polar bears in the Canadian Arctic. During the winter and spring, these large carnivores migrate to frozen coastal waters where they feed on seals. As temperatures warm during the summer, the sea ice retreats away from the mainland and the polar bears are stranded on land, no longer able to reach the ice to hunt seals. On the Hudson Bay where Stirling and his colleagues conduct their research, once the ice retreats the bears must largely rely on fat reserves for energy until the coastal waters freeze again. The amount of fat reserves stored at the end of the hunting season is particularly important for breeding females that must use this energy to give birth and rear offspring. Over recent years, the timing of sea ice breakup has occurred earlier due to global warming. Stirling and his colleagues predicted that the earlier ice breakup would reduce the number of days that polar bears have available to feed on seals, which would lead to lower stored energy reserves during the ice-free seasons and ultimately cause reduced polar bear survival.

To test this prediction, the scentists captured pregnant female polar bears during the ice-free season over a period of more than 20 years. Each year, they recorded the mean body mass, and they used satellite data to record the timing of ice breakup. You can view this data in the table.

MASS OF PREGNANT FEMALES (kg) AND DATE OF ICE BREAKUP, 1980–2007

YEAR	MASS OF PREGNANT FEMALES	DATE OF ICE BREAKUP
1980	294	JULY 5
1982	326	JULY 2
1984	248	JULY 16
1986	263	JUNE 27
1988	249	JULY 24
1990	254	JUNE 2
1992	257	JULY 21
1994	225	JUNE 27
1996	241	JULY 3
1998	254	JUNE 15
2000	267	JULY 8
2002	211	JULY 4
2004	228	JULY 9
2006	231	JUNE 24
2007	257	JUNE 22

GRAPH THE DATA

1. Create one graph with years on the *x* axis and mass of pregnant females on the *y* axis.

2. Create another graph with years on the *x* axis and date of ice breakup on the *y* axis.

3. Illustrate the relationship between the two variables on each graph by visually estimating and drawing a regression line of best fit through the distribution of data points.

INTERPRET THE DATA

1. Are the variables positively or negatively correlated in each graph?

2. Do the data in your graphs indicate that the date of ice breakup causes a decline in the mass of pregnant females? How might the date of sea ice breakup be related to polar bear survival?

3. In recent years, an increasing number of polar bears have been seen foraging for food in the urban areas around the Hudson Bay during the ice-free period in the summer. From what you have learned about polar bear feeding patterns, suggest a hypothesis that explains this behaviour.

SOURCE: Regehr, E. V., N. J. Lunn, S. C. Amstrup, and I. Stirling. 2007. Effects of earlier sea ice breakup on survival and population size of polar bears in Western Hudson Bay. *Journal of Wildlife Management* 71: 2673–2683.

CHAPTER 6

Terrestrial and Aquatic Biomes

The World of Wine

The fascinating history of winemaking dates back thousands of years. Archaeologists have found signs of winemaking in many cultures around the Mediterranean Sea, including those of the ancient Egyptians, Romans, and Greeks. More recently, Spain, Portugal, and France have become known for their wines. The entire Mediterranean region has a long tradition of cultivating wine grapes, and the production of wine has played an important role in the economic development of many societies and in religious rituals.

European explorers spread winemaking to other parts of the world. For example, in the sixteenth century, Spanish explorers brought grape vines to Chile, Argentina, and California. Grape vines accompanied the Dutch to South Africa in the seventeenth century, and the British to Australia in the nineteenth century.

Although grapes can grow in many parts of the world, specific growing conditions are required to produce grapes for the best wines. The ideal climate is a combination of hot, dry summers and wet, mild winters. The hot and dry summer climate allows the grapes to develop the right balance of sugar and acidity that provides the complex flavours of a fine wine. The dry summers also prevent various plant diseases that flourish under moister conditions. Domesticated grapes have deep roots, so they are well adapted to dry summer landscapes. Because below-freezing temperatures can harm the vines, the presence of mild, wet winters are equally important. While climate is critical, the flavour of a fine wine is also influenced by the mineral content of the soils in which the vines grow. The composition of a soil depends on the unique combination of minerals in the bedrock from which it is formed; these minerals affect how well the grape vines grow and give the grapes a distinctive flavour that characterizes the wine made from them. In short, distinctive tasting wines from around the world are the result of unique combinations of climate and soil.

Given the conditions required to make a great wine, it is perhaps not surprising that most of the major wine-producing locations around the world have the same climate—hot, dry summers followed by cool, moist winters. This is the climate of the countries surrounding most of the Mediterranean Sea. It is also the climate of most regions where wine grapes have been introduced, including Chile, Argentina, California, South Africa, and the southwestern coast of Australia. Interestingly, these regions all lie on the west side of continents and are located between 30° and 50° in the Northern and Southern Hemispheres. Not only do they have similar climates, but their landscapes also contain similar looking plants, despite being separated by thousands of kilometres. For example, while each winemaking

> "Given the conditions required to make a great wine, it is perhaps not surprising that most of the major wine-producing locations around the world have the same climate."

Growing grapes for wine during the hot, dry summer. At the Chateau Larose Trintaudon in southern France, the climate is ideal for growing the grapes that are used for wine. Photo by Grand Tour/Corbis.

region contains a large number of unique plant species, the plants are similar in their growth form. Whether in France, California, Chile, or South Africa, the plant communities are dominated by drought-adapted grasses, wildflowers, and shrubs.

In this chapter, we will explore how particular climates that are found in different locations around the world are associated with very similar looking plants and how scientists use these patterns to categorize terrestrial ecosystems. We will also examine why scientists categorize aquatic ecosystems in a different way, based on differences in salinity, flow, and depth.

SOURCE: A brief history of wine. November 5, 2007, *New York Times* http://www.nytimes.com/2007/11/05/timestopics/topics-winehistory.html

CHAPTER CONCEPTS

■ Terrestrial biomes are categorized by their major plant growth forms.

■ There are nine categories of terrestrial biomes.

■ Aquatic biomes are categorized by their flow, depth, and salinity.

As we saw in Chapter 5, climate patterns around the globe are determined by a range of factors including air currents, water currents, Coriolis forces, and local geographic features. Together, these factors cause the patterns of climate that occur in different regions of the world. Different climates provide unique seasonal temperature and precipitation conditions and these unique conditions favour different types of plants.

Terrestrial biomes are categorized by their major plant growth forms

Successful survival strategies vary with climate. In the world's deserts, for example, we find plants that are well adapted to scarce water availability. In North American deserts, many species of cacti have thick, waxy outer layers covered with hairs and spines to help reduce water loss. In Africa, we find a group of plants called euphorbs that are not closely related to the cacti of North America yet have many similar features (Figure 6.1). Though the two groups of desert-adapted plants are descended from unrelated ancestors, they look similar because they have evolved under similar selective forces, a phenomenon known as **convergent evolution.** Convergent evolution can be observed in many organisms. For example, sharks and dolphins are not closely related to each other— one is a fish and the other is a mammal—yet both have evolved fins, powerful tails, and streamlined bodies. To perform well in an aquatic environment, natural selection has favoured this set of traits because it allows both groups of animals to swim rapidly.

Convergent evolution explains why we can recognize an association between the forms of organisms and the environments in which they live. Trees found in tropical rainforests have the same general appearance no matter where they are located on Earth or their evolutionary lineage. The same can be said of shrubs inhabiting seasonally dry environments; they tend to have small, deciduous leaves and often arm their stems with spines to discourage herbivores from eating them.

Geographic regions that contain communities composed of organisms with similar adaptations are called **biomes.** Because of convergent evolution we can categorize terrestrial ecosystems by dominant plant forms that are associated with distinct patterns of seasonal temperatures and precipitation. In aquatic ecosystems, the major producers are often not plants but algae. As a result, aquatic biomes are not easily characterized by the dominant growth forms of the producers. Instead, aquatic biomes are characterized by distinct patterns of depth, flow, and salinity.

Biomes provide convenient reference points for comparing ecological processes around the globe, which makes the biome concept a useful tool that enables ecologists to understand the structure and

Convergent evolution A phenomenon in which two species descended from unrelated ancestors look similar because they have evolved under similar selective forces.

Biome A geographic region that contains communities composed of organisms with similar adaptations.

(a)

(b)

Figure 6.1 Convergent evolution. Similar conditions in the deserts of the world have selected for similar water-conserving life forms in two groups of unrelated plants: **(a)** an organ pipe cactus (*Stenocereus thurberi*) in Organ Pipe Cactus National Monument, Arizona, and **(b)** a euphorb (*Euphorbia virosa*) in Namibia, Africa. Photos by (a) All Canada Photos/Alamy; (b) imagebroker/ Alessandra Sarti/Newscom.

functioning of large ecological systems. As in all classification systems, exceptions occur. Boundaries between biomes can be unclear and not all plant growth forms correspond to climate in the same way. Australian eucalyptus trees, for example, form forests under climatic conditions that support only scrubland or grassland on other continents. Finally, plant communities reflect factors other than temperature and rainfall. Topography, soils, fire, seasonal variations in climate, and herbivory can all affect plant communities.

The overview of the major terrestrial biomes in this chapter emphasizes the distinguishing features of the physical environment and how these features are reflected in the form of the dominant plants. As a final note, although ecologists use plant forms to categorize biomes, there is generally a good association between the plant forms in a biome and the animal forms that live there. For example, deserts contain plants that are adapted to dry conditions and animals that are adapted to dry conditions.

We will use a classification system that recognizes nine major terrestrial biomes, which are listed in Figure 6.2. If we consider all combinations of average annual temperatures and average annual precipitation, as shown in the figure, we see that most places on Earth are located inside a triangular area with corners representing warm moist, warm dry, and cool dry climates. Cold regions with high rainfall are rare because water does not evaporate rapidly at low temperatures and because the atmosphere in cold regions holds little water vapour.

The nine biomes fall within three temperature ranges that we refer to often throughout this book.

Boreal forest and tundra biomes have average annual temperatures that are below 5°C. Temperate biomes—temperate rainforest, temperate seasonal forest, woodland/shrubland, and temperate grassland/cold desert—are a bit warmer, with average annual temperatures between 5°C and 20°C. Finally, tropical biomes—tropical rainforest, tropical seasonal forest/ savannah, and subtropical desert—are the warmest biomes, with average annual temperatures greater than 20°C. The global distribution of these biomes is

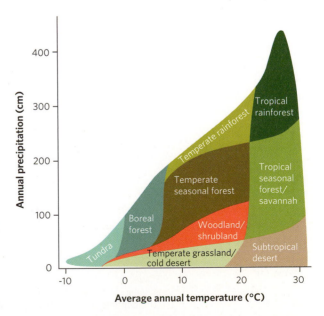

Figure 6.2 Terrestrial biomes. Distinct plant forms exist under different combinations of average annual precipitation and average annual temperature.

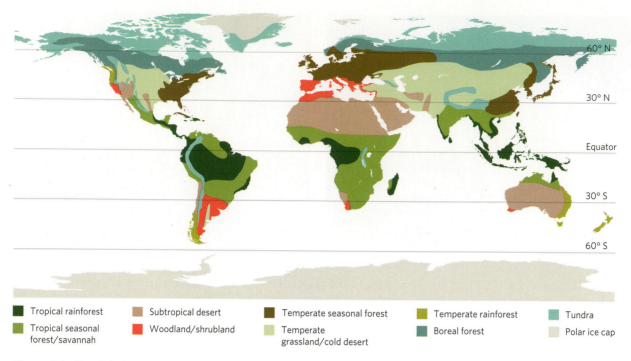

Tropical rainforest
Tropical seasonal forest/savannah
Subtropical desert
Woodland/shrubland
Temperate seasonal forest
Temperate grassland/cold desert
Temperate rainforest
Boreal forest
Tundra
Polar ice cap

Figure 6.3 The global distribution of biomes. The nine terrestrial biomes represent locations with similar average annual temperatures and precipitations and similar plant growth forms. Also shown are polar ice caps, which lack plants and therefore are not part of the biome classification system.

illustrated in Figure 6.3. As we will see, the average annual precipitation within each of these temperature categories can vary widely.

CLIMATE DIAGRAMS

To visualize the patterns of temperature and precipitation that are associated with particular biomes, scientists use **climate diagrams,** which are

graphs that plot the average monthly temperature and precipitation of a specific location on Earth. Figure 6.4 provides two sample climate diagrams. As you can see, the shaded area on the x axis indicates the

Climate diagram A graph that plots the average monthly temperature and precipitation of a specific location on Earth.

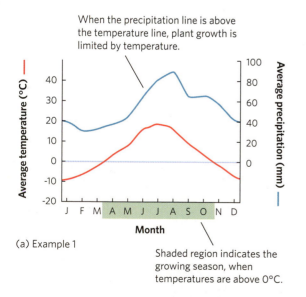

When the precipitation line is above the temperature line, plant growth is limited by temperature.

(a) Example 1

Shaded region indicates the growing season, when temperatures are above 0°C.

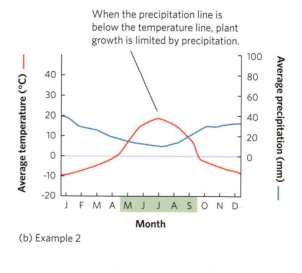

When the precipitation line is below the temperature line, plant growth is limited by precipitation.

(b) Example 2

Figure 6.4 Climate diagrams. By plotting the average monthly temperature and precipitation values over time for a particular location on Earth, we can determine how climates vary throughout the year and the length of the growing season. **(a)** In this hypothetical climate diagram, there is a seven-month growing season and plant growth is limited by temperature throughout the year. **(b)** In this example, there is a five-month growing season and plant growth is limited by precipitation.

Mean, Median, and Mode

Climate diagrams are a useful way of conveying a good deal of information about the average monthly changes in temperature and precipitation. While the climate of a particular location can vary from year to year, the climate diagrams display the typical conditions based on several years of collecting data. Using these data, one can then determine the *mean* temperature and precipitation for a particular month. The mean, or average, is calculated by summing all of the data and dividing by the total number of data points. The mean value gives you a sense of where the middle value lies in a set of data. However, this assumes that the data have a symmetrical distribution such that half of the values fall above the mean and half of the values fall below the mean.

In some sets of data, the values are not symmetrically distributed around a middle value. In such cases, a better estimate of the middle value is the *median*. The median is found by placing the data in order, from lowest to highest, and finding the number that occurs in the middle. If there is an even number of values, then there are two numbers in the middle and the median is found by taking the average of these two middle numbers. For example, consider the values:

95, 93, 90, 85, 81, 75, 63, 42, 21;

the mean = (95 + 93 + 90 + 85 + 81 + 75 + 63 + 42 + 21) ÷ 9 = 71.7;

in contrast, the median = 81.

For a set of data that contains an even number of values, the median is calculated as the mean of the middle two values. For example, consider the following values:

95, 93, 90, 85, 81, 79, 75, 63, 42, 21.

In this case, there is an even number of values and two numbers, 79 and 81, are the middle values. As a result, the median is the average of these two numbers = 80.

Sometimes scientists are not as interested in the mean or median from a set of data, but rather want to know which values occur most frequently. In this case, they determine how often each value occurs; the *mode* is the value that occurs most frequently. For example, consider the following values:

95, 93, 90, 85, 81, 81, 75, 63, 42, 21.

In this list, 81 appears more frequently than the other values, so the mode is 81. Generally, the mode is useful only for large samples where sampling of each possible value is reasonably good.

YOUR TURN For the following set of data, determine the mean, median, and mode.

12, 13, 15, 18, 17, 19, 18, 17, 12, 14, 10, 17, 19, 16, 17

Why is the mean of these values different from the median and the mode?

months in which the average temperature exceeds 0°C. These months are warm enough to allow plant growth and therefore represent the **growing season** of the biome. Climate diagrams can also indicate whether plant growth is more limited by temperature or by precipitation. For every 10°C increase in temperature, plants require an additional 2 cm of monthly precipitation to meet the increased water needs that occur under warmer temperatures. Climate diagrams set temperature and precipitation axes so that every 10°C increase in average monthly temperature corresponds to a 2 cm increase in monthly precipitation. This means that in any month in which the precipitation line goes below the temperature line, plant growth is constrained by a lack of sufficient precipitation. In contrast, any month in which the temperature line goes below the precipitation line, plant growth is constrained by a lack of sufficient temperature. Since climate is the primary force determining the plant forms of the nine different biomes, locations around the world that are from a particular biome have similar climate diagrams.

Let's take a closer look at the nine terrestrial biomes and their associated climate diagrams.

There are nine categories of terrestrial biomes

Terrestrial biomes are traditionally placed into nine categories. In this section, we will take a tour of the biomes. We start with tundras and boreal forests, which have average annual temperatures of less than 5°C. We will then examine the biomes in temperate regions, with average temperatures between 5°C and 20°C.

Growing season The months in a location that are warm enough to allow plant growth.

Finally, we will explore the biomes of tropical regions, which have average annual temperatures of more than 20°C. As we will see, seasonal patterns and amounts of precipitation can differ a great deal within a given temperature range, producing different types of biomes.

TUNDRAS

The **tundra,** shown in Figure 6.5, is the coldest biome and is characterized by a treeless expanse above permanently frozen soil, or permafrost. The soils thaw to a depth of 0.5 to 1 m during the brief summer growing season. Annual precipitation is generally less, and often much less, than 600 mm, but in low-lying areas where permafrost prevents drainage, soils

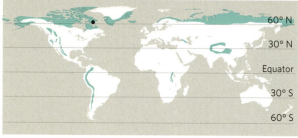

Iqaluit, Nunavut, Canada
Average annual precipitation: 404 mm

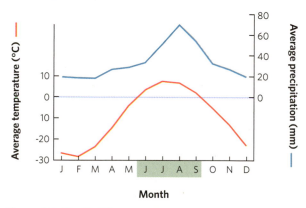

Figure 6.5 Tundra biome. Auyuittuq National Park on Baffin Island, Nunavut, is an example of a tundra biome, which is characterized by a lack of trees, and by soil that is permanently frozen. Photo by J. David Andrews / Masterfile.

may remain saturated with water throughout most of the growing season. Much of that precipitation falls as snow which, aside from watering the soil when it melts in spring, also insulates and reflects sunlight. Soils tend to be acidic because of their high content of organic matter, and they contain few nutrients. In this nutrient-poor environment, plants hold their foliage for years. Most plants are dwarf, prostrate woody shrubs, which grow low to the ground to gain protection under the winter blanket of snow and ice, since anything protruding above the surface of the snow is sheared off by blowing ice crystals. For most of the year, the tundra is an exceedingly harsh environment, but during summer days with 24 hours of sunlight, there is a rush of biological activity.

Tundra is found in the Arctic regions of Canada, Russia, Scandinavia, and Alaska, and in the Antarctic regions along the edge of Antarctica and nearby islands. At high elevations within temperate latitudes, even within the tropics, one finds vegetation resembling that of Arctic tundra, including some of the same species or their close relatives. These areas of *alpine tundra* above the tree line occur widely in the Rocky Mountains of North America, the Alps of Europe, and, especially, on the Plateau of Tibet in central Asia. In spite of their similarities, alpine and Arctic tundra have important differences. Areas of alpine tundra generally have warmer and longer growing seasons, higher precipitation, less severe winters, greater productivity, better-drained soils, and higher species diversity than Arctic tundra. Still, harsh winter conditions ultimately limit the growth of trees.

BOREAL FORESTS

Stretching in a broad belt centered at about 50° N in North America and about 60° N in Europe and Asia lies the *boreal forest.* As shown in Figure 6.6, the **boreal forest,** sometimes called **taiga,** is a biome densely populated by evergreen needle-leaved trees, with a short growing season and severe winters. The average annual temperature is below 5°C. Annual precipitation generally ranges between 40 and 1000 Iqaluit, Nunavut, Canada 0 mm, and because evaporation is low, soils are moist throughout most of the growing season. The vegetation consists of dense, seemingly endless stands of 10 to 20 m tall evergreen needle-leaved trees, mostly spruces and firs. Because of the low temperatures, plant litter decomposes very slowly and accumulates at the soil surface, forming one of the largest reservoirs of organic carbon on Earth. The needle litter produces high levels of organic acids, so the soils are acidic, strongly

Tundra The coldest biome, characterized by a treeless expanse above permanently frozen soil.

Boreal forest A biome densely populated by evergreen needle-leaved trees, with a short growing season and severe winters. *Also known as* Taiga.

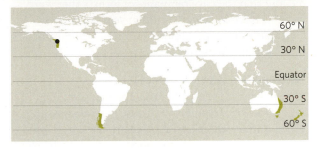

Whitehorse, Yukon, Canada
Average annual precipitation: 262 mm

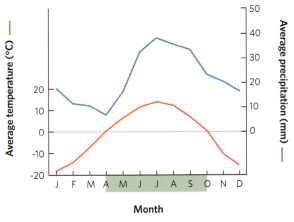

Figure 6.6 Boreal forest biome. Boreal forests, such as this one in Kluane National Park and Reserve, near Whitehorse, Yukon, typically have cold temperatures and are dominated by evergreen trees including spruces and firs. Photo by Rudolf Friederich/age fotostock.

podsolized, and generally of low fertility. Growing seasons rarely exceed 100 days, and are often half that long. The vegetation is extremely frost-tolerant; temperatures may reach –60°C during the winter. Since few species can survive in such harsh conditions, species diversity is very low. The boreal forest is not well suited for agriculture, but it serves as a source of timber products that include lumber and paper.

TEMPERATE RAINFORESTS

As we begin to move closer to the equator, we find the four temperate biomes: temperate rainforest, temperate seasonal forest, woodland/shrubland, and

Carmanah Point, British Columbia, Canada
Average annual precipitation: 2991 mm

Figure 6.7 Temperate rainforest biome. Temperate rainforests exist along the coasts of several continents, including this forest of giant Sitka spruce trees (*Picea sitchensis*) in British Columbia. They have mild temperatures and high amounts of precipitation. Photo by Radius Images/Alamy.

temperate grassland/cold desert. The **temperate rainforest** biome, shown in **Figure 6.7**, is known for mild temperatures and abundant precipitation, and is dominated by evergreen forests. These conditions are due to nearby warm ocean currents. This biome is most extensive near the Pacific coast in northwestern

Temperate rainforest A biome known for mild temperatures and abundant precipitation, and dominated by evergreen forests.

North America and in southern Chile, New Zealand, and Tasmania. The mild, rainy winters and foggy summers create conditions that support evergreen forests. In North America, these forests are dominated toward the south by coast redwood (*Sequoia sempervirens*) and toward the north by Douglas fir (*Pseudotsuga menziesii*). These trees are typically 60 to 70 m tall and may grow to over 100 m, making them very attractive for harvesting as lumber. Ecologists do not understand why these sites are dominated by needle-leaved trees, but the fossil record shows that these plant communities are very old and that they are remnants of forests that were vastly more extensive 70 million years ago. In contrast to tropical rainforests, temperate rainforests typically support few species.

TEMPERATE SEASONAL FORESTS

The **temperate seasonal forest** biome, shown in Figure 6.8, occurs under moderate temperature and precipitation conditions, and is dominated by deciduous trees. Winter temperatures can drop below freezing in this biome. The environmental conditions in this biome fluctuate much more than they do in the temperate rainforests because they do not benefit from the moderating effects of nearby warm ocean waters. In North America, the dominant plant growth form is deciduous trees, which lose their leaves each fall and include maple, beech, and oak. In North America, this biome stretches across the eastern United States and southeastern Canada; it is also widely distributed in Europe and eastern Asia. This biome is not common in the Southern Hemisphere where the larger ratio of ocean surface to land moderates winter temperatures at high altitudes and prevents frost.

In the Northern Hemisphere, the length of the growing season in this biome varies from 130 days at higher latitudes to 180 days at lower latitudes. Precipitation usually exceeds evaporation and transpiration; consequently, water tends to move downward through soils and to drain from the landscape as groundwater and as surface streams and rivers. Soils are often podsolized, tend to be slightly acidic and moderately leached, and, because of abundant organic matter, are brown in colour. The vegetation often includes a layer of smaller tree species and shrubs beneath the dominant trees, as well as herbaceous plants on the forest floor. Many of these herbaceous plants complete their growth and flower in early spring before the trees have fully leafed out.

Warmer and drier parts of the temperate seasonal forest biome, especially where soils are sandy and nutrient poor, tend to develop needle-leaved forests dominated by pines. The most important of these ecosystems in North America are the pine forests of the coastal plains of the Atlantic and Gulf coasts of the United States; pine forests also exist at higher

Bialystok, Podlaskie, Poland
Average annual precipitation: 580 mm

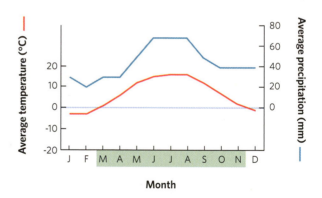

Figure 6.8 Temperate seasonal forest biome. Temperate seasonal forests have warm summers, cold winters, and a moderate amount of precipitation that favours the growth of deciduous trees. Pictured here is the Bialowieza Forest in Poland. Photo by Aleksander Bolbot/age fotostock.

elevations in the western United States. Because of the warm climate in the southeastern United States, soils there are usually nutrient poor. The low availability of nutrients and water favours evergreen, needle-leaved trees, which resist desiccation and give up nutrients slowly because they retain their needles for several years. Soils in this biome tend to be dry and fires are frequent, although most species are able to resist fire damage. The temperate seasonal forest was one of the first biomes that European settlers in North America used for agriculture.

Temperate seasonal forest A biome with moderate temperature and precipitation conditions, dominated by deciduous trees.

WOODLANDS/SHRUBLANDS

The **woodland/shrubland** biome, shown in Figure 6.9, is characterized by hot, dry summers and mild, wet winters, a combination that favours the growth of drought-tolerant grasses and shrubs. Because this type of climate is found around much of the Mediterranean Sea, it is often referred to as a Mediterranean climate regardless of where it is actually found. The woodland/shrubland biome has many different regional names including *chaparral* in southern California, *matorral* in southern South America, *fynbos* in southern Africa, and *maquis* in the area surrounding the Mediterranean Sea.

Paso Robles, California, United States
Average annual precipitation: 280 mm

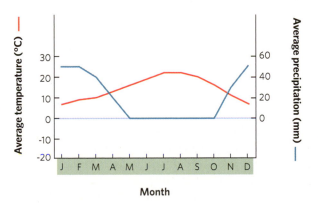

Figure 6.9 Woodland/shrubland biome. This biome is characterized by hot, dry summers and mild, wet winters, a combination that favours the growth of drought-tolerant grasses and shrubs. An example of this biome can be found in Paso Robles, California. Photo by Gary Crabbe/Enlightened Images.

As you can see in the climate diagram, although there is a 12-month growing season, plant growth is limited by dry conditions in the summer and by cold temperatures in the winter. This biome supports thick evergreen shrubby vegetation 1 to 3 m in height, with deep roots and drought-resistant foliage. The small, durable leaves of typical Mediterranean-climate plants have earned them the label of **sclerophyllous** ("hard-leaved") vegetation. Fires are frequent in the woodland/shrubland biome, and most plants have either fire-resistant seeds or root crowns that resprout soon after a fire. Traditional human use of this biome has been for grazing animals and growing deep-rooted crops such as grapes.

TEMPERATE GRASSLANDS/COLD DESERTS

The **temperate grassland/cold desert** biome, shown in Figure 6.10, is characterized by hot, dry summers and cold, harsh winters, and is dominated by grasses, nonwoody flowering plants, and drought-adapted shrubs. Plant growth is constrained by a lack of precipitation in the summer and by cold temperatures in the winter. The biome is also known by a variety of different names around the world, including *prairies* in North America, *pampas* in South America, and *steppes* in eastern Europe and central Asia.

As the biome name suggests, the dominant plant forms in temperate grasslands are grasses and nonwoody flowering plants that are well adapted to the frequent fires. Precipitation varies widely across this biome. For example, on the eastern edge of North American prairies, annual precipitation can be 1000 mm. In such areas, grasses can grow to more than 2 m high and are referred to as *tallgrass prairies*. There is even enough moisture in such areas to support the growth of trees, but the frequent fires prevent the trees from becoming a dominant component of this biome. As one moves west, annual precipitation declines to 500 mm or less. In these areas, grasses generally do not grow taller than 0.5 m and are referred to as *shortgrass prairies*. Because precipitation is infrequent, organic detritus does not decompose rapidly and this makes the soils rich in organic matter. In addition, the low acidity of the soils means that they are not heavily leached and they tend to be rich in nutrients.

Even farther west in North America, annual precipitation drops below 250 mm and the temperate grasslands grade into cold deserts, also known as

Woodland/shrubland A biome characterized by hot, dry summers and mild, wet winters, a combination that favours the growth of drought-tolerant grasses and shrubs.

Sclerophyllous Vegetation that has small, durable leaves.

Temperate grassland/cold desert A biome characterized by hot, dry summers and cold, harsh winters and dominated by grasses, nonwoody flowering plants, and drought-adapted shrubs.

Brandon, Manitoba, Canada
Average annual precipitation: 462 mm

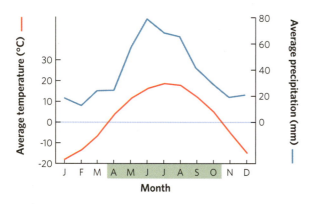

Figure 6.10 Temperate grassland/cold desert biome.
Grasslands such as this one at the Riding Mountain National Park in Alberta are characterized by hot, dry summers and very cold winters. Where moisture is more abundant, the dominant vegetation is grass. Where moisture is less abundant, in areas known as cold deserts, the dominant vegetation is composed of widely scattered shrubs. Photo by Mike Grandmaison /age fotostock.

temperate deserts. In the United States, the cold desert extends across most of the Great Basin, which sits in the rain shadow of the Sierra Nevada and Cascade Mountains. In the northern part of the region, the dominant plant is sagebrush, whereas toward the south and on somewhat moister soils, widely spaced juniper and piñon trees predominate, forming open woodlands with trees less than 10 m in stature and sparse coverings of grass. In these cold deserts, evaporation and transpiration exceed precipitation during most of the year, so soils are dry. Fires are infrequent in cold deserts because the habitat produces little organic matter to burn. However, because of the low productivity of the plant

community, grazing can exert strong pressure on the vegetation and may even favour the persistence of shrubs, which are not good forage. Indeed, many dry grasslands in the western United States and elsewhere in the world have been converted to deserts by overgrazing.

TROPICAL RAINFORESTS

Our final group of biomes is found in areas of tropical temperatures and includes tropical rainforests, tropical seasonal forests/savannahs, and subtropical deserts. **Tropical rainforests,** shown in Figure 6.11, fall within 20° N and 20° S of the equator, are warm and rainy, and are characterized by multiple layers of lush vegetation. Tropical rainforests have a continuous canopy of 30 to 40 m trees with emerging trees that occasionally reach 55 m. Shorter trees and shrubs form a layer known as the *understorey* below the canopy. The understorey also contains an abundance of epiphytes and vines. Species diversity is higher in tropical rainforests than anywhere else on Earth. This biome occurs in much of Central America, the Amazon basin, the Congo in southern West Africa, the eastern side of Madagascar, Southeast Asia, and the northeast coast of Australia. In many of these locations, however, much of the rainforest has been destroyed to harvest lumber and to make room for agriculture.

Climates that support tropical rainforests are always warm and receive at least 2000 mm of precipitation throughout the year, with rarely less than 100 mm during any single month. The tropical rainforest climate often exhibits two peaks of rainfall centered on the equinoxes, corresponding to the periods when the intertropical convergence zone passes over the equator. Rainforest soils are typically old and deeply weathered. Because they are relatively devoid of humus and clay, they take on the reddish colour of aluminum and iron oxides and they retain nutrients poorly. Despite the poor ability of these soils to hold nutrients, the biological productivity of tropical rainforests per unit area exceeds that of any other terrestrial biome. Moreover, the standing biomass exceeds that of all other biomes except temperate rainforests. This tremendous growth is possible because the continuously high temperatures and abundant moisture causes organic matter to decompose quickly, and vegetation immediately takes up the released nutrients. While rapid nutrient cycling supports the high productivity of the rainforest, it also makes the rainforest ecosystem extremely vulnerable to disturbance. When tropical rainforests are cut and burned, many of the nutrients

Tropical rainforest A warm and rainy biome, characterized by multiple layers of lush vegetation.

Sandakan, Sabah, Malaysia
Average annual precipitation: 3060 mm

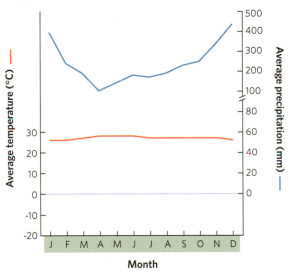

Figure 6.11 Tropical rainforest biome. Tropical rainforests, such as this site in Borneo, have very warm temperatures and very high amounts of precipitation. As a result, this biome has multiple layers of lush vegetation.
Photo by Nick Garbutt/naturepl.com.

Nairobi, Nairobi, Kenya
Average annual precipitation: 750 mm

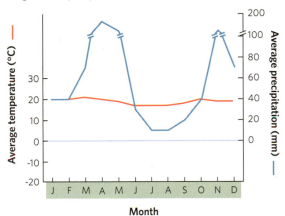

Figure 6.12 Tropical seasonal forest/savannah biome. Tropical seasonal forests and savannahs have warm temperatures like the tropical rainforests, but also distinct wet and dry seasons due to the movement of the intertropical convergence zone. As a result, this biome has trees that shed their leaves during the dry season. An example of this biome can be found in the Masai Mara National Reserve of Kenya.
Photo by Denis-Huot/ZUMAPRESS/Newscom.

are carted off in logs or go up in smoke. The vulnerable soils erode rapidly and fill the streams with silt. In many cases, the environment degrades rapidly and the landscape becomes unproductive.

TROPICAL SEASONAL FORESTS/SAVANNAHS
Tropical seasonal forests, shown in Figure 6.12, are located mostly beyond 10° N and 10° S of the equator. These regions experience warm temperatures and, as the intertropical convergence zone moves during the year, pronounced wet and dry seasons. Because tropical seasonal forests have a preponderance of

deciduous trees that shed their leaves during the dry season, this biome is sometimes referred to as a *tropical deciduous forest.* In areas where the dry season is longer and more severe, the vegetation becomes shorter and develops thorns to protect leaves from grazing animals. With even longer dry periods, the vegetation grades from dry forest into thorn forest and finally into *savannahs,* which are open landscapes

Tropical seasonal forest A biome with warm temperatures and pronounced wet and dry seasons, dominated by deciduous trees that shed their leaves during the dry season.

147

containing grasses and occasional trees, including acacia and baobab trees.

The tropical seasonal forest/savannah biome occurs in Central America, the Atlantic coast of South America, sub-Saharan Africa, southern Asia, and northwestern Australia. Fire and grazing play important roles in maintaining the character of the savannah biome. Under these conditions, grasses can persist better than other forms of vegetation. When grazing and fire are prevented within a savannah habitat, dry forest often begins to develop. As in more humid tropical environments, the soils tend to hold nutrients poorly but the warm temperatures favour rapid decomposition. Rapid decomposition provides a rapid recycling of nutrients into the soil, which trees can quickly take up and use for growth and reproduction. This rapid cycling of nutrients also makes this biome an attractive place for agriculture, including raising cattle. On the Pacific coast of Central America and on the Atlantic coast of South America, for example, over 99 percent of this biome has been converted to agriculture.

SUBTROPICAL DESERTS

Subtropical deserts, shown in Figure 6.13, are characterized by hot temperatures, scarce rainfall, long growing seasons, and sparse vegetation. Also known as *hot deserts,* subtropical deserts develop at 20° to 30° north and south of the equator, in areas associated with the dry, descending air of Hadley cells. Subtropical deserts include the Mojave Desert in North America, the Sahara Desert in Africa, the Arabian Desert in the Middle East, and the Great Victoria Desert in Australia.

Because of the low rainfall, the soils of subtropical deserts are shallow, virtually devoid of organic matter, and neutral in pH. Whereas sagebrush dominates the cold deserts of the Great Basin, creosote bush (*Larrea tridentata*) dominates in the subtropical deserts of the Americas. Moister sites support a profusion of succulent cacti, shrubs, and small trees, such as mesquite and paloverde (*Cercidium microphyllum*). Most subtropical deserts receive summer rainfall. After summer rains, many herbaceous plants sprout from dormant seeds, grow quickly, and reproduce before the soils dry out again. Few plants in subtropical deserts are frost-tolerant. Species diversity is usually much higher than in temperate arid lands.

Aquatic biomes are categorized by their flow, depth, and salinity

As we discussed earlier in this chapter, ecologists categorize aquatic biomes using a range of physical factors including water depth, water flow, and

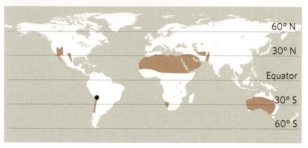

Antofagasta, Antofagasta, Chile
Average annual precipitation: 0 mm

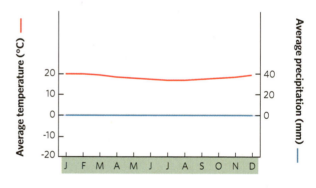

Figure 6.13 Subtropical desert biome. Subtropical deserts, such as this site in the Atacama Desert of Chile, have hot temperatures and scarce rainfall. This favours drought-resistant plants such as cactus, creosote bush, mesquite, and euphorbs. Photo by Imagebroker.net/Photoshot.

salinity. The major types of aquatic biomes include streams and rivers, lakes and ponds, freshwater wetlands, salt marshes, mangrove swamps, intertidal zones, coral reefs, and the open ocean.

STREAMS AND RIVERS

Because streams and rivers are characterized by flowing fresh water, they are often referred to as **lotic** systems. Although there is no exact

---

Subtropical desert A biome characterized by hot temperatures, scarce rainfall, long growing seasons, and sparse vegetation.

Lotic Characterized by flowing fresh water.

specification to determine classification differences between a stream and a river, in general, **streams,** also called **creeks,** are narrow channels of fast flowing fresh water whereas **rivers** are wide channels of slow flowing fresh water (Figure 6.14). As streams flow down from their headwaters, they join together with other streams and eventually grow large enough to be considered a river. Streams and some rivers are usually bordered by a **riparian zone,** which is a band of terrestrial vegetation influenced by seasonal flooding and elevated water tables.

As one moves downstream, water flows more slowly and becomes warmer and richer in nutrients. Under these conditions, ecosystems generally become more complex and more productive. In general, streams support fewer species than other aquatic biomes. Small streams are often shaded and nutrient poor, which limits the productivity of algae and other photosynthetic organisms. Much of the organic content of stream ecosystems depends on **allochthonous** inputs of organic matter, such as leaves, that come from outside the ecosystem. In large rivers, a higher proportion of the organic inputs are **autochthonous,** meaning that they are produced from inside the ecosystem by algae and aquatic plants. As rivers progress from their source, they typically become wider, slower moving, more heavily laden with nutrients, and more exposed to direct sunlight. They also accumulate sediments that are washed in from the land and carried downstream. High turbidity caused by suspended sediments in the lower reaches of silt-laden rivers can block light and reduce production.

Lotic systems are extremely sensitive to modification of their water flow by dams. In Canada, thousands of dams—built to control flooding, to provide water for irrigation, or to generate electricity—interrupt stream flow. Dams also alter water temperature and rates of sedimentation. Typically, water behind dams becomes warmer, and the original stream bottoms become filled with silt that destroys habitat for fish and other aquatic organisms. Water released downstream from large dams often has low concentrations of dissolved oxygen. Using dams for flood control changes the natural seasonal cycles of flooding that are necessary for maintaining many kinds of riparian habitats on floodplains. Dams also disrupt the natural movement of aquatic organisms upstream and downstream, fragmenting river systems and isolating populations.

PONDS AND LAKES

Ponds and **lakes** are aquatic biomes characterized by nonflowing fresh water with at least some area of water that is too deep for plants to rise above the water's surface (Figure 6.15a). Although there is no

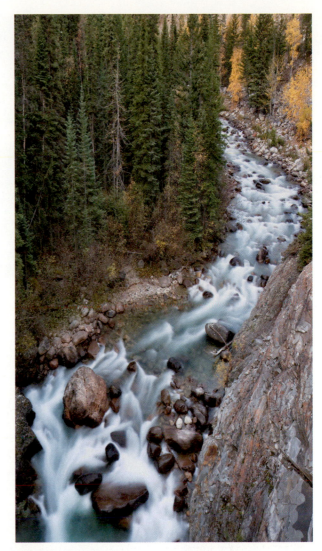

Figure 6.14 Streams and rivers. Streams and rivers are characterized by flowing fresh water. This example is on the Athabasca River in Alberta. Photo by Mcmaster Studio/Alamy.

Stream A narrow channel of fast flowing fresh water. *Also known as* **Creek**.

River A wide channel of slow flowing fresh water.

Riparian zone A band of terrestrial vegetation alongside rivers and streams that is influenced by seasonal flooding and elevated water tables.

Allochthonous Inputs of organic matter, such as leaves, that come from outside of an ecosystem.

Autochthonous Inputs of organic matter that are produced by algae and aquatic plants inside an ecosystem.

Pond An aquatic biome that is smaller than a lake and is characterized by nonflowing fresh water with some area of water that is too deep for plants to rise above the water's surface.

Lake An aquatic biome that is larger than a pond and is characterized by nonflowing fresh water with some area of water that is too deep for plants to rise above the water's surface.

(a)

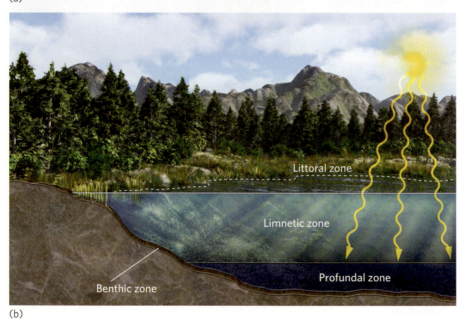

(b)

Figure 6.15 Ponds and lakes. Ponds and lakes are characterized by nonflowing fresh water with areas of water that are too deep for emergent vegetation. **(a)** Lake Louise, Alberta. **(b)** Lakes contain a variety of zones. The littoral zone exists around the edge of the lake and contains rooted, emergent plants. The limnetic zone consists of the open water in the middle of the lake where the dominant photosynthetic organisms are floating algae. Below the limnetic zone is the profundal zone, which is too deep for sunlight to penetrate enough to permit photosynthesis. The layer of sediments at the bottom of the lake is the benthic zone. Photo by Jose Moya/age fotostock.

clear-cut distinction between ponds and lakes, ponds are smaller. Many lakes and ponds were formed as glaciers retreated, gouging out basins and leaving behind glacial deposits containing blocks of ice that eventually melted. The Great Lakes of North America formed in glacial basins, overlain until 10 000 years ago by thick ice. Lakes are also formed in geologically active regions, such as the Great Rift Valley of Africa, where vertical shifting of blocks of the Earth's crust created basins in which water accumulates. Broad river valleys, such as those of the Mississippi and Amazon rivers, have *oxbow lakes,* which are broad bends of what was once the river, cut off by shifts in the main channel.

As shown in Figure 6.15b, lakes can be subdivided into several ecological zones, each with distinct physical conditions. The **littoral zone** is the shallow area around the edge of a lake or pond containing rooted vegetation, such as water lilies and

pickerelweed. The open water beyond the littoral zone is the **limnetic** (or **pelagic**) **zone,** where the dominant photosynthetic organisms are floating algae, or *phytoplankton.* Very deep lakes also have a **profundal zone** that does not receive sunlight because of its depth. The absence of photosynthesis, as well as the presence of bacteria that decompose the detritus at the bottom of the lake, causes the profundal zone to have very low concentrations of oxygen. The sediments at the bottoms of lakes and

Littoral zone The shallow area around the edge of a lake or pond containing rooted vegetation.

Limnetic zone The open water beyond the littoral zone, where the dominant photosynthetic organisms are floating algae. *Also known as* **Pelagic zone.**

Profundal zone The area in a lake that is too deep to receive sunlight.

Figure 6.16 Circulation in temperate lakes. **(a)** In the spring, seasonal winds cause the lake water to mix, which brings nutrients from the sediments to the surface water and brings oxygen from the surface water down to the deeper water. **(b)** During the summer, the surface water warms faster than the deep water, so the lake experiences thermal stratification. The zone at which water temperature changes rapidly with depth is known as the thermocline. **(c)** In autumn, the surface water cools, stratification breaks down, and autumn winds cause the surface water and deep waters to mix once again. **(d)** In winter, the surface waters are exposed to freezing temperatures, so ice forms at the surface. Because 4°C water is the most dense, the bottom of the lake does not freeze.

ponds constitute the **benthic zone,** which provides habitat for burrowing animals and microorganisms.

Circulation in Ponds and Lakes

While lakes and ponds can be separated into four separate zones based on the proximity to the shore and the amount of light penetration, the water depths can also be classified by temperature. In most lakes and ponds in temperate and polar regions, the temperature of the water forms layers. The surface water, known as the **epilimnion,** can have a different temperature than the deeper water, known as the **hypolimnion.** Between these two temperature regions is the **thermocline,** which is a middle depth of water that experiences a rapid change in temperature over a relatively short distance in depth. The thermocline serves as a barrier to a mixing between the epilimnion and hypolimnion.

Most production in a lake occurs in the epilimnion, where sunlight is most intense. Oxygen produced by photosynthesis and oxygen entering the lake at the interface of the water and atmosphere keep the epilimnion well aerated and thus suitable for animal life. Throughout a growing season, however, plants and algae often deplete the supply of dissolved mineral nutrients in the epilimnion and this curtails their growth. In the hypolimnion, which can include the lower limnetic zone and the profundal zone, bacteria continue to decompose organic material, but the reduced intensity of light causes a reduction in

photosynthesis. The result is that oxygen is used up faster than it is produced and this leads to anaerobic conditions. Oxygen is in particularly short supply deep in productive lakes that generate abundant organic matter in the epilimnion.

Lakes in the temperate zone experience changing temperatures across the seasons. These temperature changes drive changes in water density, which, in turn, causes the shallow and deep water to mix. Figure 6.16 shows this process. As you may recall from Chapter 2, water becomes more dense as it cools to 4°C and then less dense as it cools below 4°C. During the winter in cold climates, the coldest lake water (0°C) lies at the surface just beneath the ice, while the slightly warmer, denser water (4°C) sinks to the bottom of the lake. In early spring, the sun gradually warms the lake. As the surface temperature increases toward 4°C, the sun-warmed water sinks into the cooler layers immediately below and the water begins to mix. At the same time, winds drive

---

Benthic zone The area consisting of the sediments at the bottoms of lakes, ponds, and oceans.

Epilimnion The surface layer of the water in a lake or pond.

Hypolimnion The deeper layer of water in a lake or pond.

Thermocline A middle depth of water in a lake or pond that experiences a rapid change in temperature over a relatively short distance in depth.

(a)

(b)

(c)

Figure 6.17 Freshwater wetlands. This biome includes a variety of aquatic habitats. **(a)** Swamps such as this one in Campbellville, Ontario, contain emergent trees. **(b)** Marshes contain emergent nonwoody vegetation that include reeds, such as in this location near Point Pelee National Park in Leamington, Ontario. **(c)** Bogs are characterized by acidic waters and plants that are well adapted to these conditions, such as this bog in the Mingan Archipelago National Park Reserve in Quebec. Photos by (a) Wave Royalty Free/age fotostock; (b) Heeb Christian/Prisma/SuperStock; (c) © Christian GUY/imagebr/age fotostock.

surface currents that can cause deep water to rise in a manner similar to upwelling currents in the oceans. The vertical mixing of the lake water that occurs in early spring and is assisted by winds that drive the surface currents is known as the **spring overturn.** The spring overturn brings nutrients from sediments on the bottom to the surface and oxygen from the surface to the depths.

In late spring and early summer, surface layers of water gain heat faster than deeper layers. At this point, the thermocline is created. Once the thermocline is well established, the surface and deep waters no longer mix because the warmer, less dense surface water floats on the cooler, denser water below, a condition known as **stratification.** During the fall, the temperature of the surface layers of the lake drops. As this water becomes denser than the underlying water, it begins to sink. The vertical mixing that occurs in the fall and is assisted by winds that drive surface currents is called **fall overturn.** Similar to the spring overturn, fall overturn brings oxygen to deep waters and nutrients to the surface. The infusion of nutrients into surface waters in the fall may cause an explosion in the population of phytoplankton, known as the **fall bloom.** This mixing persists into late fall, until the temperature at the lake surface drops below 4°C and winter stratification ensues.

The spring and fall overturn are typical of lakes that exist in temperate climates because they experience cold winters and warm summers. The seasonality of vertical mixing is much less dramatic in lakes that are not exposed to such dramatic temperature changes. In warmer climates, water temperatures do not fall below 4°C. As a result, such lakes do not stratify in the winter, and many have only one mixing event each year following summer stratification.

FRESHWATER WETLANDS

Freshwater wetlands are aquatic biomes that contain standing fresh water, or soils saturated with fresh water for at least part of the year, and are shallow enough to have emergent vegetation throughout all depths. Most of the plants that grow in wetlands can tolerate low oxygen concentrations in the soil; many are specialized for these anoxic conditions and grow nowhere else.

Freshwater wetlands include swamps, marshes, and bogs (Figure 6.17). *Swamps* contain emergent

---

Spring overturn The vertical mixing of lake water that occurs in early spring, assisted by winds that drive the surface currents.

Stratification The condition of a lake or pond when the warmer, less dense surface water floats on the cooler, denser water below.

Fall overturn The vertical mixing of lake water that occurs in fall, assisted by winds that drive the surface currents.

Fall bloom A rapid increase in the population of phytoplankton in fall that occurs with the infusion of nutrients from fall overturn.

Freshwater wetland An aquatic biome that contains standing fresh water, or soils saturated with fresh water for at least part of the year, and which is shallow enough to have emergent vegetation throughout all depths.

trees. You can see swamps, for example, in Point Pelee National Park or the Minesing Wetland, both in Ontario. *Marshes* contain emergent nonwoody vegetation such as cattails. Some of the largest marshes in the world include the Everglades in Florida and the Pantanal of Brazil, Bolivia, and Paraguay. In contrast to swamps and marshes, *bogs* are characterized by acidic waters and contain a variety of plants including sphagnum mosses and stunted trees that are specially adapted to these conditions. Large bogs are found in northern Europe and Russia and in the Hudson Bay Lowlands, one of the largest wetland areas in the world. Other examples of protected wetland areas in Canada are Terra Nova National Park in Newfoundland and Labrador and Kejimkujik National Park in Nova Scotia.

Freshwater wetlands provide important habitat for a wide variety of animals, notably waterfowl, migrating birds, and the larval stages of many species of fish and invertebrates that are characteristic of open waters. Wetland sediments immobilize potentially toxic or polluting substances dissolved in water and therefore function as a water purification system.

SALT MARSHES/ESTUARIES

Salt marshes are a saltwater biome that contains nonwoody emergent vegetation. Salt marshes are found along the coasts of continents in temperate climates, often within *estuaries* that occur where the mouths of rivers mix with the salt water from oceans

Figure 6.18 Salt marsh. Salt marshes occur in salt water and estuaries and contain nonwoody emergent vegetation. An example of a salt marsh can be found at Judes Point, Prince Edward Island. Photo by Michael P. Gadomski/Science Source.

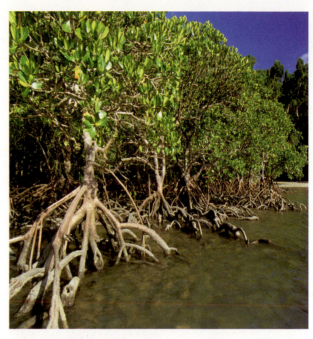

Figure 6.19 Mangrove swamps. Mangrove swamps, including this location off the coast of Australia, are saltwater biomes that contain salt-tolerant trees along tropical and subtropical coastlines. Photo by Biosphoto/T. & S. Allofs.

(Figure 6.18). Estuaries are unique because of their mix of fresh and salt water. In addition they contain an abundant supply of nutrients and sediments carried downstream by rivers. The rapid exchange of nutrients between the sediments and the surface in the shallow waters of an estuary supports extremely high biological productivity. Because estuaries tend to be areas of sediment deposition, they are often edged by extensive tidal marshes at temperate latitudes and by mangrove wetlands in the tropics. With a combination of high nutrient levels and freedom from water stress, tidal marshes are among the most productive habitats on earth. They contribute organic matter to estuarine ecosystems, which in turn support large populations of oysters, crabs, fish, and the animals that feed on them.

MANGROVE SWAMPS

Mangrove swamps are a biome that exists in salt water along tropical and subtropical coasts and contains salt-tolerant trees with roots submerged in water. This biome can also occur in estuaries where fresh water and salt water mix (Figure 6.19). Salt tolerance is a key adaptation of trees that live in mangrove swamps. By living along the coasts, these trees play important roles in preventing the erosion of coastal shorelines from constant incoming waves. The

Salt marsh A saltwater biome that contains nonwoody emergent vegetation.

Mangrove swamp A biome that occurs along tropical and subtropical coasts, and contains salt-tolerant trees with roots submerged in water.

(a)

(b)

Figure 6.20 Intertidal zone. Intertidal biomes are the coastal regions around the world that exist between the high tide and the low tide of the oceans. **(a)** Rocky coasts produce rocky intertidal habitats, such as this one in Gros Morne National Park, Newfoundland and Labrador. **(b)** Muddy coasts produce mudflat habitats around the world, including this site in the Bay of Fundy, New Brunswick. Photos by (a) © Ron Erwin/age fotostock; (b) Klaus Lang/All Canada Photos/Alamy.

swamps also provide critical habitat to many species of fish and shellfish.

INTERTIDAL ZONES

The **intertidal zone** is a biome consisting of the narrow band of coastline between the levels of high tide and low tide. As the tides come in and go out, the intertidal zone experiences widely fluctuating temperatures and salt concentrations. The species living in this biome—including crabs, barnacles, sponges, mussels, and algae—must therefore possess adaptations that allow them to tolerate such harsh conditions. Intertidal zones can occur along steep rocky coastlines, as one might find in Newfoundland, or gently sloping mudflats, as one might find in the Bay of Fundy in New Brunswick. (Figure 6.20).

CORAL REEFS

Coral reefs are a marine biome found in warm, shallow waters that remain above 20°C year-round. Coral reefs often surround volcanic islands, where they are fed by nutrients eroding from the rich volcanic soil and by deep-water currents forced upward by the profile of the island.

Corals are tiny animals—related to hydra and other cnidarians—that live in a mutualistic relationship with algae. An individual coral is a hollow tube that secretes a hard exoskeleton made of calcium carbonate. It also has tentacles that sweep detritus and plankton into the tube. As it digests these particles, the coral produces CO_2 that can be used by their symbiotic algae in photosynthesis. Some of the sugars and other organic compounds the algae produce leak into the coral tissues and further support coral growth. Although an individual coral is tiny, corals live in huge colonies. As an individual coral dies, the soft tissues decompose but the hard outer skeletons remain behind. Over time, these skeletons accumulate to form massive coral reefs.

The complex structure that corals build provides a wide variety of substrates and hiding places for algae and animals. This helps to make coral reefs among the most diverse biomes on Earth (Figure 6.21). As you may recall from our discussion of coral reefs in Chapter 2, rising sea surface temperatures in the tropics are causing the departure of the algal symbionts of corals over large areas—a phenomenon known as *coral bleaching*. Because the algal symbionts are critical to the survival of the coral, the stability of these biomes is now at risk.

THE OPEN OCEAN

The open ocean is characterized as the part of the ocean that is away from the shoreline and coral reefs. Open oceans cover the largest portion of the surface of Earth. Beneath the surface lies an immensely complex realm with large variations in temperature,

Intertidal zone A biome consisting of the narrow band of coastline between the levels of high tide and low tide.

Coral reef A marine biome found in warm, shallow waters that remain 20°C year-round.

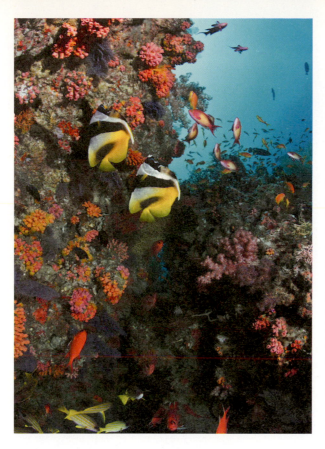

Figure 6.21 Coral reefs. The hard exoskeletons of millions of tiny corals form massive coral reefs in the ocean, which serve as home to an incredible diversity of organisms. Coral reefs can be found in shallow, warm ocean waters, such as this reef off the Maldives. Photo by Lea Lee/Getty Images.

salinity, light, pressure, and currents. Ecologists recognize a number of zones in the open ocean, shown in Figure 6.22. Beyond the range of the lowest tidal level, the **neritic zone** extends to depths of about 200 m, which correspond to the edge of the continental shelf. Because strong waves move nutrients to the sunlit surface layers from sediments below, the neritic zone is generally a region of high productivity. Beyond the neritic zone, the seafloor drops rapidly to the great depths of the **oceanic zone.** Here, nutrients are sparse, and production is strictly limited. Finally, the benthic zone consists of the seafloor underlying the neritic and oceanic zones.

Neritic zone The ocean zone beyond the range of the lowest tidal level, and which extends to depths of about 200 m.

Oceanic zone The ocean zone beyond the neritic zone.

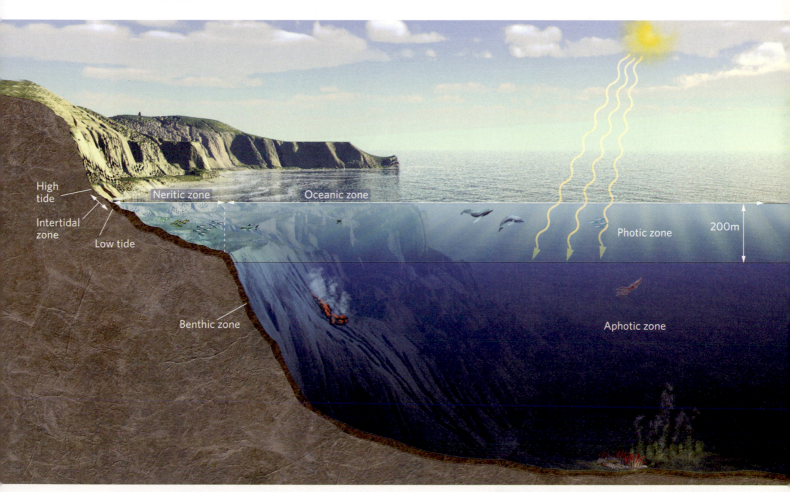

Figure 6.22 Open ocean. The open ocean is represented by water that is offshore and away from coral reefs. This biome can be broken up into several zones.

The neritic and the oceanic zones may be subdivided vertically into a *photic zone* and an *aphotic zone*. The **photic zone** is the area of the neritic and oceanic zones that contains sufficient light for photosynthesis by algae. The **aphotic zone** is the area of the neritic and oceanic zones where the water is so deep that sunlight cannot penetrate. However, as we saw in Chapter 1, bacteria in the aphotic zone use chemosynthesis to convert inorganic carbon into simple sugars. Other organisms in the aphotic zone depend on the organic material that falls from the photic zone. One of the fascinating adaptations of many organisms in the aphotic zone is the ability to generate their own source of light, known as *bioluminescence,* to help them find and consume prey. A number of jellyfish, crustaceans, squid, and fish species have independently evolved this ability.

In this chapter, we have explored how differences in climate around the world determine the types of dominant plant forms that can persist in different parts of the world, forming the basis of categorizing terrestrial biomes. In contrast, the aquatic biomes are categorized by differences in water flow, depth, and salinity. In all cases, there is a close association between the environmental conditions and the species that have evolved adaptations to live under these conditions. Of course, adaptations reflect not only the physical factors in the environment but also the many interactions with other organisms. In the next chapter, we examine the process of evolutionary adaptation and see how it has created the tremendous diversity of life on Earth.

Photic zone The area of the neritic and oceanic zones that contains sufficient light for photosynthesis by algae.

Aphotic zone The area of the neritic and oceanic zones where the water is so deep that sunlight cannot penetrate.

ECOLOGY TODAY CONNECTING THE CONCEPTS

CHANGING BIOME BOUNDARIES

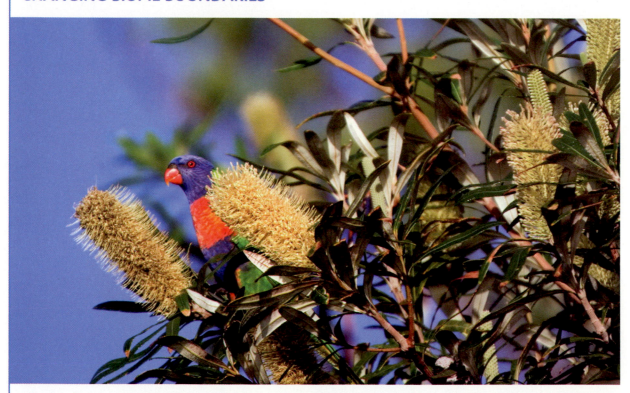

Climate change. The changing climates around the world are predicted to alter the distribution of many organisms, including plants in the genus *Banksia* in the shrublands of southwestern Australia. These changes are also affecting human agriculture, including the vineyards that are planted in the shrubland biomes around the world. Photo by Phil Morley/age fotostock.

Throughout this chapter we have seen that climate largely determines the location of the terrestrial biomes. Climatic conditions, combined with species interactions, set the edges of biome boundaries. Given our understanding of how biome boundaries form, what would happen to the biomes, and the species living in them, if the climate changed?

Records show us that during the past 130 years, temperatures of the surface of Earth have increased an average of 1°C. In fact, nine of the warmest years ever recorded happened during the 11-year period from 2001 to 2011. This small average increase in Earth's temperature conceals the fact that some regions have become 1°C to 2°C cooler during this time while others have become up to 4°C warmer. Scientists predict even greater increases in temperature and large changes in precipitation patterns for the twenty-first century. If climate determines the location of biomes and the climate is changing, it seems reasonable to predict that the boundaries of biomes are also going to change.

In some cases, scientists believe that shifting biome boundaries might occur relatively easily. Where no barriers prevent movement, plant and animal populations will be able to shift north or south over several decades without much difficulty. However, when movement is blocked, for example by mountain ranges or large highways, or suitable environmental conditions are not available, the plants and animals may not be able to survive the changing conditions. Consider the woodland/shrubland biome on the southwestern coast of Australia. This small biome is located on a relatively small area of coastal land, with an ocean to the south and west, and a desert to the north and east. Scientists predict that this biome will become hotter and drier during the current century. If this prediction is correct, organisms that cannot tolerate the increase in temperature will have nowhere

hospitable to go since the neighbouring desert biome is already too dry for them to survive. Scientists who examined one group of plants from the genus *Banksia*, which is composed of 100 species, concluded that over the next 70 years, 66 percent of these species will decline in abundance and 25 percent will go extinct.

Climate changes also affect human agriculture. Recall from the beginning of this chapter that most of the world's wine is produced in the shrubland/woodland biome. However, changes in climate have already affected the cultivation of grapes used in winemaking. In France, for example, over the past 30 years warmer growing seasons have caused grapes to ripen 16 days earlier than previously. This warmer climate alters the sugar content and acidity of the grapes—two components that must be in balance to make a fine-tasting wine. This problem is so serious for winemakers and wine consumers that in 2009 French wine growers called upon world leaders to take immediate action to try to reverse global climate change. On the other hand, locations at slightly higher latitudes that had temperatures that were historically too cool for growing quality wine grapes are now reporting increased summer temperatures that have allowed them to produce some of the best wine grapes in years. This is a good development for winemakers in countries such as England, but devastating for the French who have a long history of making some of the world's finest wines.

SOURCES: Fitzpatrick, M. C., A. D. Grove, N. J. Sanders, and R. R. Dunn. 2008. Climate change, plant migration, and range collapse in a global biodiversity hotspot: The *Banksia (Proteaceae)* of Western Australia. *Global Change Biology* 14: 1337–1352.

Iverson, J. T. How global warming could change the winemaking map. 2009. *Time Magazine*, December 3.

SUMMARY OF CHAPTER CONCEPTS

■ **Terrestrial biomes are categorized by their major plant growth forms.** Ecologists use the dominant plant forms to categorize ecosystems into terrestrial biomes because many plants have evolved convergent forms in response to similar climate conditions. Climate and dominant plant forms are similar within biomes.

■ **There are nine categories of terrestrial biomes.** The coldest biomes are the tundra and boreal forests. In temperate regions, we can find temperate rainforests, temperate seasonal forests, woodlands/shrublands, and temperate grasslands/cold deserts. In tropical latitudes, the biomes can be categorized as tropical rainforests, tropical seasonal forests/savannahs, and subtropical deserts.

■ **Aquatic biomes are categorized by their flow, depth, and salinity.** Freshwater biomes include streams and rivers, ponds and lakes, and freshwater wetlands. Saltwater biomes include salt marshes/estuaries, mangrove swamps, intertidal zones, coral reefs, and the open ocean.

REVIEW QUESTIONS

1. Why do unrelated plants often assume the same growth form in different parts of the world?

2. What information about a biome can you obtain from a climate diagram?

3. Why is the boreal forest biome found on several different continents, including North America, Europe, and Asia?

4. What types of terrestrial plants are found in each of the four biomes situated at temperate latitudes?

5. How do environmental conditions differ among the four temperate biomes?

6. Explain why tropical rainforests experience two peaks of rainfall.

7. Compare and contrast the factors used to categorize terrestrial biomes with those used to categorize aquatic biomes.

8. How do headwater streams and larger rivers differ in their major source of organic material?

9. Compare and contrast swamps, marshes, and bogs.

10. Why does productivity in the ocean differ between the photic and aphotic zones?

PRACTICING ECOLOGY: METHANE PRODUCTION IN PEATLANDS

Tim Moore, from McGill University, and his colleagues have studied how changes in precipitation influence the decomposition of peat.

Photo courtesy of Tim Moore

All things must die—and decompose. The characteristic temperature range and level of precipitation in a biome determine how fast life grows, and also how fast decomposition occurs after death. For example, the hot and wet conditions of tropical forests encourage rapid plant decomposition, whereas the relatively dry conditions of temperate grasslands foster much slower decomposition rates. The rate of decomposition is important to an ecosystem because it determines how fast nutrients are recycled for the growth of new life forms. Given that climate change threatens to alter both temperature and precipitation, we must ask an important question: How will climate change affect rates of decomposition in Earth's biomes?

Boreal forests are characterized by relatively rapid rates of plant production and slow rates of decomposition. These conditions cause partially decayed dead organic material to build up in thick peatlands. Cold temperatures reduce the activity of decomposing bacteria in the soil. High levels of precipitation often leave peatland soils so saturated with water that they contain little oxygen. Without oxygen only anaerobic bacteria can survive, and anaerobic bacteria decompose dead plants more slowly than aerobic bacteria. A common by-product of anaerobic bacteria activity is methane gas (CH_4), which is a potent greenhouse gas. As a result, environmental scientists are interested in understanding how current and future climate changes in boreal forests will affect the abundance of anaerobic bacteria. Current climate models predict less precipitation in the coming years. Since rainfall is a primary cause of soil saturation and anoxic conditions, we can hypothesize that decreased rainfall will reduce the abundance of anaerobic bacteria and the amount of CH_4 gas that peatlands emit.

Over several years, Tim Moore from McGill University and his colleagues tested this hypothesis by recording gas production of the Mer Bleue peatland in eastern Ontario. They first measured the depth of water saturation below the surface of the peat, known as the water table depth. The shallower the water table, the more saturated the soil becomes. The researchers then measured the mass of CH_4 emissions from plots of peatland soils. You can view these data in the table.

WATER TABLE DEPTH AND DAILY CH_4 EMISSIONS FROM THE MER BLEUE PEATLAND

WATER TABLE DEPTH (cm)	METHANE EMISSIONS (log [mg $CH_4/m^2/day$])
2.7	2.02
3.3	0.62
10.1	1.29
15.2	0.82
15.2	0.72
18.4	0.56
18.6	1.38
22.9	0.61
26.7	0.25
35.0	0.21
36.9	0.98
38.8	0.69
40.9	0.47
142.6	0.05
43.5	1.04

GRAPH THE DATA

1. Create a graph with water table depth on the *x* axis and mass of CH_4 on the *y* axis.

2. Visually estimate a line of best fit through the data points to illustrate the relationship between the two variables.

INTERPRET THE DATA

1. Explain how the regression line supports or rejects the hypothesis that decreased rainfall will reduce the abundance of anaerobic bacteria and the amount of CH_4 gas that peatlands emit.

2. Based on the graph, predict the change in CH_4 emission if the peatlands completely flooded and the water table was at depth zero.

3. The researchers found that the presence of beavers in peatlands is associated with higher CH_4 emissions. Why might this occur?

4. Under conditions of abundant oxygen, aerobic bacteria can decompose the peat and produce CO_2 rather than CH_4. Based on the data in the table and what you know about peatlands, what would happen to the abundance of aerobic bacteria and CO_2 production if the amount of precipitation decreased?

SOURCE: Moore, T. R., A. Young, J. L. Bubier, E. R. Humphreys, P. M. Lafleur, and N. T. Roulet. 2011. A multi-year record of methane flux at the Mer Bleue Bog, Southern Canada. *Ecosystems* 14: 646–657.

CHAPTER 7

Evolution and Adaptation

Darwin's Finches

As part of his voyage around the world, Charles Darwin spent time exploring the Galápagos Islands, where he collected birds that differed in size and shape. Darwin originally thought that he had collected different families of birds because they resembled finches, blackbirds, and grosbeaks. He shipped the birds back home to England, where ornithologist John Gould determined that the birds were all finches, but finches that looked quite different from each other. Some species ate insects; others ate the fruits and flowers of cacti; still others ate large or small seeds. Darwin hypothesized that these finches likely shared a common ancestor that originally came to the islands from South America. Darwin's hypothesis—that multiple species could descend from a common ancestor—helped him develop his theory of evolution by natural selection.

Since Darwin's time, many researchers have gone to the Galápagos Islands to examine how the process of natural selection works in the finches. In the 1970s, for example, Peter and Rosemary Grant marked all of the medium ground finches (*Geospiza fortis*) on one island and measured their beak sizes. This species of finch eats seeds, and the seed size makes a difference to these birds. Individuals with large beaks can generate the force needed to crack the largest seeds, whereas individuals with small beaks are better at handling the smallest seeds. When the island subsequently experienced a drought, overall seed abundance declined. The small, easy-to-crack seeds were quickly consumed. The large seeds that remained were much harder to crack, so only individuals with larger beaks could crack the large seeds; therefore, birds with larger beaks were more likely to survive the drought. Beak size has a genetic basis that is passed from parents to offspring, and so as the average beak size of parents increased, the beak size of offspring increased. This was evolution in action.

> "By studying how evolution occurs in wild populations, we have seen that natural selection can alter a population's traits in a relatively short time."

The evolution for large beaks that occurred during the drought was later reversed. In 1983 abundant rainfall caused small seeds to become more plentiful again. Because birds with smaller beaks handled small seeds more efficiently, individuals with smaller beaks survived better and produced more offspring than individuals with larger beaks. Consequently, the average beak size of the population declined. Once again the population evolved in response to a change in the seeds that were available.

Research on the Galápagos finches continues today. In the 1980s another species, the large ground finch (*G. magnirostris*), arrived from a neighbouring island. The large ground finch is nearly twice as big as the medium ground finch and its larger beak is better at cracking large seeds. Over 20 years, the population of large ground finches grew. In 2003, however, another drought hit the island and made seeds scarce once again. Unlike the previous drought where large seeds were the only abundant seeds remaining, now the large ground finches ate most of the large seeds and left only the scarce small seeds

The medium ground finch. This is one of more than a dozen of Darwin's finches living on the Galápagos Islands. Photo by Juergen & Christine Sohns/Animals Animals/Earth Scenes.

for the medium ground finches. With two species now competing for seeds, both suffered very high mortality. Of the medium ground finches that survived, the average beak size was smaller after the drought than before the drought. Thus, the population continued to evolve.

The decades of research on Darwin's finches have provided tremendous insight into how natural selection operates in nature. By studying how evolution occurs in wild populations, we have seen that natural selection can alter a population's traits in a relatively short time. In this chapter, we will explore the ways in which evolution causes populations to become genetically distinct and how this leads to the origin of new species.

SOURCES: P. R. Grant and B. R. Grant, Evolution of character displacement in Darwin's finches, *Science* 313 (2006): 224–226.

E. Pennisi, Competition drives big beaks out of business, *Science* 313 (2006): 156.

CHAPTER CONCEPTS

- The process of evolution depends on genetic variation.

- Evolution can occur through random processes or through selection.

- Microevolution operates at the population level.

- Macroevolution operates at the species level and higher levels of taxonomic organization.

The story of the finches of the Galápagos Islands demonstrates that evolution shapes the form and function of organisms according to properties of their environments. Evolution depends on genetic variation and this variation can arise from a number of processes. Over time, populations and species can evolve changes in traits, such as the size and shape of bird beaks. These changes can evolve due to random processes or due to the nonrandom process of selection. Some of the most important sources of natural selection include differences in physical conditions, food resources, and interactions with competitors, predators, pathogens, and individuals of the same species. In this chapter, we will examine these processes and explore how genes and the environment come together to cause the evolution of populations and new species.

The process of evolution depends on genetic variation

In Chapter 4 we discussed how the traits expressed by an individual are the outcome of genotypes and the environment interacting. As we saw in the example of finch beaks, when genetic variation is present, it allows evolution by natural selection. In this section we review the structure of DNA, the process of how genes help determine the phenotypes of organisms, and the process by which variation in genes is produced.

THE STRUCTURE OF DNA
Genetic information is contained in the molecule **deoxyribonucleic acid**, also known as **DNA**—a molecule composed of two strands that are wound together into a shape known as a double helix. Each strand is composed of subunits called *nucleotides* and each nucleotide is composed of a sugar, a phosphate group, and one of four different nitrogenous bases: adenine (A), thymine (T), cytosine (C), and guanine (G). Just as a sequence of letters signifies a particular meaning, or word, genetic information is coded in the particular order of the different nitrogenous bases. Long strands of DNA are wound around proteins into compact structures called **chromosomes**.

GENES AND ALLELES
Genes are regions of DNA that code for particular proteins, which in turn affect particular traits. Different forms of a particular gene are referred to as **alleles**. In diploid organisms—those having two sets of chromosomes—one allele comes from the mother's gamete and the other comes from the father's gamete. As you may recall, each gamete is haploid, meaning that it has just one set of chromosomes.

In many cases, a change in alleles can create differences in an organism's phenotype. The ABO blood groups in humans, for example, are determined by which allele a person inherits from each parent—

Deoxyribonucleic acid (DNA) A molecule composed of two strands of nucleotides that are wound together into a shape known as a double helix.

Chromosomes Compact structures consisting of long strands of DNA that are wound around proteins.

Alleles Different forms of a particular gene.

A, B, or O. The allele is responsible for production of the *antigens* A and B, which are molecules on the surface of our red blood cells that interact with the immune system. (Note that the O allele does not produce an antigen.) Individuals with blood type A have AA or AO genotypes; individuals with blood type B have BB or BO genotypes. All remaining individuals have either AB or OO genotypes. In this case, the link between the genotype and the phenotype is direct, and the pattern of inheritance is straightforward. For example, children of an AA father and a BB mother will all have the AB genotype.

Whereas blood type is determined by different alleles on a single gene, **polygenic** traits reflect the effects of alleles from several genes. For example, eye colour in humans is determined by at least three genes that control pigments in different parts of the iris of the eye. Patterns of phenotype inheritance that depend on interactions among multiple alleles can be quite complex.

Many phenotypes in a population can span a range of values because they are polygenic traits. Body size is a good example. Most populations exhibit a normal, or bell-shaped, distribution of body sizes, as shown in Figure 7.1. In this distribution, most individuals cluster near the middle of the range with progressively fewer located toward the upper and lower extremes. Part of this continuous variation might be due to environmental differences, such as the amount of resources available. However, much of the variation can be attributed to the actions of many genes, each with a relatively small influence on the value of the trait. If several genes influence body size, an individual's size will depend on the mix of alleles for all of these genes. The tendency for individuals to be concentrated toward the center of the distribution reflects the relative improbability of an individual inheriting mostly alleles that code for large body size or mostly alleles that code for small body size. Think of this in terms of flipping pennies. The chance of getting 10 tails in a row (about 1 in 1000) is much more remote than the chance of getting 5 heads and 5 tails (about 1 in 4).

Whereas some genes only affect a single trait, such as size, other genes affect multiple traits, an effect referred to as **pleiotropy**. For example, chickens have a gene—known as the frizzle gene—which causes feathers to curl outward rather than to lay flat along the body. However, this gene causes other variations, including faster metabolism, slower digestion, and less frequent egg laying. When a gene has pleiotropic effects, any changes in the gene can have far-reaching effects on the traits of organisms.

In some cases, the expression of one gene can be controlled by other genes. This is known as **epistasis**. In the case of mouse hair colour, for example, there is a gene that determines whether a mouse will make

Figure 7.1 Frequency distribution of polygenic traits. When a continuous trait is determined by many genes, the distribution follows a bell-shaped curve. For the heights of adult males in the United States, for example, there is a symmetrical distribution around the mean. Most individuals express an intermediate trait value whereas only a few individuals, such as professional basketball players and jockeys, express extreme trait values. Data from U.S. Census 2000.

black or brown hair pigments. However, there is a second gene that determines whether the hair will receive any pigments at all. If alleles in the second gene prevent pigments from being deposited in the hair, the alleles of the first gene become irrelevant and the mouse will have a white coat.

DOMINANT AND RECESSIVE ALLELES

Every individual has two copies of each gene, one inherited from its mother and one from its father. Exceptions to this rule include genes located on sex chromosomes, the genes of organisms that reproduce by self-fertilization, haploid organisms, and organisms such as plants that alternate between haploid and diploid generations as part of their normal life cycle. An individual with two different alleles of a particular gene is said to be **heterozygous** for that gene, as in the case of a person with the AB blood type. An individual with two identical alleles is said to be **homozygous**, for example a person with AA blood type. When an individual is heterozygous, the two different alleles may produce an intermediate phenotype, as in the case of a person with the AB blood type that expresses both

---

Polygenic When a single trait is affected by several genes.

Pleiotropy When a single gene affects multiple traits.

Epistasis When the expression of one gene is controlled by another gene.

Heterozygous When an individual has two different alleles of a particular gene.

Homozygous When an individual has two identical alleles of a particular gene.

Figure 7.2 Codominance. In snapdragons the red and white flowers are homozygous genotypes. The pink flower gets its colour from one red gene and one white gene that are codominant.
Photo by John Kaprielian/Photo Researchers, Inc.

alleles. When both alleles contribute to the phenotype, the alleles are said to be **codominant**. Codominance is also found in the flower colour of several species of plants (Figure 7.2). Alternatively, one allele may mask the expression of the other. In this case, the allele that is expressed is called **dominant** and the allele that is not expressed is called **recessive**. In domestic pigs, for example, the allele for white coat colour is dominant and the allele for black coat colour is recessive.

Fortunately, most harmful alleles are recessive and are therefore not expressed in a heterozygous individual. Any dominant harmful alleles that might arise are expressed in homozygous or heterozygous individuals. Because they reduce fitness, dominant harmful alleles are strongly selected against and removed from the population over time. In contrast, recessive harmful alleles are expressed in homozygous individuals but not in heterozygous organisms, so they can persist in a population because they are not selected against when they occur in heterozygous individuals. Examples of recessive harmful alleles in humans include those that cause sickle cell anemia and cystic fibrosis.

A **gene pool** consists of the alleles from all of the genes of every individual in a population. The gene pools of most populations that reproduce sexually contain substantial genetic variation. With the ABO blood type gene, for example, the human population of the United States includes 61 percent O alleles, 30 percent A alleles, and 9 percent B alleles. The proportions of these alleles vary among populations. People of Asian descent, for instance, tend to have higher frequencies of B alleles while people of Irish descent have higher frequencies of O alleles.

SOURCES OF GENETIC VARIATION

Now that we understand the role of genes and alleles, we need to review how we obtain genetic variation in

the traits of organisms. One of the most common ways to generate variation is through sexual reproduction. By combining a haploid sex cell of one parent with a haploid sex cell of another parent, new combinations of alleles can be produced in the offspring across many different chromosomes. The chromosomes in a haploid gamete are a **random assortment** of the chromosomes in the parent's diploid cells, meaning that they can be any combination of chromosomes that the parent received from its mother and father. When an individual produces an egg cell, for example, some chromosomes in the egg will have come from the individual's father whereas other chromosomes in the egg will have come from the individual's mother. As we will see in Chapter 9, the creation of new gene combinations through sexual reproduction represents a major strategy for species to create offspring that are resistant to rapidly evolving parasites and pathogens.

Two additional ways genetic variation arises are through *mutation* and *recombination*. **Mutation** is a random change in the sequence of nucleotides in

---

Codominant When two alleles both contribute to the phenotype.

Dominant An allele that masks the expression of the other allele.

Recessive An allele whose expression is masked by the presence of another allele.

Gene pool The collection of alleles from all individuals in a population.

Random assortment The process of making haploid gametes in which the combination of alleles that are placed into a given gamete could be any combination of those possessed by the diploid parent.

Mutation A random change in the sequence of nucleotides in regions of DNA that either comprise a gene or control the expression of a gene.

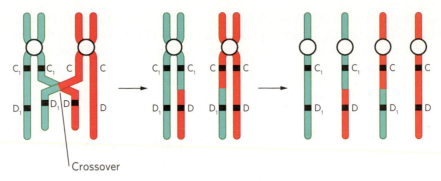

Crossover

Figure 7.3 Genetic variation through recombination. During meiosis in eukaryotic organisms, pairs of homologous chromosomes line up together. When the chromosomes exhibit crossing over, they exchange DNA and each chromosome contains a new combination of genes.

regions of DNA that either comprise a gene or control the expression of a gene. Mutations can occur anywhere along the chromosomes, although some regions of the chromosome can experience higher frequencies of mutation than others. Many mutations have no detectable effect and are referred to as *silent,* or *synonymous,* mutations. Other mutations may simply alter the appearance, physiology, or behaviour of the individual. When phenotypic changes are better suited to the environment, these phenotypes will be favoured by natural selection. Some mutations, however, can cause drastic, often lethal, changes in the phenotype. Many human genetic disorders, such as sickle cell anemia, Tay-Sachs disease, cystic fibrosis, and albinism, as well as tendencies to develop certain cancers and Alzheimer's disease, are caused by single-nucleotide mutations of individual genes.

Genetic **recombination** is the reshuffling of genes that can occur as DNA is copied during *meiosis,* the process that creates haploid gametes from diploid parent cells. During meiosis, pairs of *homologous chromosomes*—one member of which is inherited from each parent—line up next to each other. When the two chromosomes in the pair do not exchange any DNA, we end up with haploid cells that contain unaltered chromosomes. However, sometimes the two chromosomes in the pair do exchange DNA, in a process known as *crossing over,* as shown in Figure 7.3. In some cases, crossing over can also occur between *nonhomologous chromosomes.* In either case, new genes are not being created, but new combinations of alleles are produced that have the potential to produce new phenotypes.

One of the best-known examples of recombination involves the immune system of vertebrates. Vertebrates face a diversity of pathogens that continually evolve so that they become better at attacking their hosts. To combat these ever-changing pathogens, vertebrates need an ever-changing

Figure 7.4 A Kermode bear. Also called spirit bear, these bears live on islands off British Columbia's northwest coast. White fur may offer an advantage when hunting for salmon in late summer and fall. Photo by FLPA-Ignacio Yufera/age fotostock.

immune system that can identify and destroy the pathogens. Recombination provides the mechanism to create the high genetic variation in the immune system that vertebrates need to match the rapid evolution of their parasites. We will talk much more about this issue in Chapter 9.

Evolution can occur through random processes or through selection

On islands off the coast of British Columbia lives a subpopulation of black bears known as spirit bears (*Ursus americanus kermodei*) that contains individuals with either black or white fur (Figure 7.4). The spirit bears, also known as Kermode bears, are white when they are homozygous for a recessive allele of a single gene. The frequency of this recessive

Recombination The reshuffling of genes that can occur as DNA is copied during meiosis and chromosomes exchange genetic material.

allele varies among islands; for example, 21 percent of the bears on Roderick Island carry the recessive allele, while 57 percent of the bears on Gribbell Island carry it. As a result, 10 to 25 percent of the individuals on these islands are white. What has allowed this recessive allele to persist through time? It does not appear to provide a selective advantage as camouflage in the winter because the bears hibernate then. However, researchers have found that the white individuals are much more successful at catching salmon, especially on cloudy days when their white coats blend in better against a background of white clouds that make it harder for salmon to see them. This hypothesis was supported when researchers observed that salmon were twice as evasive when a human on the shoreline was dressed in black than when the person was dressed in white. Collectively, these data suggest that the recessive allele for white fur colour persists because it provides a selective advantage when it is expressed in a homozygous individual.

The story of the white bears demonstrates how evolution often happens through multiple processes. Random events, such as mutations, may or may not confer fitness advantage when they first appear. If they do, as in the case of the white bears, the mutant can persist in a population.

EVOLUTION THROUGH RANDOM PROCESSES

As we saw with the mutation that causes white deer, random processes can facilitate evolutionary change in a population. In addition to mutation, random processes include *genetic drift, bottleneck effects,* and *founder effects.*

Mutation

We have seen that mutation is one of the two primary ways that genetic variation arises in a population. Because genes often code for functions that are vital to performance and fitness, mutations that negatively impact these functions are not favoured by selection. However, a small fraction of

mutations can be beneficial. For example, Figure 7.5 illustrates what happened in the black bear population of British Columbia. A mutation for the white coat appeared, which added genetic variation to the population. After the mutation appeared and offered an advantage, the frequency of white bears increased.

Mutation rates vary a great deal in different groups of organisms, but among genes that are expressed and can be observed as altered phenotypes mutation rates range from 1 in 100 to 1 in 1 000 000 per gene per generation. The more genes that a species carries, the higher the probability that at least one gene will experience a mutation. Similarly, the larger the size of the population, the higher the probability that an individual in the population will carry a mutation.

Genetic Drift

Genetic drift, another random process, occurs when genetic variation is lost because of random variation in mating, mortality, fecundity, or inheritance. Genetic drift is more common in small populations because random events can have a disproportionately large effect on the frequencies of genes in the population. But how does one determine whether an evolved phenotype is the outcome of drift versus some other process, such as natural selection? Research on the Mexican cavefish (*Astyanax mexicanus*) provides an answer.

The Mexican cavefish is a species composed of some populations that live in cave streams and other populations that live in surface streams. Although the populations can interbreed, they look quite different. As is the case with many cave-adapted animals, the cave populations have very reduced eyes and reduced pigmentation (Figure 7.6a). However, populations living in surface streams have normal eyes and dark

---

Genetic drift A process that occurs when genetic variation is lost because of random variation in mating, mortality, fecundity, and inheritance.

Mutation ⟶   Selection for the mutant phenotype ⟶

Figure 7.5 Evolution by mutation. Mutations, such as white coat in black bears, can arise in populations. If the mutation confers a fitness benefit, the mutation can increase in frequency in the population over multiple generations.

(a)

(b)

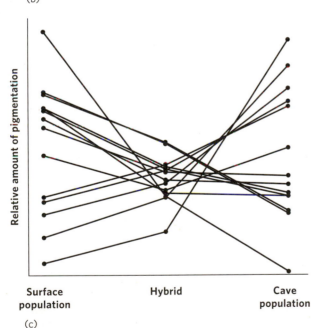

Surface population   Hybrid   Cave population

(c)

Figure 7.6 Evolution by genetic drift. (a) Mexican cavefish populations that live in surface streams have large eyes and dark pigmentation, whereas populations that live in caves have small eyes and reduced pigmentation. (b) When researchers compared how different regions of DNA that code for eye size changed between surface and cave populations, they found that all 12 regions coded for smaller eyes in the cave population than in the surface population. Because all 12 regions changed in the same direction, this suggests that natural selection selected for smaller eyes. (c) When researchers looked at changes in regions of DNA that code for pigmentation, they found 5 regions coded for increased pigmentation while 8 regions coded for decreased pigmentation. Because the 13 regions did not change in the same direction, this suggests that changes in pigmentation were due to genetic drift. Data from M. Protas et al., Regressive evolution in the Mexican cave tetra, *Astyanax mexicanus, Current Biology* 17 (2007): 452–454. Photo courtesy of Dr. Richard Borowsky.

of DNA that coded for large eyes in the surface population all coded for small eyes in the cave population. The results are shown in Figure 7.6b. This suggests that natural selection favoured all of the eye genes to evolve in a similar direction to produce smaller eyes. In contrast, when they examined the 13 regions of DNA that coded for pigmentation, they found that 5 of the regions coded for increased pigmentation in the cave populations and 8 of the regions coded for decreased pigmentation, as shown in Figure 7.6c. The lack of a consistent pattern among the 13 regions of DNA suggests that natural selection was not involved. Instead, the differences in pigmentation in the cavefish populations were likely produced by genetic drift. Given that small populations tend to experience more genetic drift than large populations, it may be the case that the cave population was initially very small.

Bottleneck Effects

A reduction in genetic variation can also occur because of a severe reduction in population size, known as the **bottleneck effect**. When a population experiences a large reduction in the number of individuals, the survivors carry only a fraction of the genetic diversity that was present in the original, larger population. Moreover, after being reduced to a small population by the bottleneck effect, the small population can then experience genetic drift. Population reductions can happen from natural causes, for example a drought that reduces the abundance of food, or anthropogenic causes, such as loss of habitat due to the construction of homes or factories.

For instance, the whooping crane (*Grus americana*) is the tallest bird in North America and the rarest of

pigmentation. To determine whether these changes were due to natural selection or to genetic drift, researchers raised individuals from the cave population, individuals from the surface population, and hybrid offspring made by interbreeding the cave and surface populations. The researchers then examined regions of the fish DNA that coded for eye size and pigmentation, which could contain one or more genes. In 2007, they reported that the 12 regions

Bottleneck effect A reduction of genetic diversity in a population due to a large reduction in population size.

Figure 7.7 The whooping crane. This is North America's tallest bird and one of its most endangered. Because of habitat destruction and hunting, populations declined sharply, which resulted in reduced genetic diversity. Photo by Larry Ditto / DanitaDelimont.com.

the crane species (Figure 7.7). Whooping cranes were once distributed across most of North America and migrated from their Arctic breeding grounds to wintering areas on the Texas coast. Cranes rely on wetlands where they feed on crustaceans, fish, berries, and grain. However, settlers in the late nineteenth and early twentieth centuries drained most of the wetlands suitable for cranes throughout North America and hunted the birds for food and sport. This dramatically reduced the population size from an estimated 1300 birds in 1869 to 88 in 1912. Only 14 individuals remained in 1938. Efforts to save the species in Canada and the United States have recently increased the whooping crane population in the wild to 382 individuals. However, the bird is still considered an endangered species by the Canadian Wildlife Service and the United States Fish and Wildlife Service.

To determine whether the dramatic decline in population size was associated with a decline in genetic diversity, researchers gathered DNA samples from museum specimens that had been collected before 1923. They defined this time period as "pre-bottleneck." The researchers also collected samples from birds after 1971, which they defined as the "post-bottleneck" period. The analyses showed that genetic diversity decreased significantly. From at least nine available combinations of alleles in the pre-bottleneck population, only three combinations remained in the current population.

The bottleneck effect is of particular interest because the subsequent reduction in genetic diversity may prevent the population from adapting to future environmental changes. This is especially true for organisms that face deadly pathogens. An inability to evolve against new strains of a pathogen could lead to

the extinction of the host organism. For example, the African cheetah (*Acinonyx jubatus*) experienced a population bottleneck approximately 10 000 years ago. Although the cause of that bottleneck is unknown, the current population contains very little genetic variation. This low genetic variation makes cheetahs more vulnerable to pathogens, including a deadly pathogen that causes the disorder known as AA amyloidosis and kills up to 70 percent of cheetahs held in captivity.

Founder Effects

The **founder effect** occurs when a small number of individuals leave a large population to colonize a new area and bring with them only a small amount of genetic variation. Following the founding by this small population, genetic drift can cause additional reductions in genetic variation. The genetic variation remains low until enough time has passed to accumulate new mutations. We can see an example of the founder effect in the Saguenay–Lac-Saint-Jean region of northeastern Quebec. More than three centuries ago, a small group of settlers, carrying only a small fraction of the genetic diversity of the European population they had left behind, founded communities in the Saguenay–Lac-Saint-Jean area. Inhabitants of this region speak French and share a distinct cultural and historical heritage. In modern time the population continues to exhibit relatively high incidences of more than a dozen rare recessive diseases compared to the general Canadian population. For example,

Founder effect When a small number of individuals leave a large population to colonize a new area and bring with them only a small amount of genetic variation.

pseudovitamin D-deficiency rickets (PDDR) is a disorder that causes the softening and weakening of bones and slows skeletal growth and development. PDDR typically occurs in 1 out of every 500 Canadians, but shows up in 1 out of every 26 inhabitants of the Saguenay–Lac-Saint-Jean region. There is even a disease that is named after the region—Charlevoix-Saguenay spastic ataxia—since it was first described in this area. What could be the reason for such a high incidence of diseases that are typically quite rare?

Many regions in Quebec have extensive population registers with information such as births, deaths, and marriages. Some of these registers reach back to the first settlers who arrived from Europe. Using these registers we can reconstruct details of the relationships among different settlers. Researchers have estimated that the approximately 30 000 current Saguenay–Lac-Saint-Jean residents are descendants from a relatively small founder population of about 400 settlers who came to the region in the seventeenth century. Molecular and genetic data indicate that the mutations responsible for the rare disorders were present in the small group of original settlers. Moreover, the average relatedness of the Saguenay–Lac-Saint-Jean population after 10 generations is approximately fourth cousin—a high degree of relatedness. In comparison, in all of Montreal, average relatedness is so low that it is indistinguishable from chance. Additional evidence of the close kinship comes from the fact that 69 percent of patients, and 85 percent of carriers of PDDR, share identical genetic markers. These patterns confirm that the small group of founders who settled in the Saguenay–Lac-Saint-Jean region contained a small proportion of the genetic diversity of the European population that they had left and, by chance, they carried a variety of recessive, disease-causing alleles.

EVOLUTION THROUGH SELECTION

The nonrandom process of *selection* also plays a substantial role in evolution. **Selection** is the process by which certain phenotypes are favoured to survive and reproduce over other phenotypes. As we saw in the story of the medium ground finches at the beginning of this chapter, selection is a powerful force that can change the phenotypes (and therefore the gene frequencies) of a population in a relatively short period of time. Depending on how the environment varies over time and space, selection can influence the distribution of traits in a population in three ways: *stabilizing selection, directional selection,* and *disruptive selection.*

Stabilizing Selection

When individuals with intermediate phenotypes have higher survival and reproductive success than those

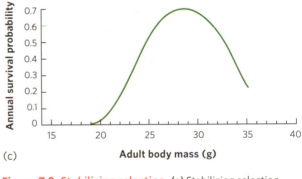

Figure 7.8 Stabilizing selection. **(a)** Stabilizing selection favours intermediate phenotypes and selects against both extremes. **(b)** In the sociable weaver bird, body size has a normal distribution. **(c)** The bird experiences stabilizing selection on body size because birds with intermediate body sizes experience high survival while birds with low and high body sizes have low survival. This selection for intermediate phenotypes would cause the next generation of birds to have a narrower distribution of phenotypes. Data from R. Covas et al., Stabilizing selection on body mass in the sociable weaver *Philetairus sociu., Proceedings of the Royal Society of London Series B 269* (2002): 1905–1909.

with extreme phenotypes, we call it **stabilizing selection**. As shown in Figure 7.8a, stabilizing selection begins with a relatively wide distribution of

Selection The process by which certain phenotypes are favoured to survive and reproduce over other phenotypes.

Stabilizing selection When individuals with intermediate phenotypes have higher survival and reproductive success than those with extreme phenotypes.

phenotypes, as illustrated by the orange line. After stabilizing selects for parents who possess intermediate phenotypes, their progeny have a more narrow distribution of phenotypes, as illustrated by the blue line. In doing so, it performs genetic housekeeping for a population, sweeping away harmful genetic variation. An example of stabilizing selection can be seen in selection for body mass in a species of bird from South Africa called the sociable weaver (*Philetairus socius*). Over an 8-year period, researchers marked nearly 1000 adult birds and examined how a bird's mass was related to its subsequent survival. As you can see in Figure 7.8b, the mass of the adults in the study follows a normal distribution with a mean of approximately 29 g. The researchers then asked how well birds of different mass survive. When mass was graphed against survival, as shown in Figure 7.8c, they found that the smallest and largest birds survived poorly, whereas birds of intermediate mass survived the best. That is, selection favoured the intermediate phenotype. When the environment of a population is relatively unchanging, stabilizing selection is the dominant type of selection. Because the average phenotype does not change, little evolutionary change takes place.

Directional Selection

At the beginning of this chapter, we saw that beak size of the medium ground finch evolved to be larger during a drought when only the large seeds were available. This is an example of **directional selection**, which occurs when an extreme phenotype experiences higher fitness than the average phenotype of the population, as shown in Figure 7.9a. In the medium ground finch, for example, Peter and Rosemary Grant quantified the distribution of beak sizes in offspring hatched in 1976, which was just prior to a drought. As you can see in Figure 7.9b, the beak sizes of these offspring followed a normal distribution with a mean beak depth of 8.9 mm. When the drought came, although all seeds became less abundant, there were proportionately more large seeds remaining. The large seeds are harder to crack, so birds with deeper beaks were better able to feed on hard seeds and they had better survival. Because beak depth is a heritable trait, the offspring that were hatched in 1978 had deeper beaks, as shown in Figure 7.9c.

Disruptive Selection

Under some circumstances, we see another type of selection known as *disruptive selection*. In **disruptive selection**, individuals with extreme phenotypes at either end of the distribution can have higher fitness

---

Directional selection When individuals with an extreme phenotype experience higher fitness than the average phenotype of the population.

Disruptive selection When individuals with either extreme phenotype experience higher fitness than individuals with an intermediate phenotype.

(a)

(c)

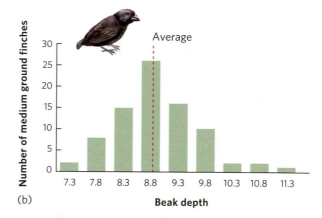

(b)

Figure 7.9 Directional selection. (a) Directional selection favours phenotypes of one extreme and selects against phenotypes of the other extreme. **(b)** Prior to a drought in 1976, the beak size in offspring of the medium ground finch had a mean depth of 8.9 mm, as indicated by the dashed red line. During the drought, when mostly large seeds were available, birds with larger beaks survived better. **(c)** Two years later, finch offspring had a mean beak depth of 9.7 mm, confirming that larger seeds caused directional selection for larger beaks. Data from R. Grant and P. Grant, What Darwin's finches can teach us about the evolutionary origin and regulation of biodiversity, *BioScience* 53 (2003): 965–975.

(a)

(c)

(b)

Figure 7.10 Disruptive selection. (a) Disruptive selection favours both extreme phenotypes and selects against intermediate phenotypes. **(b)** In the tadpoles of the spadefoot toad, an individual can have specialized mouthparts for carnivory and cannibalism (shown in this photo), omnivory, or an intermediate phenotype. **(c)** When more than 500 tadpoles of each phenotype were marked and released, about 10 percent were recovered. Of those recovered, the omnivores and carnivores experienced relatively high survival while the intermediate phenotypes experienced lower survival. Data from R. A. Martin and D. W. Pfennig, Disruptive selection in natural populations: The roles of ecological specialization and resource competition, *American Naturalist* 174 (2009): 268–281. Photo by Thomas Wiewandt/wildhorizons.com.

than individuals with intermediate phenotypes. Disruptive selection is illustrated in Figure 7.10a. For example, tadpoles of the Mexican spadefoot toad (*Spea multiplicata*) can express a range of possible phenotypes that are related to what they eat. At one extreme is the omnivorous phenotype that has small jaw muscles, numerous little teeth, and a long intestine that makes it well suited for feeding on detritus. At the other extreme is the carnivorous phenotype that has large jaw muscles, a notched mouthpart, and a short intestine that makes it well suited for feeding on freshwater shrimp and cannibalizing conspecifics. Intermediate phenotypes are not well suited to feed on either of the two food types. To test whether the tadpoles experienced disruptive selection, researchers collected more than 500 tadpoles from a desert pond, marked them to indicate their phenotype, and then returned them back to their pond. They sampled the pond again 8 days later to determine the survival of the three phenotypes. As you can see in Figure 7.10b, the omnivorous and carnivorous phenotypes survived relatively well, but the intermediate phenotypes—which had intermediate jaw muscles, tooth number, and intestine length—survived poorly. Because disruptive selection removes the intermediate phenotypes, it increases genetic and phenotypic variation within a population. In doing so, it creates a

distribution of phenotypes with peaks toward both ends of the original distribution.

Microevolution operates at the population level

The random and nonrandom processes that cause evolution can operate at a variety of levels. The evolution of populations is known as **microevolution**, and it is pervasive. It is the process responsible for producing different breeds of cats, cattle, and dogs, and for producing distinct populations of wild organisms including salmon, bears, and the flu virus. Microevolution is affected by both random process and selection. Selection at the microevolution level can be further divided into *artificial selection* and *natural selection*.

ARTIFICIAL SELECTION

In his book *On the Origin of Species,* Charles Darwin discussed the wide variety of domesticated animals that humans have bred to produce particular suites of traits. In the case of dogs, for example, humans began by domesticating grey wolves. Over time they bred individuals that possessed particular characteristics, such as body size, coat colour, and

Microevolution The evolution of populations.

ANALYZING ECOLOGY

Strength of Selection, Heritability, and Response to Selection

Researchers often wish to know exactly how much selection will move the mean phenotype in a population. For example, if a plant breeder selects for larger tomato plants, she might want to know how much larger the next generation will be. Similarly, a government agency that regulates fishing might want to know if harvesting only the largest individuals might cause the population to evolve to be smaller in the next generation.

Let's consider the case of directional selection in which one end of the phenotypic distribution is favoured. If there is selection for more extreme phenotypes and the phenotype has a genetic basis, directional selection will cause the mean phenotype to change. Can we determine exactly how much the mean phenotype will change in the next generation? To answer this question we need to know both the *strength of selection* of the phenotype and the *heritability* of the phenotype.

The **strength of selection** is the difference between the mean of the phenotypic distribution before selection and the mean after selection, measured in units of standard deviations (see Chapter 2). For example, imagine that we wanted to select for larger tomatoes. The phenotype (tomato mass) follows a normal distribution with a mean of 100 g and a standard deviation of 10 g. Now imagine that we select the upper end of the distribution and use these individuals to breed the next generation of tomatoes. If this selected group has a mean of 115 g, our selected group has a mean that is 1.5 standard deviations away from the mean of the entire population. Thus, the strength of selection is 1.5.

We also know that phenotypes are the products of genes and the environment. Since only genetic material can be passed down to the next generation, if we wish to know how much the mean phenotype will change, we have to determine what proportion of the total phenotypic variation is caused by genes. This proportion is called the **heritability** and it can range between 0 and 1. If all of the phenotypic variation that we see in a normal distribution is due to the environment, the heritability is 0. If all of the phenotypic variation is due to genetic variation, the

heritability is 1. By convention, the symbol for heritability is h^2. (This notation may be confusing because nothing is being squared.)

Using the concepts of strength of selection and heritability, we can build an equation that describes how much a population will respond to selection in the next generation. Because a population's response to selection is a function of the strength of selection and the heritability of the phenotype

$$R = S \times h^2$$

where R is the response to selection, S is the strength of selection, and h^2 is the heritability.

Using our tomato example, we can calculate how much larger the tomatoes should be in the next generation. If we select parents that are 1.5 standard deviations above the population mean and if the heritability is 0.33, then

$$R = 1.5 \times 0.33 = 0.5$$

which means that the mean phenotype of the next generation of tomatoes will be 0.5 standard deviations—or 5 g—greater than that of the parent's generation.

> **YOUR TURN** Given the following values for the strength of selection and heritability on tomato mass, calculate the expected response to selection in units of standard deviations and grams:

S	h^2
0.5	0.7
1.0	0.7
1.5	0.7
2.0	0.9
2.0	0.6
2.0	0.3
2.0	0.0

Based on your calculations, how is the response to selection affected by the strength of selection and heritability?

Strength of selection The difference between the mean of the phenotypic distribution before selection and the mean after selection, measured in units of standard deviations.

Heritability The proportion of the total phenotypic variation that is caused by genetic variation.

hunting ability. As shown in Figure 7.11, just a few centuries of breeding have created dog breeds with widely divergent phenotypes—from Saint Bernards to Chihuahuas. All of these dogs belong to the same species as the wolf and could potentially interbreed. This is an example of **artificial selection**, in which humans decide which individuals will breed and the breeding is done with a preconceived goal for the traits desired in the population. Similar artificial selection has occurred to create numerous breeds of other domesticated animals, including cattle, sheep, pigs, and chickens.

Artificial selection also has been applied to plants. One of the best-known examples is the breeding of wild mustard (*Brassica oleracea*). As you can see in Figure 7.12, the wild mustard has been bred into a diversity of vegetables by selecting for unique stem, leaf, and flower traits. Today, the wild mustard can be consumed as cabbage, Brussels sprouts, cauliflower, broccoli, kale, and kohlrabi.

Human practices may unintentionally lead to artificial selection, sometimes with detrimental consequences. For example, the widespread application of pesticides has caused resistance in more than 500 species of pests that harm food production and human health. Similarly, extensive use of antibiotics has caused many harmful human pathogens to evolve resistance to antibiotics, as described in "Drug-Resistant Tuberculosis" at the end of this chapter. In these cases, the role of evolutionary mechanisms is clear. When pesticides or antibiotics are targeted at millions of organisms, a small number of individuals commonly carry a mutation that confers resistance. Since only the mutants survive and the mutation is heritable, the next generation becomes more resistant.

Artificial selection Selection in which humans decide which individuals will breed and the breeding is done with a preconceived goal for the traits of the population.

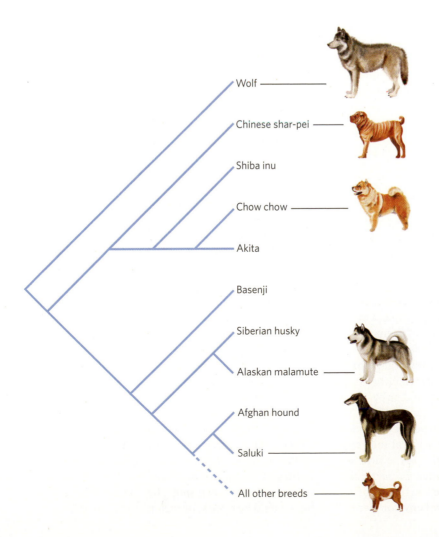

Figure 7.11 Breeds of domestic dogs. Beginning with domesticated individuals of the grey wolf, humans have created a wide diversity of dog breeds through the process of artificial selection. Data from H. G. Parker et al., Genetic structure of the purebred domestic dog, *Science* 304 (2004): 1160–1164.

Figure 7.12 Artificial selection on wild mustard. Over the years, plant breeders have produced a variety of common vegetables through artificially selecting for different traits of the wild mustard.

Darwin's case for evolution by natural selection was strengthened by his observations of how artificial selection worked. He argued that if humans could produce such a wide variety of animal and plant breeds in a few centuries through artificial selection, natural selection could certainly have had similar effects over millions of years.

NATURAL SELECTION

A person conducting artificial selection typically has a particular suite of traits in mind, for example, increased milk production in cattle. This is not the case for natural selection, which favours any trait combination that provides higher fitness to an individual. Artificial and natural selection both operate by favouring certain traits over others. Both select on traits that are heritable; the difference is in how traits are selected. There can be multiple ways to improve an individual's fitness, and all of them are favoured by natural selection, regardless of the resulting phenotype. For example, a prey organism might reduce its probability of being eaten either by hiding from predators so it is not detected or by growing spines so that it cannot be eaten. Both strategies are effective at improving the prey's fitness and both traits could be favoured by natural selection. In artificial selection, humans determine the fitness of traits and they often select attributes for particular purposes that would actually reduce the fitness of individuals if they lived in the natural environment.

Most evolutionary biologists agree that the diversification of organisms over the long history of life on Earth has occurred primarily by natural selection. Natural selection is an ecological process: It occurs because of differences in reproductive success among individuals endowed with different form or function in a particular environment. That is, as individuals interact with their environment— including physical conditions, food resources, predators, other individuals of the same species, and so on—traits that lead to greater fitness in that environment are passed on.

Evolution by natural selection is a common phenomenon in populations. For example, many predators cause selection on the traits of their prey. We can see this process in Figure 7.13 for fish that prey upon amphipods, a tiny crustacean species. The parent amphipods produce an abundance of offspring that vary in size. However, fish prefer to consume the largest amphipods because they offer the highest amount of energy per unit effort. Smaller amphipods are more likely to survive, and since body size is a heritable trait in amphipods, subsequent generations evolve smaller body sizes.

One of the most striking demonstrations of microevolution is the case of the peppered moth (*Biston betularia*). During the early nineteenth century in England, most individuals of this moth were white with black spots, but occasionally there would be a dark, or *melanistic,* individual

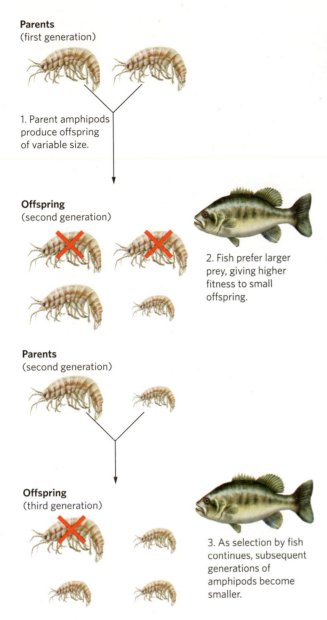

Parents
(first generation)

1. Parent amphipods produce offspring of variable size.

Offspring
(second generation)

2. Fish prefer larger prey, giving higher fitness to small offspring.

Parents
(second generation)

Offspring
(third generation)

3. As selection by fish continues, subsequent generations of amphipods become smaller.

Figure 7.13 Natural selection by predators on prey. The amphipod, a tiny crustacean, produces an abundance of offspring that vary in size. Predatory fish prefer to eat the largest amphipods and this causes selection for small body size.

(Figure 7.14a). Over the next hundred years, dark individuals became more common in forests near heavily industrialized regions, a phenomenon often referred to as **industrial melanism**. In regions that were not industrialized, the light phenotype still prevailed.

Since melanism is an inherited trait, it seemed reasonable to suppose that the environment must have been altered to give dark forms a survival advantage over light forms. The specific agent of selection was easily identified. Peppered moths rest on trees during the day. Scientists observed that air pollution in industrial areas had darkened the trees

with soot, as shown in Figure 7.14b, so they suspected that the predatory birds could see the lighter moths more easily. Because trees in nonpolluted regions were much lighter in colour, the dark moths in these regions would be more visible. To test these hypotheses, equal numbers of light and dark moths were placed on tree trunks in polluted and unpolluted woods. As you can see in Figure 7.14c, when both types of moths were attached to the light-coloured trees in unpolluted regions, birds consumed more dark moths. When both types of moths were attached to the dark-coloured trees in polluted regions, birds consumed more light moths. This confirmed that the change in phenotypes observed over time in England reflected evolution of the population in response to changing environmental conditions.

In recent years, as pollution control programs reduced the amount of soot in the air and improved the conditions of the forests, the frequencies of melanistic moths have decreased, as we would expect. Figure 7.15 shows data for the area around the industrial centre of Kirby in northwestern England. As the amount of pollution declined—measured in terms of sulphur dioxide and shown as a blue line—the bark of the trees began to turn lighter. After two decades of decreasing pollution, the trees became lighter and the frequency of the dark form of the moth decreased from more than 90 percent of the population in 1970 to about 30 percent by 1990, as shown by the black line. Similar to the medium ground finches discussed at the beginning of this chapter, the story of the melanistic moths demonstrates that microevolution can occur in a relatively short time.

Macroevolution operates at the species level and higher levels of taxonomic organization

Whereas microevolution is a process that occurs at the level of the population, **macroevolution** is a process that occurs at higher levels of organization including species, genera, families, orders, and phyla. For our purposes, we will restrict our discussion of macroevolution to the evolution of new species, a process known as **speciation**. The pattern of

Industrial melanism A phenomenon in which industrial activities cause habitats to become darker due to pollution and, as a result, individuals possessing darker phenotypes are favoured by selection.

Macroevolution Evolution at higher levels of organization including species, genera, families, orders, and phyla.

Speciation The evolution of new species.

(a) **Unpolluted woods**

(b) **Polluted woods**

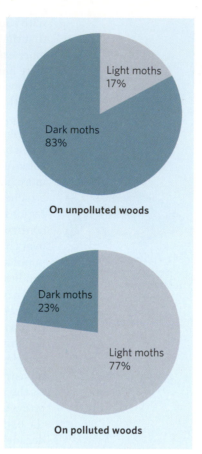

Light moths
17%

Dark moths
83%

On unpolluted woods

Dark moths
23%

Light moths
77%

On polluted woods

(c) **Moths consumed**

Figure 7.14 Selection by bird predation for different moth phenotypes. **(a)** In unpolluted forests, trees have light-coloured bark and moths with the light phenotype are better camouflaged. **(b)** In polluted forests, trees have dark-coloured bark and moths with the dark phenotype are better camouflaged. **(c)** When researchers placed both moth phenotypes on trees in polluted and unpolluted woods, fewer light-coloured moths were consumed by birds on unpolluted trees but fewer dark moths were consumed on polluted trees. Data from B. Kettlewell, Further selection experiments on industrial melanism in the Lepidoptera, *Heredity* 10 (1956): 287–301. Photos by Michael Willmer Forbes Tweedie/Photo Researchers, Inc.

speciation over time can be illustrated using *phylogenetic* trees and speciation can occur in one of two ways: *allopatric speciation* or *sympatric speciation*.

PHYLOGENETIC TREES

Scientists can often document microevolution because it can happen in a relatively short time. In some cases, scientists have monitored wild populations over time to track the evolutionary process. In other cases, we have historical documents that trace the development of domesticated plants and animals. For example, most modern dog breeds are the result of artificial selection over the last 3 centuries and records show the older breeds that gave rise to newer breeds. Understanding how macroevolution has occurred, however, is a much more daunting challenge. Because we cannot travel back in time and there are no written records from millions of years ago, the true patterns of evolution can never be known for certain, although fossils can be helpful when examining the evolution of morphological traits. In the absence of more direct evidence, scientists work

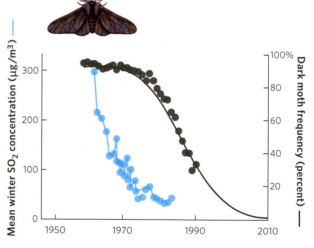

Figure 7.15 **Reversing the effects of pollution.** As industries around Kirby, England, reduced the amount of sulphur dioxide pollution they emitted into the air, the colour of the trees became lighter. After a decade of declining pollution, the frequency of the dark form of the peppered moth began to decline rapidly. Data from C. A. Clarke et al., Evolution in reverse: Clean air and the peppered moth, *Biological Journal of the Linnean Society* 26 (1985): 189–199; G. S. Mani and M. E. N. Majerus, Peppered moth revisited: Analysis of recent decreases in melanic frequency and predictions for the future, *Biological Journal of the Linnean Society* 48 (1993): 157–165.

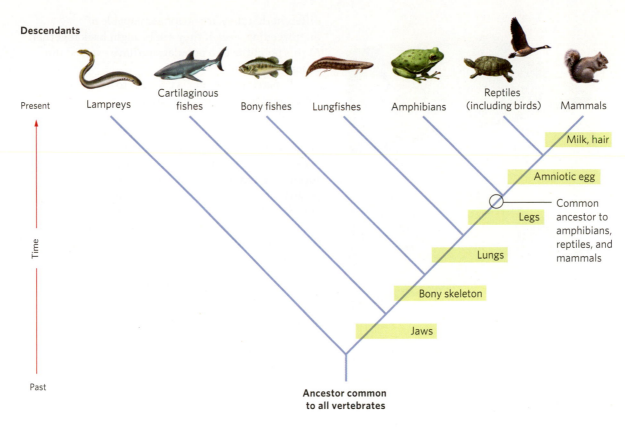

Descendants

Present

Lampreys

Cartilaginous fishes

Bony fishes

Lungfishes

Amphibians

Reptiles (including birds)

Mammals

Milk, hair

Amniotic egg

Legs — Common ancestor to amphibians, reptiles, and mammals

Lungs

Bony skeleton

Jaws

Time

Past

Ancestor common to all vertebrates

Figure 7.16 Phylogenetic tree. By looking for similarity in phenotypes and DNA, scientists can develop hypotheses about the relatedness of different groups of organisms. In this phylogenetic tree of the major vertebrate groups, the yellow boxes indicate points in time at which important new traits evolved.

from the premise that species with the greatest number of traits in common are the most closely related. These traits can include shapes and sizes of structures of living and fossilized organisms as well as the ordering of nitrogenous bases in the DNA of different organisms. To map these relationships, scientists use **phylogenetic trees**, which are hypothesized patterns of relatedness among different groups such as populations, species, or genera. In essence, phylogenetic trees are attempts to understand the order in which groups evolved from other groups. Figure 7.16 shows a phylogenetic tree for several major groups of vertebrates. From this tree, you can see that all vertebrates share a common ancestor. Over time, this ancestor has given rise to fish, amphibians, mammals, and reptiles (including birds).

ALLOPATRIC SPECIATION

Allopatric speciation is the evolution of new species through the process of geographic isolation. Imagine that we start with a single large population of a field mouse, shown as the first stage in Figure 7.17. At some point in time, a portion of the population is separated from the rest. This could occur because some

individuals colonize a new island, such as the first finches to arrive in the Galápagos Islands from South America. Alternatively, the population could be divided by a geographic barrier, such as a new river that cuts through the middle of the terrestrial habitat, a rising mountain range that cannot be crossed, or a large lake that becomes two smaller lakes. In each case, the two populations are isolated from each other, as shown in step 2 of the figure. They are no longer able to interbreed because of physical separation, so each population evolves independently. If one or both of the populations has few individuals, founder effects and genetic drift can strongly influence the direction in which that population evolves. When ecological conditions differ in the two isolated locations, natural selection will cause each population to evolve adaptations that improve fitness under local environmental conditions. Over time, as shown in step 5, the populations can become so

Phylogenetic trees Hypothesized patterns of relatedness among different groups such as populations, species, or genera.

Allopatric speciation The evolution of new species through the process of geographic isolation.

1. Original field mouse population

2. River arises, splitting the population.

3. Over many generations, each population evolves independently.

4. Eventually, the two populations become genetically distinct.

5. Later, even if the river dries up and the two populations come into contact, they may no longer be able to interbreed.

Figure 7.17 Allopatric speciation. When geographic barriers divide populations, each evolves independently. Over time, the two populations can become so different that they are not able to interbreed. At this point they have become two distinct species.

different that they are no longer capable of interbreeding, even if they are brought back together. At this point the two populations have evolved into different species.

Allopatric speciation is thought to be the most common mechanism of speciation. Figure 7.18 shows the process for Darwin's finches using a phylogenetic tree. Darwin hypothesized that the ancestor species of the finches he saw probably came from the mainland of South America. Once the ancestral finch species arrived in the Galápagos Islands, the population grew and eventually colonized many of the islands in the archipelago. The isolation and unique ecological conditions present on each island favoured the process of allopatric speciation. Researchers hypothesize that these conditions gave rise to some of the 14 currently recognized species of finches in the Galápagos Islands, although other species of finches on the islands appear to have evolved through a different process known as *sympatric speciation*.

SYMPATRIC SPECIATION

In contrast to allopatric speciation, **sympatric speciation** gives rise to new species without geographic isolation. In some cases, species evolve into a diversity of new species within a given location. An example of this is the pairs of stickleback species living in some of British Columbia's coastal lakes. Sticklebacks are small fish that live in marine, estuarine, or freshwater environments. In a few of these lakes, you can find two species that have distinct morphology, colouration, and behaviour (Figure 7.19). The benthic species is stout and feeds on gastropods and arthropods at the bottom of the lake. The limnetic species is slender, features pronounced spines, and feeds on zooplankton throughout the limnetic zone. Both species originated from the same large marine stickleback species. The benthic species reached the lakes first during a period of high sea levels about 12 500 years ago. When more marine sticklebacks gained access to these lakes during a second period of high sea levels a thousand years later, they encountered the descendants of the first invasion. Competition between these two forms led to the sympatric evolution of the benthic and limnetic species of sticklebacks in these lakes.

A common way that sympatric speciation can occur in some types of organisms is through *polyploidy*. **Polyploid** species, which contain three or more sets of

Sympatric speciation The evolution of new species without geographic isolation.

Polyploid A species that contains three or more sets of chromosomes.

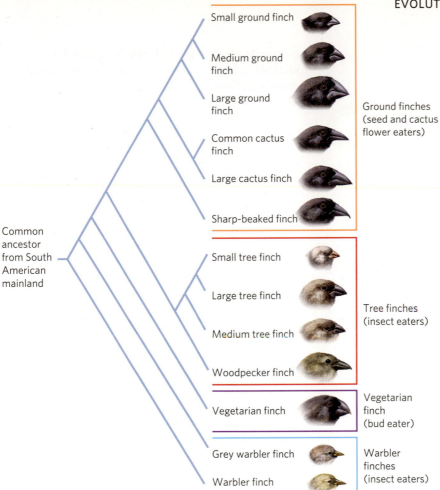

Figure 7.18 Allopatric speciation in Darwin's finches. Through allopatric speciation, a single ancestor species from the South American mainland has evolved into 14 different species of finches on the Galápagos Islands. (Note that the phylogenetic study was done on only 13 of the 14 species.)

chromosomes, arise when homologous chromosomes fail to separate properly during meiosis, with the result that gametes are diploid rather than haploid. If a diploid egg, for example, is fertilized by a haploid sperm, the resulting zygote will contain three sets of

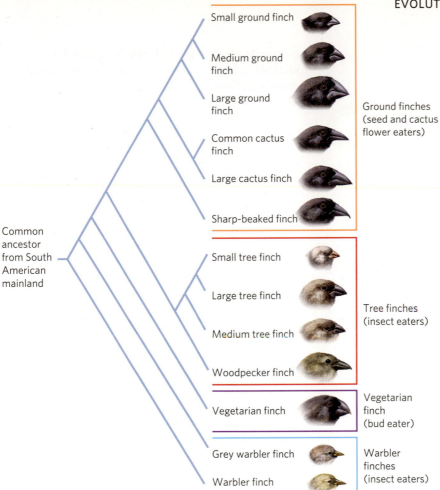

Figure 7.19 Sympatric speciation. Benthic sticklebacks differ from limnetic forms in morphology, colouration, and behaviour. A benthic gravid female is shown at top, and a limnetic gravid female below. Photo by Todd Hatfield.

chromosomes. At this point, the organism is a polyploid. Because it now has more than two sets of chromosomes, it is unable to breed with any diploid individuals. Thus, when a polyploid is formed, it instantly becomes a species that is genetically distinct from its parents. Several species of insects, snails, and salamanders are polyploid, as are 15 percent of all flowering plant species.

An interesting example of polyploidy can be found in a group of salamanders. The blue-spotted salamander (*Ambystoma laterale*) and the Jefferson salamander (*A. jeffersonianum*) are both diploid species. As Figure 7.20 illustrates, at some point in the past, a blue-spotted salamander experienced incomplete meiosis and accidentally produced a diploid gamete. It then mated with a Jefferson salamander, which produced a normal haploid gamete. The resulting offspring was a triploid salamander that instantly became a distinct species known as Tremblay's salamander (*A. tremblayi*). The Tremblay's salamander is an all-female species that produces clonal daughters. These salamanders will breed with another salamander species to stimulate their own reproduction, but they can make daughters without incorporating DNA from any of the other species. If they do incorporate the haploid sperm from

Blue-spotted
salamander

Tremblay's
salamander

Jefferson's
salamander

Figure 7.20 Polyploidy in salamanders. Triploid species can occur when one individual that experiences incomplete meiosis and produces a diploid gamete mates with another individual that experiences normal meiosis and produces a haploid gamete. The Tremblay's salamander is an all-female, triploid species that arose through sympatric speciation from the mating of the blue-spotted salamander and the Jefferson salamander.

a male of the other species, their offspring can then carry four sets of chromosomes, which would make them *tetraploids*.

Plant breeders have developed techniques that cause polyploidy to produce more desirable characteristics in plants. This is a form of artificial selection at the species level. By exposing plants to sudden cold temperatures at the time of reproduction, they can increase the chances of a plant producing gametes that are diploid rather than haploid. Polyploid plants tend to be larger and produce bigger flowers and fruits. Many beautiful flowers you see in a florist shop are the product of human-induced polyploidy. Many crops are also polyploids, including watermelons, bananas, strawberries, and wheat. As you can see in Figure 7.21, plant breeders used a species of wheat that has two sets of chromosomes to develop new species that have four or six chromosomes. The more sets of chromosomes in wheat, the larger the plant and the larger the seeds.

KEY INNOVATIONS

Sometimes evolution produces unusual traits that prove to be highly useful because they allow the species to exploit new niches and to undergo speciation at a high rate. For example, when the ancestor of bats first evolved wings, it opened up a whole new set of ecological opportunities for these mammals, including the ability to catch flying insects and to collect nectar and pollen from flowers. Today, bats are the second most species-rich order of

mammals, after rodents. Similarly, the evolution of C_4 and CAM photosynthesis allowed flowering plants to grow in regions with scarce water, as we discussed in Chapter 3.

In 2011, biologists discovered a fundamental connection between horns and wings in insects that represents a key innovation. Entomologists had long marvelled at the prominent horn-like structures present in treehopper insects, as shown in Figure 7.22. These horns serve a wide variety of functions including mimicking plant parts—such as twigs and thorns—or mimicking other insects, including aggressive ants. You may remember that insects have a pair of legs on each of the three segments of their thorax. Most insects also have a pair of wings on the second and third segments of their thorax, but not on the first segment. However, fossils indicate that ancient insects commonly had wings on their first segment. The treehopper researchers discovered that the horns of the treehoppers always emerge from the first segment of the thorax, the location where wings emerged in their ancestors. Moreover, when the horn first begins to grow, it starts as two tiny buds, similar to the beginning of wing development, and these buds later fuse to form a horn. Finally, the genes that control the development of wings in the second and third segments are also expressed in the first segment of treehoppers. Collectively, the evidence suggests that treehoppers

(a) Einkorn wheat   (b) Durum wheat   (c) Common wheat

Figure 7.21 Polyploidy in wheat. Starting with a diploid species of wheat, plant breeders have bred new species that contain four or six sets of chromosomes. **(a)** The ancestral einkorn wheat (*Triticum boeoticum*) has two sets of chromosomes and small seeds. **(b)** Durum wheat (*Triticum durum*), which is used to make pasta, was bred to have four sets of chromosomes. It has medium-sized seeds. **(c)** Common wheat (*Triticum aestivum*), which is used for bread and other baked goods, was bred to have six sets of chromosomes. It has the largest seeds.

Figure 7.22 The evolution of key innovations.
Treehoppers are a group of insects that possess distinctive horns above their bodies. New research suggests that the horns are a novel modification of ancestral wings.

took advantage of wing genes to develop a key innovation of distinctive horns that serve a wide variety of functions today. While this interpretation has been criticized by other researchers, it shows the possibility of traits changing function through evolutionary processes.

In this chapter, we have seen that the products of evolution are all around us, from our favourite breeds of animals to many of the foods that we love to eat. The processes that cause evolution at the population level are the same processes that cause evolution at the species level. Having a good understanding of evolution is not merely an academic exercise; it has real-world implications for helping humans, including in the fight against disease.

ECOLOGY TODAY CONNECTING THE CONCEPTS

DRUG-RESISTANT TUBERCULOSIS

Tuberculosis in Inuit communities. In Kangiqsualujjuaq, Quebec, more than 30 active cases of TB were reported in less than a year. Photo by Bryan and Cherry Alexander Photography/ ArcticPhoto/All Canada Photos.

Tuberculosis, or TB, is a highly infectious disease caused by a bacterium (*Mycobacterium tuberculosis*). TB has killed people for thousands of years. In 2009, for example, researchers discovered that the preserved tissues of a woman who died 2600 years ago and was mummified have genetic markers of the TB bacterium. Today, experts estimate that nearly one-third of the world's human population is infected by the bacterium, although the bacterium remains inactive and does not cause any problems for most of these people. However, the bacterium does become active in about 9 million people every year. TB causes

extensive tissue damage, weakness, night sweats, and bloody sputum. It is highly contagious; when an infected individual coughs or talks, bacteria are expelled and can survive in the air for several hours and infect other people.

Around the world each year, 2 million people die of TB. Nationally, Canada has a rate of infection comparable to other industrialized nations. Canada-born, non-native groups show an incidence of 1 TB infection per 100 000 people. However, in Inuit communities, the rate is 155.8 per 100 000. In Nunavut, the rate of infection is 62 times higher than the national average. Northern communities are isolated and less likely to receive medical treatment and supplies to treat TB infections. Incomes in most Inuit communities are low, resulting in houses that are frequently in poor condition and lack insulation. The inability to properly heat these homes can make inhabitants more vulnerable to diseases. Many homes also lack proper ventilation, which increases the exposure to airborne bacteria, and most people live in cramped conditions, which can facilitate the spread of the bacterium. Despite efforts to improve these conditions and reduce the rate of infection, researchers estimate that the goal of reducing the rate to 1 per 100 000 Inuit may still be a long way off.

Drug-resistant tuberculosis is a rapidly growing problem around the world, particularly in Africa, Russia, and China. The reason is not a mystery. Bacteria can quickly grow to incredibly high numbers and, as you may recall from our discussion of evolution, very large populations are likely to have a substantial number of individuals who possess mutations. Occasionally, a mutation makes a bacterium more resistant. Antibiotics represent a strong selective force that can quickly kill the vast majority of sensitive bacteria, thereby leaving resistant bacteria to flourish.

One of the biggest contributors to the evolution of TB resistance is thought to be the behaviour of TB patients. The typical drug treatment of TB requires taking pills daily for 1 year. Although many bacteria are killed early in the treatment, continued treatment helps eliminate every pathogen. Sometimes patients stop taking their pills because they feel better after a few months or they simply lack the money to pay for an entire year of treatment. In either case the most resistant bacteria will still survive in their bodies.

Drug-resistant TB is becoming a major problem. Researchers have developed new types of TB drugs to try to select different TB traits with the hope that evolving resistance to one drug will still make the pathogen susceptible to other drugs. However, there is now an increase in "Multiple Drug Resistant TB" or MDR-TB, a strain of the bacterium that has evolved resistance to several different drugs. In Russia, for example, nearly 20 percent of all people infected with TB are carrying the MDR-TB strain. These strains are much harder for doctors to kill and the medicines required to kill them are 100 times more expensive than the traditional drugs. Even more sobering is the discovery of what is being called "Extensively Drug Resistant TB." This version of TB has been detected in 45 countries including Russia and currently there are no drugs available to kill it.

The evolution of TB resistance is an excellent example of why we need to understand the process of evolution. Knowing the sources of genetic variation and how selection operates on this variation helps us develop drug treatment programs that are better able to control pathogens without producing strains that have multiple drug resistance.

SOURCES: Altman, L. K. Drug-resistant TB rates soar in former Soviet regions. 2008. *New York Times*, February 27. http://www.nytimes.com/2008/02/27/health/27tb.html

Goozner, M. 2008. A report from the Russian front in the global fight against drug-resistant tuberculosis. *Scientific American*, August 25. http://www.scientificamerican.com/article.cfm?id=siberia-drug-resistant-tuberculosis

Anonymous. Nunavut TB infections break record. December 17, 2010. http://www.cbc.ca/news/health/story/2010/12/17/nunavut-tuberculosis.html

Anonymous. Tuberculosis (TB). http://www.hc-sc.gc.ca/fniah-spnia/diseases-maladies/tuberculos/index-eng.php

Clark, M., P. Riben, and E. Nowgesic. 2002. The association of housing density, isolation and tuberculosis in Canadian First Nations communities. *International Journal of Epidemiology* 31: 940–945.

SUMMARY OF CHAPTER CONCEPTS

■ **The process of evolution depends on genetic variation.** Among and within populations, genetic variation is caused by the presence of different alleles, which can be dominant, codominant, or recessive. Genetic variation can be generated through mutation or recombination.

■ **Evolution can occur through random processes or through selection.** The four random processes that cause evolution are mutation, genetic drift, bottleneck effects, and founder effects. Evolution can also occur by selection, which can be stabilizing, directional, or disruptive.

Whether evolution occurs by random processes or by selection, scientists can use similarities in traits to arrange hypothesized patterns of relatedness among different groups on phylogenetic trees.

■ **Microevolution operates at the population level.** Populations can evolve due to artificial selection to produce breeds of domesticated animals and plants. Populations can also evolve due to natural selection, such as when predators selectively consume prey and when pesticides and antibiotics selectively kill the most sensitive individuals, allowing the most resistant individuals to survive and reproduce.

■ **Macroevolution operates at the species level and higher levels of taxonomic organization.** The most common process causing macroevolution is allopatric speciation in which populations become geographically isolated and independently evolve to become distinct species over time. The less common process is sympatric speciation in which species become reproductively isolated without being geographically isolated, often by forming polyploids.

REVIEW QUESTIONS

1. What is the difference between genes and alleles?

2. Why is it essential that traits be inherited for evolution to occur?

3. The insecticide DDT has been widely used to control the mosquitoes that carry malaria. How would you explain the fact that many mosquito populations are now resistant to DDT?

4. What are the three primary sources of genetic variation?

5. Compare and contrast evolution by artificial selection with evolution by natural selection.

6. Compare and contrast the processes of bottleneck effects and founder effects.

7. How should stabilizing selection and disruptive selection affect the magnitude of phenotypic variation from one generation to the next?

8. What is the assumption used when arranging species on a phylogenetic tree?

9. Distinguish between microevolution and macroevolution.

10. What is the difference between the processes involved in allopatric speciation and sympatric speciation?

PRACTICING ECOLOGY: PREDATOR PREFERENCES AND PREY PHENOTYPES

Elizabeth Boulding and her lab at the University of Guelph study the response of phenotypes to selection in intertidal and marine environments. Photo by Juan Galindo.

Predators and their prey are embroiled in an evolutionary arms race. As prey evolve adaptations to avoid being eaten, predators evolve counter-adaptations that make them better able to catch and consume prey. This arms race reflects the outcome of natural selection over many generations. Feeding preferences of predators often determine the selection on prey. For example, predators that prefer large-sized prey generate a selective force for prey to evolve smaller sizes. However, feeding preferences can change as a predator gets older. How might these changing preferences influence the selection pressure on prey through time?

The interaction between the purple shore crab (*Hemigrapsus nudus*) and its prey provide an excellent example of how natural selection by predators can change over time. The purple shore crab, which inhabits intertidal zones along the entire western coast of North America, is a predator that feeds on periwinkle snails. After capturing a snail, the crab tries to crush the snail's shell so it can consume the inner flesh. Snails with thick shells are harder to crush. Because the thickness of snail shells varies, predation by crabs should generate selection for thicker snail shells. However, the ability of crabs to crush shells also varies with age. Whereas young crabs can only crush snails with thin shells, older crabs have more powerful claws that can more easily crush snails with both thin and thick shells. As a result, older crabs are not likely to generate strong selection for snail shell thickness. We might hypothesize that the strength of selection on snail shell thickness will decrease as crabs grow older.

To test this hypothesis, Elizabeth Boulding and her lab at the University of Guelph examined the strength of selection of shore crabs on *Littoria* snails. Working on the intertidal coasts of Dixon Island, located off the southwest coast of Vancouver Island, they collected crabs that varied widely in body width, an indication of age. They placed each crab in a separate aquarium and added groups of snails. Each group of snails had an equal mean and standard deviation of snail shell thickness. As we saw in the *Analyzing Ecology* feature in this chapter, means and standard deviations allow us to calculate strength of selection. The researchers allowed each crab to feed on snails until half the snails in its aquarium were consumed. At this point, they removed the crabs and measured the means and standard deviations for shell thickness in surviving snails. By comparing the means and standard deviations of shell thickness before and after the crabs fed, the researchers could calculate the strength of selection on snail shell thickness and determine whether it changed as crab size increased. You can view their data in the table.

EFFECT OF CRAB BODY WIDTH ON STRENGTH OF SELECTION ON SNAIL SHELL THICKNESS

CRAB BODY WIDTH (mm)	STRENGTH OF SELECTION (S)
11	0.49
11	0.42
11	0.28
13	0.37
13	0.17
15	0.97
15	0.42
17	0.32
19	0.16
22	0.55
22	0.05
25	0.32
25	0.07
28	0.03
34	0.39

GRAPH THE DATA

1. Create a graph with crab shell width on the *x* axis and the strength of selection on the *y* axis.

2. Visually estimate a regression line of best fit to illustrate the relationship between the two variables.

INTERPRET THE DATA

1. Explain how the data support or refute the hypothesis that the strength of selection on snail shell thickness will decrease as crabs grow older.

2. As part of the same study, the researchers measured crab body widths on the eastern and western regions of their study site. They found that crabs were smaller on the west side of the island. Given this variation in crab size by location, would you predict the thickness of snail shells changes across the island? If so, in what way?

3. The researchers found that one snail species, *L. subrotundata*, has a thinner shell than a co-occuring species, *L. sitkana*. Given that crabs are smaller on the western side of the island, what would you predict regarding relative abundance of each snail species on the east and west coasts of the island?

SOURCE: Pakes, D., and E.G. Boulding. 2010. Changes in the selection differential exerted on a marine snail during the ontogeny of a predatory shore crab. *Journal of Evolutionary Biology* 23: 1613–1622.

CHAPTER 8

Life Histories

Live, Breed, and Die

The coho salmon (*Oncorhynchus kisutch*) is a beautiful fish that lives a dramatic life. It lays its eggs in rivers that empty into the North Pacific Ocean, from California to Alaska to eastern Russia. The hatchlings live on the energy of their yolk for several weeks and then begin to eat small prey. After growing in the river for a year, the fish swim out to the ocean where they continue to eat and grow for 1 to 3 more years. When they are ready to reproduce, they migrate back to the same river where they hatched. The females make nests in the river bottom and deposit their eggs, which the males then fertilize with their sperm. Shortly after reproducing for the first time, both the male and female salmon rapidly lose strength and physiological abilities, and die.

The coho salmon is just one of many species of salmon that live in North America. Each has a unique reproductive strategy. Several of the coho's closest relatives, including the chinook salmon (*O. tshawytscha*) and the sockeye salmon (*O. nerka*), also migrate to sea as juveniles, return for a single breeding event, and then die. Members of other species, such as the rainbow trout (*Oncorhynchus mykiss*), breed several times during their life. Some populations—known as resident rainbow trout—do not migrate out to the ocean but remain in freshwater rivers. Other populations of rainbow trout—commonly known as steelheads—migrate out to sea and return to the rivers to breed, much like the coho salmon. Unlike the coho, however, steelheads migrate several times and breed each time they return to the freshwater rivers.

> "Shortly after reproducing for the first time, the male and female salmon rapidly lose strength and physiological abilities, and die."

Researchers have looked for patterns in the breeding strategies of different species of salmon. When they examined the breeding characteristics of 12 species of salmon from North America, they found that species with larger bodies produce more and larger eggs, and that a higher fraction of their body weight is composed of gonads. When the researchers adjusted for the large differences in body size among species of salmon, they found that for a given body size, there is a trade-off between how many eggs a female can carry and the size of those eggs. A female that tries to produce more eggs can only do so if each egg is smaller. Smaller eggs produce smaller offspring that are less competitive for food and more susceptible to predators. On the other hand, larger eggs contain more yolk, which allows the offspring to hatch at a larger size. Being larger makes the hatchling fish more successful at competing for food and less vulnerable to predators.

One of the most noticeable distinctions occurs among species that breed only once compared with those that breed multiple times. On average, salmon that only breed once have larger bodies as adults and larger eggs than salmon that breed multiple times. Larger bodies allow more efficient swimming, which is important to species that migrate long distances. Based

Coho salmon. These salmon, from the Rogue River, Oregon, migrate from rivers to the ocean where they live for several years before returning to their home river to breed and then die. Photo by Mark Conlin/Alamy.

on these traits and changes over evolutionary time, it appears that larger bodies and larger eggs were key evolutionary steps that improved the survival of the offspring enough to favour the unusual strategy of breeding only once, followed immediately by death.

The story of the salmon highlights the wide variety of reproductive strategies that exist in organisms. Examining patterns of reproductive strategies among species and the selective forces experienced by species in nature helps us begin to understand why different species have evolved unique strategies. As we shall see in this chapter, organisms have evolved a wide range of alternative strategies for growth, development, and reproduction as a result of fitness trade-offs.

SOURCES: P. B. J. Crespi and R. Teo, Comparative phylogenetic analysis of the evolution of semelparity and life history in salmonid fishes, *Evolution* 56 (2002): 1008–1020.

I. A. Fleming, Pattern and variability in the breeding system of Atlantic salmon (*Salmo salar*), with comparisons to other salmonids, *Canadian Journal of Fisheries and Aquatic Sciences* 55 (S1) (1998): 59–76.

CHAPTER CONCEPTS

■ Life history traits represent the schedule of an organism's life.

■ Life history traits are shaped by trade-offs.

■ Organisms differ in the number of times that they reproduce, but they all eventually become senescent.

■ Life histories are sensitive to environmental conditions.

As we have seen in previous chapters, organisms are generally well adapted to the conditions of their environments. Their form and function are influenced by physical and biological factors. Similarly, the strategies that organisms have evolved for sexual maturation, reproduction, and longevity are also shaped by natural selection. In this chapter, we will explore the wide array of strategies that species have evolved and the trade-offs among different traits.

Life history traits represent the schedule of an organism's life

The schedule of an organism's growth, development, reproduction, and survival makes up what ecologists call the **life history** of the organism. As you can see in Figure 8.1, the life history of an organism includes the traits connected to the birth or hatching of offspring. These include the time required to reach sexual maturity; **fecundity**, which is the number of offspring produced per reproductive episode; **parity**, which is the number of episodes of reproduction; **parental investment**, which is the amount of time and energy given to offspring; and **longevity**, or **life expectancy**, which is the life span of an organism. In essence, life history traits describe an organism's strategy for obtaining the highest possible evolutionary fitness over its lifetime. In addition, life history traits represent the combined effect of many morphological, behavioural, and physiological

adaptations of organisms, all of which interact with environmental conditions to affect survival, growth, and reproduction. In this section, we will explore the wide range of life history traits that exist in nature and how these traits are often organized into strategies that allow organisms to persist under different ecological conditions.

THE SLOW-TO-FAST LIFE HISTORY CONTINUUM

Life history traits vary widely among species and among populations within a species. An organism's life history represents a solution to the problem of allocating limited time and resources to achieve maximum reproductive success. A remarkable fact about reproductive success is that the result is always nearly the same. On average, only one of the offspring that an individual produces lives to reproduce. In short, each individual only replaces itself. If this were not the case, populations would either dwindle to

Life history The schedule of an organism's growth, development, reproduction, and survival.

Fecundity The number of offspring produced by an organism per reproductive episode.

Parity The number of reproductive episodes an organism experiences.

Parental investment The amount of time and energy given to an offspring by its parents.

Longevity The life span of an organism. *Also known as Life expectancy*.

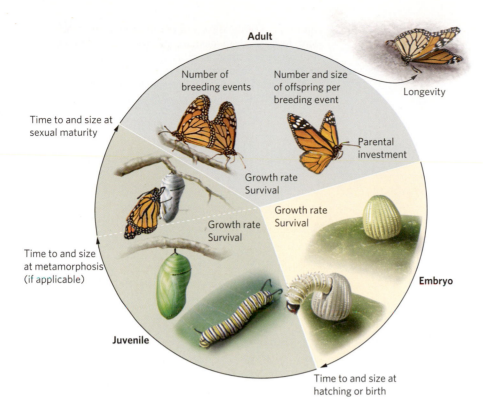

Adult

Number of breeding events

Number and size of offspring per breeding event

Longevity

Time to and size at sexual maturity

Parental investment

Growth rate Survival

Growth rate Survival

Growth rate Survival

Time to and size at metamorphosis (if applicable)

Embryo

Time to and size at hatching or birth

Juvenile

Figure 8.1 Life history traits. The schedule of an organism's life begins as an embryo. It either hatches or is born at a particular size and time. During the subsequent juvenile stage the organism grows and eventually becomes a sexually mature adult. To reach this stage, many species must first pass through metamorphosis. Adults can then reproduce in one or more breeding events with a particular level of fecundity, parental investment, and longevity. At all stages, species experience a characteristic amount of growth and some probability of surviving to the next stage.

extinction because individuals fail to replace themselves or populations would continually expand.

How organisms grow and produce offspring varies in all imaginable ways. A female sockeye salmon, after swimming up to 5000 km from her Pacific Ocean feeding ground to the mouth of a coastal river in British Columbia, still faces an upriver journey of 1000 km to her spawning ground. There she lays thousands of eggs, and promptly dies, her body wasted from the exertion. A female African elephant gives birth to a single offspring at intervals of several years, lavishing intense care on her baby until it is old enough and large enough to fend for itself in the world of elephants. Thrushes, a group of birds that includes the American robin (*Turdus migratorius*), start to reproduce when they are 1 year old, and may produce several *clutches*—which are sets of eggs—each year, with each clutch containing three or four chicks. Adult thrushes rarely live beyond 3 or 4 years. In contrast, storm petrels, which are seabirds about the size of thrushes, do not begin to reproduce until they are 4 or 5 years old and then only rear a single chick each year. They may live for 30 or 40 years. This wide variation in life history traits among species has captured the interest of researchers who wish to understand ecological conditions that favour such disparate evolutionary outcomes.

Two points can be made about this variation. First, life history traits often vary consistently with respect to life form, habitat, or conditions in the environment. Seed size, for example, is generally larger for trees than for grasses. Second, variation in one life history trait is often correlated with variation in other life history traits. For example, the number of offspring produced during a single breeding event is often negatively correlated with the size of the offspring. As a result, variations in many life history traits can be organized along a continuum of values.

We can refer to one extreme as the "slow" end of the spectrum. At this extreme, organisms such as elephants, albatrosses, giant tortoises, and oak trees require a long time to reach sexual maturity. They commonly have long life spans, low numbers of offspring, and a high parental investment in the energy given to their offspring, such as parental care, the amount of yolk in an egg, or the amount of energy stored in a seed. At the fast end of the spectrum are organisms such as fruit flies and herbaceous weedy plants, which exhibit short times to sexual maturity, high numbers of offspring, little parental investment, and short life spans.

COMBINATIONS OF LIFE HISTORY TRAITS IN PLANTS

The English ecologist J. Philip Grime conceptualized the relationship between the life history traits and environmental conditions as a triangle, with each of the three points representing an extreme condition.

Goldenrod

Wooly lousewort

Canada thistle

Figure 8.2 Combinations of life history traits in plants. Plants face the environmental challenges of competition, disturbance, and stress. Plant species living at each extreme environmental condition have evolved suites of life history traits that make the plants well suited to these environments. Data from J. P. Grime, *Plant Strategies and Vegetation Processes* (Wiley, 1979).

As shown in Figure 8.2, the environmental conditions he considered were stress, competition, and the frequency of disturbances. Grime proposed that plants functioning at the extremes of these three axes possessed combinations of traits that could be categorized as *stress tolerators, competitors,* or *ruderals.* Table 8.1 lists some of the major differences in these three plant strategies.

As their name implies, stress tolerators live under extreme environmental conditions, such as low water availability, very cold temperatures, or high salt concentrations. For example, plants living in the tundra biome, such as the wooly lousewort (*Pedicularis*

dasyantha), are typically small herbs that live for many years, grow very slowly, and achieve sexual maturity relatively late in life. Similarly, many plants that live in the desert, such as cactus, are stress tolerators because they can survive long periods of hot temperatures with no precipitation. Because growing from a seed is very difficult in such stressful environments, they put little of their energy into seeds. Instead, they rely on *vegetative reproduction,* which is a form of asexual reproduction in which new plants develop from the roots and stems of existing plants.

Where conditions for plant growth are less stressful, plants can evolve life history traits that fall

TABLE 8.1

THE LIFE HISTORY TRAITS OF PLANTS THAT ARE POSITIONED AT THE ENVIRONMENTAL EXTREMES OF STRESS, COMPETITION, AND DISTURBANCE

LIFE HISTORY TRAITS	STRESS TOLERATORS	COMPETITORS	RUDERALS
POTENTIAL GROWTH RATE	Slow	Fast	Fast
AGE OF SEXUAL MATURITY	Late	Early	Early
PROPORTION OF ENERGY USED TO MAKE SEEDS	Low	Low	High
IMPORTANCE OF VEGETATIVE REPRODUCTION	Frequently important	Often important	Rarely important

along a continuum from competitors to ruderals. With low stress and low frequency of disturbance, there is more competition. Competitors can grow relatively fast, achieve sexual maturity early in life, and devote a small proportion of their energy to seed production because they often spread by vegetative reproduction. Competitors also tend to grow large and exhibit long life spans. In forests, most species of trees fit the competitor category. In abandoned farm fields, a strong competitor is the group of tall herbaceous plants known as goldenrods.

At the other point of the triangle, with low stress and high frequency of disturbance, we find ruderals. These plants colonize disturbed patches of habitat, exhibit fast growth, early maturation, and use a high proportion of their energy to make seeds. Ruderals include many plants that we might call "weeds" in a garden, including dandelions (*Taraxacum officinale*), common ragweed (*Ambrosia artemisiifolia*), and Canada thistle (*Cirsium arvense*). Ruderals typically have seeds that are easily dispersed and that can persist in the environment for many years as they wait for favourable environmental conditions. This collection of traits allows ruderal plants to reproduce quickly and to disperse their seeds to other disturbed sites.

Life history traits are shaped by trade-offs

If we consider the many types of life history traits, it would seem that an organism could have very high fitness if it could grow fast, achieve sexual maturity at an early age, reproduce at a high rate, and have a long life span. However, no organism possesses the best of all such life history traits, which highlights the fact that organisms face trade-offs. When one life history trait is favoured, it often prevents the adoption of another advantageous life history trait. In some cases, there are physical constraints, such as the size of a mammal's uterus, which places a limit on the total volume of offspring that can be produced at one time. Thus, a female can produce several small offspring, or a few large offspring, but not many large offspring. In other cases, the trade-off reflects the genetic makeup of the organism. Because some genes have multiple effects, selection that favours genes for one life history trait can cause other life history traits to change. For instance, in the mouse-ear cress (*Arabidopsis thaliana*), artificial selection for earlier flowering also causes reduced seed production. In still other cases, the trade-offs are the result of the allocation of a finite amount of time, energy, or nutrients. Given the large number of life history traits, there are many potential trade-offs. In this section, we will discuss the *principle of allocation* and

highlight some of the most common trade-offs that have been observed.

THE PRINCIPLE OF ALLOCATION
Organisms often have limited time, energy, and nutrients at their disposal. According to the **principle of allocation**, when these resources are devoted to one body structure, physiological function, or behaviour, they cannot be allotted to another. As a result, natural selection will favour those individuals that allocate their resources in a way that achieves maximum fitness.

Selection on life history traits can be complex because when one trait is altered, it often influences several components of survival and reproduction. As a result, the evolution of a particular life history trait can be understood only by considering the entire set of consequences. For example, an increase in the number of seeds an oak tree produces may contribute to higher fitness. However, if a higher number of seeds is achieved by making each seed smaller, and if smaller seeds experience lower survival, then producing more seeds could negatively affect the tree's overall fitness. In this case, to achieve an outcome that maximizes overall fitness, evolution should favour a strategy that balances the trade-off between seed number and seed survival.

From an evolutionary point of view, individuals exist to produce as many successful progeny as possible. Doing so, however, involves many allocation problems, including the timing of sexual maturity, the number of offspring to have at any one time, and the amount of parental care to bestow upon the offspring. An optimized life history is one that resolves conflicts between the competing demands of survival and reproduction to the best advantage of the individual in terms of its fitness. Although it is widely believed that trade-offs constrain life histories, demonstrating such trade-offs has proved difficult. In some cases, trade-offs can only be exposed by using experimental manipulations.

OFFSPRING NUMBER VERSUS OFFSPRING SIZE
Most organisms face a trade-off between the number of offspring they can produce and the size of those offspring in a single reproductive event. In a low-resource environment, parents may need to invest more in each individual offspring to ensure that it has a chance of surviving to maturity at the cost of reduced fecundity. In a high-resource environment, however, even offspring with little parental investment have a

Principle of allocation The observation that when resources are devoted to one body structure, physiological function, or behaviour, they cannot be allotted to another.

good chance to survive, so parents can reduce their investment in each offspring and increase fecundity. In the Atlantic salmon (*Salmo salar*), for instance, parents do not provide any care after laying and fertilizing eggs, so the parental investment of the Atlantic salmon is reflected in the size of its eggs; larger eggs have more yolk that provides more energy for the growth of the hatchling fish, which are known as *fry*. Researchers in Nova Scotia hatched more than 30 000 salmon eggs of different sizes and released the fry into eight different streams to determine how egg size affected the survival of the fry. The researchers also quantified the quality of the streams; the highest quality streams have shallow, fast moving water, large gravel that provides cover from predation, and few trees overhead so that plenty of light is available for algae to grow and support the insect herbivores that salmon fry consume. When they surveyed the streams several months later, they discovered the best surviving fry released into the low-quality streams came from larger eggs than the fry released into the high-quality streams. When different environments favour differences in offspring size, we expect a trade-off between offspring number and offspring size. The trade-off between offspring number and offspring size for a given reproductive event can also be limited by energy and nutrients. An example of this can be seen in Figure 8.3, which illustrates the relationship between seed size and seed number in plants of the goldenrod genus (*Solidago*). Across populations and species, there is a negative correlation, demonstrating that goldenrod plants that produce more seeds also produce smaller seeds. However, although the trade-off between offspring number and offspring size can be seen in a number of species, the expected trade-off is often not observed. In many cases, while the number of offspring can be quite variable among individuals, the size of offspring can be relatively constant. This suggests that selection often favours a uniform, perhaps even optimal, offspring size and that an individual able to acquire additional energy can only use it to make greater numbers of offspring.

ANALYZING ECOLOGY

Coefficients of Determination

When ecologists want to look for life history trade-offs, they commonly plot one life history variable against another life history variable and look for a negative relationship. We saw an example of this in the case of the goldenrod seeds; the researchers used a regression to demonstrate a trade-off between seed number and seed size. As we saw in Chapter 5, a regression is a mathematical description of the relationship between two variables. For example, data points that tend to follow a straight line might be best described by using a linear regression described by the following equation:

$$Y = mX + b$$

where X and Y are the measured variables, m is the slope—which is negative in the case of the goldenrod example—and b is the intercept.

Although this equation tells us how one variable is associated with another variable, it does not tell us how strongly the two variables are related. For example, it would be valuable to know if the data points fit closely along the line or if the data points are highly variable around the line. We can answer this question using a statistic known as the *coefficient of determination*. The **coefficient of determination**, which is abbreviated as R^2, is an index that tells us how well data fit to a line. Values can range from 0 to 1, with 0 indicating a very poor fit of the data and 1 indicating a perfect fit of the data. In terms of life history trade-offs, higher R^2 values indicate that a variation in one life history trait explains a large amount of the variation in the other life history trait.

Consider the following set of hypothetical data for plant seeds that all have the same linear relationship between seed mass and seed number, as given in the equation:

Seed number = (– 4 × Seed mass) + 24

For each set of data, we can graph the relationships and include the regression line, as shown in the graphs.

SEED MASS (g)	SEED NUMBER FOR POPULATION A	SEED NUMBER FOR POPULATION B	SEED NUMBER FOR POPULATION C
1	21	22	24
1	19	18	16
3	13	14	16
3	11	10	8
5	5	6	7
5	3	2	1
	$\overline{Y} = 12$	$\overline{Y} = 12$	$\overline{Y} = 12$

Coefficient of determination (R^2) An index that tells us how well data fit to a line.

Figure 8.3 Offspring number versus offspring size. Among 14 populations and species of goldenrod plants, individuals that produce more seeds also produce smaller seeds, demonstrating that a trade-off exists between the two life history traits. Data from P. A. Werner and W. J. Platt. Ecological relationships of co-occurring goldenrods (*Solidago:* Compositae), *The American Naturalist* 110 (1976): 959-971.

Total number of seeds per goldenrod plant (log scale)

Mass of seeds (μg, log scale)

OFFSPRING NUMBER VERSUS PARENTAL CARE

The number of offspring produced in a single reproductive event can also cause a trade-off with the amount of parental care that can be provided. As the number of offspring increases, the efforts of the parents to provide food and protection will increasingly be spread thin. Presumably, there is an optimal number of offspring for parents to produce.

In a classic study of life history evolution, David Lack of Oxford University considered the number of offspring produced by songbirds. Lack observed that songbirds breeding in the tropics lay fewer eggs at a time—an average of 2 or 3 per nest—than songbirds breeding at higher latitudes, which, depending on the

To calculate R^2, we need to first calculate the mean Y value, which for all populations is a seed number of 12. We also need to determine the expected seed numbers if the data were to fall perfectly along the line. Using the line equation and the above data for seed mass, the six expected seed numbers are 20, 20, 12, 12, 4, and 4.

Next, we need to calculate the *total sums of squares,* which is the sum of the squared differences between each observed seed number (y_i) and the mean seed number ($\overline{Y}$):

$$\sum (y_i - \overline{Y})^2$$

Next we calculate the *error sums of squares,* which is the sum of the squared differences between each expected seed number (f_i) and the observed seed number (y_i):

$$\sum (y_i - f_i)^2$$

Finally, we can calculate the value of R^2 as follows:

$$R^2 = 1 - \left(\frac{\sum (y_i - f_i)^2}{\sum (y_i - \overline{Y})^2} \right)$$

For population A, we can calculate the value of R^2 as follows:

$$R^2 =$$
$$1 - \left(\frac{(21-20)^2 + (19-20)^2 + (13-12)^2 + (11-12)^2 + (5-4)^2 + (3-4)^2}{(21-12)^2 + (19-12)^2 + (13-12)^2 + (11-12)^2 + (5-12)^2 + (3-12)^2} \right) =$$
$$1 - \left(\frac{6}{262} \right) = 0.98$$

This value (0.98) tells us that a large proportion (or 98 percent) of the variation in seed number is explained by the variation in seed mass.

YOUR TURN Using the above formulas, calculate R^2 for populations B and C. Based on your three R^2 values, which set of data best fit the regression line? Which population gives you the strongest confidence that there is a negative relationship between seed size and seed number?

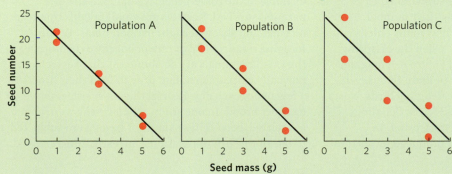

Seed number

Population A

Population B

Population C

Seed mass (g)

species, typically lay 4 to 10 eggs. In 1947, he proposed that these different reproductive strategies evolved in response to differences between tropical and temperate environments. Lack recognized that birds could improve their overall reproductive success by increasing the number of eggs, as long as a larger number of offspring did not cause a decrease in offspring survival. He hypothesized that the ability of parents to gather food for their young was limited and, if they could not provide enough, the offspring would be undernourished and therefore less likely to survive. We should therefore expect that parents produce the number of offspring that they can successfully feed. One difference between the tropics and higher latitudes is the number of daylight hours. Lack noted that parents at higher latitudes had more hours to gather food to feed their young. Therefore, assuming that the rate of food gathering is similar at low and high latitudes, he hypothesized that birds breeding at high latitudes could rear more offspring than birds breeding in the low latitudes of the tropics.

Lack made three important points. First, he stated that life history traits, such as the number of eggs laid in a nest, not only contribute to reproductive success but also influence evolutionary fitness. Second, he demonstrated that life histories vary consistently with respect to factors in the environment, such as the number of daylight hours available for gathering food for their young. This observation suggested that life history traits are moulded by natural selection. Third, he hypothesized that the number of offspring parents can successfully rear is limited by food supply. To test this idea, one could add eggs to nests to create unnaturally high numbers of hatchlings. According to Lack's hypothesis, parents should not be able to rear larger broods of chicks because they cannot gather the additional food that is required.

Such experiments have been conducted many times during the last several decades and Lack's hypothesis has commonly been supported. For example, the European magpie (*Pica pica*) typically lays seven eggs in its nest. To determine whether this is the most fit strategy for the magpie, researchers manipulated the number of eggs in magpie nests by removing one or two eggs from several nests and adding these eggs to other nests. To control for their disturbance of the nests, they also switched eggs between nests without changing their number. The researchers then waited to see how many chicks could be raised to the fledgling stage, when the offspring can leave the nest. As shown in Figure 8.4, magpies that had fewer than seven eggs or more than seven eggs ended up producing fewer fledgling birds. Nests that had eggs removed produced fewer fledglings because they started with fewer offspring. In contrast, nests with eggs added produced fewer total

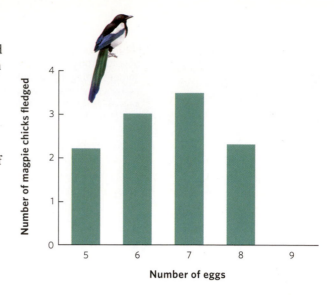

Figure 8.4 Manipulating the number of eggs in a nest. In the European magpie, the adults typically lay seven eggs. When researchers removed one or two eggs or added one or two eggs, the number of chicks that survived to fledge declined. This suggests that the typical number of eggs laid may be the optimal number for the magpie. Data from G. Högstedt, Evolution of clutch size in birds: Adaptive variation in relation to territory quality, *Science* 210 (1980): 1148–1150.

fledglings because the chicks had to share the food with more siblings. This competition with siblings caused the chicks to grow more slowly and experience higher mortality rates because the parents were unable to feed so many offspring. As Lack predicted, the number of eggs the magpie produces maximizes the number of offspring it can successfully raise.

FECUNDITY AND PARENTAL CARE VERSUS PARENTAL SURVIVAL

We saw how adding eggs to the nests of birds increased competition among the young for food that the parents brought. Consequently, parents with broods of intermediate size have the greatest fitness. Sometimes, however, having more mouths to feed stimulates parents to hunt harder for food for their chicks. In this case, an artificially enlarged brood might result in higher reproductive success in the short term. However, the additional parental effort can impose a cost on the parents that affects their subsequent fitness.

European kestrels (*Falco tinnunculus*) provide an example of the trade-off between fecundity and parental survival. Kestrels are small falcons that feed on voles and shrews caught in open fields. While foraging requires a high rate of energy expenditure, the small mammals are so abundant that kestrel pairs normally can catch enough prey to feed their brood in a few hours each day. Kestrels lay an average of five eggs per brood. In one study, when chicks were about a week old, broods in a sample of nests were subjected to one of three manipulations: investigators removed two chicks from a nest, switched chicks between nests without changing their number, or added two chicks to a nest. Investigators expected that parents of the artificially reduced and enlarged clutches would alter the amount of energy they expended looking for food for the chicks.

While parents with fewer eggs ultimately had fewer chicks fledge than the control group, parents with additional eggs had more chicks fledge, as shown in Figure 8.5. However, in spite of the

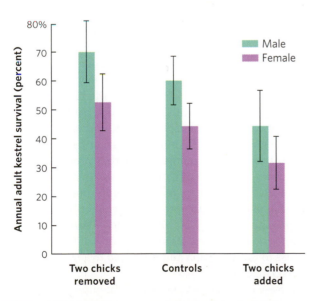

Figure 8.5 Parental care versus parental survival. When researchers removed eggs from kestrel nests, fewer chicks were fledged but male and female parents experienced increased survival. Error bars are standard deviations. When eggs were added to the nest, more chicks were fledged, but it came at the cost of decreased survival in the parents. Error bars are standard errors. Data from C. Dijkstra, et al., Brood size manipulations in the kestrel (*Falco tinnunculus*): Effects on offspring and parental survival, *Journal of Animal Ecology* 59 (1990): 269–286.

increased hunting efforts of their parents, chicks in the enlarged broods were somewhat undernourished and only 81 percent survived to fledge, compared with 98 percent survival in control nests and in nests with eggs removed. In short, adding eggs to the nest increased the total number of eggs that survived to the fledgling stage, but the percentage of surviving eggs went down. Consequently, the extra hunting effort to feed the additional two chicks only netted the parents an extra 0.8 chick, and this gain may have been diminished by the later deaths of some of the underweight fledglings. Moreover, the increased hunting efforts caused lower survival of these adults into the next breeding season. This means that increasing the number of offspring provides diminishing benefits to the parents in terms of the number of offspring that survive. Simultaneously, it causes greater adult mortality because parents must spend more energy securing food. At some point, the gains made by increasing current reproduction, which requires a large increase in parental care, are offset by the increased adult mortality, which reduces the chance of future reproduction.

GROWTH VERSUS AGE OF SEXUAL MATURITY AND LIFE SPAN

Organisms also commonly face a trade-off between putting their energy into growth or putting their energy into reproduction. In most species of birds and mammals, females grow to a particular size before they begin to reproduce. But once they initiate reproduction, females do not grow anymore—a phenomenon known as **determinate growth**. In contrast, many plants and invertebrates, as well as many fishes, reptiles, and amphibians, do not have a characteristic adult size. Rather, they continue to grow after initiating reproduction, a phenomenon known as **indeterminate growth**. Typically, indeterminate growth occurs at a decreasing rate over time.

Whether a species has determinate or indeterminate growth, the key feature shaping the trade-off between growth and reproduction is that larger females commonly produce more offspring. Because offspring production and growth draw on the same resources of assimilated energy and nutrients, increased fecundity during one year occurs at the cost of further growth that year. Moreover, for indeterminate growers, the reduced growth in one year can cause reduced fecundity in subsequent years.

Determinate growth A growth pattern in which an individual does not grow any more once it initiates reproduction.

Indeterminate growth A growth pattern in which an individual continues to grow after it initiates reproduction.

An organism with a long life expectancy should favour growth over fecundity during the early years of its life. In contrast, organisms with a short life expectancy should allocate their resources to producing eggs early in life rather than delaying reproduction and growing more. These predictions can be tested by examining the relationships among these life history traits across many different species and by conducting manipulative experiments on species in nature.

Comparisons Across Species

When we look across many different species, we see that long-lived organisms typically begin to reproduce later in life than short-lived ones. Why is this the case? If an organism has a long life span and if delaying maturity allows an organism to grow larger and produce more offspring per year once reproduction begins, natural selection will favour delayed age of maturity in these organisms. An analysis of hundreds of populations and species of animals illustrates this relationship. As you can see in Figure 8.6, as the age of sexual maturity increases, there is an associated increase in the number of years that an animal will survive after reaching sexual maturity. Moreover, different taxonomic groups fall along different regression lines. For species that are expected to live 2 years after sexual maturity, birds and mammals have the shortest times to sexual maturity whereas reptiles, fish, and shrimp have the longest times. This reflects the fact that endothermic animals can grow more rapidly than ectothermic

animals. Among species with an age of maturity of 1 year, birds have the longest expected life spans after sexual maturity. This reflects the generally lower predation risk for birds due to their ability to fly.

Comparisons Within Species

Another example occurs in pumpkinseed sunfish (*Lepomis gibbosus*), a small fish that lives in lakes in Ontario. Some of these lakes freeze over in winter, causing many fish to die. We can compare life history traits of populations that experience high adult mortality due to winterkills with populations that experience low adult mortality because they live in lakes without winterkills. As a result, the fish from populations with high adult mortality have evolved to mature earlier and invest more in their reproductive organs compared to individuals from populations with low adult mortality.

Manipulative Experiments

Another way to examine the trade-offs of growth, age of maturity, and life span is by conducting a manipulative experiment. The Trinidadian guppy (*Poecilia reticulata*), for instance, lives in the streams of Trinidad, a large island in the southern Caribbean Sea. In the lower reaches of these streams, the guppies live with a number of predatory fish species including the pike cichlid (*Crenocichla alta*), which preys on adult guppies, and the smaller killifish (*Rivulus hartii*), which preys primarily on juvenile guppies. Because of this predation, these guppies have short life expectancies. However, at higher elevations, where guppies have been able to ascend numerous small waterfalls, they live in a relatively predator-free environment and have longer life expectancies. Figure 8.7 shows the life history traits of guppy populations. In populations that face high predation risk and short life spans, males mature at smaller sizes. Females allocate a larger percentage of their body mass to reproduction, they produce more offspring, and these offspring are smaller. As predicted, researchers found the opposite to be true for populations in the predator-free sections of the streams above the waterfalls: males matured at larger sizes and females allocated less of their body mass to reproduction and produced fewer, larger offspring.

Researchers then conducted a manipulative experiment to test the hypothesis that increased mortality from predation was actually the cause of the altered life history strategies of the local guppy populations. They transplanted predators from the lower reaches of the streams to the areas above the waterfalls where predators did not historically exist. Within a few generations, the life histories of the populations above the waterfalls came to resemble

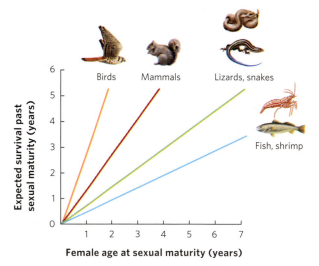

Figure 8.6 Age of sexual maturity versus life span. Using hundreds of different populations and species, we can see that different groups of animals have different relationships between these two life history traits. Data from E. L. Charnov and D. Berrigan. Dimensionless numbers and life history evolution: Age of maturity versus the adult life span, *Evolutionary Ecology* 4 (1990): 273–275.

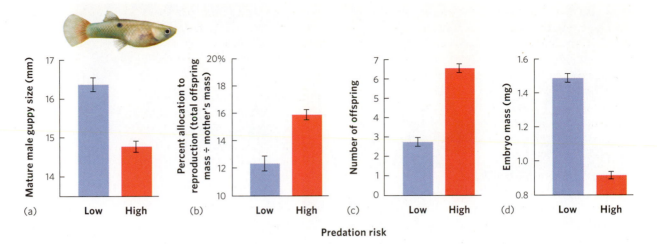

Figure 8.7 Alternate life history strategies of guppies in Trinidad. Guppy populations in streams with a high risk of predation have shorter life spans and those in streams with low predation risk have longer life spans. In response to this difference in longevity, guppies living in predator environments have evolved to **(a)** mature as smaller males, **(b)** allocate a greater fraction of energy to offspring, which is quantified as the percentage of the mother's mass that is dedicated to the production of her offspring, **(c)** produce more offspring, and **(d)** produce smaller offspring. Error bars are standard deviations. Data from D. N. Reznick et al., Life history evolution in guppies (Poecilia reticulata: Poeciliidae). 4. Convergence in life history phenotypes, *American Naturalist* 147 (1996): 319–338.

those of the populations in the lower reaches of the stream. This finding not only confirmed the basic ideas about the optimization of life history patterns, but also demonstrated the strength of predation as a selective force in evolution.

Organisms differ in the number of times that they reproduce, but they eventually become senescent

The number of times an individual reproduces during its lifetime varies a great deal among species. But in almost all species, individuals eventually experience a decline in body condition followed by death. For species that breed once, physiological decline and death follow rapidly. For species that breed many times, the decline in physiology comes more gradually. An interesting exception is in the case of bacteria, in which individuals do not always die. Instead, they undergo fission whereby a single-celled organism divides into two single-celled organisms. In this section we will examine the life history strategies of breeding once versus breeding several times, and investigate the causes of a gradual decline in physiology.

SEMELPARITY AND ITEROPARITY

Organisms can evolve to reproduce only once, a phenomenon known as **semelparity**, or multiple times throughout their life, a phenomenon known as **iteroparity**. Semelparity and iteroparity do not tell us whether an organism's life history is **annual**,

meaning a life span of one year, or **perennial**, referring to a life span of more than one year. For example, organisms that live for only one year can undertake more than one episode of reproduction, or even prolonged continuous reproduction, during that time. Iteroparity is a common life history that occurs in most species of birds, mammals, reptiles, and amphibians. Semelparity is relatively rare in vertebrate animals, but it occurs in insects and many species of plants. For example, semelparity is a common life history of crop plants including wheat and corn.

Bamboos, Agaves, and Yuccas

Although most semelparous organisms are short-lived, the best-known cases of plants occur in the long-lived bamboos and agaves, two distinctly different groups. Most bamboos are plants of tropical or warm temperate climates and they often form dense stands. Reproduction in bamboos does not appear to require substantial preparation or resources, as in the case of salmon. But bamboos probably have few opportunities for successful seed germination. Once a bamboo plant establishes itself in a disturbed habitat, it spreads for years by vegetative

---

Semelparity When organisms reproduce only once during their life.

Iteroparity When organisms reproduce multiple times during their life.

Annual An organism that has a life span of one year.

Perennial An organism that has a life span of more than one year.

(a)

(b)

Figure 8.8 Semelparous plants. **(a)** Bamboos and **(b)** agaves are two groups of plants that live for many years, reproduce one time, and then die. The flowering bamboos are from a location in Kyoto, Japan. The flowering agave is from Baja California. Photos by (a) Joi Ito; (b) A. Paul Jenkin/age fotostock.

growth, continually sending up new stalks until the habitat in which it germinated is densely packed with bamboo shoots (Figure 8.8a).

In many species of bamboo, breeding is highly synchronous over large areas, such that every individual produces flowers and makes seeds in the same year. After reproduction, the future of the entire population rests with the crop of seeds. Synchronous breeding may facilitate fertilization in this wind-pollinated plant group and also may overwhelm seed predators, which cannot consume such a large crop of seeds. Some species of bamboo, such as Chinese bamboo (*Phyllostachys bambusoides*), have a 120-year cycle of germinating, growing, and then simultaneously flowering and producing seeds.

In contrast to bamboo plants, most species of agave plants live in arid climates with sparse and erratic rainfall. Ranging from the southwestern United States through Central America, agaves grow as a rosette of leaves over several years, with the duration of this growth varying among species. When the plant is ready to reproduce, it grows a gigantic flowering stalk that produces a large number of seeds (Figure 8.8b). The growth of the stalk is so rapid that it cannot be fully supported by photosynthesis in the stalk and uptake of water by the roots. Instead, the nutrients and water necessary for stalk growth are drawn from the leaves, which die soon after the seeds are produced.

Yucca plants are close relatives of the agave plants. Although most yuccas are iteroparous, a few are not. For instance, in California the yucca species *Yucca whipplei* contains both iteroparous and semelparous

individuals. Researchers have compared the two groups—known as varieties—for total production of flower, fruits, and seeds over the lifetime of the plants. The results of this study are displayed in Figure 8.9.

Compared to the iteroparous variety, the semelparous variety produced nearly twice as many flowers and three times as many fruits. The semelparous variety also made seeds that were more likely to germinate than those of the iteroparous variety. The iteroparous variety can reproduce earlier in life, and reproduce multiple times. The semelparous variety, which takes longer to reproduce, can produce more offspring. These differences in strategies may be related to the habitats in which the two types of individuals live. Semelparous individuals live in habitats that have higher annual precipitation and are prone to fire, which may favour a strategy of a single large reproductive effort before another fire occurs. Iteroparous individuals live along the edges of the Mojave Desert where water is much more limited and there is a reduced chance of fire. This habitat favours a strategy of multiple reproductive events over time with large amounts of reproduction during years of higher precipitation.

When we consider all of the plants and animals that practice semelparity, we see that semelparity appears to arise when there is a massive amount of energy required for reproduction, such as the long migrations of salmon and the production of giant flowering stalks of agaves and yuccas. These large energy requirements make it difficult for the individuals to survive after the reproductive event.

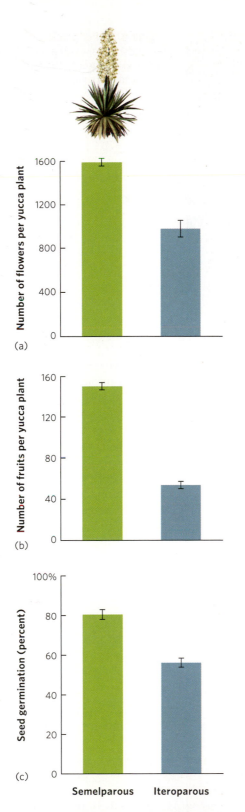

Figure 8.9 Semelparous and iteroparous yuccas. In the species *Yucca whipplei,* some varieties are semelparous and other varieties are iteroparous. Over its lifetime, the semelparous variety produces **(a)** more flowers, **(b)** more fruits, and **(c)** a higher percentage of seeds that germinate. Error bars are means ±1 standard error. Data from T. E. Huxman and M. E. Loik, Reproductive patterns of two varieties of *Yucca whipplei* (Liliaceae) with different life histories, *International Journal of Plant Sciences* 158 (1997): 778–784.

Cicadas

One of the most remarkable cases of semelparity in animals is the life cycle of periodical cicadas (Figure 8.10). Cicadas spend the first part of their life underground where they obtain nutrients from the xylem tissue of plant roots and, after some time, they emerge as adults. Their mating calls from the trees can be heard in the summer days in many parts of the Northern Hemisphere. Some cicada species have annual life cycles while other species spend several years underground with a fraction of them emerging each summer. Periodical cicadas, however, are different. They live as nymphs underground for 13 or 17 years and then emerge from the ground in synchrony to mate. The emergence of the periodical cicada is marked by nearly deafening noise as the males call to attract females during their brief mating period. The long life cycle gives the larvae time to grow to adulthood on a diet of low nutritional quality. The synchrony is probably a mechanism to overwhelm potential predators. Most of the occasional individuals that fail to emerge in synchrony come out a year earlier or later and are grabbed by predators attracted to the loud mating calls.

Scientists long wondered how periodical cicadas know when to emerge from the ground. They speculated that the cicada might count the years by the warming and cooling of the soil, or by the physiological cycles of their hosts. Researchers performed a clever experiment by growing 17-year periodical cicadas on peach trees that had been artificially selected to drop their leaves and flower twice each year. The cicadas emerged after 17 fruiting seasons had passed, rather than 17 years, demonstrating that the cicadas are sensitive to the reproductive cycles of their hosts rather than annual physical changes in their environments. How they count to 17 remains a mystery.

SENESCENCE

A few long-lived semelparous organisms die immediately after reproduction. However, iteroparous organisms experience a gradual deterioration in their physiological function over their lifetime. This leads to a gradual decrease in fecundity with an increase in the probability of mortality, a phenomenon known as **senescence**. Humans are an example of an organism that experiences senescence. Most physiological functions in humans decrease between the ages of 30 and 85 years; for example, the rate of nerve conduction and basal metabolism decrease by 15 to 20 percent, the volume of blood circulated through the kidneys

---

Senescence A gradual decrease in fecundity and an increase in the probability of mortality.

Figure 8.10 Periodical cicadas. The 13- and 17-year cicadas are semelparous insects. They spend many years underground, emerge as adults to breed, and die soon after. Pictured here is the pharaoh cicada (*Magicicada septendecim*), which is a species of 17-year cicada. Photo by ARS Information Staff.

decreases by 55 to 60 percent, and maximum breathing capacity decreases by 60 to 65 percent. Over time the function of the immune system and other repair mechanisms also declines. Using data from the Canadian population in 2009, Figure 8.11 illustrates that the incidence of death from cancers and cardiovascular diseases rises sharply with age. Birth defects in offspring and infertility also occur with increasing prevalence in women after 30 years of age, and fertility decreases dramatically in men after 60 years.

If maintaining high survival and reproduction would increase an individual's fitness at any age, why does senescence exist? Studies of aging in a variety of animals demonstrate that senescence is an inevitable consequence of natural wear and tear. It is impossible

to build a body that will not wear out eventually, just as it is impossible to build an automobile that will not wear out. Senescence might simply reflect the accumulation of molecular defects that fail to be repaired. For example, ultraviolet radiation and highly reactive forms of oxygen break chemical bonds, macromolecules become inactivated, and DNA accumulates mutations. However, this wear and tear cannot be the entire explanation for patterns of aging because maximum longevity varies widely even among species of similar size and physiology. For instance, many small insectivorous bats live 10 to 20 years in captivity, whereas mice of similar size rarely live beyond 3 to 5 years.

The rate of wear can be modified by a variety of physiological mechanisms that either prevent or repair

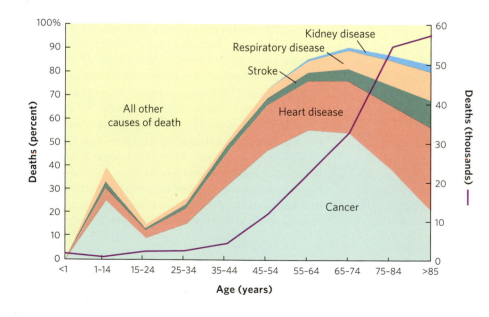

Figure 8.11 Senescence. Throughout life, humans experience a gradual decrease in physiological function and an increase in cancer and cardiovascular diseases. These contribute to increased probability of mortality. Data from Statistics Canada. Leading Causes of Deaths in Canada, 2009, CANSIM Tables 102–0561 and 102–0562.

damage. One major difference among groups that differ dramatically in life span is the cellular mechanisms for reducing the production of reactive forms of oxygen and for repairing damaged DNA and protein molecules. These appear to be better developed in long-lived animal species than in their short-lived relatives. Because these mechanisms are under genetic control, they can be modified by evolution. Mechanisms of prevention and repair require investments of time, energy, nutrients, and tissues. Therefore, the allocation of resources to these mechanisms depends on the expected life span of the individual. When a population has a low survival rate, selection should favour improvements in reproductive success at young ages and selection to delay senescence should be weak. In a population with a high survival rate, selection for delayed senescence should be strong.

This prediction is consistent with observations of natural populations. For example, because bats and birds can fly to escape predators, they lead safer lives than do rodents of a similar size. As a result, the maximum potential life spans of birds and bats are much longer than the life spans of rodents. Storm petrels, a small seabird, have a body size and metabolic rate that is similar to many rodents. However, the storm petrel can live more than 40 years whereas a rodent may only live 1 or 2 years. Because of their long life spans, birds and bats age more slowly than rodents of a similar size.

Life histories are sensitive to environmental conditions

As we saw in Chapter 4, many traits exhibit flexibility, or plasticity, in response to different environmental conditions, and life history traits are no exception. As a result, researchers have continued to discover a fascinating array of life history traits that can be altered by changing environmental conditions.

STIMULI FOR CHANGE
Many events in the life history of an organism are timed to match seasonal changes in the environment. The right timing is essential so behaviour and physiology match changing environmental conditions. For instance, flowering plants must bloom when pollinators are present and most birds must breed when there is an abundance of food to feed their chicks. To get their timing right, organisms rely on various indirect cues in the environment.

Virtually all organisms sense the amount of light that occurs each day, known as the **photoperiod**. Many can distinguish whether the photoperiod is getting shorter or longer. Within a single species, populations may be exposed to a variety of environmental conditions. Each population develops

a particular response to the photoperiod in its environment. For example, consider the grass known as sideoats grama (*Bouteloua curtipendula*). Southern populations living at 30° N flower in autumn in response to a photoperiod of 13 hours per day. In contrast, northern populations living at 47° N flower in summer in response to a photoperiod that exceeds 16 hours each day.

Another example occurs in water fleas of the genus *Daphnia*. In Ontario, water fleas enter diapause in mid-September, when the photoperiod declines to fewer than 12 hours of sunlight, but related species in Yukon enter diapause in mid-August, when the photoperiod decreases to fewer than 20 hours of sunlight. Water fleas never see 20-hour days in Ontario, but Yukon water fleas would perish from the cold if they waited for 12-hour days before entering diapause. From this we see that the critical stimulus for these organisms is the change in environmental conditions associated with a particular photoperiod. The sensitivity of individuals to these cues has been adjusted by natural selection so that the individual's response to an environmental cue is well matched to the environmental condition.

THE EFFECTS OF RESOURCES
Many types of organisms undergo dramatic life history changes during the course of their development. One of the most striking changes is the process of metamorphosis in which a larva changes into a juvenile or adult organism. Metamorphosis can be seen in many species of insects and amphibians, as in the transformation from tadpole to frog. Organisms that metamorphose have a wide range of timing options. Environmental conditions that influence timing include the amount of resources available, the temperature, and the presence of enemies.

Let's consider the different options for timing of metamorphosis by looking at the two growth curves in Figure 8.12. These curves represent the change in the mass of the barking treefrog (*Hyla gratiosa*) raised under high or low food conditions. On any particular day early in life, the individuals raised under high food conditions have a larger mass then those raised under low food conditions. As time progresses, an individual with access to high food is able to metamorphose at a relatively large mass and young age. An individual with access to low food cannot achieve the same combination of mass and age, but it can follow several alternative strategies. It might wait to mature when it achieves the same mass as individuals raised under high food conditions, though it will take longer to achieve that mass and the delay

---

Photoperiod The amount of light that occurs each day.

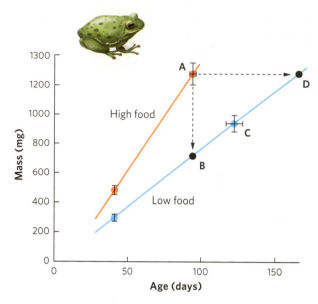

Figure 8.12 Alternative growth curves of a metamorphosing organism. The amount of food available can affect an organism's mass and age at metamorphosis. In the barking treefrog, an individual living under high food is able to metamorphose at a large mass and young age, represented by point A. An individual living under low food conditions could achieve the same age at metamorphosis if it emerges at a smaller mass, for example at point B. It could achieve the same mass at metamorphosis if it took longer to metamorphose, for example at point D. In reality, the tadpoles reach a compromise and metamorphose at a somewhat smaller mass and a somewhat later age, as indicated by point C. Error bars are standard deviations. Data from J. Travis, Anuran size at metamorphosis: Experimental test of a model based on intraspecific competition, *Ecology* 65 (1984): 1155–1160.

presence of an egg predator. Many embryos sense the chemical odours that predators emit while other embryos can detect the vibrations produced by predators. When predators are detected, the embryos can speed up their time of hatching in an attempt to leave the egg before the predator eats it. For example, the red-eyed treefrog (*Agalychis callidryas*) lives in Central America and the adults lay their eggs on leaves that hang over water. When the embryos have developed sufficiently, they hatch and drop into the water. Should a cat-eyed snake (*Leptodeira septentrionalis*) appear, however, the frog embryos sense the vibrations of the approaching snake and begin to hatch earlier than usual and drop into the water to avoid the snake (Figure 8.13). However, this response comes at the cost of hatching at a smaller size that can make the hatchling tadpoles more

in reproduction could reduce its fitness. Alternatively, it might metamorphose at the same age as individuals raised under high food conditions, even though it will be significantly smaller. The drawback of this strategy is that a smaller size at metamorphosis makes the organism more vulnerable to predation before reproduction occurs. For most metamorphosing organisms, the optimum solution is usually a compromise between these two strategies. So an organism exposed to low food typically metamorphoses at an older age and smaller mass.

THE EFFECTS OF PREDATION

The risk of predation is also an important factor in affecting the life history of organisms. As we saw in Chapter 4, predation can affect a wide range of life history traits including time to and size at hatching, time to and size at metamorphosis, and the time to and size at sexual maturity. One of the more remarkable effects of predators is their impact on the embryos of many species of aquatic organisms. In a variety of groups including fish and amphibians, the embryo that develops inside an egg can detect the

Figure 8.13 Hatching early in response to predators. As the cat-eyed snake begins to attack the eggs of the red-eyed treefrog, the embryos are stimulated to hatch early. Note the tadpole escaping the snake's attempt to eat the egg in which the tadpole lived. This photo was taken at the Corcovado National Park in Costa Rica. Photo by Karen M. Warkentin.

susceptible to predators living in the water. Therefore, when egg predators are not present, the embryo stays in the egg longer and hatches at a larger and safer size.

Studies of metamorphosing animals also find that predators commonly play an important role in affecting the size and time at which metamorphosis takes place. For example, many high-elevation streams of western Colorado contain trout, an important predator of mayfly larvae, whereas other streams lack trout. Larval mayflies living with trout metamorphose at smaller sizes and leave the streams earlier than mayflies in comparable streams that lack trout. Growth rates in the two types of streams are similar, so the difference in the time to and size at metamorphosis is entirely the result of predation risk.

Predators can also affect when organisms achieve sexual maturity. Several species of freshwater snails, for example, face higher risks of predation when they are small. As a result, when predators are present, a snail is likely to do better if it delays reproduction and uses its energy to grow. Once it has grown to a safer size, it can reproduce. Although this strategy can improve the snail's probability of survival in the presence of predators, the cost of delaying sexual maturity can be reduced fecundity. However, once the predator-induced snails begin reproducing, they can produce more eggs in each clutch because they have larger bodies. In these cases, the snails can achieve the same lifetime fecundity as snails raised without predators.

THE EFFECTS OF GLOBAL WARMING

We have seen how the life history of organisms responds to different environmental conditions found in nature. However, over the past 100 years human activity has caused a warming trend on Earth. In many regions, the difference in temperature is relatively small—a rise of 1°C or 2°C. However, even small changes in temperature can have substantial impact on an organism's physiological processes. During the past decade, researchers have begun to discover that the increase in global temperatures has caused changes in the breeding times of many animals and plants.

Animal Breeding

Researchers interested in the effect of global warming on the life histories of animals have focused on the breeding times of birds and amphibians. Figure 8.14 shows data compiled from 3450 nesting records of the North American tree swallow (*Tachycineta bicolor*) collected from 1959 to 1991. As you can see, by the end of the 4-decade period, the mean date of egg laying was 9 days earlier. Moreover, researchers found a negative correlation of the mean egg-laying date to the mean air temperature in May. Across all years, an increase in May temperature was associated with earlier egg-laying dates. From this they concluded that the variation in egg-laying dates was associated with the variation in air temperatures and that the warming temperatures over a 4-decade period might be at least part of the reason the birds are breeding earlier. Similar patterns have been found in many other species of birds across both North America and Europe.

A change in breeding dates has also been observed in several species of amphibians. In Britain researchers monitored three species of frogs and three species of salamanders for 17 years. At the end of this period, they found that two of the three frog species monitored were breeding 2 to 3 weeks earlier and all three species of salamanders were breeding 5 to 7 weeks earlier. These changes in breeding times were correlated with the average maximum temperatures that occurred just prior to breeding, which generally increased over the 17 years. A similar study of North American amphibians, however, failed

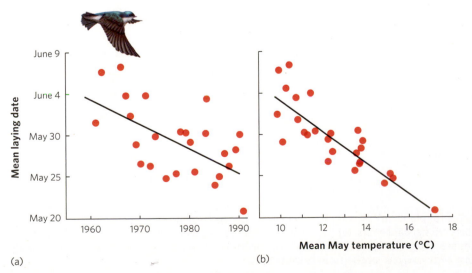

Figure 8.14 Egg laying dates in tree swallows. (a) Egg laying dates collected over a 42-year period demonstrate that tree swallows bred 9 days earlier in 1991 than in 1959. (b) The variation in egg laying date is negatively correlated with the mean air temperature in May. Data from P. O. Dunn and D. W. Winkler, Climate change has affected the breeding date of tree swallows throughout North America, *Proceedings of the Royal Society of London B*. 266 (1999): 2487–2490.

to find a relationship between changes in mean maximum air temperature over time and the initiation of breeding. At the present time, researchers do not know why amphibians in different regions of the world show different responses to global warming.

A major insight into this question came in 2012 when bird researchers reported their results on a captive population of great tits (*Parus major*). They raised the birds in temperature-controlled chambers over a period of 3 years under different simulated temperature changes. They discovered that the initiation of breeding was not in response to an increase in the mean temperature of the environment, but in response to rapid temperature increases over a period of several days in the spring. This is relevant because global climate change is predicted not only to cause an increase in mean temperatures but also to cause increases in the duration of warm spells and cold spells. The great tit results suggest that increased temperature fluctuations may be the important environmental cue that affects the life histories of many organisms.

Plant Flowering

Plants are also susceptible to change in temperature, which has the potential to alter the initiation of flower production. One of the longest studies began in the nineteenth century with the writer Henry David Thoreau, who is best known for having spent a year in a small cabin at Walden Pond in Concord, Massachusetts, and for his numerous essays about the natural world. Thoreau kept data on more than 500 species of flowering plants in Concord. From 1852 to 1858, he took notes on the dates when each plant species first began flowering in Concord.

After Thoreau's death, a local shopkeeper continued his work by observing the first flowering times of more than 700 plant species. More recently, two ecologists realized that the data could help them determine whether long-term changes in global temperatures might be associated with changes in the initial flowering times of plants. Because flowering time is sensitive to temperature as well as to photoperiod, they predicted that warmer global temperatures would cause plants to flower earlier today than in Thoreau's time. To test their hypothesis, they collected data for plant flowering times in Concord from 2003 to 2006.

In 2008, the researchers reported that over the 154-year period from 1852 to 2006, local temperatures in Concord had increased by 2.4°C. You can see these data in Figure 8.15. They also found that for the 43 most common species of plants, flowering time today is an average of 7 days earlier than in Thoreau's time. Interestingly, not all plants responded to the temperature change in the same way. In some species, initial flowering time remained unchanged, perhaps because these species use day length as their cue for flowering and day length has not changed. Other species, such as highbush blueberry (*Vaccinium corymbosum*) and yellow wood sorrel (*Oxalis europaea*), flower 3 to 4 weeks earlier now than they did in 1852. These unique data collected over a century and a half indicate that a seemingly small change in average annual temperature has been associated with dramatic changes in initial flowering time.

Consequences of Altered Breeding Events

The changing breeding seasons of plants and animals in response to global warming does not by itself cause any problems to the species that are responding. Problems can arise, however, when a species depends on the environment to provide the necessary

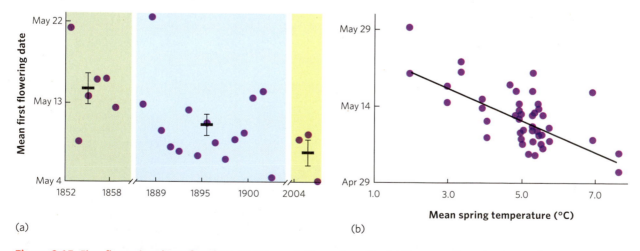

(a)

(b)

Figure 8.15 First flowering dates for plants in Concord, Massachusetts. (a) The mean flowering time today is 7 days earlier than in the 1850s. Error bars are standard errors. **(b)** The variation in first flowering time is associated with the mean temperature of the 1 or 2 months preceding each species' flowering time. Data from A. J. Miller-Rushing and R. B. Primack, Global warming and flowering times in Thoreau's Concord: A community perspective, *Ecology* 89 (2008): 332–341.

resources with an altered breeding season. The pied flycatcher (Ficedula hypoleuca), for example, is a bird that breeds in Europe each spring. In 1980, researchers in the Netherlands found that the date of egg hatching of the flycatcher began just a few days before the peak abundance of caterpillars, which are a major prey item for the flycatcher chicks. As spring temperatures warmed over the next 2 decades, however, tree leaves appeared 2 weeks earlier and the caterpillars reached their peak abundance 2 weeks earlier. The pied flycatcher, however, retained its normal date of egg hatching, which was 2 weeks later than the new time for peak caterpillar abundance. As a result, the chicks of the flycatcher no longer have a major source of food and the pied flycatcher population has declined by 90 percent.

From our discussion of life history traits, you can see that natural selection has favoured a wide variety of strategies for the life history of different species. Different life histories evolve as a result of different selection pressures on traits such as mortality, fecundity, and longevity combined with a considerable number of potential trade-offs among life history traits. As is true for other traits, the genes that code for these traits interact with the environments that organisms experience, ultimately to produce the life history traits of individuals.

ECOLOGY TODAY CONNECTING THE CONCEPTS

SELECTING ON LIFE HISTORIES WITH COMMERCIAL FISHING

Human selection on the life history of fish. For several decades, commercial fishing vessels, like this boat in the north Atlantic Ocean, have harvested the largest individuals and thereby caused unintentional selection for smaller fish. Because of this selection, some species of fish now achieve sexual maturity earlier. Photo by Jeff Rotman/Alamy.

Throughout this chapter we have seen how natural selection has shaped the evolution of life histories by favouring individuals that are best suited to their environment. But what if a change in the selection process favoured individuals with different suites of traits? This is precisely what happens in many commercial fisheries because they harvest only the largest individuals. For many years this seemed like a wise way to manage the exploitation of wild populations because it protected the small individuals, allowing them a chance to grow. It is also a common practice for state agencies to set a minimum size for fish, such as bass and salmon, that anglers are permitted to keep.

Based on our discussion in this chapter, you should be able to predict what will happen to the life history traits of species that experience a great deal of fishing pressure, particularly from large commercial fishing boats that can collect thousands of fish. When the smallest fish are either not caught or are thrown back into the water, we impose a high mortality rate on large adults and leave the smaller, younger fish behind for breeding. As we have seen in this chapter, high adult mortality favours the evolution of smaller adult sizes, earlier times to maturity, higher fecundity, and shorter life span.

During the past 2 decades, researchers have begun to investigate whether large-scale fishing can cause unintended evolution of fish life histories and they have confirmed that commercial fishing imposes substantial selection on fished populations. Atlantic cod (*Gadus morhua*) has been fished off Canada's east coast for more than 5 centuries. In the beginning, populations were so large that they seemed inexhaustible. For instance, in 1497 John Cabot wrote that cod on the Grand Banks of Newfoundland were so plentiful that they could be caught by simply lowering a bucket into the water. However, the number of fishing vessels increased from around 650 vessels in Cabot's time to more than 1600 vessels 2 centuries later. The number of fish caught per year rose from 190 000 tonnes in the last half of the eighteenth century to almost a million tonnes around 1900. With new technology in ship and fishing gear design, catches rose to almost 2 million tonnes by the mid-1960s. Over the next 2 decades, cod populations collapsed and the fishery was finally closed in 1992. While cod fishing methods changed over the centuries, one aspect remained constant: Fishermen focused on catching the largest fish. Over 3 decades starting in the late 1970s, the length at maturity fell by 14 to 20 percent and age at maturity decreased by 16 to 30 percent. Researchers have also modelled that the observed reductions in size and age at maturity in these cod populations more than double the chance that populations will decline due to higher costs of reproduction and lower hatching success in younger, smaller fish that reproduce for the first time.

The data collected by commercial fishing boats does not typically contain information on other life history traits such as fecundity and longevity. However, based on our knowledge of common life history trade-offs it is reasonable to assume that an increase in adult mortality and a decline in age at maturity imposed by fishing practices coincide with increases in size-adjusted fecundity and declines in longevity.

One challenge in determining the effect of fishing pressure on life history is that it can cause other significant changes. As we noted earlier in this chapter, environmental changes—such as resource levels—can affect life history traits. For example, fishing from commercial fishing boats could reduce competition among the remaining fish. We have seen that reduced competition can allow a faster time to sexual maturity. In some cases, researchers cannot differentiate between the environmental induction of life history changes from reduced competition and the evolution of life history changes. In other cases, however, scientists have been able to document that a fish population held at low numbers for several decades continues to exhibit life history changes over time. In such cases, the changes in life history are more likely to be the result of evolution by artificial selection.

The impact of human selection on natural populations is not limited to fish; similar impacts have been found in hunted mammals and some plants. In all of these cases, identifying the factors that naturally cause the evolution of life histories and how various life history traits trade off with each other has helped fisheries managers understand how human harvest of wild populations can have unintended consequences.

SOURCES: Hutchings, J. A. 2005. Life history consequences of overexploitation to population recovery in Northwest Atlantic cod (*Gadus morhua*). *Canadian Journal of Fisheries and Aquatic Sciences* 62: 824-1/n-832.

Darimont, C. T., S. M. Carlson, M. T. Kinnison, P. C. Paquet, T. E. Reimchen, and C. C. Wilmers. 2009. Human predators outpace other agents of trait change in the wild. *Proceedings of the National Academy of Science* 106: 952–954.

SUMMARY OF CHAPTER CONCEPTS

■ **Life history traits represent the schedule of an organism's life.** Species differ in a wide range of traits that help determine their fitness throughout their life including the birth or hatching of offspring, the time required to reach sexual maturity, fecundity, parity, and longevity. These traits are under the influence of natural selection and often evolve in particular combinations.

■ **Life history traits are shaped by trade-offs.** Trade-offs can occur because of physical constraints, time or energetic constraints that affect allocation, or genetic correlations that cause selection favouring one trait to come at the cost of another trait. Common trade-offs include offspring number versus offspring size, and growth versus reproduction.

■ **Organisms differ in the number of times that they reproduce, but they all eventually become senescent.** Semelparous organisms breed once in their life whereas iteroparous organisms breed more than once. Regardless of how many times an organism breeds, it ultimately experiences a decay in physiological function followed by death. In semelparous organisms, this decay in function occurs rapidly after reproduction. In iteroparous organisms, the decay in function can be very gradual.

■ **Life histories are sensitive to environmental conditions.** As with all phenotypes, life history traits are the product of genes and environments interacting. Some of the most common environmental influences on life history traits include variation in resources and predators, both of which can induce substantial changes in the life history of organisms. Current anthropogenic changes in the environment can also affect life history traits by altering environmental cues—such as temperature—that induce life history changes.

REVIEW QUESTIONS

1. Why are trade-offs among life history traits so commonly observed?

2. Using Grime's categorization of plant life history traits, why might ruderals spread via easily dispersed seeds whereas stress tolerators spread vegetatively?

3. Why should organisms with low annual survival rates begin to reproduce at an early age?

4. Explain why organisms face a fundamental trade-off between growth and fecundity?

5. Compare and contrast semelparous and iteroparous life history strategies.

6. Why might natural selection act more strongly on traits that improve reproductive success early in life rather than later in life?

7. Give two reasons why an individual could face a trade-off between offspring number and offspring size.

8. What is the mechanism by which parental care can increase current fitness but decrease later fitness?

9. Why might an organism use cues such as photoperiod to predict the future state of its environment?

10. In terms of environmental cues, explain why many species of plants flower earlier in association with warmer spring temperatures whereas other species of spring plants do not?

PRACTICING ECOLOGY: TROPHY HUNTING FOR BIGHORN SHEEP

David Coltman from the University of Alberta and his colleagues have studied how sport hunting has altered the evolutionary trajectory of bighorn rams. Photo by Pierrick Blanchard, courtesy of David Coltman.

For many species of large mammals such as deer, antelope, and wild sheep, bigger is often better. Males with larger body sizes often gain dominance and therefore produce more offspring. Males with large defensive structures such as horns and antlers also have an advantage. They are more likely to win fights with other males to obtain females and can protect themselves better against enemies. However, large males with large horns or antlers are more vulnerable to trophy hunters who prefer them to males with smaller defensive structures. In *Ecology Today* in this chapter we saw that commercial fishing for large fish can affect the life history of the Atlantic cod. Is it possible that sport hunting can similarly alter the life history of large mammals with large horns or antlers?

In the mountains of western Alberta, bighorn rams *(Ovis canadensis)* are grazers that live on steep slopes. The breeding success of males increases with age, body mass, and horn length—all of which help determine a male's dominance ranking in a herd. Dominant males are usually the only ones within a population that breed and pass their genes on to future generations. Natural predation on rams is rare, but sport hunters prize large rams with long horns. Hunter selection for horn size leads us to hypothesize that sport hunting artificially selects for individuals with smaller body mass and shorter horns.

To test this hypothesis, David Coltman at the University of Alberta and his colleagues have monitored the bighorn ram population of Ram Mountain in southern Alberta for over 30 years. During this time, Coltman and his team have captured, tagged, and measured the horn lengths of 894 rams. Using large scales, they have also measured body mass of each ram. You can view a sampling of their data from 1980–2002 in the table.

HORN LENGTH AND BODY MASS OF BIGHORN RAMS ON RAM MOUNTAIN, 1980–2002

YEAR	HORN LENGTH (cm)	BODY MASS (kg)
1980	59.5	75.9
1982	64.4	80.1
1984	64.5	76.5
1986	73.2	84.7
1988	61.7	75.9
1990	59.5	80.3
1992	53.8	63.7
1994	59.5	73.9
1996	45.5	60.7
1998	46.9	64.5
2000	47.4	65.1
2002	53.4	55.7

GRAPH THE DATA

1. Create one scatterplot with year on the *x* axis and horn length on the *y* axis.

2. Create a second scatterplot with year on the *x* axis and body mass on the *y* axis.

3. Visually estimate a regression line for both graphs.

INTERPRET THE DATA

1. Explain how the data support or refute the hypothesis that sport hunting artificially selects individuals with smaller body mass and shorter horns for breeding.

2. Do the regressions indicate that hunter preferences have artificially selected on horn and body mass phenotypes over time?

3. In a later study, it was found that most of the rams that hunters harvested were not of breeding age, but experienced rapid horn growth during their early years of development. Based on this information, what phenotypes are sport hunters artificially selecting for in ram populations?

SOURCES: Coltman, D. W., P. O'Donoghue, J. T. Jorgenson, J. T. Hogg, C. Strobeck, and M. Festa-Bianchet. 2003. Undesirable evolutionary consequences of trophy hunting. *Nature* 426: 655–658.

Hengeveld, P. E. and M. Festa-Bianchet. 2011. Harvest regulations and artificial selection on horn size in male bighorn sheep. *Journal of Wildlife Management* 75: 189–197.

CHAPTER 9

Reproductive Strategies

The Sex Life of Honeybees

Honeybees (*Apis mellifera*) have a complicated sex life. They live in hives that may contain tens of thousands of bees, usually progeny of the same mother, known as the *queen*. Like many organisms, the queen bee produces sons and daughters, but she does so in a rather unique way. Early in her life, the queen bee flies out of the hive and mates in the air with a group of male bees. Male bees, known as *drones,* are smaller than the queen. A queen mates with several drones, but the larger drones contribute more sperm than the smaller drones. The queen stores sperm from this single mating event in a special organ in her body, known as the *spermatheca,* where it remains viable for several years. She uses these sperm to fertilize her eggs and make diploid daughters, known as *workers*. In contrast, she creates drones by laying unfertilized haploid eggs. The drones a queen produces rarely mate with her, but instead mate with other queens outside the hive. After mating, the drones die.

A key issue for a successful hive is the proper sex ratio of drones and workers. In a typical beehive, the queen bee may produce a few dozen drones but tens of thousands of workers. Since the workers do the vast majority of the work in the hive, producing many more workers than drones is in the best interest of the queen. Since workers only live for 4 to 7 weeks, the queen must constantly produce more of them.

> **"The Cape honeybee can lay diploid eggs without ever mating."**

Genetically, the workers and the queen are quite similar. They are both female and they both arise from a fertilized, diploid egg. What makes them different is the food they are given as larvae. For the first few days of life, all larvae are fed royal jelly, a liquid produced by the worker bees, but after this the larvae destined to be workers are switched to a diet of honey and pollen. These workers are not capable of mating with a drone, but they can lay unfertilized eggs. Larvae destined to be queens continue to be fed royal jelly, which allows the future queen to grow very large. The queen's size allows her to eventually lay as many as 2000 eggs per day.

When a beehive experiences the decline of its queen, new larval queens are often already being produced as replacements. Sometimes, however, the death or departure of a queen happens unexpectedly and there are no replacements. In such cases, some of the workers lay eggs but, since the workers cannot mate, their eggs are haploid and destined to turn into drones. Without a queen to lay fertilized eggs, the colony will eventually die.

Scientists have recently discovered an exception to this scenario in a subspecies of the honeybee known as the Cape honeybee (*Apis mellifera capensis*) that is found in southern Africa. Cape honeybee workers can lay diploid eggs without ever mating and, therefore, can ensure the persistence of the beehive. Researchers recently discovered that a single recessive gene controls this ability. As a result, depending on which alleles they carry, individual workers can lay either haploid eggs that give rise to drones or diploid eggs that give rise to workers.

A honeybee hive. In most populations of honeybees, a queen can reproduce by laying either haploid eggs that produce sons or fertilized diploid eggs that produce daughters. Photo by Franz Christoph Robil/age fotostock.

The complexity of honeybee reproduction serves as an excellent example of the variety of reproductive strategies that have evolved. These options include reproducing with or without a sexual partner, choosing the number of mating partners, selecting the best traits of a sexual partner, varying the way the sex of offspring is determined, and controlling the number of sons versus daughters. In this chapter, we will explore the wide range of reproductive strategies in a diversity of organisms.

SOURCE: H. M. G. Lattorff, R. F. A. Moritz, and S. Fuchs, A single locus determines thelytokous parthenogenesis of laying honeybee workers (*Apis mellifera capensis*), *Heredity* 94 (2005): 533–537.

CHAPTER CONCEPTS

- Reproduction can be sexual or asexual.

- Organisms can evolve as separate sexes or as hermaphrodites.

- Sex ratios of offspring are typically balanced, but they can be modified by natural selection.

- Mating systems describe the pattern of mating between males and females.

- Sexual selection favours traits that facilitate reproduction.

The evolution of reproductive strategies involves a large number of different factors, many of which are influenced by ecological conditions. For example, organisms can evolve to reproduce sexually or asexually; each strategy has unique costs and benefits, particularly, as we shall see, when species interact with parasites and pathogens. Organisms can evolve to reproduce as separate sexes or as hermaphrodites that possess both female and male sexual organs. If they reproduce as hermaphrodites, they must also evolve solutions to the problems associated with self-fertilization. In many species the sex ratio of offspring can be altered to respond to changing ecological conditions. Finally, we see many different mating strategies to improve fitness, including the number of mates and the preference for certain traits in the opposite sex. This chapter explores how ecological conditions affect the evolution of sex and the strategies that organisms have evolved to increase their fitness.

Reproduction can be sexual or asexual

All organisms reproduce but they have a variety of ways in which they accomplish this task. In plants, animals, fungi, and protists, reproduction may be accomplished through either *sexual reproduction* or *asexual reproduction*. In this section, we will look at the two processes and compare their costs and benefits.

SEXUAL REPRODUCTION

As we discussed in Chapter 7, reproductive function in most animals and plants is divided between two sexes. When progeny inherit DNA from two parents, we say they are the result of **sexual reproduction**. Sexual gametes are produced through meiosis within the primary sexual organs, known in animals as **gonads**. Meiosis produces haploid cells, each of which contains a single full set of chromosomes. In animals, these haploid cells can immediately act as gametes. In plants and many protists, the haploid cells develop into multicellular, haploid stages of the life cycle that eventually produce gametes. Of the two sets of chromosomes inherited by an organism's parents, the distribution of the chromosomes into the haploid cells is generally random and the mixing of chromosomes from the two parents results in new combinations of genes in the offspring. Ultimately, two gametes join together in an act of fertilization to produce a diploid zygote.

ASEXUAL REPRODUCTION

In contrast to sexual reproduction, progeny produced by **asexual reproduction** inherit DNA from a single parent. This can be accomplished through either *vegetative reproduction* or *parthenogenesis*.

Vegetative Reproduction

Vegetative reproduction occurs when an individual is produced from the nonsexual tissues of a parent. Many plants can reproduce by developing new shoots that sprout from leaves, roots, or rhizomes (i.e., underground shoots). Figure 9.1 shows an example of this in Ross's sandwort (*Minuartia rossii*), an Arctic

Sexual reproduction A reproduction mechanism in which progeny inherit DNA from two parents.

Gonads The primary sexual organs in animals.

Asexual reproduction A reproduction mechanism in which progeny inherit DNA from a single parent.

Vegetative reproduction A form of asexual reproduction in which an individual is produced from the nonsexual tissues of a parent.

Figure 9.1 Vegetative reproduction. Organisms that use vegetative reproduction produce offspring from nonsexual tissues. This Arctic Ross's sandwort rarely flowers. Instead, leaf buds fall off easily, are transported by wind or water, and quickly develop roots.

plant that reproduces mainly through the dispersal of leaf buds. If you have ever placed a plant clipping into a glass of water and watched it grow roots to become a new plant, you have witnessed this type of reproduction. Individuals that descend asexually from the same parent and bear the same genotype are known as **clones**. Many simple animals, such as hydras, corals, and their relatives, also reproduce in this way; they produce buds along their body that develop into new individuals. Bacteria and some species of protists reproduce by duplicating their genes and then dividing the cell into two identical cells, a process known as **binary fission**.

Parthenogenesis

In contrast to vegetative reproduction, some organisms reproduce asexually by producing an embryo without fertilization, a process known as **parthenogenesis**. In most cases, parthenogenetically produced offspring arise from diploid eggs, which do not require any genetic contribution from sperm. Parthenogenesis has evolved in plants and several groups of invertebrates such as the Cape honeybees mentioned at the beginning of this chapter. Many Arctic insects, such as mayflies and midges, are also parthenogenetic. This may reduce costs associated with mating in cold and

variable conditions. It also allows Arctic insects to be less affected by an occasional cool summer that can make it more difficult for these ectotherms to move about and locate mates. Animal species that reproduce only by parthenogenesis are typically composed entirely of females.

Parthenogenesis is relatively rare in vertebrates. It has never been observed as a natural occurrence in mammals but has been confirmed in a few species of lizards, amphibians, birds, and fishes. For a long time it was thought that snakes and sharks were not capable of parthenogenesis. In 2007, however, researchers confirmed that a virgin female hammerhead shark (*Sphyrna tiburo*) gave birth to daughters that were genetically identical to the mother. In 2011, researchers discovered that a female boa constrictor gave birth to two litters of daughters through parthenogenesis (Figure 9.2). The growing

Clones Individuals that descend asexually from the same parent and bear the same genotype.

Binary fission Reproduction through duplication of genes followed by division of the cell into two identical cells.

Parthenogenesis A form of asexual reproduction in which an embryo is produced without fertilization.

Figure 9.2 A fatherless boa constrictor. This female boa constrictor was the product of parthenogenesis. As a result, the recessive caramel colouration of the mother was passed on to all of her clonal daughters. Photo by Warren Booth.

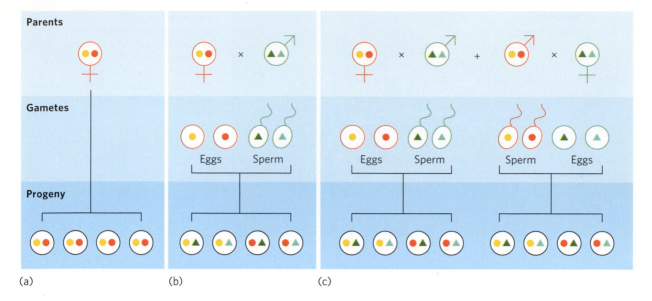

Figure 9.3 The cost of meiosis. If two hypothetical female organisms can only produce four eggs, **(a)** the female that uses asexual reproduction contributes eight copies of her genes. **(b)** In contrast, the female that uses sexual reproduction only contributes four copies of her genes. **(c)** A hermaphrodite can pass on eight copies of its genes by producing four eggs that are fertilized by another individual's sperm and four copies of its genes when it uses its sperm to fertilize another individual's eggs.

evidence suggests that parthenogenesis may be more common than we currently appreciate and that some species can reproduce through both sexual reproduction and parthenogenesis.

Parthenogenesis can produce offspring that are clones of the parent or offspring that are genetically variable. Clones are produced when germ cells develop directly into egg cells without going through meiosis. In contrast, genetically variable offspring are produced when germ cells proceed partially or entirely through meiosis. In partial meiosis, germ cells pass through the first meiotic division, but suppression of the second meiotic division results in diploid egg cells. Although a sexual union is not involved, these eggs differ from one another genetically because of recombination between pairs of homologous chromosomes and the independent assortment of chromosomes during the first meiotic division. When the germ cells experience complete meiosis, the gamete-forming cells of the female are haploid and then fuse to form a diploid embryo.

COSTS OF SEXUAL REPRODUCTION

Sexual and asexual reproduction are both viable strategies, but sexual reproduction comes with a number of costs. For example, sexual organs require considerable energy and use resources that could be devoted to other purposes. In addition, mating itself can be a substantial task. Many plants must produce floral displays to attract pollinators and most animals conduct elaborate courtship rituals to attract mates. These activities require time and resources. They can

also elevate the risk of herbivory, predation, and parasitism.

For organisms in which the sexes are separate—that is, in which an individual is either male or female—sexual reproduction has an additional cost of reduced fitness. To understand this cost, we need to remember that every parent's goal is to leave as many copies of its genes as possible in the next generation; this maximizes the parent's fitness. In the case of asexual reproduction, as shown in Figure 9.3a, a parent contributes two sets of chromosomes to each of its offspring. In the case of sexual reproduction, each parent contributes only one set of chromosomes to each offspring because the gametes produced by meiosis are haploid, as shown in Figure 9.3b. Females using either mode of reproduction produce the same number of offspring, but the female parent using sexual reproduction leaves behind half as many copies of its genes as a female using asexual reproduction. This 50 percent reduction due to sexual versus asexual reproduction is known as the **cost of meiosis**.

The cost of meiosis can be counterbalanced by hermaphroditism, a reproductive strategy that is found in most plants and in many invertebrates. Consider an individual that uses sexual reproduction and possesses both male and female function. Such an individual can contribute one set of its genes to

Cost of meiosis The 50 percent reduction in the number of a parent's genes passed on to the next generation via sexual reproduction versus asexual reproduction.

offspring produced through female function and another set to offspring produced through male function. As shown in Figure 9.3c, this allows a hermaphrodite to contribute twice as many copies of its genes to its offspring than is possible for an individual that can be only male or female.

The cost of meiosis can also be offset when the sexes are separate and the male helps the female take care of the offspring. Consider the case in which a male's parental assistance allows the female to raise twice as many offspring as she could care for alone. In this case, the 50 percent cost of meiosis is offset by doubling the number of offspring that can be raised when both parents care for them.

BENEFITS OF SEXUAL REPRODUCTION

If sexual reproduction is so costly, then it must persist because it provides substantial benefits. These benefits include purging harmful mutations and creating genetic variation that helps offspring deal with future environmental variation, including the existence of rapidly evolving parasites and pathogens.

Purging Mutations

Mutations occur in all organisms, and most are harmful. In asexually reproducing organisms there is no way to lose mutations from one generation to the next, so mutations continue to accumulate over generations, especially if the asexual parents produce clonal offspring. In contrast, sexually reproducing organisms can lose deleterious mutations during meiosis, which involves the random assortment of genes, or after fertilization of the gametes. Of all the gametes that are produced, those that form zygotes may not contain the mutation. Alternatively, if some of the gametes used to make offspring do contain the mutation, the union of two gametes that both contain the recessive mutation produces offspring that are homozygous recessive for the harmful mutation. A homozygous recessive offspring will express the harmful mutation and, as a result, it is likely that the offspring will not be viable and will fail to pass the mutation to the next generation.

Because species that only reproduce asexually do not have any means of purging mutations, deleterious mutations slowly accumulate over many generations. In time we would expect individuals of such a species to experience poor growth, survival, and reproduction, which would lead to eventual extinction. If this hypothesis is correct, then species that use asexual reproduction today would have adopted this mode of reproduction only recently. Asexually reproducing species that arose long ago would probably have gone extinct by now.

To test the hypothesis that asexual reproducing species do not persist in nature as long as sexual reproducing species, we can look at the patterns of asexual reproduction within a phylogeny. If the hypothesis is correct, we should observe that asexual reproduction has evolved relatively recently. For example, most vertebrate species that reproduce asexually belong to genera that have a sexual ancestor and contain mostly sexual species, with asexual species that are relatively recently evolved. We observe this pattern with salamanders in the genus *Ambystoma,* fishes in the genus *Poeciliopsis,* and lizards in the genus *Cnemidophorus.* This observation suggests that purely asexual species typically do not have long evolutionary histories. If they did, we would expect to see large groups of related species—such as most species within a genus—all using asexual reproduction. Such a pattern would suggest that their common ancestor used asexual reproduction. In fact, the long-term evolutionary persistence of asexual populations appears to be low. This fits with the explanation that accumulation of mutations and lack of genetic variation cause species that reproduce asexually to go extinct.

However, not all asexual species fit this pattern. For example in the bdelloid rotifers, an ancient group of more than 300 species of freshwater and soil organisms, all species are asexual and entirely female. Similarly, some groups of protists have existed for hundreds of millions of years and appear to lack sexual reproduction. One way that such species could avoid extinction is by producing offspring more rapidly than new deleterious mutations arise, so that some individuals would always retain the nonmutated parental genotype and produce the next generation, a process known as clonal selection. However, groups such as these continue to challenge our efforts to understand the full range of costs and benefits that favour the evolution of sexual or asexual reproduction.

Genetic Variation and Future Environmental Variation

A second benefit of sexual reproduction is the production of offspring with greater genetic variation. If the environment were homogeneous across time and space, parents that are well adapted to the environment could use asexual reproduction to produce clonal offspring that are also well adapted. However, as we have discussed in earlier chapters, environmental conditions typically change across time and space. As a result, offspring are likely to encounter different environmental conditions than their parents did. Because environmental conditions vary, offspring with genetic variation have an increased probability of possessing gene combinations that will help them adapt to different environmental conditions. Most theoretical models that examine the importance of abiotic environmental variation conclude that temporal and spatial variation in the physical environment does not produce a large enough advantage to offset the cost of meiosis. A promising alternative explanation is that temporal and spatial variation in the biotic environment—particularly variation in pathogens—provides a large advantage to sexual reproduction.

Genetic Variation and Evolving Parasites and Pathogens

To understand why sexual reproduction provides an evolutionary benefit when species experience variation in pathogens, we first must realize that pathogens have much shorter generation times and much larger population sizes than most of the host species they infect. Because pathogens have the potential to evolve at a much faster rate than their hosts, they can evolve ways to get past host defences. Without rapid host evolution, pathogens could drive their hosts to low numbers or even to extinction. For example, in 1998 researchers described a newly discovered species of pathogen that was causing widespread amphibian deaths in Central America. The pathogen is a type of chytrid fungus (*Batrachochytrium dendrobatidis*) that can infect a wide variety of amphibian species. By 2012, the fungus had been detected on every continent inhabited by amphibians. In some parts of the world, including Central America, it appears that this deadly pathogen may have been recently introduced and that many species of frogs in the region have no adaptations to combat it. As a result, scientists now believe that dozens of species have gone extinct.

The harmful effects of pathogens place a premium on hosts to evolve new defences rapidly. As we have seen, sexual reproduction produces offspring with a greater range of genetic combinations, and some of these combinations might be better suited to combat the pathogen. In short, there is an evolutionary race between hosts, which are trying to evolve adaptations rapidly enough to combat the pathogen, and pathogens, which are trying to evolve adaptations rapidly enough to elude host defences. The hypothesis that sexual selection allows hosts to evolve at a rate sufficient to counter the rapid evolution of parasites is called the **Red Queen hypothesis**, after the famous passage in Lewis Carroll's *Through the Looking-Glass, and What Alice Found There,* in which the Red Queen tells Alice, "Now, here, you see, it takes all the running you can do, to keep in the same place."

Testing the Red Queen Hypothesis

One of the most compelling tests of the Red Queen hypothesis focuses on a species of freshwater snail (*Potamopyrgus antipodarum*) that is a common inhabitant of lakes and streams in New Zealand. The snails can become infected by parasites, including trematode worms in the genus *Microphallus*. The life cycle of the pathogen is shown in Figure 9.4. The worm's life cycle begins when the snails ingest worm eggs. The eggs hatch into larvae that form cysts in the sexual organs and cause the snails to become sterile. Ducks then eat the infected snails and the pathogens mature sexually inside the intestines of the ducks, where they produce eggs asexually. These eggs exit the duck when it defecates in the water, thereby completing the cycle. Not surprisingly, *Microphallus* is most abundant in shallow waters of lakes where ducks feed.

The snail's mode of reproduction depends on the depth of water in which it lives. In the shallower regions of the lake, where the parasitic worm is more common, a higher proportion of the snails use sexual reproduction. Such populations contain about 13 percent males—enough to maintain some genetic diversity through sexual reproduction. In deeper regions, where the parasite is rare, a higher proportion of the snails use asexual reproduction. Although asexual populations reproduce faster than sexual populations, asexual clones cannot persist in the face of high rates of parasitism. As a result, the asexual snails do not survive well in the shallower regions of the lake where they are more likely to encounter the parasitic worm.

If the parasitic worms evolve to specialize on the snails with which they coexist, then parasites living in the shallow water should be good at infecting snail populations that live in shallow water. Similarly, other parasites that live in deep water should be good at infecting snail populations in that region. Researchers tested this hypothesis with parasites and snails from several different lakes in New Zealand. As shown in Figure 9.5, shallow-water snails from several different lakes were infected most readily by shallow-water parasites, and deep-water snails were infected most readily by deep-water parasites. Averaged across all parasite sources, infection rates were relatively low in deep-water snails because few parasites live in this region and, therefore, have had less opportunity to evolve an ability to infect the deep-water snails. However, because deep-water habitats contain fewer parasites, asexual lineages of snails have a reproductive advantage over sexual lineages because of their more rapid reproduction.

Recent studies have continued to support the Red Queen hypothesis. In the roundworm (*Caenorhabditis*

Red Queen hypothesis The hypothesis that sexual selection allows hosts to evolve at a rate that can counter the rapid evolution of parasites.

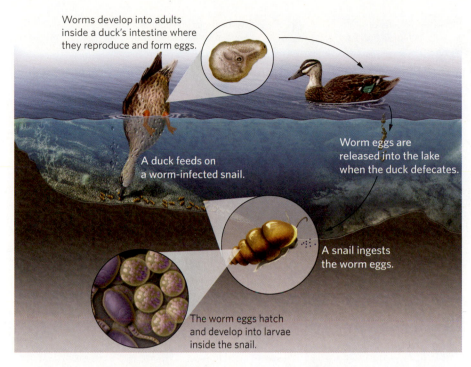

Figure 9.4 A pathogen's life cycle through snails and ducks. Eggs of the pathogenic worm *Microphallus* are inadvertently consumed by snails. The eggs hatch into larvae and form cysts in the snails, causing the snails to become sterile. When the snails are consumed by ducks in the shallow waters, the worm turns into an adult and reproduces in the ducks' intestines. When the ducks defecate, the eggs are deposited back into the water.

elegans), for example, researchers raised individuals in the laboratory that were genetically destined to reproduce either sexually or asexually and then exposed populations containing two different types of worms to a bacterial parasite. In 2011, the researchers reported that when they allowed the bacteria to evolve to infect the worms, the parasite quickly drove the asexual worms to extinction. In contrast, the sexual worms continually evolved resistance against the parasite and persisted. When the researchers prevented the bacteria from evolving, the asexual individuals came to dominate the population.

Organisms can evolve as separate sexes or as hermaphrodites

Species in nature have evolved an incredible array of sexual strategies for male and female function, as you

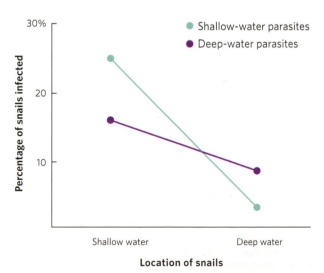

Figure 9.5 Snail infection by a pathogen. Shallow water holds more parasites than deep water because ducks defecate and release pathogen eggs more frequently in shallow water. The large number of pathogens allows them to evolve rapidly in response to any changes in the snails. Snails in shallow water are challenged by a need to evolve defences rapidly against the pathogens that respond quickly. As a result, a much higher percentage of shallow-water snails get infected than deep-water snails. The pathogens also show an increased ability to infect the populations of snails with which they coexist. Pathogens from shallow water are better able to infect shallow-water snails than deep-water snails. Similarly, pathogens from deep water are better able to infect deep-water snails than shallow-water snails. Data from C. M. Lively and J. Jokela, Clinal variation for local adaptation in a host-parasite interaction, *Proceedings of the Royal Society of London B* 263 (1996): 891–897.

Figure 9.6 Plant mating strategies. Hermaphroditic plants, such as the Saint-John's-wort wildflower, possess perfect flowers that contain male and female structures within a single flower. Monoecious plants, such as hazelnut shrubs, have separate male and female flowers, but a single plant possesses both types of flowers. Dioecious plants, such as the white campion wildflower, are comprised of some individuals that only contain male flowers and other individuals that only contain female flowers.

can see in Figure 9.6. Most vertebrates and some plants have separate sexes, whereas most plants and some animals are hermaphrodites. Hermaphroditic plants, such as the wildflower known as Saint-John's-wort (*Hypericum perforatum*), possess male and female functions in the same flower. Flowers that have both male and female parts within each flower are known as **perfect flowers**. When the two functions are produced at the same time, we call it a **simultaneous hermaphrodite**. Examples of simultaneous hermaphrodites include many species of mollusks, worms, and plants. When an individual possesses one sexual function and then switches to the other, we call it a **sequential hermaphrodite**. Some species of plants are sequential hermaphrodites, as are some mollusks, echinoderms, and fishes.

Some plants possess separate male and female flowers. Plants that have separate male and female flowers on the same individual plant are known as

monoecious. For example, every hazelnut tree (*Corylus americana*) has separate male flowers and female flowers. When an individual plant contains only male flowers or only female flowers, the species is known as **dioecious**. For example, the white campion (*Silene latifolia*) is a wildflower comprised of some individuals that produce only male flowers

Perfect flowers Flowers that contain both male and female parts.

Simultaneous hermaphrodites Individuals that possess male and female reproductive functions at the same time.

Sequential hermaphrodites Individuals that possess male or female reproductive function and then switch to possess the other function.

Monoecious Plants that have separate male and female flowers on the same individual.

Dioecious Plants that contain either only male flowers or only female flowers on a single individual.

and other individuals that produce only female flowers. Although perfect-flowered hermaphrodites account for more than two-thirds of flowering plant species, nearly all imaginable sexual patterns are known. Populations of some plant species can be composed of a complex mixture of hermaphrodites, males, females, and monoecious individuals. In other species, individual plants produce both perfect flowers and flowers that are only male or only female.

COMPARING STRATEGIES

We would expect natural selection to favour the reproductive strategy with the highest fitness. For instance, in organisms such as flowering plants, a plant could evolve to make male flowers, female flowers, or hermaphrodite flowers. To determine whether evolution should favour separate sexes or hermaphrodites, we need to compare the amount of fitness an individual would gain by investing in only male or female reproduction versus the amount of fitness it would gain by investing in both male and female reproduction. As depicted in Figure 9.7, if a male individual can invest in female function and gain a lot of female fitness while only giving up a small amount of male fitness, then selection will favour the evolution of hermaphrodites. A similar

Figure 9.8 When separate sexes have a fitness advantage over hermaphrodites. When male or female individuals add the other sexual function and experience a large decline in original sexual function, they can achieve a higher total fitness by staying as separate sexes rather than being a hermaphrodite. In this example, the fitness of a hermaphrodite equals the fitness derived from male function (H_M) plus the fitness derived from female function (H_F).

scenario can be considered for a female individual that adds male function. This occurs because the total fitness as a hermaphrodite through male plus female function exceeds the fitness of being only a male or only a female. In the case of flowers, the basic flower structure and the floral display necessary to attract pollinators—for those species that rely on them—are already in place in male flowers and female flowers. This should make the fitness costs of adding a sexual function relatively small while providing large fitness benefits. As we noted earlier, about two-thirds of all flowering plant species are hermaphrodites.

In some cases, the fitness cost of investing in a second sexual function is too high to be offset by the fitness benefits of being a hermaphrodite. Figure 9.8 illustrates how a reduction in female function can allow an investment in male function, but the total fitness is less than males and females would experience if they were to retain a single sexual function. For instance, sexual function in complex animals requires gonads, ducts, and other structures for transmitting gametes. Moreover, in many animals, being male requires large expenditures of time and energy for attracting mates and fighting with other males, while being female requires specializations for egg production or time and energy needed to care for the offspring. Because these costs can be quite high,

Figure 9.7 When hermaphrodites have a fitness advantage over separate sexes. When male or female individuals can add the other sexual function with little decline in original sexual function, they can achieve a higher total fitness by being a hermaphrodite than by being only male or female. In this example, the fitness of a hermaphrodite equals the fitness derived from male function (H_M) plus the fitness derived from female function (H_F).

we would predict that hermaphroditism should occur only rarely among animal species that actively seek mates and engage in brood care. In contrast, we would predict that hermaphroditism should occur commonly among sedentary aquatic animals that mate by simply shedding their gametes into the water. Researchers have found evidence that supports both of these predictions.

SELFING VERSUS OUTCROSSING OF HERMAPHRODITES

One of the challenges for individuals that possess both male and female function is the problem of *self-fertilization*. Self-fertilization, also known as selfing, occurs when an individual uses its male gametes to fertilize its own female gametes. As we discussed in Chapter 4, self-fertilization poses a fitness cost due to inbreeding depression. Therefore, selection should favour individuals that do not use selfing when they have an opportunity to breed with other individuals, a strategy known as *outcrossing*. Some species avoid the problems of selfing by being sequential hermaphrodites. For example, the blue-headed wrasse (*Thalassoma bifasciatum*), a fish species common in coral reefs, can be functionally female when it is a small adult but then becomes functionally male later in life when it has grown larger. Similarly, if a plant releases the pollen in its anthers before making its stigma receptive to pollen, the flower will not be able to pollinate itself. Other species have self-incompatibility genes. Individuals with the same self-incompatibility genotype—including an individual mating with itself—cannot produce offspring.

MIXED MATING STRATEGIES

As we discussed in Chapter 4, some hermaphrodites use a mixture of mating strategies. When a mate can be found, the individual prefers to breed by outcrossing to avoid the costs of inbreeding. When a willing mate cannot be found, the individual self-fertilizes. Self-fertilizing will not provide as many viable offspring as outcrossing, but it might be better than not reproducing at all.

In some cases, using a mixture of outcrossing and selfing is a response to a lack of resources. Attracting mates can be energetically expensive, as is the case with plants that must produce nectar to attract pollinators. For example, in orange jewelweed the production of outcrossing flowers is more energetically expensive than the production of selfing flowers, which do not need to invest in nectar to attract pollinators. Plants that experience herbivory of their leaves have less energy to make outcrossing flowers. As a result, these plants produce a higher proportion of selfing flowers, as shown by the data for orange jewelweed in **Figure 9.9**.

Figure 9.9 Mixed mating strategies in orange jewelweed. In a survey of populations, those populations receiving higher amounts of herbivory also produced higher proportions of selfing flowers and, therefore, a lower proportion of outcrossing flowers. Error bars are standard errors. Data from A. A. Steets and T.-L. Ashman, Herbivory alters the expression of a mixed-mating system, *American Journal of Botany* 91 (2004): 1046–1051.

Sex ratios of offspring are typically balanced, but they can be modified by natural selection

In organisms with separate sexes, the sex ratio of male to female offspring is often one to one. The exceptions offer interesting insights into evolutionary forces that favour particular sex ratios in an individual's offspring. In this section, we will examine the mechanisms that determine whether an offspring will be male or female. We will then examine the underlying reasons for the wide range of sex ratios that are observed in nature.

MECHANISMS OF SEX DETERMINATION

In previous chapters, we learned that the phenotypes of organisms are often determined by a combination of genetics and the environment. The sex of offspring is no different, although the influence of genetics and the environment on the sex of offspring differs among species.

Genetic Sex Determination

In mammals, birds, and many other organisms, sex is determined by inheritance of sex-specific chromosomes. In most mammals, females have two X chromosomes whereas males have one X chromosome and one Y chromosome. Birds have the opposite pattern of genetic sex determination: males have two copies of the Z chromosome whereas females have one Z chromosome and one W chromosome. In both cases, the sex that possesses two different chromosomes—male mammals and female birds—produces an approximately equal number of gametes containing each sex chromosome. On average, half the progeny in these populations will be female and half will be male.

In insects, genetic sex determination is accomplished in a variety of ways. In grasshoppers and crickets, for instance, all individuals are diploid, but females have two sex chromosomes whereas males have only one. In honeybees and other members of their order including other bees, ants, and wasps, sex is determined by whether an egg is fertilized. Fertilized eggs, which receive two sets of chromosomes, become females whereas unfertilized eggs become males.

Environmental Sex Determination

In some species, sex is largely determined by the environment in a process known as **environmental sex determination**. In reptiles, including several species of turtles, lizards, and alligators, the sex of an individual is determined by the temperature at which the egg develops. In turtles, embryos incubated at lower temperatures typically produce males whereas those incubated at higher temperatures produce females. The reverse is generally true in alligators and lizards. This type of environmental sex determination is known as *temperature-dependent sex determination*. Because the genotype has the ability to produce multiple phenotypes, temperature-dependent sex determination is a type of phenotypic plasticity.

For decades, biologists have wondered whether temperature-dependent sex determination is adaptive. We can test this hypothesis by investigating how the importance of temperature-dependent sex determination varies with environmental conditions. For example, the Atlantic silverside (*Menidia menidia*) is a common fish that lives along the Atlantic coast from Florida to the Gulf of St. Lawrence. In this species, the relative influence of genes and the environment in determining sex varies among populations. Some populations show the one-to-one sex ratio we expect from genetic sex determination. In other populations, sex ratios differ from the one-to-one ratio and temperature plays a more important role than genes in determining sex. Which conditions favour environmental sex determination in silverside populations? In populations with long growing seasons, as found in the southern part of the silverside's range, females that hatch at the beginning of the season, when temperatures are low, gain a distinct advantage over females that hatch later in the growing season. These early-hatching females have more time to grow larger and when they become sexually mature they are able to produce more or larger eggs. In contrast, the reproductive success of males does not depend on

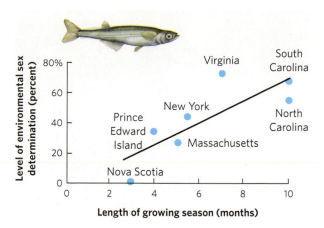

Figure 9.10 Environmental sex determination in the Atlantic silverside. The importance of environmental sex determination in the Atlantic silverside increases with the length of the growing season. The longer the growing season, the larger the proportion of fish whose sex was determined by temperature. This ensures female fish can grow large. Data from D. O. Conover and S. W. Heins, Adaptive variation in environmental and genetic sex determination in a fish, *Nature* 326 (1987): 496–498.

size, so there is no benefit of early hatching for males. The advantage of temperature-controlled sex determination diminishes in the northern part of the silverside's range where they experience a shorter growing season. We would therefore expect a positive correlation between the length of the growing season and the degree to which sex is determined by temperature. This is what researchers found when they plotted the length of the growing season against the proportion of fish with temperature-controlled sex determination, as shown in Figure 9.10. In the areas with long growing seasons, such as South Carolina or Virginia, sex determination occurs by temperature for almost 80 percent of Atlantic silverside. On the other hand, in the Nova Scotia population none of the fish exhibits environmental sex determination. Nova Scotia's Bay of Fundy features such low summer temperatures and such a short growing season in the Atlantic silverside's range that environmental sex determination is not adaptive in this population.

Not all cases of environmental sex determination are driven by temperature. In some species, sex determination is driven by the social environment in which an individual lives. As mentioned in our discussion of sequential hermaphrodites, the blue-headed wrasse is a coral reef fish that is able to change from female to male as it gets older

---

Environmental sex determination A process in which sex is determined largely by the environment.

Figure 9.11 Blue-headed wrasse. Young fish are typically female and live in schools with a dominant male. If the male leaves or dies, the largest female changes into a new dominant male. Photo by Barry B. Brown/wildhorizons.com.

(Figure 9.11). Blue-headed wrasses typically begin life as females and live in large schools along with one or a few dominant males. If for some reason the dominant male leaves or dies, the largest female will then change sex and become the new dominant male. Being large is important for males, because they must defend a territory from other males that attempt to breed with the females in the school.

OFFSPRING SEX RATIO

Now that we understand the mechanisms that help determine the sex of offspring, we can consider the factors that favour particular sex ratios of offspring. Biologists increasingly appreciate that a female can have a large influence on the sex ratios of her offspring. In species where the males have two different types of sex chromosomes, as in mammals, researchers have found that the females of some species can control whether X-chromosome or Y-chromosome sperm are allowed to fertilize their eggs. In species where the females contain two different types of sex chromosomes, as in birds, the females of some species can determine the sex ratio of their offspring by controlling the fraction of her eggs that receive the Z versus the W chromosome during meiosis. In the hymenopteran insects—bees, wasps, and ants—the female determines the sex of her offspring by whether or not she fertilizes the eggs.

A different approach to controlling offspring sex ratio is through selective abortion. In red deer (*Cervus elaphus*), for example, adult females breed in early autumn and give birth the following spring. Sons are often larger at birth and require more of the mother's milk than daughters. As a result, sons require a larger investment from the mother than daughters. Researchers in Spain examined the fetuses of 221 harvested red deer to determine whether the sex ratio of the fetuses was affected by the mother's

age—categorized as adult, subadult, or yearling—and whether the sex ratio changed through the gestation period. As shown in Figure 9.12a, adult mothers produced a nearly even offspring sex ratio. In contrast, yearling mothers, which are smaller and have less energy, were much more likely to produce the less energetically expensive daughters. Averaged across all months of the deer harvest, yearling mothers carried about 25 percent male fetuses and 75 percent female fetuses. To determine how the yearling mothers accomplish this uneven sex ratio, researchers examined yearling mothers that were harvested at different months throughout the winter,

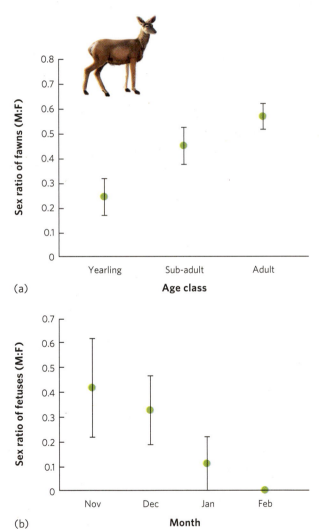

Figure 9.12 Offspring sex ratios in red deer. (a) Yearling deer produce a low proportion of male fawns, but this proportion increases in older deer. (b) Among yearling deer, the proportion of male fetuses is initially high but then declines as gestation continues throughout the winter as female deer selectively abort male fetuses. Error bars are standard errors. Data from T. Landete-Castillejos et al., Age-related foetal sex ratio bias in Iberian red deer (*Cervus elaphus hispanicus*): Are male calves too expensive for growing mothers? *Behavioral Ecology and Sociobiology* 56 (2004): 1–8.

illustrated in Figure 9.12b. They found that the yearling mothers initially had a nearly even sex ratio of offspring. As winter progressed, however, the proportion of male fetuses declined sharply, suggesting that the yearling mothers were capable of selectively aborting the more expensive male fetuses as gestation progressed. This phenomenon is not restricted to red deer; it has also been observed in other mammals including rats and mice.

Regardless of how the sex ratio is controlled, in most species the sex ratio of male to female offspring is nearly even. Is this a ratio produced by default, or are there adaptive reasons for it? To answer this question we can compare the conditions that favour a one-to-one sex ratio with the conditions that favour a deviation from that ratio.

Frequency-Dependent Selection

To understand how selection results in a one-to-one sex ratio in most species, we have to consider the best strategy for an individual whenever the sex ratio might deviate from being balanced. Let's imagine, for instance, that a population had more females than males and each male bred with only one female. In this situation, some females would remain unbred. A parent that produced all sons would have higher fitness than a parent that produced an even number of

sons and daughters, because some of the daughters might not find mates. Similarly, if a population had a surplus of sons and one male bred with only one female, a parent that produced all daughters would have higher fitness than a parent that produced an even number of sons and daughters, because some of the sons might not find mates. As you can see in this example, individuals of the less abundant sex enjoy greater reproductive success because they compete with fewer individuals of the same sex for breeding. Thus, whenever the population has an abundance of one sex, natural selection will favour any parents that produce offspring of the less abundant sex. Over time, as the less abundant sex becomes more common and the common sex becomes less abundant, populations tend to be balanced at an approximate one-to-one ratio. Therefore, the best sex ratio strategy for parents depends on the frequencies of males and females in a population. The evolution of the sex ratio is said to be the product of **frequency-dependent selection**, which occurs when natural selection favours the rarer phenotype in a population.

Frequency-dependent selection When the rarer phenotype in a population is favoured by natural selection.

ANALYZING ECOLOGY

Frequency-Dependent Selection

We can better understand frequency-dependent selection if we work though an example with real numbers. The American black vulture (*Coragyps atratus*) is a large bird that feeds on animal carcasses throughout much of North and South America. The female vulture typically lays two eggs in its nest.

If a population is composed of five females and two males, and we assume that a male can breed with more than one female, how many copies of genes, on average, does each male and female contribute to the next generation?

Total number of eggs = 10

Average female fitness = 10 female gene copies ÷ 5 females = 2 gene copies/female

Average male fitness = 10 male gene copies ÷ 2 males = 5 gene copies/male

If a population is composed of five females and eight males, how many copies of genes, on average,

does each male and female contribute to the next generation?

Total number of eggs = 10

Average female fitness = 10 female gene copies ÷ 5 females = 2 gene copies/female

Average male fitness = 10 male gene copies ÷ 8 males = 1.25 gene copies/male

In this example, you can see that when males are the less abundant sex, males have the higher fitness. When females are the less abundant sex, females have the higher fitness.

> **YOUR TURN** Using the same assumptions above, calculate the number of gene copies per male and per female under the following two scenarios:
>
> 1. Four males and five females
> 2. Six males and five females

Given the four scenarios that we have explored, what sex ratio will be favoured by natural selection over the long term?

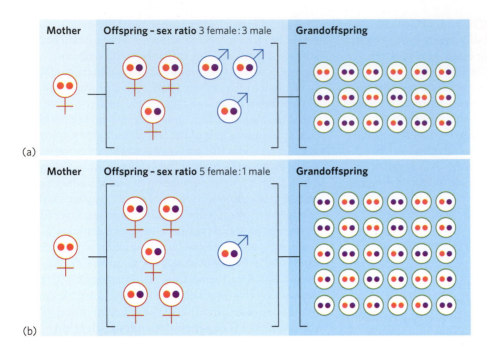

Figure 9.13 The effect of different sex ratios on a mother's fitness. When one son can breed with multiple daughters in an isolated population, the sex ratio of the mother's offspring will affect the number of grandoffspring that can be produced. In this example, we assume each female can produce six offspring. **(a)** If a mother produces three daughters and three sons, 18 grandoffspring can be produced. **(b)** If a mother produces five daughters and one son, 30 grandoffspring can be produced.

Highly Skewed Sex Ratios

We have seen that natural selection favours a one-to-one sex ratio. In some cases, however, we can observe highly skewed sex ratios. Fig wasps (*Pegoscapus assuetus*) offer a good example of this phenomenon. The female fig wasp arrives at a flower of a fig tree carrying pollen from another fig tree. She crawls in through a small hole in the flower and pollinates the flower. Once inside, the female lays its eggs in the developing fruit and then dies. It is fascinating that her eggs can be comprised of up to 90 percent daughters. Similar to the bees that we discussed at the beginning of this chapter, wasps can easily adjust the sex ratio of their offspring because fertilized eggs become daughters whereas unfertilized eggs become sons. Once the eggs hatch, the larvae feed on the fig fruit and seeds. The larvae metamorphose into adult wasps and breed with their siblings while still inside the flower. The young males breed with the young females, chew a hole in the side of the flower, and then die. The fertilized young females then escape through the hole and fly off to pollinate a new fig flower.

The high proportion of daughters among fig wasps occurs because male competition for females is among brothers. This phenomenon, known as **local mate competition**, occurs when competition for mates takes place in a limited area and only a few males are required to fertilize all the females. When only one fertilized mother enters a fig flower, the only males available to fertilize her daughters are her sons. From the mother's perspective, it does not matter which son passes along her genes to her grandoffspring. One son can fertilize multiple daughters, which means there is no fitness benefit to having multiple sons. The mother's fitness is therefore

a function of how many eggs can be produced by her daughters, so it is in the mother's best interest to produce many daughters and just enough sons to fertilize those daughters. As you can see in Figure 9.13, when mating options are restricted such that the only mates available for daughters are their brothers, mothers that produce a higher proportion of daughters than sons will have more grandoffspring and therefore greater evolutionary fitness.

Sometimes, however, two or more females lay their eggs in the same flower. In this situation, a mother's male offspring can mate either with their sisters or with the daughters of the other mothers. When this occurs, a mother will obtain greater fitness if she produces extra sons so she has enough sons to fertilize all the females in the flower. As expected, researchers observe that when a fig flower contains more than one mother, the mothers lay a higher proportion of male eggs.

Mating systems describe the pattern of mating between males and females

Although many species of algae and fungi have gametes that are of similar size, animals and plants produce sperm that are a fraction of the size of eggs. This smaller sperm takes much less energy to produce than an egg. Because of this difference, a female's reproductive success depends both on the number of eggs she can make and on the quality of the mates that she finds. Because most males can produce

Local mate competition When competition for mates occurs in a very limited area and only a few males are required to fertilize all of the females.

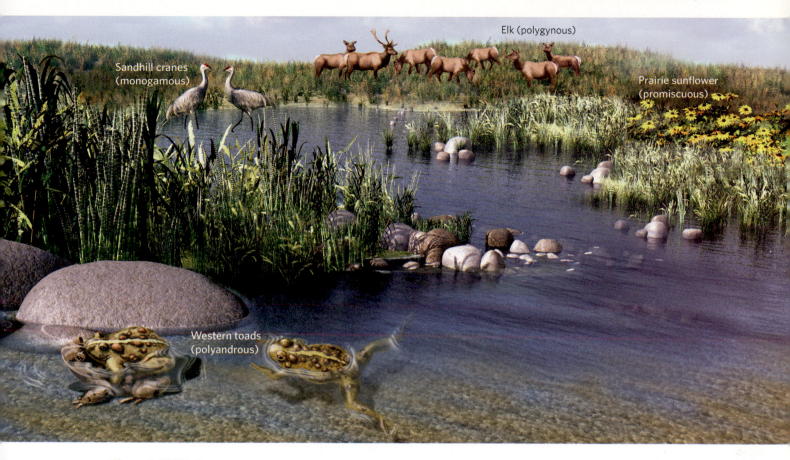

Elk (polygynous)

Sandhill cranes
(monogamous)

Prairie sunflower
(promiscuous)

Western toads
(polyandrous)

Figure 9.14 Mating systems. Promiscuity occurs in species of outcrossing plants such as the prairie sunflower (*Helianthus petiolaris*). Polygyny exists when a male mates with several females, as in elk (*Cervus canadensis*). Polyandry occurs when a female mates with multiple males, as is the case for the western toad (*Anaxyrus boreas*). Monogamy for life was thought to be the rule in 90 percent of all birds, as in the sandhill crane (*Grus canadensis*). However, recent genetic analyses have confirmed that most bird species actually participate in extra-pair copulations.

millions of sperm, a male's reproductive success generally depends on how many females he can fertilize. In this section, we will discuss the **mating system** of species, which describes the number of mates each individual has and the permanence of the relationship among mates. Like sex ratios, the mating system of a population is subject to natural selection. Consequently, mating systems are often a product of the ecological conditions under which species live. Figure 9.14 illustrates the four mating systems: promiscuity, polyandry, polygyny, and monogamy.

PROMISCUITY

Promiscuity is a mating system in which individuals mate with multiple partners and do not create a lasting social bond. Among animal taxa as a whole, promiscuity is by far the most common mating system. Promiscuity is universal among outcrossing plants because they send pollen out to fertilize the eggs of multiple individuals and receive pollen from multiple individuals.

When animals release eggs and sperm directly into the water or when pollen is released into the wind, much of the variation in mating success is

random. Whether a particular sperm is the first to find an egg is largely a matter of chance. When males attract or compete for mates, however, reproductive success can be influenced by factors such as body size and the quality of courtship displays. Even when fertilization is random, males that produce the most sperm or pollen and males that produce the most competitive sperm or pollen are bound to father the most offspring.

POLYGAMY

Polygamy is a mating system in which a single individual of one sex forms long-term social bonds with more than one individual of the opposite sex. Most

---

Mating system The number of mates each individual has and the permanence of the relationship with those mates.

Promiscuity A mating system in which males mate with multiple females and females mate with multiple males and do not create a lasting social bond.

Polygamy A mating system in which a single individual of one sex forms long-term social bonds with more than one individual of the opposite sex.

often, a male mates with more than one female, which is known as **polygyny**. In some species, polygyny evolves when males compete for females and the females all prefer only the few best males. In this case, the largest and healthiest males may mate with the vast majority of the females. Polygyny also can evolve when a male is able to defend a group of females from other males or when a male can defend a patchy resource that is attractive to multiple females. The guanaco, a relative of the camel that lives in the Patagonian region of South America, provides an example. The land in Patagonia is generally dry, with patches of wet habitat that grow nutritious plants that guanacos prefer to eat. When a male guanaco defends a wet patch from other males, he is able to mate with many of the females who come to the wet patch to eat the plants.

The opposite of polygyny is **polyandry**, a mating system in which a single female breeds with multiple males. The account of the queen honeybee at the beginning of this chapter is an example of polyandry because she mates with multiple drones. Polyandry commonly occurs when the female is in search of genetically superior sperm or has received material benefits from each male with whom she mates. For instance, in some species of insects, including some butterflies, the female receives a nutritional package of food— known as a *spermatophore*—from a male. The female uses the protein in the spermatophore to make her eggs. In polyandrous species, the more spermatophores that a female can collect, the more eggs she can produce.

MONOGAMY

Monogamy is a mating system in which a social bond between one male and one female persists through the period that is required for them to rear their offspring. In some cases, the bond may endure until one of the pair dies. Monogamy is favoured when males make important contributions in raising the offspring, such as when a male bird's help enables his chicks to grow and survive much better than they would without his help. Therefore his parenting improves his fitness. Having a social bond with only one other individual occurs in about 90 percent of bird species because males can offer much of the same care to offspring as females, such as incubating eggs, gathering food for the chicks, and protecting the chicks from predators. In mammals, however, fewer than 10 percent of species have a social bond with only one other individual. Because male mammals cannot provide the same offspring care as females, particularly because females lactate, the growth and survival of mammal offspring is less reliant on a male's presence.

For decades, ecologists observed that many species of birds have a single partner with which they breed and raise offspring. This male represents their single social bond. With the development of new DNA technologies, scientists began to examine the genetic relationships between parents and offspring and discovered that offspring in a nest often have different fathers. This means that the mother copulated with other males. While these females had a social bond with only one male, they were actually breeding with other males, a behaviour known as **extra-pair copulation**. In some monogamous species, a third or more of the clutches contain offspring sired by another male—usually a male in a neighbouring territory. Since neighbouring males also have a social bond with their own partner, both males and females use extra-pair copulations. Thanks to developing DNA technologies, we now know that 90 percent of bird species thought to be monogamous actually engage in extra-pair copulations.

Extra-pair copulations surely increase the fitness of the neighbouring males, but how does this behaviour increase the fitness of the female? One way would be if her reproductive success is improved by creating greater genetic variation among her offspring so at least one might be better suited to future environmental conditions. She could also benefit if the neighbouring males have better genotypes than her mate. In a bird known as the bluethroat (*Luscinia svecica*), for example, extra-pair copulations are common. Researchers in Norway examined the immune response of chicks whose father was the social partner of the female versus the extra-pair partner of the female. To do this, the researchers inject a small amount of foreign material—which is extracted from kidney beans—into a bird's wing and then they measure the amount of swelling that occurs at the site of the injection; the greater the swelling, the stronger the bird's immune response. As you can see in Figure 9.15, the amount of swelling in the wing was greater in offspring produced by extra-pair copulations, which demonstrates that they had a stronger immune response. This suggests that females seek extra-pair copulations as a way to obtain superior genotypes from another male and produce offspring that have superior immune systems.

Polygyny A mating system in which a male mates with more than one female.

Polyandry A mating system in which a female mates with more than one male.

Monogamy A mating system in which a social bond between one male and one female persists through the period that is required for them to rear their offspring.

Extra-pair copulations When an individual that has a social bond with a mate also breeds with other individuals.

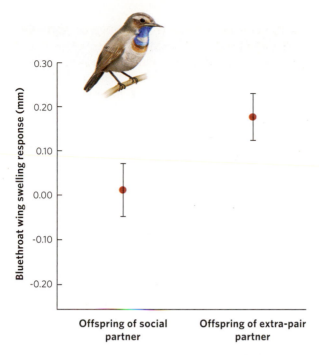

Figure 9.15 Female benefits of extra-pair copulations. In bluethroats, the strength of the immune system can be measured in terms of a wing-swelling response when the wing is injected with a chemical derived from plants. Larger wing-swelling responses indicate stronger immune systems. Error bars are standard deviations. Data from A. Johnsen, V. Andersen, C. Sunding, and J. T. Lifjeld, Female bluethroats enhance offspring immunocompetence through extra-pair copulations, *Nature* 406 (2000): 296–299.

The male participating in a monogamous social bond does not benefit if his mate breeds with other males. The threat that she will participate in extra-pair copulations has selected for **mate guarding**, a behaviour in which one partner prevents the other partner from participating in extra-pair copulations. A variety of mate guarding behaviours have evolved. In some species, an individual simply stays near its mate and keeps other potential suitors away, while in other cases the mate makes future breeding physically impossible. In 2004, for instance, researchers discovered that the golden orb-weaving spider (*Argiope aurantia*) inserts its sperm-transferring appendages into the two openings of the female's reproductive system. Within a few minutes of transferring his sperm, the male's heart stops and he dies with his appendages still stuck inside the female, making it impossible for the female to mate again. Such adaptations are effective at reducing the chance that an individual's partner will breed with another mate.

The various mating systems are the product of natural selection and are shaped by the ecological conditions under which each species lives. Because the mating systems often involve attracting a mate or defending against other members of the same sex,

natural selection has also caused the evolution of many sex-specific traits, as we will see in the next section.

Sexual selection favours traits that facilitate reproduction

We have seen that a male's reproductive success is typically determined by the quantity of females he can breed, whereas a female's reproductive success is typically determined by the quality of males that can fertilize her limited number of eggs. This means that females should generally be the choosier sex. But what exactly should she be choosing? In broad terms, females should select those males that improve her fitness by the greatest amount. She might select the male with the best genotype or the male with the most resources for her and her offspring.

Because females are choosy, males should compete strongly with other males for the opportunity to breed. This intense competition among males for mates has resulted in the evolution of male traits that are either used to attract females or used in contests and combat among males. Natural selection for sex-specific traits that are related to reproduction is referred to as **sexual selection**. In this section, we will explore how males and females have evolved different traits as a result of sexual selection, what traits females prefer in their mates, and how the fitness interests of males and females can cause conflict between the sexes.

SEXUAL DIMORPHISM

One result of sexual selection is **sexual dimorphism**, which is a difference in the phenotype between males and females of the same species, such as that observed in the honeybees at the beginning of this chapter. Sexual dimorphism includes differences in body size, ornaments, colour, and courtship behaviour. Traits related to fertilization—such as gonads—are referred to as **primary sexual characteristics** while traits related to differences in body size, ornaments, colour, and courtship are known as **secondary sexual characteristics**.

Mate guarding A behaviour in which one partner prevents the other partner from participating in extra-pair copulations.

Sexual selection Natural selection for sex-specific traits that are related to reproduction.

Sexual dimorphism The difference in the phenotype between males and females of the same species.

Primary sexual characteristics Traits related to fertilization.

Secondary sexual characteristics Traits related to differences between the sexes in terms of body size, ornaments, colour, and courtship.

Figure 9.16 Sexual size differences. In the golden silk spider (*Nephila clavipes*), females are much larger than males. Photo by Millard H. Sharp/Photo Researchers, Inc.

Figure 9.17 Male weapons. In some species, such as these caribou, males spar with each other for the right to breed females. Repeated contests cause selection for large weapons, including large antlers. Photo by Stephen J. Krasemann/www.agefotostock.com.

Sexual dimorphism can evolve due to differences in life history between the sexes, contests between males, or mate choice by females. Body size differences are common between the sexes of many animals because there has been selection for an increased number of gametes produced or for an increase in parental care by one sex. In fish and spiders, for instance, egg production is directly related to body size. This selection for greater gamete production in females without concurrent selection for greater sperm production in males could be the underlying cause of the larger size of females than males in many species of spiders and fish (Figure 9.16).

Sexual dimorphism can also occur when males compete for mates. In this case, selection will favour the evolution of weapons for combat. Such weapons include the horns of mountain sheep, the leg spurs of chickens and turkeys, and the antlers of caribou. Male caribou spar with each other (Figure 9.17), and the male with the smaller set of antlers usually abandons the competition. This favours the evolution of larger antlers and body size in males. Males that win such contests are more likely to gain access to females. Sexual dimorphism also commonly arises when one sex is choosy when selecting a mating partner. Since the females are the choosy sex in most species, female selection of males with particular traits can cause the sexual selection for extreme traits.

THE EVOLUTION OF FEMALE CHOICE

A female's preference for particular male traits should be tied to those features that improve her fitness. In terms of broad categories, we can consider two types of female preferences: *material benefits* and *nonmaterial benefits*.

Material benefits are those physical items that a male can provide a female including a site for raising offspring, a high-quality territory, or abundant food. In these cases, the benefit to the female is

straightforward—a site for raising offspring and resources for producing eggs and feeding the offspring should improve a female's fitness.

It has been more of a challenge for scientists to understand female choices when the female does not receive any material benefits from males. If the female choice is an adaptation, then there must be some benefit. One of the first demonstrations of female choice in nature for male traits that provided no material benefit came from a study of tail length in male long-tailed widowbirds (*Euplectes progne*), a small, polygynous species that inhabits the open grasslands of central Africa. The females are mottled brown, short-tailed, and drab. In contrast, during the breeding season males are jet black with a red shoulder patch, and they sport a half-metre-long tail that is conspicuously displayed in courtship flights (Figure 9.18a). The most successful males may attract up to a half dozen females to nest in their territories, but the males provide no care for their offspring. To determine what the females are choosing, researchers cut the tail feathers of some males to shorten them and glued the clipped feather ends onto the feathers of the tails of other males to lengthen them. As you can see in Figure 9.18b, males with experimentally elongated tails attracted significantly more females than those with shortened or unaltered tails. This suggests that even though tail length has no effect on a male's ability to maintain a territory, females choose mates on the basis of tail length.

Why would females choose a male based on a trait such as the length of his tail? According to the **good genes hypothesis**, an individual chooses a mate that possesses a superior genotype. In the grey treefrog, for example, females prefer males that can produce the

---

Good genes hypothesis The hypothesis that an individual chooses a mate that possesses a superior genotype.

(a)

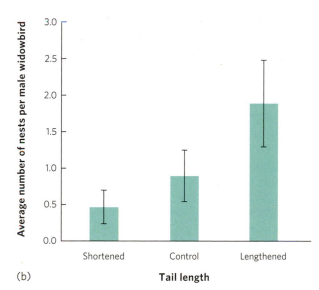

(b)

Figure 9.18 Long-tailed widowbird. (a) During the breeding season, males have exceptionally long tails. **(b)** When researchers made male tails shorter, the same length, or longer than typical males, females preferred males with longer tails. Error bars are standard errors. Photo by FLPA/Dickie Duckett/age fotostock. Data from M. Andersson, Female choice selects for extreme tail length in a widowbird, *Nature* 299 (1982): 818–820.

longest mating calls. Long calls can only be produced by the largest and healthiest male frogs. If male size and health have a large genetic component, choosing this trait might benefit the female's offspring. Indeed, when researchers forced females to breed with both long- and short-calling males, the offspring of long-calling males grew faster than the offspring of short-calling males. Subsequent research, illustrated in Figure 9.19, found that the offspring of long-calling males grew faster because they spend more time feeding compared to the offspring of short-calling males.

According to a second hypothesis, known as the **good health hypothesis**, individuals choose the

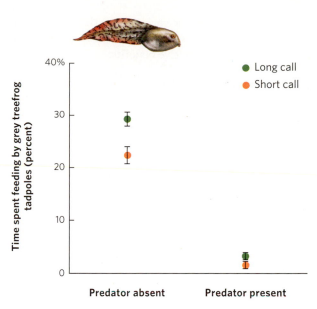

Figure 9.19 Good genes. In the grey treefrog, females who chose long-calling males produced offspring that spent more of their time eating than did the offspring of short-calling males, regardless of whether predators were present or absent. Error bars are standard errors. Data from G. V. Doty and A. M. Welch, Advertisement call duration indicates good genes for offspring feeding rate in gray treefrogs, *Behavioral Ecology and Sociobiology* 49 (2001): 150–156.

healthiest mates. Good health could be the outcome of either superior genetics or a superior upbringing with abundant resources. As a result, the good genes hypothesis and the good health hypothesis are not mutually exclusive. Females might prefer healthy males because such males may be both genetically superior and pose a lower risk of passing on a number of different parasites and diseases.

RUNAWAY SEXUAL SELECTION

Once female preference for a male trait has evolved, the trait may continue to evolve over time. For example, if females prefer longer tails in their mates, and there is genetic variation to select from, longer tails will continue to evolve in males. When selection for preference of a sexual trait and selection for that trait continue to reinforce each other, the result can be **runaway sexual selection**. Runaway sexual selection is thought to have favoured the evolution of such extreme traits as the half-metre-long tails of the male widowbird, the giant tail fans of the peacock, and other large male ornaments such as horns, tusks, and antlers. Runaway selection continues until males

Good health hypothesis The hypothesis that an individual chooses the healthiest mates.

Runaway sexual selection When selection for preference of a sexual trait and selection for that trait continue to reinforce each other.

either run out of genetic variation for the trait or until the fitness costs of possessing extreme traits begin to outweigh the reproductive benefits.

THE HANDICAP PRINCIPLE

If sexually selected traits indicate intrinsic attributes of male quality—at least initially, before runaway sexual selection occurs—we are then faced with a paradox. Presumably, extreme traits burden males by requiring energy and resources to maintain them, and they make males more conspicuous to predators. If this is the case, it is hard to imagine how extreme traits indicate a higher quality mate.

One intriguing possibility is that elaborate male secondary sexual characteristics act as handicaps. If a male can survive with sexual traits that require extra energy to build or that increase the risk of predation, these traits might signal a superior genotype. This idea, known as the **handicap principle**, argues that the greater the handicap an individual carries, the greater its ability to offset that handicap with other superior qualities.

One factor that might attract females to certain males is a genetically based, high resistance to parasites and pathogens. As you know, parasites evolve rapidly and thereby continually apply selection for genetic resistance in the host. Because parasites and pathogens can impair the construction of secondary sex characteristics, a male bird that has an elaborate and showy plumage might communicate to females that because he has the energy to build elaborate feathers, he can resist parasites and pathogens. If this resistance can be inherited by the offspring of the males, then the secondary sex characteristics are *honest signals* of the genetic superiority of the males and indicate that only individuals with superior genes could resist parasites and maintain a bright and showy plumage.

Numerous studies have found that parasites and pathogens affect male attractiveness. In rock doves (*Columba livia*)—also known as pigeons—hatchling birds can be infected by mites (*Dermanyssus gallinae*) that live in the nest. To determine the effect of the mites on the young doves, researchers fumigated one set of infected nests to kill the mites but did not fumigate a second set of infected nests. The nestlings living in the nests with mites experienced lower survival and slower growth as a nestling (Figure 9.20). Other researchers who have examined the effects of lice on rock doves have found that females prefer louse-free males to louse-infested males by a ratio of three to one. In ring-necked pheasants (*Phasianus colchicus*), females prefer males with longer spurs on their lower legs. Longer spurs are linked genetically to major histocompatibility complex (MHC) genes that influence susceptibility to disease. Males with longer spurs have MHC alleles that are linked to longer life spans. Therefore, females

Figure 9.20 The effect of mites on pigeons. The pigeons on the left were raised in nests that were fumigated to reduce the mite population. The pigeons on the right were raised at the same time, but in unfumigated nests that had high mite populations. The mite infections cause lower survival, slower growth, and areas of missing feathers. Photo by Dale H. Clayton.

that choose to mate with long-spurred males should produce offspring with a higher chance of surviving to reproduce as adults.

SEXUAL CONFLICT

Mating decisions were once thought to serve the mutual interest of both participants. More recently, scientists have come to appreciate that mating partners often behave according to their own self-interest. In lions, for example, when a new dominant male takes over the pride, he often kills newborn cubs that have been fathered by the previous dominant male so that the female loses her entire reproductive effort. The male obtains a fitness benefit because females without newborn cubs come into breeding condition more rapidly, which allows the new dominant male to father offspring sooner.

In some species of water striders sexual conflict can be intense (Figure 9.21). When males attempt to seize females, females struggle to displace the males. These struggles expose females to increased predation and energy expenditure. In some species, such as *Gerris incognatus*, females have pronounced abdominal spines. Researchers at the University of Toronto hypothesized that these spines help females dislodge males attempting to mate with them. To test this, the researchers created three groups of females: females with elongated spines made by adding glue to the spine tips, females with shortened spines, and females with unaltered spines. The researchers then measured the time females struggled before they dislodged a male and the number of matings each female allowed. Females with elongated spines dislodged males sooner than females with shortened and unaltered spines; they also mated less frequently. These results support the hypothesis that long abdominal spines in this species is a morphological adaptation to reduce the cost of male mating attempts. Examples such as these demonstrate that sexual interactions can reflect different decisions that represent the self-interest of males and females.

The handicap principle The principle that the greater the handicap an individual carries, the greater its ability must be to offset that handicap.

Figure 9.21 Sexual conflict in water striders. Male mating attempts are costly for females and they usually struggle to displace the males. In some species, females have evolved spines that allow the females to dislodge the males quicker and thus reduce the number of matings. Photo by A. Hartl/ www.agefotostock.com.

We have seen the ecological conditions that favour sexual versus asexual reproduction as well as the factors that favour balanced versus biased sex ratios. We have also seen how the different investments into eggs versus sperm, as well as the need to care for offspring, have shaped the evolution of different mating systems and the evolution of secondary sex characteristics. The evolution of sex remains an active area of research and there is still much to be learned. Today, researchers continue to make new discoveries and our understanding of sex evolution continues to grow.

ECOLOGY TODAY CONNECTING THE CONCEPTS

MALE-HATING MICROBES

(a)

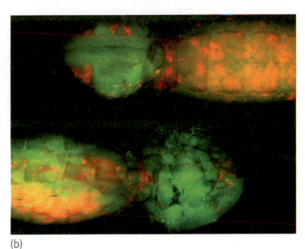

(b)

Infected with Wolbachia. The Wolbachia bacteria can infect a diversity of insect species, such as this mosquito (*Plasmodium berghei*). **(a)** The mosquito is seen under normal light. **(b)** When the mosquito and bacteria are treated to glow in different colours, the bacterial cells show up as red spots inside the mosquito's body. Photo (a) by CDC/Jim Gathany. Photo (b) from *Applied and Environmental Microbiology*, (2012), 78, 1497–1495, doi: 10.1128/AEM.06751-11. Reproduced with permission from American Society for Microbiology. Image courtesy of Jason Rasgon and Ryuichi Koga.

Throughout this chapter, we have seen how organisms have evolved a wide range of reproductive strategies. These strategies are typically adaptive; they improve an individual's fitness under particular ecological conditions. We have considered the different mating strategies as the result of the individual's genes and its environment. However, there is a growing realization that bacteria and other microorganisms living inside an organism can control its reproductive strategies.

One of the most widespread groups of sex-altering bacteria in insects belongs to the genus *Wolbachia*. Though first discovered nearly a century ago living in mosquitoes, it was not until 1990 that scientists realized that the bacteria could fundamentally change an individual's reproductive strategies. *Wolbachia* infects a wide variety of invertebrates including spiders, crustaceans, nematodes, and insects. In fact, it is currently estimated that 70 percent of all insect

species experience infections by this bacteria, which can live throughout the tissues of its host. In 2007 researchers discovered that the parasite's entire genome has even been incorporated into the genome of the tropical fruit fly (*Drosophila ananassae*). An important aspect of *Wolbachia's* life history is that it is only passed down to offspring via the mother's infected egg. From the microbe's perspective, therefore, the host's female offspring are important for improving the microbe's fitness whereas the host's male offspring are useless.

To improve its fitness, *Wolbachia* has evolved ways to exploit or eliminate these useless males. In mosquitoes, *Wolbachia* alters sperm to prevent them from fertilizing the eggs of uninfected females, which ensures that subsequent generations will be dominated by offspring of infected females that carry the bacteria. When researchers treated wasps with an antibiotic, they discovered that *Wolbachia* causes parthenogenesis and prevents some species from producing males. Once the bacteria were gone, the wasps no longer underwent parthenogenesis. Scientists now wonder if parthenogenesis in other species might also be the result of a similar bacterium rather than an adaptive strategy.

In woodlice (*Armadillidium vulgare*), *Wolbachia* converts males into females by suppressing male hormones. This ability for hormones to determine sex in the embryos is similar to the effects of hormone manipulations by scientists investigating temperature-dependent sex determination in Jacky dragons that we saw earlier. In other species of invertebrates, the bacterium simply kills young males and prevents them from entering the adult population. In species with adults of both sexes, such as the Ugandan butterfly (*Acraea encedon*), the increase in females relative to males has caused a reversal in the typical sex roles; females now court males and compete with other females for breeding opportunities. Note that none of these sex-determining mechanisms, changes in offspring sex ratio, or mating systems has been selected as adaptive strategies of the host. Rather, they are in the best interest of the bacteria.

Wolbachia are not the only bacteria with such effects. Bacteria in the genus *Rickettsia* are well-known for causing diseases such as typhus and Rocky Mountain spotted fever, but other species in this genus commonly occur in invertebrates without causing disease. In 2011, researchers reported that they found a species of *Rickettsia* living in the sweet potato whitefly (*Bemisia tabaci*). In the southwestern United States, infection rates increased from 1 percent in 2000 to nearly 100 percent by 2009. This incredibly rapid spread of the bacterium occurred because infected female flies nearly doubled their reproductive output, the survival rate of their offspring increased, and the offspring sex ratio was altered from 50 percent daughters to nearly 85 percent daughters. This is interesting from the standpoint of host–parasite interactions and the evolution of reproductive strategies. It is also important for agriculture. The sweet potato white fly is a pest that sucks the sap out of the leaves of many crops including cotton, vegetables, and ornamental plants. Scientists worry that the massive increase in infection rates by *Rickettsia* will lead to a massive population increase of this pest.

Collectively these studies suggest that there is much more to the evolution of sex than we currently know, and that understanding the evolution of sex in these organisms can have important implications for humans, including the number of pests attacking our food supply.

SOURCES: Himler, A. G., et al. 2011. Rapid spread of a bacterial symbiont in an invasive whitefly is driven by fitness benefits and female bias. *Science* 332: 254–256.

Knight, J. 2011. Meet the Herod bug. *Nature* 412: 12–14.

SUMMARY OF CHAPTER CONCEPTS

■ **Reproduction can be sexual or asexual.** Asexual reproduction can either be through vegetative reproduction or through parthenogenesis. Compared to asexual reproduction, sexual reproduction results in fewer copies of a parent's genes in the next generation. This can be offset by adopting a hermaphrodite sexual strategy or by providing parental care that results in raising twice as many offspring. The benefits of sexual reproduction include purging harmful mutations and creating genetic variation to help offspring deal with future environmental variation.

■ **Organisms can evolve as separate sexes or as hermaphrodites.** If an individual possessing only male or female function can add a large amount of the other sexual function while giving up only a small amount of its current sexual function, selection will favour the evolution of hermaphrodites. If not, selection will favour the evolution of separate sexes. To avoid inbreeding depression, hermaphrodites have evolved adaptations to prevent selfing and mixed strategies for when selfing is the best option.

■ **Sex ratios of offspring are typically balanced, but they can be modified by natural selection.** Depending on the species, sex may be largely determined by genetics or by the environment. In many species, females have the ability to manipulate the sex ratio by controlling which sperm are used to fertilize the eggs or which sex chromosomes end up in the eggs, or by selective abortion of the fertilized embryos. In most organisms, the sex ratio is approximately one to one due to frequency-dependent selection. When offspring are isolated from the rest of the population and are subjected to local mate competition, highly skewed offspring sex ratios can be adaptive.

■ **Mating systems describe the pattern of mating between males and females.** While many species are socially monogamous, recent studies have demonstrated that many individuals participate in extra-pair copulations. As a result of this infidelity, species have evolved a variety of mate guarding behaviours to prevent a reduction in their fitness.

■ **Sexual selection favours traits that facilitate reproduction.** The difference in the energetic costs of gametes and the costs of parental care typically causes female fitness to be a function of mate quality and male fitness to be a function of mate quantity. As a result, females are typically selective in choosing mates whereas males compete strongly with each other to mate as often as possible. Male competition for mates has favoured the evolution of sexually dimorphic traits including body size, ornaments, colouration, and courtship behaviours. Females choose particular males to obtain material benefits, such as nesting sites or food, or for nonmaterial benefits, such as good genes or good health. The best reproductive choices of males and females are often not reciprocal, which can cause conflicts between the sexes.

REVIEW QUESTIONS

1. Compare and contrast the costs and benefits associated with sexual versus asexual reproduction.

2. How does the Red Queen hypothesis help us understand the fitness benefits of sexual reproduction?

3. When the fitness increment of increased male function results in a larger cost in fitness through female function, why should a population evolve separate sexes rather than hermaphrodites?

4. Given that self-fertilization leads to inbreeding depression, under what conditions should a hermaphrodite use self-fertilization?

5. When a population is composed of two sexes, why does the rarer sex have a fitness advantage?

6. How does local mate competition favour the production of female-biased sex ratios in offspring?

7. Explain how environmental sex determination can be an adaptive strategy.

8. Compare and contrast monogamy, polygyny, polyandry, and promiscuity.

9. Explain how extra-pair copulation has favoured the evolution of mate guarding.

10. Why might exaggerated secondary sexual characteristics in males demonstrate a superior genotype to females?

PRACTICING ECOLOGY: HOW TO HANDLE SNEAKY MALES

Bryan Neff and his lab at the University of Western Ontario study behavioural ecology and the evolution of mating systems. Photo by John W. Heath. Courtesy of Bryan Neff.

When there is intense competition among males for mates, females are likely to ignore inferior males in favour of superior males. Consequently, natural selection should favour strategies that allow inferior males to fertilize the eggs of a female before the superior male drives them away. However, any eggs fertilized by an inferior male lower the number fertilized by the superior male, which reduces his fitness. What can a superior male do to improve his fitness when inferior males fertilize some of the eggs?

The bluegill sunfish (*Lepomis macrochirus*) provides a good example of this dilemma. The bluegill is found in lakes throughout North America. Superior males—which are the largest males—create depressions in the benthos to serve as nests. When the male attracts a female, she releases her eggs into the nest and he releases a cloud of sperm. However, inferior males frequently sneak into the nest, often mimicking the behaviour of a nonthreatening female. The inferior male quickly casts his sperm onto the eggs before the superior male chases him away. After mating, the superior males guard the nest, but they are frequently observed eating some of the eggs and hatched fry. Scientists have wondered whether superior males are removing offspring fertilized by inferior males so they do not waste energy protecting unrelated offspring. If so, we would hypothesize that cannibalism occurs more in nests with eggs and fry that have been fertilized by inferior males.

To test this hypothesis, Bryan Neff and his lab at the University of Western Ontario have examined bluegill colonies of Lake Opinicon, which is located just north of Lake Ontario. During the mating season, they observed 40 superior males in their nests after eggs were fertilized. The researchers removed some of the fertilized eggs from each nest, extracted DNA from the eggs, and determined the percentage of the eggs that were fertilized by inferior males. As the eggs developed, the researchers also quantified cannibalistic behaviour the superior males exhibited toward the eggs and, later, toward the newly hatched fry. This behaviour was quantified by counting the number of pecking motions the superior males made toward eggs and fry, which are indicative of eating. You can view their data in the table.

INFERIOR MALE PATERNITY IN BLUEGILL NESTS AND CANNIBALISM BY SUPERIOR MALES ON EGGS AND FRY

INFERIOR MALE PATERNITY (%)	PECKS TOWARD EGGS	PECKS TOWARD FRY
8.23	12	3
25.12	13	16
13.64	17	3
3.97	11	1
2.11	8	0
12.40	7	7
2.01	23	3
0.00	0	0
27.46	18	11
14.96	2	12

GRAPH THE DATA

1. Create a scatterplot graph with inferior male paternity on the *x* axis and the number of pecks toward eggs on the *y* axis, and visually estimate a regression line.

2. Create a second scatterplot graph with inferior male paternity on the *x* axis and the number of pecks toward fry on the *y* axis, and visually estimate a regression line.

INTERPRET THE DATA

1. Compare the spread of data points around the mean regression line in both scatterplot graphs. Which graph should have a larger coefficient of determination (R^2)? What does this comparison tell you about how well each *y* variable correlates with percentage of inferior male paternity?

2. Explain how the data support or refute the hypothesis that cannibalism of eggs and fry occurs more in nests containing eggs that have been fertilized by inferior males.

3. In another study, Neff and his colleagues found that fry emit olfactory cues that superior males can use to assess paternity, but eggs do not emit these cues. Does this finding support the hypothesis that cannibalism occurs more in nests containing eggs that have been fertilized by inferior males?

SOURCES: Neff, B. 2003. Paternity and condition affect cannibalistic behavior in nest-tending bluegill sunfish. *Behavioral Ecology and Sociobiology* 54: 377–384.

Neff, B., and Sherman, P. W. 2003. Nestling recognition via direct cues by parental male bluegill sunfish (*Lepomis macrochirus*). *Animal Cognition* 6: 87–92.

CHAPTER 10

Social Behaviours

The Life of a Fungus Farmer

The leaf-cutter ant is an extraordinary farmer. Living in colonies of several million individuals, these ants leave the colony each day to harvest leaves from the surrounding forest. Using their sharp mandibles, they slice through leaves to cut off pieces that are many times larger than their bodies. They then carry the pieces back to the nest, which can rise several metres out of the ground and sink tens of metres below the ground. Back at the nest, the ants consume the sap in the leaves, but they do not eat the leaves. Instead, they use the leaves to grow a specialized species of fungus that they consume.

There are more than 40 species of leaf-cutter ants, which live primarily in Mexico, Central America, and South America. Like honeybees, the leaf-cutter ants form enormous societies of co-operating individuals. An ant colony normally has a single queen that can live for 10 to 20 years. Early in her life, the queen participates in a mating flight with males. The sperm she receives then are held inside her body and remain viable for the rest of her life. She uses them sparingly when fertilizing eggs to make daughters; in some species, the queen releases only one or two sperm per egg. Occasionally she lays unfertilized eggs to make sons whose only function is to mate with other queens. The millions of individuals in the nest are daughters of the same queen and sisters to each other. They all forgo reproduction.

> "There are nearly 30 different jobs for the worker ants and different workers are suited to different jobs."

Daughters in this ant society are the workers in the colony and the division of labour among the workers is amazingly complex. Scientists estimate that there are nearly 30 different jobs for the worker ants and they have determined that different workers are suited to different jobs. The worker *caste* is composed of several subcastes, known as *minims, minors, mediae,* and *majors.* Ants in each subcaste differ dramatically in size and shape. The largest workers, the majors, can be 200 times more massive than the smallest workers, the minims. The differences in subcastes are thought to be a phenotypically plastic response to different diets the ants receive as larvae. In addition to size and shape differences, the jobs of a given worker can also change during its lifetime. For example, when workers are young, they spend most of their time inside the nest where they build tunnels, air condition the nest, and raise the larvae. When large leaf pieces arrive at the nest, other workers cut them into smaller pieces and then even smaller workers mash them into tiny bits. The smallest workers bring a strand of fungus to the mashed bits and tend the fungus gardens, a job that includes removing undesirable species of fungi that can be pathogenic to the ants.

The oldest and largest ants act as soldiers and go out of the nest to collect leaves. The process of leaf collection is complex. Some ants climb into trees and cut large pieces of leaves that drop to the ground, while others relay leaf pieces back to the nest. When workers first begin leaf cutting, they possess razor-sharp jaws, known as mandibles, that are very

Leaf-cutter ants. Through an extensive division of labour, leaf-cutter ants work together to bring pieces of leaves back to their nest. In this photo, a large worker ant carries a cut piece of leaf while a much smaller worker ant rides on top to discourage parasitoid flies from attacking the worker. Photo by Gavriel Jecan/age fotostock.

effective at cutting tough leaves. Researchers have recently discovered that these mandibles become dull over time, causing older workers to take twice as long to cut a leaf. When an ant's cutting performance declines, it shifts to the task of leaf carrying. This change in job allows older individuals to continue contributing to the ant society. The division of labour also helps the group to maintain a high foraging efficiency.

While the ants benefit from living in a very large group, the large colonies and conspicuous foraging behaviour make them quite noticeable to their enemies. For example, a species of parasitoid fly that is specialized to hunt large leaf-cutter ants lays its eggs inside the necks of foraging ants. To reduce the risk of these attacks, leaf-cutter ants have evolved several tactics. Smaller individuals, which are less attractive to the fly, forage during the day. At night, when the flies do not hunt, the larger, more effective foragers go out to collect leaves. Small workers also serve as guards by riding on leaves that larger ants carry. When a fly approaches, these tiny guards prevent the fly from laying its eggs on the neck of the large ant that is carrying the leaf. When not repelling attacks, these hitchhiking individuals clean the leaf of undesirable micro-organisms.

The leaf-cutter ants illustrate an extreme case of social behaviour and living in groups. As we will see in this chapter, being social is a common occurrence in the animal world and the ecological conditions under which species live have important consequences for the costs and benefits of social living.

SOURCES: B. Hölldobler and E. O. Wilson, *The Leafcutter Ants* (Norton, 2011).

R. M. S. Schofield et al., Leaf-cutter ants with worn mandibles cut half as fast, spend twice the energy, and tend to carry instead of cut, *Behavioral Ecology and Sociobiology* 65 (2011): 969–982.

CHAPTER CONCEPTS

- Living in groups has costs and benefits.

- There are many types of social interactions.

- Eusocial species take social interactions to the extreme.

In the course of a lifetime, an individual typically interacts with many members of its species. Interactions with mates, offspring, other relatives, and unrelated individuals in one's species are known as **social behaviours**. Like most behaviours, social behaviours have a genetic basis and are therefore subject to natural selection. As a result, many types of social behaviours have evolved to favour the cohesiveness of family groups and populations and to constrain antagonism among conspecifics.

Although the study of social behaviour typically focuses on animals, many other organisms interact with conspecifics in ways that might be considered to be social. For example, bacteria and protists can sense the presence of individuals of the same species, often through chemical secretions, and react in "friendly" or "aggressive" ways. During parts of their life cycles, free-living slime moulds respond to others when they aggregate to form large fruiting bodies. Even plants communicate with one another. When one plant is damaged by herbivores, it emits volatile compounds. Other individuals detect the compounds and respond to them by producing chemical or structural defences against future herbivore attacks.

In this chapter, we will focus on the social behaviours of animals, which play important roles in the ecology of individuals, populations, communities, and ecosystems. Social interactions pervade nature and form an important part of the environment to which populations adapt. We will explore some of the implications of interacting within social groups, and describe various ways that individuals manage social relationships. We will also examine how different ecological conditions affect the evolution of social behaviours.

Living in groups has costs and benefits

Animals are social for a variety of reasons. In some cases, offspring remain with their parents to form family groups and do not disperse. In other cases,

Social behaviours Interactions with members of one's own species, including mates, offspring, other relatives, and unrelated individuals.

Figure 10.1 Group defence. Adult muskox (*Ovibos moschatus*), like these from Victoria Island, Canada, form an outward-facing circle and place the calves inside the circle where they are safe from approaching predators. Photo by Eric Pierre/NHPA/Science Source.

individuals are attracted to each other for breeding. Individuals can also aggregate because they are independently attracted to the same habitat or resource. For instance, vultures gather around a carcass and dung flies congregate on cow pies. In this section, we will examine the costs and benefits of living in social groups and then discuss how animals use *territories* and *dominance hierarchies* in social interactions.

BENEFITS OF LIVING IN GROUPS
Animals generally form groups to increase their survival, rate of feeding, or success in finding mates.

Survival
While an individual might not be able to fend off the attack of a predator alone, a group of individuals can be quite effective at doing so (**Figure 10.1**). Another survival mechanism available to social groups is a phenomenon known as the *dilution effect*. The **dilution effect** refers to the reduced, or diluted, probability of predation to a single animal when it is in a group. In an aggregation of prey, the predator has many prey choices, so the individual living in a group has a lower probability of being caught. The dilution effect is an important benefit of large groups such as herds of mammals, flocks of birds, and schools of fish.

A lower probability of predation in groups also allows individual prey animals to spend less time watching for predators. Consider the case of the European goldfinch (*Carduelis carduelis*), a small bird that feeds on the seed heads of plants in open fields and hedgerows. If you watch closely as the birds feed, you will notice that they raise their heads and look around for predators. The total number of head raises conducted by the group increases with flock size; as you can see in **Figure 10.2a**, the larger the group, the more eyes are on the lookout for predators. As flock size increases, however, each individual can raise its

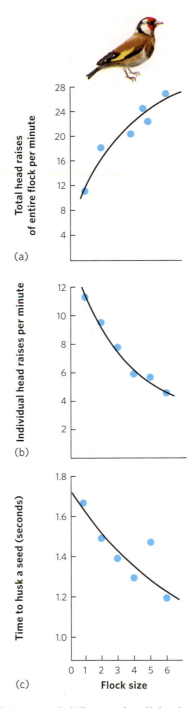

Figure 10.2 Increased vigilance when living in a group. In the European goldfinch, an increase in flock size results in **(a)** an increase in the total number of head raises performed by the flock, **(b)** a decrease in the number of head raises performed by an individual, and **(c)** a decrease in the time required to husk a seed. Data from E. Glück, Benefits and costs of social foraging and optimal flock size in goldfinches (*Carduelis carduelis*), *Ethology* 74 (1987): 65–79.

head less frequently, as shown in Figure 10.2b. Because each individual spends less time looking for predators, it can spend more time feeding. The data in Figure 10.2c show that when a goldfinch spends less

Dilution effect The reduced, or diluted, probability of predation to a single animal when it is in a group.

time looking for predators, it can husk a seed much faster and therefore consume seeds more quickly. Vancouver Island marmots (*Marmota vancouverensis*) show similar patterns. Today's population of 200 individuals is only 80 to 90 percent of the population of the 1980s. Marmots today spend 10 times longer on vigilance behaviour and their feeding rate declined by 86 percent compared to marmots in the larger population of three decades ago.

Feeding

Living in groups can also help animals locate and consume resources. Having many conspecifics all searching for food means that there are many more sets of eyes that might find food when it is rare. In some cases, animals may find food easily but have difficulty in capturing and killing it when they are alone. In lions, for example, a lone female has a low probability of capturing and killing a zebra, but if she hunts with many other lions, the probability of a capture goes up dramatically.

Mating

Socializing can also provide mating benefits since being social makes it easier to locate potential mates. An extreme example of socializing for mating benefits occurs when animals aggregate in large groups to attract members of the opposite sex by making calls or displaying in ways that capture the attention of potential mates. The location of the aggregation, known as a **lek**, is used only for displaying. The site has no other value either to the displaying sex or to the attracted sex. For example, males of the ruff (*Philomachus pugnax*)—a medium-sized wading bird that lives in northern Europe and Asia—come together at a lek and participate in mating displays to attract females. On the island of Gotland in Sweden, researchers observed ruff leks to determine whether lek size affected ruff mating. As you can see in Figure 10.3a, males in larger leks were more successful at attracting females. In addition, as illustrated in Figure 10.3b, males in larger leks experienced a higher percentage of successful copulations with females, which confirms that forming social groups provides fitness benefits to the male birds.

COSTS OF LIVING IN GROUPS

The benefits of group living can certainly be substantial for many species, but group living can also come with costs that include predation and competition.

Predation

Groups of animals are much more conspicuous to predators than are individual animals. In a grassland, for example, it is easier for a predator to spot a herd of antelopes than an individual antelope. Given the propensity of antelopes to live in herds, this cost of being detected is outweighed by the benefits of the dilution effect and of more eyes to detect approaching predators.

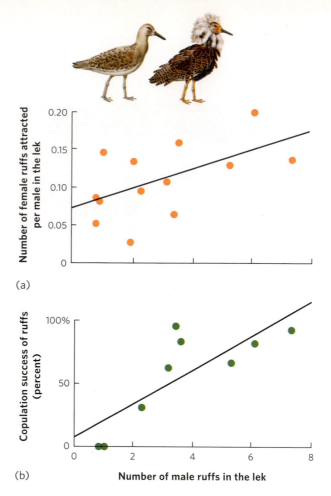

(a)

(b)

Figure 10.3 Breeding benefits in a lek. Among displaying ruff males, those displaying to females in larger groups **(a)** are more likely to attract females and **(b)** have a higher probability of successfully copulating. Data from J. Högland, R. Montgomerie, and F. Widemo, Costs and consequences of variation in the size of ruff leks, *Behavioral Ecology and Sociobiology* 32 (1993): 31–39.

The risk of parasites and pathogens can also increase when living with conspecifics. Many species of parasites and pathogens spread from one host to another. High population density can increase the rate that disease spreads and can lead to epidemics. For instance, coral reefs that experience fishing pressure generally have fewer fish compared to reefs that are protected from fishing. In 2008, researchers reported the results of a study on fish parasites from a protected and unprotected coral reef in the central Pacific Ocean. As shown in Figure 10.4, they found that fish on the protected reef were infested with a higher number of parasite species than the same species living in a coral reef that was fished. In addition, fish living in the protected reef frequently carried higher numbers of each species of parasite.

The parasite and disease costs of group living can also be readily observed in modern aquaculture operations, which farm aquatic species for human consumption. These operations raise oysters, salmon, catfish, shrimp, and other edible species at very high densities. Under such conditions, a single infected

Lek The location of an animal aggregation to put on a display to attract the opposite sex.

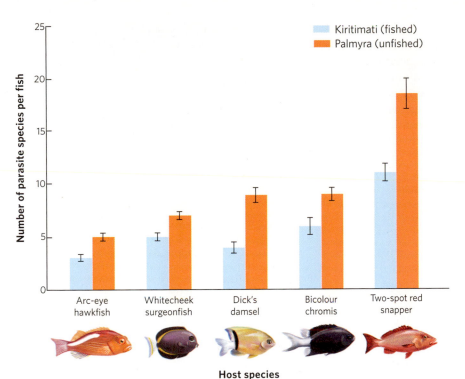

Figure 10.4 Parasite occurrence in coral reef fishes. Coral reefs protected from fishing have higher fish densities. A survey of five different species of fish found that fish living in a reef that is subject to fishing pressure contained fewer species of parasites than the same species living in a reef that is protected from fishing. Error bars are 95% confidence intervals. Data from K. D. Lafferty, J. C. Shaw, and A. M. Kuris, Reef fishes have higher parasite richness at unfished Palmyra Atoll compared to fished Kiritmati Island, *EcoHealth* 5 (2008): 338–345.

individual can rapidly spread parasites and pathogens to the rest of the group.

Increased transmission of parasites and diseases to groups that live in higher densities makes it undesirable for people to feed wild animals such as deer. When there is a steady, easily available source of food, deer form large aggregations around the food. This aggregating behaviour makes it more likely that the animals will experience outbreaks of parasites compared to when they live in smaller family groups. Similar concerns exist for livestock operations in which animals are raised under very high densities. In this situation, diseases can jump to native wildlife populations and have dramatic effects with such diseases as rinderpest, avian influenza, and West Nile virus.

Competition

Another major cost of living in groups is competition for food. While larger groups are better at locating food, the food must then be shared among all of the individuals in the group. For example, transient killer whales off British Columbia feed exclusively on other marine mammals, such as porpoises and seals, which they hunt as a group. As you can see in Figure 10.5 single whales or pairs of whales have a smaller energy intake per day than groups of three. Trios are more likely to detect prey and will be more successful in capturing the prey than single individuals or pairs. As groups become larger than three, daily energy intake decreases again. While such large groups are more likely to detect prey than smaller groups, more individuals have to share the captured prey. Groups of three individuals offer the best balance between the ability to detect and capture prey, and the individual

share of the prey each whale receives. For this reason, the most frequently observed group size of transient killer whales is three whales.

Each species that has evolved to live in groups faces different costs and benefits, which depend on the ecological conditions under which it lives. Assuming a genetic component for such social behaviour, we expect natural selection to favour the evolution of group sizes that balance the costs and benefits for each species.

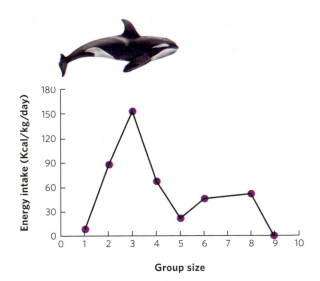

Figure 10.5 Energy intake and group size in transient killer whales. In small groups, whales are unlikely to detect and capture prey. In large groups, an increasing number of individuals has to share the prey. A group of three whales provides the best balance and provides the most energy for each individual. Data from R.W. Baird and L.M. Dill, Ecological and social determinants of group size in transient killer whales, *Behavioral Ecology* 7 (1996): 408–416.

TERRITORIES

Many species of animals have evolved to manage living near other conspecifics by establishing a *territory* or a *dominance hierarchy*. A **territory** is any area defended by one or more individuals against the intrusion of others. Territories can be either transient or relatively permanent, depending on the stability of the resources in the territory and how long an individual needs those resources. For example, many migratory species establish summer breeding territories and defend them for several months. Defending a high-quality territory generally assures greater resources, such as abundant food or nest sites. This typically improves the attractiveness of a territory holder as a mate, and therefore its fitness. When breeding is complete for the season, migratory species move on to their wintering grounds where they establish new territories. Shorebirds that stop at several points along the way of their long migration defend feeding areas for a few hours or days and then continue their migratory trip. Hummingbirds and other nectar feeders defend individual flowering bushes and abandon them when those bushes cease producing flowers. As long as a resource can be defended and the benefits of defending the resource outweigh the costs, animals are likely to maintain territories.

DOMINANCE HIERARCHIES

In some situations, defending a territory is impractical. This can occur when an individual is surrounded by so many conspecifics that it becomes impractical to defend against them all, when resources are available for only short periods of time, or when the benefits of living in a group override the benefits of defending a territory. In such circumstances, individuals of many species form dominance hierarchies. A **dominance hierarchy** is a social ranking among individuals in a group, typically determined through fighting or other contests of strength or skill. Once individuals order themselves into a dominance hierarchy, subsequent contests among them are resolved quickly in favour of higher-ranking individuals. In a linear dominance hierarchy, the first-ranked member dominates all others, the second-ranked dominates all but the first-ranked, and so on down the line to the last-ranked individual, who dominates no one else in the group.

There are many types of social interactions

Most social interactions can be considered as an action by one individual, the **donor** of the behaviour, directed toward another individual, the **recipient** of the behaviour. One individual delivers food, the other receives it; one individual attacks, the other is attacked. When the attacked individual responds—by standing its ground or by fleeing—it becomes the donor of this subsequent behaviour. Every interaction between two individuals has the potential to affect

Figure 10.6 The four types of social interactions. Co-operation occurs when the donor and recipient of a behaviour both obtain a positive fitness effect. Selfishness occurs when the donor obtains a positive fitness effect while the recipient obtains a negative fitness effect. Altruism occurs when the donor obtains a negative fitness effect while the recipient obtains a positive fitness effect. Spitefulness occurs when the donor and recipient both obtain a negative fitness effect.

the fitness of both individuals, either in a positive or negative way. To understand how an interaction affects both participants, it can be useful to categorize the interactions. In this section, we will explore the four types of social interactions between donors and recipients and then examine the conditions that favour a donor helping or harming a recipient.

THE TYPES OF SOCIAL INTERACTIONS

Social behaviours can be placed into one of four categories, as illustrated in Figure 10.6: *co-operation, selfishness, spitefulness,* and *altruism*. When the donor and the recipient both experience increased fitness from the interaction, we call it **co-operation**. When one lion helps another kill a gazelle, for instance, both individuals experience a fitness benefit. When the donor experiences increased fitness and the recipient experiences decreased fitness, we call it **selfishness**.

Territory Any area defended by one or more individuals against the intrusion of others.

Dominance hierarchy A social ranking among individuals in a group, typically determined through contests such as fighting or other contests of strength or skill.

Donor The individual who directs a behaviour toward another individual as part of a social interaction.

Recipient The individual who receives the behaviour of a donor in a social interaction.

Co-operation When the donor and the recipient of a social behaviour both experience increased fitness from an interaction.

Selfishness When the donor of a social behaviour experiences increased fitness and the recipient experiences decreased fitness.

Selfishness is a common interaction between two conspecifics that compete for a resource such as food. The winner of the competition receives a fitness benefit while the loser experiences a fitness loss. **Spitefulness** occurs when a social interaction reduces the fitness of both donor and recipient. The fourth type of interaction, **altruism**, increases the fitness of the recipient but decreases the fitness of the donor.

Under what conditions would natural selection favour each type of social interaction? For both co-operation and selfishness, the interactions benefit the donor. We would therefore expect natural selection to favour any donors that engage in either co-operation or selfishness. In contrast, spitefulness cannot be favoured by natural selection under any circumstance given that both participants experience lower fitness. Consistent with this prediction, spitefulness is not known to occur in natural populations. Explaining the evolution of altruism presents a unique challenge because it requires natural selection to favour individuals who improve the fitness of others while reducing their own fitness. We will explore this challenge in the next section.

ALTRUISM AND KIN SELECTION

Altruistic behaviour presents an evolutionary problem because it does not lead to an increase in *direct fitness*. **Direct fitness** is the fitness that an individual gains by passing on copies of its genes to its offspring. We would expect selfish individuals to prevail over altruistic individuals because selfishness increases the fitness of the donor. Despite this expectation, altruism has evolved in many species. For example, some of the most extreme cases of altruism occur in colonial species, such as leaf-cutter ants and honeybees, in which workers forgo personal reproduction to rear the offspring of the dominant female.

We can explain altruistic behaviour by looking beyond direct fitness. When an individual has an altruistic interaction with a relative, it increases the fitness of the relative with which it shares genes through a common ancestor. When you help a relative improve its fitness, you are indirectly passing on more copies of your genes, which gives you **indirect fitness**. An individual's **inclusive fitness** is the sum of its direct fitness and indirect fitness. When considering how selection operates, we say that direct fitness is favoured by **direct selection**. Indirect fitness through relatives is favoured by **indirect selection**, also known as **kin selection**.

As we have noted, indirect or kin selection occurs because an individual and its relatives carry copies of some of the same genes inherited from a recent common ancestor. The probability that copies of a particular gene are shared by relatives is known as the **coefficient of relatedness**. As depicted in Figure 10.7,

Spitefulness When a social interaction reduces the fitness of both donor and recipient.

Altruism A social interaction that increases the fitness of the recipient and decreases the fitness of the donor.

Direct fitness The fitness that an individual gains by passing on copies of its genes to its offspring.

Indirect fitness The fitness that an individual gains by helping relatives pass on copies of their genes.

Inclusive fitness The sum of direct fitness and indirect fitness.

Direct selection Selection that favours direct fitness.

Indirect selection Selection that favours indirect fitness. *Also known as* Kin selection.

Coefficient of relatedness The numerical probability of an individual and its relatives carrying copies of the same genes from a recent common ancestor.

Figure 10.7 Coefficients of relatedness. The coefficient of relatedness is the probability that one individual possesses the same copy of a gene as another individual through a shared relative. In this family tree, we see that the individual in the red box has a 0.5 coefficient of relatedness with its parents, siblings, and offspring. More distant relatives have lower coefficients of relatedness, as indicated by the boxes with lighter shades of red. The coefficients of relatedness are based on the assumption that none of the mates are related to the highlighted individual.

its value for diploid organisms depends on the degree of relationship between two individuals. If we focus on the individual in the red box of the family tree, we see that the coefficient between this individual and its offspring is 0.5 because the individual has two sets of genes but gives only one set to its offspring. As a result, they only have half of their genes in common. This means that the coefficient of relatedness between our focal individual and its parent is also 0.5. If we next consider the focal individual and its siblings, we see that these two individuals have a 0.5 probability of receiving copies of the same gene from a parent. In the case of two cousins, the probability drops to 0.125 (one in eight) of inheriting copies of the same gene from one of their grandparents, which are their closest shared ancestors. Using these coefficients of relatedness, we can calculate the indirect fitness as the benefit given to a recipient relative (B) multiplied by the coefficient of relatedness between the donor and the recipient relative (r):

Indirect fitness benefit = $B \times r$

In the case of nonrelatives, there is a zero probability that an individual carries the same genes from a recent common ancestor. In examining these different coefficients of relatedness, we can see that an individual has a higher probability of leaving more copies of its genes in the next generation by promoting the fitness of its closest relatives and gains nothing by promoting the fitness of nonrelatives.

Understanding the role of kin selection and coefficients of relatedness helps resolve the puzzle of how altruistic social interactions can evolve. Whereas selfish interactions provide direct fitness benefits to the donor, altruistic interactions provide indirect fitness benefits to the donor, weighted by the coefficient of relatedness between the donor and the recipient. If the inclusive fitness of altruistic behaviours exceeds the inclusive fitness of selfish behaviours, then altruism will be favoured by natural selection.

The evolution of altruistic behaviour becomes clear when we examine the costs and benefits in an equation. Genes for altruistic behaviour will be favoured in a population when the benefit to the recipient (B) times the recipient's coefficient of relatedness to the donor (r) is greater than the direct fitness cost to the donor (C):

$B \times r > C$

If we rearrange this equation, we can show that for altruism to evolve, the cost-benefit ratio must be less than the coefficient of relatedness between donor and recipient:

$C/B < r$

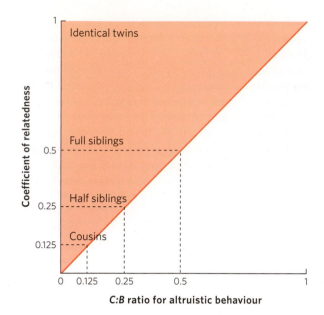

Figure 10.8 Conditions that favour the evolution of altruistic behaviours. An altruistic behaviour will evolve whenever the ratio of donor costs and recipient benefits (*C/B*) is less than the coefficient of relatedness between the donor and recipient. The region in red indicates the conditions that favour the evolution of altruistic behaviours.

Figure 10.8 shows this relationship graphically. Based on this equation and the figure, we can see that altruism is favoured when the cost to the donor is low, the benefit to the relative is high, and the donor and its relative are closely related.

A study of wild turkeys (*Meleagris gallopavo*) in California has shown how altruistic behaviour can be maintained through kin selection. Male turkeys display at leks by puffing up their feathers and strutting back and forth to attract females. Males may either display alone or as part of a coalition with other males (Figure 10.9). When a coalition of males

Figure 10.9 Wild turkey coalition. When two or more male turkeys display together to attract a female, only the dominant male sires the offspring. These individuals live in Texas. Photo by Larry Ditto/age fotostock.

displays together, only the dominant male copulates with the females that they attract. This raises the question of why the subordinate males in a coalition spend their time and energy displaying when they do not produce any offspring. Researchers obtained their first clue by using genetic data. Males in a coalition were more closely related than males drawn at random from the population. Indeed, the average coefficient of relatedness between groups of displaying males was 0.42, which suggests that paired

males represent a mixture of full brothers ($r = 0.5$) and half brothers ($r = 0.25$).

The researchers then determined the average number of offspring sired by the different types of males. Dominant males in a coalition sired an average of 6.1 offspring; subordinate males in a coalition sired 0 offspring; and males that displayed alone sired an average of 0.9 offspring. With these data, we can evaluate the fitness of the subordinate male in the coalition. By being a part of the coalition, the

---

ANALYZING ECOLOGY

Calculating Inclusive Fitness

In the pied kingfisher (*Ceryle rudis*), a fish-eating bird from Africa and Asia, adult males often forgo their own reproduction to help their parents raise offspring. Researchers have identified *primary helpers* and *secondary helpers*. Primary helpers are sons of the parents and they work hard to protect the nest and to bring food to the chicks. In some cases, one of the parents disappears and is replaced by an unrelated mate, so the son ends up helping a parent and a step-parent. Secondary helpers—which are unmated males from other families—are not related to the parents and do not work as hard to feed and protect the offspring. After helping for one year, both types of helpers set up their own nest the following year. A third group of males, known as *delayers,* do not help but simply delay reproduction until their second year.

Researchers followed several nests of kingfishers and determined how much each helper improved the fitness of the parents in the first year (B_1), the probability of surviving and finding a mate in the subsequent year (P_{sm}), and the fitness of the helper when he bred independently in the second year (B_2). They also quantified the coefficients of relatedness between the helpers and the parents being helped in the first year (r_1) and between the

helpers and their own offspring produced in the second year (r_2). The coefficient of relatedness for the primary helpers was 0.32 in year 1. This was the result of some nests retaining both parents of the helper ($r = 0.5$) and other nests having one parent of the helper plus a step-parent ($r = 0.25$).

Based on these data, the researchers calculated the inclusive fitness of the primary helpers, secondary helpers, and delayers.

As you can see, the primary helper had inclusive fitness that was a bit higher after 2 years. The primary helpers obtain about half of their inclusive fitness by helping their parents raise their siblings in year 1 and the other half by having their own offspring in year 2. In contrast, the secondary helpers did not gain any indirect fitness in year 1, but had a higher probability of surviving and finding a mate in year 2, which improved their direct fitness. The delayers obtained no indirect fitness in year 1 and had a poor ability to attract mates in year 2, leading to a low inclusive fitness.

YOUR TURN Use the table below to calculate the change in inclusive fitness if the primary helpers only improved the fitness of their parents by 1.0 rather than 1.8. Under this scenario, which strategy of helping would be most favoured by natural selection?

MALE ROLE	YEAR 1			YEAR 2				INCLUSIVE FITNESS
	B_1	r_1	INDIRECT FITNESS	B_2	r_2	P_{sm}	DIRECT FITNESS	
PRIMARY HELPER	1.8	× 0.32	= 0.58	2.5	× 0.5	× 0.32	= 0.41	0.58 + 0.41 = 0.99
SECONDARY HELPER	1.3	× 0.00	= 0.00	2.5	× 0.5	× 0.67	= 0.84	0.00 + 0.84 = 0.84
DELAYER	0.0	× 0.00	= 0.00	2.5	× 0.5	× 0.23	= 0.29	0.00 + 0.29 = 0.29

SOURCE: Data from H.-U. Reyer, Investment and relatedness: A cost/benefit approach of breeding and helping in the pied kingfisher, *Animal Behaviour* 32 (1984): 1163–1178.

subordinate male forgoes the ability to breed as a solo male, which would have allowed him to sire 0.9 offspring. Therefore, his cost of being altruistic is 0.9 offspring. By helping his brother or half brother to become highly successful at attracting females, he allows his brother to sire 6.1 offspring. As we have learned, the subordinate male's average coefficient of relatedness to his brother is 0.42. Therefore, the indirect fitness benefit to the subordinate male can be calculated as:

$$\text{Indirect fitness benefit} = B \times r = 6.1 \times 0.42 = 2.6$$

This means that a subordinate male obtains greater inclusive fitness by helping his brother than by going out on his own to attract females.

The concept of kin selection has given ecologists a better understanding of the evolutionary reasons underlying a wide variety of altruistic and selfish behaviours in animals. In the next section, we will explore the evolution of an extreme form of altruistic behaviour in which individuals completely forgo their reproduction to help others.

Eusocial species take social interactions to the extreme

We have seen that many animals are social and interact with conspecifics in a variety of ways. Some animals are so extremely social that we call them *eusocial*. **Eusocial** species are distinguished by four characteristics:

1. Several adults living together in a group
2. Overlapping generations of parents and offspring living together in the same group
3. Co-operation in nest building and brood care
4. Reproductive dominance by one or a few individuals, and the presence of sterile individuals

Figure 10.10 Eusocial species. These Texas leaf-cutter ants (*Atta texana*) are one of many highly social species. Pictured here is a queen and her young daughters that are tending a fungus garden deep inside their underground nest. Photo by Alexander Wild.

Among the insects, eusocial species are limited to the order Hymenoptera, which includes bees, ants, and wasps, and the order Isoptera, which includes the termites (Figure 10.10). Not only are these social insects of evolutionary interest, they are also major players in ecosystem processes. Eusocial insects pollinate plants, consume plant and animal material on a large scale, and recycle wood and organic detritus. Their dominance in the world is due in large part to the immense success of eusociality. Other than insects, the only animals known to be eusocial are two species of mammals that live in underground tunnels in Africa: the naked mole rat (*Heterocephalus glaber*) and the Damaraland mole rat (*Fukomys damarensis*).

Eusocial species are fascinating to ecologists because most individuals in a eusocial group forgo sexual maturation and reproduction. Instead, they specialize at tasks that include defending the group, foraging for the group, or taking care of the subsequent offspring of their parents. Because nonreproductive individuals are specialized for these tasks, they are known as sterile *castes*. A **caste** consists of individuals within a social group who share a specialized form of behaviour.

How can natural selection produce individuals that lack any reproductive output and therefore lack any direct fitness? To help answer this question, we will start by examining the unique breeding habits of the hymenopteran insects.

EUSOCIALITY IN ANTS, BEES, AND WASPS

In Chapter 9 we talked about the unique reproductive behaviour of the hymenopteran insects. You may recall that this group of insects makes sons by laying unfertilized eggs and makes daughters by laying fertilized eggs. This method of reproduction is detailed in Figure 10.11. Because one sex is haploid and the other sex is diploid, the hymenopterans are said to have a **haplodiploid** sex-determination system. As we will see, this sex-determination system helps to favour the evolution of eusociality.

Social Organization

Eusocial insect societies are dominated by one egg-laying female or a few egg-laying females, referred to as **queens**. The queens in colonies of ants, bees, and wasps mate only once during their lives and store

---

Eusocial A type of social animal in which individuals live in large groups with overlapping generations, co-operation in nest building and brood care, and reproductive dominance by one or a few individuals.

Caste Individuals within a social group sharing a specialized form of behaviour.

Haplodiploid A sex-determination system in which one sex is haploid and the other sex is diploid.

Queen The dominant, egg-laying female in eusocial insect societies.

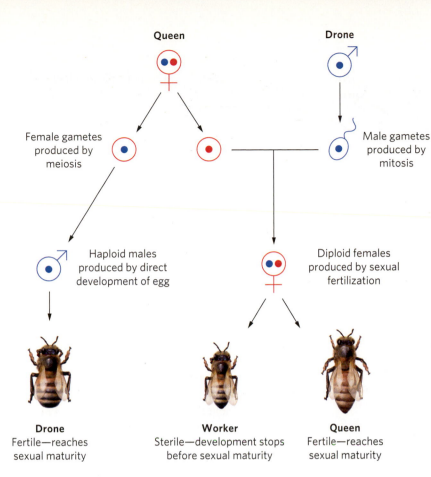

Queen

Drone

Female gametes produced by meiosis

Male gametes produced by mitosis

Haploid males produced by direct development of egg

Diploid females produced by sexual fertilization

Drone
Fertile—reaches sexual maturity

Worker
Sterile—development stops before sexual maturity

Queen
Fertile—reaches sexual maturity

Figure 10.11 The haplodiploid mating system of hymenopteran insects. In ants, bees, and wasps, sons are produced when a queen's haploid gamete remains unfertilized and it develops into a haploid son. Daughters are produced when a queen's haploid gamete is fertilized by a drone's haploid gamete and it develops into a diploid daughter.

enough sperm to produce all their offspring—up to a million or more offspring over a period of 10 to 15 years in some army ants. The nonreproductive progeny of a queen gather food and care for developing brothers and sisters, some of which become sexually mature, leaving the colony to mate and to establish new colonies.

As we discussed in Chapter 9, honeybee societies have a simple organization. The offspring of a queen are divided among a sterile worker caste that is female and a reproductive caste that consists of drones and future queens. Whether an individual becomes a sterile worker or a fertile female is determined by the length of time the larva is fed royal jelly. As a result, the worker caste represents an arrested stage in the development of reproductive females; they stop developing before sexual maturity. Eusocial behaviour appears in a large number of hymenopteran species.

Coefficients of Relatedness

Haplodiploidy is important to the evolution of eusocial animals because it creates strong asymmetries in coefficients of relatedness. The queen has the same genetic relationship to sons as she has to daughters ($r = 0.50$), so she can be relatively indifferent to the sex of her reproductive offspring—the drones and new queens that leave the nest—especially when the sex ratio among reproductive individuals in other local hives is nearly equal. However, the relatedness among siblings is quite unique in hymenopterans. If a queen has mated with a single drone, then all females have the same set of genes from their haploid father. This causes females to be more closely related to each other than is typical in a diploid mating system. In the haplodiploid mating system, the coefficient of relatedness between a given female and her sisters is 0.75. In contrast, her brothers are haploid and therefore do not have any genes from the drone, so the average coefficient of relatedness between a given female and her brothers is 0.25. This skewed relatedness favours the evolution of social groups that contain nonreproductive female workers.

To understand how being haplodiploid favours eusociality, we can compare the different options for obtaining fitness in these organisms. For example, a female worker that raises a fertile sister receives the fitness benefit of that individual multiplied by the coefficient of relatedness ($r = 0.75$). When a female raises its own offspring, it receives the same fitness benefit multiplied by the coefficient of relatedness between a mother and daughter ($r = 0.50$). This means that the indirect fitness obtained from caring for a sister exceeds the direct fitness that could be obtained from caring for a daughter. Thus, co-operation is likely to be greater among all-female castes than among male castes or, especially, among mixed castes. This may explain why workers in hymenopteran societies are all female.

The fact that workers are three times more closely related to their sisters than to their brothers may also explain why broods of reproductive individuals usually favour females, by about three to one on a weight basis, in spite of the queen's indifference to the sex ratio. When a female worker can help to rear more female than male reproductive individuals, her own inclusive fitness may actually be higher than it would be if she raised a brood of her own consisting of an equal number of males and females. Under these circumstances, it is not surprising that sterile castes have evolved.

EUSOCIALITY IN OTHER SPECIES

Although the hymenopterans are among the best known of the eusocial insects, there are a few other species that are also eusocial, including termites and a group of mammals known as naked mole rats. Unlike the haplodiploid hymenopterans, these latter two groups are both diploid.

Termites

Termite colonies can be massive structures that are dominated by a mated pair called the king and queen (Figure 10.12a). Because the queen's role is almost entirely limited to egg production, her abdomen is grossly distended to allow her to produce thousands of eggs (Figure 10.12b). The king and queen produce sons and daughters by sexual reproduction. With a few exceptions, these sons and daughters can both act as workers in the termite society but they remain sexually immature unless either the king or the queen dies. Many species of termites also have a second caste of nonreproductive individuals known as *soldiers*. As their name implies, soldiers help to defend the nest against intruders such as ants. Soldiers typically have a very large head that can be used to block the openings of a termite nest to prevent ants from entering.

Mole Rats

There are dozens of species of mole rats, but only two are known to be eusocial: the naked mole rat and the Damaraland mole rat. These rodents live in underground tunnels in African grasslands in colonies of up to 200 individuals. In naked mole rats, a single queen and several kings are responsible for all of the reproduction in the colony and all individuals are diploid (Figure 10.13). Although the workers in the group are capable of reproducing, they forgo reproduction in favour of caring for their younger siblings and taking care of the colony. Current research suggests that the subordinate males and females do not do this willingly. Instead, the dominant female harasses them, which causes them to be stressed. Stress reduces the levels of sex hormones in the subordinates and makes them less motivated to breed.

THE ORIGINS OF EUSOCIALITY

From its distribution across a variety of distantly related species ranging from hymenopterans to mammals, it is clear that eusociality has evolved independently many times. Even within the hymenopterans, eusociality appears to have evolved several times. This indicates that eusociality was not due to a single mutation in an ancestor that caused all of the descendant species to be eusocial today. It is tempting to conclude that eusociality is caused by a haplodiploid sex-determination system, but many haplodiploid species are not eusocial, and some eusocial species—such as termites and naked mole rats—are not haplodiploid. Based on these observations, we are left to conclude that while being haplodiploid appears to favour the evolution of eusocial behaviour by providing large indirect fitness effects when workers forgo breeding to help their sisters, it is not required for the evolution of eusocial behaviour.

Throughout our discussion, we have focused on the importance of the coefficient of relatedness with the idea that the coefficient needs to be relatively high to favour eusociality. However, this equation could also be satisfied if the cost of forgoing personal reproduction is very low. If individuals that chose to leave a colony had a very low likelihood of surviving and setting up a new colony, then direct fitness would be very low. In naked mole rats, for example, some individuals leave the home colony to form new colonies, but most of these small new colonies do not persist beyond one year. As a result, the cost of forgoing reproduction—in terms of lost direct fitness—would be very small. When this cost is small, a large coefficient of relatedness is no longer required to favour eusocial behaviour.

(a)

(b)

Figure 10.12 Termite colonies. (a) A massive termite colony located in the Northern Territory, Australia. **(b)** A queen termite (*Macrotermes gilvus*) living deep inside a colony surrounded by workers. Because her primary role is egg production, the queen's abdomen is much larger than that of workers and soldiers. Photos by (a) Thomas P. Widmann/ picture-alliance/DUMONT Bildar/Newscom and (b) Anthony Bannister/NHPA/Science Source.

Figure 10.13 Naked mole rats. The naked mole rat is one of two species of mammals that are eusocial. In this photo, a queen is laying on top of the workers. Photo by Raymond Mendez/Animals Animals/Earth Scenes.

The origins of eusocial behaviour are still actively debated. Many researchers have argued that it has been favoured by the inclusive fitness of the altruistic sterile offspring. Others have argued that it has evolved to enhance the direct fitness of parents that force the offspring to forgo reproduction. Like many biological questions, continued research will likely provide new insights that will help us to understand eusocial societies.

Behavioural relationships among the social insects represent one extreme along a continuum of social organization, from animals that live alone except to breed to those that aggregate in large groups organized by complex behaviour. Regardless of their complexity, all social behaviours balance costs and benefits to the individual, and the magnitude of these costs and benefits is often determined by the ecological conditions in which these behaviours exist. Like morphology and physiology, behaviour is strongly influenced by genetic factors and therefore is subject to evolutionary modification by natural selection. The evolution of behaviour becomes complicated when individuals interact within a social setting, and the interests of individuals within a population may either coincide or conflict. Understanding the evolutionary resolution of social conflict in animal societies continues to be one of the most challenging and important concerns of biology.

ECOLOGY TODAY CONNECTING THE CONCEPTS

HEN-PECKED CHICKENS

Domesticated chickens. Chicken survival and egg production depend not only on the traits of each chicken but also on the way each chicken interacts with other chickens in its cage. Photo by Phillip Hayson/Photo Researchers, Inc.

The social behaviour of animals in the wild is inherently interesting, but understanding social behaviour also has practical applications for domesticated animals. There are costs and benefits of raising domesticated animals in large groups. Raising animals such as pigs and chickens in high densities is more cost effective because it requires less space. However, as we have seen, living at high densities has downsides that include increased risks of disease transmission, increased competition, and increased fighting. For farmers who raise domesticated animals, raising animals in high densities translates into fewer healthy animals and smaller profits. However, during the past 2 decades, researchers with insights into group living have helped breed more productive domesticated animals. Some of the leading research has been conducted with chickens (*Gallus gallus*).

Chickens are well known for fighting when raised under crowded conditions. For example, if one chicken receives a small injury that causes a spot of blood on its feathers, other chickens will begin to peck at the spot, causing greater injury. In the process, some chickens will become dotted with blood spots and other chickens, in turn, will also peck them. This incessant behaviour is the origin of the phrase "hen-pecked." Fighting among chickens can reduce overall health and egg production, and can even cause death of the chickens. In fact, fighting can cause so much lost profit to poultry farms that there is a long-standing practice of cutting off the ends of chicken beaks to reduce injuries. Although beak trimming is effective at reducing fighting and cannibalism, it is a controversial practice and was banned in the European Union in 2011. This has motivated animal breeders to find alternative ways to reduce fighting among crowded domesticated animals.

The traditional method of breeding animals has been to select for those individuals with the best survival, growth, or other traits such as increased egg production. After decades of selection on traits, the amount of heritable variation in these traits is quite low, which makes it difficult to conduct further selection. In the mid 1990s, however, researchers discovered that they could cause large responses to selection if they based their selection on the best performing social groups of chickens living together rather than on just the best performing individual chickens. Groups of hens were caged together and cages with the highest survival were used to generate offspring for the next generation. At the start of the experiment, annual hen mortality was 68 percent. After six generations of selection for the best surviving groups, annual hen mortality dropped to 9 percent,

which was similar to the mortality rate of hens that were raised alone with no opportunities to fight. Moreover, lifetime egg production increased dramatically, from 91 eggs per hen in the first generation to 237 eggs per hen by the sixth generation. The increase in egg production was caused by the hens living longer and the hens being in better physical condition, which allowed them to lay more eggs per day. These data suggest that selection for the best social groups could reduce fighting to a level that would eliminate the need to trim chicken beaks.

During the past 5 years, researchers have begun to understand why selecting on social groups of animals could cause such dramatic increases in animal production. Biologists have traditionally focused only on the heritability of traits for an isolated individual. However, the performance of an individual depends not only on the individual but also on the way it interacts with other individuals. For example, if an individual is particularly aggressive and lives in a group of other aggressive individuals, the group will spend more time fighting and more deaths will occur. In contrast, a social group with individuals who fight less with each other will have lower stress, better health, and a reduced mortality rate. Because social animals interact with conspecifics, we need to consider the heritability of traits not just when an individual lives alone but when an individual lives in a social group. In chickens, for example, the heritability of survival for an individual can be quite low, but the heritability of the trait when social interactions are included can be two to three times higher because social interactions affect survival. Similar results have been observed in other farmed animals including domesticated pigs (*Sus scrofa*) and mussels (*Mytilus galloprovincialis*). All of this suggests that by considering the social interactions of animals we can improve agricultural production of domesticated animals.

SOURCES: Muir, W. M. 1996. Group selection for adaptation to multiple-hen cages: Selection program and direct responses. *Poultry Science* 75: 447–458.

Wade, M. J., et al. 2010. Group selection and social evolution in domesticated animals. *Evolutionary Applications* 3: 453–465.

SUMMARY OF CHAPTER CONCEPTS

■ **Living in groups has costs and benefits.** The benefits of social living include the dilution effect in which large groups of prey have a reduced likelihood of being killed by a predator, reduced need for personal vigilance, and increased ability to find food and mates. The costs include an increase in visibility to predators, increased risk of parasite and pathogen transmission, and increased competition for food. In response to social living, many species have evolved the ability to establish territories and dominance hierarchies to manage individual interactions.

■ **There are many types of social interactions.** When we envision interactions in terms of donors and recipients, we can devise four types of social interactions: co-operation, selfishness, spitefulness, and altruism. Co-operation and selfishness of donors should be favoured by natural selection whereas spitefulness should not. Altruism can be favoured when the recipient of an altruistic act is closely related to the donor, as measured by the coefficient of relatedness.

■ **Eusocial species take social interactions to the extreme.** Eusocial animals consist of many individuals living together with dominant individuals reproducing and subordinate individuals forgoing reproduction. Eusocial species are common among the haplodiploid species of bees, ants, and wasps, but also exist in diploid species of termites and at least two species of mammals. A high coefficient of relatedness favours the evolution of eusocial behaviour, but it is not required. Equally important may be the presence of a low cost of lost fitness in species that have a low probability of leaving the group and reproducing on their own.

REVIEW QUESTIONS

1. If living in large groups has costs and benefits, under what conditions would natural selection favour group living?

2. Why might individuals give up defending territories if the density of their population increases?

3. Explain the costs and benefits that might influence the optimal flock size in birds.

4. Why should natural selection not favour spiteful behaviour?

5. How could helping a relative improve the helper's fitness?

6. In the kin selection explanation for the evolution of altruism, why is the benefit to the recipient weighted by its coefficient of relatedness to the donor?

7. Why are selfish behaviours less favoured when the donor and recipient are related to each other?

8. Compare the coefficient of relatedness between brothers and sisters in diploid organisms with haplodiploid organisms.

9. How does a haplodiploid sex-determination system favour the evolution of eusociality?

10. What evidence is there that a haplodiploid sex-determination system is not required for the evolution of eusociality?

PRACTICING ECOLOGY: PLANTS THAT KNOW THEIR KIN

As a researcher at the University of Toronto, Jay Biernaskie studied the fitness consequences of kin recognition in plants. Photo by Jen Perry. Courtesy of Jay Biernaskie.

Like all organisms, plants compete for resources. But competition comes at a cost. For example, plants compete for light by growing taller than their neighbours, and they compete for soil nutrients by extending their roots deeper than the roots of other plants. Although these strategies allow an individual to grow and develop in the face of competition, they also reduce the energy the plant has left for reproduction. As a result, plants that experience high competition are not likely to produce as many seeds as plants that experience low competition. However, if plants grow near siblings and are capable of kin selection, they might reduce investment in competition—which could increase investment in reproduction for all neighbouring siblings. Are plants capable of recognizing and responding to the relatedness of their neighbours?

The ivyleaf morning glory (*Ipomoea hederacea*) is a climbing vine common throughout most of North America. Its flowers are mostly self-pollinated, which leads to offspring with high levels of relatedness. The morphology of the morning glory is phenotypically plastic; roots grow deeper when they experience limiting nutrients in the soil, and stems grow taller when they detect low levels of sunlight. These investments of energy in plant growth come at the expense of investments in flower and seed production. If the morning glory experiences kin selection, we can hypothesize that individuals that grow near siblings will invest less in competitive traits, such as root and stem growth, and will invest more in reproduction compared to individuals grown near nonsiblings.

To test this hypothesis, Jay Biernaskie conducted an experiment at the University of Toronto in which he manipulated the relatedness of three ivyleaf morning glory individuals grown together in clay pots. He collected seeds from three genotypes: A, B, and C. Using these genotypes, he established three treatments containing three siblings of a particular genotype and one treatment containing a mixture of nonsiblings. Once plants flowered and were pollinated, he measured the total mean number of seeds produced in each treatment. You can view a sample of his data in the table.

MEAN NUMBER OF SEEDS FROM VINES GROWING WITH SIBLINGS OF THE SAME GENOTYPE OR MIXED GENOTYPES

SIBLINGS OF GENOTYPE A	SIBLINGS OF GENOTYPE B	SIBLINGS OF GENOTYPE C	MIXED NON-SIBLINGS
32.2	40.0	24.0	20.5
30.5	47.6	31.0	22.7
26.9	28.8	34.0	26.6
26.3	27.5	37.2	29.7
32.6	27.8	40.9	30.6
32.7	24.2	30.4	32.9
29.2	36.3	39.0	23.4
32.2	33.3	40.5	21.6
24.9	33.6	33.5	19.0
28.8	30.6	30.0	42.3

GRAPH THE DATA

1. For each of the four treatments calculate the means and the standard errors of the means.

2. Create a bar graph that displays the means and standard errors for the four treatment groups.

INTERPRET THE DATA

1. Explain how the data support or refute the hypothesis that siblings grown together will show increased seed production relative to nonsiblings grown together.

2. Compare the differences between the means of each sibling treatment. What do these differences suggest about genetic variation in the ability of the plant to recognize kin?

3. Given the data, is natural selection likely to favour increased or decreased kin recognition in the ivyleaf morning glory?

4. In the same study, Biernaskie measured vine height and root mass in each treatment replicate. Although he found no difference in vine height among treatments, he did find that root mass was significantly greater in the mixed-sibling treatment. How does this result relate to your observations?

SOURCE: Biernaskie, J.M. 2011. Evidence for competition and cooperation among climbing plants. *Proceedings of the Royal Society B-Biological Sciences*, 278: 1989–1996.

CHAPTER 11

Population Distributions

Bringing Back
the Mountain Boomer

When Alan Templeton was a Boy Scout, he encountered his first collared lizards (*Crotaphytus collaris*), also known as mountain boomers, in the Ozark Mountains of Missouri. He was struck by the brightly coloured males that ran around in the forest openings that dotted the mountains. Two decades later, as a biology professor at Washington University in St. Louis, Missouri he returned to the Ozark Mountains and was shocked to find that most of the lizards were gone. He began a course of research to identify the causes of the decline and to determine the steps that could restore the lizard population.

The collared lizard is a fascinating animal that feeds on insects such as grasshoppers. It prefers to live in dry, open areas and has a geographic range that spans the central United States from Kansas to Mexico; Missouri lies on the eastern edge of this range. Although much of the Ozark Mountains are forested, there are openings in the forest, known as glades, that contain exposed bedrock with the hot and dry conditions that provided suitable habitat for the lizards. These small, patchy habitats were once surrounded by savannahs and forests with open understoreys. Over time these habitats began to change.

> "Over time, lizard populations in the glades declined and the geographic range shrank; only a few glades continued to support lizard populations."

In the Ozark Mountains, forest fires were historically common. These fires removed the small understorey trees and accumulating leaves scattered across the forest floor. Forest fires were also important for maintaining the glades as open, sunny areas. However, beginning in the 1940s, a national campaign to suppress forest fires caused the glade habitats to be invaded by eastern red cedar trees (*Juniperus virginiana*). The cedar trees shaded the glades and made them cooler, which is not conducive to the growth of an ectotherm such as the collared lizard. The shade also reduced the number of insect prey, including grasshoppers. As a result, the glades became less suitable for the lizard. Over time, lizard populations in the glades declined and the geographic range shrank. Only a few glades continued to support lizard populations. Moreover, there appeared to be no movement of the lizards from these glades to other glades that were unoccupied.

Templeton and his colleagues began their research in the 1980s. They discovered that many glades once occupied by lizards no longer had any lizard population. They took several steps to restore those populations. First, they cut down the cedar trees in several glades to make the habitat more suitable. Next, they reintroduced lizards to three sites. They marked the animals to estimate the abundance and density

A male collared lizard. Males of this species have a bright orange throat that intimidates other males and attracts females. Photo by Francois Gohier/Photo Researchers, Inc.

of each population and to determine whether the lizards they introduced would disperse to colonize the neighbouring glades. Dispersal events were rare and no neighbouring glades were colonized, despite the fact that some were only 50 m away. Like the cedar trees that shaded the glades, the thick forest understorey and accumulating leaf litter shaded much of the ground, which made the forest cooler and caused lower abundances of insect prey. At this point the researchers realized that they had two tasks. They needed to improve both the habitat of the glades where the lizards spent most of their time and the forested habitat through which the lizards travelled when dispersing from one glade to another.

In cooperation with the Missouri Department of Conservation, the team used controlled fires to burn away the cedars in many glades as well as the small understorey trees and leaves in the forest habitat. The lizards responded so well that 2 months after the burn, they began to colonize new glades. In 2011, Templeton and colleagues reported that from the original three glades where lizards were reintroduced, lizards eventually spread to more than 140 glades. Burning the forested habitat allowed more sunlight to come in and dramatically increased the number of grasshoppers for the lizards to eat when they travelled between glades.

The great success in restoring the collared lizard required researchers to understand its habitat needs and to recognize that the regional population was actually composed of many small populations interconnected over a large spatial area. In this chapter, we will explore how spatial distribution and the movement of individuals among habitats influence the long-term persistence of species.

SOURCES: A. R. Templeton, The transition from isolated patches to a metapopulation in the eastern collared lizard in response to prescribed fires, *Ecology* 92 (2011): 1736–1747.

Restoration as science: The case of the collared lizard, *Science Daily*, August 22, 2011, http://www.sciencedaily.com/releases/2011/08/110822091918.htm

CHAPTER CONCEPTS

- The distribution of populations is limited to ecologically suitable habitats.

- Population distributions have five important characteristics.

- The distribution properties of populations can be estimated.

- Population abundance and density are related to geographic range and adult body size.

- Dispersal is essential to colonizing new areas.

- Many populations live in distinct patches of habitat.

In previous chapters we have focused on ecology at the individual level. The story of the collared lizard demonstrates that studying ecology at the population level is also fascinating and can have real-world applications. In this chapter, we will focus on the **spatial structure** of populations, which is defined as the pattern of density and spacing of individuals. We will begin by examining how the distribution of suitable habitat affects the distribution of populations, including pest populations that are new to a region. We will then discuss the many properties of population distributions and examine how we estimate these properties. Next, we will investigate the importance of individuals moving between patches of habitats. Finally, we will learn how providing strips of favourable habitat can facilitate the movement of individuals between habitats to help ensure the persistence of populations over time.

Spatial structure The pattern of density and spacing of individuals in a population.

The distribution of populations is limited to ecologically suitable habitats

As we have seen in previous chapters, the natural world varies from place to place. At the beginning of this chapter we saw that collared lizards inhabited habitat patches, known as glades, interspersed within a forest. As these glades were invaded by cedar trees, however, they became less suitable for the collared lizards. In this section, we will explore how ecologists determine the suitability of habitats. We will see that understanding habitat suitability is critical for understanding where a species is capable of living and the extent to which a species can expand its geographic range.

DETERMINING SUITABLE HABITATS

In Chapter 1, we mentioned that a species niche includes the range of abiotic and biotic conditions it can tolerate. We are now ready to elaborate on this point. To explore the concept of niche, scientists find it useful to draw a distinction between the *fundamental niche* and the *realized niche* of a species.

The **fundamental niche** of a species is the range of abiotic conditions under which species can persist. This includes the range of temperature, humidity, and salinity conditions that allow a population to survive, grow, and reproduce.

Although a species can potentially live under the conditions of its fundamental niche, many favourable locations can remain unoccupied because of other species in those locations. For example, the presence of competitors, predators, and pathogens can often prevent a population from persisting in an area, despite the existence of favourable abiotic conditions. The range of abiotic and biotic conditions under which a species persists is its **realized niche**. The realized niche determines the *geographic range* of a species or of various populations that compose a species. The **geographic range** is a measure of the total area covered by a population. For example, the American chestnut tree (*Castanea dentata*) was once a very common tree in the eastern forest of North America because it grew and reproduced well under the abiotic and biotic conditions that existed in this region for thousands of years. Around 1900, however, a fungus (*Cryphonectria parasitica*) introduced from Asia caused a deadly disease known as chestnut blight. The fungus spread rapidly throughout the eastern forests and killed billions of trees. As a result of this biotic interaction between the chestnut trees and the fungus, few places remain where adult trees can persist. In short, the fungus has caused a major reduction in the realized niche of the American chestnut.

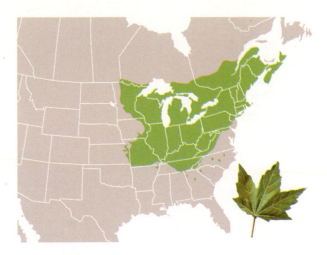

Figure 11.1 The geographic range of the sugar maple. The range is limited in the north by low winter temperatures, to the south by high summer temperatures, and to the west by droughts.

When we think about the geographic range of a species or population, we need to realize that individuals often do not occupy every location within the range. This is because climate, topography, soils, vegetation structure, and other factors influence the abundance of individuals. Consider the case of the sugar maple tree. As shown in Figure 11.1, its geographic range includes the midwestern United States, the northeastern United States, and southeastern Canada. Its distribution is limited by cold winter temperatures at the northern extent of its range, by summer droughts at the western extent, and by hot summer temperatures at the southern extent. Throughout this entire range, however, the sugar maple does not live everywhere. For instance, it cannot live in marshes, newly formed sand dunes, recently burned areas, and a variety of other habitats that lie outside its fundamental niche. As a result, the geographic range of the sugar maple is actually composed of a patchwork of smaller occupied and unoccupied areas.

The distribution of a plant known as horse gentian (*Triosteum aurantiacum*) provides an excellent example of how small-scale variation in the environment can create geographic ranges that are composed of small patches of suitable habitat.

Fundamental niche The range of abiotic conditions under which species can persist.

Realized niche The range of abiotic and biotic conditions under which a species persists.

Geographic range A measure of the total area covered by a population.

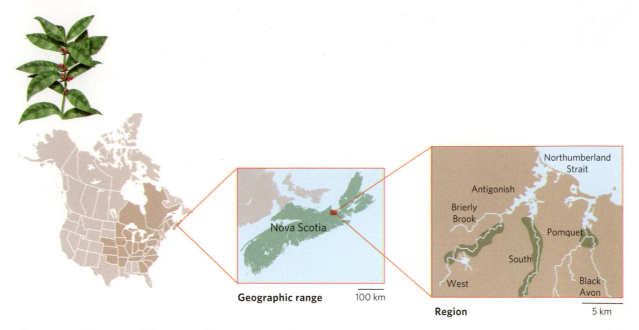

Figure 11.2 The geographic range of horse gentian plant. A 2006 survey recorded this rare plant for the first time in three river valleys in Nova Scotia. In these valleys, horse gentian grew best in clearings and in areas of low acidity, indicated by the green shading. Data from B. R. Taylor and J. C. Tam, Local distribution of the rare plant *Triosteum aurantiacum* Subsp. *aurantiacum* in Northeastern Nova Scotia, Canada, *Rhodora* 114 (2012): 366–382.

Although the plant's geographic range spans eastern North America, its patchy regional distribution in rich forests and thickets makes it a rare plant, and it is considered endangered or vulnerable. In Nova Scotia, which is the northern limit of the distribution, horse gentian grows mainly in floodplains of rivers and streams. Variations in canopy cover and soil acidity further restrict the plant's range in this area. You can examine this patchy distribution in Figure 11.2, which shows previously unknown locations in three river valleys in northeastern Nova Scotia.

Although patterns of distribution would suggest that only certain habitats are suitable, we can test whether this is the case. In a recent experiment, researchers observed that four species of moss occur in habitats along a gradient from wet to dry in an Alberta peatland. Among the three species of peatmoss, *Sphagnum angustifolium* lives in wet areas whereas *S. magellanicum* and *S. fuscum* live in increasingly drier areas. The fourth species, red-stemmed feathermoss (*Pleurozium schreberi*), lives in the driest habitat. To determine if environmental conditions cause these different distributions, researchers collected samples from each of the four species. They then planted five samples from each species in the habitat where they are typically found. These samples served as a control. Researchers also transplanted five samples from each species in each of the other three habitats. After 1 year the researchers measured plant growth and observed how healthy the plants looked. Plants that were all green and without abnormal growth were given a score of 5. Plants that featured abnormal growth and black tissues were given a score of 0. The results of this experiment are shown in Figure 11.3. *S. angustifolium* grew well in its own habitat and in the habitat of *S. magellanicum*, but did not grow well in the drier areas. *S. magellanicum* grew better in the habitat of *S. angustifolium* than its own, while *S. fuscum* grew equally well in all habitats. *P. schreberi* grew well only in its own dry habitat. All species were healthiest in their own habitats. When grown outside their habitats, most species experienced slower growth, and all experienced reduced health.

ECOLOGICAL NICHE MODELLING

As a general rule, the more suitable the habitat, the larger a population can grow. This fundamental relationship between the environment and the distribution of populations allows ecologists to predict the actual or potential distributions of species. This predictive ability has a number of important applications. For example, when biologists wish to bring species back from the brink of extinction, they need to know the habitat conditions that the species requires. With this knowledge, biologists can determine the locations that would provide the highest probability of successful reintroductions. Similarly, if a new pest species is accidentally introduced to a continent, biologists can assess what habitat might be

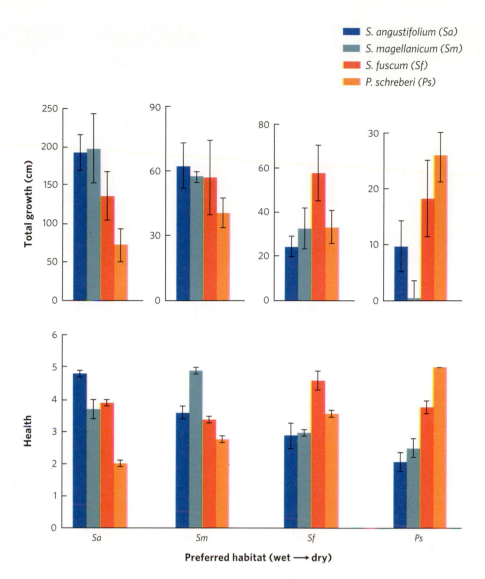

Figure 11.3 The distribution of four moss species. Three species of peatmoss and one species of feather moss grow in different areas of an Alberta peatland. *Sphagnum angustifolium*, *S. magellanicum*, *S. fuscum*, and *Pleurozium schreberi* prefer increasingly drier habitats. When transplanted into the habitats of the other three species, most mosses grew slower, measured here as total growth of all shoots in each transplant. All mosses suffered reduced health. Error bars are standard errors. Data from R. C. Mulligan and L. D. Gignac, Bryophyte community structure in a boreal poor fen: Reciprocal transplants, *Canadian Journal of Botany* 79: 404–411.

suitable and predict the area over which it might spread. This allows biologists to predict the extent of the damage it might cause.

Predicting the potential geographic range of a single population or of all populations of a given species is a major challenge when there are few individuals living in the wild. One way researchers have overcome this challenge is by using historic data on the distributions of populations. Such data are often available from collections of preserved organisms stored in museums and herbariums. In addition, when species have been introduced from other continents, researchers can try to determine the suitable habitat conditions found on the originating continent.

The process of determining the suitable habitat conditions for a species is known as **ecological niche modelling**. Because temperature and precipitation have a dominant influence on the distribution of biomes, modellers often begin by mapping the known locations of a species and then by quantifying the ecological conditions at the locations where the species has been recorded. The modeller can potentially include many additional variables such as different soil types and the presence of potential predators, competitors, and pathogens that might limit the

Ecological niche modelling The process of determining the suitable habitat conditions for a species.

(a) (b)

Figure 11.4 Ecological modelling of an invasive species. (a) Researchers collected data on the environmental conditions where the Chinese bushclover lives in Asia and used these data to predict the entire native range of the plant in Asia. **(b)** They then used the data to predict the future geographic range of Chinese bushclover in North America, where the plant was introduced 100 years ago and has continued to spread slowly. Data from A. T. Peterson et al., Predicting the potential invasive distributions of four alien plant species in North America, *Weed Science* 51 (2003): 863–868.

population's distribution. The range of ecological conditions that are predicted to be suitable for a species is the **ecological envelope** of the species. The concept of the ecological envelope is similar to the realized niche, but the realized niche includes the conditions under which a species currently lives whereas the ecological envelope is a prediction of where a species could potentially live. Using the conditions in the ecological envelope, the modeller can determine what other locations in a region have the same ecological conditions and therefore might serve as additional suitable habitat.

Modelling the Spread of Invasive Species

Ecological niche modelling can be a useful way to predict the expansion of pest species introduced to a continent where they have not previously lived. One example is the Chinese bushclover (*Lespedeza cuneata*), which is native to eastern Asia. In the late 1800s, this species of bushclover was brought to North Carolina to help control erosion, reclaim land that had been mined, and provide food for cattle. Over time, though, the plant quickly spread into the Great Smoky Mountains National Park and throughout the grasslands of the United States where it displaces native plants. To determine the likely extent of its future spread, ecologists gathered data on the environmental conditions under which the bushclover lived in 28 locations in Asia. As illustrated in Figure 11.4a, they then quantified the ecological envelope for the clover to predict the entire geographic range in Asia. They used this data to successfully predict the potential geographic range in North America. As you can see in Figure 11.4b, they

predicted all locations where the clover has already spread. The model also predicts that the clover has the potential to live in many other locations, including southern Canada, with widespread negative effects on the plant communities. The fact that not all of these predicted areas have Chinese bushclover in them yet suggests either that there are additional ecological factors important to the clover and not included in the model or that the clover simply has not had enough time to disperse to these more distant locations.

HABITAT SUITABILITY AND GLOBAL WARMING

Knowledge of the environmental conditions that make a habitat suitable can also be used to understand the shift in the geographic ranges of species as the environmental conditions of the world change. During the past century, for example, the average temperature of Earth has increased by 0.8°C. Some regions of the world—such as Alaska and northern Canada—have warmed as much as 4°C. In the relatively shallow North Sea, situated between the United Kingdom and Norway, temperatures in the bottom waters have increased 0.7°C per decade since the 1970s. As shown in Figure 11.5a, this translates to a 2°C increase in temperature from 1977 to 2003. Given that most species of fish have optimal temperature ranges, we might predict that warming ocean waters could cause southern fish species, which live in warm waters, to move north.

Ecological envelope The range of ecological conditions that are predicted to be suitable for a species.

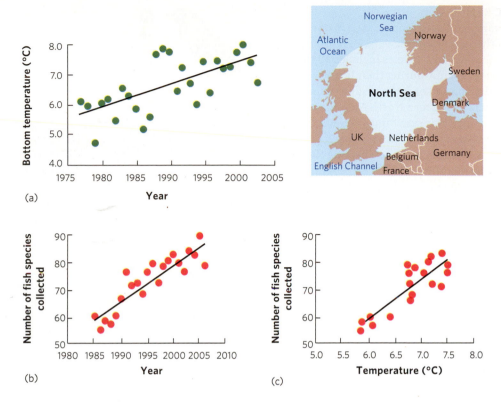

Figure 11.5 Changes in the distribution of fish in the North Sea. (a) From 1977 to 2003, the temperature of the water at the bottom of the North Sea increased by 2°C. **(b)** Surveys of fish trawls from 1985 to 2006 found that the total number of fish species collected per year in the North Sea increased from 60 to approximately 90. **(c)** The total number of fish species collected per year was positively correlated with the average sea temperatures during the previous 5 years. Data from J. G. Hiddink and R. ter Hofstede, Climate induced increases in species richness of marine fishes, *Global Change Biology* 14 (2008): 453–460.

During the same period that temperatures in the North Sea were being monitored, the International Council for the Exploration of the Sea (ICES) compiled data on fish distributions by pulling large nets (called "trawls") along the ocean floor. From 1985 to 2006, scientists from six countries worked together to fish the bottom of the ocean at 300 locations distributed throughout the North Sea. Based on 7000 trawl samples, the researchers reported in 2008 that fish species richness in the North Sea had increased steadily over 22 years. As you can see in Figure 11.5b, there were about 60 species in the mid-1980s but this grew to nearly 90 species 2 decades later. This list of species included dozens of more southerly species that had expanded their ranges northward.

The increase in species richness was positively correlated with the increase in bottom-water temperatures in the North Sea. This correlation— shown in Figure 11.5c—suggests that warmer temperatures are more hospitable to a greater variety of species, and that the warming of the North Sea has allowed more southerly species to expand the northern edges of their ranges into the area. Hence, this is a case of global warming increasing the diversity of species in a region.

The increase in the diversity of species in the North Sea not only is dramatically changing the fish community but also may affect the important commercial fisheries that depend on this community. For example, the three species whose ranges have contracted as the North Sea became warmer—the wolffish (*Anarhichas lupus*), the spiny dogfish (*Squalus acanthias*), and the ling (*Molva molva*)—are all commercially important, whereas more than half of the southerly species with expanded ranges have little or no commercial value. As a result, while these shifts in species distributions with warming temperatures may increase fish diversity, it could actually decrease the value of North Sea commercial fisheries.

Population distributions have five important characteristics

When ecologists study the distribution of a population like the collared lizard in the Ozark Mountains, they typically need to know the characteristics of the population, including geographic range, *abundance, density, dispersion,* and *dispersal.* As we will see, each of these properties tells us something important about how individuals are distributed.

GEOGRAPHIC RANGE

We saw an example of geographic range in our discussion of the sugar maple tree (see Figure 11.1). The geographic range of a species includes all the areas its members occupy during their life. For example, the distribution of sockeye salmon includes not only the rivers of western North America and eastern Asia that are their spawning grounds, but also vast areas of the North Pacific Ocean where individuals grow to maturity before making the long migration back to their birthplace. The geographic range is an important measure because it tells us how large an area a population occupies. If a population is restricted to a small area, for example, it may be very susceptible to a natural disaster such as a hurricane or a fire that can wipe it out. This is a serious challenge for **endemic** species, which live in a single, often isolated, location, such as the Galápagos finches on islands off the coast of South America. Populations with a larger geographic range are less vulnerable to such events because much of the population would remain unaffected. Species with very large geographic ranges that can span several continents are known as **cosmopolitan** species.

ABUNDANCE

The **abundance** of a population is the total number of individuals that exist within a defined area. For example, we might count the total number of lizards on a mountain, the number of sunfish in a lake, or the number of coconut trees on an island. The total abundance of a population is important to ecologists because it provides a measure of whether a population is thriving or on the brink of extinction.

DENSITY

The **density** of a population is the number of individuals per unit area or volume. If we know the abundance of a population in a given area and we know the size of the area, then we can calculate density by dividing the abundance by the area. Examples of density include the number of bears per km² in British Columbia, the number of cattails per m² in a pond, or the number of bacteria per ml of water. Density is a valuable measure because it tells ecologists how many individuals are packed into a particular area. If a habitat can support a higher density than currently exists, the population can continue to grow in the area. If the population density is greater than what the habitat can support, either some individuals will have to leave the area or the population will experience lower growth and survival.

Although individuals only live in suitable habitats, not all habitats are equal in quality because the environment is inherently variable. Some habitat patches have abundant resources that support a large

■	Above 100
■	>30–100
■	>10–30
■	>3–10
■	>1–3
■	>0.05–1
	None counted

Figure 11.6 Densities across a geographic range. In the red-winged blackbird, most areas with highest densities are near the centre of its geographic range and the lowest densities are near the periphery of its geographic range. Data from http://www.mbr-pwrc.usgs.gov/bbs/ra2011/ra04980.htm.

number of individuals while others have scarce resources and can only support a few individuals. Across a large geographic area, the highest concentrations of individuals are typically near the center of a population's geographic range. As one moves closer to the limits of the geographic range, biotic and abiotic conditions become less ideal. As a result, the margins of the geographic range typically contain fewer individuals. Consider, for example, the geographic range of the red-winged blackbird (*Agelaius phoeniceus*), a common songbird associated with marshes and water bodies across North America. As illustrated in Figure 11.6, this bird has most of its highest densities in the centre of its geographic range and its lowest densities near the periphery. However, because environmental conditions do not vary

Endemic Species that live in a single, often isolated, location.

Cosmopolitan Species with very large geographic ranges that can span several continents.

Abundance The total number of individuals in a population that exist within a defined area.

Density In a population, the number of individuals per unit area or volume.

Clustered

(a)

Evenly spaced

(b)

Random

(c)

Figure 11.7 Three types of dispersion patterns. (a) Clustered dispersion is characterized by individuals that are aggregated. For example, sulphur tuft fungi (*Hypholoma fasciculare*) grow in clusters anywhere there is a rotting stump, such as this site in the United Kingdom. **(b)** Evenly spaced dispersion is characterized by each individual maintaining a minimum distance between neighbours, as is the case for these blue-eyed cormorants (*Phalacrocorax atriceps*) nesting in Patagonia, Argentina. **(c)** Random dispersion is characterized by each individual's position being independent of the location of other individuals, such as these dandelions growing in a pasture in Bulgaria. Photos by (a) Gary K. Smith/naturepl.com; (b) Juan Carlos Munoz/naturepl.com; (c) FLPA/Bob Gibbons/age fotostock.

smoothly, the pattern of the red-winged blackbird's preferred habitat is highly irregular.

DISPERSION

Dispersion of a population describes the spacing of individuals with respect to one another within the geographic range of a population. As shown in Figure 11.7, dispersion patterns can either be *clustered, evenly spaced,* or *random.*

Clustered Dispersion

In **clustered dispersion**, individuals are aggregated in discrete groups. Some clustered dispersions result from individuals living in social groups, as we

discussed in Chapter 10. For instance, some birds live in flocks and some fish live in schools. Other clustered dispersions occur because individuals stay near clustered resources. For example, salamanders and sow bugs aggregate under logs because individuals of both species are attracted to dark, moist places. Another cause of clustered dispersions is

Dispersion The spacing of individuals with respect to one another within the geographic range of a population.

Clustered dispersion A pattern of population dispersion in which individuals are aggregated in discrete groups.

offspring that remain close to their parents. Some species of trees, such as the quaking aspen, form clusters of stems because a parent tree gives rise to offspring by sending up new stems from its roots, a form of vegetative reproduction. As a result, we commonly see clusters of aspen trees that are composed of a parent tree surrounded by its offspring. Within such a cluster, however, the stems tend to be *evenly spaced*.

Evenly Spaced Dispersion

In **evenly spaced dispersion**, each individual maintains a uniform distance between itself and its neighbours. In agricultural settings we can observe even spacing in crops such as corn or apple trees because farmers want each plant to have sufficient resources to maximize crop production. In natural settings, even spacing most commonly arises from direct interactions between individuals. For example, plants positioned too close to larger neighbours often suffer from shading and root competition. In addition, some plants can emit chemicals from their leaves and roots that inhibit the growth of other plants around them. As these crowded individuals die, the remaining individuals become more evenly spaced. We can also observe evenly spaced distributions in animals that defend territories, such as birds and lizards. Because territory size commonly depends on the amount of resources available, neighbouring territory holders commonly defend areas that are similar in size, which causes the territory holders to be evenly spaced.

Random Dispersion

In **random dispersion**, the position of each individual is independent of the position of other individuals in the population. Random dispersions are not common in nature, largely because abiotic conditions, resources, and interactions with other species are not randomly distributed. However, if we wished to know whether the dispersion pattern of a population is clustered or evenly spaced, we would have to demonstrate statistically that its distribution was significantly different from random.

DISPERSAL

Dispersal is the movement of individuals from one area to another. Dispersal is distinct from migration, which is the seasonal movement of individuals back and forth between habitats, such as birds flying north and south with the changing seasons. In contrast, dispersal involves individuals leaving their habitat of origin—where a seed was made or where a squirrel was born—and typically not returning. As we saw at the beginning of this chapter, dispersal is of great interest to ecologists because it is the mechanism by which individuals can move between suitable habitats and, in some cases, colonize suitable habitats that are

not already inhabited by the species. Dispersal can also be a way to avoid areas of high competition or high predation risk. For example, when a fish arrives in a section of a stream, many of the aquatic insects in the stream disperse by floating downstream at night to avoid being eaten.

The distribution properties of populations can be estimated

Although we can easily define the distribution properties of populations, how do we actually quantify them? Four of the properties—abundance, density, geographic range, and dispersion—require that we determine where individuals are located and how many are located in particular areas. The fifth property, dispersal, is somewhat unique because it requires that we quantify the movement of individuals.

QUANTIFYING THE LOCATION AND NUMBER OF INDIVIDUALS

One way to determine the number of individuals in an area is to conduct a **census**, which means counting every individual in a population. Every 5 years, for example, the Canadian government attempts a census with the goal of counting every person living in the country. For most species, however, it is not feasible to count every individual in the population. As a result, scientists must conduct a **survey**, in which they count a subset of the population. Using these samples, they estimate the abundance, density, and distribution of the population. Scientists have developed a variety of ways to make these estimates, including *area-* and *volume-based surveys, line-transect surveys,* and *mark-recapture surveys* (Figure 11.8).

Area- and Volume-Based Surveys

Area- and volume-based surveys define the boundaries of an area or volume and then count all of the individuals within that space. The size of the defined space is typically related to the abundance and density of the population. For example,

---

Evenly spaced dispersion A pattern of dispersion of a population in which each individual maintains a uniform distance between itself and its neighbours.

Random dispersion A pattern of dispersion of a population in which the position of each individual is independent of the position of other individuals in the population.

Dispersal The movement of individuals from one area to another.

Census Counting every individual in a population.

Survey Counting a subset of the population.

Area- and volume-based surveys Surveys that define the boundaries of an area or volume and then count all of the individuals in the space.

(a)   (b)   (c)

Figure 11.8 Estimating population abundance, density, and distribution. (a) Area-based studies count the number of individuals within a fixed area, such as this researcher working on the Dendles National Natural Reserve in England. **(b)** Linear-transect studies, including this one at the Great Barrier Reef in Australia, count the number of individuals that are observed along a predefined line. **(c)** Mark-recapture studies collect a sample of the population, mark them, and then return them. For example, horseshoe crabs (*Limulus polyphemus*) in the U.S. state of Delaware are marked with small circular tags. A short time later, a second sample is collected to determine the proportion of marked animals in the population, which can be used to estimate the size of the total population. Photos by (a) Paul Glendell/Alamy; (b) Suzanne Long/Alamy; (c) Patrick W. Grace/Photo Researchers, Inc.

researchers who wish to know the number of bacteria in the soil might collect samples of soil that are only a few cubic centimetres in volume. In contrast, researchers who wish to know the number of individual corals on a coral reef might sample areas that are 1 m². At the most extreme, researchers who want to estimate the abundance, density, and distribution of large mammals might count the number of individuals in aerial photos that cover hundreds of square metres. By taking multiple samples, scientists can determine how many individuals are present in an average sample area of land or volume of soil or water.

Line-Transect Surveys
Line-transect surveys count the number of individuals observed as one moves along a line. There are many variations on this technique. For example, researchers examining small plants in a field or forest might tie a long string between two fixed points and count the number of individuals the string crosses. Alternatively, researchers might count all individuals that are observed within a fixed distance of a line, such as the number of trees on a savannah located within 100 m of a line. A similar approach has been used in surveys of amphibians. In this case, observers count the number of frogs that can be heard along a predetermined path. If we know how far, on average,

a person can hear a frog call, we can estimate the number of frogs calling in an area that includes both sides of the path. Such line-transect data can be converted into area estimates.

One of the most famous line-transect studies is the annual Christmas bird count. The bird count began in 1900 when 27 volunteers from the Audubon Society positioned themselves at different locations in North America and counted all the birds they saw in one day. Today, tens of thousands of volunteers go outside during their winter holidays and follow a predetermined path that covers a 24-km circle. Throughout the day, the volunteers count the number of individuals of every bird species they can see or hear within this circle. This long-term survey of birds in North America has provided incredibly valuable data that has helped scientists determine which species of birds have populations that are increasing, stable, or declining.

Mark-Recapture Surveys
Area- and volume-based and line-transect studies are very useful for organisms that do not move—such as plants and corals—or animals that are not easily

Line-transect surveys Surveys that count the number of individuals observed as one moves along a line.

Mark-Recapture Surveys

To estimate the number of individuals in a population using mark-recapture surveys, we need to know how many individuals were initially sampled and marked. For example, let's imagine that crayfish researchers collected 20 crayfish from a 300-m^2 stretch of stream and marked them with a dot of red fingernail polish. Once the fingernail polish was dry, the crayfish were returned to the stream. After waiting one day for the marked crayfish to move around the stream, the researchers collected another sample of crayfish. In this second sample, they captured 30 crayfish, 12 of which were marked. Based on these data, how many crayfish were in the 300 m^2 of stream? To find out, we can use an equation that considers how many individuals are marked and—after the marked individuals are released back into the population—the ratio of total individuals to marked individuals in the entire population.

Let's look at how we estimate the size of the population. First, note that that we capture and mark a number of individuals (M) from an entire population whose size is defined as N. Therefore, the fraction of marked individuals in the entire population is $\frac{M}{N}$.

When we go back and capture individuals the second time, we record the number of individuals in the second capture (C) and the number of marked individuals that are recaptured (R). The fraction of marked individuals in the recaptured population is $\frac{R}{C}$.

The fraction $\frac{R}{C}$ and our first fraction, $\frac{M}{N}$, actually represent the same fraction, which is the proportion of marked individuals in a sample. Given that these two fractions should represent the same number, we can set the two ratios equal to each other and solve for the unknown variable (N), which is the total size of the population

$$\frac{M}{N} = \frac{R}{C}$$

$$N = M \times C \div R$$

Applying this equation to our crayfish data, the estimated number of crayfish in the stream is

$$N = 20 \times 30 \div 12$$

$$N = 50$$

YOUR TURN What is the estimated size of the crayfish population if the second sample collected 48 crayfish and the number recaptured in that sample was 24?

Based on your estimate of crayfish abundance and the data provided on stream area, what is your estimate of crayfish density?

---

disturbed—such as snails—and therefore less likely to leave the area while a location is being sampled. Some animals, however, are very sensitive to the presence of researchers and will leave the area while other species are well camouflaged and difficult to find. Both situations can cause us to underestimate the number of individuals in a population, so for these situations we need a different type of sampling technique. One effective method is the use of **mark-recapture surveys**. As the name implies, mark-recapture surveys collect a number of individuals from a population and mark them. These individuals are then returned to the population. Once enough time has passed for the marked individuals to mingle throughout the population, a second sample of the population is made. Based on the number originally marked, the total number collected the second time, and the number of marked animals collected the second time, we can estimate the size of the population. Mark-recapture studies are commonly conducted on birds, fish, mammals, and highly mobile invertebrates. The actual calculations for arriving at this estimate are discussed above in "Analyzing Ecology: Mark-Recapture Surveys."

QUANTIFYING THE DISPERSAL OF INDIVIDUALS
Quantifying the dispersal of individuals from a population requires identifying the source of individuals. As we will see, this can be done by ensuring that there is only one possible source of individuals and then determining how far individuals disperse from this single location. In other cases, individuals are marked and then observed or recaptured at some later time to determine how far they moved from the location where they were

---

Mark-recapture survey A method of population estimation in which researchers capture and mark a subset of a population from an area, return it to the area, and then capture a second sample of the population after some time has passed.

Figure 11.9 Measuring dispersal. This blue dasher dragonfly (*Pachydiplax longipennis*) has been marked with a number on its wings for a mark-recapture study. Photo by S. J. McCauley.

marked (Figure 11.9). In animal studies, possible marks include ear tags, radio transmitters, or leg bands. In plant studies, researchers can mark pollen with fluorescent powders and then examine surrounding flowers to determine how far the pollen grains have been moved either by the wind or by pollinators.

A common measure of dispersal is the **lifetime dispersal distance**, which is the average distance that an individual moves from where it was hatched or born to where it reproduces. By knowing the lifetime dispersal distance, we can also estimate how rapidly a growing population can increase its geographic range. For example, when researchers marked eight species of songbirds with leg bands, they found that lifetime dispersal distances averaged between 344 and 1681 m. A lifetime dispersal

distance of about 1 km per generation is therefore not unusual for populations of songbirds. At this rate, the descendants of an average individual might traverse an entire continent in a thousand generations or so.

These calculations suggest that it will take an average species of songbird more than 1000 generations of dispersal to travel across a continent. However, it can actually happen much faster because a few individuals in a population can disperse much farther than the average bird in the population. An excellent example of this occurred following the introduction of the European starling to the United States. In 1890 and 1891, 160 European starlings were released in the vicinity of New York City. Within 60 years, the population had spread 4000 km from New York to California, at an average rate of about 67 km per year. You can see this rapid spread of starlings in Figure 11.10. The expansion occurred rapidly because a few individuals dispersed much longer distances than the average, and established new populations beyond the range boundary of the species. Although the few individuals that might move over such long distances are rare, they can have large effects on population distributions. Today, the starling lives throughout most of the United States and throughout southern Canada.

Population abundance and density are related to geographic range and adult body size

Given the many properties of population distributions, all of which can span a wide range of values, ecologists often search for relationships that might

---

Lifetime dispersal distance The average distance an individual moves from where it was hatched or born to where it reproduces.

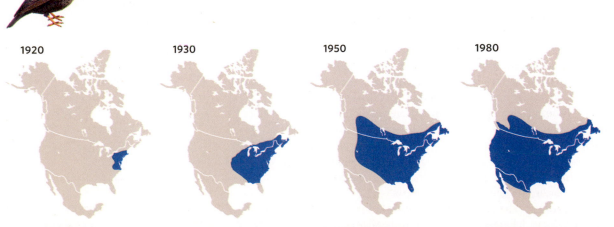

Figure 11.10 The rapid dispersal of European starlings across North America. After being introduced near New York City in 1890 and 1891, the bird rapidly spread across the continent over a 60-year period. The blue-shaded area represents the range of breeding populations. The blue dots indicate detections of starlings that exhibited unusually long dispersal, thereby facilitating the rapid spread of the species. Today, the starling population covers more than 7 million km^2 of North America from coast to coast. Data from http://www.bio.georgiasouthern.edu/bio-home/harvey/lect/lectures.html?flnm=bgin&ttl=Patterns%20of%20 Distribution&ccode=el&mda=scrn

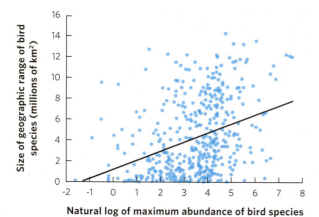

Figure 11.11 Abundance and range size of birds. Data from 457 species of North American birds show that bird species with higher abundances are generally more widely distributed. However, this relationship contains a great deal of variation around the line of best fit due, in part, to annual variations in abundance and geographic ranges. Data from B. McGill and C. Collins, A unified theory for macroecology based on spatial patterns of abundance, *Evolutionary Ecology Research* 5 (2003): 469–492.

explain the underlying causes of variation in these values. Common patterns include the relationship between population abundance and geographic range and the relationship between population density and adult body size.

POPULATION ABUNDANCE AND GEOGRAPHIC RANGE

Ecologists have consistently found that populations with high abundance also have a large geographic range. This pattern has been observed in plants, mammals, birds, and protists. Consider the data on the birds of North America in Figure 11.11. We can see a positive relationship; species with the highest abundances also are the species with the largest range sizes. This same pattern was observed in collared lizards; as their abundance declined, fewer glades were occupied and the geographic range of the population shrank. In the bird data, however, there is a great deal of variation around the regression line. If we were to quantify the coefficient of determination, we would find that abundance of the populations only explained 13 percent of the variation in the geographic ranges of the populations. To review the concept of coefficient of determination, see "Analyzing Ecology: Coefficients of Determination" on page 190 in Chapter 8.

The causes of the range-abundance relationship are still widely debated. There is general consensus that resource availability plays an important role. For example, if a species relies on resources that are only available in a small geographic area, the species will only inhabit a small geographic range. If resources are abundant over a large geographic area, we might

expect that species to cover a large geographic range and to be abundant. In short, the distribution of resources should cause a positive relationship between abundance and geographic range.

Although we do see this positive relationship, we are left to wonder about all of the unexplained variation in the relationship shown in Figure 11.11. In some cases, we might be observing fluctuations within a geographic range. For example, individuals might disperse to more marginal habitats during years that favour especially high reproduction and survival. The range would then contract during years that have low reproduction and survival. Such year-to-year fluctuations can cause the data on the abundance and geographic range of a population to vary a great deal. This can account for some of the variability that we see.

The relationship between population abundance and distribution suggests that reducing the geographic range of a population—for example by converting habitat to agricultural purposes or to housing—will also reduce the size of the population. Similarly, factors that reduce the overall size of the population will simultaneously reduce the geographic range of the population because marginal habitats will no longer receive as many dispersing individuals.

POPULATION DENSITY AND ADULT BODY SIZE

Another common pattern across species is the relationship between population density and body size. Generally, the density of a population is negatively correlated to the body size of the species. Figure 11.12 shows that in herbivorous animals, the

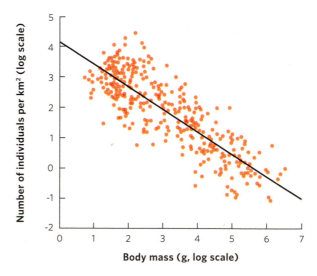

Figure 11.12 Body mass and population density. Across 200 species of herbivorous mammals, population density declines as a function of adult body mass. Data from J. Damuth, Interspecific allometry of population density in mammals and other animals: The independence of body mass and population energy-use, *Biological Journal of the Linnean Society* 31 (1987): 193–246.

smallest-bodied species—such as mice—live at the highest densities, and the largest-bodied species—such as elephants—live at the lowest densities. Part of this relationship is a matter of body size relative to space. A square metre of soil harbours hundreds of thousands of small arthropods, whereas a single elephant simply would not fit into this space. Even if it could be squeezed in, a square metre of soil would not produce enough food to sustain the elephant. A large individual requires more food and other resources than a small one. As a result, we expect a given plot of habitat to support fewer large individuals than it does small individuals.

Dispersal is essential to colonizing new areas

As we mentioned earlier in this chapter, dispersal plays a key role in allowing individuals to move between patches of habitat and to colonize suitable habitats that are not inhabited. In this section, we will discuss how suitable habitats may remain unoccupied due to *dispersal limitation;* we will also examine how *habitat corridors* can play an important role in facilitating dispersal.

DISPERSAL LIMITATION

In some cases, as we saw with the Chinese bushclover, not enough time has passed since the introduction of a species for individuals to spread to every suitable habitat. In other cases, such as the collared lizards, there are substantial barriers that prevent dispersal. A common barrier to dispersal is the presence of large expanses of inhospitable habitat that an organism cannot cross, such as an ocean that a plant seed cannot cross or a large desert that an amphibian cannot cross. Recall that the sugar maple has a preferred range of temperatures and precipitation that favours its presence in the northern temperate forests of North America. Although there are no other suitable areas for sugar maples to grow in North America, there is an abundance of suitable habitat in Europe and Asia. Indeed, several other species of the maple genus (*Acer*) live in habitats that should be suitable for the sugar maple. Sugar maple does not exist in Europe and Asia because its seeds are not capable of dispersing from North America across the oceans to these distant regions. Sometimes the inhospitable habitat is not particularly expansive but still serves as an effective barrier to dispersal, as was the case with fire-suppressed forests that surrounded glades and prevented collared lizard dispersal. The absence of a population from suitable habitat because of barriers to dispersal is called **dispersal limitation**.

Occasionally, individuals cross formidable barriers and disperse long distances without assistance from humans. We know this because many species of plants and animals have populated remote islands, such as the Hawaiian Islands, before humans arrived. However, humans have affected the dispersal of many species. For example, human activities such as road building and forest clearing have created barriers to dispersal for some species. Humans have also assisted in the dispersal of plants and animals for thousands of years. For instance, aboriginal peoples brought dogs to Australia and Polynesians distributed pigs and rats throughout the small islands of the Pacific. In more recent times, foresters have transplanted fast-growing eucalyptus trees from Australia and pines from California to points all over the world for timber and fuel wood. Other species have been intentionally moved to assist them in dispersing over barriers created by humans, such as roads. Still other species have been accidentally moved over large distances by riding along in the ballast water of cargo ships or by attaching themselves to the outside of ship hulls. In most cases, individuals are introduced into a new area but never establish a viable population. In some cases, however, the introduced individuals are able to develop into a population that can grow and expand its geographic range over time. We'll have much more to say about such invasive organisms later in this book, but their success in many places outside their native ranges emphasizes the role of barriers to dispersal in limiting species' distributions.

HABITAT CORRIDORS

In some landscapes, dispersal is facilitated by narrow strips of favourable habitat known as **habitat corridors** that are located between large patches of habitat. For example, two forests might be separated by an open field with a stream that has a narrow band of trees along its banks. This narrow band of trees along the stream can serve as a corridor of suitable habitat between the forests and allow individuals to easily disperse between them.

In recent years, ecologists have conducted large manipulative experiments to test the importance of habitat corridors. For example, in the pine forests of the U.S. state of South Carolina, storms and fires often create cleared areas that are suitable habitat for many species of understorey plants and animals. To test the importance of habitat corridors in facilitating dispersal between these cleared areas, researchers cleared large (1.375 ha) patches of forest. Using a

---

Dispersal limitation The absence of a population from suitable habitat because of barriers to dispersal.

Habitat corridor A strip of favourable habitat located between two large patches of habitat that facilitates dispersal.

Figure 11.13 Manipulating habitat corridors. Researchers cleared patches in the pine forests of South Carolina. Using groups of five patches, the central patch served as a source of dispersers. The outer patches included one patch that was connected to the central patch by a cleared habitat corridor, while the other three were not connected. One unconnected patch was a simple rectangle. The other two patches had cleared corridor paths, termed "wings," that were not connected to another patch. Photo by Ellen Damschen and the Forest Service.

group of five cleared areas, the central cleared area served as the source of dispersers in the experiment. Of the remaining four patches, one was connected to the central patch with a cleared path that could serve as a corridor. The other three outer patches included a simple rectangular-shaped patch and two patches with "wings," meaning that they were patches with corridors that did not connect to other patches (Figure 11.13). The researchers replicated this experimental design eight times.

The researchers used a number of methods for measuring dispersal. They followed movements of the common buckeye butterfly (*Junonia coenia*) by marking individuals in the central patch and recaptured them in the four outer patches. To track pollen movement, the investigators planted eight male winterberry plants (*Ilex verticillata*) in each central patch and then planted three mature female plants in each of the four outer patches. Fertilized flowers producing fruits in the outer patches would indicate that pollen had moved from the central patch. To track seed and fruit movements, they planted individuals of a local holly (*Ilex vomitoria*) and wax myrtle (*Myrica cerifera*) in the central patch. In some cases, the fruits in the central patch were dusted with coloured fluorescent powder. The researchers then collected samples of bird droppings in traps placed under artificial perches in each outer patch. Fecal matter that fluoresced under ultraviolet light indicated that birds had consumed fruits in the central patch and dispersed them to the outer patches.

The results of these experiments are shown in Figure 11.14. The outer patch connected by a habitat

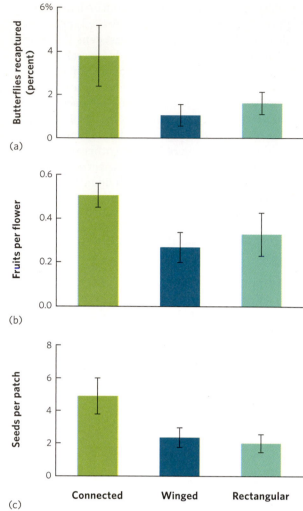

Figure 11.14 The effects of manipulated corridors on dispersal. Compared to unconnected patches that were either winged or rectangular in shape, patches connected by a habitat corridor experienced (a) a higher number of recaptured butterflies, (b) greater pollination of plants, leading to more fruits per flower, and (c) greater seed dispersal by birds. Error bars are standard errors. Data from J. J. Tewksbury et al., Corridors affect plants, animals, and their interactions in fragmented landscapes, *Proceedings of the National Academy of Sciences USA* 99 (2002): 12923–12926.

corridor had much more dispersal from the central patch. The researchers observed more dispersal by butterflies, more traffic by insect pollinators—which caused more fruits to be produced—and greater movement of fruits and seeds found in bird droppings. Indeed, the movement of fruits and seeds through the corridors was so frequent that the number of species of herbs and shrubs increased faster in the connected patches. Experiments such as these confirm the importance of habitat corridors. Indeed, conservation

Figure 11.15 Conservation of habitat corridors. Numerous groups and organizations in Canada and the United States are working to create a protected corridor connecting mountainous regions in the Yukon with similar regions in Yellowstone National Park.

efforts have increasingly considered the preservation of corridor habitats. For example, nongovernmental groups, government agencies, local communities, First Nations, scientists, and a range of interest groups in Canada and the United States are collaborating to create a corridor that will span 1.3 million square kilometres from the Yukon to Yellowstone National Park. The corridor is shown in Figure 11.15 and will connect a number of protected areas in five U.S. states and four Canadian provinces and territories. The project is intended to protect large natural areas important for grizzly bears and numerous bird and fish species.

Many populations live in distinct patches of habitat

In our discussion of the collared lizard, we saw that individuals in the region live in many small groups, with each group restricted to a particular glade. The lizards also move between habitats, but such movement depends on the quality of the forested habitat connecting the glades. When ecologists want to understand the movement of individuals among patches of habitats and how this affects the abundance of animals in each patch of habitat, they need to consider the quality of the habitat. They also have to consider the difficulty in reaching the habitat, which is determined by the distance to the habitat and the difficulty in crossing the area between habitat patches. In this section, we will explore how habitat quality affects the distribution of individuals among habitats and how ecologists have conceptualized the distribution and movement of individuals among suitable patches of habitat.

THE IDEAL FREE DISTRIBUTION AMONG HABITATS
When habitats differ in quality and individuals can easily move among habitat patches, natural selection should favour those individuals that can choose the habitat that provides them with the most energy; this would improve their fitness. If all individuals are able to distinguish between high- and low-quality habitats, and assess the benefit of living in either habitat, then all individuals should initially move to the high-quality habitats, which are represented by a

green line in Figure 11.16. However, as more and more individuals choose the high-quality habitat, the resources available must be divided among many individuals. This reduces the amount of resources available to each individual, also known as the *per capita benefit*. At some point, the per capita benefit in the high-quality habitat falls so low that an individual would be better off if it moved over to the low-quality habitat, which is indicated by the orange line in Figure 11.16. Continued increases in the population size would continue to add individuals to both the high- and low-quality habitats such that individuals in both habitats would experience the same per capita benefit. When all individuals have perfect knowledge of habitat variation and they distribute themselves in a way that allows them to have the same per capita benefit, we call it the **ideal free distribution**.

An example of individuals exhibiting an ideal free distribution can be seen in an experiment in which stickleback fish (*Gasterosteus aculeatus*) were presented with high- and low-quality habitats. Using water fleas (*Daphnia magna*), which are readily consumed by stickleback fish, the researchers added different numbers of water fleas to opposite ends of an aquarium so that each end of the aquarium represented a habitat of different quality. When the first fish were initially placed into the aquarium with no water fleas present, the fish were distributed equally between the two halves of the aquarium. In one experiment, researchers added 30 water fleas per minute to one end

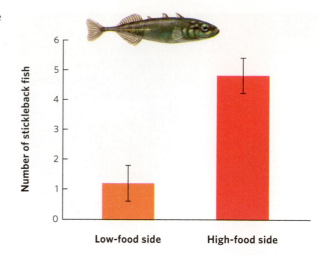

Figure 11.17 The ideal free distribution in stickleback fish. When six fish were placed in an aquarium with the two ends receiving a five-to-one difference in food rations, the fish distributed themselves in a four-to-one ratio. As a result, each fish received a similar per capita amount of food. Error bars are standard deviations. Data from M. Milinski, An evolutionarily stable feeding strategy in sticklebacks, *Zeitschrift für Tierpsychologie* 51 (1979): 36–40.

of an aquarium and six per minute to the other, a ratio of five to one. The results of this experiment are shown in Figure 11.17. Within 5 minutes, the fish had distributed themselves between the two halves in a ratio that was approximately four to one, which is close to the five to one ratio predicted by an ideal free distribution. When the provisioning ratio was changed—that is, the high- and low-quality patches were switched between ends of the aquarium—the fish quickly adjusted their distribution ratio. How they achieved this ideal free distribution was not determined, but they may have used the rate at which they encountered food items, or perhaps the number of other fish close by, as cues to patch quality.

The ideal free distribution tells us how individuals should distribute themselves among habitats of differing quality, but individuals in nature rarely match the ideal expectations. In some cases individuals may not be aware that other habitats exist. Also, the fitness of an individual is not solely determined by maximizing its resources. Other factors that influence an individual's use of a particular habitat include the presence of predators or of a territory owner that precludes moving into a high-quality habitat.

When reproductive success has been measured in the field, ecologists commonly find that individuals living in high-quality habitats have higher reproductive success. Often dominant individuals have monopolized

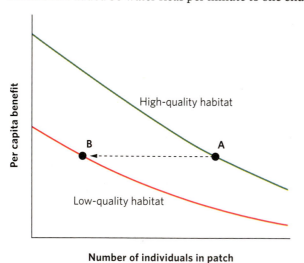

Figure 11.16 The ideal free distribution. Given the existence of high- and low-quality habitats, the first individual to arrive should select the high-quality habitat. As more individuals choose the high-quality habitat, however, the per capita benefit of the habitat declines. At some point (point A), it is equally beneficial for an individual to move to the low-quality habitat (point B) because the per capita benefit equals that experienced in the high-quality habitat. Although the low-quality habitat will have fewer individuals than the high-quality habitat, all individuals will experience the same per capita benefits.

Ideal free distribution When individuals distribute themselves among different habitats in a way that allows them to have the same per capita benefit.

the best resources, and they tend to produce surplus offspring while individuals in the poorest habitats do not produce enough offspring to replace themselves. As a result, the populations living in the high-quality habitats serve as a source of dispersing offspring that move to low-quality habitats, which allows populations in low-quality habitats to persist.

A study of the blue tit (*Parus caeruleus*), a small songbird from southern Europe, helps to illustrate this point. The blue tit breeds in two kinds of forest habitat, one dominated by the deciduous downy oak (*Quercus pubescens*) and the other by the evergreen holm oak (*Quercus ilex*). Long-term studies in southern France have revealed that the downy oak habitat produces more caterpillars, an important food item for the blue tits. This difference in caterpillar availability is reflected in the population densities of the birds. As you can see in Figure 11.18, the downy oak forest supports more than six times as many adult breeding pairs as the holm oak forest. Moreover, each breeding pair produces about 60 percent more offspring per year in the downy oak forest than in the holm oak forest. If we assume that juvenile survival is 20 percent in the first year and that annual adult survival is 50 percent— values that are typical of temperate songbirds— the population in the downy oak habitat would experience an annual growth of 9 percent per year if the surplus individuals did not disperse out of the habitat. At the same time, the population in the low-quality, holm oak habitat would experience an annual decline of 13 percent per year if no birds emigrated from the high-quality habitat. This would cause the population in the low-quality habitat to rapidly go extinct. In reality, the populations persist over time because the low-quality habitat receives surplus individuals from the high-quality habitat.

CONCEPTUAL MODELS OF SPATIAL STRUCTURE

The ideal free distribution examines how individuals should distribute themselves when they know about the quality of different habitats and can freely move among them. In many cases, however, it is not easy to move from one patch of habitat to another, either because the other patches are relatively far away or because the region between the patches is fairly inhospitable. In such scenarios, most individuals in the patch stay put, only occasionally dispersing to other habitat patches. As a result, the larger population is broken up into smaller groups of conspecifics that live in isolated patches, which we call **subpopulations**.

When individuals frequently disperse among subpopulations, the whole population functions as a single structure and they all increase and decrease in abundance synchronously. When dispersal is infrequent, however, the abundance of individuals in each subpopulation can fluctuate independently of

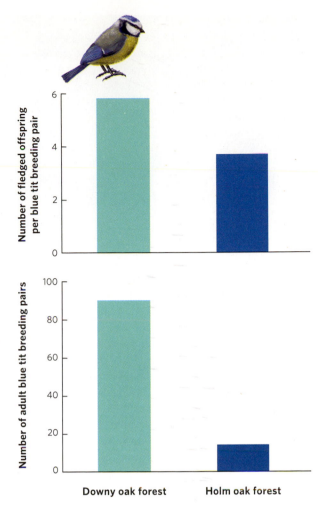

Figure 11.18 The effects of habitat quality. In blue tit songbirds, deciduous downy oak forests support many more adult breeding pairs than evergreen holm oak forests. In addition, birds living in the downy oak forest produce about 60 percent more fledglings per nest. Data from J. Blondel, Habitat heterogeneity and life-history variation of Mediterranean blue tits (*Parus caeruleus*), The Auk 110 (1993): 511–520.

one another. In considering the spatial structure of subpopulations, ecologists have devised three models: the *basic metapopulation model,* the *source–sink metapopulation model,* and the *landscape metapopulation model.* These three models range from less to more complex. We shall explore the dynamics of these models in more detail in Chapter 13.

The Basic Metapopulation Model

The **basic metapopulation model** describes a scenario in which there are patches of suitable habitat embedded within a matrix of unsuitable habitat. All

---

Subpopulations When a larger population is broken up into smaller groups that live in isolated patches.

Basic metapopulation model A model that describes a scenario in which there are patches of suitable habitat embedded within a matrix of unsuitable habitat.

Figure 11.19 The landscape metapopulation model. The landscape model contains the most realistic conditions of habitat patches that differ in quality and a surrounding matrix that is variable in its suitability for dispersal and the presence of dispersal barriers.

suitable patches are assumed to be of equal quality. Some suitable patches are occupied while others are not, although the unoccupied patches can be colonized by dispersers from occupied patches. The basic metapopulation models emphasize how colonization and extinction events can affect the proportion of total suitable habitats that are occupied.

The Source–Sink Metapopulation Model
The **source–sink metapopulation model** builds upon the basic metapopulation model and adds the reality that different patches of suitable habitat are not of equal quality. As we saw in the study of blue tits, it is common that occupants of high-quality habitats serve as a source of dispersers. Such populations are referred to as **source subpopulations**. At the same time, there can be low-quality habitats that rarely produce enough surplus offspring to produce any dispersers. These habitats depend on outside dispersers to maintain the subpopulation. These subpopulations are known as **sink subpopulations**.

The Landscape Metapopulation Model
The **landscape metapopulation model** is even more realistic than the source–sink model because it considers both differences in the quality of the suitable patches and the quality of the surrounding matrix. As we discussed earlier in this chapter, the habitat in the surrounding matrix can vary in quality for dispersing organisms. Consider, for example, the challenge that metamorphosing frogs have when they move from their natal pond. They face risks of both predation and desiccation. Moving through a grassy field poses a much higher risk of predation and drying out than moving through a humid forest. Although

neither field nor forest is a suitable habitat for the frog to reproduce, each presents different magnitudes of dispersal barriers. For a regional population of collared lizards to persist and grow requires both high-quality glade habitats and a high-quality matrix of open forest that facilitates dispersal. The landscape model, which is illustrated in Figure 11.19, represents the most realistic, yet also the most complex, spatial structure of populations.

Throughout this chapter we have examined the spatial structure of populations by considering the characteristics of abundance, density, and geographic range. We have also focused on the importance of suitable habitats and dispersal and how ecological niche modelling can help us predict the distributions of populations in the future. In doing so, we have largely ignored the fact that populations increase and decrease in abundance, but in the next chapter, we will take an in-depth exploration of how populations grow and how this growth is regulated.

Source–sink metapopulation model A population model that builds upon the basic metapopulation model and accounts for the fact that not all patches of suitable habitat are of equal quality.

Source subpopulations In high-quality habitats, subpopulations that serve as a source of dispersers within a metapopulation.

Sink subpopulations In low-quality habitats, subpopulations that rely on outside dispersers to maintain the subpopulation within a metapopulation.

Landscape metapopulation model A population model that considers both differences in the quality of the suitable patches and the quality of the surrounding matrix.

REVIEW QUESTIONS

1. What is the difference between population distribution and population structure?

2. Why is the realized niche considered a subset of the fundamental niche?

3. The American bullfrog is native to eastern North America, but it has been moved by humans and now thrives in western North America. What does this suggest about the cause of the bullfrog's historical range limit?

4. How can ecological niche modelling be used to predict the spread of introduced species?

5. What mechanisms could cause evenly spaced versus clustered distributions of individuals within populations?

6. Suppose that 100 cattle were allowed to graze in either one of two pastures. If the grass was three times more productive in pasture A than in pasture B, how many cattle would be in each pasture if they followed an ideal free distribution? What might prevent this distribution of cattle from happening?

7. What realities of nature do source–sink models and landscape models include that basic metapopulation models do not include?

8. How would lifetime dispersal distance and neighbourhood size differ for a plant whose seeds are carried by the wind and a plant whose seeds drop near the parent plant?

9. Why do most species have their highest densities near the centre of their geographic range?

10. Explain why researchers commonly find a relationship between adult body size and population density.

PRACTICING ECOLOGY: BLOWING ACROSS THE GEOGRAPHIC RANGE

Chris Eckert and his lab at Queen's University in Ontario study evolutionary adaptations that enable plants to disperse across a landscape. Photo by Lawrence Harder.

Life on the edge is difficult— particularly when you are a plant. At the edges of a plant's geographic range, individuals often face less than optimal ecological conditions for growth, survival, and reproduction. Since offspring are likely to experience similarly difficult conditions, individual fitness is likely to increase if plants on the edge widely disperse their seeds. Wider dispersal increases the probability that some of their offspring will end up in suitable habitats. We might therefore expect natural selection to favour adaptations that enable plants on the edge of their geographic range to disperse more widely than plants in the interior of a range.

The pink sand verbena (*Abronia umbellata*) is a perennial herb found along coastal areas spanning latitudes from southern British Columbia to Baja California, Mexico. After bunches of pink flowers blossom, the plants produce fruits with three or four papery wings. When the fruits mature, they break off and disperse as the wind catches their wings. Wing sizes among the fruits differ and are positively correlated with dispersal distance. Given the variation in wing size among verbena populations, we might hypothesize that selection would favour larger wing size at the edges of the pink sand verbena's geographic range.

To test this hypothesis, Chris Eckert and his lab from Queen's University in Ontario sampled fruits from 36 pink sand verbena populations at different latitudes between British Columbia and Baja California. In each population, they sampled a fruit from 12 different plants. For each fruit, they ranked wing size from 1 to 5, with 1 being the smallest. They then summed the ranks for all wings on a fruit. Next, they calculated the mean wing size for each pink sand verbena population. Because fruits that require a lower wind velocity to move will disperse farther, the researchers also considered the wind velocity needed to move a single fruit. They measured the velocity by placing each fruit in front of a fan and gradually increased the wind velocity until the fruit moved from a resting position. You can view data for 14 of the pink sand verbena populations in the table.

MEAN WING SIZE AND WIND VELOCITY AT MOVEMENT FOR PINK SAND VERBENA FRUIT AT DIFFERENT LATITUDES

LATITUDE (°N)	MEAN WING SIZE	WIND VELOCITY (m/s)
30.5	8.16	1.75
31.8	12.34	1.64
33.6	6.01	2.25
34.0	4.84	2.48
34.0	2.64	2.96
34.9	1.65	2.99
35.0	2.43	3.47
36.8	3.27	2.87
37.0	2.95	2.61
37.9	3.08	2.65
39.3	7.35	2.58
40.8	6.86	2.07
41.3	10.67	1.85
41.4	9.94	1.77

GRAPH THE DATA

1. Create a scatterplot graph with latitude on the *x* axis and wing size index on the *y* axis.

2. Create a second scatterplot graph with latitude on the *y* axis and wind velocity needed to move a fruit on the *x* axis.

3. Visually estimate and draw a regression line of best fit through the data on both graphs. Note that a curvilinear regression line may fit the data better than a linear regression line.

INTERPRET THE DATA

1. Explain how the data support or refute the hypothesis that wing size and dispersal distance is greater at the edges of geographic range.

2. Assume there is a trade-off between investing energy in wings and increasing the number of seeds in a fruit. Hypothesize what the relationship might be between latitude and seed number per fruit.

SOURCE: Darling, E., K. E. Samis, and C. G. Eckert. 2008. Increased seed dispersal potential towards geographic range limits in a Pacific coast dune plant. *New Phytologist* 178: 424—435.

CHAPTER 12

Population Growth and Regulation

The Human Population Explosion

The number of people living on Earth has been increasing rapidly for thousands of years. Although estimates are understandably crude, experts believe that a million years ago the ancestors of modern humans numbered a million individuals. With the advent of agriculture 10 000 years ago, greater food production allowed further population growth, so that by the year 1700 the human population had grown to about 600 million people. During the Industrial Revolution, increased wealth brought better nutrition, sanitation, and improvements in medical treatments. As a result, the mortality rate of children fell and the life span of adults increased. An exponential increase in population growth followed these developments. According to the United Nations, the world population hit 1 billion in 1804. Although it took more than

> **"By 2012, our planet held 7 billion people."**

1 million years to reach this mark, it only took 123 years to reach the 2 billion mark in 1927. In recent decades, a billion people have been added every 12 to 13 years, and the population has doubled in just 40 years. By 2012, our planet held 7 billion people.

The rapid growth of the human population in the past has led many scientists to predict population sizes in the future. In the eighteenth century, for example, the British economist Thomas Malthus examined data on human population growth and concluded that the rapid rate of growth would cause humans to outstrip the food supply quickly. The logic of Malthus was simple; as the density of humans increased, each person would have less food to eat and the rates of survival and reproduction would go down. One reason that this did not occur during the nineteenth century is that new technologies increased food production. Food production has continued to grow throughout the twenty-first century and this has supported further population growth.

Scientists have developed mathematical models of population growth that predict how much more the human population will grow. One of the challenges in making these predictions is the differing population growth rates among developed and developing countries. Developed countries that formerly had rapid growth, including the United States and most European countries, now have populations that are growing slowly or even declining. In contrast, developing countries still have rapid growth rates because they have a disproportionate number of young people who will reproduce in the coming decades, leading to further growth in these populations. When modellers of population growth take all of this into consideration, they estimate that the world population will eventually stabilize at 8 to 12 billion people by 2050. An important question, however, is whether our planet can sustain such a large population.

The explosive growth of the human population is a clear example of how populations can grow when resources are abundant.

Human population growth. During the past 2 centuries, the number of people on Earth has been growing at an unprecedented rate. In this photo, a crowd gathers as part of the Oktoberfest celebration in Munich, Germany. Photo by Stefan Auth/age footstock.

Mathematical models help us predict future increases. Although it is interesting from the perspective of population growth, as we shall see in this chapter, human population increases are also of great interest to ecologists because nearly every major ecological concern is connected to the growing number of humans on Earth.

SOURCES: United Nations Population Fund, *State of the World Population 2011*, http://www.unfpa.org/webdav/site/global/shared/documents/publications/2011/EN-SWOP2011-FINAL.pdf.

"Population seven billion: UN sets out challenges," *BBC News*, October 26, 2011.

CHAPTER CONCEPTS

■ Under ideal conditions, populations can grow rapidly.

■ Populations have growth limits.

■ Population growth rate is influenced by the proportions of individuals in different age, size, and life history classes.

Studying ecology at the population level can take a number of different perspectives. In the previous chapter, for example, we examined the spatial distribution of populations. In this chapter we will examine changes in population size, known as population dynamics. We will look at how populations grow and we will explore the factors that regulate their growth.

Historically, studies of population growth focused on human populations. As mentioned at the beginning of this chapter, Thomas Malthus examined data on human populations in the eighteenth century and concluded that it was growing so rapidly that there would soon be insufficient food to support its size. This insight influenced nineteenth-century scientists, including Charles Darwin who realized that this reasoning could apply to every organism on Earth.

Understanding population growth and regulation today is important because it permits us to predict future population growth. This predictive ability helps us manage the population sizes of species that are harvested by humans, species that are declining and need to be saved from extinction, and pest species that invade new regions and need to be controlled. The study of populations is known as **demography**. In this chapter we will explore key demographic models of how populations grow.

Under ideal conditions, populations can grow rapidly

Populations of any species can grow at an incredibly rapid rate. Some of the earliest studies of population growth came from researchers who examined human populations. In this section we will explore how ecologists use population growth models to predict how populations will grow over time.

THE EXPONENTIAL GROWTH MODEL

To understand how populations grow, we must first understand the concept of population *growth rate*.

The **growth rate** of a population is the number of new individuals that are produced in a given amount of time minus the number of individuals that die. We typically consider the growth rate on an individual, or per capita, basis. Under ideal conditions, individuals can experience maximum reproductive rates and minimum death rates. When this happens, a population achieves its highest possible per capita growth rate, which is called the **intrinsic growth rate**, denoted as *r*. For example, abundant resources allow a white-tailed deer to have twins, a western mosquito (*Aedes sierrensis*) to lay up to 200 eggs in a single clutch, and a white oak tree (*Quercus alba*) to produce more than 20 000 acorns in one reproductive season. In addition to facilitating high birth rates, abundant resources also lead to lower death rates. When conditions are less than ideal, for example in times of reduced resources or increased disease, a population's growth rate is less than its intrinsic growth rate.

Once we know the intrinsic growth rate for a population, we can estimate how a population will grow over time under ideal conditions. To do this, we can use the **exponential growth model**, so named because it uses a formula that has an exponent:

$$N_t = N_0 e^{rt}$$

where e is the base of the natural log (e = 2.72). This equation tells us that when conditions are ideal the

---

Demography The study of populations.

Growth rate In a population, the number of new individuals that are produced in a given amount of time minus the number of individuals that die.

Intrinsic growth rate (r) The highest possible per capita growth rate for a population.

Exponential growth model A model of population growth in which the population increases continuously at an exponential rate.

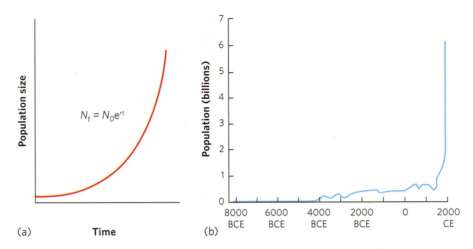

Figure 12.1 The exponential growth model. (a) Over time, a population living under ideal conditions can experience a rapid increase in population size. This produces a J-shaped curve. **(b)** Human population growth is an excellent example of a J-shaped curve. Data from http://www2.uvawise.edu/pww8y/Supplement/ConceptsSup/PopulationSup/ChartWldPopGro.html.

size of the population in the future (N_t) depends on the current size of the population (N_0), the population's intrinsic growth rate (r), and the amount of time over which the population grows (t).

The exponential growth model demonstrates that populations living under ideal conditions can experience rapid growth that, when graphed, produces a **J-shaped curve** as shown in Figure 12.1a. Figure 12.1b illustrates that the human population has been growing exponentially during the past 300 years.

Using the exponential model, we can also determine the rate of growth at any point in time. To do this, we can take the derivative of the exponential growth equation and obtain the following equation:

$$\frac{dN}{dt} = rN$$

where $\frac{dN}{dt}$ represents the change in population size per unit of time. In words, this equation tells us that the rate of change in population size at any particular point in time depends on the population's intrinsic growth rate and the population's size at that point in time. Populations with higher intrinsic growth rates or a larger number of reproductive individuals will experience a greater rate of increase in population size. Another way to think about this equation is that it tells us the slope of the line relating population size to time at any given point. Looking at the graphs in Figure 12.1, for example, the slope of the line is very shallow early in time and becomes very steep later in time. This confirms that the population initially grows more slowly because there is a small number of reproductive individuals, but then grows much faster as the number of reproductive individuals increases.

The exponential growth model for a population resembles the way money grows when it is earning interest in a bank account. The number of individuals in the population is analogous to the number of dollars in your bank account, and the intrinsic rate of growth is analogous to the annual rate of interest. Imagine that you had $1000 in your account, you received 5 percent annual interest, you never added or withdrew money from the account, and any interest earned would be deposited directly into the account at the end of the year. After 1 year the balance would be $1050, which is an annual growth of $50. The balance after the second year would be $1102.50—an annual growth of $52.50. In the tenth year, the annual growth would be $77.57, and in the twentieth year the annual growth would be $126.35. As you can see, a constant interest rate results in an ever-increasing balance. Similarly, a constant intrinsic growth rate results in ever-increasing numbers of individuals in a population and the population experiences exponential growth.

THE GEOMETRIC GROWTH MODEL

The exponential growth model can be applied to species that reproduce throughout the year, such as humans. However, most species of plants and animals have discrete breeding seasons. For example, most birds and mammals reproduce in the spring and summer when there are abundant resources available for their offspring. As an example, let's look at the California quail (*Callipepla californica*), a bird from western North America that lays one or two clutches of eggs in the spring. The quail population experiences a large boost in its population size in the spring but then the population slowly declines over the summer, autumn, and winter due to deaths. You

J-shaped curve The shape of exponential growth when graphed.

Figure 12.2 The discrete breeding events of the California quail. Each generation of new quail produced each spring is represented by a different colour. Because births only happen in the spring, the growth of the California quail is better modelled by a geometric growth model than by an exponential growth model. Data from J. T. Emlen, Jr., Sex and age ratios in survival of California quail, *Journal of Wildlife Management* 4 (1940): 92–99.

can see the changes in population sizes in Figure 12.2, where each colour in the graph represents the new generation of young quail that are produced each spring. When we examine the pattern of population abundance over several years, we can see that the exponential growth model, which assumes continuous births and deaths throughout the year, does not describe animals such as quail that have a distinct breeding period. For these species, ecologists use the **geometric growth model** because it compares population sizes at regular time intervals. In the case of the California quail, for instance, the geometric growth model allows us to compare population sizes at yearly intervals.

The geometric growth model is expressed as a ratio of a population's size in 1 year to its size in the preceding year (or some other time interval). This ratio is assigned the symbol λ, which is the lowercase Greek letter lambda. A λ value greater than 1 means the population size has increased from 1 year to the next because there have been more births than deaths. When λ is less than 1, the population size has decreased from 1 year to the next because there have been more deaths than births. Because there cannot be a negative number of individuals, the value of λ is always positive.

If N_0 is the size of a population at time 0, then its size 1 time interval later would be:

$$N_1 = N_0 \lambda$$

Next, we can examine how we predict population size after more than 1 time interval. For example, if we wanted to estimate population size after 2 time intervals:

$$N_2 = (N_0 \lambda)\lambda$$

which can be rearranged to

$$N_2 = N_0 \lambda^2$$

In more general terms, then, we can express the geometric growth model as

$$N_t = N_0 \lambda^t$$

where t equals time. Imagine that we start with a population of 100 quail and we have an annual growth rate of $\lambda = 1.5$. After 5 years, the size of the population is

$$N_5 = N_0 \times \lambda^5$$

$$N_5 = 100 \times 1.5^5$$

$$N_5 = 759$$

Using the geometric growth model, we can also calculate how much the size of a population changes between time intervals. The change in population size from 1 time interval to the next is referred to as ΔN, where Δ, the capital Greek letter delta, means "a change in." The change in population size is the number of individuals at the end of the interval minus the number of individuals at the beginning of the interval

$$\Delta N = N_0 \lambda^t - N_0$$

which can be rearranged to

$$\Delta N = N_0 (\lambda^t - 1)$$

If we return to our example in which the initial population size is 100 and the annual rate of growth is 1.5, the change in population size after 5 years will be

$$\Delta N = 100 (1.5^5 - 1)$$

$$\Delta N = 659$$

COMPARING THE EXPONENTIAL AND GEOMETRIC GROWTH MODELS

Notice that the equation for exponential growth is identical to the equation for geometric growth, except that e^r takes the place of λ. Thus, geometric and exponential growth are related by

$$\lambda = e^r$$

Geometric growth model A model of population growth that compares population sizes at regular time intervals.

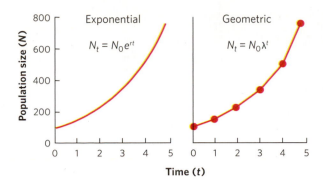

Figure 12.3 Comparing the growth of populations using the exponential and geometric growth models. The exponential model uses continuous data, whereas the geometric model uses discrete data that are calculated at each time point. However, the two models reveal the same increase in population size over time.

which can be rearranged to

$$\log_e \lambda = r$$

From this relationship, we can see that λ and r are directly related to each other. Indeed, if we were to graph the growth of populations over time using both models and setting $\lambda = e^r$, we would find identical growth curves. We see this comparison in **Figure 12.3.** Although the exponential model has continuous data points whereas the geometric model has discrete data points, the two models show the same pattern in population growth. **Figure 12.4** compares the relationships between λ and r by examining the values of λ and r when populations are decreasing, constant, or increasing. When a population is decreasing, $\lambda < 1$ and $r < 0$. When a population is constant, $\lambda = 1$ and $r = 0$. When a population is increasing, $\lambda > 1$ and $r > 0$.

POPULATION DOUBLING TIME

We can appreciate the capacity of a population for growth by observing the rapid increase of organisms introduced into a new region with a suitable environment. For example, after sea otters were reintroduced into British Columbia waters in the 1970s, their population grew from 70 to 345 individuals within 7 years. This means that the population experienced an annual growth rate of $\lambda = 1.26$. In other words, the population grew, on average, by almost one third each year. Populations maintained under optimal conditions in laboratories can have very high growth rates. Under ideal conditions, the value of λ can be as high as 24 for field voles (*Microtus agrestis*), a small mouse-like mammal, 10 billion (10^{10}) for flour beetles, and 10^{30} for water fleas.

One way of appreciating the potential growth rate of populations is to estimate the time required for a population to double in abundance, known as the **doubling time**. To understand how we determine doubling time, we can start by rearranging the exponential growth equation:

$$N_t = N_0 e^{rt}$$

$$e^{rt} = N_t \div N_0$$

When a population doubles, its size is twice its original size at time 0. As a result, we can replace $(N_t \div N_0)$ with the value 2 and determine the time required (t_2) for a population to double in size:

$$e^{rt} = 2$$

$$rt = \log_e 2$$

$$t = \log_e 2 \div r$$

For the geometric model, the equation is nearly the same, except r is replaced by $\log_e \lambda$

$$t_2 = \log_e 2 \div \log_e \lambda$$

Given that the value of $\log_e 2$ is 0.69, the doubling time for the ring-necked pheasant, which

---

Doubling time The time required for a population to double in size.

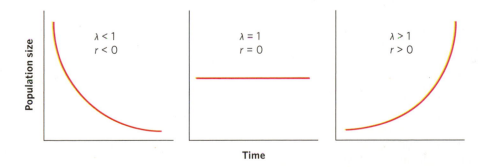

Figure 12.4 A comparison of λ and r values when populations are decreasing, constant, or increasing. When the population is decreasing, $\lambda < 1$ and $r < 0$. When the population is constant, $\lambda = 1$ and $r = 0$. When the population is increasing, $\lambda > 1$ and $r > 0$.

has an annual growth rate of $\lambda = 2.78$, can be calculated as

$$t_2 = 0.69 \div \log_e 2.78$$

$$t_2 = 0.67 \text{ years}$$

$$t_2 = 246 \text{ days}$$

The same calculation gives doubling times of 79 days for the field vole, 11 days for the flour beetle, and only 3.6 days for the water flea. Using this calculation, demographers have determined that the doubling time for the human population is currently 40 years.

The exponential and geometric growth models are excellent starting points for understanding how populations grow. There also is abundant evidence that real populations do initially grow rapidly, just as these models suggest. As we will see in the next section, however, no population can sustain exponential growth indefinitely. As populations become more abundant, they are limited by other factors such as competition, predation, and pathogens.

Populations have growth limits

In nature we commonly observe that there are limits on how large a population can grow. Ecologists categorize these limits as being either *density independent* or *density dependent*. In this section we will review these two types of limits on population growth and discuss how we can incorporate them into population growth models.

DENSITY-INDEPENDENT FACTORS

As the name suggests, **density-independent** factors limit population size regardless of the population's density. Common density-independent factors include natural disasters such as tornadoes, hurricanes, floods, and fires. However, other less dramatic changes in the environment, including extreme temperatures and droughts, can also limit populations. In all of these cases, the impact on the population is not related to the number of individuals in the population. Consider, for example, what happens when a hurricane hits a coastal forest. The number of trees that survive the hurricane is independent of the number of trees that were present before the hurricane arrived.

A classic study on density-independent effects was conducted by James Davidson and Herbert Andrewartha in Australia during the 1930s and 1940s. The apple thrip (*Thrips imaginis*) was a common insect pest in Australia that would occasionally increase to very large numbers and devastate apple trees and rose bushes throughout large regions of the country. From 1932 to 1946, the researchers surveyed thrip populations on rose bush flowers in an attempt to understand the causes of the large variations in

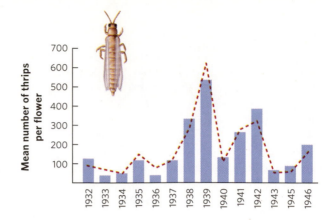

Figure 12.5 Predicting density-independent growth of the apple thrip. Scientists surveyed thrip populations on the flowers of rose bushes from 1932 to 1946. The dashed line represents the predicted number of thrips each year based on a model that included temperature and rainfall variables that cause density-independent effects on thrips. The bars represent the number of thrips actually observed each year. Data from J. Davidson and H. G. Andrewartha, The influence of rainfall, evaporation and atmospheric temperature on fluctuations in the size of a natural population of *Thrips imaginis* (Thysanoptera), *The Journal of Animal Ecology* 17 (1948): 200–222.

population size. They suspected that these population changes were the result of density-independent factors such as seasonal variation in temperature and rainfall. Figure 12.5 shows the population sizes the researchers predicted and the population sizes they observed. As you can see, the predicted changes in thrip abundance closely matched the actual number of thrips they counted in their surveys.

Understanding the role of density-independent factors remains an important goal of ecologists today. For example, in 2011 researchers from the U.S. Forest Service and the Canadian Forest Service reported on the impacts of bark beetles in western North America. Bark beetles are native insects that consume the phloem tissues of trees, causing them to die. During the past several decades, bark beetles have killed billions of coniferous trees covering millions of hectares from Mexico to Alaska (Figure 12.6). Under warmer temperatures, bark beetles develop faster and have more generations per year, but they can be killed by unusually cold temperatures. As a result, temperature is a major determinant of population size. Concerned about the possible effect of climate change on the population of bark beetles, researchers looked at expectations for increased climate warming and applied their knowledge of bark beetle ecology to project future growth. They concluded that bark beetle populations are likely to increase dramatically and cause more damage to conifer trees in the future.

Density independent Factors that limit population size regardless of the population's density.

DENSITY-DEPENDENT FACTORS

Density-dependent factors affect population size in relation to the population's density. Ecologists find it useful to break these factors down into two categories. **Negative density dependence** occurs when the rate of population growth decreases as population density increases. **Positive density dependence**, also known as **inverse density dependence**, occurs when the rate of population growth increases as population density increases. Because positive density dependence was first proposed by the ecologist Warder Allee in 1931, it is also known as the **Allee effect**.

Some of the most common negative density-dependent factors include resources that are available in limited supply, such as food, nesting sites, and physical space. When a population is small, there is an abundance of resources for all individuals, but as the population increases, such resources are divided among more individuals. As a result, the per capita amount of resources declines and at some point it reaches a level at which individuals find it difficult to grow, reproduce, and survive. This was the concern Malthus expressed in the eighteenth century about the rapid growth of the human population.

Although competition for food is a common challenge for populations that grow to large sizes, it is not the only challenge. Crowded populations also can have higher levels of stress, transmit diseases at a greater rate, and attract the attention of predators. All

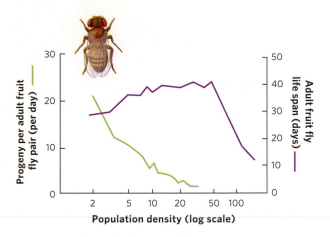

Figure 12.7 Limits on population growth. Larger populations cause increased competition. This graph shows that as competition among fruit flies increases, they experience a decline in the number of progeny produced per adult breeding pair per day. Increased competition also leads to a decline in the life span of adult flies. Data from R. Pearl, The growth of populations, *Quarterly Review of Biology* 2 (1927): 532–548.

of these factors can act to slow, and finally halt, population growth.

Negative Density Dependence in Animals

In a classic experiment that investigated density-dependent regulation of populations, Raymond Pearl introduced pairs of fruit flies (*Drosophila melanogaster*) at different densities into laboratory bottles containing identical amounts of food. As the number of individuals in the bottles increased, competition for food became more intense and the number of progeny produced per pair of adult flies decreased, as shown in Figure 12.7. With even larger populations, the life span of the adults also sharply declined, demonstrating that both the young and old fruit flies were affected by the competition for food.

Although negative density dependence is easily demonstrated in the laboratory, a number of natural experiments have confirmed the existence of the phenomenon in nature. For example, deer mice (*Permyscus maniculatus*) occur most commonly in mature forests and grasslands, but also in houses and barns. While deer mice can eat insects and bird eggs, their main food supply is seeds. Depending on seed availability, populations can vary considerably from year to year. In a study spanning 36 years, researchers

Figure 12.6 Bark beetles. Many of these lodgepole pines in central British Columbia, Canada, have been severely damaged by bark beetles, which consume the phloem tissues of the pine trees. The mountain pine beetle (*Dendroctonus ponderosae*) is one of several species of bark beetles that is predicted to become more abundant with warmer global temperatures in the future. Photos by Tom Nevesely/age fotostock and (inset) Spectrum Photofile.

Density dependent Factors that affect population size in relation to the population's density.

Negative density dependence When the rate of population growth decreases as population density increases.

Positive density dependence When the rate of population growth increases as population density increases. *Also known as inverse density dependence or the Allee effect.*

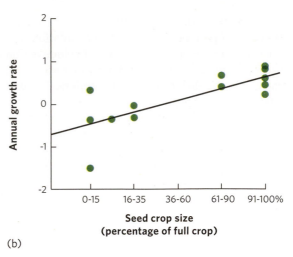

(a)

(b)

Figure 12.8 Negative density dependence in deer mice. **(a)** Deer mice live in forests and grasslands, but are also common around human settlements. A trapping study spanning 36 years in Algonquin Park, Ontario, revealed that the annual intrinsic growth rate of a population of deer mice declined with increasing mouse abundance. **(b)** Growth rate in this population depends on seed availability. Years with high seed availability are associated with higher growth rates than years with low seed availability.
Data from J. B. Falls, E. A. Falls and J. M. Fryxell. Fluctuations of deer mice in Ontario in relation to seed crops. *Ecological Monographs* 77(2007): 19-32

placed live traps in Algonquin Park, Ontario, and counted the number of deer mice caught. In Figure 12.8 we can see a negative density dependence in the deer mouse population: As the population size increased, annual growth rate declined. The researchers argued that this was due to the limited seed availability; as Figure 12.8b illustrates, growth rate increased with increasing seed availability. These data show that the population growth rate can depend on the size of the population and food availability.

Negative Density Dependence in Plants

Plants also experience reduced fecundity and increased mortality at high densities. When plants are grown at high densities, each plant has access to fewer resources such as sunlight, water, and soil nutrients. As a result, it is harder for each plant to grow. We can see this outcome in a study of flax plants (*Linum usitatissimum*) that were grown at a wide range of densities and then dried to determine their mass. The data from this experiment are shown in Figure 12.9. When seeds were sown at a density of 60 seeds per m², the average dry weight of individuals was between 0.5 and 1 g. When seeds were sown at densities of 1440 and 3600 per m², most of the individuals weighed less than 0.5 g. Plants with reduced growth typically experience reduced fecundity. As a result, the high competition caused by high plant densities will cause a population of flax plants to grow more slowly.

Under very high densities, competition among conspecifics can cause plants to die. In an experiment using horseweed (*Erigeron canadensis*), seeds were sown at a density of 100 000 per m². Competition among the tiny seedlings was intense; as you can see in Figure 12.10a, over an 8-month period most of them died. A close examination of the two *y* axes in the

figure illustrates that there was a hundredfold decrease in population density but a thousandfold increase in the average weight of the surviving individuals. As a result, the total mass of the plants increased tenfold over time.

In Figure 12.10b, you can see the same data, but now we have graphed the density of surviving plants over time against the average dry mass per plant, using a log scale. When we graph these data, we see that the data fall on a line with a slope of approximately $-\frac{3}{2}$. Plant ecologists call this relationship a **self-thinning curve**. This relationship has been observed frequently across a wide variety of species. The insights that emerge from the self-thinning curve have a number of practical uses. For example, the curve can be used to predict survival and growth of crop plants that might be sown at different densities in agricultural fields or the growth and survival of tree seedlings that might be planted at different densities in forest plantations.

In plants, animals, and other taxonomic groups, negative density dependence tends to bring populations under control and to maintain their abundances at a level close to the maximum number that can be supported by the environment. In many cases, populations are affected by density-independent and density-dependent factors. Next we consider the fascinating process of positive density dependence.

POSITIVE DENSITY DEPENDENCE

Whereas negative density dependence causes population growth to decrease as population density

---

Self-thinning curve A graphical relationship that shows how decreases in population density over time lead to increases in the size of each individual in the population.

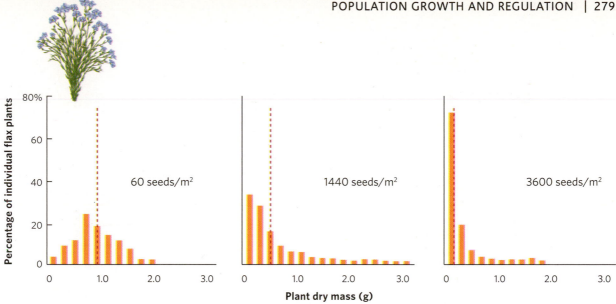

Figure 12.9 Negative density dependence in flax. When flax seeds are sown at higher densities, the average plant is smaller. Smaller plants are less fecund, so high densities cause plant populations to increase at a slower rate. The dashed red line in each panel represents the mean dry mass. Data from J. Harper, A Darwinian approach to plant ecology, *Journal of Ecology* 55 (1967): 247–270.

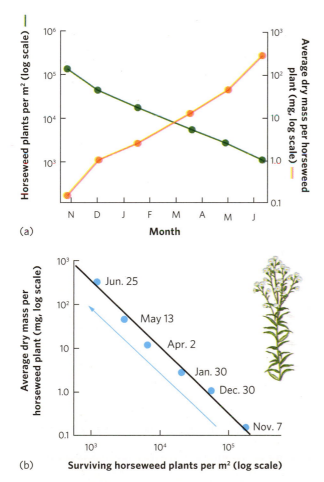

Figure 12.10 Self-thinning curve. Horseweed seeds were sown at a density of 100 000 per m². **(a)** Over time, the number of survivors declined by 100 while the average mass of the surviving plants increased by 1000. **(b)** When we plot plant density against average mass per plant on log scales, we see it follows a line with a slope of approximately $-\frac{3}{2}$. Data from J. Harper, A Darwinian approach to plant ecology, *Journal of Ecology* 55 (1967): 247–270.

increases, positive density dependence causes population growth to increase as population density increases. Positive density dependence typically occurs when population densities are very low and can be caused by several different mechanisms. In some cases, very low densities make it hard for individuals to find mates or, in the case of flowering plants, to obtain pollen, which can reduce reproductive success and slow population growth. Very low densities can also lead to the harmful effects of inbreeding, as discussed in Chapter 4. There can also be problems associated with small population sizes that, by chance, have uneven sex ratios. Small populations with a low proportion of females can suffer low population growth rates. Finally, individuals living in smaller populations can face a higher predation risk than those living in large populations, as we discussed in Chapter 10. In short, while negative density dependence causes slow population growth due to overcrowding, positive density dependence causes slow population growth due to undercrowding.

Positive density dependence has been demonstrated in a wide variety of species. Many plant species, for example, avoid the costs of inbreeding through a number of mechanisms that prevent self-fertilization. They depend on receiving pollen from other plants, but this can be difficult when individuals live at low densities and are widely spread over an area. In a study of cowslip (*Primula veris*), a plant of European grasslands, researchers were interested in determining why many of the smaller populations were declining. When they looked at reproduction in populations of different sizes, they found that populations of fewer than 100 individuals produced fewer seeds per plant, as can be seen

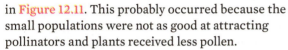

Figure 12.11 Positive density dependence in cowslip plants. At low densities, plants appear to be pollen-limited, which causes them to produce fewer seeds. At higher densities, each plant produces many more seeds. Data from M. Kéry et al., Reduced fecundity and offspring performance in small populations of the declining grassland plants *Primula veris and Gentiana lutea, Journal of Ecology* 88 (2000): 17–30.

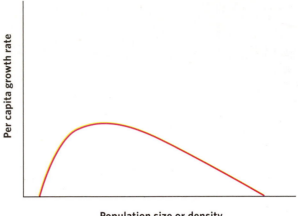

Figure 12.12 Positive and negative density dependence. When populations are very small, an increase in density can cause an increase in the population's per capita growth rate. However, when populations are large and resources start to become limiting, further increases in density can cause a decrease in the population's per capita growth rate.

in Figure 12.11. This probably occurred because the small populations were not as good at attracting pollinators and plants received less pollen.

A recent study found that positive density dependence also occurs in parasites. In 2012, researchers reported on reproduction in salmon lice (*Lepeophtheirus salmonis*), which feed on the skin of salmon. When breeding, male and female lice form a mating pair. Researchers in British Columbia counted the number of mating pairs on each fish and found that the probability of an individual forming a mating pair was positively correlated with the number of lice of the opposite sex. Individuals living at low densities have a hard time finding a mate and therefore the lice experienced positive density dependence.

Although we often think of positive and negative density dependence as isolated phenomena, populations can be regulated by both processes, as shown in Figure 12.12. As we move from low densities to intermediate densities, the effects of positive dependence play an important role. Increased densities provide more individuals for breeding and the growth rate of the population can improve. Above some intermediate density, resources start to become limiting and negative density dependence begins to play a role. As densities continue to increase, negative density dependence continues to slow population growth and, eventually, the population growth rate falls to zero.

An example of the dynamic between positive and negative effects of density on population growth can be found in a population of herring (*Clupea harengus*)

that spawns near Iceland. As you can see in Figure 12.13, the herring experience low population growth at low densities, high population growth at

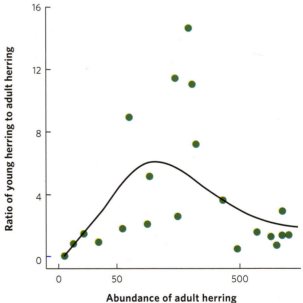

Figure 12.13 Positive and negative density dependence in herring. The highest population growth is measured as a ratio between the number of young fish and the number of adult fish. The highest growth rate of the population occurs at intermediate densities. At higher densities, there is negative density dependence. At lower densities, there is positive density dependence. Data from R. A. Myers et al., Population dynamics of exploited fish stocks at low population levels, *Science* 269 (1995): 1106–1108.

intermediate densities, and then a return to low population growth at high densities. The existence of positive density dependence in the herring population concerns fisheries managers because it indicates that if the herring were driven to low densities by fishing, it would be difficult for the population to bounce back. In fact, the populations could experience negative growth rates and become extinct. Fortunately, most fish populations do not appear to experience positive density dependence.

The herring study highlights why those who manage populations in the wild need to understand positive density dependence. For example, when a species has declined to very low numbers and we wish to improve its numbers, management strategies need to consider how to avoid inbreeding and how to ensure that each female can encounter a sufficient number of mates to fertilize all of her eggs. The concept of positive density dependence also offers an opportunity for ecologists to help control undesirable pest species. For example, several pest insects have been controlled by releasing sterile males into the population. This skews the sex ratio and causes the females to breed with sterile males, leading to a lower growth rate in the pest population. In other cases, researchers have simply reduced the size of an undesirable population to the point that individuals have a hard time finding mates.

THE LOGISTIC GROWTH MODEL

Although positive and negative density dependence both occur in nature, the effects of negative density dependence have received much more attention from population modellers. Researchers have developed population growth models that mimic the behaviour of many natural populations: rapid initial growth followed by slower growth as populations grow toward their maximum size. The maximum population size that can be supported by the environment is known as the **carrying capacity** of the population, denoted as **K**.

To model the slowing growth of populations at high densities, we use the **logistic growth model**. The logistic growth model builds on the exponential growth model, but we add a term in parentheses that accounts for a decline in growth rate as the population approaches its carrying capacity:

$$\frac{dN}{dt} = rN\left(1 - \frac{N}{K}\right)$$

When the number of individuals in the population is small relative to the carrying capacity, the fraction $\frac{N}{K}$ is close to 0 and the term inside the parentheses approaches 1. When this happens, the equation becomes nearly identical to the exponential

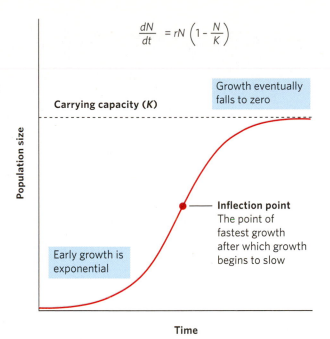

Figure 12.14 The logistic growth model. When the population is small, growth is exponential. When the population is large, growth slows. The inflection point indicates the point at which the population experiences its fastest rate of growth. Above the inflection point, growth begins to slow. When the population reaches its carrying capacity, *K*, growth is zero. As a result of this growth pattern, the logistic growth model is represented by an S-shaped growth curve.

growth model. However, when the number of individuals in the population approaches the carrying capacity, the term inside the parentheses approaches 0. As a result, the population's rate of growth approaches 0. When we plot the growth of a population over time using the logistic growth curve, we obtain a *sigmoidal*, or **S-shaped curve**, which is shown in Figure 12.14. The middle point on the curve, where the population experiences its highest growth rate, is known as the **inflection point**. Above the inflection point, the population growth begins to slow.

We can further understand how the logistic growth model behaves by examining the effect of population size on the rate of increase of

---

Carrying capacity (K) The maximum population size that can be supported by the environment.

Logistic growth model A growth model that describes slowing growth of populations at high densities.

S-shaped curve The shape of the curve when a population is graphed over time using the logistic growth model.

Inflection point The point on a sigmoidal growth curve at which the population achieves its highest growth rate.

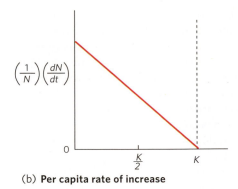

(a) **Overall rate of population increase**

(b) **Per capita rate of increase**

Figure 12.15 The effect of population size on the rate of increase and the per capita rate of increase. **(a)** As populations grow from a small number to their carrying capacity, the rate of population growth increases until it reaches the inflection point, after which it decreases. **(b)** On a per capita basis, the rate of increase continually declines.

the population $\left(\frac{dN}{dt}\right)$. As shown in **Figure 12.15a**, as the population increases from a very small size, the rate of increase grows because the number of reproductive individuals increases. After reaching one-half of the carrying capacity, which corresponds to the inflection point of the S-shaped curve, the rate of increase begins to slow because the reproductive individuals are each obtaining fewer resources.

We can also examine how population size affects the rate of increase on a per capita basis, which is:

$$\left(\frac{1}{N}\right)\left(\frac{dN}{dt}\right)$$

As you can see in Figure 12.15b, individuals in the population continually decline in their ability to contribute to the growth of the population. These insights help us understand that the logistic growth curve shows an initial rapid increase in growth due to the increasing number of individuals in the population, followed by a slowing rate of growth as a large number of individuals experience limited per capita resources.

The logistic equation describes the growth of many populations. One of the classic studies was conducted by Russian biologist Georgyi Gause in 1934. Gause raised two species of protists, *Paramecium aurelia* and *Paramecium caudatum,* in test tubes and added a fixed amount of food each day. Both species initially experienced exponential growth, which then slowed until they reached carrying capacity. Because the two species differ in many ways, their populations stabilized at different carrying capacities, which are illustrated as blue lines in **Figure 12.16**. Gause suspected that the cause of this maximum population size was the amount of available food. To test this hypothesis, he did the experiment again, but this time doubled the amount of food. As you can see from the orange line in each

graph, the two species once again experienced rapid initial growth followed by slower growth that stabilized. With twice as much food available, however, the two species stabilized at population

(a)

(b)

Figure 12.16 Logistic growth in two species of *Paramecium.* **(a)** When *Paramecium aurelia* was raised under low or high food supply, the population size—measured in terms of the total volume of protist cells—grew rapidly at first and slowed as they reached their carrying capacity. When twice as much food was provided, the species achieved a population size that was twice as large. **(b)** A similar pattern was found for another species of protist, *Paramecium caudatum.* Data from G. Gause, *The Struggle for Existence* (Zoological Institute of the University of Moscow, 1934).

sizes that were twice as large as in the first experiment. This early experiment confirmed that logistic growth can happen in real organisms and that the increased availability of a limiting resource can increase the carrying capacity of a population.

PREDICTING HUMAN POPULATION GROWTH WITH THE LOGISTIC EQUATION

The logistic growth model was first developed in 1838 by the Belgian mathematician Pierre François Verhulst. Verhulst had read Thomas Malthus's 1798 essay and sought to formulate a natural law governing the growth of populations. Nearly a century later, in 1920, Raymond Pearl and Lowell Reed independently developed the logistic growth model when they examined human population growth in the United States.

Using census data that had been collected every 10 years, Pearl and Reed confirmed the observation by Malthus that the population in the United States initially grew exponentially from 1790 to 1910. They noticed, however, that by 1910 this growth began to slow, as illustrated in Figure 12.17, and that the rate of growth appeared to be declining over time. Based on these data, Pearl and Reed applied their logistic growth equation to the census data and predicted that the population of the United States, which was 91 million in 1910, had a carrying capacity of 197 million. However, they were careful to note that this prediction depended on the population supporting itself using only the land of the United States.

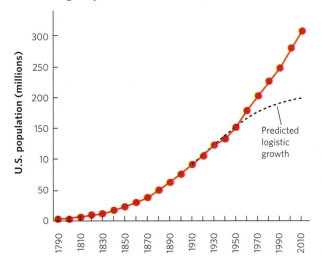

Figure 12.17 The growing human population in the United States. A census conducted every 10 years from 1790 to 1910 indicated that the U.S. population was continuing to increase. However, the rate of increase started to decline by 1910. Based on these data, Pearl and Reed predicted that the population would reach a maximum of 197 million people. In reality, the U.S. population has continued to grow. Data from R. Pearl and L. J. Reed, On the rate of growth of the population of the United States since 1790 and its mathematical representation, *Proceedings of the National Academy of Sciences* 6 (1920): 275–288; and U.S. Census Bureau (http://2010.census.gov/2010census/data/apportionment-pop-text.php).

As you can see in the figure, the U.S. population has greatly exceeded this prediction; by 2012, the population was more than 310 million people. There are several reasons for this. One important reason is that technological advances have allowed farmers in the United States to produce food much more efficiently. Moreover, a substantial amount of the food consumed by the U.S. population comes from other regions of the world. In addition to the increases in available food, we have had major improvements in public health and medical treatment that have increased survival rates substantially, particularly for infants and children. Finally, the logistic equation did not incorporate the millions of immigrants that came to the United States after 1910.

Population growth rate is influenced by the proportions of individuals in different age, size, and life history classes

In our discussion of population models so far, we have assumed that all individuals in the population have an identical intrinsic growth rate—that is, they have the same birth rates and death rates. While this assumption is helpful to generate simple models of population growth, we know that survival and fecundity commonly vary with an individual's age, size, or life history stage. In terms of age, individuals cannot reproduce until they have achieved reproductive maturity. Individual mass often plays a large role in determining the individual's fecundity, with larger individuals having higher fecundity (see Chapter 8). We can also consider different life history stages. For example, tadpoles experience low survival and no fecundity, whereas later in life, as frogs, they experience high survival and high fecundity. Many perennial plants alternate between years in which they are reproductive and years in which they are nonreproductive. In this section, we will examine how changes in survival and fecundity among different classes in a population affect population growth. In doing so, we will focus on age classes while recognizing that the same principles apply to size classes and life history classes.

AGE STRUCTURE

The **age structure** of a population is the proportion of individuals that occurs in different age classes. A population's age structure can tell us a great deal about its past and its potential for future growth. Consider the different patterns of age structure

---

Age structure In a population, the proportion of individuals that occurs in different age classes.

Figure 12.18 **Age structure for human populations in India, the United States, and Canada in 2012.** The age structure of India has a high proportion of young people, which indicates a growing population. The age structure of the United States has very similar numbers of people from 0 to 50 years of age, which indicates a stable population. Canada's age structure has fewer young people than middle-aged people. This is an indication that its population may decline, assuming that factors such as fertility and immigration remain constant. Recent increases in fertility and immigration may contribute to continued growth. Data from http://www.census.gov/population/international/data/idb/informationGateway.php.

age classes. If the young age classes experience good nutrition and health care, they will survive well and the number of reproductive individuals 2 decades from now will be much greater than it is today, which will cause the population to grow.

We can contrast this with the age structure pyramid of the United States. The number of people in the younger age classes is very similar to those in the middle age classes. In this case, we would predict that 2 decades from now we would have a similar number of reproductive individuals, so the population should remain relatively stable.

The other extreme is represented by Canada. In Canada, the current reproductive age classes have experienced reduced reproduction such that the number of people in the younger age classes is less than the number in the middle age classes. If we project ahead 2 decades, we would predict that there will be fewer reproductive individuals in the future and, as a result, the population in Canada should decline. However, Statistics Canada predicts that while population growth will slow due to this aging population, recent increases in fertility and immigration will contribute to continued population growth.

among human populations in 2012 that are displayed in Figure 12.18. In these figures, known as *age structure pyramids,* we see that the nations of India, the United States, and Canada all have declining numbers of people after 50 years of age, due to senescence. But each has a very different pyramid shape below 50 years of age. Age structure pyramids with broad bases reflect growing populations, pyramids with straight sides reflect stable populations, and pyramids with narrow bases reflect declining populations.

In the case of India, for example, people in the younger age classes far outnumber those in the middle

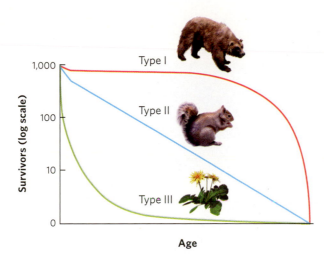

Figure 12.19 Survivorship curves. Individuals that have a type I curve experience high survival until later in life. Individuals with a type II curve experience a steady decline in survival throughout life. Individuals that have a type III curve experience low survival when young and high survival until later in life.

SURVIVORSHIP CURVES

To understand the differences in survival among different age classes we can graph survival over time, as illustrated in Figure 12.19. Using this technique we find that species can be categorized according to type I, type II, or type III survivorship curves. A type I survivorship curve depicts a population that experiences very little mortality early in life and high mortality late in life. Examples include humans, elephants, and whales. A type II survivorship curve occurs when a population experiences relatively constant mortality throughout its life span. Organisms with type II curves include squirrels, corals, and some species of songbirds. A type III survivorship curve depicts a population with high mortality early in life and high survival later in life. This is a common pattern in many species of insects and in plants such as dandelions and oak trees that produce hundreds or thousands of seeds. In reality, most populations have survivorship curves that combine features of type I and type III curves, with high mortality early and late in life.

LIFE TABLES

Because both birth and death rates can differ among age, size, and life history classes, we need a way to incorporate this information into our estimates of population growth rates. If we can find a way to do this, we can better predict how a population will change in abundance over time, which can be important for managing populations that humans consume, populations of species that we wish to conserve, and populations of pests that we wish to control. We can achieve this goal if we know the

proportions of individuals in different classes and the birth and death rates of each age class. Collectively, such data can tell us whether a population is expected to increase, decrease, or remain unchanged.

To determine how age, size, or life history classes affect the growth of a population, ecologists use *life tables*. **Life tables** contain class-specific survival and fecundity data that help us understand and predict the growth of the population. Because it can be hard to ascertain paternity in many species, life tables are typically based on females and fecundity is defined as the number of female offspring per reproductive female. For some populations with highly skewed sex ratios or unusual mating systems (see Chapter 9), using only females can pose problems, but in most cases a female-based life table provides a useful model of population growth.

To understand how ecologists use life tables, let's consider a hypothetical population composed of 100 individuals distributed into different age classes, as illustrated in Table 12.1. In this table, the age classes are denoted by x and the number of individuals in each age class is denoted by n_x. The value of n_x represents the number of individuals present immediately after the population has produced offspring. In the table, survival rate from one age class to the next age class is denoted as s_x. (You can think of "s" for survival.) In our example, the column labelled as s_x indicates that the new offspring have a 50 percent survival rate, 1-year-olds have an 80 percent survival rate, 2-year-olds have a 50 percent survival rate, and no 3-year-olds survive to become 4-year-olds. The fecundity of each age class is denoted by b_x. (You can think of "b" for birth.) In the table, the column labelled as b_x indicates that the new offspring cannot reproduce, but 1-year-olds can each produce one offspring, 2-year-olds can each produce three offspring, and 3-year-olds can each produce two offspring.

TABLE 12.1
A HYPOTHETICAL POPULATION OF 100 INDIVIDUALS WITH AGE-SPECIFIC RATES OF SURVIVAL AND FECUNDITY

AGE (x)	NUMBER OF INDIVIDUALS (n_x)	SURVIVAL RATE (s_x)	FECUNDITY (b_x)
0	20	0.5	0
1	10	0.8	1
2	40	0.5	3
3	30	0.0	2
4	0	—	—
TOTAL	100		

Life tables Tables that contain class-specific survival and fecundity data.

TABLE 12.2
THE NUMBER OF SURVIVORS AND NEW OFFSPRING AFTER 1 YEAR FOR THE HYPOTHETICAL POPULATION IN TABLE 12.1

AGE (x)	NUMBER OF INDIVIDUALS (n_x)	SURVIVAL RATE (s_x)	NUMBER SURVIVING TO NEXT AGE CLASS $(n_x) \times (s_x)$	FECUNDITY (b_x)	NUMBER OF NEW OFFSPRING PRODUCED $(n_x) \times (s_x) \times (b_x)$	CENSUS OF POPULATION 1 YEAR LATER
0	20 ×	0.5 =		0	0	74
1	10	0.8	10 ×	1 =	10	10
2	40	0.5	8	3	24	8
3	30	0.0	20	2	40	20
4	0		0			0
TOTAL	$N_0 = 100$		38	+	74 =	$N_1 = 112$

The data in this life table allows us to calculate the expected size of the population after one year. Table 12.2 shows us the calculations for the population in Table 12.1. To calculate the number of individuals that survive to the next age class we multiply the number of individuals in an age class by the annual survival rate of that age class, as shown in the red-shaded region of the table. For example, if we start with 0-year-olds and their survival rate is 50 percent, then we can calculate the number of 1-year-olds that we will have in the following year:

Number surviving to next age class = $(n_x) \times (s_x)$

Number surviving to next age class = 20×0.5

Number surviving to next age class = 10

To determine the number of new offspring that will be produced next year, we multiply the number of individuals that survive to the next age class—as shown in the orange-shaded region—by the fecundity of that age class. For example, if 10 individuals will survive to be 1-year-olds and each 1-year-old can produce one offspring, then we can calculate the number of new offspring that this age class will produce, as shown in the yellow-shaded region of the table:

Number of new offspring produced = $(n_x) \times (s_x) \times (b_x)$

Number of new offspring produced = $20 \times 0.5 \times 1$

Number of new offspring produced = 10

If we do these calculations for every age class, we find that next year we will have a total of 74 new offspring. In addition, we will have 10 individuals that are now 1 year old, 24 individuals that are 2 years old, and 40 individuals that are now 3 years old. You can see a summary of this in Table 12.3.

When we add up all of the age classes, we find that these survival and fecundity rates have caused the population to grow from a total size of 100 to a new

TABLE 12.3
POPULATION SIZE INITIALLY AND 1 YEAR LATER BASED ON THE HYPOTHETICAL POPULATION IN TABLE 12.1

AGE (x)	NUMBER OF INDIVIDUALS (n_x)	NUMBER OF INDIVIDUALS 1 YEAR LATER $(n_x + 1)$
0	20	74
1	10	10
2	40	8
3	30	20
4	0	0
TOTAL	$N_0 = 100$	$N_1 = 112$

total size of 112 individuals 1 year later. As a result, the geometric growth rate of the population is

$$\lambda = \frac{N_1}{N_0} = \frac{112}{100} = 1.12$$

We can continue these calculations to estimate the size of the population for many years into the future. In Table 12.4, for example, you can see the results of these same calculations for 8 years into the future. We can also continue to calculate the geometric rate of growth of the population for each year. If we examine these values of λ, we can see that they initially fluctuate widely between 1.05 and 1.69. As the years pass, however, the values of λ settle down to 1.49. Provided that survival and fecundity of each class stays constant over time, λ will stabilize at a single value and the proportion of individuals in each age class also will stabilize. When the structure of a population is not changing over time, we say it has a **stable age distribution**. If the survival and

---

Stable age distribution When the age structure of a population does not change over time.

TABLE 12.4
PROJECTING POPULATION GROWTH OVER 8 YEARS BASED ON THE HYPOTHETICAL POPULATION IN TABLE 12.1

	YEAR 0	YEAR 1	YEAR 2	YEAR 3	YEAR 4	YEAR 5	YEAR 6	YEAR 7	YEAR 8
n_0	20	74	69	132	175	274	399	599	889
n_1	10	10	37	34	61	87	137	199	299
n_2	40	8	8	30	28	53	70	110	160
n_3	30	20	4	4	15	14	26	35	55
N	100	112	118	200	279	428	632	943	1403
λ		1.12	1.05	1.69	1.40	1.53	1.48	1.49	1.49

fecundity rates were to change, the value of λ and the proportion of individuals in each age class would be altered as well.

Calculating Survivorship

We have examined life tables using the survival rate (s_x) from one age class to the next age class. Another useful measure of survival is the probability of surviving from birth to any later age class, which we call survivorship and denote as l_x. (You can think of "l" for living.) Survivorship in the first age class is always set at 1 because all individuals in the population are initially alive. Survivorship at any given age class is the product of the prior year's survivorship and the prior year's survival rate. For example, survivorship to the second year is calculated as

$$l_2 = l_1 s_1$$

and survivorship to the third year is calculated as

$$l_3 = l_2 s_2$$

The calculated survivorships in our hypothetical population are shown in Table 12.5.

Calculating the Net Reproductive Rate

The **net reproductive rate** is the total number of female offspring that we expect an average female to produce over the course of her life. If this value is greater than 1, then each female more than replaces

TABLE 12.5
CALCULATING THE SURVIVORSHIP (l_x) OF THE HYPOTHETICAL POPULATION IN TABLE 12.1

AGE (x)	NUMBER OF INDIVIDUALS (n_x)	SURVIVAL RATE (s_x)	SURVIVORSHIP (l_x)
0	20	0.5	1.0
1	10	0.8	0.5
2	40	0.5	0.4
3	30	0.0	0.2
4	0	—	0.0
TOTAL	100		

herself in the population and the population will grow. The net reproductive rate, denoted as R_0, is calculated in two steps. First, we multiply the probability of living to each age class (l_x) by the fecundity of the respective age classes (b_x). Second, we sum these products, as shown in the following equation

$$R_0 = \sum l_x b_x$$

An example is shown in Table 12.6 for our hypothetical population. In this case, the net

Net reproductive rate The total number of female offspring that we expect an average female to produce over the course of her life.

TABLE 12.6
CALCULATING THE NET REPRODUCTIVE RATE BASED ON THE HYPOTHETICAL POPULATION IN TABLE 12.1

AGE (x)	SURVIVAL RATE (s_x)	SURVIVORSHIP (l_x)	FECUNDITY (b_x)	(l_x) × (b_x)
0	0.5	1.0	0	
1	0.8	0.5	1	0.5
2	0.5	0.4	3	1.2
3	0.0	0.2	2	0.4
4	—	0.0	—	0.0
NET REPRODUCTIVE RATE (R_0) = $\sum l_x b_x$ = 2.1				

TABLE 12.7
CALCULATING THE GENERATION TIME, *T*, FROM THE HYPOTHETICAL POPULATION IN TABLE 12.1

AGE (x)	SURVIVAL RATE (s_x)	SURVIVORSHIP (l_x)	FECUNDITY (b_x)	$(l_x) \times (b_x)$	$(x) \times (l_x) \times (b_x)$
0	0.5	1.0	0		0.0
1	0.8	0.5	1	0.5	0.5
2	0.5	0.4	3	1.2	2.4
3	0.0	0.2	2	0.4	1.2
4	—	0.0	—	0.0	0.0
NET REPRODUCTIVE RATE				$(R_0) = \Sigma\, l_x b_x = 2.1$	$\Sigma\, x l_x b_x = 4.1$

reproductive rate is 2.1, which means that each female is producing 2.1 daughters on average, so the population will grow.

Calculating the Generation Time

We can also use life tables to calculate the **generation time (T)** of a population, which is the average time between the birth of an individual and the birth of its offspring. To calculate generation time, we must make three sets of calculations, as shown in Table 12.7. First, within each age class we multiply the age (*x*), survivorship (l_x), and fecundity (b_x). Second, we take the sum of these products. This provides us with the expected number of births for a female, weighted by the ages at which she produced the offspring.

Using these calculations, we can now calculate generation time as

$$T = \frac{\Sigma x l_x b_x}{\Sigma l_x b_x}$$

$$T = \frac{4.1}{2.1}$$

$$T = 1.95 \text{ years}$$

This tells us that in our hypothetical population, the average time between the birth of an individual and the birth of its offspring is about 2 years.

Calculating the Intrinsic Rate of Increase

We can make connections between life tables and our earlier population growth models by using life table data to estimate the population's intrinsic rate of increase (λ or *r*). When an intrinsic rate of increase is estimated from a life table, we assume that the life table has a stable age distribution. However, stable age distributions rarely occur in nature because the environment varies from year to year in ways that can affect survival and fecundity. As a result, any approximation of λ or *r* is necessarily restricted to the set of environmental conditions that the population

experiences. There are complicated equations to estimate λ and *r*. However, we can provide close approximations, denoted as λ_a and r_a (where the letter "*a*" indicates an approximation), based on our estimates of net reproductive rate (R_0) and generation time (*T*)

$$\lambda_a = R_0^{\frac{1}{T}}$$

which given our earlier calculations simplifies to

$$\lambda_a = 2.1^{\frac{1}{1.95}}$$

$$\lambda_a = 1.46$$

Note that this value of 1.46 is close to our observed l of about 1.49 after the population achieved a stable age distribution (see Table 12.3). If we wanted to calculate ra , we could do so using the following equation

$$r_a = \frac{\log_e R_0}{T}$$

$$r_a = \frac{\log_e 2.1}{1.95}$$

$$r_a = 0.38$$

You can see that a population's intrinsic rate of increase depends on both the net reproductive rate (R_0) and the generation time (*T*). A population grows when R_0 exceeds 1, which is the replacement level of reproduction for a population. In contrast, a population declines when $R_0 < 1$. The rate at which a population can change increases with shorter generation times.

COLLECTING DATA FOR LIFE TABLES

To determine the age structure of a population and predict future population growth, we need to collect data or organisms from different ages. This objective

Generation time (T) The average time between the birth of an individual and the birth of its offspring.

ANALYZING ECOLOGY

Calculating Life Table Values

Life table data are collected for many different species. For example, a group of researchers spent a decade following a population of grey squirrels (*Sciurus carolinensis*) in North Carolina to quantify the survival and fecundity of the squirrels. Below is a table of the age-specific survival and fecundity.

YOUR TURN Using the data in the table, calculate survivorship (l_x), net reproductive rate (R_0), generation time (T), and the intrinsic rate of increase (λ).

GREY SQUIRREL LIFE TABLE

(x)	(n_x)	(l_x)	(b_x)	(l_x) × (b_x)	(x) × (l_x) × (b_x)
0	530	1.000	0.05		
1	134	0.253	1.28		
2	56	0.116	2.28		
3	39	0.089	2.28		
4	23	0.058	2.28		
5	12	0.039	2.28		
6	5	0.025	2.28		
7	2	0.022	2.28		

SOURCE: Data from Barkalow, F. S., et al. 1970. The vital statistics of an unexploited gray squirrel population, *Journal of Wildlife Management* 34: 489–500.

can be met by constructing either a *cohort life table* or a *static life table*. A **cohort life table** follows a group of individuals born at the same time from birth to the death of the last individual. In contrast, a **static life table** quantifies the survival and fecundity of all individuals in a population during a single time interval. As we will see, the two types of life tables are used in different situations and have different advantages and disadvantages.

Cohort Life Tables

Cohort life tables are readily applied to populations of plants and sessile animals in which marked individuals can be continually tracked over the course of their entire life. The cohort life table does not work well for species that are highly mobile or for species with very long life spans, such as trees. One of the problems in using a cohort life table is that a change in the environment during one year can affect survival and fecundity of the cohort that year. This makes it difficult to disentangle the effects of age from the effects of changing environmental conditions. For example, imagine that you mark a population of plants that lives for several years. However, in the middle of this life span the population experiences a severe drought that causes sharp reductions in survival and fecundity. The following year is wet, which causes high survival and fecundity. When you construct the life table, you cannot determine whether the age-specific changes in survival and fecundity are due to a change in age or a change in the environment.

An excellent example of a cohort life table comes from Peter and Rosemary Grant, who have studied several species of ground finches in the Galápagos Islands off the coast of Ecuador. On the small island of Daphne Major, the Grants were able to capture all the birds on the island and mark them with uniquely coloured plastic leg bands. Because this island is so isolated, few birds left from the island or arrived from elsewhere. Using 210 cactus finches (*Geospiza scandens*) that fledged in 1978, the researchers constructed a cohort life table for the population. In

Cohort life table A life table that follows a group of individuals born at the same time from birth to the death of the last individual.

Static life table A life table that quantifies the survival and fecundity of all individuals in a population during a single time interval.

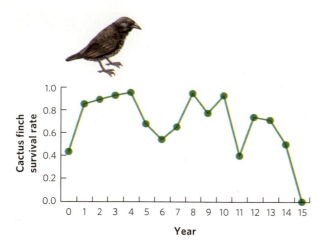

Figure 12.20 Annual survival rate of cactus finches.
Using a cohort of cactus finches, researchers were able to calculate the annual survival over a 15-year period. Data from P. R. Grant and B. R. Grant, Demography and the genetically effective sizes of two populations of Darwin's finches, *Ecology* 73 (1992): 766–784.

Figure 12.20 you can see the annual survival rates for these birds. As is the case for many species, survival rate was low in the population's first year and then remained high for several subsequent years. However, survival was quite variable throughout the life of the birds. This variation in survival reflects variation in the environment due to climatic changes during El Niño years (see Chapter 5). Because El Niño years are wet years, the vegetation produces abundant food for the finches and this causes high survival. Following El Niño years are periods of several dry years. During this time, food becomes scarce for the finches and this causes low survival. These data emphasize a disadvantage of cohort life tables. As mentioned above, it is difficult to determine whether low survival at a particular age is due to the age of the individuals or due to the environmental conditions that occur during that year.

Static Life Tables

Using a static life table avoids many of the problems of the cohort life table. By considering the survival and fecundity of individuals of all ages during a single time interval, differences among the age classes are quantified under the same environmental conditions, so age is not confounded with time. In addition, static life tables allow us to look at the survival and fecundity of all individuals during a snapshot in time, which means we can examine both species that are highly mobile as well as species that have long life spans.

To construct a static life table, we must be able to age all individuals since we do not mark them at birth. A number of techniques are available to do this. For example, we can count annual growth rings in the trunks of trees, in the teeth of mammals, and in the ear bones of fish. The one major concern when using static life tables is that the age-specific data on survival and fecundity only apply to the environmental conditions that existed at the time the data were collected. Because the life table may not be representative of years in which the environmental conditions are quite different, it is helpful to construct static life tables for multiple years to assess how much environmental variation affects the predicted population growth.

A classic study using a static life table was conducted by Olaus Murie, who examined the survival of Dall mountain sheep (Ovis dalli) at a site in Alaska during the 1930s. Murie knew that the horns of the sheep contained annual growth rings that could be used to age the sheep. He also knew that following a cohort of highly mobile sheep for 15 years in Alaska was not feasible, so he took a static approach by conducting a search of all sheep skeletons in the area. He found a total of 608 skeletons from sheep that had died in recent years and assigned an age to each individual based on its horns. Using these estimates, he determined how many of the original 608 sheep died in each age class. His results, depicted in Figure 12.21, show a low

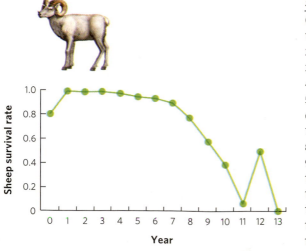

x	n_x	s_x	l_x
0	608	0.801	1.00
1	487	0.986	0.801
2	480	0.983	0.789
3	472	0.985	0.776
4	465	0.961	0.765
5	447	0.937	0.735
6	419	0.931	0.689
7	390	0.892	0.641
8	348	0.770	0.572
9	268	0.575	0.441
10	154	0.383	0.253
11	59	0.068	0.097
12	4	0.500	0.007
13	2	0.000	0.003
14	0		0.000

Figure 12.21 A static life table and survival rate curve for Dall sheep.
Using the ages of Dall sheep skeletons found in Alaska, researchers calculated the survival and survivorship of different age classes. When graphed, these data show that the survival rate of the sheep remains quite high for the first 7 years and then rapidly declines. There were only four animals in the 12-year age class and this resulted in an unreliable high estimate of survival rate. Based on data from O. Murie, The Wolves of Mt. McKinley (U.S. Department of the Interior, National Park Service, Fauna Series No. 5, Washington, D.C., 1944), as quoted by E. S. Deevey, Jr., Life tables of natural populations of animals, *Quarterly Review of Biology* 22 (1947): 223–314.

survival rate during the first year, followed by high survival for the next 7 years. After 7 years of age, the annual survival rate began to drop, with the exception of those at 12 years of age. However, only four animals made it to 12 years, which gives us a poor estimate of the typical survival rate for 12-year-old sheep. Because these data were taken from the skeletons of sheep that presumably died over a relatively short period, we do not see the large variation in survival rate that we saw in the cohort life table of the cactus finch.

Throughout this chapter we have examined how ecologists use their understanding of populations to develop mathematical models that mimic natural patterns of population growth. Although populations commonly grow rapidly when at low densities, they become limited as the populations grow larger. The simplest population models are helpful starting points, but they are not sufficient for most species that have survival and fecundity rates that vary with age, size, or life history stage. Using life tables helps to incorporate this complexity. As we will see below in "Ecology Today: Connecting the Concepts," the analysis of life tables can be very helpful in setting priorities for management strategies to save species from extinction.

ECOLOGY TODAY CONNECTING THE CONCEPTS

SAVING THE SEA TURTLES

Sea turtles. Population models have identified the stages in life that need the most protection to allow turtle populations to recover. These models are relevant to all species of sea turtles, including this green sea turtle (*Chelonia mydas*) off the coast of Hawaii. Photo by Masa Ushioda/Image Quest Marine.

Under the cover of darkness, the swimmers come to shore. After spending a year out in the ocean, the female sea turtles crawl up the shore and dig a nest to lay their eggs. After about 2 months, these eggs will hatch and the hatchlings will scramble toward the ocean where they will face a gauntlet of predators. If they survive, they will need 2 or 3 decades to become reproductively mature. They may live for 80 years.

Not long ago, many more females came to shore. Of the six species of sea turtles, all have declined over the past several centuries. For example, scientists estimate that there were once 91 million green turtles (*Chelonia mydas*) in the Caribbean; today there are about 300 000 which is a decline of more than 99 percent. Sea turtles, which live in temperate and tropical waters, are declining worldwide. Historically, people hunted them for food and used turtle shells for decorations; more recently, land development has caused the loss of beach habitat for egg laying. Commercial fishing operations accidentally catch turtles. Across all six species, more than 300 000 were caught along the U.S. coast in the 1980s, and more than 70 000 of these died before being returned to water. About 98 percent of these accidental catches were from shrimp trawlers pulling large nets through the water.

For many years, ecologists struggled to find a solution that would reverse the decline of the sea turtles. They felt one important step might be to protect the remaining nesting areas and to incubate large numbers of eggs artificially. They released the hatchlings into the oceans to improve their probability of survival, but after spending a great deal of time and money on this effort, population modellers had an interesting insight. When they began constructing life tables for turtles—a real challenge, given that the turtles spend nearly all of their life at sea—they discovered that improving the survival of the hatchlings would have little positive impact on the growth of the population. Because so few hatchlings survive to adulthood naturally, adding even thousands more would have little impact on the turtle population. Rather, the life tables indicated that improving the survivorship of the adult turtles was the key to growing the population.

This insight completely changed the emphasis of sea turtle conservation. The focus shifted to the commercial fishing operations that were accidentally killing adult sea turtles. Pressure mounted for commercial shrimp fishermen to install turtle excluder devices on their trawling nets. These devices work like a sieve; small shrimp pass through the sieve and get captured in the net while sea turtles, which are too big to pass through the sieve, are shunted out of the net. The installation of these devices has been a tremendous success. In 2011, researchers reported that the number of turtles that had been accidentally netted had been reduced by 60 percent and the number of turtles that were killed had declined by 94 percent.

Scientists continue to work on obtaining age-specific survival rates on the six species of sea turtles, which is still a challenge since the animals can migrate thousands of miles and males do not return to nesting beaches. Indeed, the most recent population models indicate that improving the survival of the hatchlings does have a small positive effect on the population, although improving the survival of older animals continues to have a much greater effect. The story of the sea turtle makes it clear that being able to model population growth can be a critical step in bringing species back from the brink of extinction.

SOURCES: Finkbeiner, E. M., et al. 2011. Cumulative estimates of sea turtle bycatch and mortality in USA fisheries between 1990 and 2007. *Biological Conservation* 144: 2719–2727.

Marzaris, A. D., et al. 2006. An individual based model of a sea turtle population to analyze effects of age dependent mortality. *Ecological Modelling* 198: 174–182.

SUMMARY OF CHAPTER CONCEPTS

■ **Under ideal conditions, populations can grow rapidly.** When populations grow at their intrinsic growth rate, they can initially increase at exponential rates, which can be modelled using either the exponential growth model or the geometric growth model.

■ **Populations have growth limits.** The limits can be due to density-independent factors, which regulate population sizes regardless of the population's density. The limits can also be due to density-dependent factors, which affect population growth in a way that is related to the population's density. Negative density dependence causes populations to grow more slowly as they become larger

whereas positive density dependence causes populations to grow faster as they become larger. Ecologists use the logistic growth model to demonstrate negative density dependence. The logistic model mimics rapid population growth when populations are small and slow population growth when populations approach their carrying capacity. The logistic growth model has been used to predict human population growth, but human populations have exceeded these predictions due to improvements in food production, international trade, and public health.

■ **Population growth rate is influenced by the proportions of individuals in**

different age, size, and life history classes. Most organisms have rates of survival and fecundity that change over their lifetime, as illustrated by survivorship curves. Life tables were developed to incorporate age-, size-, or life history-specific rates of survival and fecundity. Using life tables, we can determine survivorship (l_x), net reproductive rates (R_0), generation times (T), and approximations of the intrinsic growth rates (r_a and λ_a). The data needed for life tables can be collected by following a cohort and building a cohort life table or by examining all individuals during a snapshot in time and developing static life tables.

REVIEW QUESTIONS

1. What is the difference in approach between the geometric growth equation and the exponential growth equation?

2. Given the relationship between λ and r in the geometric and exponential growth equations, can you demonstrate mathematically why λ must be 1 when r is 0?

3. Why might a model that contains age structure better reflect the reality of population growth than a model that lacks age structure?

4. What is the difference between a stable population and a stable age distribution?

5. If a life table projects a population size of 100 females and the sex ratio of the population is 1 : 1, how large is the entire population?

6. In a life table, what is the fundamental difference between survival rate (s_x) and survivorship (l_x)?

7. Compare and contrast a cohort life table and a static life table.

8. What is the relationship between generation time and the rate of population growth?

9. What are the different causes of slow population growth at low population sizes versus high population sizes?

10. What evidence would you need to determine whether a population experiences negative density dependence or positive density dependence?

PRACTICING ECOLOGY: ACIDIFICATION AND ZOOPLANKTON

Shelley Arnott from Queen's University in Ontario studies factors that determine distribution and abundance of freshwater invertebrates. Photo by Celia Symons. Courtesy of Shelley Arnott

Understanding the regulation of population growth can offer insight into the recovery of species following major disturbances. For example, in Chapter 2 we discussed how decades of acid precipitation have caused many lakes and streams to have much lower pH. In some cases, pH levels in the 1970s fell so low that many aquatic species could not survive. Although acid precipitation has declined substantially since 1980, many species are still recovering. Researchers are concerned that the recovery of some species has been slow because they experience positive density dependence. How can we predict which species will experience positive density dependence?

Zooplankton in freshwater ecosystems are particularly sensitive to acid precipitation. These tiny organisms are important to the ecosystems of rivers and lakes because they consume algae and are prey for many carnivores. Researchers studying zooplankton in the lakes of Killarney Provincial Park, Ontario, found that two previously dominant species of zooplankton—*Daphnia mendotae* and *Epischura lacustris*—nearly disappeared as lakes acidified. Although both species are recovering with reduced acid precipitation, *D. mendotae* is recovering faster. According to one hypothesis, *E. lacustris* is recovering more slowly because it is more sensitive to low pH conditions. A second hypothesis considers the way the two species reproduce. *D. mendotae* is parthogenic whereas *E. lacustris* uses sexual breeding, which means that it has to find mates. Therefore, we might also hypothesize that *E. lacustris* is recovering more slowly because it experiences positive density dependence when living at low densities.

Shelley Arnott and her lab at Queen's University in Ontario tested these hypotheses by subjecting the two species of zooplankton to different pH levels and population densities. Working in Carlyle Lake, they set up floating plastic bags, each containing 1500 L of lake water. Half of the bags contained *D. mendota* at pH 5, 6, or 7, and at low, medium, or high starting population densities. The remaining bags contained *E. lacustris* raised under the same treatments. Over 8 weeks, the researchers measured changes in population sizes and calculated the mean intrinsic growth rate of each treatment. You can view these data in the tables.

INTRINSIC GROWTH RATES OF *D. MENDOTAE* AT THREE pH LEVELS AND THREE STARTING DENSITIES

STARTING DENSITY	pH 5	pH 6	pH 7
LOW	−0.019	0.029	0.032
MEDIUM	−0.025	0.043	0.034
HIGH	−0.034	0.023	0.028

INTRINSIC GROWTH RATES OF *E. LACUSTRIS* AT THREE pH LEVELS AND THREE STARTING DENSITIES

STARTING DENSITY	pH 5	pH 6	pH 7
LOW	−0.025	0.001	0.000
MEDIUM	−0.039	0.036	0.032
HIGH	−0.032	0.032	0.041

GRAPH THE DATA

1. Create one bar graph showing the treatment means of intrinsic growth rate for *D. mendotae* at each starting density and for each pH level.
2. Create a second bar graph showing the treatment means of intrinsic growth rate for *E. lacustris* at each starting density and for each pH level.

INTERPRET THE DATA

1. Do the data support the hypothesis that *E. lacustris* is recovering more slowly because it is more sensitive to acidic water?
2. Do the data support the hypothesis that *E. lacustris* is recovering more slowly because it experience positive density dependence when living at low densities?
3. During the period of lake acidification, an acid-tolerant zooplankton species—*Bythotrephes longimanus*—dominated Carlyle Lake. This species preys on other zooplankton such as *E. lacustris* and *D. mendotae*. However, as pH levels rose, the abundance of *B. longimanus* decreased because it persists well in water with neutral pH. How might the fluctuating abundance of *B. longimanus* affect the decline and subsequent recovery of *E. lacustris* and *D. mendotae*?

SOURCE: Gray, D. K. and S. E. Arnott. 2011. The interplay between environmental conditions and Allee effects during the recovery of stressed zooplankton communities. *Ecological Applications* 21: 2652–2663.

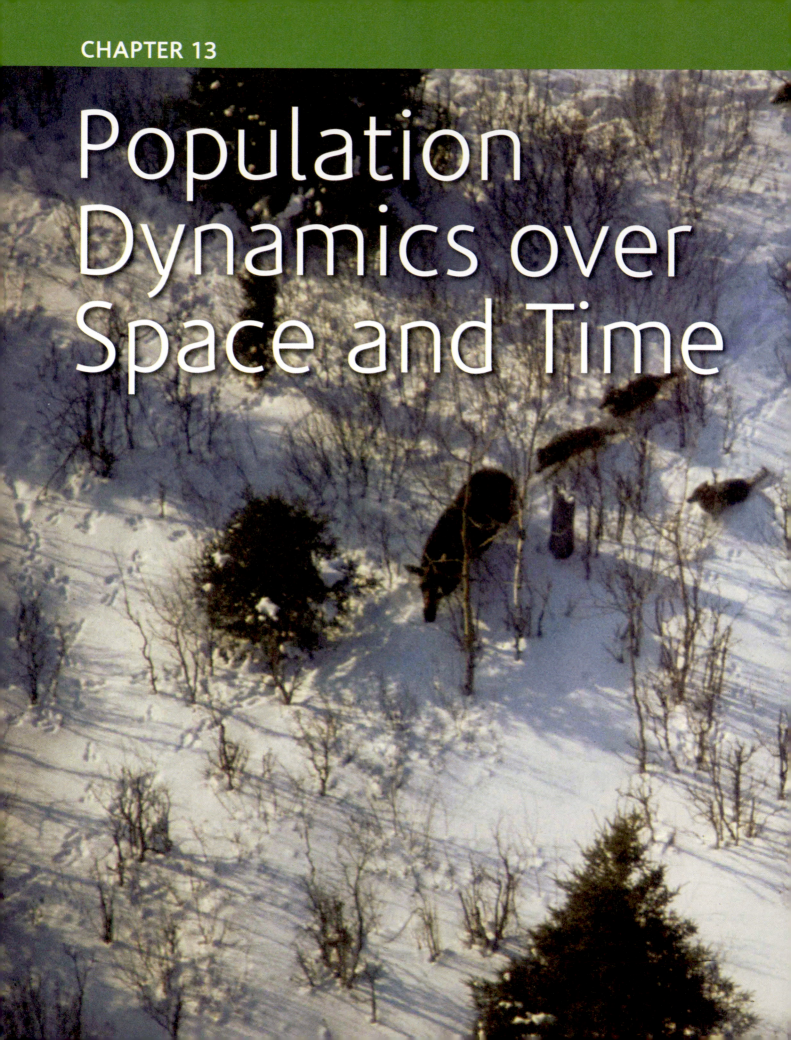

CHAPTER 13

Population Dynamics over Space and Time

Monitoring Moose in Michigan

Isle Royale, which is part of the state of Michigan and located off the northern shore of Lake Superior, has long served as a natural laboratory for ecologists. In the early 1900s, the island was colonized by moose from the mainland. This unlikely feat probably occurred during winter when the lake was frozen and the moose could travel the 40 km from the shores of the province of Ontario. The moose found an abundance of food on the island, and during subsequent decades the population grew rapidly. For more than 100 years, ecologists have watched the number of moose fluctuate widely due to changes in food, predators, and pathogens.

> "For more than 100 years, ecologists have watched the number of moose fluctuate widely due to changes in food, predators, and pathogens."

The initial changes in the population size of moose were driven by changes in the abundance of food. The population grew to approximately 3000 animals by the early 1930s but in 1934, when the population finally exceeded the island's carrying capacity, the moose suffered starvation and many died. In 1936 the island experienced extensive forest fires that stimulated new plant growth and, as a result, the moose population began to increase. By the late 1940s, the moose once again exceeded their carrying capacity and experienced another decline due to starvation.

The population would probably have continued this cycle of growth and decline, but sometime in the late 1940s grey wolves (*Canis lupus*) arrived on Isle Royale by crossing the ice during the winter. Beginning in 1958, ecologist David Mech and his colleagues began estimating the number of moose and the number of wolves on the island. This was a unique research opportunity. Because both immigration and emigration of the wolves and moose were extremely rare, the researchers could watch the dynamics of the two populations unfold over time. They have estimated the wolf and moose populations every year since 1958 and have continued to make new discoveries.

During the 1960s and early 1970s, there were about 24 wolves on the island while the moose population increased from about 600 to 1500. In the 1970s, there was a series of winters with deep snow that made it difficult for the moose to move around and locate food. At this time the population of wolves grew to 50 animals. These two factors caused a sharp drop in the moose population.

In 1981, a lethal virus (canine parvovirus) arrived on the island, probably carried to the island by a visitor's domestic dog. The virus caused the wolf population to experience a major decline, dropping from approximately 50 to 14 animals. With fewer wolves on the island, the moose population rebounded and it continued to grow. In 1996, the moose once again exceeded their carrying capacity and experienced widespread starvation, causing the population to drop from approximately 2400 to 500 individuals. From 1997 to 2011, the number of moose has fluctuated between 500 and 1000 while the number of wolves has fluctuated between 15 and 30.

The wolves and moose of Isle Royale in Lake Superior. For more than 50 years, researchers have tracked large swings in moose and wolf population sizes. Photo by Rolf Peterson.

The story of the moose and wolf populations on Isle Royale during the past 100 years highlights the fact that populations in nature often experience large fluctuations in size. These fluctuations can have many different causes that operate either separately or together. In this chapter, we will examine how populations fluctuate over space and time. Understanding the causes of such fluctuations can help us understand the natural fluctuations of organisms in nature so we are able to predict the future abundances of populations and the species with which they interact.

SOURCES: R. O. Peterson, Wolf-moose interaction on Isle Royale: The end of natural regulation? *Ecological Applications* 9 (1999): 10–16.

J. A. Vucetich and R. O. Peterson, The influence of top-down, bottom-up and abiotic factors on the moose (*Alces alces*) population of Isle Royale, *Proceedings of the Royal Society of London*-Series B 271 (2004): 183–189.

J. A. Vucetich and R. O. Peterson, *Ecological Studies of Wolves on Isle Royale: Annual Report 2010–2011*, School of Forest Resources and Environmental Science, Michigan Technological University, 2011.

CHAPTER CONCEPTS

- Populations fluctuate naturally over time.

- Density dependence with time delays can cause populations to be inherently cyclic.

- Chance events can cause small populations to go extinct.

- Metapopulations are composed of subpopulations that can experience independent population dynamics across space.

In the previous chapter, we saw that populations can be regulated by density-dependent and density-independent factors. In this chapter, we look at the causes of variation in population size, an issue that has become increasingly relevant as many species dwindle toward extinction and human activities fragment habitats into ever smaller, more isolated patches. Because these fluctuations include both random and cyclical changes over time, we will incorporate them into population models. We will also investigate how population fluctuations can cause small populations to be more prone to extinction. With this understanding of population variation over time, we will move on to examine population fluctuations over space by expanding our discussion of metapopulations from Chapter 12 to include metapopulation models and current research in metapopulations.

Populations fluctuate naturally over time

All populations experience fluctuations in size over time. There are many reasons for these fluctuations, such as changes in the availability of food and nesting sites, predation, competition, disease, parasites, and weather or climate. Despite these fluctuations, some populations tend to remain relatively stable over long periods. For example, researchers have studied the wolverine (*Gulo gulo*) hunt as a measure of population size across Canada for more than a decade. During that time, harvests have been relatively stable,

fluctuating between 350 and 700 individuals, as shown in Figure 13.1.

Other populations exhibit much wider fluctuations. In Lake Erie, for example, researchers monitored the number of waterborne algal cells, known as phytoplankton, throughout the year. As you can see in

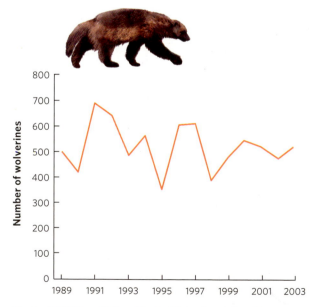

Figure 13.1 Population fluctuations in wolverines across Canada. From 1989 to 2003, wolverine harvests, which relate to total population size, have been relatively stable, ranging between 350 and 700 individuals. Data from B. G. Slough, Status of the wolverine *Gulo gulo* in Canada, *Wildlife Biology* 13 (2007): 76–82.

Figure 13.2 Population fluctuations in the algae of Lake Erie. The number of algal cells in the water has fluctuated widely over time. The populations of green algae and diatoms that compose the phytoplankton varied by several orders of magnitude on a scale of days, weeks, and months. Data from C. C. Davis, Evidence for the eutrophication of Lake Erie from phytoplankton records, *Limnology and Oceanography* 9 (1964): 275–283.

Figure 13.2, the algae fluctuated widely from nearly 0 cells per cm^3 in June to more than 7000 cells per cm^3 in September. These wide fluctuations occurred across scales of days, weeks, and months.

Wolverine populations are inherently more stable than algae. We can explain this by looking at differences in body size, population response time, and sensitivity to environmental change. Small organisms such as algae can reproduce in a matter of hours, which means that their populations respond very quickly to favourable and unfavourable environmental conditions. Their small bodies and the associated high surface-area-to-volume ratio cause them to be much more affected by unfavourable environmental changes, including abiotic conditions such as temperature. As a result, their rates of survival and reproduction can decline quickly. With larger animals that live for several years, a population at a given time includes individuals born over a long period, which tends to even out the effects of short-term fluctuations in birth rate. In addition, organisms with large body sizes can maintain homeostasis in the face of unfavourable environmental changes and therefore exhibit higher survival.

AGE STRUCTURE FLUCTUATIONS

We have discussed population fluctuations over time by looking at surveys of population size. In some cases, however, we can detect fluctuations over time by examining a population's age structure. When a certain age group contains an unusually high or low number of individuals, it suggests that the population experienced unusually high birth or death rates in the past.

A classic example of age structure as an indication of population fluctuations comes from data of the commercial harvest of whitefish (*Coregonus clupeaformis*) in Lake Erie from 1945 to 1951. Biologists

are able to determine the age of each fish by examining the scales of the harvested fish. As you can see in Figure 13.3, in 1947 there was a particularly large number of 3-year-old whitefish. This suggests that in 1944 the whitefish population experienced a very high amount of reproduction. This cohort continued to dominate the population's age structure in subsequent years, with high numbers of 4-year-old fish in 1948 and 5-year-old fish in 1949. Even in 1950, the 6-year-old age class contained more individuals than had been observed in that age class in previous years.

The age structure of a forest is also readily analyzed. As you may know, the age of a tree can be determined by counting the number of rings in its trunk; under most circumstances, one ring is added each year. To examine how fluctuations affect the age structure of forest populations, researchers work at sites that have not been logged, because logging alters the age structure of the current tree population.

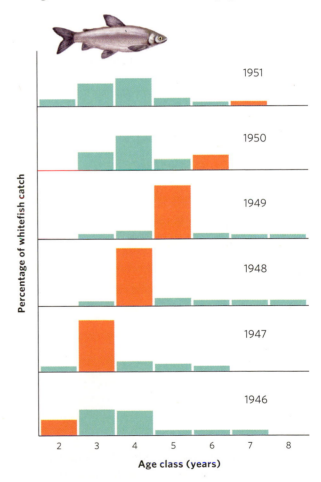

Figure 13.3 Whitefish age structure. The whitefish population in Lake Erie experienced an unusually high amount of reproduction in 1944. However, young fish are not captured by fishing nets until they are 2 years old, so the large 1944 cohort was not detected until 1946. This large reproduction event led to age structures that were dominated by that cohort of fish in subsequent years. Data from G. H. Lawler, Fluctuations in the success of year-classes of whitefish populations with special reference to Lake Erie, *Journal of the Fisheries Research Board Canada* 22 (1965): 1197–1227.

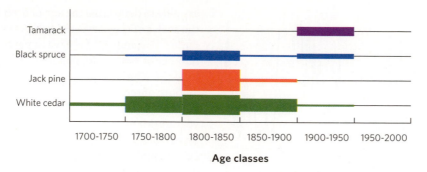

Figure 13.4 Age structure in an old-growth forest. By dating the ages of existing trees in an old-growth forest in Manitoba, researchers could determine how age structures of populations changed over time. The thickness of the bars indicates the relative abundance of each species. Data from K.L. Grotte, D.K. Heinrichs, and J.C. Tardif, Old-growth characteristics of disjunct *Thuja occidentalis* stands at their northwestern distribution limit, Central Canada, *Natural Areas Journal* 32 (2012): 270–282.

In one study, researchers investigated an area of old-growth forest in Manitoba that had not been logged for more than 300 years. Researchers determined the age of each tree as well as environmental factors, such as droughts and fire, from ring samples they took by drilling into the tree trunks. Figure 13.4 shows data from their survey. The researchers found that the forest in the 1700s was composed of white cedar (*Thuja occidentalis*) and some black spruce (*Picea mariana*). A fire in the 1800s created openings in much of the forest, after which cedar and spruce re-established. There was also an increase in the proportion of new jack pines (*Pinus banksiana*), which grow well in high light conditions. As the jack pines matured, they cast so much shade that new jack pine seedlings did not survive well, which lead to a decline in jack pine during the late 1800s. Today, most jack pine trees in this forest are about 150 years old. As the jack pines came to dominate and shade the forest, white cedar and black spruce—which are very tolerant of shady environments—began to grow. Cedar and spruce continued to recruit new individuals over time, which has created a population age structure that is more evenly distributed than that of jack pine. Tamarack is most often found in pure stands or mixed forests with black spruce and white cedar. It began populating the area when jack pines declined and spruce and cedar began to dominate again.

OVERSHOOTS AND DIE-OFFS

In Chapter 12, we saw that modelled populations experiencing density dependence initially grow quite rapidly but that this growth rate eventually slows as the population reaches its carrying capacity. However, populations in nature rarely follow a smooth approach to their carrying capacity. In many cases, they grow beyond their carrying capacity, a phenomenon ecologists refer to as an **overshoot**. An overshoot can occur when the carrying capacity of a

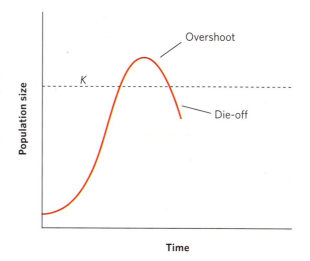

Figure 13.5 Population overshoots and die-offs. Some populations can overshoot their carrying capacity either because the carrying capacity has decreased or because the population can increase by large amounts in a single breeding season. Populations that overshoot their carrying capacity subsequently experience a die-off that causes a rapid decline in the population.

habitat decreases from one year to the next. For example, if one year has abundant rainfall and the next year has a drought, the habitat will produce less plant mass in the second year. As a result, the carrying capacity for herbivores that rely on the plants for food will be reduced. A population that overshoots its carrying capacity is living at a density that cannot be supported by the habitat. Such populations experience a **die-off**, which is a substantial decline in density that typically goes well below the carrying capacity. Figure 13.5 illustrates an

Overshoot When a population grows beyond its carrying capacity.

Die-off A substantial decline in density that typically goes well below the carrying capacity.

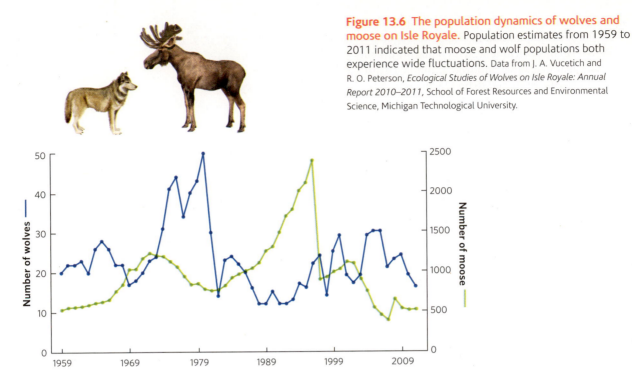

Figure 13.6 The population dynamics of wolves and moose on Isle Royale. Population estimates from 1959 to 2011 indicated that moose and wolf populations both experience wide fluctuations. Data from J. A. Vucetich and R. O. Peterson, *Ecological Studies of Wolves on Isle Royale: Annual Report 2010–2011*, School of Forest Resources and Environmental Science, Michigan Technological University.

overshoot and die-off in a population of a hypothetical organism. The population begins small but grows at such a rapid rate that it exceeds the carrying capacity. Following the overshoot, the population experiences a rapid die-off to a point that is below the carrying capacity. We saw an example of this in the story at the beginning of this chapter that described wolves and moose on Isle Royale. Over time the moose exceeded their carrying capacity and the population experienced massive die-offs from starvation. You can see these data in Figure 13.6.

A natural experiment with caribou in Alaska provides another good example of overshoots and die-offs. In 1911, the United States government introduced 25 caribou to St. Paul Island, Alaska, to provide a source of meat for the local population. The island contained no predators of caribou and the caribou population quickly began to reproduce. The caribou fed on a variety of items during the spring, summer, and fall, but they relied on lichens to get them through the winter. By 1938, the caribou population had swelled to more than 2000 individuals. As you can see in Figure 13.7, this rate of growth followed a J-shaped growth curve that we associate with exponential growth models. As the population grew in the early years, the lichens they consumed in the winter remained abundant. However, with continued growth of the caribou population, the lichens became rare, which suggests that the caribou had far exceeded the carrying capacity of the island. Following a peak in 1938, the caribou population began to experience a massive die-off, probably from

a combination of scarce winter food and unusually cold winters. In 1940 and 1941, the government culled several hundred caribou in an attempt to reduce the size of the herd to get it closer to the island's now much-reduced carrying capacity. Despite this effort, the population continued to decline and, by 1950, only 8 individuals remained. In 1951, 31 new caribou were

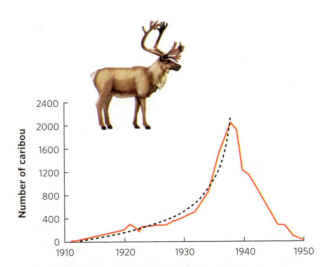

Figure 13.7 The overshoot and die-off of a caribou population. A herd of 25 caribou was introduced to St. Paul Island in Alaska in 1911. The population experienced a rapid increase in size that approximates an exponential J-shaped growth curve, shown by the dashed black line. After growing to nearly 2000 animals in 1938, the population crashed, probably because the animals exhausted the food supply. Data from V. B. Scheffer, The rise and fall of a reindeer herd, *Scientific Monthly* 73 (1951): 356–362.

brought to the island to supplement the population. Since 1980, the population has been managed by the St. Paul Tribal Government, which maintains the population at just a few hundred animals. In 2012, the population was estimated to be 125 caribou. The caribou, maintained at a sustainable population size, can now continue to provide valuable and affordable meat to the local residents.

CYCLIC POPULATION FLUCTUATIONS

As we have just seen, populations can experience large fluctuations over time, including overshoots and die-offs. Over longer periods, some populations oscillate up and down in regular patterns known as **population cycles**.

An interesting example of a population cycle comes from long-term records on the gyrfalcon. Falconry was a popular pastime among European nobility during the seventeenth and eighteenth centuries, and the gyrfalcon was especially prized because it is one of the largest and most beautiful falcons. During this time, Danish royalty imported gyrfalcons from Danish territories in Iceland where the falcons were trapped and then transported to Copenhagen. The Danish royalty would then present the falcons as diplomatic gifts to the royal courts of Europe. The governor of Iceland wrote the export permits, which provide a detailed historical record of gyrfalcon exports over several decades. As shown in Figure 13.8, the number of gyrfalcons exported from Iceland between 1731 and 1793 reveals 10-year cycles of abundance. After 1770, falconry became less popular and exports declined. Exports no longer exhibited regular fluctuations because the low demand for the birds could be satisfied even during years of low abundance.

In some cases, cyclic population fluctuations occur among species and across large geographic areas. In Finland, for example, biologists conducted annual surveys in 11 provinces to determine the abundance of three species of grouse: capercaillie (*Tetrao urogallus*), black grouse (*Tetrao tetrix*), and hazel grouse (*Bonasa bonasia*). After monitoring the birds for 20 years and graphing the data—as shown in Figure 13.9—they found that all three species experienced population cycles every 6 to 7 years. Moreover, these three species appear to exhibit high and low populations at the same time and this synchrony occurred across all of the provinces.

These examples demonstrate that some populations can exhibit remarkably regular fluctuations. The cause of such cycles, and of their synchrony, has been an interesting and persistent question in ecology. In subsequent chapters we will look at how species interactions also influence population cycles.

Figure 13.8 Cyclic fluctuations in gyrfalcons. From 1730 to 1770, falconry was popular in Denmark and the number of falcons captured and exported from Iceland showed regular cycles that occurred approximately every 10 years. After 1770, the popularity of falconry declined and the low demand for falcons could be met even in low-population years. Data from Ó. K. Nielsen and G. Pétursson, Population fluctuations of gyrfalcon and rock ptarmigan: Analysis of export figures from Iceland, *Wildlife Biology* 1 (1995): 65–71.

Density dependence with time delays can cause populations to be inherently cyclic

We have seen that some species have regular cycles of population increases and decreases. The challenge for us is to figure out the causes of these cycles. Some types of environmental variation are cyclical, including daily, lunar or tidal, and seasonal cycles. However, most environmental variation occurs irregularly over time. Prolonged periods of abundant rainfall or drought, extreme heat or cold, or natural disasters such as fires and hurricanes are not predictable. We must therefore conclude that irregular variation in the environment is not likely to be the underlying cause of regular population cycles. In this section, we will examine the inherent cycling behaviour of populations and apply a modified version of the logistic growth equation to help us understand some of the causes.

THE INHERENT CYCLING BEHAVIOUR OF POPULATIONS

In the 1920s and 1930s, population modellers discovered that populations can naturally oscillate above and below their carrying capacity when they are subjected to environmental variation, even if the environmental variation is random over time. Populations have an inherent periodicity and tend to fluctuate up and down, although the time required to complete a cycle differs among species.

Population cycles Regular oscillation of a population over a longer period of time.

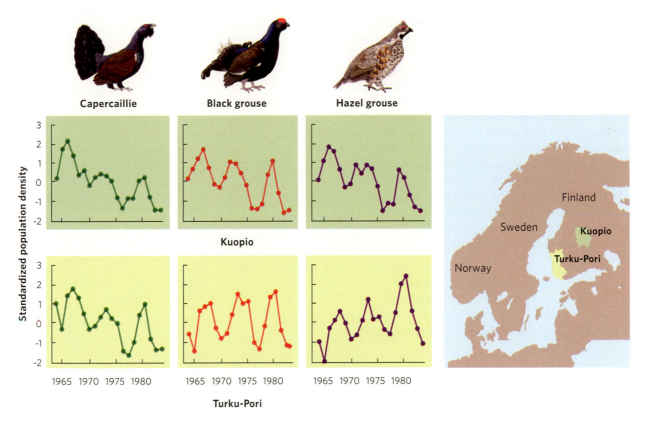

Figure 13.9 Cyclical fluctuations of grouse species in two different provinces of Finland. The three species of grouse experience cycles every 6 to 7 years and appear to fluctuate in synchrony with each other and across the provinces. Data from J. Lindström et al., The clockwork of Finnish tetraonid population dynamic, *Oikos* 74 (1995): 185–194.

To help us understand such behaviour, we can think of a population as being analogous to a swinging pendulum. We know that a pendulum is stable when hanging straight up and down. The pull of gravity will cause any movement of the pendulum to the right or left to move back toward the centre. However, since this movement back to the centre has momentum, the pendulum overshoots the stable, centre position and swings to the other side. Gravity then pulls it back toward the centre where momentum once again causes it to overshoot the stable, centre position.

Populations behave like the pendulum; the momentum of increases and decreases in a population causes it to oscillate. Populations are stable at their carrying capacity. When reductions in the population's size occur—because of events such as predation, disease, or a density-independent event—the population responds by growing. If the growth is sufficiently rapid, the population can grow beyond its carrying capacity. We see this phenomenon when there is a delay between the initiation of breeding and the time that offspring are added to the population. Populations that overshoot their carrying capacity subsequently experience a die-off that causes the population to swing back toward its carrying capacity.

Because death rates are high and birth rates are low, the population can experience a large reduction and undershoot its carrying capacity.

DELAYED DENSITY DEPENDENCE

We can model population cycles by starting with the logistic growth model that we introduced in Chapter 12. You may recall that this model incorporates density dependence, which causes population growth rates to slow down as the population increases in size. The key to making density-dependent populations cycle in these models is to incorporate a delay between a change in environmental conditions and the time the population reproduces. When density dependence occurs based on a population density at some time in the past, we call it **delayed density dependence**.

Delayed density dependence can be caused by any number of factors. For example, large herbivores such as moose often breed in the fall but do not give birth until the following spring. If food is abundant in the fall, the carrying capacity is high, but by the time the

---

Delayed density dependence When density dependence occurs based on a population density at some time in the past.

offspring are born in the spring, the carrying capacity of the habitat could be much lower. The offspring will still be born, but the population will now exceed the carrying capacity of the habitat because the amount of reproduction was based on the earlier environmental conditions. We can also think about time delays for predators. When predators experience an increase of prey, their carrying capacity increases. However, it may take weeks or months for the predators to convert abundant prey into higher reproductive rates. By this time, the prey may no longer be abundant. The lack of prey will cause the carrying capacity of the predator to decline just as the predator population is increasing. In both scenarios, the population experiences a time delay in density dependence.

Modelling Delayed Density Dependence

We can modify the logistic growth model to demonstrate delayed density dependence. As you may recall, the logistic model uses the following equation:

$$\frac{dN}{dt} = rN\left(1 - \frac{N}{K}\right)$$

where $\frac{dN}{dt}$ is the rate of change in the population size, r is the intrinsic growth rate, N is the current size of the population at time t, and K is the carrying capacity.

To incorporate a time delay, we begin by defining the amount of time delay as τ, which is the Greek letter tau. Now we can rewrite the logistic growth equation by making the density-dependent part of the equation based on the population's size at τ time units in the past:

$$\frac{dN}{dt} = rN\left(1 - \frac{N_{t-\tau}}{K}\right)$$

In words, this equation tells us that the population slows its growth when the population's size, at τ time units in the past, approaches the carrying capacity.

Whether a population cycles above or below the carrying capacity depends both on the magnitude of the time delay and on the magnitude of the intrinsic growth rate. As the time delays increase, density dependence is further delayed, making the population more prone to both overshooting and undershooting the carrying capacity. In addition, having a high intrinsic rate of growth allows a population to grow more rapidly in a given amount of time, making an overshoot of the carrying capacity more likely.

Population modellers have determined that the amount of cycling in a population experiencing delayed density dependence depends on the product of r and τ, as illustrated in Figure 13.10. As you can see in Figure 13.10a, when this product is a low value ($r\tau < 0.37$), the

(a)

(b)

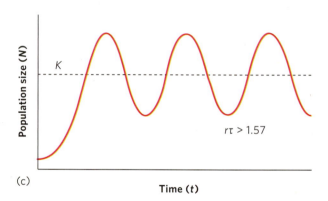

(c)

Figure 13.10 Population cycles in models containing delayed density dependence. (a) In population models where the product of $r\tau$ is a low value ($r\tau < 0.37$), the population approaches the carrying capacity without any oscillations. **(b)** When the product of $r\tau$ is an intermediate value ($0.37 < r\tau < 1.57$), the population will exhibit damped oscillations. **(c)** When the product is a high value ($r\tau > 1.57$), the population will oscillate over time as a stable limit cycle.

population approaches the carrying capacity without any oscillations. If this product is an intermediate value ($0.37 < r\tau < 1.57$), as shown in Figure 13.10b, the population initially oscillates but the magnitude of the oscillations declines over time, a pattern known as **damped oscillations**. When the product is a high value ($r\tau > 1.57$), as shown in Figure 13.10c, the

---

Damped oscillations A pattern of population growth in which the population initially oscillates but the magnitude of the oscillations declines over time.

Delayed Density Dependence in the Flixweed

Flixweed (*Descurania sophia*) is a weed that is native of Europe but has been introduced into North America. Researchers who have studied this weed have found that the number of plants per m² of soil fluctuates in a cyclical manner over time. The population grows according to a delayed density dependence model where $K = 100$, $r = 1.1$, and $\tau = 1$.

The flixweed, a common plant in Europe and North America. Photo by Nigel Cattlin/Alamy.

From plant surveys, we know that there were 10 plants per m² in year 1 and 20 plants per m² in year 2. Based on these data, we can calculate the expected change in population size in year 3:

$$\frac{dN}{dt} = rN_2 \left(1 - \frac{N_1}{K}\right)$$

$$\frac{dN}{dt} = (1.1)(20)\left(1 - \frac{10}{100}\right)$$

$$\frac{dN}{dt} = 20$$

Rounded off to the nearest whole number, the flixweed will add 20 individuals to the population in year 3. Given that the population in year 2 is 20 individuals, adding 20 more individuals in year 3 will produce a total population size of 40 individuals.

We can continue the calculations to determine the population size in year 4:

$$\frac{dN}{dt} = rN_3 \left(1 - \frac{N_2}{K}\right)$$

$$\frac{dN}{dt} = (1.1)(40)\left(1 - \frac{20}{100}\right)$$

$$\frac{dN}{dt} = 35$$

As we can see, the flixweed population will increase by another 35 individuals in year 4, producing a total population size of 75 individuals.

YOUR TURN Using the data provided, calculate the population sizes of the plant from year 5 through year 15. Based on the product of $r\tau$, what type of oscillating behaviour do you expect to see in this population even before doing the calculations? Graph the results to confirm your prediction.

population continues to exhibit large oscillations over time, a pattern known as a **stable limit cycle**.

CYCLES IN LABORATORY POPULATIONS

Population models that incorporate delayed density dependence help us understand how time delays cause populations to oscillate in regular cycles. Because models do not identify specific mechanisms by which time delays occur, ecologists have investigated real populations using laboratory experiments to discover these mechanisms.

In some cases, delayed density dependence occurs because the organism can store energy and nutrient reserves. The water flea *Daphnia galeata,* for example, is a tiny zooplankton species that lives in lakes throughout the Northern Hemisphere. When the population is low and there is an abundance of food, individuals can store surplus energy in the form of lipid droplets. As the population grows over time to the carrying capacity and food becomes scarce, adults with stored energy can continue to reproduce. *Daphnia* mothers can also transfer some of these lipid droplets to their eggs, which allows their offspring to grow well

Stable limit cycle A pattern of population growth in which the population continues to exhibit large oscillations over time.

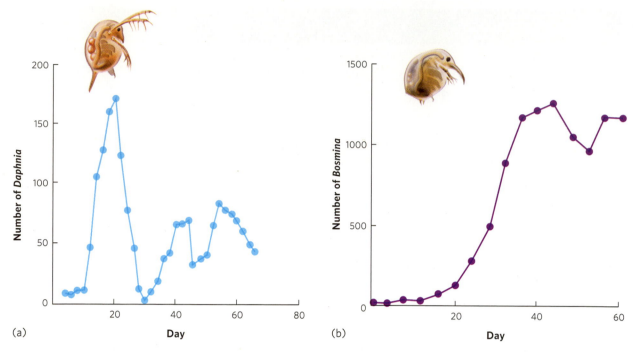

Figure 13.11 The importance of energy reserves in causing population cycling. (a) *Daphnia galeata* water fleas can store high amounts of energy, which allows them to survive and reproduce even after reaching carrying capacity. When energy reserves run out, the population crashes to very low numbers and then rebounds and continues to oscillate. (b) *Bosmina longirostris* water fleas can only store a low amount of energy, so as the population nears carrying capacity, they experience reduced survival and reproduction. As a result, the population remains near its carrying capacity and oscillates much less. Data from C. E. Goulden and L. L. Hornig, Population oscillations and energy reserves in planktonic cladocera and their consequences to competition, *Proceedings of the National Academy of Sciences* 77 (1980): 1716–1720.

even if the carrying capacity of the lake has been exceeded. Eventually, the stored energy is used up and the *Daphnia* population crashes to low numbers. When the population is low, the food can once again become abundant and the cycle begins once more. You can see these oscillations in Figure 13.11a.

We can contrast the story of the *Daphnia* water flea with the story of another species of water flea, *Bosmina longirostris. Bosmina* do not store as many lipid droplets as *Daphnia,* so when their population approaches the lake's carrying capacity, they have little energy to buffer the reduction in food. As a result, *Bosmina* populations do not exhibit large oscillations. Instead, as illustrated in Figure 13.11b, they grow to their carrying capacity and remain there.

Delayed density dependence can also occur when there is a time delay in development from one life stage to another. In a classic study of developmental delays, A. J. Nicholson examined the effect of time delays between the larval and adult stages of the sheep blowfly (*Lucilia cuprina*), an insect that feeds on the flesh of domestic sheep.

In his first experiment, Nicholson fed the larvae a fixed amount of food—thereby setting a carrying capacity for the larvae—but he fed the adults an unlimited amount of food. At the start of the experiment, the larvae began to metamorphose into the first set of adults, as shown in the orange line in Figure 13.12a. These adults then laid eggs that hatched into more larvae that eventually became adults. The adult fly population rapidly increased to more than 4000 individuals.

As the adult population increased, the unlimited food supply allowed them to continue laying eggs. The large number of eggs hatched into a large number of larvae, but, because the larvae had a limited food supply, they did not grow well enough to metamorphose into adults. The larvae died and no new adults were produced. Eventually the adult population crashed. However, before the last few adults died, they laid a small number of eggs. When these eggs hatched, the fixed food supply provided an abundance of food for the low number of larvae. As a result, the larvae experienced a high rate of survival and most of them metamorphosed into adults. These new adults then laid a large number of eggs and the cycle began again. In short, there was a delay between the time that the adults produced a large number of eggs and the time these eggs hatched into larvae, died from high larval competition, and failed to produce

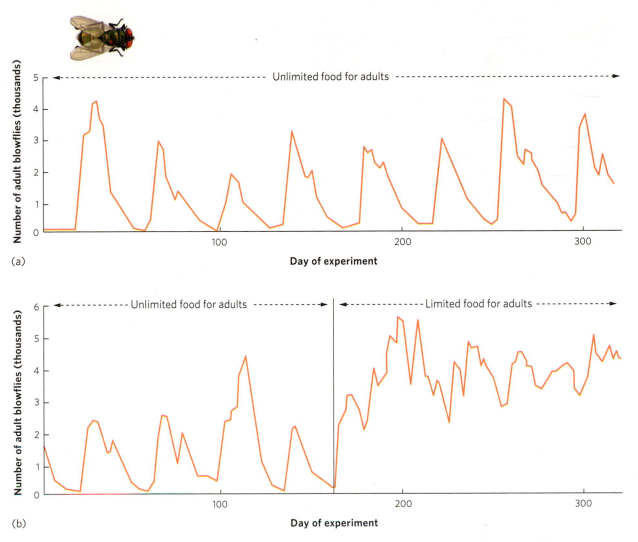

(a)

(b)

Figure 13.12 Population cycling in sheep blowflies. (a) When researchers limited food for larvae but not adults, they observed a delay between the time that the adults produced a large number of eggs and the time these eggs hatched into larvae, died from high larval competition, and failed to produce new adults. As a result, the adult population experienced regular cycles. **(b)** When adults were initially raised with unlimited food, but then given limited food halfway through the experiment, they began to experience density dependence without a time delay. As a result, the adult population still fluctuated but no longer experienced regular cycles. Data from A. J. Nicholson, The self-adjustment of populations to change, *Cold Spring Harbor Symposia on Quantitative Biology* 22 (1958): 153–173.

new adults. This time delay appeared to cause the adult population cycles.

Nicholson then reasoned that if limited food for offspring caused the time delay, limiting the adult food should eliminate the time delay and reduce the extreme fluctuations in the adult population. To test this hypothesis, he ran the experiment again by starting with unlimited adult food, but halfway through the experiment he limited the adult food. Under these conditions, the adults experienced density dependence without delay. As you can see in Figure 13.12b, although the abundance of the adult population still fluctuated, it no longer exhibited regular population cycles. These laboratory studies

confirmed that time delays between life stages caused population cycles.

Chance events can cause small populations to go extinct

In our models of population growth, we have seen that when populations are large, density-dependent factors cause slower growth, and when populations are small, density-dependent factors cause faster growth. Given these outcomes, it is hard to see how populations could go extinct, yet we know that extinction does occur in real populations in nature. In this section, we will explore the relationship between population size and

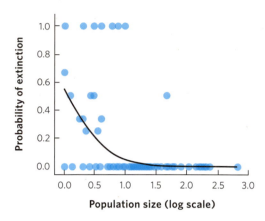

Figure 13.13 Smaller populations and the likelihood of extinction. Researchers surveyed small islands in three archipelagos off the Labrador coast for evidence of eider duck nests. Populations on islands with larger population sizes had a reduced probability of going extinct during the 6-year period of the study. Data from K.G. Chaulk, G.J. Robertson, and W.A. Montevecchi, Extinction, colonization, and distribution patterns of common eider populations nesting in a naturally fragmented landscape, *Canadian Journal of Zoology* 84 (2006): 1402–1408.

the probability of extinction. We will then examine the underlying causes of this relationship.

EXTINCTION IN SMALL POPULATIONS

In nature we find that smaller populations are more vulnerable to extinction than larger populations. Biologists described this phenomenon after studying populations of eider ducks (*Somateria mollisssima*) living on small islands in three archipelagos off Labrador. Over 6 years, researchers surveyed the islands for evidence of eider duck nests. The researchers categorized population size by the number of nests found and recorded whether each population contained nests the following year, or whether it had gone extinct. They then modelled the probability of a population going extinct over 6 years for populations of each size category. As shown in Figure 13.13, the smallest populations experienced the highest probability of extinction and the largest populations experienced the lowest probability of extinction. Similar patterns have been observed for many other groups of animals, including mammals, reptiles, and amphibians.

The increased extinction rate of small populations is also seen in plants. For example, researchers in Germany examined the persistence of 359 populations of plants that spanned eight different species. In 1986, they counted the number of individuals in each population and 10 years later found that 27 percent of

the populations had gone extinct. As you can see in Figure 13.14, when the researchers placed the populations into one of six population sizes and

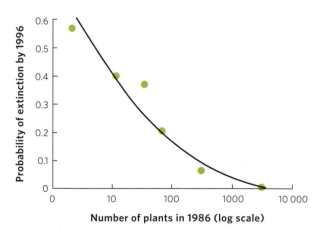

Figure 13.14 The probability of plant populations going extinct. Researchers surveyed 359 plant populations in Germany in 1986 and then again in 1996. They then placed the populations into six size categories and averaged extinction rates across the eight species. They found that populations with the lowest number of individuals experienced the highest probability of extinction over a 10-year period. Data from D. Matthies et al., Population size and the risk of local extinction: Empirical evidence from rare plants, *Oikos* 105 (2004): 481–488.

averaged the extinction rates across the eight species, they found that the probability of extinction was high for smaller population sizes and low for larger populations.

EXTINCTION DUE TO VARIATION IN POPULATION GROWTH RATES

We have seen that populations are more likely to go extinct when they are small. However, the density-dependent population models we have examined show that small populations have more rapid growth rates than large populations. This suggests that a small population should quickly rebound and grow into a larger population. By this reasoning, small populations should be resistant to extinction. How do we resolve the fundamental difference between these models and our observations of actual populations?

The models that we have studied to this point have assumed a single birth rate and death rate for each individual in the population. When a model is designed to predict a result without accounting for random variation in population growth rate, we say it is a **deterministic model**. Although deterministic models are simpler to work with, not every individual in the real world has the same birth rate or the same probability of dying. Rather, because we see random variation among individuals in a population, the growth rate is not constant; it can vary over time. Models that incorporate random variation in population growth rate are known as **stochastic models**. When random variation in birth rates and death rates is due to differences among individuals and not due to changes in the environment, we call it **demographic stochasticity**. In contrast, when random variation in birth rates and death rates is due to changes in the environmental conditions, we call it **environmental stochasticity**. Examples of environmental stochasticity include changes in the weather, which can cause small changes in the growth rate of a population, and natural disasters, which can cause large changes in the growth rate of a population.

When we use stochastic models, there is an average growth rate with some variation around that average. The actual growth rate experienced by the modelled population can take any of the values within the range of this variation. As a result, a population's growth rate may be above or below the average growth rate. If a population experiences a string of years with above average growth rates, it will have faster growth. If the population happens to have a string of years with below average growth rates, it will have slower growth. Which outcome actually occurs is determined strictly by chance.

We can begin to account for stochasticity by adding it to the exponential growth model we discussed in

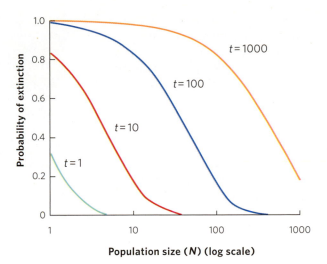

Figure 13.15 The probability of extinction in a population model. Using an exponential growth model and stochastic rates of birth and death, the average probability of extinction increases with decreasing population sizes (N) and longer times into the future (t).

Chapter 12. For example, we can set the birth rate and death rate to 0.5, which are reasonable values for adult mortality and recruitment in a population of terrestrial vertebrates. We can let populations vary above and below these average birth and death rates at random. When a population randomly experiences years of low birth rates or high death rates, it is more likely to go extinct. A string of bad years can drive the population extinct faster in small populations than in large populations. Moreover, at a given population size, as time passes there is an increased chance that a population will have a string of bad growth years and go extinct. You can see the effect of population size and time on the average probability of extinction in Figure 13.15. For example, in a population with 10 individuals, the average probability of extinction within 10 years is 0.16, the average probability of extinction within 100 years is 0.82, and extinction becomes virtually certain (0.98) within 1000 years. In contrast, a population with an initial size of 1000 has an average probability of extinction of only 0.18 within 1000 years. By adding the reality of stochasticity to

Deterministic model A model that is designed to predict a result without accounting for random variation in population growth rate.

Stochastic model A model that incorporates random variation in population growth rate.

Demographic stochasticity Variation in birth rates and death rates due to random differences among individuals.

Environmental stochasticity Variation in birth rates and death rates due to random changes in the environmental conditions.

our population models, we see that although populations of any size eventually have some chance of going extinct, the smallest populations face the greatest risk of extinction.

Metapopulations are composed of subpopulations that can experience independent population dynamics across space

We have considered population dynamics over time, but population dynamics also occur over space. Whether on land or in water, species that have a particular geographic range are commonly subdivided into smaller subpopulations. This happens because preferred habitat is not continuous but occurs as patches of suitable habitat that are surrounded by a matrix of unsuitable habitat. When individuals are able to move between patches, we consider the collection of subpopulations to be a metapopulation, a concept that we first introduced in Chapter 11. As you may recall, the example of the collared lizards in Chapter 11 involved a metapopulation that did not live in the entire forest, but rather in glades, where it experienced more favourable temperatures and a more abundant food supply. In this section, we will examine why populations are broken up into metapopulations and review the three types of metapopulation models. We will then examine studies of real metapopulations in nature.

THE FRAGMENTED NATURE OF HABITATS

Metapopulations occur when a habitat is fragmented. A good example is the existence of wetlands that are dotted across the landscape of many parts of North America (Figure 13.16). Each wetland can contain populations of many different species of amphibians, crustaceans, snails, and aquatic plants. The terrestrial habitat located between these wetlands is generally inhospitable to these organisms, yet individuals of many species are able to disperse across these intervening regions to arrive at another wetland. As you might guess, the amount of movement among wetlands depends on the distance between neighbouring wetlands and how far and fast an individual can move.

Metapopulations also occur as a result of human activities such as clearing forests, draining large wetlands, and constructing roads, housing developments, and commercial properties—all of which contribute to breaking up large habitats into a number

Figure 13.16 Wetlands within a terrestrial landscape. In many regions of the world, such as this site in the Canadian prairies, wetlands provide patches of habitat that are hospitable to aquatic organisms such as amphibians, crustaceans, snails, and aquatic plants. However, the terrestrial habitat between the wetlands is often inhospitable to these organisms. Photo ©Ducks Unlimited Canada/www.ducks.ca.

of smaller habitats (Figure 13.17). Because the small habitats represent only fragments of the original, we refer to this process as **habitat fragmentation**. Small habitats typically support small populations, which, as we have seen, are prone to extinction. However, a group of small populations that are interconnected by occasional dispersal have a unique dynamic. While some habitat fragments experience extinctions of their subpopulations, others become colonized by dispersers that form new subpopulations. This combination of extinctions balanced by colonizations allows the metapopulation to persist over time.

In Chapter 11, we examined three different types of metapopulation models. The basic metapopulation model assumes that all habitat patches are equal in quality and that the habitat matrix between patches is inhospitable. The source–sink metapopulation model builds upon the basic model but incorporates the fact that habitat patches differ in quality. High-quality patches are known as *sources* because they produce high numbers of individuals that can disperse to other patches. In contrast, low-quality patches are known as *sinks* because they produce few individuals and rely on dispersers coming in from other patches to keep the subpopulation from going extinct. The landscape metapopulation model is even more realistic because it

---

Habitat fragmentation The process of breaking up large habitats into a number of smaller habitats.

Figure 13.17 Habitat fragmentation due to human activities. Activities such as construction, harvesting of lumber, and agriculture have created fragmented habitats throughout the world. For example, the originally forested landscape on this site in Exmoor, England, is currently composed of cleared fields with fragments of forests between the fields. Photo by Adam Burton/age fotostock.

acknowledges that both the patches and the habitat matrix can vary in quality.

The dynamics of a metapopulation can fall along a range of possible outcomes. If the subpopulations rarely exchange individuals, the fluctuations in abundance will be independent among subpopulations. Some subpopulations will increase over a period of time while others will decrease or stay relatively constant. At the other extreme, if subpopulations are highly connected by individuals frequently moving among habitat patches, the subpopulations will act as one large population, with all experiencing the same fluctuations. Between these two extremes is the scenario in which individuals occasionally move between habitat patches, such as when juvenile animals disperse away from their family to find a mate. In this case, fluctuations in abundance in one subpopulation can influence the abundance of other subpopulations.

THE BASIC MODEL OF METAPOPULATION DYNAMICS

Having reviewed the spatial structure of metapopulations, we can now examine the population dynamics that occur in metapopulations. We will start with the basic metapopulation model that contains a number of simplifying assumptions. Although these assumptions are not realistic, they can help us understand the basic dynamics of a metapopulation.

In this model, we begin with a population that is divided into several subpopulations, each occupying a distinct patch of habitat. We assume that these habitat patches are of equal quality, and that because each occupied patch has the same subpopulation size, each subpopulation supplies the same number of dispersers to other habitat patches.

For the entire collection of habitat patches that exist, we will assume that some fraction of them is occupied, which we denote as p. We will also assume that there is a fixed probability of each patch becoming unoccupied—that is, its subpopulation going extinct—in a given amount of time, which we shall refer to as e. Finally, we will assume that there is a fixed probability that each unoccupied patch could be colonized, which we will denote as c. Using these variables, the proportion of occupied patches when colonization and extinction achieve an equilibrium, as indicated by $\hat{p}$, is given by the following equation:

$$\hat{p} = 1 - \frac{e}{c}$$

This basic model indicates how we might increase the number of occupied habitat patches, which would increase the total number of individuals in the metapopulation given that each occupied patch is assumed to have an equal number of individuals. One way would be to provide corridors between neighbouring populations, thereby increasing the rate of colonization, as we saw in the case of the collared lizards. A second way to increase the number of occupied habitat patches would be to decrease the rates of extinction by reducing the major causes of population decline in subpopulations.

The concept of the metapopulation is important for the conservation of species around the world as humans continue to fragment terrestrial and aquatic habitats. The models and empirical research inform us that a key to preserving populations is by maintaining large fragments of habitat whenever possible because populations in these large habitats are less likely to go extinct. When we can only preserve small fragments, we must ensure that individuals can disperse to and from them so they can be colonized, which will help prevent small, declining subpopulations from going extinct.

OBSERVING METAPOPULATION DYNAMICS IN NATURE

The basic metapopulation model indicates that a metapopulation persists because of a balance between extinction of some habitat patches and colonization of others. Although these are the predictions of the model, we need to see if these processes occur in nature. One of the most extensive studies of metapopulations has been conducted on the Glanville fritillary butterfly (*Melitaea cinxia*) by ecologist Illka Hanski and colleagues. On the Åland Islands of Finland, this butterfly lives in isolated patches of dry meadows (**Figure 13.18a**). The researchers found 1600 suitable meadows on the islands, but found that only 12 to 39 percent of these patches were occupied in any given year. Over a 9-year period, they observed that more than 100 of the occupied patches experienced extinction each year and more than 100 of the unoccupied patches experienced colonization. Based on their observations, it became clear to the researchers that no single patch was safe from extinction, but the metapopulation persisted because the extinctions were offset by continued colonizations.

To closely examine extinctions and colonizations over time, Hanski and colleagues introduced populations of the Glanville fritillary to the small island of Sottungia off the coast of Finland. The island lacked this species of butterfly, but it contained 20 meadows that offered suitable butterfly habitat. The researchers introduced the Glanville fritillary to 10 meadows. They surveyed all 20 meadows for a decade to observe the dynamics of the metapopulation they had created. From their starting point of 10 subpopulations, they observed large fluctuations in the occupation of habitat patches over time, as shown in Figure 13.18b. During the course of their study, there were as few as 3 subpopulations and as many as 14 subpopulations. The number of extinctions ranged between 0 and 12 per year and the number of colonizations ranged between 0 and 9 per year. None of the original 10 subpopulations persisted for the entire decade. Instead, each subpopulation went extinct at

(a)

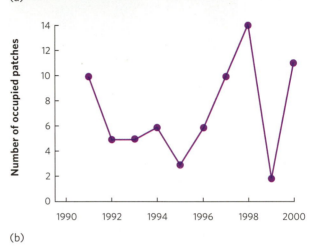

(b)

Figure 13.18 Metapopulation dynamics of a butterfly. **(a)** The Glanville fritillary butterfly lives in isolated meadows on the Åland Islands of Finland. Here, the butterflies exist as a metapopulation. **(b)** On the Finnish island of Sottungia, researchers placed butterflies in 10 of the 20 available meadows. Over time, the number of occupied patches varied from 3 to 14 with all of the original patches going extinct and some of them being recolonized. After a decade, the metapopulation was composed of 11 occupied patches, a number that is very close to its original starting point. Data from I. Hanksi et al., Checkerspots and conservation biology, in *On the Wings of Checkerspots*, ed. P. Ehrlich and I. Hansk, Oxford University Press, 2004. Photo by Robert Thompson/naturepl.com.

different times and dispersers from other patches subsequently recolonized many of the habitat patches that experienced extinctions of their butterfly subpopulations. At the end of the decade-long experiment, the metapopulation consisted of 11 subpopulations, which is quite close to the original 10 subpopulations that were started at the beginning of the experiment. So, while there was a great deal of change going on at the subpopulation level, the metapopulation persisted.

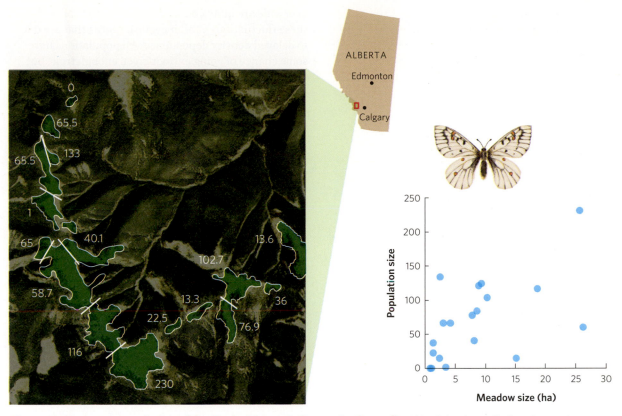

Figure 13.19 A metapopulation of the Rocky Mountain Parnassian butterfly. This alpine butterfly lives in meadows on the eastern slopes of the Canadian Rocky Mountains, such as in the Kananaskis region of Alberta. The numbers next to each patch indicate how many butterflies biologists counted. Data from J. Roland, N. Keyghobadi, and S. Fownes, Alpine *Parnassius* butterfly dispersals: Effects of landscape and population size, *Ecology* 81 (2000): 1642–1653.

THE IMPORTANCE OF PATCH SIZE AND PATCH ISOLATION

We have noted that the assumptions of the basic model of metapopulation dynamics do not allow for variations found in nature. For example, habitat patches are rarely equal in quality. Some patches are larger or contain a higher density of resources needed by a population. On the eastern slopes of the Canadian Rocky Mountains, for example, the Rocky Mountain Parnassian butterfly (*Parnassius smintheus*) lives in small fragmented meadows. As you can see in Figure 13.19, each fragment is a different size and the number of butterflies in each meadow ranges from 1 to 230. As you can also see, not all patches are equally distant from all other patches.

When each patch in a metapopulation supports a different number of individuals, we expect the small patches to experience higher rates of extinction. As a result, small patches are less likely to be occupied than large patches. At the same time, we might also predict that more distant patches will have a lower probability of being occupied than closer patches. We make this prediction because successful dispersal is a function of the distance an individual has to disperse.

Therefore, unoccupied patches that are close to occupied patches have a better chance of colonization. Moreover, subpopulations on the brink of extinction can be supplemented by the arrival of dispersers from other subpopulations. When dispersers supplement a declining subpopulation and thereby prevent the subpopulation from going extinct, we call this phenomenon the **rescue effect**. These two mechanisms of colonization and the rescue effect should cause less isolated patches to experience a higher probability of being occupied in a metapopulation.

The effect of patch size and patch isolation on patch colonization has been tested using a series of islands in two lakes in Finland. Researchers surveyed a large number of islands to determine how many individuals of the common shrew (*Sorex araneus*) lived on each. The islands varied in size from about 0.1 to 1000 ha, and they varied in isolation from less than 0.1 to more than 2 km from other islands or from the shore of the lake. The researchers evaluated their

Rescue effect When dispersers supplement a declining subpopulation and thereby prevent the subpopulation from going extinct.

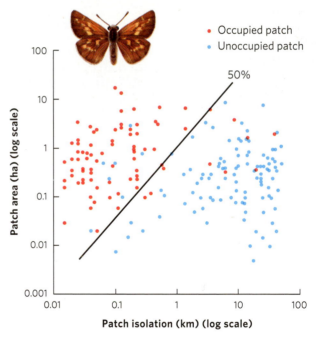

Figure 13.20 Effects of patch size and patch isolation on patch occupancy by the common shrew. The shrew was much less likely to occupy patches that were small or more isolated. Data from A. Peltonen and I. Hanski, Patterns of island occupancy explained by colonization and extinction rates in shrews, *Ecology* 72 (1991): 1698–1708.

over time are quite common in nature and some of these fluctuations exhibit cyclic patterns that are due to delayed density dependence. All populations have some chance of going extinct given enough time, but the smallest populations face a much higher risk of extinction due to demographic and environmental stochasticity. These population dynamics over time can be offset by population dynamics over space as observed in metapopulations in which habitat fragments can experience extinctions and colonizations over time in a manner that allows the entire metapopulation to persist over time. As we will see in "The Recovery of the Black-footed Ferret" on the next page, understanding this interplay of population dynamics over time and space can be critical in bringing species back from the brink of extinction.

data in terms of patch size and patch isolation. As shown in Figure 13.20, they found that shrews were much less likely to occupy the smaller islands and those that were more isolated.

Similar patterns have been found in other species and other areas of the world. In Britain, for example, rabbits feed heavily on calcareous grasslands and the skipper butterflies (*Hesperia comma*) prefer to live in these heavily grazed areas. The grassland patches varied from about 0.01 to 10 ha in area and the distances between patches ranged from 0.02 to 100 km. The researchers examined which patches the butterflies occupied. Examining their data in Figure 13.21, you can see that the largest and least isolated patches were occupied whereas the smallest and most isolated patches were not occupied. As we will see in Chapter 22, patch size and isolation are not only important to the metapopulation of one species, but also to the total number of species that we find on habitat islands.

Throughout this chapter, we have examined how populations vary over time and space. Fluctuations

Figure 13.21 Butterfly occupation of habitat patches that differ in size and isolation. The skipper butterfly in Britain occupies the largest patches of grasslands and those that are the least isolated. The line indicates the combinations of patch area and patch isolation that correspond to 50 percent probability of patch occupancy. Data from C. D. Thomas and T. M. Jones, Partial recovery of a skipper butterfly (*Hesperia comma*) from population refuges: Lessons for conservation in a fragmented landscape, *Journal of Animal Ecology* 62 (1993): 472–481.

ECOLOGY TODAY CONNECTING THE CONCEPTS

THE RECOVERY OF THE BLACK-FOOTED FERRET

Black-footed ferrets. An understanding of how populations fluctuate over time and space has allowed scientists to facilitate the recovery of the black-footed ferret. Photo by Shattil & Rozinski/naturepl.com.

The black-footed ferret (*Mustela nigripes*) is an elusive animal of the North American West. As a nocturnal predator and a member of the weasel family, it feeds almost entirely on prairie dogs. In fact, an average ferret can consume 125 to 150 prairie dogs per year. The ferret's life is so tightly connected to prairie dogs that it lives in old prairie dog burrows within aggregations of prairie dogs known as "prairie dog towns." Historically, the ferret is thought to have possessed a geographic range that spanned from Texas to Arizona and north all the way up to southern Alberta and Saskatchewan. No data exist on historic population sizes, but their numbers were probably in the tens of thousands and likely varied in relation to the fluctuating abundances of prairie dogs. By the 1980s, however, the ferret was thought to have gone extinct in the wild.

The decline of the black-footed ferret occurred over the course of more than a century. With settlement of the West in the 1800s, farmers converted a large portion of the land occupied by prairie dogs to cropland. This loss of habitat substantially reduced the carrying capacity of rangelands for prairie dogs, which in turn caused a large decline in the carrying capacity of ferrets. By the 1920s, a campaign of poisoning prairie dog towns was initiated to reduce the number of prairie dogs, which were thought to compete with cattle and sheep for rangeland plants. The poisons also killed the ferrets, but they also died from canine distemper and sylvatic plague. Canine distemper is a viral disease that is native to North America and deadly to members of the weasel family. Sylvatic plague is a disease introduced to North America from Asia in 1900 and is highly lethal to ferrets. A combination of habitat loss, poisoning, and disease caused prairie dog and ferret populations to decline rapidly.

The situation of the ferret started to receive attention in the 1960s when a single, small population was discovered in South Dakota. In a scramble to save the species from extinction, it was classified as an endangered species in 1967, which meant that efforts would be taken to promote the recovery of the species. Nine animals were brought into captivity for a breeding program. This initial attempt at captive breeding was not very successful; a few offspring were born but none survived beyond a few days. At the same time, the habitat at the South Dakota site continued to become more fragmented. As the ferret populations became smaller, the ferret faced a higher risk of extinction due to stochastic factors. Indeed, the small ferret population in nature went extinct in 1974 and the last ferret in captivity died in 1979. Everyone thought the ferret was extinct.

Two years later, a report that a rancher's dog killed a ferret near the town of Meeteetse, Wyoming, led to the discovery of a population of more than 120 ferrets living in a nearby prairie dog town. However, in 1985, this population was infected by canine distemper and sylvatic plague and began a rapid decline. Because small populations are prone to extinction from stochastic factors, biologists captured the remaining 24 ferrets and brought them into a new captive breeding program in Wyoming. Although only 18 of the 24 ferrets survived, this time the captive breeding program was so successful that additional programs were started in other locations. These locations included several zoos to insure against a stochastic catastrophe at the Wyoming facility.

Within 6 years of bringing ferrets into captivity, biologists began to reintroduce animals into the wild where prairie dogs were abundant. By this time, because the concept of the metapopulation was much better understood, they chose to conduct several reintroductions throughout western North America. This strategy accounted for the possibility that should some subpopulations go extinct due to habitat loss, disease, or other stochastic factors, other subpopulations could persist and serve as future sources of new reintroductions. The recovery program has been an incredible success. Ferrets have been introduced to 19 sites throughout the United States, Canada, and Mexico and their number has grown from the original 18 animals that survived in captivity to more than 1000 animals today. Its success reflects a tremendous effort by scientists working to identify and improve the factors that determine ferret population abundance over time and space.

SOURCES: Roelle, J. E., et al., eds. 2006, Recovery of the black-footed ferret—progress and continuing challenges: U.S.G.S. Scientific Investigations Report 2005–5293, 288 pp.

Robbins, J. 2008. Efforts on 2 fronts to save a population of ferrets. *New York Times*, July 15.

Black-footed ferret recovery program. http://blackfootedferret.org/

SUMMARY OF CHAPTER CONCEPTS

■ **Populations fluctuate naturally over time.** This occurs because density-dependent and density-independent factors can change from year to year and from place to place. The magnitudes of the fluctuations are often related to the ability of a species to resist changes in the environment and to differences in their life histories, including reproductive rates and life span. In some species, the population can overshoot its carrying capacity and then experience a rapid die-off. In populations that have an age structure, fluctuations in population size over time can be detected by disproportionate numbers of individuals in particular age classes. Many species experience cyclic fluctuations in population size.

■ **Density dependence with time delays can cause populations to be inherently cyclic.** Delayed density dependence allows populations to fluctuate above and below their carrying capacity. Delayed density dependence can be incorporated into our population models by having the population's growth rate depend on the population density that occurred at some time in the past. Using these models, we find that the magnitude of the fluctuations depends on the product of the intrinsic growth rate (r) and the time delay (τ). Increasing values of this product causes the population to shift from experiencing no oscillations to damped oscillations to a stable limit cycle. Experiments have confirmed that time delays due to energy reserves or development times between life stages can cause cyclic fluctuations.

■ **Chance events can cause small populations to go extinct.** Smaller populations are more likely to go extinct than large populations. This occurs due to demographic and environmental stochasticity.

■ **Metapopulations are composed of subpopulations that can experience independent population dynamics across space.** Metapopulations exist when a habitat exists in small fragments, either naturally or from human activities. The basic model of metapopulation dynamics informs us that metapopulations persist due to a balance between extinctions in some habitat patches and colonizations that occur in other patches. Although the basic metapopulation model assumes that all patches are equal, in reality, larger patches generally contain larger subpopulations and patches that are less isolated are more likely to be occupied as a result of both the rescue effect and higher rates of recolonization.

REVIEW QUESTIONS

1. Explain why energy reserves can make a population experience delayed density dependence.

2. How does the age structure of a population inform us about population fluctuations over time?

3. In models of delayed density dependence, why does the product $r\tau$ determine the magnitude of population oscillations?

4. Explain why populations can be inherently cyclic.

5. Why does the probability of extinction due to stochastic processes decline as population size increases?

6. Compare and contrast demographic stochasticity and environmental stochasticity.

7. Why is extinction less likely in models of stochastic extinction that include density dependence than in models that exclude density dependence?

8. In a metapopulation, what is the relationship between the amount of dispersal among subpopulations and

the synchrony of fluctuations among subpopulations?

9. If you were trying to save an endangered species that lived in a metapopulation, how might you try to increase the proportion of occupied patches?

10. In the basic model of metapopulation dynamics, how might the rescue effect alter both the probability of colonization and the probability of extinction?

PRACTICING ECOLOGY: VOLES AND BERRIES

Charles Krebs at the University of British Columbia and his colleagues study the causes and consequences of rodent population fluctuations in North American forests. Photo by Alice Kenney.

Small rodents such as mice, squirrels, and moles are a familiar sight in the forest. They serve as prey for a variety of predators, and they consume berries, nuts, and other seeds. They also serve as vectors for seed dispersal since they frequently hide the food they collect in the ground where it will germinate if left for too long. Because small rodents are vital to forest ecosystems, ecologists want to understand the factors that determine the year-to-year variation in rodent populations. What factors are responsible for variation in the major species of North American rodents?

The northern red-backed vole (*Myodes rutilus*) is a small rodent with a wide distribution throughout northern Canada, Alaska, Scandinavia, and Russia. This vole lives for one year and produces multiple litters of offspring during the spring and summer. While the northern red-backed vole consumes many different food items, it prefers berries. During the winter, the vole relies on stored energy from the food it consumes during the summer and fall. We might therefore hypothesize that the abundance of berries consumed during the summer and fall determines vole population size during the following year.

Charles Krebs from the University of British Columbia and his colleagues tested this hypothesis by tracking the abundance of berries and voles in the southwestern Yukon between 1997 and 2009. During July of each year—before berries ripened and were consumed—the researchers estimated the abundance of several berry species including black crowberry (*Empetrum nigrum*), lingonberry (*Vaccinium vitis-idaea*), and kinnikinnick (*Arctostaphylos uva-ursi*). They estimated berry abundance by counting all berries in 0.32 m² quadrats placed on the ground. Each August they estimated the abundance of voles by spending 3 days catching voles at four different sites around Kluane Lake. At each site, they placed 100 traps in a grid and baited them with oats. You can view their data in the table.

THE MEAN NUMBER OF BERRIES PER QUADRAT AND THE MEAN NUMBER OF VOLES PER HECTARE, FROM 1997 TO 2009

YEAR	CROWBERRY	LINGONBERRY	KINNIKINNICK	VOLES
1997	34.3	0.0	4.7	6.2
1998	13.2	1.0	3.5	13.6
1999	24.0	2.4	6.1	6.1
2000	13.5	16.2	4.7	2.6
2001	52.7	48.5	14.8	11.1
2002	30.5	29.4	11.7	28.9
2003	30.5	17.8	18.7	0.1
2004	47.6	15.2	14.8	1.0
2005	18.2	22.9	14.2	18.0
2006	17.3	24.6	22.1	4.2
2007	43.3	20.1	16.5	2.9
2008	75.0	17.9	14.9	15.2
2009	39.3	14.5	7.2	28.6

GRAPH THE DATA

1. Create a line graph with year on the *x* axis and abundance of each berry species on the *y* axis.

2. Create a second line graph with year on the *x* axis and vole abundance on the *y* axis.

INTERPRET THE DATA

1. Which berry species exhibits year-to-year variation in abundance that follows the pattern of vole abundance?

2. Does the abundance of voles appear to have a time-lag response to the abundance of berries? Given what you know of the vole breeding patterns, explain why delayed-density dependence may or may not be present.

3. Does the data support the hypothesis that food resources in the fall predict vole population size during the following summer?

SOURCES: Gilbert, B.S. and C.J. Krebs. 1991. Population dynamics of Clethrionomys and Peromyscus in southwestern Yukon 1973–1989. *Holarctic Ecology* 14: 250–259.

Krebs, C.J., K. Cowcill, R. Boonstra, and A.J. Kenney. 2010. Do changes in berry crops drive fluctuations in small rodents in the southwestern Yukon? *Journal of Mammalogy* 91: 500–509.

CHAPTER 14

Predation and Herbivory

A Century-long Mystery of the Lynx and the Hare

For centuries, naturalists, hunters, and trappers have noticed that populations of many species often experience large fluctuations, and that some species fluctuate at regular intervals. In 1924, the ecologist Charles Elton drew attention to regular population fluctuations in many species of high-latitude animals in Canada, Scandinavia, and Siberia. In particular, he focused on snowshoe hares (*Lepus americanus*) and Canada lynx (*Lynx canadensis*). Elton examined data that had been compiled from the Hudson's Bay Company, a Canadian firm that had been purchasing pelts from trappers for more than 70 years. He assumed that the number of purchased pelts over time reflected the abundance of the two species. Elton and his fellow ecologists were fascinated by regular cycles of high and low density among lynx and hare populations that occurred approximately every 10 years. The 10-year cycles in the abundance of lynx and hares were clear, but the mechanisms causing these cycles have been debated for nearly a century.

> "The 10-year cycles in the abundance of lynx and hares were clear, but the mechanisms causing these cycles have been debated for nearly a century."

There have been numerous hypotheses for these 10-year cycles. When Elton wrote his classic paper, some biologists hypothesized that animals possessed a "physiological rhythm" that caused both the lynx and the hare to reproduce abundantly in some years and sparingly in others. Elton rejected this hypothesis because it was very unlikely that such a rhythm would be synchronized for all individuals of different ages and for all individuals across large regions. Instead, he favoured an explanation related to the 9- to 13-year cycle of sunspots, which are periods of increased solar activity. If the cycle of sunspots could substantially affect the climate and, therefore, the growing conditions of the plants that hares consumed, it could explain the cycles of the hares. The lynx cycle, which occurred about 2 years later than the hare cycle, was thought to reflect the fact that lynx primarily consume hares. When hares are abundant, lynx have more food and therefore reproduce more in subsequent years, but when hares are rare, the lynx reproduce poorly and many of them starve, which causes the lynx population to decline.

Since Elton's original work, ecologists have determined that although the sunspot cycle is similar in length to the hare cycle, it has never closely matched the timing of the hare cycle. Nor have they been able to find a climate-driven mechanism that connects the sunspot cycle to the hare cycle. With these hypotheses eliminated, researchers turned their attention to competition and predation. For many decades a lively debate focused on the possibility that the hare cycles were caused by the hares exceeding their carrying capacity, which could explain the observation that hare reproduction declines as the hare population grows. Another hypothesis suggested that lynx predation caused the cycles. When lynx were rare, the hares survived better. As lynx became more numerous, they began consuming hares faster than the hares could reproduce, thereby causing the hare population to decline.

Canada lynx and snowshoe hare. For nearly 100 years, ecologists have been examining the regular fluctuations in the populations of these species to determine the causes. Photo by Tom & Pat Leeson.

It seemed impossible to determine the answer without conducting some experiments. From 1976 to 1985, a large experiment manipulated the presence or absence of supplemental food for the hares. Although the supplemental food increased the carrying capacity for the hare population, it still cycled in synchrony with populations that were not fed. This suggested that the hare population does not decline due to a lack of food. However, if the hares do not experience a lack of food, why does their rate of reproduction decline at higher densities?

In a subsequent experiment, researchers built large fences so they could manipulate the presence or absence of supplemental food and the presence or absence of lynx predation. Both excluding lynx and adding food increased the peak population size of the hares, but the hare populations still cycled. However, the fence that excluded the lynx did not exclude other predators, including owls and hawks, which continued to kill hares. Among the hares that died, more than 90 percent died from predation and few from starvation, which further confirmed that predation rather than food availability contributed to the cyclical decline of hares.

A new insight came in 2009 when researchers discovered that the declining rate of hare reproduction under high hare densities is caused by high densities of predators, which induce high levels of stress in the hares. Eventually, the stress of the predation threat becomes so high that the hares experience reduced reproduction. Once the hare population declines, there are many fewer predators so the hare stress level is much reduced and hare reproduction returns to a high level. In short, while the abundance of food can affect the number of hares in the population, it appears that the lynx-hare population cycles can be attributed to a combination of direct predation and the indirect effects of predator stress that cause reduced hare reproduction.

The century-long investigation of the lynx-hare cycles illustrates that consumers and the resources that they consume can interact in complex and interesting ways. In this chapter, we will examine how predators and herbivores can affect the populations of the species they consume, including how the abundance of predators and prey populations can cycle over time, how consumers catch their prey, and how prey defend themselves.

SOURCES: C. S. Elton, Periodic fluctuations in the numbers of animals: Their causes and effects, *British Journal of Experimental Biology* 2 (1924): 119–163.

M. J. Sheriff, The sensitive hare: Sublethal effects of predator stress on reproduction in snowshoe hares, *Journal of Animal Ecology* 78 (2009): 1249–1258.

C. J. Krebs, Of lemmings and showshoe hares: The ecology of northern Canada, *Proceedings of the Royal Society B.* 278 (2011): 481–489.

CHAPTER CONCEPTS

■ Predators and herbivores can limit the abundance of populations.

■ Populations of consumers and consumed populations fluctuate in regular cycles.

■ Predation and herbivory favour the evolution of defences.

Most species consume resources and serve as a resource for other species to consume. For example, plants and algae consume nutrients, water, and light; these resources allow plants and algae to photosynthesize and grow. While plants and algae are alive, they are consumed by herbivores, parasites, and pathogens. After they die, these producers are consumed by detritivores and decomposers. Similarly, animals consume a wide variety of other organisms at the same time that they are subject to consumption by carnivores, parasites, and pathogens. After they die, animals are consumed by scavengers, detritivores, and decomposers. As you can see, a tremendous number of interactions occur among species in nature. These interactions, critical to the composition of species in different communities, are the subjects of the next four chapters.

In this chapter we will focus on interactions between predators and their prey and between herbivores and producers. We will examine the

Figure 14.1 Lizard predation on spiders. (a) In the Bahamas, researchers introduced a total of 20 spiders to each of five small islands with predatory lizards and to five small islands without lizards. **(b)** Spiders were introduced to each island on two dates, indicated by the red arrows. The spiders remained rare or absent on islands with lizards but increased 10-fold on islands without lizards. Error bars are standard errors. Data from T. W. Schoener and D. A. Spiller, Effect of predators and area on invasion: An experiment with island spiders, *Science* 267 (1995): 1811–1813. Photo by Jason J. Kolbe.

(a)

(b)

conditions under which predators and herbivores can limit the population sizes of the species they consume. We will also look at models of predators and herbivores to help us understand how populations of these consumers fluctuate in relation to the populations of the species they consume. We will conclude the chapter by exploring how predators and herbivores have favoured the evolution of defences in prey and plants.

Predators and herbivores can limit the abundance of populations

Are populations limited primarily by what they eat or by what eats them? Studies of predation and herbivory attempt to answer this question by looking at whether or not predators and herbivores reduce the size of prey and producer populations below the carrying capacities set by resources. Understanding these relationships is of great practical concern to those interested in the management of crop pests, game populations, and endangered species. It also has far-reaching implications about the interactions among species that share resources—knowledge that helps us to understand the structure of ecological communities.

PREDATORS

As we noted with lynx and hares at the beginning of this chapter, ecologists have long been fascinated by predator–prey interactions. For the past century,

researchers have repeatedly observed large effects of predators on prey populations in nature and in manipulative experiments. For example, researchers conducted a survey of 93 Caribbean islands and found that the smaller islands contained spiders whereas the larger islands contained spiders and lizards, which are predators of the spiders. When they compared the density of spiders on islands with and without the predatory lizards, they found that spiders were about 10 times more abundant on islands without the lizards.

To test whether lizard predation was the cause of the lower spider densities, the researchers conducted a manipulative experiment on islands in the Bahamas that ranged in size from about 200 to 4000 m². They selected five islands with brown anole lizards (*Anolis sagrei*) and five islands without lizards. On these 10 islands, they introduced a species of orb-weaving spider (*Metepeira datona*) as a prey that was not originally present on the islands. They introduced the spiders on two occasions, for a total of 20 spiders per island. Then they surveyed the spider populations 4 days after each introduction and detected lower spider densities on the islands containing lizards. This suggested that predation from lizards had already started to affect the spider populations. Over the next 5 years, the spider populations that lived on islands without lizards became 10 times more abundant, as you can see in Figure 14.1. In contrast, the spider

Figure 14.2 Brown tree snake. The introduction of this snake to the island of Guam led to the decline and local extinction of birds, bats, and lizards. Photo by John Mitchell/Science Source.

populations living with lizards became either rare or extinct, which confirmed that predatory lizards reduce the densities of spiders.

Sometimes the introduction of a predator is the result of an accident. When a species is introduced to a region of the world where it has not historically existed, we say that the species is *introduced, exotic,* or *non-native.* If the introduced species spreads rapidly and negatively affects other species, we say that the species is *invasive.* On the island of Guam, for example, many species of birds, bats, and lizards lived without predatory snakes for thousands of years. Sometime after World War II, however, the brown tree snake (*Boiga irregularis*) was accidentally introduced to Guam by ships carrying supplies (Figure 14.2). While the snake was native to the South Pacific, the animals on Guam had no evolutionary history of living with snakes and many had no defences against them. Over the next 20 years, as the population of the brown tree snake grew exponentially, it had a significant and devastating effect on the island fauna. Today, Guam has experienced sharp declines or extinctions in nine species of forest-dwelling birds, all three species of bats, and several species of lizards.

As we mentioned in Chapter 1, parasitoids represent a unique type of predator that lives within and consumes the tissues of a living host, eventually killing it. Like many other predators, parasitoids also can limit the abundance of their prey. This can be seen in an example of wasps and scale insects. The California red scale insect (*Aonidiella aurantii*) is a worldwide pest in citrus orchards. It feeds on the leaves and fruits of the trees, which causes a great deal of damage and makes the fruit unmarketable. Fortunately, a small species of parasitoid wasp (*Aphytis melinus*) is able to control the abundance of the red scale insect by laying eggs inside the scale insects and ultimately killing them.

To demonstrate the magnitude of this control, researchers in California mimicked an insect outbreak by adding large numbers of scale insects to four trees. Ten other trees with typical low numbers of scale insects served as controls. You can see the results of this experiment in Figure 14.3. Shortly after adding large numbers of scale insects, there was a significant increase in the number of juvenile and adult wasps. As the wasps became more abundant, the population of scale insects began a rapid decline. Within a few months, the large population of scale insects was reduced to the same level found on trees that had never received an experimental addition of scale insects. Although the wasps were not able to completely eliminate the scale insects, they held the population of scale insects to a level that minimally affected the citrus crops.

MESOPREDATORS

Two levels of predators often exist in ecological communities: *mesopredators* and *top predators.* **Mesopredators** such as coyotes, weasels, and feral cats are relatively small carnivores that consume herbivores. In contrast, **top predators** such as wolves, mountain lions, and sharks typically consume both herbivores and mesopredators. Throughout history, top predators have interfered with human activities such as ranching, farming, hunting, and fishing. To protect our livelihood, we have reduced or eliminated the world's top predators, but this has caused unintended consequences. In 2009 researchers estimated that the decline of the top predators in North America had allowed 60 percent of all mesopredators to expand their geographic ranges.

Mesopredators Relatively small carnivores that consume herbivores.

Top predators Predators that typically consume both herbivores and mesopredators.

California red
scale insects

Parasitoid wasp

(a)

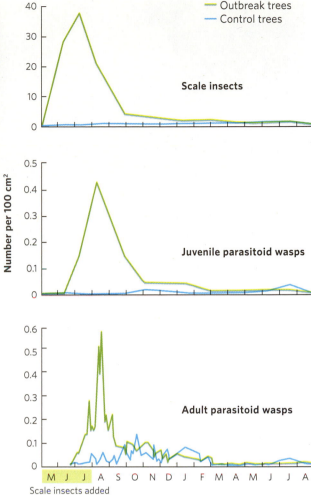

Scale insects added

(b)

Month

Figure 14.3 Effects of a parasitoid on populations of red scale insects. (a) The California red scale, a tiny insect with a hard shell, feeds on the sap of citrus trees and other plants. The parasitoid wasp lays its eggs under the hard shell of the scale insect. When the eggs hatch, the larvae slowly consume the scale insect. (b) After introducing large numbers of scale insects to trees—during the months indicated by yellow shading on the *x* axis—the scale insect population grew rapidly. Soon after the addition of the scale insects, the number of juvenile and adult parasitoid wasps began to rapidly increase. This increase in parasitoid wasps caused a subsequent rapid decline in scale insects. Data from W. Murdoch et al., Host suppression and stability in a parasitoid-host system: Experimental demonstration, *Science* 309 (2005): 610–613.

The expansion in the range and abundance of mesopredators has had dramatic effects on the prey they consume. For example, the reduction of shark populations in the Atlantic Ocean from overharvesting has led to an increase in the cownose ray (*Rhinoptera bonasus*), a major mesopredator that consumes bay scallops (*Argopecten irridians*). This has caused a large reduction in bay scallops. In some cases, researchers have found that the benefits to humans of removing a top predator are much smaller than the damage inflicted by a mesopredator that becomes more abundant after the top predator declines. In Australia, for example, there has long been a campaign to remove dingoes and feral dogs because they kill sheep. However, removing these top predators has caused an increase in red foxes (*Vulpes vulpes*), which also eat sheep. This increase in red foxes has led to more than triple the loss in sheep that occurred when dingoes and feral dogs helped to control the red fox population.

HERBIVORES

Like predators, herbivores can have substantial effects on the species they consume. One of the classic examples is the control of the prickly pear cactus in Australia. The prickly pear cactus is a group comprised of several different species that are native to North and South America. In the nineteenth century, many of these species were brought to Australia for a variety of reasons, including use as ornamental plants and as "living fences" for pastures. The cacti rapidly spread across the continent to the point that they dominated thousands of hectares of pasture and rangeland. To combat the spread of the cactus, biologists in the 1920s collected cactus moths (*Cactoblastis cactorum*) in South America and introduced the moths to Australia. These moths are natural herbivores of the cactus; the caterpillar stage of the moth consumes a portion of the cactus and the injuries it causes allow pathogens to infect the plants. As you can see in the before and after photographs in

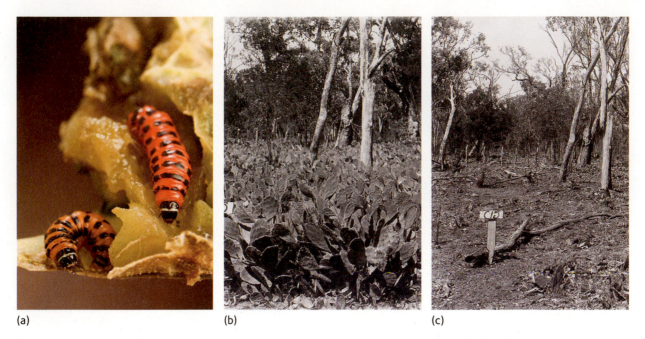

(a) (b) (c)

Figure 14.4 The prickly pear cactus in Australia. Following the import of the cactus from South America to Australia, its numbers increased dramatically. **(a)** To reduce the abundance of the cactus, the cactus moth was introduced to Australia from South America. **(b)** A site in Queensland, Australia, in 1926 prior to the introduction of the moth. **(c)** The same site 3 years after the introduction of the moth. Photos courtesy of (a) USDA/ARS, photographer Peggy Greb; (b) The Alan Fletcher Research Station, Department of Lands, Queensland; and (c) the Queensland Government.

Figure 14.4, the moths quickly reduced the cactus to very low abundances. While the moth has not completely eradicated the cactus because the cactus is able to disperse to moth-free areas, today the prickly pear cactus exists only in small pockets around Australia.

Herbivorous insects have been used in similar ways in North America. In California, a plant known as Saint-John's-wort, which is toxic to livestock, was accidentally introduced from Europe in the early 1900s. It then spread to almost all mainland U.S. states and to all Canadian provinces except Alberta and Saskatchewan. In the 1950s, biologists decided to introduce a leaf-feeding beetle (*Chrysolina quadrigemina*) that consumed Saint-John's-wort in Europe. After introduction of the beetle, the weed rapidly declined in abundance, as you can see in Figure 14.5. Biologists estimate that the beetle has now eliminated 99 percent of the Saint-John's-wort population in North America.

The effects of herbivores are easily observed. For example, when domesticated animals such as cattle or sheep are raised at very high densities, little plant life remains. Similarly, deer and geese can consume large amounts of plants. Their impact can be demonstrated by fencing off areas to prevent grazing; the fenced areas typically have a greater total mass, or *biomass*, of plants and a higher composition of plants that are

preferred by the herbivores (Figure 14.6). The plant species that remain in the areas where grazing occurs are those that the herbivores prefer not to eat.

Herbivore effects also occur in aquatic habitats. In rocky shore communities, for example, sea urchins

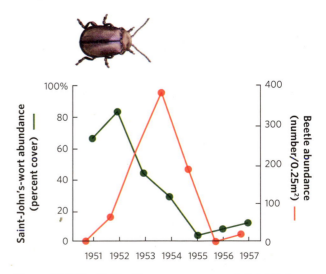

Figure 14.5 Beetle herbivory on common Saint-John's-wort. Following the introduction of leaf-feeding beetles, the beetle population initially increased and the plant began a rapid decline. Once the plant population declined, the beetle population also declined. Data from C. B. Huffaker and C. E. Kennett, A ten-year study of vegetational changes associated with biological control of Klamath weed, *Journal of Range Management* 12 (1959): 69–82.

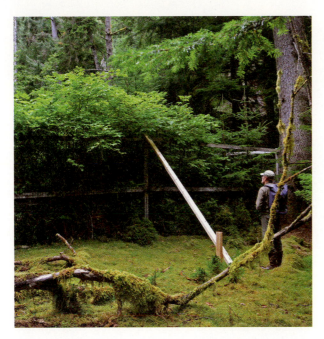

Figure 14.6 Fencing out deer. In this deer enclosure in Gwaii Haanas National Park Reserve and Heritage Site, Haida Gwaii, B.C., the long-term fencing of an area to prevent deer herbivory has allowed a much greater abundance of plants to grow. Photo by Jean-Louis Martin RGIS/CEFE-CNRS.

control populations of algae. When researchers conduct experiments in which they remove sea urchins from an area, the biomass of algae increases rapidly while the species composition of the algae also changes. In the presence of sea urchins, the remaining algae consist mostly of those species the sea urchins do not like and those that can withstand intense grazing. When sea urchins are removed, however, more palatable species such as the large brown algae become more abundant and can shade

out the other species of algae. Such studies demonstrate that the influence of herbivores on the abundance of producers affects the species composition of the entire community.

Populations of consumers and consumed populations fluctuate in regular cycles

In the previous chapter, we discussed how populations can fluctuate over time and space, and we saw that some populations experience population cycles. At the beginning of this chapter we introduced a study of population cycles for lynx and snowshoe hares. You can see the data from this study in Figure 14.7. The hare and lynx populations both experienced cycles of 9 to 10 years, with the lynx cycles lagging about 2 years behind the hare cycles.

It turns out that other large herbivores in the boreal and tundra regions of Canada—such as muskrat, ruffed grouse, and ptarmigan—also have population cycles of 9 to 10 years. Smaller herbivores, such as voles, mice, and lemmings, tend to have 4-year cycles. Studies of predators in these regions have revealed that some predators—including red foxes, marten, mink, goshawks, and horned owls— feed on larger herbivores and have long population cycles. In contrast, other predators— including Arctic foxes (*Vulpes lagopus*), rough-legged hawks (*Buteo lagopus*), and snowy owls (*Bubo scandiacus*)—feed on small herbivores and have short population cycles. The close synchrony of population cycles between consumers and the populations they consume suggests that these oscillations are the result of interactions between them.

Figure 14.7 Cyclic fluctuations in the populations of snowshoe hares and lynx. Records from the Hudson's Bay Company of pelts purchased from trappers demonstrate that the populations of hares and lynx went through cycles approximately every 10 years. Data from D. A. MacLulich, *Fluctuations in the number of the varying hare* (*Lepus americanus*), University of Toronto Studies, Biological Series No. 43 (1937).

To understand the mechanisms underlying the predator–prey cycles, it is useful to examine them in the context of population models.

CREATING PREDATOR–PREY CYCLES IN THE LABORATORY

During the early twentieth century, biologists became interested in using predators and pathogens to control populations of crop and forest pests. One of the leading researchers in this effort was Carl Huffaker, a biologist at the University of California at Berkeley who pioneered the biological control of crop pests. Huffaker sought to understand the conditions that cause predator and prey populations to fluctuate. He chose two species of mites that lived on citrus trees: the six-spotted mite (*Eotetranychus sexmaculatus*) was the prey and the western predatory mite (*Typhlodromus occidentalis*) was the predator. In a series of experiments, he established populations on large trays that contained oranges—which were both habitat and food for the prey—with rubber balls interspersed among the oranges, as illustrated in Figure 14.8. On each tray, he varied the number and distribution of the oranges.

In most experiments, Huffaker initially added 20 female prey per tray and then introduced two female predators 11 days later; because both species reproduce parthenogenetically, no males were added. When the prey were introduced to trays without predators, the prey population levelled off at between 5500 and 8000 mites. When predators were added, the predator population increased rapidly and soon wiped out the prey population. Without prey to feed on, though, predators quickly became extinct. However, it took longer for predators and prey to go extinct if the oranges were spread far apart from each other because it took longer for the predators to locate the prey. In short, predators and prey could not coexist over time.

Huffaker reasoned that if predator dispersal could be further impeded, the two species might be able to coexist. To accomplish this, he introduced barriers to predator dispersal. The predatory mites disperse by walking, but the prey mites use a silk line that they spin to float on wind currents. Based on these differences, Huffaker modified his trays to give the prey a dispersal advantage by placing a mazelike pattern of Vaseline barriers among the oranges to slow the dispersal of the walking predators. He also placed vertical wooden pegs throughout the trays for the six-spotted mites to use as jumping-off points. This arrangement produced a series of three population cycles during the 8-month experiment, as depicted in Figure 14.9. The distribution of predators and prey throughout the trays continually shifted as the prey, which were on their way to extermination on one orange, recolonized another orange and stayed one step ahead of their predators. In short, Huffaker created a metapopulation in the laboratory.

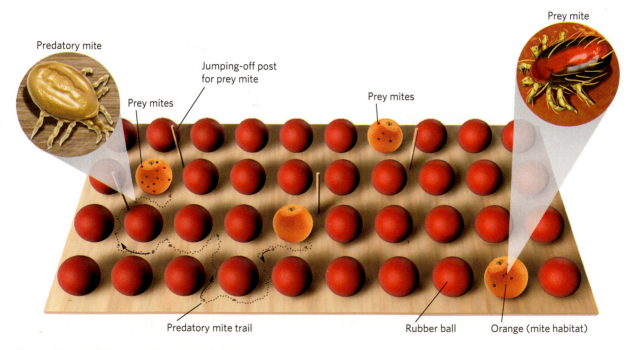

Figure 14.8 Huffaker's predator–prey lab experiment. To determine the factors that cause predator–prey cycles, Huffaker manipulated the number and distribution of oranges in a tray filled with oranges and rubber balls. He also added posts that only the prey species were able to climb. The prey drifted from the posts to colonize new oranges. Because the predators had to walk from one orange to another and avoid Vaseline barriers, the prey stayed one step ahead of the predators.

Figure 14.9 Predator–prey cycles in laboratory mites. When the prey species was given a dispersal advantage, it could colonize predator-free oranges and avoid going extinct. The cycling of the predator population lagged behind the prey population because it required more time to disperse to oranges containing prey and it required time to reproduce after finding prey. Data from C. B. Huffaker, Experimental studies on predation: Dispersion factors and predator-prey oscillations, *Hilgardia* 27 (1958): 343–383.

Huffaker's experiment demonstrated that predators and prey could not coexist in the absence of suitable refuges for the prey. However, the predator and prey populations were able to coexist through time in a spatial mosaic of suitable habitats that provided a dispersal advantage to the prey. Two kinds of time delays caused the populations to cycle: one was the result of predators dispersing more slowly between food patches than their prey, and the other was the result of time needed for predator numbers to increase through reproduction. From this we can conclude that stable population cycles can be achieved when the environment is complex enough that predators cannot easily find scarce prey.

MATHEMATICAL MODELS OF PREDATOR–PREY CYCLES

Even before Huffaker's laboratory experiments on predator–prey cycles, mathematicians Alfred Lotka and Vito Volterra were developing models of predator–prey interactions. The **Lotka-Volterra model** incorporates oscillations in the abundances of predator and prey populations and shows predator numbers lagging behind those of their prey. It does this by calculating both the rate of change in the prey population and the rate of change in the predator population.

Let's begin with the prey population. Following the convention of our population models in Chapter 12, we will designate the number of prey as N. We will then designate the number of predators as P. In words, the growth rate of the prey population depends on the rate of individuals being added to the prey population minus the rate of individuals being killed by predators:

$$\frac{dN}{dt} = rN - cNP$$

The first term in this equation (rN) represents the exponential growth of a prey population based on the

intrinsic growth rate (r), as we saw in Chapter 12. For simplicity, this term does not include density dependence. The second term (cNP) represents the loss of individuals due to predation. The model assumes that the predation rate is determined by the probability of a random encounter between predators and prey (NP) and the probability of such an encounter leading to the prey's capture (c) (think of c for "capture efficiency").

Next we turn to the predator population. The equation for the predator population is similar to that for the prey population in that it has two terms, one term that represents the birth rate of the predator population and a second term that represents the death rate of the predator population:

$$\frac{dP}{dt} = acNP - mP$$

The first term in the equation ($acNP$) represents the birth rate of the predator population. It is determined by the number of prey consumed by the predator population (cNP) multiplied by the efficiency of converting consumed prey into predator offspring (a). The second term (mP) represents the death rate of the predator population and is determined by the per capita mortality rate of predators (m) multiplied by the number of predators (P).

Changes in the Prey Population

We can use the Lotka-Volterra equations to determine the conditions that allow the prey population to be stable. By definition, a population is stable when its rate of change is zero. We can write this as:

$$0 = rN - cNP$$

Lotka-Volterra model A model of predator–prey interactions that incorporates oscillations in the abundances of predator and prey populations and shows predator numbers lagging behind those of their prey.

We can rearrange this as:

$$rN = cNP$$

This indicates that the prey population becomes stable when the addition of prey (rN) is balanced by the consumption of prey (cNP). We can further simplify this equation as:

$$P = r \div c$$

This indicates that the prey population will be stable when the number of predators equals the ratio of the prey's growth rate and the predator's capture efficiency.

Now that we know the conditions that make the prey population stable, we can also explore the conditions that cause the prey population to increase or decrease. The prey population will increase when the addition of prey (rN) exceeds the consumption of prey (cNP), which we can write as:

$$rN > cNP$$

We can rearrange this inequality as:

$$P < r \div c$$

This inequality represents the number of predators that the population of prey can support and still increase. This number is higher when the growth potential of the prey population (r) is higher or when the predators are less efficient at capturing prey (c). Using the same logic, the prey population will decrease whenever

$$P > r \div c$$

Changes in the Predator Population

We can also use these equations to understand the conditions that make the predator population remain stable, increase, or decrease. Once again, the population is stable when the rate of change is zero:

$$0 = acNP - mP$$

We can rearrange this as:

$$acNP = mP$$

This indicates that the predator population becomes stable when the addition of predators ($acNP$) is balanced by the mortality of predators (mP). We can further simplify this equation as:

$$N = m \div ac$$

Given that this is the condition needed for a stable predator population, we can also determine the conditions required for predator populations either to increase or to decrease. The predator population will increase when the addition of predators exceeds the mortality of predators, which we can write as:

$$acNP > mP$$

We can rearrange this as:

$$N > m \div ac$$

This inequality represents the number of prey required to support the growth of the predator population. This number is higher when the death rate of predators (m) is higher, and it is lower when predators are more efficient at capturing prey (c) and converting them into offspring (a). Using the same logic, the predator population will decrease whenever

$$N < m \div ac$$

Trajectories of Predator and Prey Populations

Knowing the conditions under which predator and prey populations increase, decrease, or remain stable helps us understand why predator and prey populations sometimes cycle. Figure 14.10a plots the abundance of both populations. We can draw a horizontal line at the point where $P = r \div c$, which is the number of predators associated with a stable prey population. This line is called the **equilibrium isocline**, or **zero growth isocline**, for the prey because it indicates the points at which a population is stable. At any combination of predator and prey numbers in the region below the equilibrium isocline, there are relatively few predators and the prey population increases. In the region above this line, the prey population decreases because predators remove them faster than they can reproduce.

We can also plot the equilibrium isocline for the predator population, as shown in Figure 14.10b. In this graph, we draw a vertical line at the point where $N = m \div ac$, which is the number of prey that causes the predator population to be stable. Any combination of predator and prey numbers that lies in the region to the right of this line allows the predator population to increase because there is an increased abundance of prey to consume. In the region to the left, the predator population decreases because there are not enough prey available.

We can now combine our understanding of how predator and prey populations change in abundance by considering the trajectory of both populations

---

Equilibrium isocline The population size of one species that causes the population of another species to be stable. *Also known as* Zero growth isocline.

(a)

(b)

Figure 14.10 Equilibrium isoclines for prey and predator populations. **(a)** The prey population is stable when the number of predators is equal to $r \div c$. A higher number of predators causes the prey population to decrease, whereas a lower number of predators causes the prey population to increase. **(b)** The predator population is stable when the number of prey is equal to $m \div ac$. A higher number of prey causes the predator population to increase, whereas a lower number of prey causes the predator population to decrease.

simultaneously, which we call a **joint population trajectory**.

Figure 14.11 traces the path of the joint population trajectory. Beginning with the lower right region, predators and prey both increase, and their joint

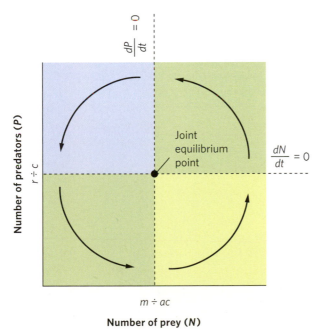

Figure 14.11 A joint population trajectory of predator and prey populations. In the upper left region, predator and prey populations are both declining. In the lower left region, predators have declined, which allows the prey to increase. In the lower right region, a further increase in prey allows the predator population to begin to increase. In the upper right region, predators have increased to a point that they cause a decline in the prey population.

population trajectory moves up and to the right. In the upper right region, prey are still abundant enough that predators can increase, but the increasing number of predators depresses the prey population. Accordingly, the joint population trajectory moves up and to the left. In the upper left region, the continued decline in prey causes the predator population to decline, so the trajectory moves down and to the left. In the lower left region, the continued decline in predators allows the prey population to start increasing, which causes the trajectory to move down and to the right and completes the cycle. Together, the trajectories in the four regions define a counterclockwise cycling of predator and prey populations.

At the centre of Figure 14.11, you can see the **joint equilibrium point**, which is the point at which the equilibrium isoclines for predator and prey populations cross. The joint equilibrium point represents the combination of predator and prey population sizes that falls exactly at this point and will not change over time. If either of the populations strays from the joint equilibrium point, the populations will oscillate around the joint equilibrium point rather than returning to it.

Based on the joint population trajectory of the Lotka-Volterra model, we can plot the changes in

Joint population trajectory The simultaneous trajectory of predator and prey populations.

Joint equilibrium point The point at which the equilibrium isoclines for predator and prey populations cross.

sizes of the two populations over time, as shown in Figure 14.12. This visualizes how the two populations cycle and reveals that the prey population remains one-fourth of a phase ahead of the predator population. Although some populations cycle due to time delays, as we discussed in Chapter 13, the predator–prey equations do not; the predator–prey equations are continuous equations in which population changes are immediate. Instead, the cycling is the result of each population responding to changes in the size of the other population. This pattern is reminiscent of the lynx population discussed at the beginning of this chapter that follows the fluctuations in the hare population.

FUNCTIONAL AND NUMERICAL RESPONSES

The Lotka-Volterra model provides an explanation for population cycles that relies on a very simplified version of nature. As we have already discussed, it does not include time delays or density dependence, and it does not incorporate the real foraging behaviour of most predators. To get a more realistic

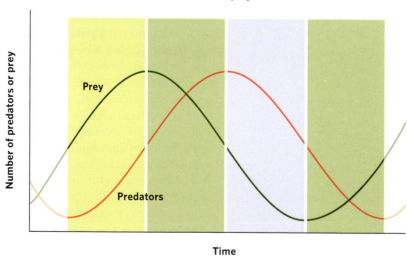

Figure 14.12 Predator–prey oscillations. The Lotka-Volterra model illustrates how the abundance of predator and prey populations cycle. With a decline in predators, prey can increase in abundance. The increase in prey then provides more food for predators; this allows predators to increase their population size through reproduction. As predators grow more numerous, they start to kill prey at such a high rate that the prey population begins to decline. A decline in available prey reduces the survival and reproduction of the predator population, which starts to decline. Over time, both populations cycle up and down in abundance, with predator abundance lagging behind prey abundance.

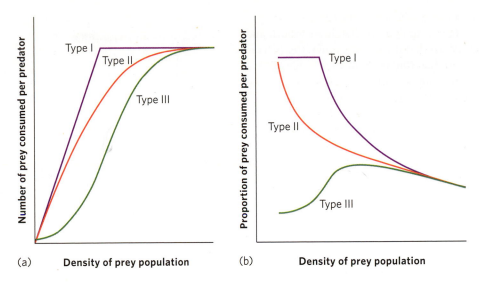

Figure 14.13 Functional responses of predators. (a) When we consider the number of prey consumed per predator, we see that a type I response exhibits a linear relationship between the density of the prey population and the number of prey a predator consumes until eventually the predator is satiated. In contrast, a type II response exhibits a slowing rate of prey consumption as prey density increases; the slowing rate is due to an increase in the time spent handling the additional prey. Eventually, the predator achieves satiation. A type III response also shows a slowing rate of consumption with increased prey density. In addition, this response exhibits the effect of a predator learning a search image for prey as the prey density increases. **(b)** When we consider the proportion of prey consumed, we see that the type I response results in a constant proportion of prey being consumed—prior to satiation—as prey density increases. A type II response results in a decreasing proportion of prey being consumed. Finally, a type III response initially causes an increased proportion of prey consumed, followed by a decrease.

picture of predator and prey relationships, we have to consider the *functional response* and the *numerical response* of predators. As we will see, both of these responses help to stabilize the cycling of predator and prey populations.

The Functional Response

The relationship between the density of prey and an individual predator's rate of food consumption is known as the **functional response** of the predator. There are three potential categories of functional responses, as illustrated in Figure 14.13. For each of these categories, we can examine the *number of prey* consumed by each predator, shown in Figure 14.13a, or the *proportion of prey* consumed by each predator, shown in Figure 14.13b. A key point to remember is that whenever prey density increases and a predator can consume a higher proportion of those prey, the predator has the ability to regulate the growth of the prey population.

A **type I functional response**, indicated by the purple line, occurs when a predator's rate of prey consumption increases in a linear fashion with an increase in prey density until satiation occurs. As shown in Figure 14.13a, this means that increases in prey density result in an ever-increasing number of prey consumed by a predator until the predator becomes satiated and can consume no additional prey.

Some species of predators, such as web-building spiders that catch an increasing number of prey as the density of prey increases, exhibit a type I functional response. As you can see in Figure 14.13b, this means that as prey density increases, each predator continues to consume a constant proportion of the prey up until the point of predator satiation. Once the prey become so dense that predator satiation occurs, the predators consume a continuously decreasing proportion of the prey. This is the functional response used by the Lotka-Volterra model that we discussed.

The **type II functional response**, shown by the orange line, occurs when the number of prey consumed slows as prey density increases and then plateaus when satiation occurs. The number of prey consumed slows because as predators consume more prey, they must spend more time handling the prey.

Functional response The relationship between the density of prey and an individual predator's rate of food consumption.

Type I functional response A functional response in which a predator's rate of prey consumption increases in a linear fashion with an increase in prey density until satiation occurs.

Type II functional response A functional response in which a predator's rate of prey consumption begins to slow down as prey density increases and then plateaus when satiation occurs.

For example, when a pelican catches a fish, it must take the time to manipulate the fish in its mouth and position the fish so that it will slide down the pelican's throat. The more fish that the pelican catches, the more time it must spend handling fish, which leaves less time available to hunt for fish. Eventually, because the predator is spending so much of its time handling large numbers of fish and has little time remaining to catch additional fish, its predation rate levels off. In Figure 14.13b we can see that in a type II functional response any increases in prey density are associated with a slowing rate of prey consumption. As this occurs, the predators consume a smaller proportion of prey.

In a **type III functional response**, illustrated by the green line in Figure 14.13a, a predator exhibits low prey consumption under low prey densities, rapid consumption under moderate prey densities, and slowing prey consumption under high prey densities. Figure 14.13b shows how this type of functional response affects the proportion of prey consumed. As prey density increases there is an initial increase in the proportion of prey consumed. However, as the predators spend more time handling prey and become satiated, this proportion subsequently declines—just as we saw in the type II response.

Low consumption at low prey densities can be the result of three factors. First, at very low prey densities the prey can hide in refuges where they are safe from predators. A predator can consume prey only after the prey become so numerous that some individuals are unable to find a refuge.

Second, at low prey densities predators have less practice locating and catching the prey, and therefore are relatively poor at doing it. As prey densities increase, however, the predators learn to locate and identify a particular species of prey, a phenomenon known as a *search image*. A **search image** is a learned mental image that helps the predator locate and capture food, just as a person might locate a can of cola in a grocery store by searching for a small, red cylinder.

A third factor that can cause relatively low consumption of prey at low densities is the phenomenon of prey switching. This occurs when one prey species is rare and a predator changes its preference to another prey species that is more abundant. If the first species then increases in density, the predator can switch back to it.

Many laboratory and field studies have demonstrated type III functional responses. For example, researchers examined the feeding preferences of a predatory insect known as a backswimmer (*Notonecta glauca*). Figure 14.14 shows

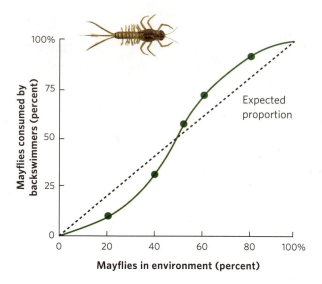

Figure 14.14 A type III functional response. Using predatory insects known as backswimmers, researchers manipulated the proportion of available prey: larval mayflies (shown above the graph) and isopods. The dashed black line indicates the expected proportion of mayflies consumed if the predator's preference was not affected by the proportional abundance of mayflies. When mayflies were rare, the predator consumed fewer mayflies than expected at random. When mayflies were common, the predator consumed more mayflies than expected at random. Data from M. Begon and M. Mortimer, *Population Ecology*, 2nd ed. (Blackwell Scientific Publications, 1981), after J. H. Lawton, J. R. Beddington, and R. Bonser in M. B. Usher and M. H. Williamson (eds.), *Ecological Stability*, pp. 141–158, (Chapman & Hall, 1974).

what happened when the researchers presented the predator with two types of prey—isopods (*Asellus aquaticus*) and larval mayflies (*Cloeon dipterum*)—and manipulated the proportions of each prey species. If the predator could not develop a search image, it would consume the two prey species in proportion to their availability. If the predator could develop a search image, it would eat fewer than expected of the rare prey and more than expected of abundant prey. As predicted, when the mayflies were rare, the backswimmer consumed fewer mayflies than expected, given their abundance. When mayflies were abundant, however, the backswimmers consumed more mayflies than expected, based on their abundance.

Type III functional response A functional response in which a predator exhibits low prey consumption under low prey densities, rapid consumption under moderate prey densities, and slowing prey consumption under high prey densities.

Search image A learned mental image that helps the predator locate and capture food.

The reason that the predator switched to the more abundant prey was because its success rate was higher with the prey that was present at high densities. For example, when larval mayflies were rare, predator attacks were successful less than 10 percent of the time. When the mayflies were common, predator attacks were successful nearly 30 percent of the time. Researchers attributed the improved capture rate to practice; the backswimmer became more proficient at catching the mayflies when it had more opportunities to catch them. The backswimmer exhibited no innate preference for either prey species, only a preference for the one that was more abundant.

The Numerical Response

As we have seen, the functional response considers changes in the number of prey consumed by each predator. A predator's functional response helps us understand how many prey can be consumed by predators, and therefore the conditions under which predators can regulate prey populations. The **numerical response** is a change in the number of predators through population growth or population movement due to immigration or emigration. Populations of most predators usually grow slowly relative to populations of their prey, although the movement of mobile predators from surrounding areas can occur quite rapidly when the density of prey changes.

An example of numerical response occurs in local populations of the bay-breasted warbler (*Dendroica castanea*), a small insectivorous bird of eastern North America. During outbreaks of the spruce budworm, populations of the warbler increase dramatically. In most years, the warbler populations live at densities of about 25 breeding pairs per km². However, during outbreak years, the warblers congregate in areas where the spruce budworm is living in high densities and warbler population densities can reach 300 breeding pairs per km². As a result of this rapid numerical response, the predator has the potential to reduce prey densities quickly and to regulate the prey's abundance.

Predation and herbivory favour the evolution of defences

Given the large effects that predators can have on their prey and herbivores can have on producers, it is not surprising that many species have evolved strategies to defend themselves. In this section, we will review the types of defences that have evolved and how they function. We will also look at how some

predators and herbivores have evolved counterdefences.

DEFENCES AGAINST PREDATORS

To understand the defences prey use against predators, we first need to understand predator hunting strategies. Predator hunting strategies can be categorized as either *active hunting* or *ambush hunting*, also known as *sit-and-wait hunting*. A predator that uses active hunting spends most of its time moving around looking for potential prey. For example, American robins actively hunt when they move around a lawn searching for earthworms. In contrast, a predator that uses ambush hunting lies in wait for a prey to pass by. Chameleons can sit very still as they wait for an insect to pass. When the insect is close enough, the chameleon shoots out its long tongue and the sticky, prehensile tip grabs the unsuspecting prey.

Regardless of hunting mode, we can think of hunting by predators as a series of events: detecting the prey, pursuing the prey, catching the prey, handling the prey, and consuming the prey. As we will see, prey have evolved defences to thwart the predator at different points in this series of events.

Behavioural Defences

Some of the most common behavioural defences against predators include alarm calling, spatial avoidance, and reduced activity. Alarm calling is used by many species of birds and mammals to warn their relatives that predators are approaching. Prey that use spatial avoidance move away from the predator. If the predator moves toward them, they continue to move away from the danger. Prey that follow a strategy of reduced activity are less likely to come into contact with a predator. When the prey detect the presence of a predator, they reduce their movements, which minimizes the chance of being detected. In a study of six species of tadpoles, a researcher placed animals in tubs of water with one of two treatments. The first contained a caged predator that could not kill the tadpoles but could emit chemical cues to scare the tadpoles. The second was a control treatment that consisted of an empty cage. Once the experiment was set up, the researcher watched the tadpoles to determine their activity level, which was defined as the percent of the time that the animals spent moving. As shown in

---

Numerical response A change in the number of predators through population growth or population movement due to immigration or emigration.

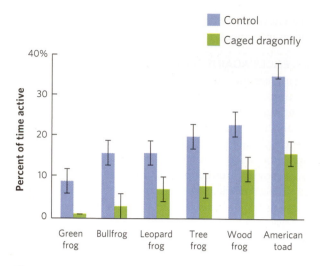

Figure 14.15 Behavioural defences. Tadpoles commonly avoid predation by becoming less active, which is measured as the portion of the day that they spend moving. Being less active reduces the probability of being detected by a predator. Error bars are standard errors. Data from R. A. Relyea, The relationship between predation risk and antipredator responses in larval anurans, *Ecology* 82 (2001): 541–544.

Figure 14.15, each species exhibited a different level of activity when predators were absent and all species reduced their level of activity in the presence of the predator.

Crypsis

Another way to reduce the probability of being detected by a predator is through camouflage that either matches the environment or breaks up the outline of an individual to blend in better with the background environment, a phenomenon known as **crypsis**. Various animals resemble sticks, leaves, flower parts, or even bird droppings. These organisms are not so much concealed but mistaken for inedible objects and passed over by predators. Common species that use crypsis include stick insects, katydids, and horned lizards (Figure 14.16). Some species have a fixed colour pattern that aids in crypsis while other species, such as the octopus, are able to rapidly change colour in ways that make themselves match their background.

Structural Defences

Although some prey use behaviour and crypsis to avoid being detected, other species employ mechanical defences that reduce the predator's ability to capture, attack, or handle the prey. One of the best-known examples of a mechanical defence is the barbed quills of the porcupine; more than 30 000 quills cover the porcupine's body and can penetrate the flesh of an attacking predator. In other species, the structural defences are phenotypically plastic and therefore only induced when the prey detects a predator in the environment. For example, water fleas—tiny freshwater crustaceans—that detect the chemical cues of predators early in their life can develop spines along different parts of their body to deter predators from consuming them. Other

Crypsis Camouflage that either allows an individual to match its environment or breaks up the outline of an individual to blend in better with the background environment.

(a)

(b)

Figure 14.16 Cryptic prey. Some prey avoid detection by blending into the environment. Crypsis is a strategy used by a number of animals including those shown here: **(a)** a katydid and **(b)** a willow ptarmigan (*Lagopus lagopus*) in winter plumage. (a) Photo by Dr. James L. Castner/Visuals Unlimited, Inc. (b) Photos by Ron Erwin/age fotostock.

(a)

(b)

No predator   **Predator**

Figure 14.17 Structural defences. In the crucian carp, individuals living with predatory fish develop a deep, hump-shaped body that allows the carp to escape predation. These predator-induced fish have **(a)** greater muscle mass, which allows **(b)** greater acceleration while swimming away from predators. Error bars are standard errors. Data from P. Domenici et al., Predator-induced morphology enhances escape locomotion in crucian carp, *Proceedings of the Royal Society B* 275 (2008): 195–201.

mechanical defences involve changes in the overall shape of the body. For example, the crucian carp (*Carassius carassius*), a species of fish that lives in Europe and Asia, develops a deep, hump-shaped body over a period of many weeks when the carp smells a predatory fish in the water. As you can see in Figure 14.17, carp with a hump-shaped body have greater muscle mass and can accelerate more quickly away from a predator.

Chemical Defences

Prey can also use chemical defences to deter a predator. Skunks, well-known for using this strategy, spray potential threats with foul-smelling chemicals from posterior glands. Many insects also use chemical defences. For example, when monarch butterfly caterpillars feed on milkweed, they store some of the milkweed toxins in their body, which make the butterfly very distasteful to predatory birds. Bombardier beetles take a different approach. Their

abdomens contain two glands, each of which makes a distinct chemical. When agitated, the beetle mixes these two chemicals, causing a reaction that makes the liquid approach 100° C. They shoot the boiling hot liquid out of the abdomen, causing pain or death to small predators, as illustrated in Figure 14.18.

While chemical defences are often effective in deterring predation, these defences are even more effective if the prey can communicate their distastefulness to the predator before an attack occurs. In many species, distastefulness has evolved in association with very conspicuous colours and patterns, a strategy known as **warning colouration**, or **aposematism**. Predators quickly learn to avoid markings such as the black and orange stripes of monarch butterflies; the insect tastes so bitter that a single experience is well-remembered. Conspicuous combinations of black, red, and yellow adorn such diverse animals as bombardier beetles, yellow-jacket wasps, and coral snakes. These colour combinations consistently advertise noxiousness; some predators have evolved innate aversions to them and do not need to go through the process of learning to avoid a particular prey.

Mimicry of Chemical Defences

When predators avoid aposematic species, any individuals of palatable species that resemble the distasteful, aposematic species would also be favoured by selection. Over generations, the palatable species can evolve to more closely resemble the

Warning colouration A strategy in which distastefulness evolves in association with very conspicuous colours and patterns. *Also known as* Aposematism.

Figure 14.18 Chemical defences. The bombardier beetle (*Stenaptinus insignis*) mixes two chemicals inside its abdomen that react to release a boiling hot fluid to deter predators.

(a)

(b)

(c)

Figure 14.19 Batesian mimicry. **(a)** The common wasp (*Vespula vulgaris*) possesses aposematic colouration as a warning to predators that an attack could result in being stung and injected with painful chemicals. Other harmless species that do not possess any stinging ability have evolved to resemble the colour patterns of the wasp. These include **(b)** a species of hover fly (*Helophilus pendulus*) and **(c)** the hornet clearwing (*Sesia apiformis*). Their resemblance to the wasp reduces their risk of predation. Photos by (a) Nick Upton/naturepl.com; (b) Geoff Dore/naturepl.com; and (c) FLPA/Gianpiero Ferrar/age fotostock.

aposematic species. When palatable species evolve warning colouration that resembles unpalatable species, we call it **Batesian mimicry**, named for Henry Bates, the nineteenth-century English naturalist who first described it. In his journeys to the Amazon region of South America, Bates found numerous cases of palatable insects that did not retain the cryptic patterns of their close relatives but instead had evolved to resemble brightly coloured, unpalatable species (Figure 14.19).

Experimental studies have demonstrated that mimicry does confer an advantage to the mimics. For example, toads that consumed live bees and received a sting on the tongue subsequently avoided palatable drone flies, which mimic the appearance of bees. In contrast, when naive toads consumed dead bees with stingers removed, they consumed the bees and the drone flies. This result indicated that toads learned to associate the conspicuous and distinctive colour patterns of live bees with an unpleasant experience.

Another type of mimicry, called **Müllerian mimicry**, occurs when several unpalatable species evolve a similar pattern of warning colouration. Müllerian mimicry is named after its discoverer, the nineteenth-century German zoologist Fritz Müller. When multiple species of prey have conspicuous colour patterns and all are unpalatable, a predator that learns to avoid one prey species will later avoid all prey species with a similar appearance. For example, most of the bumblebees and wasps that co-occur in mountain meadows share a pattern of black and yellow stripes and they all have the ability to sting a predator. Similarly, in Peru several species of poison dart frogs, all in the genus *Ranitomeya*, closely resemble each other. In four regions of that country researchers have found three species that vary in colouration according to location, including

one species (*R. variabilis*) that looks very different across two locations. A fourth species, *R. imitator*, is also unpalatable and has populations in each of these four locations that closely resemble the other species that is present (Figure 14.20).

Costs of Defences Against Predators

Many types of defences against predators can be costly, as we discussed in our coverage of predator- and herbivore-induced defences in Chapter 4. For example, because behavioural defences frequently result in reduced feeding activity or increased crowding in locations away from predators, behavioural defences often come at the cost of reduced growth and development. Similarly, most mechanical defences are energetically expensive to produce, such as the 30 000 quills of a porcupine. In some cases, the costs of defence are so high that they can come at the cost of growth and reproduction. When this happens, the presence of predators can cause smaller population sizes even when they do not consume the prey.

Less is known about the costs of chemical defences in prey, but recent studies suggest that these defences are also energetically costly to produce. In ladybugs, which are technically known as ladybird beetles, many species are red with black spots. These warning colours communicate to predators that the ladybugs taste bad because of chemicals in their bodies known as alkaloids. However, there is a large amount of variation in the concentration of alkaloids that each ladybug can produce. In 2012, researchers reported that only beetles that consumed large amounts of food had sufficient energy to produce high

Batesian mimicry When palatable species evolve warning colouration that resembles unpalatable species.

Müllerian mimicry When several unpalatable species evolve a similar pattern of warning colouration.

Figure 14.20 Müllerian mimicry. Müllerian mimics are a collection of unpalatable species that share a pattern of warning colouration. At four locations in Peru, researchers have found pairs of species of poison dart frogs that are unpalatable and closely resemble each other. For each location, the frog on the left is a particular species of unpalatable frog. In all four locations, the frog on the right is the unpalatable *R. imitator,* so named because in each location it has evolved to imitate the colouration of the other frog. Photos by Evan Twomey, East Carolina University.

concentrations of carotenoid chemicals in their bodies, which give the ladybugs a more intense red colour, as shown in Figure 14.21a. Moreover, as illustrated in Figure 14.21b, beetles that produced more carotenoids also produced higher concentrations of alkaloids. As a result, ladybugs with the highest energy diet can better advertise their level of toxicity to predators and thereby reduce their chances of being attacked.

Counter-Adaptations of Predators

If predation can select for prey to evolve a wide range of defences, then prey defences should favour the selection for counter-adaptation in predators. In this way, predators and prey experience an evolutionary arms race between prey defences and predator offences. When two or more species affect each other's evolution, we call it **coevolution**. In the case of the porcupine, for example, the spines deter most predators. However, bobcats (*Lynx rufus*) and wolverines have an effective solution. When these

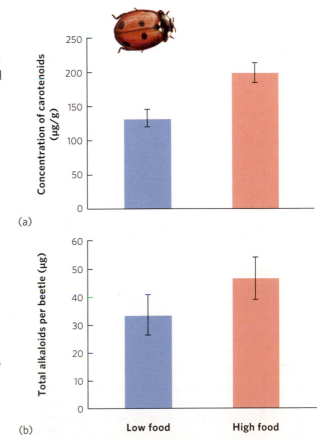

Figure 14.21 Costly chemical defences in ladybugs. In male ladybugs, individuals given a high amount of food produced **(a)** a higher concentration of carotenoids that make the ladybug's body more intensely red, and **(b)** a greater amount of defensive alkaloid chemicals. Error bars are standard errors. Data from J. D. Blount et al., How the ladybird got its spots: Effects of resource limitation on the honesty of aposematic signals, *Functional Ecology* 26 (2012): 334–342.

predators find a porcupine, they flip the porcupine on its back and attack the belly, which is not defended by spines. Other common predator adaptations include high-speed locomotion to catch their prey and camouflage that allows them to ambush their prey.

Some predators can also evolve to handle the toxic chemicals produced by prey. For example, the cane toad (*Bufo marinus*) is a species that was introduced into Australia in 1935. Like other species of toads, the cane toad contains toxins in its skin that can cause predators to become sick or die. As a result, predators in the native range of cane toads do not attack cane toads even though these same predators regularly consume other species of amphibians. When the toads were introduced to Australia, predators of amphibians, such as black snakes (*Pseudechis porphyriacus*), had no evolutionary experience with cane toads. When they ate the toads, most snakes had no resistance to the toxins and they died. In 2006, however, researchers reported that some populations of the black snakes were consuming cane toads, which suggests that while most black snakes that consumed cane toads died, some snakes must have been resistant to the toad's toxins and survived. Over time, the selection for snake resistance to the toad toxin must have resulted in the evolution of resistant populations. To test this hypothesis, the researchers fed samples of toad skin to snakes from different populations around Australia. They then examined how much the toad skin reduced the swimming speed of each snake, which allowed the researchers to determine the susceptibility of each snake to the toxin. As you can see in Figure 14.22, the researchers found that snake populations that coexisted with cane toads for the longest amount of time evolved the lowest susceptibility to the cane toad toxin. This evolution would have occurred in less than 70 years, which is a remarkably short time.

DEFENCES AGAINST HERBIVORES

The selective pressure from predators has caused the evolution of prey defences. Similarly, continued selective pressure from herbivores has caused the evolution of defences against herbivory. In some cases, these defences are induced by an herbivore attack and are therefore phenotypically plastic. In other cases, plant defences are fixed and therefore are expressed whether or not an herbivore has attacked the plant. In both cases, some herbivore species have evolved counter-adaptations. In fact, some species of herbivores are so specialized in countering the defences of a particular plant species that they do not consume any other plants.

Structural Defences

When it comes to structural defences, plants have evolved a variety of traits to deter herbivores from consuming their leaves, stems, flowers, or fruit.

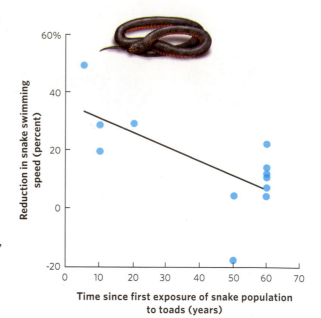

Figure 14.22 Counter-adaptations of predators to prey chemical defences. Researchers measured the swimming speed of black snakes before and after being fed the skin of cane toads, which contains toxins. They found that snake populations that had coexisted for the longest times with the toad had the lowest susceptibility to the toad toxin. Data from B. L. Phillips and R. Shine, An invasive species induces rapid change in a native predator: Cane toads and black snakes in Australia, *Proceedings of the Royal Society B* 273 (2006): 1545–1550.

Some plants, such as cacti, roses, and blackberries, have sharp spines and prickles that inflict pain on the mouths of herbivores. Other plants grow a wooly layer of hair over their leaf surfaces to make it difficult for herbivorous insects to penetrate them.

Chemical Defences

A wide variety of chemical defences have evolved in plants. For example, western red cedar (*Thuja plicata*) produce chemicals called terpenes that are thought to repel deer with their smell or to have toxic effects on the deer's digestive and nervous systems. Researchers studied red cedar on Haida Gwaii, British Columbia, and found that cedars with higher terpene concentration suffered less browsing from Sitka black-tailed deer (*Odocoileus hemionus sitkensis*) than trees with lower terpene concentration. Other plant chemicals include sticky resins and latex compounds that are hard to consume. Some plants also produce alkaloids—including caffeine, nicotine, and morphine—that have a wide range of toxic effects on herbivores. Other chemicals in plants such as tannins are difficult for herbivores to digest. It is generally thought that the chemicals produced by plants are the by-products of a plant's physiology.

Coevolution When two or more species affect each other's evolution.

ANALYZING ECOLOGY

Understanding Statistical Significance

In examining adaptations of prey, counter-adaptations of predators, or any other ecological measurements, we often look at experiments in which researchers find differences among experimental manipulations. Until now, we have not explored how ecologists assess when such differences are meaningful versus when the differences are due to chance.

For any group of measurements, such as the concentration of toxins in ladybugs fed high and low amounts of food, there will be variation among the individuals of each group. If we were to sample ladybugs at random from the high- and low-food treatment, we would find that the mean toxin concentration is higher in the high-food treatment. However, the measurements taken on some of the individuals in the high-food treatment could overlap with the measurements taken on some of the individuals in the low-food treatment. When the means are similar and the distribution of the data from two groups are almost entirely overlapping, we would have to conclude that the two groups are nearly identical in whatever we are measuring, as shown in the figure below.

In contrast, when the means are very far apart and the distribution of the data from two groups exhibit no overlap, as in the case below, we would feel confident that we have two completely different groups of individuals.

Although these two extreme alternatives demonstrate when two groups have completely overlapping or completely nonoverlapping distributions, we need to know how much overlap between the two groups is acceptable in order to conclude that the groups are different from each other.

Scientists agree that two distributions can be considered "significantly different" if we can sample the two distributions many times and find that the means of those distributions overlap less than 5 percent of the time. This somewhat arbitrary, but widely accepted, cutoff value is known as alpha (α). Thus, we say that our cutoff for statistical significance is $\alpha < 0.05$. Determining that something has statistical significance is not the same as stating that a difference between two means is large, substantial, or important. In other words, the everyday use of "significant" is not synonymous with the scientific use of "significantly different."

YOUR TURN In Chapter 2 "Analyzing Ecology: Standard Deviation and Standard Error," we mentioned that when data have a normal distribution, about 68 percent of the data fall within one standard deviation of the mean, 95 percent of the data fall within two standard deviations of the mean, and 99.7 percent of the data fall within three standard deviations of the mean. Based on this information, if you had two groups with identically shaped distributions of data, approximately how many standard deviations apart would they have to be in order for the two groups to be considered significantly different?

For many of these chemical defences, one or more species of herbivore has evolved tolerance. For example, a shrub in Polynesia called the Tahitian noni (*Morinda citrifolia*) produces toxic chemicals with such a foul smell that the plant has the nickname "vomit fruit" (Figure 14.23). Most species of fruit flies avoid this plant because if they were to land on it, they would

die. However, one species of fruit fly (*Drosophila sechellia*) can tolerate the defensive chemicals. It lays its eggs on the vomit flower and they have a distinct fitness advantage because they experience no competition from other species of fruit flies.

A similar situation exists for the monarch butterfly. As mentioned in our discussion of prey defences, the

Figure 14.23 Chemical plant defences and counter-adaptations of specialized herbivores. The Tahitian noni, also known as the vomit flower, is a shrub that produces such a foul smell that most species of fruit flies avoid it. However, one species of fruit fly has evolved that tolerates the chemicals and therefore lays its eggs on the plant. Photo by US National Park Service/Bryan Harry.

monarch caterpillar specializes in feeding on milkweed plants, which produce toxic chemicals. The monarch caterpillar readily feeds on milkweed plants because it has evolved to resist the effects of a group of chemicals known as *cardiac glycosides,* which can stop the heart of many other herbivores. It also sequesters some of the chemicals to serve as a defence against its own predators. In 2012, researchers made a striking discovery about the evolution of insects' ability to consume the plants containing cardiac glycosides. They found that a diversity of insects from different orders—including flies, beetles, true bugs, and butterflies—independently evolved the same changes in a gene that offers resistance to the effects of the toxin. In short, the different groups of insects exhibited convergent evolution.

Tolerance

Some plants that have not evolved extensive defences against herbivores take an alternative strategy of tolerating herbivory. Plants taking this strategy are able to grow new tissues rapidly to replace those that are consumed. For example, in the wetlands of the Hudson Bay Lowlands in northern Manitoba, the lesser snow goose (*Anser caerulescens caerulescens*) is the major herbivore. Researchers studied the effects of goose herbivory by comparing plant mass and growth in areas that were fenced to exclude geese and similar areas where geese were allowed to graze. Over two growing seasons, they found that plant mass and growth were higher in fenced than in unfenced areas. Plants that were grazed experienced relatively high fitness despite being partially consumed.

Costs of Herbivore Defences

For decades, researchers have investigated whether plant defences come at the cost of reduced fitness. When defence traits are phenotypically plastic,

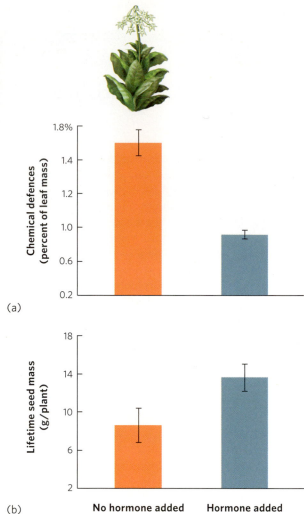

Figure 14.24 Costs of defences against herbivores in tobacco plants. Researchers damaged tobacco plants to simulate the act of herbivory. They then prevented a chemical response in some individuals by applying plant hormones to the damaged area of the leaves. **(a)** Damaged plants treated with the plant hormone produce a lower amount of chemical defence than damaged plants not treated with the hormone. **(b)** Damaged plants treated with the plant hormone also experienced higher lifetime fitness, as measured by the mass of seeds produced. Error bars are standard errors. Data from I. T. Baldwin et al., The reproductive consequences associated with inducible alkaloid responses in wild tobacco, *Ecology* 71 (1990): 252–262.

researchers can compare the fitness of individuals with induced defences against the fitness of noninduced individuals. For example, tobacco plants (*Nicotiana sylvestris*) respond to herbivory by producing chemical defences including nicotine. Researchers damaged one group of tobacco plants to induce an increase in chemical defences. As a control, they also damaged a second group of plants and then treated the damaged areas with a plant hormone that prevented the chemical defences from responding. When they later counted the number of seeds produced by the two groups, they found that the group with increased chemical defences produced fewer seeds, as you can see in Figure 14.24.

A second approach to quantifying the costs of defences against herbivores is to make the genes responsible for defences nonfunctional. For instance, researchers have examined the growth of different genotypes of mouse-ear cress, a tiny plant that is native to Europe and Asia. In 2011, they reported that individuals with intact defence genes commonly grew more slowly than individuals with nonfunctional defence genes. This confirmed that plants pay a cost for fixed defences against herbivores.

ECOLOGY TODAY CONNECTING THE CONCEPTS

THE TROUBLE WITH CATS AND RABBITS

Introduced species on Macquarie Island. Introduced rabbits have caused devastating effects on the plants of the island, which do not have an evolutionary history of living with rabbits. Attempts to reduce the rabbit populations have caused feral cats to switch from eating rabbits to eating the island's native birds. Photo by James Doust, courtesy Australian Antarctic Division.

Islands throughout the world often contain a variety of endemic species that have evolved together for millions of years. As we have seen in this chapter, new species that are introduced to islands can have devastating effects on the native plants and animals. Because the native species do not share an evolutionary history with the introduced species, they have not evolved defences against introduced predators and herbivores. These introductions happen commonly on islands. One typical response is an attempt to remove the intruders and thereby reverse the harmful effects, although the actual outcomes of these efforts can lead to unintended consequences.

An example of this occurred on Macquarie Island, a small island located halfway between Australia and Antarctica. Island plants and animals existed together for eons in a tundra biome that contained a large number of seabirds and land birds, as well as a diversity of plants that included tall grasses. Humans began to visit the island in the 1800s and they brought several species of animals with them.

In the early 1800s, the island was used by seal hunters as a place to rest and to resupply their ships. These visitors introduced house cats (*Felis catus*), which soon became feral. While there are no data on the impact of the feral cats, it is generally assumed that they acted as a mesopredator and preyed upon the abundant island birds. In 1878, the seal hunters introduced European rabbits (*Oryctolagus cuniculus*) to the island to serve as a food source whenever the sailors returned to the island. Despite the fact that the cats fed on the rabbits, over time the rabbit population grew to very large numbers.

Data collected from the 1950s through the 1970s suggest that the rabbit population experienced large fluctuations approximately every 10 years, similar to those seen in hares and other animals that live in high northern latitudes. Because the plants on the island had no evolutionary history with the rabbits, periods of high rabbit numbers caused devastating effects on the abundance of palatable plant species; when the rabbit population experienced large declines, the vegetation rebounded.

Because periods of high rabbit populations caused such adverse effects on the island's vegetation, scientists introduced the European rabbit flea (*Spilopsyllus cuniculi*) in 1968. The fleas can carry a *Myxoma* virus that causes the disease myxomatosis, which is fatal to rabbits. The scientists quickly learned that the virus did not persist well on the island, so beginning in 1978 they reintroduced it every year. The virus had its desired effect and caused the rabbit populations to plunge from a high of 130 000 individuals in 1978 to fewer than 20 000 by the 1980s. As the rabbit population declined, the vegetation of the island began to recover. However, the changes to the island community did not end there.

With the rapid decline of the rabbits, the cats of the island exhibited prey-switching behaviour—described in our discussion of predator functional responses—and added more birds to their diet. They consumed an estimated 60 000 seabirds per year and caused the extinction of two species of endemic land birds. To help save the island's undefended birds, government officials decided to eradicate all of the cats from the island.

Starting in 1985, scientists removed 100 to 200 cats from the island each year. An analysis of the cat stomachs in 1997 indicated that while the cat population was killing birds, the cat population was still consuming approximately 4000 rabbits per year, so the cats—in combination with the *Myxoma* virus—were probably still regulating the rabbit population.

When the last cat was removed in 2000, the island community responded rapidly. For example, five species of seabirds started breeding on the island, some of which had not bred there for more than 80 years. The rabbits also responded. From 1985 to 2000, the rabbit population increased from approximately 10 000 to more than 100 000 individuals. This increase was probably caused by a combination of fewer predatory cats, a rebounding abundance of plants, and a decline in the amount of virus distributed each year. By 2006, there were 130 000 rabbits and these herbivores had consumed so much vegetation that parts of the island resembled a closely cropped lawn.

These changes on the island underscored the importance of understanding all the interactions that introduced species can have on a community, including interactions among the introduced species themselves. In the case of Macquarie Island, it has become clear to scientists that removing the introduced cats was not enough to return the island community to its former condition. The introduced rabbits had to be removed as well. As a result, the Australian government agreed to spend $24 million to remove the rabbits from the island as well as several other introduced mammals, including mice and rats. Rabbit poison, distributed around the island in 2011, was very successful in killing the vast majority of the rabbits. In 2012 hunters and rabbit-hunting dogs were brought to the island to eliminate the few remaining rabbits. The success of this latest effort will determine whether an island that has been invaded by introduced predators and herbivores can ever return to its original state.

SOURCES: Bergstrom, D. M., et al. 2009. Indirect effects of invasive species removal devastate World Heritage Island. *Journal of Applied Ecology* 46: 83–81.

Dowding, J. E., et al. 2009. Cats, rabbits, *Mxyoma* virus, and vegetation on Macquarie Island: A comment on Bergstrom et al. (2009). *Journal of Applied Ecology* 46: 1129–1132.

Bergstrom, D. M., et al. 2009. Management implications of the Macquarie Island trophic cascade revisited: A reply to Dowding et al. (2009). *Journal of Applied Ecology* 46: 1133–1136.

http://www.abc.net.au/news/2012-04-19/rabbit-hunters-head-to-macquarie-island/3961270

SUMMARY OF CHAPTER CONCEPTS

■ **Predators and herbivores can limit the abundance of populations.** Using observations in nature and manipulative experiments, ecologists have found that predators commonly limit the abundance of prey and herbivores commonly limit the abundance of producers.

■ **Populations of consumers and consumed populations fluctuate in regular cycles.** Cycling populations have been observed frequently in nature and recreated in laboratory experiments. Lags in the response times of predator movement and reproduction linked to changes in the abundance of prey cause these cycles. Mathematical models have been developed to mimic the cycling behaviour of predator and prey populations.

■ **Predation and herbivory favour the evolution of defences.** Prey have evolved a wide variety of defences including behavioural defences, mechanical defences, chemical defences, crypsis, and mimicry. Producers have evolved defences against herbivores including mechanical defences, chemical defences, and tolerance. Evolved defences are commonly costly and can sometimes be countered by subsequent adaptations in predators.

REVIEW QUESTIONS

1. How could you experimentally test whether herds of African antelope affect the abundance of plants on which they graze?

2. Explain why an herbivore that consumes many different species of plant might be less successful at regulating the abundance of a well-defended plant species compared to an herbivore that specializes on eating a single species of plant.

3. In the classic experiments of C. F. Huffaker with mites and oranges, what mechanisms allowed the predator and prey populations to persist?

4. Which factors determine the duration of population cycles and which determine the magnitude of change in population sizes?

5. Explain in words the equations of the Lotka-Volterra model for the change in the population sizes of prey and predators.

6. According to the Lotka-Volterra model of predator–prey interactions, why do predator and prey populations cycle?

7. How do search image formation and prey switching behaviour lead to a type III functional response in predators?

8. Compare and contrast a predator's numerical response and functional response.

9. How might a type II functional response prevent a predator from controlling a large prey population?

10. Explain the causes of an evolutionary arms race between consumers and the species that they consume.

PRACTICING ECOLOGY: INDIRECT EFFECTS OF DEER HERBIVORY

Peter Arcese from the University of British Columbia and his colleagues study the effects of animal management and conservation on ecosystems. Photo by Owen Jongejan.

Human colonization of North America has led to the elimination of top predators and a rapid increase in the herbivore populations that the predators would normally consume. Overabundance of herbivores has a direct effect on plant community composition; the abundance of edible plant species quickly decreases while the abundance of inedible plants increases. In addition, increased herbivory can remove many of the understorey plants that live on the forest floor, which reduces food and habitat for other forest dwelling species. How does the expansion of herbivore populations affect other species that live in North American forests?

The black-tailed deer, a subspecies of the mule deer (*Odocoileus hemionus*), is endemic to the Gulf Islands and San Juan Islands to the west of Washington State and Vancouver Island. Population sizes were once controlled by cougar and wolf predation and hunting by First Nations. Over the past century, humans have eliminated both predator species from the area, and removal of First Nations from their native land has severely reduced hunting pressure on the deer. During this time, the average height of forest plants has changed; the understorey plants that were easily accessible and palatable to deer have disappeared, while the inaccessible larger plants that extend into the midstorey and canopy of the forest have remained intact. Scientists have considered how this change may affect other species in the forest. For example, many bird species depend on understorey plants to provide resources for insect prey and nesting habitat. Hence, we might hypothesize that deer grazing activity in the understorey results in a negative relationship between deer abundance and the abundance of understorey bird species but has no effect on the abundance of canopy bird species.

Peter Arcese from the University of British Columbia and his colleagues tested this hypothesis by examining bird community composition on 18 of the Gulf Islands and San Juan Islands that were ranked as having low or high deer abundance. They then determined the abundance of all bird species on each island, with each bird species identified as being either an understorey or canopy species. You can view their data in the table. For simplification, bird species are identified by numbers instead of taxonomic names.

ABUNDANCE OF UNDERSTOREY AND CANOPY BIRD SPECIES (INDIVIDUALS/HA) ON ISLANDS WITH LOW AND HIGH DEER DENSITIES

UNDERSTOREY BIRDS			CANOPY BIRDS		
	Low deer	High deer		Low deer	High deer
SPECIES 1	0.66	0.07	SPECIES 10	0.05	1.38
SPECIES 2	2.13	0.09	SPECIES 11	1.34	0.41
SPECIES 3	0.66	0.38	SPECIES 12	1.82	2.3
SPECIES 4	0.87	0.38	SPECIES 13	0.45	0.39
SPECIES 5	4.39	1.24	SPECIES 14	1.44	0.21
SPECIES 6	2.23	0.26	SPECIES 15	1.78	0.19
SPECIES 7	4.35	1.04	SPECIES 16	0.27	0.3
SPECIES 8	2.05	0.38	SPECIES 17	1.73	0.21
SPECIES 9	0.94	0.57	SPECIES 18	0.1	0.74

GRAPH THE DATA

1. Calculate the means and standard errors of the means for each of the four columns.

2. Create a bar graph showing means and standard errors for the abundance of understorey bird abundances at low and high deer densities.

3. Create a second bar graph showing means and standard errors for the abundance of canopy bird abundances at low and high deer densities.

INTERPRET THE DATA

1. Explain how the data support or refute the hypothesis that deer abundance is negatively related to the abundance of understorey bird species, but has no effect on the abundance of canopy bird species.

2. In the metapopulation of islands represented in this study, are islands with high deer densities likely to be serving as source or sink populations for understorey bird species? If deer are removed from an island, what is likely to happen to the composition of the bird community?

3. Predict what will happen to understorey bird populations if both nest predators and deer are removed from an island.

SOURCES: Martin, T. G., P. Arcese, and N. Scheerder. 2011. Browsing down our natural heritage: Deer impacts on vegetation structure and songbird populations across an island archipelago. *Biological Conservation*,. 144: 459–469.

CHAPTER 15

Parasitism and Infectious Diseases

The Life of Zombies

Zombie films scare us because they depict zombies as the walking dead in search of victims they will turn into more zombies. But something similar to this happens commonly in nature; some species of parasites infect a host and take control of its life for their own benefit.

Consider the case of the amber snail (*Succinea putris*) from Europe. This animal lives along the edges of streams and ponds. It normally spends its time in the shade of terrestrial vegetation where it chews on leaves and remains hidden from the eyes of predatory birds. An amber snail will occasionally consume bird feces, which sometimes contain the eggs of a parasitic flatworm (*Leucochloridium paradoxum*). These eggs hatch inside the snail and grow, but to reproduce, the parasites must spend the next stage of their lives inside a bird. To achieve this goal, the larvae slowly make their way into the snail's eyestalks. These eyestalks are normally pale and slender, but the parasitic infection causes the eyestalks to become enlarged and banded with colours that pulsate in a way that resembles a moving caterpillar. The parasites also take control of the snail's brain and force the snail to move up a plant's stem, which is not something an amber snail normally does. Snails that move up plant stems are more easily noticed by predatory birds and, because the eyestalks look like caterpillars, birds consume the infected snails. The parasite completes the second stage of its life inside the bird and the cycle continues. The flatworm reproduces, its eggs leave the bird through feces, and snails consume them.

> "The ability of parasites to act as puppet masters over the behaviour of their victims is just one way that parasites have evolved to improve their fitness."

Parasites can control the behaviour of many different animals. In Thailand, for example, carpenter ants (*Camponotus leonardi*) normally spend their time living in nests in the hot, dry rainforest canopy. Occasionally the carpenter ants travel from the canopy to the ground where they can be exposed to the spores of the fungus *Ophiocordyceps unilateralis,* the so-called zombie-ant fungus, before heading back up to their nests. An ant that is infected subsequently moves down to the humid, understorey vegetation. It stops at about 25 cm above the ground in a place that many previously infected ants have also gone, known to researchers as the "ant graveyard." Here it bites the underside of a leaf and maintains a death grip on a leaf vein as it dies. After the ant dies, the fungus grows a spore-producing structure out of the ant's head and releases its spores into the environment.

This unusual ant behaviour does not benefit the ant, but it greatly benefits the fungus. The higher humidity found lower in the rainforest is more suited to fungal growth than the drier conditions high in the canopy. Moreover, if the infected ant were to die in its nest in the

A parasitized amber snail. The snail on the right has one normal eye stalk that is pale and slender and another that is infected by a parasitic flatworm, which causes the eye stalk to become enlarged and colourful. It also pulsates in a way that is attractive to predatory birds. Photo by Alex Teo Khek Teck via Flickr.

An infected carpenter ant. In Thailand, carpenter ants that become infected by a fungus crawl down from the canopy and attach themselves to the underside of a leaf by biting the leaf vein and then dying. After death, a spore-producing stalk grows out of the ant's head and releases its spores into the environment. Photo by David Hughes/Penn State University.

canopy, its nest mates would remove the corpse before the fungus could grow the spore-producing structure that is critical to its reproduction.

The ability of parasites to act as puppet masters over the behaviour of their victims is just one way that parasites have evolved to improve their fitness. As we will discover in this chapter, parasites come in a wide variety of forms and their effects on hosts can range from mild to lethal. Adaptations that enable parasites to infect hosts and adaptions that help hosts to resist infections by parasites offer intriguing insights into the strategies of parasite–host interactions.

SOURCES: S. B. Andersen et al., The life of a dead ant: The expression of an adaptive extended phenotype, *American Naturalist* 174 (2009): 424–433.

D. P. Hughes et al., Behavioural mechanisms and morphological symptoms of zombie ants dying from fungal infection, *BMC Ecology* 11 (2011): 13.

T. Lefevre, and F. Thomas, Behind the scene, something else is pulling the strings: Emphasizing parasitic manipulation in vector-borne diseases, *Infection, Genetics and Evolution* 8 (2008): 504–519.

CHAPTER CONCEPTS

- Many different types of parasites affect the abundance of host species.

- Parasite and host dynamics are determined by the parasite's ability to infect the host.

- Parasite and host populations commonly fluctuate in regular cycles.

- Parasites have evolved offensive strategies while hosts have evolved defensive strategies.

The struggle between parasites and hosts has produced many fascinating examples of ecological interactions and evolutionary adaptations. In Chapter 1, we defined a parasite as an organism that lives in or on another organism, called the host, and causes harmful effects as it consumes resources from the host. Some hosts have **infection resistance**, which is the ability of a host to prevent the infection from occurring, while other hosts have **infection tolerance**, which is the ability of a host to minimize the harm that an infection can cause. The number of parasites of a given species that an individual host can harbour is known as the host's **parasite load**. A parasite typically has only one or a few species of hosts, although a given host could contain dozens of species of parasites. In fact, it is estimated that approximately half of all species on Earth are parasites.

When parasites cause an infectious disease, we call them *pathogens*. However, infection by a pathogen does not always result in a disease. For example, humans can be infected with human immunodeficiency virus (HIV) but they may never experience the disease symptoms known as Acquired Immune Deficiency Syndrome (AIDS). In many cases, it is not known what causes a host to transition from being infected by a pathogen to experiencing the disease.

Infectious diseases take a large toll on people; the World Health Organization estimates that more than 25 percent of all human deaths are caused by infectious diseases. Note that only infectious diseases are caused by pathogens; there are many noninfectious diseases in which pathogens do not play a role.

In this chapter, we will focus on the interaction between parasites and their hosts. In later chapters, we will discuss the larger role that parasites can play in communities and ecosystems. We will begin by looking at the many different types of parasites that exist, including those that have large effects on ecological communities, crops, domesticated animals, and human health. We will then examine the factors that determine whether parasites can infect hosts, spread rapidly through a population, and cause widespread harmful effects. Because mathematical

Infection resistance The ability of a host to prevent an infection from occurring.

Infection tolerance The ability of a host to minimize the harm that an infection can cause.

Parasite load The number of parasites of a given species that an individual host can harbour.

models can help us understand the population dynamics of interacting species, we will also discuss parasite–host models. Finally, we will consider how parasites have evolved to increase their chances of infecting hosts, and how hosts have evolved to combat the risk of infection.

Many different types of parasites affect the abundance of host species

Parasites typically have specific habitat needs and, as a result, often live in particular places on a host organism. In humans, for instance, parasites find certain parts of the body to be highly suitable habitats. As illustrated in Figure 15.1, head lice live in the hair, liver flukes reside in the liver, the fungus that causes athlete's foot resides in the feet, and so on.

We can categorize the wide variety of parasites as either *ectoparasites* or *endoparasites*. Both are common in plants and animals. **Ectoparasites** are

---

Ectoparasite A parasite that lives on the outside of an organism.

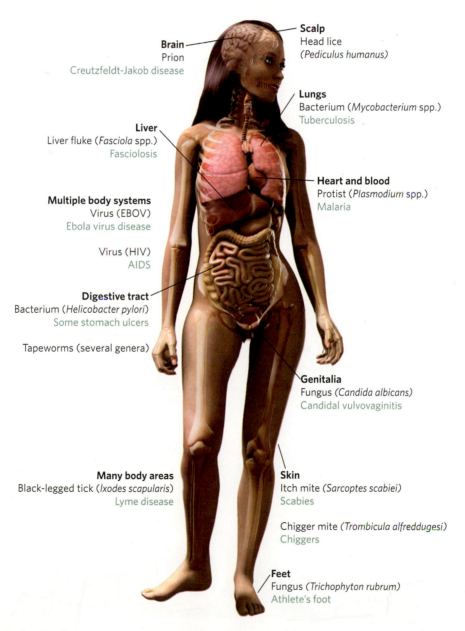

Scalp
Head lice
(*Pediculus humanus*)

Brain
Prion
Creutzfeldt-Jakob disease

Lungs
Bacterium (*Mycobacterium* spp.)
Tuberculosis

Liver
Liver fluke (*Fasciola* spp.)
Fasciolosis

Heart and blood
Protist (*Plasmodium* spp.)
Malaria

Multiple body systems
Virus (EBOV)
Ebola virus disease

Virus (HIV)
AIDS

Digestive tract
Bacterium (*Helicobacter pylori*)
Some stomach ulcers

Tapeworms (several genera)

Genitalia
Fungus (*Candida albicans*)
Candidal vulvovaginitis

Many body areas
Black-legged tick (*Ixodes scapularis*)
Lyme disease

Skin
Itch mite (*Sarcoptes scabiei*)
Scabies

Chigger mite (*Trombicula alfreddugesi*)
Chiggers

Feet
Fungus (*Trichophyton rubrum*)
Athlete's foot

Figure 15.1 Preferred habitats. Parasites have preferred habitats on their host. The human body, for example, offers a wide range of habitats for parasites.

TABLE 15.1
COMPARING THE CONSEQUENCES OF ENDOPARASITE AND ECTOPARASITE LIFESTYLES

FACTOR	ECTOPARASITES	ENDOPARASITES
Exposure to natural enemies	High	Low
Exposure to external environment	High	Low
Difficulty of movement to and from host, for parasite or its offspring	Low	High
Exposure to host's immune system	Low	High
Ease of feeding on host	Low	High

parasites that live on the outside of organisms whereas **endoparasites** are parasites that live inside organisms. Each lifestyle has advantages and disadvantages, as summarized in Table 15.1. For example, because ectoparasites live on the outside of their hosts, they do not have to combat the immune system of a host and they can easily move on and off a host. In addition, they do not directly cause diseases in their host. In contrast, because endoparasites live inside their hosts, they must contend with the immune system of their host, and they may have a difficult time getting in and out of their host's body. Endoparasites have the advantage of being protected from the external environment and are therefore not exposed to most of their enemies. In addition, living inside the host gives endoparasites easy access to the host's body fluids on which they feed. In contrast, ectoparasites are exposed to the variable conditions of the external environment, including natural enemies, and they must find a way to pierce the flesh of their host to feed. In this section, we will discuss some of the most common types of ectoparasites and endoparasites and the effects they have on their hosts.

ECTOPARASITES

A variety of organisms live as ectoparasites, as shown in Figure 15.2. Most ectoparasites that attack animals are arthropods. Among the best known ectoparasites are two groups of arachnids—ticks and mites—and two groups of insects—lice and fleas. There are many other groups of ectoparasites, including nematodes–also known as roundworms–and some species of

Endoparasite A parasite that lives inside an organism.

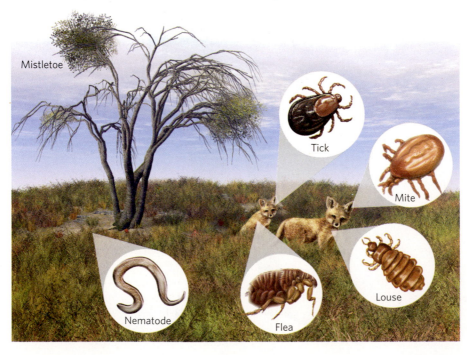

Figure 15.2 Ectoparasites. Ectoparasites are parasites that live attached to the outside of their host. Common ectoparasites include ticks, fleas, lice, mites, mistletoes, and nematodes.

lamprey fish. Each of these animals attaches to a host and consumes the host's blood and other body fluids.

Plants also have ectoparasites, but they are most commonly either nematodes or other species of plants. Nematodes that act as plant ectoparasites live in the soil and feed on plant roots. The 1-mm-long worms attach to a plant root, inject digestive enzymes that break down root cells, and then consume the resulting slurry. This parasitic behaviour can reduce the growth, reproduction, and survival of the host.

Approximately 4000 species of plants make their living as ectoparasites on other plants. For example, mistletoes embed rootlike organs into the branches of trees and shrubs. Because such plants have their roots inside the host and their shoots outside the host, they are sometimes categorized as *hemiparasites*. The mistletoe leaves conduct photosynthesis, but the plant obtains water and minerals from the host plants. Some species, such as dwarf mistletoes (in the genus *Arceuthobium*) cause death in conifer trees—particularly under drought conditions—by extracting too much of the host's water and nutrients. In some species of mistletoe, the plant's fruit is consumed by birds, which inadvertently disperse the seeds from the fruit when they defecate onto the branches of other trees and shrubs. Other species of mistletoe are able to project seeds laterally up to 15 m.

ENDOPARASITES

Endoparasites can be categorized as either *intracellular* or *intercellular*. As the names imply, intracellular parasites live inside the cells of a host, whereas intercellular parasites live in the spaces between cells that include the cavities of a host's body. Intracellular parasites are very small; examples include viruses, small bits of proteins known as *prions,* and some types of bacteria and protists. Intercellular parasites are much larger; examples include some types of protozoa and bacteria, fungi, and a group of worms known as *helminths,* which include the nematodes. Each of these groups of endoparasites can have substantial effects on their host's survival, growth, and reproduction. Because endoparasites often cause diseases, they can alter the abundance of host species and change the composition of ecological communities.

Viruses

Late in the nineteenth century, researchers struggled to find the cause of a disease striking tobacco plants. They recognized that the pathogen was too small to be a bacterium. After many years of research, they were able to identify it as a virus, which they named tobacco mosaic virus. Plants that are infected exhibit a mosaic of colours in their leaves and may also have growths that resemble blisters (**Figure 15.3**). Although this virus was first discovered in tobacco, we now

Figure 15.3 Tobacco mosaic virus. Plants infected with the tobacco mosaic virus, such as this tobacco plant, develop blisters and light areas on their leaves. These effects can cause substantial reductions in plant growth. Photo by Nigel Cattlin/ Science Source.

know that it can infect more than 150 species of plants. Viruses that infect plants can have a devastating effect on food production, which makes them a matter of great concern to farmers and consumers.

Animals can also be infected by many different pathogenic viruses. For example, there are different genera of the pox virus that can infect mammals: cowpox in cattle, monkeypox in primates and rodents, and smallpox in humans. Hoofed animals such as deer, cattle, and sheep are susceptible to bluetongue virus, which can cause high mortality rates. Similarly, the West Nile virus—named for the location of the first known human case in the West Nile region of Uganda in 1937—can be highly lethal to some species of birds. The mosquitoes that carry this virus usually transmit the disease to birds, but occasionally to horses and humans, both of which can die from the infection. The virus arrived in the United States in 1999 and in Canada in 2001, and spread rapidly. Several species of North American birds experienced high rates of death, including blue jays (*Cyanocitta cristata*), American robins, and American crows (*Corvus brachyrhynchos*). At the same time, horses and people started to become infected. In the decade following the introduction of the virus into North America more than 1000 people died. As a result, public health officials in Canada and the United States have made great efforts to reduce mosquito populations that spread the disease. These efforts, possibly combined with the evolution of reduced virulence by the pathogen, have caused the number of infections and deaths to decline dramatically from 2003 to 2011, as illustrated in Figure 15.4. However, in 2012 there was a sharp increase in the number of infections and deaths in the United States. The sharp increase is thought to be due to unusually warm and humid conditions combined with abundant rain that created a lot of standing water where mosquitoes can breed and subsequently build up a large adult population.

Both mammals and birds are susceptible to strains of influenza virus. For example, a strain of influenza known as the *Spanish flu* is caused by the H1N1 virus and normally infects only birds, but in 1918 it also infected humans. Researchers hypothesize that the virus experienced a mutation that allowed it to survive in humans and to infect other humans directly. They believe that humans initially contracted the disease by handling domesticated ducks and chickens. However, once the virus found its way into humans, it rapidly spread around the world and killed up to 100 million people. More recently, the H1N1 virus has been found on pig farms, which is why the pathogen also has the name *swine flu*.

In 2006, a similar strain of virus known as *bird flu*, or H5N1, infected and killed large numbers of birds and also jumped from domesticated birds to humans. Hundreds of thousands of domesticated birds were killed in an attempt to stop the spread of the virus (Figure 15.5). By 2012, the World Health Organization reported that 595 people had been infected with bird flu and more than half of them had died. The ability of pathogens to jump from traditional hosts to human hosts, as we will discuss in the next section, causes a great deal of concern for human health in the future.

Figure 15.5 Controlling the spread of bird flu. In an attempt to stop the spread of bird flu, millions of domesticated birds in Asia were killed and their bodies burned. Data from http://www.cdc.gov/ncidod/dvbid/westnile/surv&control.htm#maps. Photo by SONNY TUMBELAKA/AFP/Getty Images.

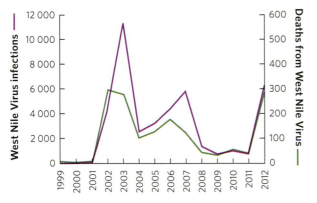

Figure 15.4 Human infections and deaths from West Nile virus in Canada and the United States. Following the introduction of the virus to North America in 1999, the number of human infections and deaths rose rapidly through 2003. Efforts to control mosquito populations, which carry the virus, caused a sharp decline in these effects on humans through 2011. In 2012, however, unusually warm temperatures with high precipitation in some regions resulted in large mosquito populations and a new surge in West Nile infections and deaths. Note that the two *y* axes have different scales. Data from http://www.cdc.gov/ncidod/dvbid/westnile/surv&control.htm#maps.

Prions

Prions are a recently discovered category of pathogenic parasite. All prions begin as a beneficial protein in the brain of an animal, but occasionally a protein folds into an incorrect shape and becomes pathogenic. Prions do not contain any RNA or DNA; instead, they replicate by coming into contact with normal proteins and causing them to fold incorrectly, with the first prion serving as a template.

One of the best-known diseases caused by prions is bovine spongiform encephalopathy, commonly known as mad cow disease. This disease, which is always fatal, gets its name from the way that infected cows lose control of their body, as if they have gone mad. While cows cannot transmit prions to each other, in the 1980s it was a common practice to feed cows the ground remains of other cows. Cows that consumed the mutated prions from the dead cattle became infected and ultimately passed the infection on to humans. More than 150 people have died from mad cow disease. Today, new rules forbid feeding dead cattle to other cattle and the incidence of mad cow disease has declined.

Chronic wasting disease is a more common prion disease. It infects members of the deer family, including white-tailed deer, mule deer (*Odocoileus hemionus*), elk, and moose. Individuals infected with chronic wasting disease excrete prions that are later consumed inadvertently by others. Infected individuals begin to lose weight and eventually die (Figure 15.6). Currently, chronic wasting disease is concentrated in Colorado and Wyoming, but small numbers of infected deer have been discovered in other regions of Canada and the United States.

Protozoans

Protozoans are a group of parasites that can cause a variety of diseases. For example, several protozoans cause diarrhea in humans and other animals. They are also the source of malaria in humans and avian malaria in birds. Malaria is carried by mosquitoes that acquire the protists when they feed on an infected individual and then transfer the protists to uninfected individuals they subsequently feed on. In Hawaii, as well as in some other parts of the world, there is no historic presence of avian malaria, so the birds of the Hawaiian Islands have not evolved defences against the protists. Mosquitoes were accidentally introduced into Hawaii in the early 1800s and by the early 1900s the protists that cause avian malaria had also arrived, which has contributed to the decline and extinction of many native bird species in these islands.

Bacteria

Bacterial parasites can cause a wide variety of plant and animal diseases. In plants, bacterial infections can cause spotted leaves, wilted stems, scabbed surfaces, and large abnormal tissue growths known as *galls* (Figure 15.7). Because bacteria must enter a

Figure 15.6 Chronic wasting disease. Deer infected with prions that cause chronic wasting disease lose weight and eventually die. Photo courtesy of Mike Hopper, Game Warden, Kansas Department of Wildlife, Parks & Tourism.

plant through a wound in the plant's tissues, they commonly require the assistance of herbivores that pierce a plant's tissues.

Bacterial infection in animals can also be quite harmful to the host. One of the earliest discovered species of deadly bacteria was anthrax (*Bacillus anthracis*). Anthrax, originally isolated from cows and sheep, is also highly lethal to humans. After an incubation period of about 1 week, the growing bacterial population begins to release toxic compounds that cause internal bleeding, which often brings rapid death. Other common bacterial pathogens include those that cause plague, pneumonia, salmonella, leprosy, and many sexually transmitted diseases, all of which occur in a wide variety of animals.

(a)   (b)

Figure 15.7 Bacterial disease of plants. Common plant diseases that are caused by bacteria include **(a)** shot-hole such as this *Pseudomonas syringae mors-prunorum* on a cherry tree and **(b)** Crown gall (*Rhizobium radiobacter*) shown on a birch tree in England. Photos by (a) Nigel Cattlin/Science Source; and (b) Caroline Morgan.

Fungi

Fungal parasites have large ecological impacts on a wide range of plants and animals. Fungal diseases have devastated many dominant plant species, including critical food sources for humans. A number of tree species in North America have suffered tremendous declines as a result of pathogenic fungi that have been introduced from Europe and Asia. The American chestnut tree was once one of the tallest tree species in the temperate forests of the eastern United States and southern Ontario and comprised up to 50 percent of all trees in the forest. It was also one of the most prized trees for lumber as well as for the crop of edible nuts that it produced every autumn. Around 1900, however, species of Asian chestnut trees imported to New York inadvertently carried a fungus (*Cryphonectria parasitica*). While Asian chestnut trees had a long evolutionary history with the fungus and were resistant to its harmful effects, American chestnut trees had no such history. As a result, the American chestnut rapidly succumbed to the disease known as chestnut blight that was caused by the fungus. The decline of the chestnut trees also likely had widespread effects on the many animal species that relied on the tree's large crop of nuts for food. Today, the American chestnut tree is quite rare; young seedlings can emerge from seeds but they ultimately become infected with the fungus and die early in life.

Similar problems have occurred with other trees, including the American elm tree (*Ulmus americana*). The elm was widely planted along North American streets because of its attractive arching branches that provided shade (Figure 15.8a). In the 1930s, a fungus from Asia was introduced into Europe and North America that caused a rapid decline in elm trees. Because the fungal disease was first described by researchers in Holland, it is known as Dutch elm disease. The American elm had no evolutionary history with the Asian fungus, and 95 percent of the elm trees across North America have succumbed (Figure 15.8b). For instance, up to 80 percent of elm trees died in a single year in Toronto, where the proportion of these trees has now declined from approximately 12 percent to around 4 percent. Winnipeg spends over $2 million to keep its loss in elm trees to around 3 percent. The small number of remaining trees appears to be resistant to the fungal disease, and researchers are currently working to plant these resistant genotypes back into eastern forests.

Some fungal parasites also damage crop plants. One of the best-known examples is a group of fungi that causes a disease known as *rust*. Throughout human history, rust diseases have affected many of our most important crops, including wheat, corn, rice, coffee, and apples. The effect of rust can range from reduced food production to a complete loss of the crop

(a)

(b)

Figure 15.8 Declining trees in North America. The introduction of a fungus that causes Dutch elm disease killed many American elms, which used to be common along North American streets. **(a)** A street in Detroit, Michigan in 1974 before Dutch elm disease. **(b)** The same street in 1981 after Dutch Elm disease killed all of the elm trees. Photos by Jack H. Barger, U.S. Forest Service.

costing millions of dollars of damage. Widespread rust infections, such as wheat rust, have caused the starvation of millions of people over the past 2 centuries.

Animals can also be infected by fungi. One high profile animal fungal disease is caused by a species of chytrid fungus that infects amphibians (*Batrachochytrium dendrobatidis*). During the 1990s, scientists working in Central America began to notice massive die-offs of amphibians and subsequently determined that the dead animals were infected with the fungus. The fungus, which lives in the outer layers of the amphibian's skin, causes an imbalance of ions in the body that ultimately causes the animal's heart to stop.

During 2 decades of surveying amphibians from Costa Rica to Panama, researchers found that the fungus was moving from the northwest to the southeast, as shown in Figure 15.9a. Between 2000 and July 2004, at a site in El Copé, Panama, a team of researchers sampled more than 1500 amphibians

(a)

Figure 15.9 The deadly chytrid fungus in Central America. **(a)** Surveys of amphibians in Central America demonstrated that the fungus was spreading from the northwest to the southeast. **(b)** Surveys of amphibians at the El Copé site detected a sharp decline in amphibians, both in species that are active during the day and species that are active at night. Data from K. Lips et al., Emerging infectious disease and the loss of biodiversity in a Neotropical amphibian community, *Proceedings of the National Academy of Sciences, USA* 103 (2006): 3165–3170.

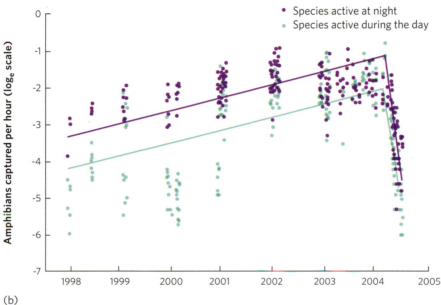

(b)

before the fungus arrived; not one individual tested positive. By October 2004, however, 21 of 27 sampled species were infected in at least 10 percent of their populations and by December 2004, 40 species tested positive for the fungus.

The researchers also estimated the sizes of amphibian populations in El Copé with transect surveys of amphibians that were active either during the day or during the night. After the fungus arrived in 2004, the number of live amphibians declined sharply (Figure 15.9b). The dead amphibians included 38 different frog species. Moreover, 99 percent of the 318 dead individuals collected had moderate to severe chytrid infections. In 2010, researchers reported that of the 63 species of amphibians that were present at the site before the fungus arrived, 30 species were gone. Relatively little is known about this fungus, although the massive die-offs suggest that amphibians have few defences against it, which indicates that the fungus may have been introduced to Central America. The fungus is now suspected to have caused the extinction of dozens of species of amphibians around the world.

Helminths

Helminths include several groups of roundworms and flatworms that can cause serious diseases. We have already discussed the impact of nematodes as ectoparasites on plant roots. Nematodes can also live as endoparasites within the tissues of plants, where they damage the plant's growth, reproduction, and survival. In animals, disease-causing helminths include hookworms that feed on the blood of intestines, lungworms that live in the lungs, and echinostome worms that live in the kidneys. In livestock, infection by liver flukes has been a problem for centuries. Livestock inadvertently consume these parasites when they drink water containing fluke-infected snails or when they eat grass that contains a stage of the parasite that is excreted by snails. Liver flukes are a particular problem in sheep where they cause liver damage, hemorrhaging, and sudden death. Left untreated, liver flukes can kill up to 10 percent of a sheep population; fortunately, drugs can cure infected animals.

Emerging Infectious Diseases

Many infectious diseases have been infecting hosts for thousands of years while others have only emerged recently. When a new disease is discovered, or a formerly common disease that has become rare suddenly increases in occurrence, it is called an **emerging infectious disease**. New diseases typically emerge when a mutation allows a pathogen to jump to a new host species. We have already mentioned several emerging infectious diseases, including the chytrid fungus that has decimated amphibians around the world, the H5N1 bird flu that jumped from birds to humans, and mad cow disease that infected cattle, sheep, and humans. Since the 1970s, the world has experienced an average of one new emerging infectious disease each year.

In 2006 researchers identified an emerging disease infecting bats in a cave near Albany, New York. These bats had a white-coloured fungus (*Geomyces destructans*) growing on their noses, and they were dying in large numbers (Figure 15.10). Researchers named the disease white-nose syndrome. They hypothesized that the fungus causes bats hibernating in caves to come out of their normal torpor and to use up their fat reserves until they become emaciated and die. By 2012, the U.S. Fish and Wildlife Service reported that the fungus had spread to 4 provinces and 16 U.S. states, and killed more than 5 million bats. In caves containing infected bats, the mortality rate can reach 100 percent. This is important because bats provide many important services, including the consumption of a large number of insects. Therefore, it is plausible that the death of so many bats will have substantial effects on many communities and ecosystems.

An interesting detail in the story of white-nose syndrome is that bats in Europe also carry the fungus but do not experience widespread death. This raises the

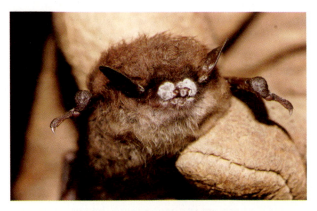

Figure 15.10 White-nose fungus in bats. This species, known as the little brown bat, is infected by the fungus *Geomyces destructans*. One symptom of the infection is that the bat's nose turns white. When a bat colony becomes infected, a large proportion of the bats die. Photo by AP Photo/ New York Department of Environmental Conservation, Ryan von Linden, File.

Figure 15.11 Testing North American versus European strains of white-nose fungus. Bats exposed to the control treatment did not die. In contrast, bats exposed to the North American fungus strain or the European fungus strain died in large numbers. However, bats died faster when exposed to the European strain than the American strain. Data from L. Wernicke et al., Inoculation of bats with European *Geomyces destructans* supports the novel pathogen hypothesis for the origin of white-nose syndrome, *PNAS* 109 (2012): 6999–7003.

possibility that the fungus is native to Europe and has only recently been introduced to North America. Alternatively, it could be that the fungus has always been in North America but recently mutated to become much more lethal. To test these alternative hypotheses, researchers isolated strains of the fungus from European and North American bats. They then exposed the little brown bat (*Myotis lucifugus*), a species from North America, to fungal spores from the European strain, fungal spores from the North American strain, or a control treatment containing no fungal spores. After the exposures, they monitored bat survival for 120 days. In 2012, the researchers reported their results, which you can see in Figure 15.11. In the control treatment, no bats died, but bats exposed to either fungal strain experienced decreasing rates of survival, which suggests that the fungus is a recent arrival from Europe and not a mutated North American native. Moreover, the bats died more slowly when exposed to the North American strain of fungus than the European strain, which may suggest that North American bats were already beginning to evolve some level of resistance to their local strain of fungus. Unfortunately, there is currently no treatment to prevent the continued death of bats from white-nose syndrome.

In early 2003, another emerging infectious disease made the news around the world and caused the World Health Organization to issue a travel advisory for Toronto. Severe Acute Respiratory Syndrome (SARS) originated in China from where it spread in a matter of months. Once it arrived in Toronto, it infected approximately 400 people and killed 44 between

Emerging infectious disease A disease that is newly discovered, or has been rare and then suddenly increases in occurrence.

Vertical transmission from parent to offspring (e.g., *Chlamydia*)

Horizontal transmission through a vector (e.g., West Nile virus)

Horizontal transmission directly between conspecifics (e.g., bird flu)

Horizontal transmission through multiple hosts (e.g., some helminths)

Figure 15.12 Mechanisms of parasite transmission. Parasites can be transmitted vertically or horizontally. When transmission is horizontal, the parasite can be transmitted through a vector such as a mosquito, transmitted directly between two conspecifics, or transmitted to other species. When transmission is vertical, a parent host transmits the parasite to its offspring, such as when a mother bird transmits lice to its hatchlings at the nest.

February and June of that year—the highest infection and death rates outside of Asia. Globally, over 8000 people were diagnosed with SARS and more than 900 people died from it. The virus that causes SARS occurs in a number of mammal species, such as palm civets, raccoon dogs, and bats. Infected individuals of those species were sold on Chinese markets for food, which allowed the virus to jump to humans.

Parasite and host dynamics are determined by the parasite's ability to infect the host

Parasites and their hosts have populations that fluctuate over time, much like the dynamics of predator and prey populations that we discussed in Chapter 14. Unlike predators, however, parasites generally have a higher reproductive rate than their hosts and often do not kill them. If we know what causes fluctuations in parasite and host populations, we can predict when parasitism will be prevalent in a host population. This allows us to anticipate population changes in hosts and, in some cases, to intervene and reduce the harmful effects of a parasite on a species of concern. The probability that a host will become infected by a parasite depends on numerous factors that include the parasite's mechanism of transmission, its mode of entry

into a host's body, its ability to jump between species, the existence of reservoir species, and the response of the host's immune system.

MECHANISMS OF PARASITE TRANSMISSION

The first factor that determines the risk of infection by a parasite is the mechanism of transmission. As illustrated in Figure 15.12, there are many different mechanisms. Parasites can move between hosts through *horizontal transmission* or *vertical transmission*. **Horizontal transmission** occurs when a parasite moves between individuals other than parents and their offspring. For example, horizontal transmission occurs when helminths are transmitted from snails to frogs or from frogs to birds. Horizontal transmission can also occur between conspecifics, such as the transmission of bird flu from one bird to another. Some parasites, such as the prions that cause mad cow disease, cannot be naturally transmitted from one individual to another. The risk of transmission is essentially zero in these cases. As we have seen with mad cow disease, dead infected cows were fed to other cows, which significantly raised the risk of infection. At the other extreme of transmission are influenza viruses, which easily pass between

Horizontal transmission When a parasite moves between individuals other than parents and their offspring.

individuals. You may have experienced the rapid transmission of the flu virus if you or someone in your family became ill with flu; most likely, other members of your family quickly became infected. For most hosts, the risk of becoming infected by a parasite generally increases with host population density because higher densities mean that individuals are likely to come into contact with the parasite more often.

Some parasites require another organism, known as a **vector**, to disperse from one host to another. For example, the West Nile virus is transmitted from one host to another by a mosquito that picks up the virus from one infected bird and transmits it to an uninfected bird or other animal that it bites. In this case, the mosquito serves as the vector.

Still other parasites require transmission to multiple host species to complete their life cycle, as we saw in the story of the amber snail. Some species of helminths, for example, spend the first stage of their life in a snail, the second stage of their life in an amphibian, and the final stage of their life in a bird. This requirement for multiple hosts poses a substantial challenge for the parasite, which must find all of its required hosts during its lifetime.

Vertical transmission occurs when a parasite is transmitted from a parent to its offspring. In this case, the parasite must evolve in such a way that it does not cause the death of its host until after the host has reproduced and passed the parasite to its offspring. Many sexually transmitted diseases can be passed by vertical transmission. For example, chlamydia is a disease caused by several different species of bacteria within two genera, *Chlamydia* and *Chlamydophila*, that infect mammals, birds, and reptiles. In humans, chlamydia causes inflammation of the urethra and cervix, but is typically not lethal. While the bacterium is often horizontally transmitted between individuals, it can also be transmitted vertically from an infected mother to her fetus. When this happens in humans, the newborn baby can suffer eye infections and pneumonia.

Researchers continue to learn about the wide range of pathways that parasites can take to infect a host. The existence of vertical transmission and various methods of horizontal transmission can make it very challenging to predict and control the spread of infectious diseases in humans, crops, domesticated animals, and wild organisms.

MODES OF ENTERING THE HOST

The mode of entry into the host's body also affects the ability of the parasite to infect the host. As we have discussed, some species of parasites, such as leeches, are able to pierce the tissues of the host. Other parasites, such as some viruses, bacteria, and protists, rely on another organism to penetrate the host's tissues and use the damaged tissue as an entry point. The protist that causes avian malaria, for example, depends on mosquitoes to inject the protist into a bird's body after the protist completes a critical life stage inside the mosquito's body. Without the mosquito, the malaria life cycle could not be completed.

JUMPING BETWEEN SPECIES

If a parasite specializes on only one host species and is only able to cause a lethal disease in that host species, then it might eventually run out of hosts and face extinction. One solution sometimes favoured by natural selection is for the parasite to be nonlethal to the host; this allows a host population to persist. We will talk more about this strategy later in the chapter.

Alternatively, the parasite might evolve the ability to infect other species. We saw an example of this with bird flu, which infected several species of birds before a mutation occurred that allowed the virus to infect humans. A similar scenario occurred with HIV. For many years it was hypothesized that human HIV originated in chimpanzees, and in 2006 researchers identified a population of chimps in the West African nation of Cameroon that carried a genetically similar strain of the virus. The researchers suspect that the virus jumped from chimpanzees to humans when local hunters consumed the chimpanzees as food. Other examples of parasites jumping between species include the chytrid fungus, which can jump between amphibian species, and the canine parvovirus, which can jump from cats to dogs.

RESERVOIR SPECIES

One way that parasite populations persist in nature is through the use of *reservoir* species. **Reservoir species** carry a parasite but do not succumb to the disease that the parasite causes in other species. Because reservoir species do not die from the infection, they serve as a continuous source of parasites as other susceptible host species become rare. For example, some species of birds can be infected with the protist that causes avian malaria, but they are resistant to developing the disease. However, mosquitoes that feed on these resistant species of birds can pick up the protist and transfer it to other species of susceptible birds that become infected and die. In the case of SARS, which we discussed above, bats serve as the reservoir species. In contrast to palm civets and raccoon dogs, which are susceptible to the virus, bats can carry the virus without developing the

Vector An organism that a parasite uses to disperse from one host to another.

Vertical transmission When a parasite is transmitted from a parent to its offspring.

Reservoir species Species that can carry a parasite but do not succumb to the disease that the parasite causes in other species.

disease. In this way, the reservoir species ensure the persistence of the parasite population over time.

THE HOST'S IMMUNE SYSTEM

A host's immune system can play a role in combatting an infection from endoparasites. As a result, some parasites have evolved the ability to escape the immune system by making themselves undetectable. For example, when HIV enters a human cell, it can hide from the body's immune system by living in the cytoplasm or by incorporating itself into the chromosomes of the cell. Because the body's immune system searches for infections on the outside of cells, it cannot detect the virus. Other parasites have evolved additional strategies to escape the immune system. For instance, parasitic worms known as schistosomes produce a protective outer layer around their bodies that prevents them from being detected by the host's immune system. Still others, such as the protists that cause African sleeping sickness, are able to continually change the compounds present on their outer surface so they become a moving target that stays one step ahead of the immune system as it tries to respond to the infection.

Parasite and host populations commonly fluctuate in regular cycles

In our discussion of predators and prey in Chapter 14, we saw that population fluctuations are common and sometimes occur in regular cycles. Because parasites and hosts represent consumers and resources, they show similar population dynamics. In this section, we will examine how parasites and hosts fluctuate over time. We will also look at a mathematical model that helps us describe and understand the behaviour of infectious pathogens and hosts.

POPULATION FLUCTUATIONS IN NATURE

As we discussed in the previous chapter, the density of hosts can affect how easily parasites are transmitted from one host to another. An excellent example of this can be seen in the dynamic between the forest tent caterpillar (*Malacosoma disstria*)—a major herbivore of broadleaf trees in Canada and the United States—and a group of viruses that can infect and kill the caterpillar. The forest tent caterpillar can defoliate trees over thousands of square kilometres. In years of high caterpillar densities, they remove the majority of a tree's leaves and tree growth is reduced up to 90 percent. In 2012, researchers in Canada reported that caterpillar populations had cycles that last 10 to 15 years and that these fluctuations were fairly synchronous across large geographic areas. In three different locations in Ontario, the fluctuations in caterpillar populations exhibit the same pattern of growth and decline over time, as shown in Figure 15.13. Although the caterpillar is susceptible to predators and parasitoids, viruses have the greatest ability to reduce its abundance. When the tent caterpillar population is

Figure 15.13 Cyclic population fluctuations of forest tent caterpillars. At three sites in Ontario, researchers quantified the population sizes of tent caterpillars by measuring the intensity of leaf removal on the trees. Over a 60-year period, the caterpillars exhibited large population fluctuations every 10 to 15 years. The rapidly growing populations ultimately succumb to an outbreak of infection by a virus and the populations quickly die back. Data from B. J. Cooke et al., The dynamics of forest tent caterpillar outbreaks across east-central Canada, *Ecography* 35 (2012): 422–435.

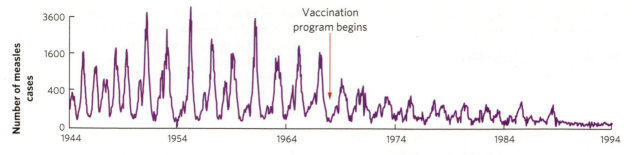

Figure 15.14 The cyclical occurrence of measles in a human population. Before vaccinations became available in 1968, the human population in London, England, experienced cycles of measles every 2 years. Once vaccination became available and the number of people getting vaccinated increased, the number of measles cases declined and the fluctuations no longer occurred. Data from P. Rohani et al., Opposite patterns of synchrony in sympatric disease metapopulations, *Science* 286 (1999): 968–971.

high, the virus comes into contact with new hosts more frequently and each caterpillar is more likely to get the virus. Under these conditions, large numbers of caterpillars die. As host population density decreases, it becomes harder for the viruses to find a new host and the prevalence of the disease declines. Since fewer caterpillars become infected and die, the caterpillar population starts to increase again.

Fluctuations in parasite and host populations can also be caused by changes in the proportion of the host that has achieved immunity. When a host species possesses the ability to become immune to a parasite after an initial infection, continued infection in the population causes an increasing proportion of the population to develop immunity. When a high proportion of the population is no longer susceptible to the parasite, its spread is slowed. Measles, for example, is a highly contagious viral disease that stimulates lifelong immunity in humans. In unvaccinated populations, measles typically produces epidemics at 2-year intervals. Once most of the population is infected and develops immunity, the number of new measles cases declines sharply. As humans continue to reproduce, however, there are enough new children born without immunity to initiate another measles outbreak after 2 years. The number of measles cases in London, England, from 1944 to 1968—before vaccinations became available—shows a distinct pattern of 2-year cycles, as you can see in Figure 15.14. Once vaccinations became available and a higher proportion of susceptible individuals were immunized, the number of new cases dropped sharply and the fluctuations declined and eventually disappeared.

MODELLING PARASITE AND HOST POPULATIONS

For pathogenic parasites, we can understand the dynamics of the parasite and the host using models that are similar to the Lotka-Volterra predator–prey model. However, the parasite–host model differs from the predator–prey model in two key ways. Parasites, unlike predators, do not always remove host individuals from a population, and hosts may develop

immune responses that make some individuals resistant to the pathogen.

The simplest model of infectious disease transmission that incorporates immunity is the **Susceptible-Infected-Resistant (S-I-R) model**. In this model, all individuals begin as susceptible to the pathogen (S). Of those, some number becomes infected (I). Of the infected individuals, some number develops resistance via immunity (R). We can use this model to examine the conditions that favour an epidemic versus the conditions that cause a disease to decline. The proportions of S, I, and R individuals in a population are determined by rates of transmission of the disease and acquisition of immunity, as well as the birth of new, susceptible individuals.

In a population of hosts, the first individual to be infected by a pathogen is known as the *primary case* of the disease. Any individuals infected from this first individual are known as *secondary cases*. The rate at which the pathogen spreads through the population depends on two opposing factors. One factor is the rate of transmission between individuals (b), which includes both the rate of contact of susceptible individuals with an infectious individual and the probability of infection when there is contact. The other factor is the rate of recovery (g), which determines the period of time from when an individual is infected and can transmit the infection to when the individual's immune system clears the infection and the individual becomes resistant to any future infections.

The rate of infection and the rate of recovery can be used to determine whether an infectious disease will spread through a population. A disease will spread whenever the number of newly infected individuals is greater than the number of recovered individuals. To determine the number of newly infected individuals, we need to know the probability that an infected individual and a susceptible

Susceptible-Infected-Resistant (S-I-R) model The simplest model of infectious disease transmission that incorporates immunity.

individual will come into contact with each other. If we assume that individuals in the population move around randomly, the probability of contact is the product of their proportions in the population:

Probability of contact between susceptible and infected individuals $= S \times I$

Once infected and susceptible individuals come into contact, we also have to consider the rate of infection between them (g):

Rate of infection between susceptible and infected individuals $= S \times I \times g$

Next we have to determine how many individuals are recovering from the infection. We can do this by knowing the proportion of infected individuals (I) and the rate of recovery from an infection (b):

Rate of recovery of infected individuals $= I \times b$

Now we can determine whether an infection will spread through a population. To do this, we need to calculate the reproductive ratio of the infection (R_0), which is the number of secondary cases produced by a primary case during its period of infectiousness. The reproductive ratio of the infection is the ratio of new infections to recoveries:

$$R_0 = (S \times I \times b) \div (I \times g)$$
$$R_0 = S \times (b \div g)$$

If $R_0 > 1$, the infection will continue to spread through the population and an epidemic will occur. This happens because each infected individual infects more than one other individual before it recovers from the disease and becomes resistant. When $R_0 < 1$, the infection fails to take hold in the host population. This happens because each infected individual fails to infect another individual, on average, before it recovers and becomes resistant.

Figure 15.15 illustrates the dynamics of a typical disease. Imagine that we start with a population that is composed entirely of uninfected individuals. When an infectious disease arrives, there is initially a rapid increase in the number of infected individuals but, over time, infected individuals subsequently recover and become resistant (R). At this point, the number of susceptible individuals (S) decreases, so the value of R_0 decreases. When R_0 declines to < 1, the growing epidemic can no longer sustain itself. As a result, the number of infected individuals achieves a peak in abundance and then declines.

With this understanding of how R_0 affects the spread of diseases, we can now look at the R_0 values for different diseases that are caused by parasites. HIV is transmitted through rather limited mechanisms including direct sexual contact, blood transfusion, or

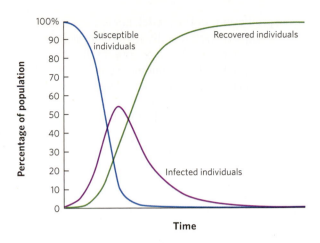

Figure 15.15 The dynamics of an infection over time. In the basic S-I-R model, all individuals in a population are initially susceptible. When the infection is introduced at the beginning of the time period, there is an initial rapid growth in the number of infected individuals. As some infected individuals recover and become resistant, there are fewer susceptible individuals left to infect, so the number of infected individuals declines.

perinatal transmission from mother to offspring. HIV has a relatively low range of R_0 values, from 2 to 5. Typical values for R_0 in childhood diseases of humans—measles, chicken pox, and mumps, among others—range from 5 to 18 at the time a population is initially infected. At the extreme, malaria, which is transmitted by mosquitoes, has an R_0 value greater than 100 in crowded human populations. This high value occurs because mosquitoes are excellent vectors for transmitting the parasite and infected people remain infectious for long periods of time.

Assumptions of the S-I-R Model

The basic S-I-R model has several important assumptions. For example, it assumes there are no births of new susceptible individuals and that individuals retain any resistance they develop. In such a model, the result is an epidemic that runs its course until all individuals in the population have become resistant or there are too few susceptible individuals remaining to sustain the spread of the disease. Some pathogens, such as influenza viruses, fit these assumptions. The basic model also explains why vaccinations slow or stop the spread of diseases such as influenza; by vaccinating individuals, we reduce the size of the susceptible population (S), which reduces the value of R_0. This makes it harder for an epidemic to sustain itself.

Other factors can be added to the model, including births of susceptible infants, lag times between when an individual is infected and becomes infectious to others, host mortality, host population dynamics, and transmission of disease from parent to offspring. These additional factors can have substantial effects on the predictions of the model. For example, the birth of new susceptible individuals can cause cyclic fluctuations in the model, as we observed in the case of the measles data from London (Figure 15.14).

If a pathogen can kill the host, the pathogen should increase in abundance until the hosts begin to die. As the host population declines, the pathogen population will subsequently decline and as the pathogen population declines, the host population should subsequently recover. This is analogous to the predator–prey cycles we discussed in Chapter 14. However, some pathogens do not follow these dynamics because they do not attack a single host species. For instance, the chytrid fungus that we discussed earlier in this chapter infects dozens of species of amphibians. A host species can decline all the way to extinction, yet the fungus does not decline in abundance because it can infect other species of amphibians. Some of these newly infected species will become ill and die while others never develop the disease and serve as reservoirs for the disease. When a pathogen is not restricted to a single host species, it has the ability to persist and spread even after it causes one of its hosts to go extinct.

Parasites have evolved offensive strategies while hosts have evolved defensive strategies

Parasites gain fitness when they find a suitable host and reproduce. Hosts will experience higher fitness if they avoid being parasitized. Therefore, natural selection has favoured the evolution of parasite offences and host defences. In this section, we will examine some of these strategies as well as look at evolutionary relationships between parasites and hosts.

PARASITE ADAPTATIONS

As we saw at the beginning of the chapter, parasites have evolved a wide array of strategies for finding and infecting hosts. Some of the most fascinating examples of parasite adaptation involve strategies that increase the probability of parasite transmission. For example, the pathogenic fungus *Entomorpha muscae* infects both houseflies (*Musca domestica*) and yellow dungflies (*Scatophaga stercoraria*). To improve its chance of transmitting its spores, it has two different strategies. When the fungus infects and kills female houseflies, the bodies become attractive to male houseflies looking for a mate. Although it is unclear what attractive cues the fungus produces, when a male housefly tries to mate with a dead female, spores of the fungus are transferred to the male's body.

When the same fungus infects the yellow dungfly, the fly climbs up nearby vegetation, moves to the upwind side of the plant, and hangs upside down on the underside of a leaf. The fly then moves its wings toward the leaf, and moves its abdomen—which is swollen with fungal spores—away from the leaf. Once it reaches this position, the dungfly dies and the fungal spores erupt from its abdomen. The erupting spores can then potentially infect dungflies that pass below, as illustrated in Figure 15.16. When researchers compared the behaviour of infected dungflies to uninfected

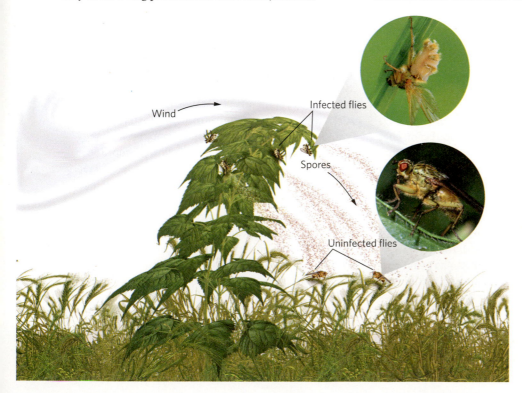

Figure 15.16 Transmission of fungal spores in the yellow dungfly. Infected flies climb up a plant, position themselves upside down on the underside of a leaf, and die as the spores erupt from their abdomen. In contrast, uninfected flies position themselves lower on plants and sit upright on the top sides of leaves. Based on D. P. Maitland, A parasitic fungus infecting yellow dungflies manipulates host perching behaviour, *Proceedings of the Royal Society of London* Series B 258 (1994): 187–193. Photos by (top) Biosphoto/Julien Boisard; and (bottom) Michael Hutchinson/ naturepl.com.

Wind

Infected flies

Spores

Uninfected flies

(a)

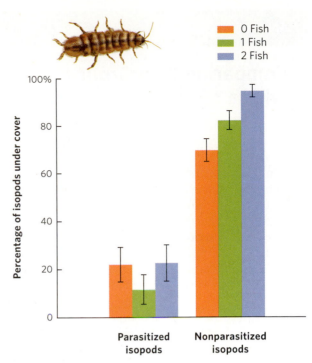

Figure 15.17 Fungal control of the perching behaviour in the yellow dungfly. **(a)** Compared to uninfected dungflies, infected dungflies perch higher on plants. Error bars are 1 standard deviation. **(b)** Uninfected dungflies never perch on the underside of a leaf whereas infected flies do so 91 percent of the time. Data from D. P. Maitland, A parasitic fungus infecting yellow dungflies manipulates host perching behaviour, *Proceedings of the Royal Society of London* Series B 258 (1994): 187–193.

Figure 15.18 Parasite-induced changes in isopod behaviour. When isopods are not infected by parasites, they generally stay under cover whether there are fish present or not. When they are infected, the isopods move out from under the cover, which makes them susceptible to predatory fish. Error bars are standard errors. Data from L. J. Hetchel et al., Modification of antipredator behaviour of *Caecidotea intermedius* by its parasite *Acanthocephalus dirus, Ecology* 74 (1993): 710–713.

dungflies, they found that uninfected flies perched lower on plants and never perched on the underside. You can view these data in Figure 15.17.

For parasites that require a series of different hosts to complete their life stages, it is challenging to find a way to get transmitted from one host to another. Some parasites—such as trematodes that first infect snails and subsequently infect tadpoles—have evolved a simple strategy of leaving the body of the first host and then searching for the second host. Other parasites have evolved ways of manipulating the behaviour of the first host to ensure that the second host consumes it. We saw an example of this with the amber snail discussed at the beginning of the chapter. There are

many examples of parasites taking control of the host's behaviour. For example, mice can inadvertently consume a parasitic protist, *Toxoplasma gondii,* when feeding near cat feces. Once the mouse becomes infected, it no longer avoids an area marked by the smell of bobcat (*Lynx rufus*) urine but, instead, is mildly attracted to it. As a result, the mouse is more likely to be eaten by the bobcat, which is a valuable outcome for the protist since it can only reproduce in the gut of a cat.

Similarly, small crustaceans known as isopods (*Caecidotea intermedius*) normally hide in refuges from predatory fish. However, if the isopods are infected by a parasitic worm (*Acanthocephalus dirus*), they spend less time in the refuge and more time out in the open where the fish are more likely to notice them, as you can see in Figure 15.18. This manipulation of the isopod's behaviour is beneficial to the parasite because fish serve as the parasite's second host.

HOST ADAPTATIONS

Host species have evolved a range of defences to combat parasite invaders. Earlier in the chapter we noted that the immune system detects parasites that attempt to enter the body and tries to kill them. Other elements of the immune system are more general and

Comparing Two Groups with a *t*-Test

When the dungfly researchers examined the perching height of infected and uninfected flies, they concluded that the infected flies perched significantly higher on vegetation. How did they come to that conclusion? They compared the mean heights for the infected and uninfected flies and used a statistical test to determine if the two means were different.

In Chapter 14, we discussed how scientists typically consider two means to be different if we can sample their distributions many times and find that the means of those sampled distributions overlap less than 5 percent of the time. We commonly call this critical cutoff *alpha*—abbreviated as α. Therefore we say that our critical cutoff is $α = 0.05$. To determine if the means of the two groups are significantly different, we need to know three things from each group: the mean, the variance, and the sample size.

With these three parameters, we can calculate a **t-test**, which is a test that determines if the distributions of data from two groups are significantly different. An important assumption of the *t*-test is that the values from both groups are normally distributed.

We begin by calculating the *differences between the means* of the two groups:

$$\overline{X}_1 - \overline{X}_2$$

where $\overline{X}_1$ is the mean of group 1 and $\overline{X}_2$ is the mean of group 2.

Next we calculate the *standard error of the difference in the two means*:

$$\sqrt{\frac{s_1^2}{n_1} + \frac{s_2^2}{n_2}}$$

where s_1^2 is the sample variance of group 1, s_2^2 is the sample variance of group 2, n_1 is the sample size of group 1, and n_2 is the sample size of group 2 (you can review the concept of sample variances in Chapter 2).

The next step is to divide the first calculated value by the second calculated value:

$$t = \frac{\overline{X}_1 - \overline{X}_2}{\sqrt{\frac{s_1^2}{n_1} + \frac{s_2^2}{n_2}}}$$

t-test A statistical test that determines if the distributions of data from two groups are significantly different.

Figure 15.19 Self-medicating chimpanzees. When chimpanzees become ill from intestinal parasites, they swallow whole leaves and chew bitter twigs of plants not normally in their diet. These leaves and twigs reduce the number of parasites in their body and help them recover. Photo by NHPA/Photoshot.

do not recognize particular species of pathogens as foreign organisms. Some species of plants and animals can produce antibacterial and antifungal chemicals that kill bacterial and fungal parasites. For example, researchers have recently reported that many species of amphibians naturally release antimicrobial peptides onto their skin, which inhibit the growth of the deadly chytrid fungus.

Host species have also evolved mechanical and biochemical defences to combat parasites. We see this with chimpanzees living in Tanzania that become infected with intestinal worms. Instead of eating their normal food diet, chimps that are ill with the intestinal worms pick a few leaves from plants in the genus *Aspilia* and swallow them whole (Figure 15.19). The leaves of these plants are covered in tiny hooks and, as the leaves pass through the chimp's digestive system, the hooks pull nematode parasites (*Oesophagostomum stephanostomum*) out of the digestive tract so they can be expelled from the chimp's body with feces. Infected chimps also chew on bitter twigs from the *Vernonia* plant, something that healthy chimps do not do. The sick chimps become well within 24 hours because the twigs contain chemicals that kill a variety of different parasites. These plants are also consumed by people in

As you can see from this equation, the value of t becomes larger as the difference between the means becomes larger, or the standard error of the difference in the means becomes smaller.

For example, consider two groups of data with the following parameters:

GROUP	MEAN	SAMPLE VARIANCE	SAMPLE SIZE
1	20	10	5
2	10	10	5

In this case:

$$t = \frac{20-10}{\sqrt{\frac{10}{5}+\frac{10}{5}}} = \frac{10}{\sqrt{2+2}} = 5$$

The last thing we need to calculate is known as the degrees of freedom, which is defined as the sum of the two sample sizes minus 2. In this example, the degrees of freedom is:

$$5+5-2=8$$

With this value of t calculated, we can then determine if our value exceeds a critical level of t and therefore determine whether the means are significantly different. We can find the critical value of t from a statistical table. If you look at the t-table in the "Statistical Tables" section at the back of this book, you can find the column that uses an alpha value of 0.05 and then find the row that contains 8 degrees of freedom. The critical t-value is the number found in this row and column. In this case, the critical t-value is 2.3. Because our calculated t-value exceeds the critical t-value, we can conclude that the two means differ significantly.

YOUR TURN For the following data, calculate the t-value.

GROUP	MEAN	SAMPLE VARIANCE	SAMPLE SIZE
1	10	4	8
2	5	4	8

Next calculate the degrees of freedom.

Based on your calculated t-value and the critical t-value using $\alpha = 0.05$, determine if the two groups are significantly different.

the region when they experience symptoms of parasite infections. In short, both chimps and humans use plants to medicate themselves against parasitic infections.

COEVOLUTION

When we consider the evolutionary forces that affect adaptations of parasites and hosts, we often find that as one species in the interaction adapts, the other species responds by adapting as well. As we discussed in Chapter 14, when two or more species continue to evolve in response to each other's evolution we call it coevolution. When all species involved in an interaction coevolve together, no one species is likely to get an upper hand. For example, when a parasite evolves an advantage that makes it more effective at infecting hosts, the hosts will then be under stronger selection to evolve defences against that parasite.

We can see an example of coevolution between parasites and hosts in the case of rabbits in Australia. In the 1800s European rabbits were introduced to many regions of the world where they had not lived historically. In Australia, for instance, rabbits were first introduced from England in 1859. Within a few years, local ranchers were erecting fences to keep

them out and organizing shooting parties to help control the size of the population. Eventually, hundreds of millions of rabbits ranged throughout most of the continent, destroying sheep pasture lands and thereby threatening wool production (Figure 15.20). The Australian government tried

Figure 15.20 A growing rabbit population in Australia. After being introduced to Australia in 1859, the rabbit population exploded and consumed nearly all of the vegetation that was needed by ranchers for raising sheep. Photo © John Carnemolla/Corbis.

poisons, predators, and other control measures, all without success.

In 1950 the government tried a new approach. A virus known as *Myxoma* had been discovered in South American rabbits. The virus—which was carried by mosquitoes—causes a disease known as myxomatosis. The disease had minor effects on South American rabbits, but it turned out to be highly lethal to European rabbits; an infected rabbit dies within 48 hours. The Australian government introduced the virus and initially it had a devastating effect on the rabbits. Among infected rabbits, 99.8 percent died and the rabbit population declined to very low numbers. However, 0.2 percent of the infected rabbits proved to be resistant and passed on their resistant genes. Before the introduction of the *Myxoma* virus, a few rabbits possessed resistant genes, but they had never been favoured. In a second outbreak of the disease, only 90 percent of infected rabbits died. By the third outbreak, only 40 to 60 percent of the rabbits died and the rabbit population of Australia started to increase. You can see the decline in rabbit mortality over time in Figure 15.21.

The declining percentage of rabbits that died was due partly to the evolution of increased resistance in the rabbit population, but because the rabbit population was initially killed in large numbers, the rapidly dwindling host population also favoured any virus strains that infected the rabbits without killing them. Dead rabbits represent a dead end for the virus, especially if the rabbit dies before it is bitten by a mosquito, which is the only way the virus can be transmitted.

Over time, the interaction between the rabbits and the *Myxoma* virus resulted in a more resistant host population and a less lethal pathogen. This allowed the rabbit population to rebound and the virus to

Figure 15.21 Coevolution of Australian rabbits and the *Myxoma* virus. In rabbit populations that had never been exposed to the virus, nearly all infected rabbits died. However, a small percentage of resistant rabbits survived and passed on their resistant genes to the next generation. At the same time, selection favoured less lethal virus strains, which could reproduce and get transmitted by mosquitoes that only feed on live rabbits. Since rabbit populations evolved higher resistance and viruses evolved to be nonlethal, rabbit populations that experienced more epidemics of the disease experienced less mortality. Data from F. Fenner and F. N. Ratcliffe, *Myxomatosis* (Cambridge University Press, 1965).

persist. This dynamic is probably what researchers observed in South America, where the rabbit population and *Myxoma* virus persisted together for a long time. To meet the Australian government's goal of controlling the rabbit population in Australia, scientists continue to introduce new strains of the virus that are highly lethal to rabbits because the new strains have no history of coevolution with the Australian rabbits. In this way, they maintain the effectiveness of the *Myxoma* virus as a pest control agent.

ECOLOGY TODAY CONNECTING THE CONCEPTS

OF MICE AND MEN ... AND LYME DISEASE

The black-legged tick. The tick, also known as a deer tick, is the vector of the bacteria that causes Lyme disease in humans. Photo by Scott Bauer/USDA-ARS.

Back in the 1970s, a number of children in a Connecticut community experienced a suite of mysterious symptoms. Many had a bull's-eye–shaped rash on their skin followed by flulike symptoms that developed into other symptoms similar to arthritis and various neurological disorders. Physicians discovered that the children had been infected by a pathogenic bacterium (*Borrelia burgdorferi*). Because the community was in the town of Lyme, Connecticut, the disease came to be known as Lyme disease.

The parasite that causes Lyme disease is carried by ticks that feed on wild animals. Occasionally these ticks latch onto people and transmit the parasite. In North America, the primary vector is the black-legged

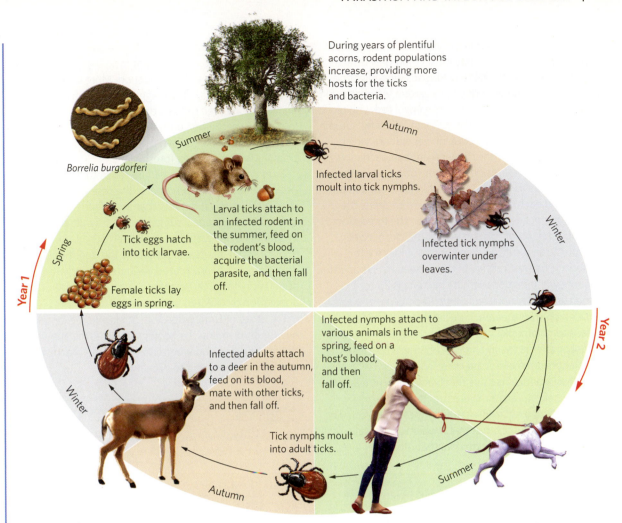

During years of plentiful acorns, rodent populations increase, providing more hosts for the ticks and bacteria.

Borrelia burgdorferi

Summer

Autumn

Infected larval ticks moult into tick nymphs.

Winter

Infected tick nymphs overwinter under leaves.

Larval ticks attach to an infected rodent in the summer, feed on the rodent's blood, acquire the bacterial parasite, and then fall off.

Spring

Tick eggs hatch into tick larvae.

Female ticks lay eggs in spring.

Year 1

Infected nymphs attach to various animals in the spring, feed on a host's blood, and then fall off.

Year 2

Infected adults attach to a deer in the autumn, feed on its blood, mate with other ticks, and then fall off.

Winter

Tick nymphs moult into adult ticks.

Summer

Autumn

The cycle of Lyme disease. A diverse array of species affects the prevalence of the bacterium that causes Lyme disease. The cycle begins when ticks attach themselves to infected rodents and become infected with the bacterium. After they drop from the rodents, they moult into nymphs and spend the winter under leaves. In the spring the nymphs attach themselves to hosts again and then drop and moult into adults. As adults, most ticks attach themselves to deer, search for mates, and then lay eggs. During years of high acorn production by oak trees, there is an abundance of food for rodents that produces large rodent populations. Large rodent populations lead to large tick populations that can infect a variety of animals including deer and humans.

tick (*Ixodes scapularis*), also known as the deer tick. The ticks and the parasite probably existed in North America for thousands of years before the disease was identified. According to the Centers for Disease Control and Prevention, currently 20 000 to 30 000 cases of Lyme disease occur every year in the United States, most in the Northeast and Midwest. In Canada—where incidences of Lyme disease have been recorded since 2009—approximately 260 cases were reported in 2011.

Over 4 decades, ecologists and medical researchers have determined that to understand Lyme disease requires a knowledge of the ecological community in which the parasite lives. First they had to determine how ticks obtain the parasite. They found that 99 percent of newly hatched ticks do not carry the parasite, so vertical transmission is not likely. Instead, they discovered that the tick obtains the parasite from infected hosts. But because the black-legged tick can live on a wide variety

of hosts, the researchers needed to determine which hosts affected the abundance of infected ticks.

By examining patterns of abundance in nature and conducting manipulative experiments, the researchers were able to work out the life cycle of the black-legged ticks. The ticks are predominantly creatures of the forest. Larval ticks hatch from eggs each summer but since they are unable to climb very high on vegetation, they can only infect small animals living close to the ground, such as birds and rodents. The most commonly infected rodents are chipmunks (*Tamias striatus*) and white-footed mice (*Peromyscus leucopis*). These two species can be infected without exhibiting any signs of disease, so they act as reservoir species that maintain the bacterial population. After feeding on the rodents for a few days, the ticks drop off their host and moult into nymphs.

As nymphs, the ticks spend the winter living under the leaves on the forest floor. The following spring, they climb a short distance up in the vegetation and wait for another host to pass by. They most commonly attach to birds, small mammals, and any humans who might be taking a walk in the woods or in a garden. After attaching themselves to a host and becoming engorged with blood, the ticks drop off again and moult into adults by autumn. Adult ticks can climb higher up on vegetation, which allows them to jump onto larger mammals such as white-tailed deer. The deer's body not only offers a blood meal but also serves as a location for adult male and female ticks to find each other and to mate. From there, they drop off the deer and lay their eggs on the forest floor each autumn. All of this movement of ticks between multiple hosts results in many opportunities for the bacteria to be transmitted. In fact, researchers found that frequency of infection with the bacteria is 25 to 35 percent for nymphs and 50 to 70 percent for adults.

Based on this research, it became clear that deer, mice, and chipmunks were keys to the tick's life cycle. But what affects the abundances of these species? The researchers discovered one more key player: oak trees. Oak trees produce massive numbers of acorns every few years and acorns are a major food item for deer, chipmunks, and mice. In high-acorn years, tick-carrying deer gather under the oak trees, which causes an aggregation of reproducing ticks that drop their eggs under the trees. Chipmunks and mice are also attracted to the acorns since the abundant food permits higher survival and increased reproduction. More rodents and an increased density of tick eggs cause an increase in the number of infected rodents the following summer. In low-acorn years when both deer and rodents spend more time in maple forests, the prevalence of infected ticks shifts from the oak forests.

To examine what steps can be taken to reduce the risk of Lyme disease to humans, the researchers also used mathematical models that included all of the key species. These models suggested that, while reductions in deer densities would have little effect on the population of infected ticks unless the deer population were completely eradicated, reductions in the rodent population could cause a large reduction in the population of infected ticks.

The story of Lyme disease is an excellent example of how understanding the ecology of parasites allows us to predict when and where they pose the greatest risk to wild organisms and to humans.

SOURCES: Ostfeld, R. 1997. The ecology of Lyme-disease risk. *American Scientist* 85: 338–346.

Ostfeld, R., et al. 2006. Climate, deer, rodents, and acorns as determinants of variation in Lyme-disease risk. *PLOS Biology* 4: 1058–1068.

SUMMARY OF CHAPTER CONCEPTS

■ **Many different types of parasites affect the abundance of host species.** Ectoparasites live on host organisms whereas endoparasites live in host organisms. As a group, parasites include a wide range of species that include plants, fungi, protozoa, helminths, bacteria, viruses, and prions. Among parasites that cause diseases—known as pathogens—those that have recently become abundant are called emerging infectious diseases.

■ **Parasite and host dynamics are determined by the parasite's ability to infect the host.** The transmission of parasites can be horizontal— either through direct transmission or transmission by a vector—or vertical from parent to offspring. The ability to infect a host also depends on the parasite's mode of entering the host, its ability to infect reservoir species, its ability to jump to new host species, and its ability to avoid the host's immune system.

■ **Parasite and host populations commonly fluctuate in regular cycles.** These fluctuations occur because transmission increases with host density but decreases as an increased proportion of the host population develops immunity. These fluctuations can be modelled using the S-I-R model.

■ **Parasites have evolved offensive strategies while hosts have evolved defensive strategies.** Natural selection has favoured parasites that can improve their probability of transmission, including manipulations of host behaviour. Hosts have evolved both specific and general immune responses to combat host infection. Hosts also can employ mechanical and biochemical defences against parasites. Coevolution occurs when the parasite and host continually evolve in response to each other.

REVIEW QUESTIONS

1. Compare and contrast the advantages and disadvantages of life as an ectoparasite versus an endoparasite.

2. Why are parasites often not very harmful to hosts in their native range but highly detrimental to hosts in an introduced range?

3. What allows a parasite to jump to a new species of host?

4. If prions are not normally transmittable among cattle, what procedure caused the spread of mad cow disease?

5. What constitutes an emerging infectious disease?

6. Contrast horizontal versus vertical transmission of a parasite.

7. Using the basic S-I-R model of parasite and host dynamics, explain why the proportion of infected individuals in the population declines over time.

8. In the S-I-R model of parasite and host dynamics, how does the outcome change if we allow new susceptible individuals to be born into the population?

9. When using a *t*-test, what is an important assumption regarding the distribution of the data?

10. Explain why pathogens that are highly lethal to their hosts might become less lethal over time.

PRACTICING ECOLOGY: THE COST OF HAVING A BODYGUARD

Jacques Brodeur from Université de Montréal and his colleagues study interactions between herbivores and their enemies. Photo by Anne Audy.

As we have seen in this chapter, many parasites complete part of their life cycle by manipulating the body of a host. However, in order to control its host, a parasite must expend energy. Often, a host ceases to feed once infected, so the energy that the parasite uses must come from stored energy that the parasite could otherwise use for its own development. In addition, the body manipulation necessarily requires that the parasite transfer chemical compounds to the host's brain, and these chemicals are likely expensive to produce. Do these costs detract from the fitness of the parasite?

Dinocampus coccinellae is a parasitic wasp that lays a single egg inside the spotted lady beetle (*Coleomegilla maculata*). After feeding on the beetle's eggs, gonads, and fat reserves, the larva paralyzes the beetle and emerges from the anal gland as a pupa. The pupa then spins a cocoon around the legs of the beetle and develops for an average of 7 days until it emerges as an adult wasp. Only 25 percent of beetles survive during the time that the pupa is in its cocoon. A beetle that does survive exhibits bodyguard behaviour over the cocoon; the pupa makes the beetle twitch when provoked by an external stimulus and this behaviour reduces predation on the pupa. However, the pupa must use energy to control the beetle's behaviour. We might hypothesize that this energy expense comes at a cost to the future fitness of the parasitic wasp.

Jacques Brodeur from the Université de Montréal and his colleagues tested this hypothesis by observing parasitized beetles in the laboratory. They collected wasps and beetles from the field and allowed the wasps to parasitize the beetles in the lab. After pupae emerged from the beetles and spun cocoons, the researchers recorded the number of days each beetle exhibited twitching behaviour, which corresponds to the amount of time the beetle survived and the pupa was able to control its movement. Once the wasps emerged from their cocoons, the researchers counted the number of eggs in female wasps as a measure of wasp fitness. You can view their data in the table.

DURATION OF BEETLE TWITCHING BEHAVIOUR (DAYS) AND NUMBER OF EGGS IN EMERGENT FEMALE WASPS

DURATION OF TWITCHING BEHAVIOUR	NUMBER OF EGGS
2.9	103
4.9	102
4	102
7.9	101
8.9	99
9	98
7	100
5.9	100
6.1	101
7	103
4.9	101
8	99

GRAPH THE DATA

1. Create a scatterplot graph with duration of twitching behaviour on the *x* axis and number of eggs in emergent wasps on the *y* axis.

2. Visually estimate and draw a regression line of best fit through the data.

INTERPRET THE DATA

1. Explain how the data support or refute the hypothesis that the energy expense of controlling a lady beetle host comes at a cost to future fitness of the developing wasp.

2. Unlike animals that are born with a fixed number of eggs, female *D. coccinellae* wasps are able to produce eggs throughout their lifetime. How might the later production of eggs compensate for the fitness cost of controlling the behaviour of a host beetle?

3. In the same study, the researchers found no relationship between duration of beetle twitching behaviour and the emergent wasp's lifespan. However, the amount of food after emergence positively correlated with wasp lifespan. What does this suggest about the allocation of energy for fecundity versus energy allocated to survival in *D. coccinellae*?

SOURCE: Maure, F., J. Brodeur, N. Ponlet, J. Doyon, A. Firlej, E. Elguero, and F. Thomas. 2011. The cost of a bodyguard. *Biology Letters*, 7: 843–846.

CHAPTER 16

Competition

The Complexity of Competition

In the deserts of the American Southwest, many species of herbivores make their living in unique ways. These include large mammals such as deer and antelope, several species of rodents, and numerous species of ants. For a long time, no one considered the possibility of strong competition among animal groups of very different sizes because ecologists assumed that when competition existed, only animals of similar size and type would compete with each other. In the 1970s, ecologist James Brown and his colleagues questioned this assumption. Their research revealed that herbivores of different sizes compete with each other, and that this competition can create a complex cascade of effects.

> "While rodents and ants compete over the short term, the ants rely on the rodents over the long term."

Brown and his colleagues conducted field experiments in which they manipulated the presence of seed-eating rodents and seed-eating ants. They built large fences around several plots of land that enclosed rodents alone, ants alone, rodents and ants together, or neither ants nor rodents. This allowed researchers to determine whether rodents and ants competed for their primary food resource, seeds.

The enclosure experiment provided clear evidence that rodents and ants compete for seeds. For example, when ants were excluded, the number of rodents increased by 18 percent and individual rodent growth increased by 24 percent. When the rodents were excluded, the number of ant colonies increased by 71 percent. The researchers also found that when either group was excluded, seed abundance in the enclosures increased substantially but when both groups were excluded, the number of seeds increased even more. These results provided further evidence that the two groups compete for seeds. The researchers also quantified the seed preferences of each group; although rodents preferred larger seeds and ants preferred smaller seeds, the size of seeds that both consumed overlapped a good deal.

The interactions between rodents and ants also shaped competition among plants in the experimental plots. When either rodents or ants were excluded, more seeds were left to germinate. This increased the plant density and led to more competition among plants. Although more competition did not reduce plant survival, it did reduce plant growth and reproduction.

For 4 decades researchers continued to observe the original plots as well as additional plots they created later. Over this period they saw a number of surprising effects. For example, they found that over the long term rodents actually helped provide more seeds for the ants. During the first year, eliminating rodents from the enclosures caused an increase in the number of ants, but several years later the number of ants in these enclosures began to decline. To determine the reason for the decline in

Kangaroo rat. Experiments on seed-eating kangaroo rats and ants in Arizona have revealed that the two groups of herbivores compete with each other in complex ways. Pictured here is a bannertail kangaroo rat (*Dipodomys spectabilis*) with its cheek pouches stuffed with seeds. Photo by Mary McDonald/naturepl.com.

ants, the researchers looked carefully at the seed diets of the two competing groups and found that the combination of rodents and ants collectively consumed the entire range of seed sizes. Competition for seeds kept rodent and ant densities sufficiently low so that neither species could eliminate its seed sources. It also promoted a diversity of plants that continually produced a full range of seed sizes. However, when rodents were excluded from the enclosures for several years, ants continued to consume the smallest seeds. With few small seeds left to germinate, the plant species that make small seeds dwindled. In addition, small-seed plant species were competitively inferior to the large-seed plant species. Because of these factors, plants with small seeds declined so much that ants no longer had enough food, and the ant population declined. In short, while rodents and ants compete over the short term, the ants rely on the rodents over the long term to maintain an adequate supply of plants that have the smaller seeds they can eat.

The interactions between ants and rodents in the desert highlights the fact that limited resources can control the abundance of competing populations. Such competition is pervasive and the interactions between competing species can be complex. In this chapter, we will see that competition can occur in a wide variety of ways that shape the composition of communities.

SOURCES: J. H. Brown and D. W. Davidson, Competition between seed-eating rodents and ants in desert ecosystems, *Science* 196 (1977): 880–882.

D. W. Davidson et al., Granivory in a desert ecosystem: Experimental evidence for indirect facilitation of ants by rodents, *Ecology* 65 (1984): 1780–1786.

K. M. Thibault et al., Long-term insights into the influence of precipitation on community dynamics in desert rodents, *Journal of Mammalogy* 91 (2010): 787–797.

CHAPTER CONCEPTS

- Competition occurs when individuals experience limited resources.

- The theory of competition is an extension of logistic growth models.

- The outcome of competition can be altered by abiotic conditions, disturbances, and interactions with other species.

- Competition can occur through exploitation or direct interference, or it may be apparent competition.

In Chapter 1, we defined competition as a negative interaction between two species that depend on the same limiting resource to survive, grow, and reproduce. Competition can help determine where a species can live in nature and how abundant a population can become. It can occur among many groups of organisms including predators, herbivores, and parasites.

In this chapter, we will explore different types of competition and the resources for which species compete. We will then examine models of competition, which extend the Lotka-Volterra models introduced in Chapter 14. Competition models allow us to predict the conditions that determine when a species will win a competitive interaction. Knowing how species compete for a resource is important, but we must also consider how other interactions might alter or even reverse the expected outcomes of competition, including abiotic effects, disturbances, and interactions with other species. At the end of the

chapter we will examine a variety of cases that appear to describe competition but actually reflect other processes, including predation and herbivory.

Competition occurs when individuals experience limited resources

When we study competition, we differentiate between **intraspecific competition**, which is competition among individuals of the same species, and **interspecific competition**, which is competition among individuals of different species. We considered intraspecific competition in our discussion of negative density dependence in Chapter 12, where we

Intraspecific competition Competition among individuals of the same species.

Interspecific competition Competition among individuals of different species.

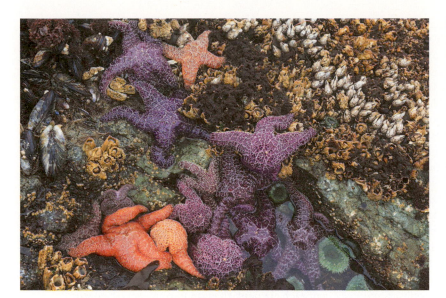

Figure 16.1 Competition for space. Sessile organisms that live in the rocky intertidal biome compete for open space on the rocks as a point of attachment. This photo from Whaler Islets off the coast of Vancouver Island, British Columbia, shows a number of sessile organisms at low tide including mussels, giant green sea anemones (*Anthopleura xanthogrammica*), goose barnacles (*Lepas anserifera*), and ochre sea stars (*Pisaster ochraceus*). Photo by Ron Watts/ age fotostock.

saw that an increase in a population's density causes a decline in the growth rate of the population. We have discussed interspecific competition much less. Interspecific competition can cause the population of one species to decline and eventually die out. Both intraspecific and interspecific competition play substantial roles in determining the distribution and abundance of species on Earth.

In this section, we will examine how competition for a limited resource can cause one species to outcompete another. We will explore the wide variety of resources that are available to organisms, including resources that are fixed in abundance and those that are renewable. With our understanding of resources, we will then investigate the importance of the most limiting resource in a population and examine patterns of competition between closely and distantly related species.

THE ROLE OF RESOURCES

A **resource** is anything an organism consumes or uses that causes an increase in the growth rate of a population when it becomes more available. For plants, resources generally include sunlight, water, and soil nutrients such as nitrogen and phosphorus. Each is used by most plants and plays a role in the growth of plant populations. Resources for animals generally include food, water, and space. For example, animals such as mussels and barnacles spend most of their lives attached to rocks in the intertidal zone along the ocean shores (Figure 16.1). Open space is a critical resource for them because as the rocks become more crowded there is less space to grow. As a result, the growth and fecundity of adults decline, and there are few places on the rocks for offspring to settle. Similarly, many birds may compete for a

limited number of nest sites or cavities, and many prey species compete for a limited number of holes and crevices in which they can hide from predators. In each of these cases, when more space becomes available, the populations can increase.

Ecological factors that cannot be consumed or used are not considered resources. Temperature, for instance, plays a major role in the growth and reproduction of organisms in nature, but temperature is not a resource because it is not consumed or used. The same is true for other abiotic factors, including pH and salinity. These can affect the growth and reproduction of organisms, but because organisms do not consume or use them, they are not resources and species do not compete for them.

Renewable versus Nonrenewable Resources

We can categorize resources as either *renewable* or *nonrenewable*. **Renewable resources** are constantly regenerated. For example, the rodents and ants described at the beginning of the chapter compete for seeds, and every year new plants grow and renew the seed supply. Similarly, sunlight is continually generated by the Sun. In contrast, **nonrenewable resources** are not regenerated. For instance, space is a resource that typically has a fixed availability. If we think of the rocky intertidal habitat, there is a fixed

---

Resource Anything an organism consumes or uses that causes an increase in the growth rate of a population when it becomes more available.

Renewable resources Resources that are constantly regenerated.

Nonrenewable resources Resources that are not regenerated.

number of rocks to which algae and animals can attach; space becomes available only when a competitor leaves or dies.

Renewable resources can originate from either inside or outside the ecosystem in which the competitors live. For example, dead leaves that fall into streams from the surrounding forest serve as food for stream insects. Because these resources come from outside the system, however, competition can reduce resource abundance but cannot affect the rate of the resource supply. Moreover, resources originating from outside the system do not respond to the rate of resource consumption. Sunlight continually strikes the surface of Earth regardless of the rate at which plants and algae consume it, and the amount of local precipitation is largely independent of the rate at which plants use water.

Competitors can affect both the supply of and demand for resources that originate within the ecosystem. When competing herbivores or competing predators consume another species in the system, they reduce the supply of the species that they consumed. However, reducing the abundance of a consumed species affects its growth and reproduction, so consumption also affects the future supply of the resource. We saw an example of this with the competition between rodents and ants. When rodents were eliminated from the enclosures, the increased ant population continually ate the small seeds, and after a few years there were fewer plants that produced small seeds.

In some cases, the supply rate of a renewable resource generated within an ecosystem is only indirectly affected by competitors. For example, plants take up nitrate from the soil and use it to grow. Competing herbivores that consume the plants return large amounts of nitrogen compounds back to the soil when they defecate and after death when their bodies decompose. These nitrogen compounds are further broken down by microorganisms, which release the nitrogen as nitrate, a form plants can use. The herbivores can affect the supply rate of nitrate to the plants, but the chain of events has many links and takes such a long time that the herbivores do not have an immediate effect on the rate of future plant population size through this indirect pathway.

Leibig's Law of the Minimum

Although consumers can reduce the abundance of both renewable and nonrenewable resources, not all resources limit consumer populations. For example, all terrestrial animals require oxygen, but an increase in the population of an animal species does not depress the concentration of oxygen in the atmosphere to the point that population growth is limited. Well before the concentration of oxygen can become limiting, some other resource, such as food supply, will decline in abundance to the point that it limits the population's growth.

At one time, ecologists thought that populations were limited by the single resource that was most scarce relative to demand. This is known as *Liebig's law of the minimum,* after Justus von Liebig, a German chemist who articulated the idea in 1840. According to **Liebig's law of the minimum**, each population increases until the supply of the most limiting resource prevents it from increasing further.

The amount of a resource that limits a population's growth depends on the resource. For example, the microscopic, glass-shelled algae known as diatoms require both silica and phosphate to grow and reproduce. When one diatom species, *Cyclotella meneghiniana,* is grown under different concentrations of each element, the population growth ceases whenever the concentration of silica is reduced to 0.6 micromolar (μM) or whenever the concentration of phosphate is reduced below 0.2 μM. According to Liebig's law of the minimum, whichever resource reaches its limiting value first will be the resource that regulates the growth of the diatom population.

If we know the minimum amount of a resource that is required for populations to grow, we should be able to predict which species is the best competitor for the resource. An experiment with two species of diatoms, shown in Figure 16.2, demonstrates how competition for silica reduces the availability of silica and affects the outcome of competition. When the two species are raised separately, *Asterionella formosa* and *Synedra ulna* both experience rapid population growth followed by a plateau as they reach their carrying capacity, shown in Figure 16.2a. When *Asterionella formosa* reaches its carrying capacity, it drives down the abundance of silica to 1 μM. However, when *Synedra ulna* reaches its carrying capacity, it drives down the abundance of silica to 0.4 μM, which is not enough silica to support the other diatom's population. As a result, *Synedra ulna* should outcompete *Asterionella formosa.* When a researcher put both species together, that is exactly what happened. As illustrated in Figure 16.2b, the two species drove the abundance of silica down to a level that allowed *Synedra ulna* to persist but caused *Asterionella formosa* to decline to extinction. The outcome between these two diatoms is one that is commonly observed in nature: when two species compete for a single limiting resource, the species

---

Liebig's law of the minimum Law stating that a population increases until the supply of the most limiting resource prevents it from increasing further.

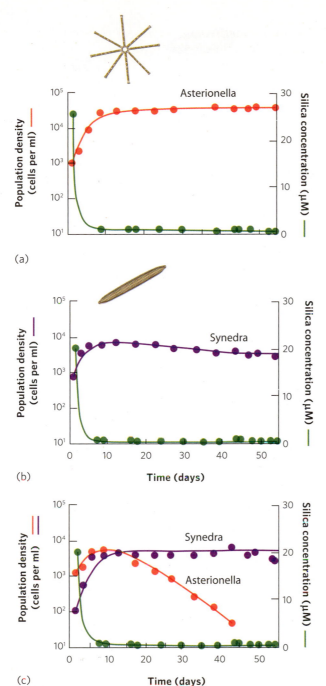

(a)

(b)

(c)

Figure 16.2 Competition for the most limiting resource. When raised separately, **(a)** *Asterionella* and **(b)** *Synedra* diatoms grow rapidly and then persist with population sizes near their carrying capacity. At their carrying capacity, both species dramatically reduce the abundance of silica, although *Synedra* causes the silica to be driven down to a lower abundance than *Asterionella*. **(c)** When combined, the two species again drive down the abundance of silica to the point that there is still enough to meet the minimum requirements of the *Synedra* population but not enough to meet the minimum requirements of the *Asterionella* population. As a result, the *Asterionella* population declines. Data from D. Tilman et al., Competition and nutrient kinetics along a temperature gradient: An experimental test of a mechanistic approach to niche theory, *Limnology and Oceanography* 26 (1981): 1020–1033.

that persists is the one that can drive down the abundance to the lowest level.

Interactions Among Resources

Leibig's law of the minimum assumes that each resource has an independent effect on the growth of a population. In other words, it assumes that if a given resource limits the growth of individuals and populations, increasing the availability of other resources will not improve this growth. However, this is not always the case, as seen in research on small balsam (*Impatiens parviflora*), a plant that is common in English woodlands. Researchers wanted to know if this plant would respond independently to two different resources: fertilizer and light. They sowed seeds of the plants in pots filled with soil. Control pots received only water while fertilized pots received a solution containing both water and fertilizer. These two groups of pots were then grown under one of four different light intensities for 5 weeks. As you can see in Figure 16.3, plants grown in low-nutrient soil experienced only a small increase in growth when the amount of light was increased. Similarly, plants grown at a low light intensity experienced a small growth increase when soil nutrients were increased.

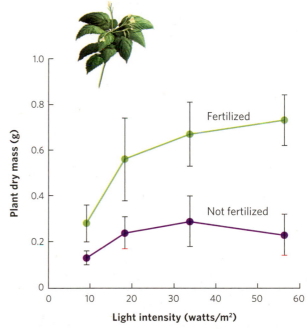

Figure 16.3 Interactions among resources. The small balsam plant experienced little growth when light intensity was increased under low soil fertility. However, when light intensity and fertility were both increased, the large amount of growth that followed was much higher than the sum of the two separate effects. Error bars are 95 percent confidence intervals. Data from W. J. H. Peace, and P. J. Grubb, Interaction of light and mineral nutrient supply in the growth of *Impatiens parviflora*, *New Phytologist* 90 (1982): 127–150.

However, plants grown under high soil fertility and high light intensity experienced an increase in growth that was much larger than the sum of the separate effects of fertility and light intensity. This means that an increase in one resource can have a much larger effect on a population when there is a simultaneous increase in a second resource.

THE COMPETITIVE EXCLUSION PRINCIPLE

A classic study of competition between species was conducted by Russian biologist Georgyi Gause in the 1930s. As you may recall from Chapter 12, Gause conducted laboratory experiments that explored how populations of protists in the genus *Paramecium* grew when living alone or together with limited resources. He began by growing two species—*P. aurelia* and *P. caudatum*—separately in test tubes. The test tubes contained a fixed amount of food—the bacteria *Bacillus pyocyaneus*. When grown separately, the population of each paramecium species initially experienced rapid growth and then began to plateau as it reached its carrying capacity. You can see these data in the first two graphs of Figure 16.4. However, when the two species were grown together in the test tube, Gause observed a different outcome. As you can see in Figure 16.4c, the population of *P. aurelia* persisted in the test tube but the population of *P. caudatum* declined to very low levels by the time the experiment ended. In short, *P. aurelia* was the superior competitor, presumably by driving down the abundance of the bacteria to such a low point that *P. caudatum* could not persist.

Similar competition experiments have been conducted hundreds of times using a wide variety of organisms. These experiments frequently produce the same result: one species persists and the other dies out. This common pattern led to the development of the **competitive exclusion principle**, which states that two species cannot coexist indefinitely when they are both limited by the same resource. Researchers commonly find that when two species are limited by the same resource, one species either is better at obtaining the resource or is better able to survive when the resource is scarce.

Competition Among Closely Related Species

As we saw in the case of ants and rodents, competition can occur between both closely and distantly related organisms. Charles Darwin believed that competition is most intense between closely related species because, he reasoned, they possess similar traits, probably consume similar resources, and, therefore, have the potential to compete strongly.

After Darwin, naturalists observed that closely related species living in the same region often grow in different habitats. They hypothesized that because these species would compete strongly for

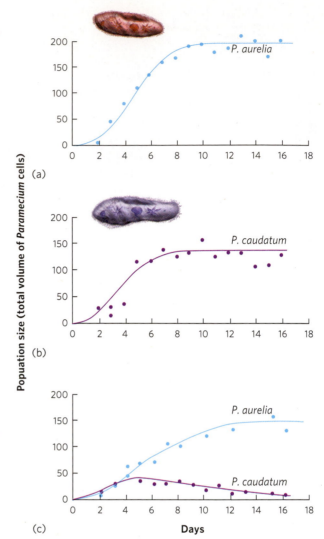

Figure 16.4 Competition between two species of protists. Gause raised different species of protists and measured the populations in terms of the total volume of cells in the container. When he raised each species separately on a fixed daily ration of bacteria, the population of **(a)** *P. aurelia* and **(b)** *P. caudatum* grew rapidly until they reached their carrying capacity. **(c)** When the two species were grown together, *P. aurelia* continued to grow well but *P. caudatum* was reduced to a very low population density. Because *P. caudatum* was still present in low numbers, *P. aurelia* did not attain the same population size as when it was raised alone. Data from G. F. Gause, *The Struggle for Existence* (Williams & Wilkins, 1934).

the same resources, natural selection would favour differences in habitat use. These differences in habitat use would allow each species to have a competitive advantage within its preferred habitat and to have a competitive disadvantage in habitats preferred by closely related species.

Competitive exclusion principle The principle that two species cannot coexist indefinitely when they are both limited by the same resource.

The first experimental test of this hypothesis was conducted in 1917 by the British botanist Arthur Tansley, who worked with two species of small perennial plants known as bedstraw. Heath bedstraw (*Galium saxatile*) typically lives on acidic soils whereas white bedstraw (*G. sylvestre*) typically lives on alkaline soils. To determine whether each species competed best under the conditions where it naturally occurred, Tansley and his colleagues planted them separately and together in deep boxes containing either acidic or alkaline soils, as illustrated in Figure 16.5. Because the boxes were located at a single site—the Botanic Garden in Cambridge, England—any differences in growth would be due to the types of soils in which the bedstraw was planted.

When planted separately, each species germinated and grew in both types of soil. However, each species presented more vigorous germination and growth in the soil type characteristic of its natural

Heath bedstraw grown alone

White bedstraw grown alone

Both species grown together

Alkaline soil   **Acidic soil**

Figure 16.5 Tansley's bedstraw experiment. White bedstraw naturally lives in alkaline soils and heath bedstraw naturally lives in acidic soils. Each species was planted either in its preferred soil or in the preferred soil of the other species; each species grew and survived best in its own preferred soil. This experiment showed that although closely related species can be intense competitors, the competition is reduced when competitors evolve to perform best in different habitats.

habitat. When grown together on alkaline soils, the white bedstraw overgrew and shaded the heath bedstraw. When grown together on acidic soils, heath bedstraw outcompeted the white bedstraw. Tansley concluded that although both species are capable of living in both soil types when they are grown separately, intense competition from a closely related species restricts their distribution in nature to the soil type that gives them a competitive advantage. Tansley's work has inspired researchers to investigate competition between closely related species in many other groups of organisms. For example, mosses also compete with each other. Researchers investigating the competitive abilities of sphagnum and feather mosses in a boreal peat land in northeastern Alberta found that sphagnum mosses outcompete feather mosses in wooded and open areas. The scientists found competitive differences even within the same genus. *Sphagnum magellanicum* outcompeted two other sphagnum species in both types of areas.

Competition Among Distantly Related Species

While competition can be quite intense among closely related species, it can also be intense among distantly related species that consume a common resource. As we saw in Figure 16.1, open space on rocks in the intertidal biome is a resource used by numerous distantly related species such as barnacles, mussels, algae, sponges, and others, all of which compete intensely for this limited space. As we will see later in this chapter, the outcome of this competition among so many competitors depends on their abilities to compete as well as their abilities to tolerate different abiotic conditions and predation.

Another example of competition among different species occurs among animals that eat krill (*Euphausia superba*). Krill are shrimp-like crustaceans that live in the oceans surrounding Antarctica, and are consumed by virtually every type of large marine animal, including fish, squid, penguins, seals, and whales. Commercial exploitation of whales in the Southern Hemisphere has caused a decline in whale populations while penguin and seal populations have increased. This suggests that reducing the number of whales has eased competition for krill and allowed penguin and seal populations to grow.

Intense competition can also occur among distantly related species in terrestrial ecosystems. On the forest floor, spiders, ground beetles, salamanders, and birds all consume the invertebrates that live in the leaf litter. In desert ecosystems, birds and lizards eat many of the same insect species. As we discussed at the beginning of the chapter, desert ants and rodents compete for seeds. When we examine the patterns of seed sizes consumed by ants and rodents as shown in

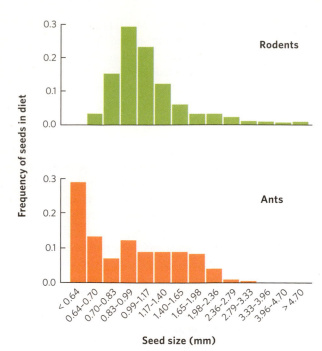

Figure 16.6 Seed size competition in rodents and ants. The two groups of herbivores in the deserts of Arizona consume seeds of similar size, although rodents prefer somewhat larger seeds and ants prefer somewhat smaller seeds. This overlap in seed consumption means that these two distantly related species can compete with each other. Data from J. H. Brown and D. W. Davidson, Competition between seed-eating rodents and ants in desert ecosystems, *Science* 196 (1977): 880–882.

Figure 16.6, we can see that the two groups both consume a wide range of seed sizes, but the ants tend to consume more of the small seeds whereas the rodents tend to consume more of the large seeds.

The theory of competition is an extension of logistic growth models

You may recall from Chapter 12 that population growth and regulation can be modeled using the logistic growth equation. Extensions of this equation can help us understand the population dynamics of two species that are competing with each other. In this section, we will begin by examining Lotka-Volterra competition models when there is a single limiting resource. We will then expand our perspective and consider models that include multiple limiting resources. Although these models represent conditions that are simplified from nature, they provide a useful foundation for exploring competitive interactions and the conditions under which species can coexist.

COMPETITION FOR A SINGLE RESOURCE

To understand how the populations of two species compete for a single resource, let's begin with the logistic growth equation introduced in Chapter 12:

$$\frac{dN}{dt} = rN\left(1 - \frac{N}{K}\right)$$

As you may recall, the part of the equation in parentheses represents intraspecific competition for resources. As the size of the population, N, approaches the carrying capacity of the environment, K, the term in parentheses approaches zero. At this point, the population's growth rate is zero, which means that the population has achieved a stable equilibrium.

If we wish to use this equation to model competition between two species that compete for a single resource, we need to consider the carrying capacity of the environment relative to the number of individuals from both species that are being supported. For example, imagine an environment that has a carrying capacity of 100 rabbits. Let's also assume that the food needed to support 100 rabbits could instead support 200 squirrels. The environment could also support many different combinations of rabbits and squirrels, such as 90 rabbits and 20 squirrels or 80 rabbits and 40 squirrels.

To include a second species in our equation, we need to add two pieces of information: the number of individuals of the second species and how much each individual of the second species affects the carrying capacity of the first species. For example, if we wanted to know the growth rate of a rabbit population, we would need to know the number of rabbits and squirrels in the area, and—in terms of resource consumption—how many squirrels equal one rabbit. Given that we want to model the changes in population for two species, we will use one equation for each. In these equations, we denote species 1 and 2 using the subscripts 1 and 2, respectively.

$$\frac{dN_1}{dt} = r_1 N_1 \left(1 - \frac{N_1 + \alpha N_2}{K_1}\right)$$

$$\frac{dN_2}{dt} = r_2 N_2 \left(1 - \frac{N_2 + \beta N_1}{K_2}\right)$$

The first equation tells us that the rate of change in the population of species 1 depends on its intrinsic rate of growth (r), the number of individuals of species 1 that are present (N_1), and—in parentheses—the combined effects of species 1 and species 2 in consuming the resource relative to the carrying capacity. The second equation tells us the same thing for species 2.

In these equations, we use two variables, the terms α and β, which are called *competition coefficients*. **Competition coefficients** are variables that convert

Competition coefficients Variables that convert between the number of individuals of one species and the number of individuals of the other species.

between the number of individuals of one species and the number of individuals of the other species. In the preceding equations, α converts individuals of species 2 into the equivalent number of individuals of species 1. Similarly, β converts individuals of species 1 into the equivalent number of individuals of species 2. In our example of rabbits and squirrels competing for a common food resource, we can define rabbits as species 1 and squirrels as species 2. In that case, the value of α would be 0.5 because one squirrel is equivalent to 0.5 rabbits. Similarly, the value of β would be 2.0 because one rabbit is equivalent to two squirrels.

Graphing the Populations of Each Species at Equilibrium

Using these two equations, we can create graphs that help us understand when each of the populations will reach the point at which it no longer increases or decreases. This point is the population's equilibrium point. We can start by determining the conditions under which species 1 will have a population at equilibrium. This will happen whenever the change in population size per unit time is zero:

$$\frac{dN_1}{dt} = r_1 N_1 \left(1 - \frac{N_1 + \alpha N_2}{K_1}\right) = 0$$

$$r_1 N_1 \left(1 - \frac{N_1 + \alpha N_2}{K_1}\right) = 0$$

There are two conditions that can make this equation be at equilibrium. One condition occurs when $N_1 = 0$. Not surprisingly, if we have no individuals of species 1, there will be no growth of species 1. When N_1 is not

zero, we can find equilibrium by rearranging the equation:

$$r_1 N_1 \left(1 - \frac{N_1 + \alpha N_2}{K_1}\right) = 0$$

$$\left(1 - \frac{N_1 + \alpha N_2}{K_1}\right) = 0$$

$$\frac{N_1 + \alpha N_2}{K_1} = 1$$

$$N_1 + \alpha N_2 = K_1$$

$$N_1 = K_1 - \alpha N_2$$

In words, this equation tells us that the number of N_1 individuals depends on the total carrying capacity for species 1 minus the amount of resources consumed by some number of N_2 individuals. Based on this equation, we can draw a line that represents all combinations of N_1 and N_2 that would exist when the N_1 population is at equilibrium. We can plot this line on a graph that has N_1 on the x axis and N_2 on the y axis, as shown in Figure 16.7a. To draw this line, we need to identify the x and y intercepts on the graph.

The x intercept can be found by setting N_2 equal to zero. If we do this, the equation simplifies to

$$N_1 = K_1 - \alpha(0)$$
$$N_1 = K_1$$

As you can see from this equation, when there are no N_2 individuals, the N_1 population achieves equilibrium when it reaches its carrying capacity, K_1.

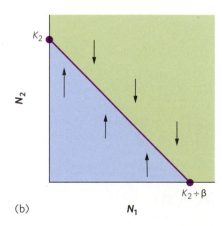

Figure 16.7 Equilibrium population sizes of two species. (a) The equilibrium isocline for species 1, shown as a green line, represents all combinations of species 1's population sizes (N_1) and species 2's population sizes (N_2) that cause species 1 to experience zero growth. When species 2 has an abundance of zero, species 1 will have an abundance of K_1, which is the carrying capacity of species 1. **(b)** The equilibrium isocline for species 2, shown as a purple line, represents all combinations of species 1's population sizes (N_1) and species 2's population sizes (N_2) that cause species 2 to experience zero growth. When species 1 has an abundance of zero, species 2 will have an abundance of K_2, which is the carrying capacity of species 2. The arrows indicate how each population either increases or decreases as it moves away from the equilibrium line.

The y intercept can be found by setting N_1 equal to zero. If we do this, the equation simplifies to

$$0 = K_1 - \alpha N_2$$
$$\alpha N_2 = K_1$$
$$N_2 = K_1 \div \alpha$$

As you can see from this equation, when there are no N_1 individuals, equilibrium occurs when the N_2 population reaches $K_1 \div \alpha$ individuals.

Now that we have identified the x and y intercepts, we can construct the line on the graph that represents all combinations of N_1 and N_2 that cause the population of N_1 to be at equilibrium, as illustrated by the green line in Figure 16.7a. As you can see in this figure, as we increase N_2, the equilibrium population size of N_1 becomes smaller because species 2 is now consuming some of the resource that species 1 needs. Because this equation of a straight line represents population sizes at which a population experiences zero growth, it is called a **zero population growth isocline**.

The zero population growth isocline for species 1 represents the values of N_1 that are in equilibrium across a range of N_2 values. As a result, when the N_1 population is to the left of the isocline, as indicated by the blue-shaded region, it will experience population growth, and move to the right until it reaches the isocline. In contrast, when the N_1 population is to the right of the isocline, as indicated by the green-shaded region, it will experience a population decline, thereby moving to the left until it reaches the isocline.

Now that we have graphically represented the zero population growth isocline for species 1, we need to do the same for species 2. To do this, we set the second equation equal to zero and then rearrange the equation:

$$\frac{dN_2}{dt} = r_2 N_2 \left(1 - \frac{N_2 + \beta N_1}{K_2}\right) = 0$$

$$r_2 N_2 \left(1 - \frac{N_2 + \beta N_1}{K_2}\right) = 0$$

$$\left(1 - \frac{N_2 + \beta N_1}{K_2}\right) = 0$$

$$\frac{N_2 + \beta N_1}{K_2} = 1$$

$$N_2 + \beta N_1 = K_2$$

$$N_2 = K_2 - \beta N_1$$

You might notice that this equation looks very similar to the equation we derived for N_1. To determine the x and y intercepts for the equilibrium line of species 2, we can go through the same mathematical steps that we used for species 1. For example, when we set $N_1 = 0$, equilibrium will occur for species 2 when it reaches its carrying capacity:

$$N_2 = K_2 - \beta(0)$$
$$N_2 = K_2$$

If we then set $N_2 = 0$, equilibrium will occur when

$$0 = K_2 - \beta N_1$$
$$\beta N_1 = K_2$$
$$N_1 = K_2 \div \beta$$

Using these two intercepts, we can plot a zero population growth isocline for species 2, as illustrated by the purple line in Figure 16.7b. As you can see in this figure, as N_1 increases, the equilibrium population size of N_2 decreases because species 1 is consuming the resource that species 2 needs. Because this is a zero population growth isocline for species 2, when the N_2 population is above the isocline, as indicated by the green-shaded region, it will experience a population decline until it reaches the isocline. In contrast, when the population is below the isocline, as indicated by the blue-shaded region, it will experience population growth until it hits the isocline.

In drawing both of these population isoclines, you might notice that the growth rates of the two populations (r_1 and r_2) have no effect on the position of the isocline. While the growth rates determine how quickly a population can reach an equilibrium, they do not affect the location of the equilibrium.

Predicting the Outcome of Competition

Now that we know the conditions under which each of two species reaches an equilibrium population size, we can determine if one of the two species will win the competitive interaction or if the two species will coexist. We do this by overlapping the two zero population growth isoclines for species 1 and species 2. As illustrated in Figure 16.8, overlaying the two lines can produce four possible outcomes. In all four cases, when both species start with small population sizes, as indicated by the yellow-shaded region, both species will experience population growth. In contrast, when both species start with large population sizes, as indicated by the green-shaded region, both species will experience a population decline. However, it is the region that lies between these two extremes where we can determine the outcome of competition.

In Figure 16.8a, the isocline for species 1 lies farther out than the isocline for species 2. As a result, any combination of the two species numbers that falls

---

Zero population growth isocline Population sizes at which a population experiences zero growth.

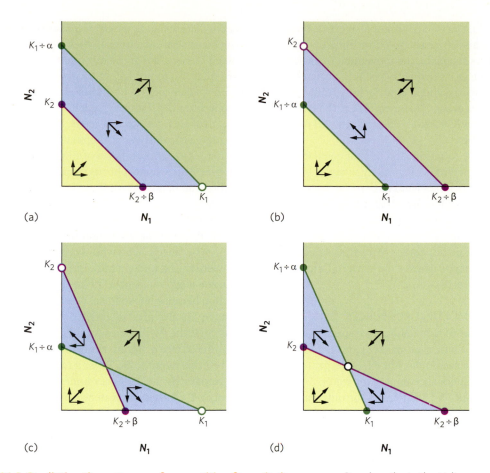

Figure 16.8 Predicting the outcome of competition for a single resource. Based on the Lotka-Volterra competition equations, there are four possible outcomes of competition. **(a)** If the zero population growth isocline for species 1 is farther out than species 2, species 1 will increase to its carrying capacity and species 2 will go extinct. **(b)** If the isocline for species 2 is farther out than species 1, species 2 will increase to its carrying capacity and species 1 will go extinct. **(c)** If the isoclines cross and the two carrying capacities are the outermost points on the two axes, the winner depends on the initial number of each species. **(d)** If the isoclines cross and the two carrying capacities are the innermost points on the two axes, then the two species will coexist. As in Figure 16.7, the green line represents the population isocline for species 1 and the purple line represents the population isocline for species 2. The thin black arrows represent increases and decreases in population sizes for each species whereas the thick black arrows represent the net effects of population changes in both species.

within the blue-shaded region means that species 1 is below its isocline and its population will grow whereas species 2 is above its isocline and its population will decline. The net effect of species 1 moving to the right and species 2 moving down—both indicated by thin arrows—is that the combination of the two species moves down and to the right, as indicated by the thick arrow. This combination will continue to move until eventually it reaches an equilibrium. At this point, N_1 reaches its carrying capacity and N_2 goes extinct.

In Figure 16.8b, we see the opposite situation; the isocline for species 2 lies farther out than the isocline for species 1. In this case, any combination of the two species that falls within the blue-shaded region means that species 1 is above its isocline and its population will decline whereas species 2 is below its isocline and its population will grow. With species 1

moving to the left and species 2 moving up, the net effect is that the combination of the two species moves up and to the left. As this process continues, the combination will reach an equilibrium; N_2 reaches its carrying capacity and N_1 goes extinct.

In Figure 16.8c, the two isoclines cross, with K_1 and K_2 as the two most extreme points on the axes. In this scenario, there are two possible equilibria and the outcome depends on the initial number of each species. If the combination of the two species falls within the blue-shaded region, the outcome of competition depends on the combination of N_1 and N_2 that are present. If the combination of N_1 and N_2 fall within the upper left blue region, the net movement is toward K_2 and species 2 wins the competitive interaction. If the combination of N_1 and N_2 fall within the lower right blue region, the net movement is toward K_1 and species 1 wins.

In Figure 16.8d, the two isoclines also cross, but this time with K_1 and K_2 as the two innermost points on the axes. In this scenario, the movement of each population relative to its respective isocline causes a net movement toward a single equilibrium point where the two lines cross. This means that the two competing species will coexist; one species will not outcompete the other.

In general, these equations tell us that coexistence of two competing species is most likely when interspecific competition is weaker than intraspecific competition—that is, when the competition coefficients α and β are less than 1. In other words, coexistence of two competing species occurs when individuals of each species compete more strongly with themselves (conspecifics) than with individuals of other species (heterospecifics).

COMPETITION FOR MULTIPLE RESOURCES

Although it is simpler to think about two species competing for one resource, in reality species in nature compete for multiple resources. For example, grassland plants might simultaneously compete for water, nitrogen, and sunlight. We can predict the outcome of this more complex competition by thinking about a case in which two species compete for two different resources. If species 1 is better able to sustain itself at low levels of both resources than species 2, then species 1 will win the competitive interaction. However, the outcome changes when species 1 is better at persisting at a low level of one resource but species 2 is better at persisting at a low level of another resource. In this case, because neither species can drive the other to extinction, the two species should coexist.

This principle was first demonstrated by David Tilman in a series of experiments conducted with two species of diatoms: *Asterionella formosa* and *Cyclotella meneghiniana*. We encountered these species when we discussed Liebig's law of the minimum. Both species require silica (Si) to produce their glassy outer shells and both species require phosphorus (P) for their metabolic activities. However, each species differs in how much of each resource it needs. *Cyclotella* uses silicon more efficiently, so its population can persist when there is a low abundance of silicon. In contrast, *Asterionella* utilizes phosphorus more efficiently, so its population can persist when there is a low abundance of phosphorus. Based on these differences, Tilman predicted that if the diatoms were raised under a low ratio of Si/P, *Cyclotella* would dominate the competitive interaction because silicon would be low in abundance. If they were raised under a high ratio of Si/P, *Asterionella* would dominate the competitive interaction because phosphorus would be low in abundance. He also predicted that at intermediate ratios of Si/P, the two species would coexist because each is limited by a different resource.

To test this hypothesis, Tilman raised the two species separately and together under a range of different Si/P ratios in laboratory containers, as shown in Figure 16.9. When raised separately, both species were able to survive under all of the Si/P ratios. However, *Cyclotella* dominated the containers when they were raised together at a low Si/P ratio of 0.6, while *Asterionella* dominated the containers when they were raised at a high Si/P ratio of 455. In contrast, when they were raised at an intermediate Si/P ratio of 38, the two species coexisted and neither became dominant. These were the first results to demonstrate that competing species can coexist when there are multiple resources and each species is limited by a different resource.

Figure 16.9 Species competing for multiple resources. In the experiment, two species of diatoms compete for silicon and phosphorus. *Cyclotella* is more limited by a low abundance of phosphorus whereas *Asterionella* is more limited by a low abundance of silicon. **(a)** When raised under a low ratio of Si/P, *Cyclotella* dominates the competitive interaction. **(b)** When raised under an intermediate ratio of Si/P, neither species dominates. Instead, the two species coexist. **(c)** When raised under a high ratio of Si/P, *Asterionella* dominates the competitive interaction. Data from D. Tilman, Resource competition between planktonic algae: An experimental and theoretical approach, *Ecology* 58 (1977): 338–348.

Figure 16.10 Competition between barnacles. In Great Britain, Poli's stellate barnacle lives in the upper intertidal zone because it is more resistant to desiccation and it cannot compete successfully in the lower intertidal zone. In contrast, the rock barnacle lives in the lower intertidal zone because it is the superior competitor but is not resistant to the desiccating environment of the upper intertidal zone.

The outcome of competition can be altered by abiotic conditions, disturbances, and interactions with other species

We have considered what happens when two species compete for shared resources. However, in nature, competition occurs in the context of variable environmental conditions and in the presence of other types of species interactions, including predation and herbivory. In this section, we will examine how competitive outcomes can be altered by abiotic conditions, disturbance, and interactions with other species.

ABIOTIC CONDITIONS

The ability to persist when resources are scarce is important to winning in a competitive situation, but it can be overwhelmed by the ability to persist under harsh abiotic conditions. We can see an example of this in a classic study of barnacles, conducted by Joseph Connell in Great Britain. Barnacles begin their lives as larvae that float on the ocean currents and then settle on open, rocky spaces of coastal intertidal zones. They spend the rest of their lives as sessile adults, commonly living in dense groups. They feed on plankton that they filter from the water that washes over them. Plankton are not a limiting resource because the barnacles cannot substantially reduce the vast amount of plankton that exists in coastal waters. However, they are often limited by open spaces on rocks in the intertidal habitat.

Connell observed that the larvae of two barnacle species on the coast of Scotland were distributed broadly along the upper and lower intertidal zones. However, adults were distributed in two separate zones according to species; Poli's stellate barnacles (*Chthamalus stellatus*) live in the upper intertidal zone, whereas rock barnacles (*Semibalanus balanoides*, formerly *Balanus balanoides*) live in the lower intertidal zone. The two species have only a small overlap in their distributions, as shown in Figure 16.10.

To understand the reason for this distribution, Connell conducted a series of experiments in which he removed one or the other species. He discovered that the stellate barnacles were able to live in the upper intertidal zone because they are resistant to desiccation, which happens during periods of low tide. Though they could live in the lower intertidal zone, competition from the rock barnacle prevented it. The rock barnacle has a rapid growth rate and heavier shell than the stellate barnacle, which allows the rock barnacle to expand in size, pry the stellate barnacle off the rock, and take over the newly opened space. Connell discovered this when he manually removed the rock barnacles from the lower part of the intertidal zone and found that the stellate barnacles quickly moved down and thrived. However, when he removed the stellate barnacle from rocks in the upper intertidal zone, the rock barnacles did not move up because they could not survive the desiccation that occurs in the upper intertidal zone. In short, the distribution of the two species is the result of a trade-off that many species experience—the trade-off between competitive ability and the ability to tolerate a challenging abiotic factor.

DISTURBANCES

Competitive interactions among species can also be altered by disturbances. For example, fescue prairie (Figure 16.11), characterized by fescue grasses and the most northerly prairie in North America, was once widespread in Alberta and Saskatchewan. Today, however, only between 5 and 15 percent of fescue prairie remain in those provinces. Fire once burned this prairie every 12 to 26 years, which maintained high species diversity, but, with the beginning of fire suppression in the late 1800s, species diversity began to decrease. Without fire, one fescue species, the plains rough fescue (*Festuca hallii*), and the tree trembling aspen (*Populus tremuloides*) both became more abundant. Fire suppression altered the outcome of plant competition in fescue prairie because it allowed the superior competitors previously removed by fire to persist.

Figure 16.11 Fescue prairie in Saskatchewan and Alberta. This type of grass prairie depends on regular fires to maintain its species diversity. Fire eliminates many species that would compete with plants that have roots or seeds able to survive fire. Photo by John Foster/Radius Images/Masterfile.

PREDATION AND HERBIVORY

The outcome of competition can also be altered by the presence of predators and herbivores. We frequently find a trade-off between competitive ability and resistance to predators or herbivores; the most competitive plants are often the most susceptible to herbivores and the most competitive animals are typically the most susceptible to predators. For example, many animals are good competitors because they move a lot in search of food and they have high feeding rates. This activity, though, makes the animals very noticeable to predators and, as a result, animals that are more competitive are more likely to be killed by predators.

Studies of aquatic communities have shown that predators can reverse the outcome of competition. For example, researchers conducted a competition experiment using large, outdoor tanks to simulate ponds. Each pond was supplied with 200 tadpoles of the spadefoot toad (*Scaphiopus holbrookii*), 300 tadpoles of the spring peeper (*Pseudacris crucifer*), and 300 tadpoles of the southern toad (*Anaxyrus terrestris*). After setting up a series of identical tanks, each one was randomly assigned a treatment of 0, 2, 4, or 8 predatory newts (*Notophthalmus viridescens*).

In the absence of newts, the spadefoot toad's superior competitive ability was clear. As depicted in Figure 16.12, survival was high for spadefoot toad tadpoles, moderate for southern toad tadpoles, and low for spring peeper tadpoles. Spadefoot toad tadpoles had high survival because they are very active and rapidly consume whatever algae are available. In contrast, southern toad tadpoles are somewhat less active and spring peeper tadpoles are very inactive and therefore grow slowly. When the predatory newts were added, the outcome changed dramatically. As the number of newts increased, the survival of the spadefoot toads dropped precipitously and the survival of the southern toads dropped modestly, which is consistent with the southern toad's higher activity levels. Compared to the control treatment, the addition of eight newts reduced the survival of spadefoot toads by 75 percent and caused the survival of southern toad tadpoles to decrease by 32 percent. However, the survival of spring peepers increased by 26 percent. Because the newts consumed the superior competitors, the outcome of competition among the three species of tadpoles was reversed.

Herbivores can have a similar effect on competition. For example, several species of goldenrod dominate fallow fields throughout North America. These plants can grow more than a metre high and cast shade upon their shorter competitors. A species of beetle (*Microrhopala vittata*) specializes in consuming the goldenrods. Every 5 to 15 years the

Figure 16.12 Altering competition with predators. In the absence of predatory newts, the spadefoot toads and southern toads survive well because they are superior competitors. With the addition of predatory newts, the survival of the two toad species decreases while the survival of the spring peepers increases. Error bars are standard errors. Data from P. J. Morin, Predatory salamanders reverse the outcome of competition among three species of anuran tadpoles, *Science* 212 (1981): 1284–1286.

Figure 16.13 Altering competition with herbivores. Over a 10-year period, the plot on the right was sprayed with an insecticide to prevent beetles from consuming the goldenrods—the tall plants with yellow flowers—while the plot on the left was left unsprayed. The unsprayed plot experienced an outbreak of beetles and this greatly reduced the density and height of the goldenrod plants. As a result, many species of plants that are inferior competitors became more numerous. Photo by W. Carson.

beetle population achieves very high densities, known as an insect outbreak, during which the beetles consume large amounts of the goldenrod plants.

To determine the impact of the beetles on the plant community, researchers conducted a long-term experiment in which some plots were sprayed with an insecticide to kill the goldenrod-consuming beetles while other plots were left unsprayed. The results were dramatic (Figure 16.13). In the sprayed plots that contained few herbivorous beetles, the goldenrods grew tall and cast shade on the other species of plants. The goldenrods dominated the community because of their superior competitive ability. In the unsprayed plots, the beetles experienced an outbreak midway through the experiment, which significantly reduced the abundance of goldenrod plants. Moreover, the goldenrods that remained were substantially shorter, so they cast less shade on the inferior competitors. With fewer large goldenrods present, the inferior competitors became much more abundant. After 10 years, the researchers confirmed that the beetle was able to reverse the outcome of competition among plants. Another example of herbivory involves the tent caterpillar, whose population fluctuations were discussed in Chapter 15. Tent caterpillars cause changes in populations of arthropods living in the leaves the caterpillars eat. Researchers sampled trembling aspen leaves in Alberta and recorded the incidence of caterpillar damage and arthropod presence. Out of 10 arthropod species, abundance of 7 species declined with increasing caterpillar density. Abundance of 2 species remained unchanged, and 1 species became more abundant with increasing tent caterpillar density.

Competition can occur through exploitation or direct interference, or it may be apparent competition

Competition can be categorized in several ways. We have seen numerous examples of **exploitative competition**, in which individuals consume and drive down the abundance of a resource to a point that other individuals cannot persist. For example, rodents and ants that compete for seeds in the desert experience exploitative competition. Sometimes competitors do not immediately consume resources but defend them, a type of competition known as **interference competition**. While interference competition is a direct interaction between two individuals, exploitative competition is considered indirect competition because it operates through a resource. Finally, two species can appear to compete for a shared resource, but in fact experience a negative effect from *apparent competition*. In this section, we will explore the concepts of interference competition and apparent competition.

Exploitative competition Competition in which individuals consume and drive down the abundance of a resource to a point that other individuals cannot persist.

Interference competition When competitors do not immediately consume resources but defend them.

INTERFERENCE COMPETITION: AGGRESSIVE INTERACTIONS

Aggressive interactions are an effective form of interference competition that occurs between species of animals. For example, in the deserts of New Mexico, two species of ants—the long-legged ants (*Novomessor cockerelli*) and the red harvester ants (*Pogonomyrmex barbatus*)—compete for seeds and for any insects they can catch and subdue. However, the long-legged ants have a unique method of competing. Early in the morning, they emerge from their nests, find the nests of the red harvester ants, and plug the nest entrances with stones and soil. As you can see in Figure 16.14, when researchers examined the effects of this aggression, they found that harvester ants with plugged entrances required several hours to unplug their nests before they could begin foraging for seeds. This direct aggression by the long-legged ants allows them to forage in the morning with minimal competition from the harvester ants.

INTERFERENCE COMPETITION: ALLELOPATHY

Another type of interference, known as **allelopathy**, occurs when organisms use chemicals to harm their competitors. For example, black walnut trees (*Juglans nigra*) produce *juglone,* an aromatic organic compound that inhibits certain enzymes in other plants. The compound—found in the leaves, bark,

Figure 16.15 The common reed. The common reed, a wetland plant, can be found in many locations around the world, including this site on a coastal wetland in Somerset, England. In North America, some genetic strains are native whereas others have been introduced from Europe. Photo by Martin Fowler/Alamy.

roots, and seed husks of the black walnut tree—makes it difficult for most other species of plants to germinate and grow under a walnut tree. The competition between mosses discussed above may also be due to allelopathy. Sphagnum mosses increase the acidity of their surrounding area and make it less suitable for other plants to grow.

Allelopathy can also be an effective strategy for invasive plants, which invade and dominate a community. The wetland plant known as the common reed (*Phragmites australis*) is found throughout the world (Figure 16.15). In North America, some genetic strains are native and other strains have been introduced from Europe. In locations such as the Great Lakes region of North America, the introduced strain is spreading rapidly and has displaced many other species of native wetland plants. For many years, researchers hypothesized that the common reed was successful at spreading due to allelopathy, but they lacked evidence. In 2007, researchers reported that the root of the common reed produced a chemical known as *gallic acid,* which is highly toxic to the roots of many other plants. By harming the roots of other species, the common reed is able to impair the growth of its interspecific competitors. The researchers also discovered that the chemicals produced by the introduced strains of the reed were much more lethal to some plants than those produced by the native strain, as shown in Figure 16.16. This result is consistent with the observation that the introduced strains of the reed are able to invade a wetland and spread much more rapidly than the native strains.

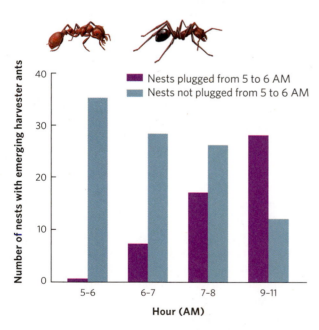

Figure 16.14 Interference competition in ants. Red harvester ants and long-legged ants compete for seeds and insects. When the harvester ant nest is not plugged by the long-legged ant, a large number of colonies emerge by 6 AM. However, if the nest entrances are plugged by the long-legged ant, it can take several hours for the harvester ants to clear the openings and start foraging. Data from D. M. Gordon, Nest-plugging: Interference competition in desert ants (*Novomessor cockerelli* and *Pogonomyrmex barbatus*), *Oecologia* 75 (1988): 114–118.

Allelopathy A type of interference that occurs when organisms use chemicals to harm their competitors.

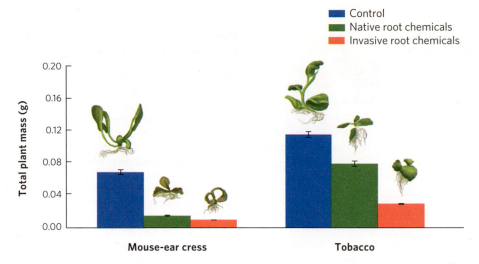

Figure 16.16 **Allelopathy in the common reed.** Root chemicals from native or invasive strains of the common reed were extracted and then added to pots of mouse-ear cress or tobacco plants. The addition of the root chemicals caused both plant species to grow poorly compared to the control. In the case of the tobacco plant, chemicals from the invasive strain of the reed inhibited growth more than the native strain. Error bars are standard deviations. Data from T. Rudrappa et al., Root-secreted allelochemical in the noxious weed *Phragmites australis* deploys a reactive oxygen species response and microtubule assembly disruption to execute rhizotoxicity, *Journal of Chemical Ecology* 33 (2007): 1898–1918.

Allelopathy does not always take the form of toxic effects on interspecific competitors. In Australia, for example, researchers hypothesize that flammable oils in the leaves of some eucalyptus could be an adaptation that promotes frequent fires in the leaf litter of the forest floor, which kill seedlings of competitors.

Regardless of whether organisms use direct aggressive interactions or allelopathic chemicals, it is clear that interference competition can be an important mechanism in determining the abundance and distribution of species.

APPARENT COMPETITION

Throughout this chapter, we have defined competition as a negative interaction between two individuals for a limited resource. Sometimes, however, two species can share a resource and have negative effects on each other without competing for a resource. When two species have a negative effect on each other through an enemy—including a predator, parasite, or herbivore—we call it **apparent competition**. As its name suggests, apparent competition causes an outcome that looks like competition, but the underlying mechanism is different.

Apparent competition can be observed in a variety of communities. For example, the ring-necked pheasant and the grey partridge (*Perdix perdix*) live in many of the same habitats in the United Kingdom. For 50 years, a large decline in the number of grey partridges has been attributed to increased agriculture, although researchers suspected that a parasitic nematode (*Heterakis gallinarum*) might also play a role. Infected pheasants experience few harmful effects from the parasite, but infected partridges can suffer weight loss, reduced fecundity, and death.

To determine the susceptibility of each species to infection by parasites in the environment, researchers allowed both bird species to feed either in pens with parasite eggs scattered across the soil or in pens without the eggs. After 50 days of living in the pens without added parasite eggs, they found that both species contained low parasite infections, as shown in Figure 16.17. However, in the pens that contained added parasite eggs, the pheasants became infected at rates that were more than 20 times higher than the rate of infection in the partridges. Moreover, when both birds were fed the same number of nematode eggs over a 100-day period, the pheasants excreted over 80 times more nematode eggs than the partridges. Pheasants can carry a large number of parasites without being harmed and they excrete a large number of parasites that subsequently infect and harm partridges. The presence of pheasants is associated with partridges that gain less weight and have a low rate of survival. The relationship between the birds has the appearance of competitors for a shared resource, but the real cause is a shared parasite.

There are many examples of pathogens that cause apparent competition. For instance, when Europeans colonized other continents, particularly North and South America, they brought with them diseases such as smallpox. The Europeans could tolerate many of these diseases because they had a long evolutionary history of living with them, but the Native Americans lacked this evolutionary history and the introduced diseases devastated their populations. While the Europeans appeared to cause competition, the reduction of the Native American population occurred primarily because the Europeans brought new pathogens to the Western Hemisphere.

Apparent competition When two species have a negative effect on each other through an enemy—including a predator, parasite, or herbivore.

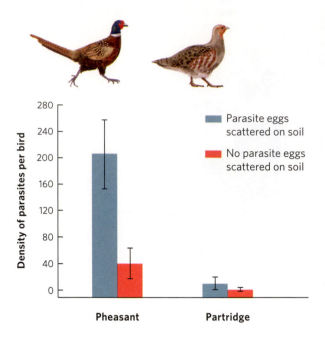

Figure 16.17 Apparent competition in pheasants and partridges. Pheasants (left) and partridges (right) were held in pens with no parasite eggs, or with parasitic eggs of nematodes scattered across the soil. In the absence of parasite eggs, some birds exhibited a low level of background infection. In the presence of parasite eggs, pheasants had a much higher number of parasites in their bodies after 50 days. Pheasants are not harmed by the parasites, but they serve as a major reservoir of new parasite eggs, which they excrete onto the soil. Partridges, which experience harmful effects from the parasite, had a low number of parasites and therefore excreted few parasite eggs into the environment. The presence of pheasants causes apparent competition because they release large numbers of parasite eggs, which leads to a decline in the partridges. Error bars are standard errors. Data from D. M. Tompkins et al., The role of shared parasites in the exclusion of wildlife hosts: *Heterakis gallinarum* in the ring-necked pheasant and the grey partridge, *Journal of Animal Ecology* 69 (2000): 829–840.

In some cases, species involved in apparent competition can vary over time due to factors such as habitat change. For example, woodland caribou west of the Athabasca River in northeastern Alberta began to decline in the 2000s after having been stable throughout the 1990s. This decline coincided with increasing forestry and oil industry activities that removed old-growth forest and peat lands. This habitat change made the area more suitable for moose. Researchers hypothesized that a growing moose population had attracted more wolves, which in turn increased predation on caribou. However, when the scientists collected data on population sizes and wolf predation, they found that the declining caribou population was not due to apparent competition with moose.

The data showed that the wolf population almost doubled from approximately 6 wolves per 1000 km^2 during the 1990s to over 11 wolves per 1000 km^2 in the 2000s. While the moose population remained unchanged, white-tailed deer (*Odocoileus virginianus*) had invaded the region. The deer population increased more than 17-fold, from 0.2 deer per 100 km transect line in the 1990s to 3.5 deer per 100 km transect line in the 2000s.

Wolves responded by changing their diet, as you can see in Figure 16.18a. Fewer scats contained moose hair, and scats with deer and caribou hair became

(a)

(b)

Figure 16.18 Apparent competition between moose and woodland caribou in northeastern Alberta. Researchers compared the composition of wolf scat and wolf pack movements before industrial activity in the 1990s and after forestry and oil industry activities increased in the 2000s. **(a)** The proportion of wolf scats containing moose hair decreased in the 2000s. In contrast, scat containing white-tailed deer, woodland caribou, and beaver hair increased. **(b)** In the 1990s, wolves remained in the highlands around the river (dark green areas). This changed in the 2000s, when wolves increasingly moved into peat lands and old-growth forest (light green areas), which are the main habitat for woodland caribou.

Chi-square Tests

When the researchers in California were trying to determine the causes of the bare zone around coastal shrubs, they captured 28 mice in the bare zone near the shrubs and 1 mouse in the grass. Although this is a substantial difference, how do we know if this difference is statistically significant? In Chapter 14, we discussed the importance of testing hypotheses using a criterion of $\alpha < 0.05$ for statistical significance. In Chapter 15 we discussed testing hypotheses using t-tests when we have estimates of mean values and standard deviations. Sometimes, however, our data do not consist of means and standard deviations, but rather of individuals that are counted, such as the number of mice in the bare zone versus those in the grass.

When our data consist of counts, we can test whether the data are significantly different from an expected distribution using a *Chi-square test*. A **Chi-square test** is a statistical test that determines whether the number of observed events in different categories differs from an expected number of events, which is based on a particular hypothesis. In the case of the mice, the researchers sampled both habitats equally and they captured a total of 29 mice from both areas. If the mice exhibited no habitat preference, we would have expected to find 14.5 mice in the bare zone and 14.5 mice in the grass. As a result, we want to statistically compare the observed distribution of mice—28 mice in the bare zone and 1 mouse in the grass—to an even distribution between the two habitats. Now that we know the observed distribution and the expected distribution for mice that exhibited no preference, we can create a small table and conduct the Chi-square test.

	OBSERVED	EXPECTED
BARE ZONE	28	14.5
GRASS	1	14.5

To calculate the Chi-square value, we use an equation that includes our observed (O) and expected (E) values to calculate the value of Chi-square (χ^2).

$$\chi^2 = \sum_{i=1}^{n} \frac{(O_i - E_i)^2}{(E_i)}$$

$$\chi^2 = \frac{(28 - 14.5)^2 + (1 - 14.5)^2}{14.5}$$

$$\chi^2 = 25.1$$

As was the case when we did a t-test in Chapter 15, we have to compare our calculated value to a table of values to determine if our observed distribution is significantly different from the expected distribution. The first step in that process is determining the degrees of freedom:

Degrees of freedom = (number of observed categories − 1) = (2 − 1) = 1

Now we can examine the Chi-square table provided in the Statistical Tables in the appendix. Using these tables, we can compare our calculated value (25.1) to the critical Chi-square value when we have 1 degree of freedom and we set $\alpha = 0.05$. In this case, the critical Chi-square value is 3.841. Since our calculated value is greater than the critical value, we can conclude that the observed distribution of mice differed significantly from an equal distribution of mice between the two habitats.

YOUR TURN Using a Chi-square test, determine whether an observed distribution of 12 mice in the bare zone and 8 mice in the grass differs significantly from an expected distribution of 10 mice in the bare zone and 10 mice in the grass.

Chi-square test A statistical test that determines whether the number of observed events in different categories differs from an expected number of events, which is based on a particular hypothesis.

---

more frequent. Predation on beavers (*Castor canadensis*) also increased, although it is not clear why. The researchers suspected that increasing deer populations may have brought wolves into areas where beavers are more common, or that the beaver population may have been growing since their reintroduction to this area in the mid 1900s. The larger wolf population also changed its distribution patterns. Figure 16.18b shows that wolves remained close to the river uplands in the 1990s. A decade later, wolves were increasingly seen in caribou habitat, peat lands, and old growth forests.

Taken together, these results demonstrate that white-tailed deer had replaced moose in an apparent competition with woodland caribou. Following industrial activities, deer invaded the area, which allowed wolf populations to increase. The growth of the wolf population, and its enlarged overlap with caribou habitat, caused the increase in caribou predation by wolves and the following decline in the caribou population.

Throughout this chapter, we have seen that competition can be a major force affecting the distribution and abundance of organisms. This competition can occur by exploitation or interference and it can occur among species both closely and distantly related. As we will see in "Ecology Today: Finding the Forest in the Ferns," understanding the role of competition and how it interacts with disturbance, herbivory, and apparent competition is important for understanding how the natural world works, including biomes that provide humans with valuable natural resources.

ECOLOGY TODAY CONNECTING THE CONCEPTS

FINDING THE FOREST IN THE FERNS

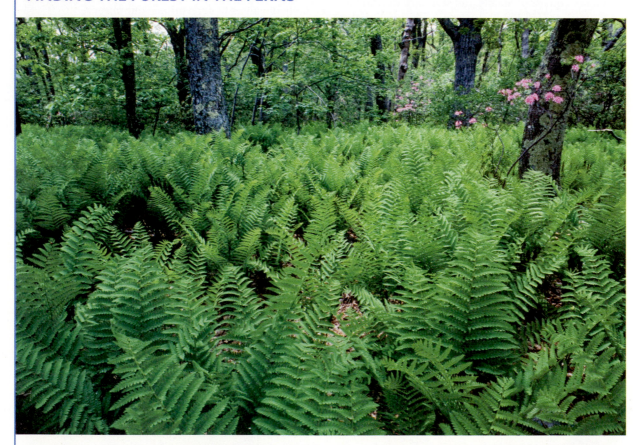

A mature temperate forest carpeted with hay-scented fern. The dense fern layer is associated with a low number of tree seedlings, as evidenced by the lack of small trees growing in the understorey of this forest in Shenandoah National Park, Virginia. Photo by Greg Dale/National Geographic Stock.

In forests around the world, adult trees produce offspring that can become the next generation of trees. This next generation will be available to support the other species that live in the forest and will be available for such human uses as lumber and paper products. For a long time, it was assumed that the composition of trees in the next generation would be determined by the composition of seedlings produced by adult trees, the ability of each species to tolerate shade as a seedling, and the amount of shade that is present. For example, because seedlings of quaking aspen trees (*Populus tremuloides*) cannot tolerate shade from adult trees, they regenerate only when there are large openings in the forest that allow high amounts of sunlight. In contrast, seedlings of American beech trees are highly tolerant of shade and can grow in forests even under a dense canopy of leaves. While these processes are important determinants of the composition of the next generation of forest, they can be overwhelmed by processes that promote the dense growth of a single plant species that dominates the understorey.

An excellent example of this phenomenon has occurred in the forests of central Pennsylvania. Historically, these forests had a variety of tree species in the upper canopy and a variety of younger trees in the understorey, but over decades as the forests were logged, an abundance of plant growth occurred. In addition, top predators such as mountain lions and wolves were removed, which allowed a dramatic increase in the white-tailed deer population. Deer consumed large numbers of seedlings but not an unpalatable plant known as the hay-scented fern (*Dennstaedtia punctilobula*). Although the fern historically comprised less than 3 percent of the understorey vegetation, today it is the dominant plant in more than one-third of Pennsylvania forests. Few seedlings can emerge out of the dense fern layer, so there are many large trees in the forest canopy but few young trees to serve as the next generation.

For many years, foresters hypothesized that ferns were causing competitive exclusion by casting high amounts of shade over the seeds and seedlings. It was thought that ferns prevented the germination and growth of the tree seeds and seedlings by consuming nearly all the light and making it a limiting resource. Competition as a mechanism between these distantly related species was completely consistent with the low number of tree seedlings that were emerging from the fern layer. However, an alternative explanation was that the fern layer provided an attractive refuge for rodents from predatory hawks and owls and that these rodents were consuming tree seeds and tree seedlings.

In 2008, researchers reported the results of an experiment to determine which of these two hypotheses was correct: whether the negative effect of the ferns was caused through competition for resources or through apparent competition by facilitating the rodents. They began by selecting plots of ferns within a forest, each with an area of 4 m². For each plot, they either fenced out rodents or let rodents roam throughout the plot and they either removed the ferns or left them in place. After conducting the four different manipulations, the researchers began collecting data by first examining how many rodents were captured in each type of plot. The most common rodent species were deer mice,

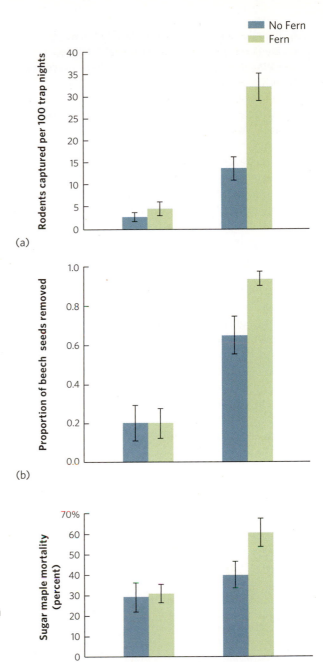

Apparent competition from ferns in a forest. To determine whether the effect of ferns was due to competition or apparent competition, the researchers used fencing to exclude rodents and manipulated the presence of ferns. **(a)** Traps caught very few rodents inside fenced plots, but they caught many rodents in unfenced plots, especially when ferns were present. **(b)** Few seeds of beech trees were removed from fenced plots, but many seeds were removed from unfenced plots, especially when ferns were present. **(c)** Similarly, sugar maple seedlings experienced low rates of mortality in fenced plots, but high rates of mortality in unfenced plots in which ferns remained in place. Error bars are standard errors. Data from A. Royo and W. C. Carson, Direct and indirect effects of a dense understory on tree seedling recruitment in temperate forests: Habitat-mediated predation versus competition, *Canadian Journal of Forest Research* 38 (2008): 1634–1645.

chipmunks, and red-backed voles (*Clethrionomys gapperi*). As you can see in part (a) of the graph, rodents rarely got inside the fenced plots, regardless of fern coverage. However, rodents were frequently captured in the unfenced plots and more than twice as many rodents were captured when the ferns remained in place. When the researchers added beech seeds to the plots, they found that only a small percentage of seeds were removed in the fenced plots—regardless of fern coverage—as shown in part (b) of the graph. However, a large percentage of seeds were removed in the unfenced plots, and even more if the ferns remained in place. Similar results were seen for many other species of seeds, including black cherry (*Prunus serotina*) and sugar maple (*Acer saccharum*). Finally, when the researchers set out seedlings of sugar maple, they observed low mortality in the fenced plots— regardless of whether ferns were removed—as shown in part (c) of the graph. In the unfenced plots, however, seedling mortality was higher, but only when ferns were present. These results confirmed that the primary effect of the ferns on tree seedlings was not as a competitor for light, but rather that they served as a refuge for rodents that eat the seeds and seedlings of beech, black cherry, and sugar maple. It was clearly a case of apparent competition.

A different outcome occurred when the researchers examined red maple (*A. rubrum*). Although seed predation occurred more often on red maple when ferns were present, many red maple seeds were present on the forest floor. In the unfenced plots, the presence of ferns caused fewer red maple seedlings to germinate. However, ferns had the same effect in the fenced plots, which indicated that the overall effect of ferns on red maple was not due to seed predation but to resource competition.

The results confirmed that, while fern effects on red maple were due to resource competition, the fern effects on American beech, black cherry, and sugar maple were due to apparent competition mediated through the fern's protection of rodents. These results confirm that both competition and apparent competition can play substantial roles in determining the species that will compose the next generation of a forest and therefore determine the diversity of species and the future species that will be available for harvest.

SOURCES: Royo, A, and W. C. Carson. 2006. On the formation of dense understory layers in forests worldwide: Consequences and implications for forest dynamics, biodiversity, and succession. *Canadian Journal of Forest Research* 36: 1345–1362.

Royo, A, and W. C. Carson. 2008. Direct and indirect effects of a dense understory on tree seedling recruitment in temperate forests: Habitat-mediated predation versus competition. *Canadian Journal of Forest Research* 38: 1634–1645.

SUMMARY OF CHAPTER CONCEPTS

■ **Competition occurs when individuals experience limited resources.** Competition can either be intraspecific or interspecific and occurs when there is a limited resource. Resources can either be renewable or nonrenewable and they can be generated either from within an ecosystem or from outside an ecosystem. Leibig's law of the minimum states that a population will increase until the most limiting resource prevents further growth, although we now appreciate that different resources can have interactive effects on population growth. The competitive exclusion principle states that two species cannot coexist indefinitely when they are both limited by the same resource.

■ **The theory of competition is an extension of logistic growth models.** The simplest models consider competition for a single resource and consider the zero population growth isoclines of two competing species. Using these models, we can make predictions regarding the conditions under which two species can win a competitive outcome or coexist. Under the more realistic situation of multiple limiting resources, we can have coexistence of multiple species of competitors when each species is limited by a different resource.

■ **The outcome of competition can be altered by abiotic conditions, disturbances, and interactions with other species.** If a species is competitively superior but not tolerant of extreme abiotic conditions, it will not be able to dominate areas that experience such conditions. Similarly, competitively superior species that cannot persist with frequent disturbances, such as fire, cannot come to dominate inferior competitors. In the same way, superior competitors that are more vulnerable to herbivores or predators cannot outcompete inferior competitors because they are preferentially harmed or killed.

■ **Competition can occur through exploitation or direct interference, or it may be apparent competition.** Exploitative competition occurs when one species consumes enough of a resource that another species can no longer persist. In contrast, interference competition occurs when a species defends a resource and prevents other individuals from consuming it. Interference competition includes aggressive interactions among species and allelopathy. Sometimes species appear to be competing because the presence of one species has a negative effect on the population of the other. In cases of apparent competition, the underlying mechanism is not competition, but another type of interaction such as predation, herbivory, or parasitism.

REVIEW QUESTIONS

1. What is the difference between renewable and nonrenewable resources?

2. How does Liebig's law of the minimum explain limits on a population's growth?

3. If two species require the same limiting resource, what would you predict about their ability to coexist?

4. Why are population growth rates at zero when two species are at equilibrium?

5. Explain why ecologists predict that closely related species will often experience more intense competition than distantly related species.

6. Compare and contrast exploitative competition and interference competition.

7. Explain the concept of the "competition coefficients" that are used in the Lotka-Volterra competition equations.

8. Why is allelopathy considered a form of interference competition?

9. Under what conditions can two species competing for two resources coexist?

10. Explain the difference between competition and apparent competition.

PRACTICING ECOLOGY: THE IMPORTANCE OF DIVERSITY

Marc Cadotte at the University of Toronto studies how plant diversity alters competition and primary productivity. Photo by Robin Marushia.

Humans rely on plant productivity for timber, fuel, agriculture, and feed for livestock. Because of these benefits, we often try to maximize the density of plants in forests, grazing lands, and croplands. However, maximizing density generates competition among plants and can reduce individual plant growth. Fertilizing soils to reduce belowground competition for nutrients is a costly solution and is not feasible in many areas such as wild grasslands and forests. Given that plant resources are limited, what is a suitable management strategy to increase plant productivity in these ecosystems?

Grasslands are particularly important grazing habitats for many species of wild and domesticated herbivores. Because plants grow at a high density, resources are almost always limiting and competition among plants can be intense. However, each grassland plant species has a unique suite of traits that allow it to acquire resources such as water, nutrients, and light in often unique ways. For example, some plants grow deep roots to obtain nutrients that plant species with shallow roots cannot access. Maintaining a unique suite of traits may reduce the intensity of resource competition between plant species. Hence, we might hypothesize that total plant production in a grassland will increase with the diversity of plant traits represented in a grassland community.

Marc Cadotte from the University of Toronto tested this hypothesis in an experimental grassland on the Koffler Scientific Reserve near King City, Ontario. He collected seeds from 17 different plant species that naturally occur around the reserve, including grasses, forbs, and legumes. Each species differs in traits such as root depth and plant height; traits of taxonomically related species differed less than traits of taxonomically unrelated species. After ploughing the field, he established 4 plots of a square metre each and planted a similar density of seeds in each plot. He established plots that contained species with similar or dissimilar traits by planting seeds of species that were either taxonomically related or taxonomically unrelated. Then, 2 years later, he removed and measured the total aboveground biomass of plants in each plot. You can view his data in the table.

ABOVEGROUND BIOMASS (g) IN PLOTS WITH SIMILAR AND DISSIMILAR TRAITS

SIMILAR TRAITS	DISSIMILAR TRAITS
145	706
189	769
235	912
352	1108
491	1205
554	1387
657	1654
598	429
216	1289
379	1156

GRAPH THE DATA

1. Calculate the means and standard errors of the means for each column.

2. Create a bar graph showing the mean and standard error for each treatment.

3. Calculate a *t* value (see "Analyzing Ecology" in Chapter 15) to statistically compare the two treatment means.

INTERPRETING THE DATA

1. Based on the *t*-test—and setting $\alpha = 0.05$—are the two treatment means significantly different from each other?

2. Explain how the data support or refute the hypothesis that total plant production in a grassland will increase with the diversity of traits.

3. In the same experiment, Cadotte found that some plants grew larger in the presence of other plant species than they did when they were grown alone in a greenhouse. This is a common phenomenon, particularly when one species grows next to a nitrogen-fixing species. Develop a hypothesis to explain why nitrogen-fixing plants might facilitate the growth of other plant species.

SOURCE: Cadotte, M. W. 2013. Experimental evidence that evolutionarily diverse assemblages result in higher productivity. *Proceedings of the National Academy of Science*, 110: 8996–9000.

CHAPTER 17

Mutualism

Living with Crabs

Life as a tiny crab has its challenges. There is a continuous need to find lunch and, at the same time, to avoid becoming someone else's lunch. Consider the red-ridged clinging crab (*Mithrax forceps*), which reaches an adult size of only 2.5 cm. The red-ridged crab lives on the reefs off the coast of North Carolina and spends much of its time among the many branches of ivory bush coral (*Oculina arbuscula*) where it hides from predators. Researchers find that crab survival is much higher in areas where the coral is present than in areas without the coral.

Ivory bush coral, however, is a poor competitor on the shallow-water reefs. Large species of algae, commonly known as seaweeds, grow very fast and smother the coral. As we discussed in Chapter 2, corals are animals that have a mutualistic relationship with tiny algae, known as zooxanthellae, that live inside them. When the corals become shaded by the smothering seaweeds, the zooxanthellae cannot photosynthesize and both the corals and zooxanthellae die. So how do corals, and therefore crabs, persist on the reefs?

> **"Crabs assist corals by reducing competition from seaweeds while corals assist crabs by providing safe places to hide from predators."**

Fortunately for the coral, the red-ridged clinging crab mows down any seaweed that tries to establish itself near the coral. It is able to eat all species of shallow-water seaweed, even those with chemical defences that repel many species of reef fish. To confirm the importance of the crab to the coral, researchers removed crabs from areas containing coral and found that the rate of death among corals increased fourfold. In short, crabs assist corals by reducing competition from seaweeds while corals assist crabs by providing safe places to hide from predators.

To determine whether corals offer crabs something more than a useful place to hide, researchers gave crabs a choice between living among the branches of live corals or of dead corals. The crabs overwhelmingly chose live corals because, as researchers discovered, the crabs feed on a layer of coral mucus that only exists on live corals. This mucus provides a high-energy food for crabs.

Researchers also discovered that the mutualistic relationship between crabs and corals changed in deep-water reefs where lower light and colder temperatures cause seaweed to grow more slowly. Moreover, one of the common species of seaweed on the deep-water reef is unpalatable to crabs. As a result, crabs are not able to offer any benefit to the coral, but the crabs still use coral for protection and may still consume the high-energy mucus of the coral. This means that the mutually beneficial interaction in the shallow reefs changes to a one-sided benefit in the deep reefs.

The red-ridged clinging crab. In shallow-water reefs off the coast of North Carolina, red-ridged crabs remove seaweed from around corals. The corals provide the crabs with a place to live and a source of food. Photo by Azure Computer & Photo Services/Animals Animals/Earth Scenes.

The interaction of crabs and corals illustrates how species commonly depend on one another to exist in nature and how a mutual reliance can change under different environmental conditions. In this chapter, we will see a wide variety of ways in which species rely on each other to improve their growth, survival, and reproduction.

SOURCES: J. J. Stachowicz and M. E. Hay, Mutualism and coral persistence: The role of herbivore resistance to algal chemical defense, *Ecology* 80 (1999): 2085–2101.

M. E. S. Bracken, et al., Whole-community mutualism: Associated invertebrates facilitate a dominant habitat-forming seaweed, *Ecology* 88 (2007): 2211–2219.

CHAPTER CONCEPTS

- Mutualisms can improve the acquisition of water, nutrients, and places to live.

- Mutualisms can aid in defence against enemies.

- Mutualisms can facilitate pollination and seed dispersal.

- Mutualisms can change when conditions change.

- Mutualisms can affect communities.

In Chapter 1, we defined a mutualism as a positive interaction between two species in which each species receives benefits that only the other species can provide. Mutualisms are common in nature: Corals live with their symbiotic algae, crabs assist corals by removing competing seaweed, and, as we saw in Chapter 10, leaf-cutter ants farm a fungus. In this chapter, we will see how species have evolved to participate in mutualistic interactions for a wide variety of benefits. Some species benefit by obtaining resources while others benefit by obtaining a place to live, aid in defence, and assistance in pollination or seed dispersal. When we consider mutualistic relationships, it is tempting to think that each species is trying to help the other. However, as you might recall from our review of natural selection in Chapter 1, selection favours any strategy that increases the fitness of the individual. Regardless of the specific benefit, there are requirements for the evolution of mutualistic interactions and conditions under which a positive, mutualistic relationship can change to a neutral or negative interaction. Finally, we are interested in mutualisms not only because they are a common interaction in nature, but because they can affect the abundance of populations, the distribution of species, the diversity of communities, and the functioning of ecosystems.

provide fitness benefits to each other and require each other to persist, we call them **obligate mutualists**. We saw an example of this in Chapter 1 when we discussed the tubeworms and chemosynthetic bacteria that live together near deep-sea hydrothermal vents. Tubeworms provide a place for bacteria to live and bacteria provide food for the tubeworms; neither species can survive without the other. In contrast, **facultative mutualists** provide fitness benefits to each other, but the interaction is not critical to the persistence of either species. For example, a group of tiny insects known as aphids suck the sap from plants and produce a droplet rich in carbohydrates that is consumed by several species of ants. The ants gain a source of food and in exchange they protect the aphids from predators. Although both groups benefit, it is a facultative mutualism because each can persist without the other. A mutualism between two species can be composed of two obligate mutualists, one obligate and one facultative mutualist, or two facultative mutualists.

One of the most common functions of mutualisms is to help species acquire resources they need, such as water, nutrients, and a place to live. In previous chapters we have discussed a few examples of such mutualisms. In Chapter 1 we saw that lichens are composed of a fungus living with either green algal cells or cyanobacteria (see Figure 1.13). This fungus

Mutualisms can improve the acquisition of water, nutrients, and places to live

We can categorize mutualisms in several ways. For example, some mutualists are **generalists**, which means that one species interacts with many other species. Other mutualists are **specialists**, which means that one species interacts with either one other species or a few closely related species. When two species

Generalist A species that interacts with many other species.

Specialist A species that interacts with one other species or a few closely related species.

Obligate mutualists Two species that provide fitness benefits to each other and require each other to persist.

Facultative mutualists Two species that provide fitness benefits to each other, but the interaction is not critical to the persistence of either species.

Figure 17.1 A mutualism between coral and zooxanthellae. A coral grabs food with its tentacles, which contain stinging cells, and pulls the food into its mouth. Zooxanthellae algae live along the surface of a coral's tentacles where they obtain sunlight for photosynthesis. Corals provide a place for algae to live and emit CO_2 that algae use in photosynthesis. As algae photosynthesize, they provide sugars and O_2 that corals consume.

provides the algae with water, CO_2 from fungal respiration, and nutrients and, in exchange, the algae provide the fungus with carbohydrates from photosynthesis. Similarly, in Chapter 2 we discussed how corals provide a home for photosynthetic algae known as zooxanthellae. As you can see in Figure 17.1, the coral catches bits of food with its tentacles and during digestion the coral emits CO_2, which the algae use during photosynthesis. The algae then produce sugars and O_2, some of which can be consumed by the coral. Other animals also incorporate symbiotic algae into their bodies. Similarly, in Chapter 2 we discussed how the eggs of spotted salamanders incorporate algae in the tissues of the embryo. Similarly, when the green sea slug (*Elysia chlorotica*) consumes algae, it stores the chloroplasts from the algae inside its tissues and thus gains most of its energy through photosynthesis (Figure 17.2). In this section, we will review how numerous species of plants, animals, fungi, and bacteria interact in mutualisms to gain water, nutrients, and places to live.

ACQUISITION OF RESOURCES IN PLANTS

Although plants obtain water and soil minerals through their root systems, ecologists have learned that many plants also rely on mutualisms with fungi and bacteria to help them obtain nutrients.

Figure 17.2 The green sea slug. When the slug first hatches from an egg, it is brown. However, as it begins to eat algae, it stores the chloroplasts from the algae in its own tissues. As the slug accumulates a large number of chloroplasts, its body turns green and it is able to acquire most of its energy through photosynthesis rather than through herbivory. Photo by Nicholas E. Curtis and Ray Martinez, University of South Florida.

Plants and Fungi

Fungi that surround plant roots and help plants obtain water and minerals are known as **mycorrhizal fungi**. The network of fungal hyphae provide plants with minerals such as nitrogen and phosphorus and with water from the surrounding soil. Plants provide the fungi with the sugars produced by photosynthesis. Because fungi can increase the amount of minerals obtained by the plants, they are able to increase the plants' tolerance to both drought and salt stress. They can also help plants combat infections from pathogens. This mutualism is valuable to many species of plants, including plants that humans use for food and materials.

Mycorrhizal fungi can be divided into either *endomycorrhizal fungi* or *ectomycorrhizal fungi*. **Endomycorrhizal fungi**, illustrated in Figure 17.3a, are characterized by hyphal threads that extend far out into the soil and penetrate root cells between the cell wall and the cell membrane. One fungal species can commonly infect multiple plant species. There

---

Mycorrhizal fungi Fungi that surround plant roots and help plants obtain water and minerals.

Endomycorrhizal fungi Fungi characterized by hyphal threads that extend far out into the soil and penetrate root cells between the cell wall and the cell membrane.

Fungal hyphae grow into root cells between the cell wall and cell membrane.

Fungal hyphae grow between root cells.

Root epidermis

Fungal sheath

Fungal hyphae extend into soil.

(a) Endomycorrhizal fungus

(b) Ectomycorrhizal fungus

Figure 17.3 Mycorrhizal fungi. Mycorrhizal fungi can be categorized as endomycorrhizal fungi or ectomycorrhizal fungi. **(a)** Endomycorrhizal fungi have hyphae that penetrate the root cells of plants and reside between the cell wall and cell membrane. **(b)** Ectomycorrhizal fungi have hyphae that do not penetrate the root cells but instead grow between the root cells of plants.

are several types of endomycorrhizal fungi. The most common type is **arbuscular mycorrhizal fungi**, which comprise a group of mycorrhizae that infects a tremendous number of plants including grasses and apple, peach, and coffee trees. Arbuscules are branching hyphal structures found within plant cells that help the fungus provide nutrients to the plant.

Ectomycorrhizal fungi, illustrated in Figure 17.3b, are characterized by hyphae that surround the roots of plants and enter between root cells but rarely enter the cells. These fungi are currently known to live only in mutualistic relationships with trees and shrubs. Species of ectomycorrhizal fungi also tend to form mutualistic relationships with fewer plant species than do endomycorrhizal fungi.

The mutualistic relationship between plants and mycorrhizal fungi goes back more than 450 mya to the time when plants first evolved to live on land. This ancient interaction between the ancestral plants and fungi probably explains why so many modern species of plants and fungi continue to interact as mutualists. Researchers currently estimate that mutualism between plants and fungi involves more than 6 000 species of mycorrhizal fungi and 200 000 species of plants, which is about two-thirds of all plant species.

Plants and Bacteria

In some cases, mutualistic interactions between plants and bacteria convert unusable forms of minerals into forms that plants can use. One of the best-known examples of this is the group of bacteria in the genus *Rhizobium* that live in a mutualistic relationship with numerous species of legumes, including important crops such as beans, peas (*Pisum sativum*), and alfalfa (*Medicago sativa*). When legumes detect the presence of *Rhizobium* bacteria in the soil or when bacteria enter the plant through an opening in the root, the plant develops small nodules that surround the bacteria on the roots and provide them with a place to live (Figure 17.4). Plants also provide the bacteria with the products of photosynthesis. In exchange, the bacteria do something the plant cannot do; they convert atmospheric nitrogen—a form of nitrogen plants cannot use—into ammonia, a form of nitrogen plants can readily use. This mutualism can be quite valuable to plants, especially when they are living in areas of low soil fertility. We will discuss this phenomenon in more detail in Chapter 21.

ACQUISITION OF RESOURCES IN ANIMALS

Animals also use a variety of organisms to help them obtain food and habitat. These interactions range from protozoans living in animals to mutualisms between animals.

Figure 17.4 Root nodules that contain *Rhizobium* bacteria. Legumes such as this bean plant can enter into a mutualistic relationship with *Rhizobium* bacteria. The plant provides the bacteria with sugars from photosynthesis and a root nodule in which the bacteria can grow. In return, the bacteria convert atmospheric nitrogen to ammonia, a form of nitrogen that the plant can use. Photo by Nigel Cattlin/Science Source.

Animals and Protozoans

Termites are a group of insects that consume wood, which is difficult to digest since it is composed largely of lignin and cellulose. To assist in this effort, species of protozoa that are able to consume lignin and cellulose live in termite guts. The protozoa receive a home in the termite gut with a constant source of food from wood that the termite consumes and, in exchange, the termite receives nutrients from the waste products of protozoan digestion. Many other animals also contain microbes in their digestive system. Humans, for example, host hundreds of species of microbes—bacteria, fungi, and protozoa—that largely seem to be beneficial. In fact, a person's digestive system contains 10 times more bacterial cells—from more than 500 bacterial species—than the total number of human cells in that person's body.

Arbuscular mycorrhizal fungi A type of endomycorrhizal fungi that infects a tremendous number of plants including apple trees, peach trees, coffee trees, and grasses.

Ectomycorrhizal fungi Fungi characterized by hyphae that surround the roots of plants and enter between root cells but rarely enter the cells.

Mutualisms Between Animals

Mutualisms for acquiring resources can also occur between two species of animals. A fascinating example occurs between humans and a bird known as the greater honeyguide (*Indicator indicator*). For centuries, people in Africa have consumed the honey produced by bees, but locating the beehives is a challenge. While the greater honeyguide likes to consume bee larvae and bee's wax, it has a hard time getting into beehives. The local people and the honeyguide both obtain resources by working together. Over time, local people have learned to use calls to attract the attention of the bird and then follow the bird to the beehive. Along the way, the bird stops to perch in nearby trees. As the bird gets closer to the beehive, it flies shorter distances from one perch to the next and perches lower in trees, as shown in Figure 17.5. Local people have learned how to interpret the bird's behaviour and they follow it to the hive. When they find the hive, they scoop out the honey and leave pieces of the honeycomb with beeswax and bee larvae on the ground for the honeyguide to consume. It is thought that the honeyguide may have originally evolved this behaviour as a mutualism with other honey-consuming mammals, such as the honey badger (*Mellivora capensis*).

Some animals provide a habitat for other animals in exchange for reciprocal benefits. For example, alpheid shrimp live in the ocean and have very poor vision. They burrow into the sand and allow a group of fish known as gobies to share their burrows. In contrast to shrimp, gobies have excellent vision and are able to see shrimp predators. In exchange for receiving a burrow, a goby allows a shrimp to stay in close contact by permitting the shrimp to place an antenna on it once the goby leaves the burrow. If the goby sees a predator, it warns the shrimp by

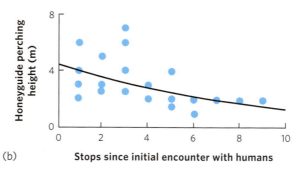

(a)

(b)

Figure 17.5 The greater honeyguide. When the greater honeyguide leads humans or other honey-consuming mammals to a beehive, the bird benefits by consuming the discarded honeycombs that contain beeswax and bee larvae. Humans have learned that as the greater honeyguide gets close to a beehive, **(a)** the bird flies shorter distances between stops and **(b)** at each stop it perches lower on trees. Error bars are standard deviations. Data from H. A. Isack and H.-U. Reyer, Honeyguides and honey gatherers: Interspecific communication in a symbiotic relationship, *Science* 243 (1989): 1343–1246.

twitching. The shrimp detects the twitching via its antenna and heads back into the burrow for protection (Figure 17.6).

Figure 17.6 Habitat mutualist. The alpheid shrimp (*Alpheus randalli*) digs a burrow that it shares with a fish known as the pinkbar goby (*Amblyeleotris aurora*) in the waters off Indonesia. In exchange for this habitat, the goby allows the shrimp to place an antenna on its body so that the shrimp can feel the goby twitching when it sees an approaching predator. Photo by Jim Greenfield/imagequestmarine.com.

(a) (b)

Figure 17.7 A mutualism between ants and acacia trees. (a) Acacia trees, such as this swollen thorn acacia in Panama, have large brown thorns that the ants can hollow out for nests where they can raise their larvae. The plants also have green nectaries that provide nectar the ants consume. In exchange for these benefits, the ants attack herbivores that try to eat the acacia tree and they attack vines and other plants that grow near the acacia tree. **(b)** Because the ants attack encroaching plants, acacia trees with ants are typically surrounded by bare ground. Photos by Alexander Wild.

Mutualisms can aid in defence against enemies

As we saw with the alpheid shrimp, mutualisms can increase the ability of a species to defend itself against enemies. To obtain a defence from a mutualistic partner, an organism must provide some type of benefit in return. In this section, we will examine a number of ways that mutualisms have evolved to benefit both the species being defended and the species providing the defence.

PLANT DEFENCE

Plants are involved in a number of mutualisms that help defend them from enemies. Two well-known examples are the mutualisms between ants and acacia trees and the mutualisms between fungi and plants.

Acacia trees are found in tropical forests throughout the world; they face a variety of herbivores and numerous competitors including vines that try to wrap around the acacia's branches. In Central America, acacia trees are commonly inhabited by ants in the genus *Pseudomyrmex* that constantly patrol the tree branches. The acacia trees provide large thorns with pulpy centres that ants hollow out and convert into nests. The trees also contain nectaries, which produce nectar that the ants consume (Figure 17.7a). In exchange, these ants bite and sting any herbivores—from small insects to large mammals—that attempt to consume the leaves. The ants also eliminate plants that attempt to grow near their home tree by chewing on them until they die (Figure 17.7b).

In a classic study, Dan Janzen compared acacia trees that had ants living on them and acacia trees from which he removed the ants. As you can see in Figure 17.8a, he found that trees with ants had a lower percentage of herbivorous insects than trees from which the ants had been removed. In addition, trees

with ants grew to be 14 times heavier than trees without ants. When Janzen monitored the trees over a 10-month period, he found that trees with ants had

(a)

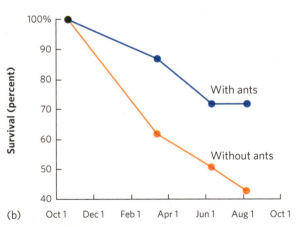

(b)

Figure 17.8 The effect of ants on acacia trees. (a) When ants were removed from acacia trees, the trees experienced a large increase in the number of herbivorous insects feeding on them. **(b)** As a result, the removal of ants caused much lower survival of the acacia trees. Data from D. H. Janzen, Coevolution of mutualism between ants and acacias in Central America, *Evolution* 20 (1966): 249–275.

much higher survival rates than trees without ants, as shown in Figure 17.8b. In short, Janzen demonstrated that the ants are critical to the survival and growth of the acacia trees.

More recently, researchers have discovered that the ant-acacia mutualism has additional benefits. In 2010, researchers reported that the chemicals in the nectaries contain numerous proteins with antibacterial properties. To determine if ants helped to distribute these chemicals on the leaves, the researchers used two sets of trees: one group had the ants present and the other group had the ants removed for 2 weeks. Researchers also examined the effect of one ant species that acted as a mutualist (*P. ferrugineus*) and another ant species that did not (*P. gracilis*). You can view the results of the experiment in Figure 17.9. When *P. ferrugineus* was present, the plant leaves were nearly free of bacteria, but when *P. gracilis* was present, there was no significant reduction in bacteria. The researchers do not yet know how the *P. ferrugineus* causes a reduction in leaf bacteria, but it seems likely that *P. ferrugineus* is distributing some of the antibacterial nectar to the leaves whereas *P. gracilis* does not. These results demonstrated that ants in a mutualistic relationship with trees not only defend the tree against herbivores and competitors but appear to also defend the tree against bacteria that act as plant pathogens.

Figure 17.10 Cleaner wrasse. The small cleaner wrasse, *Labroides dimidiatus*, removes large numbers of ectoparasites from larger fish, such as this yellowfin goatfish (*Mulloidichthys vanicolensis*) in the Red Sea. The cleaner wrasse benefits from a reliable food source while the larger fish benefits from having a lower number of parasites. Borat Furlan/Getty Images.

Some plants can defend themselves from herbivores through mutualisms with fungi known as **endophytic fungi** that live within the plant's tissues. These fungi produce chemicals that can repel insect herbivores and also provide drought resistance by increasing the concentration of minerals in plant tissues, which increases the plant's ability to absorb and retain water from the soil. In exchange, the plant supplies the fungi with the products of photosynthesis. While endophytic fungi can be beneficial to plants, some of the chemicals they produce can be quite harmful to herbivores on which humans depend. For example, when the grass known as tall fescue (*Festuca arundinacea*) contains endophytic fungi, the fungi produce chemicals that are highly toxic to cattle, sheep, goats, and horses. Researchers are currently working to identify alternative strains of fungi that can offer defence against insect herbivores and drought resistance without being toxic to livestock.

ANIMAL DEFENCE

Animals also participate in mutualisms that aid their defence. An excellent example occurs in a group of fish known as cleaner wrasse. These tiny fish spend their life consuming ectoparasites that are attached to other, much larger fish (Figure 17.10). As the cleaner wrasse approaches, the larger fish opens its mouth and flares its gills to permit access to the many parasites that are attached to its body. The number of parasites removed can be substantial; a single cleaner wrasse can consume more than 1200 parasites per day. The tiny cleaner wrasse benefits from having a

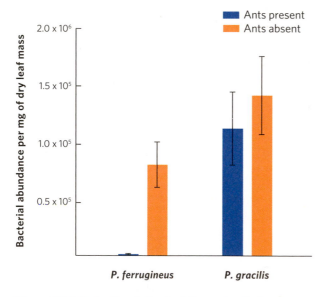

Figure 17.9 Reduction in bacteria by mutualistic ants. The presence of the mutualistic ant (*P. ferrugineus*) causes a large reduction in the abundance of bacteria on the leaves of the acacia tree. In contrast, the presence of a nonmutualistic ant (*P. gracilis*) has no effect on bacterial abundance. Error bars are standard errors. Data from M. González and M. Heil, *Pseudomyrmex* ants and *Acacia* host plants join efforts to protect their mutualism from microbial threats, *Plant Signaling and Behavior* 5 (2010): 890–892.

Endophytic fungi Fungi that live inside a plant's tissues.

Figure 17.11 Oxpeckers. On the African savannah oxpeckers, such as this red-billed oxpecker in Kruger National Park in South Africa, remove ticks from a variety of grazing mammals, such as this impala (*Aepyceros melampus*). Photo by Robert Harding Picture Library/SuperStock.

grazing mammals in Africa. The researchers quantified the oxpecker preferences for different species of mammals by observing grazing mammals of several species and dividing the number of oxpeckers on a given species by the total number of species grazing. Then they quantified the abundance of ticks on an individual of each mammal species. Although they found no relationship between the species preferences of oxpeckers and the thickness of the animals' hides, they found positive correlations between the species preferences of oxpeckers and tick abundance, as you can see in Figure 17.12. These results suggest that the birds are acting primarily as mutualists whose preferences are geared toward feeding on ticks.

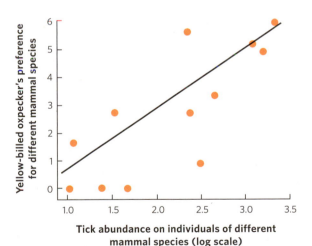

Figure 17.12 Oxpecker preferences for different species of mammals. The preferences of red-billed oxpeckers and yellow-billed oxpeckers are positively correlated to the abundance of ticks on the different mammals. Neither bird showed a preference for mammals with thin hides. These data suggest that oxpeckers seek out particular species of mammals primarily to consume ticks as a mutualist rather than to consume bits of mammal flesh as a parasite. Data from C. L. Nunn et al., Mutualism or parasitism? Using a phylogenetic approach to characterize the oxpecker-ungulate relationship, *Evolution* 65 (2011): 1297–1304.

large source of food and the larger fish benefit by having fewer parasites.

A similar situation exists for large terrestrial animals in Africa. Two species of birds, the red-billed oxpecker (*Buphagus erythrorhynchus*) and the yellow-billed oxpecker (*B. africanus*), perch on the backs of grazing animals such as rhinos and antelopes (Figure 17.11). While perched on the backs of the grazing animals, the birds consume ticks that are attached to the mammals. However, because the birds also peck at the wounds caused by the ticks, scientists wondered whether the birds were mutualists or parasites. If oxpeckers act primarily as mutualists, their preferences for certain species of grazing mammals should be related to the number of ticks carried by each species. Alternatively, if the birds act primarily as parasites that seek to peck at mammal flesh, they should prefer mammals with thinner hides, which their beaks can penetrate more easily. In a 2011 study, researchers examined these relationships in both species of oxpeckers using up to 15 species of

Mutualisms can facilitate pollination and seed dispersal

In addition to providing resources and defence, mutualisms can also provide the valuable services of plant pollination and the dispersal of plant seeds. Indeed, without these services many species could not reproduce or colonize areas throughout their geographic range. In this section, we will examine the various mutualisms that serve these functions.

POLLINATION

To make seeds, flowering plants need pollen to fertilize their ovules. Some plants, such as grasses, commonly rely on the wind to blow pollen from one plant to another. Other plants rely on animals to carry the pollen. Over evolutionary time, plants have evolved a number of reward mechanisms to entice pollinators to visit their flowers. For example, the common honeybee visits flowers that offer both nectar and pollen. The bees consume both items, but the bees also inadvertently transfer some of the pollen that is attached to their bodies as they move from flower to flower. Many plants appear to have evolved flowers that are specialized in attracting one particular type of pollinator. For example, flowers pollinated by hummingbirds tend to be red—a colour hummingbirds favour. Plants pollinated by hummingbirds also tend to possess long tubular flowers inaccessible to most insects. However, because hummingbirds have long tongues, they can easily reach the nectar inside. Similarly, many bat-pollinated flowers are large, are often open only at night, and contain high volumes of nectar to attract the large pollinators. In some plant species, the volume of nectar increases during the night when bats are most likely to be foraging (Figure 17.13).

Although flowers in many species of plants can be pollinated by several different pollinator species, some plants, such as the group known as yuccas, have evolved very specific mutualisms with their pollinators. In fact, most species of yucca plants rely on a single species of yucca moth for pollination. Unlike most pollinators, however, a visit from the yucca moth is not a brief affair. A female yucca moth arrives at a yucca flower and lays eggs in the ovaries of the flower (Figure 17.14). To ensure that the flower makes seeds for her offspring to eat, the female moth climbs to the top of the flower and adds pollen to the flower's stigma. This pollen produces a long tube that grows down through the style and into the ovary where it deposits the male gametes, which then fuse with the female gametes to produce seeds. The moth eggs hatch into caterpillars that feed on the seeds that the flower forms. As a result, the plant gains a very effective pollinator at the cost of a few seeds.

Despite the consumption of seeds by the caterpillars, enough seeds remain for the yucca to

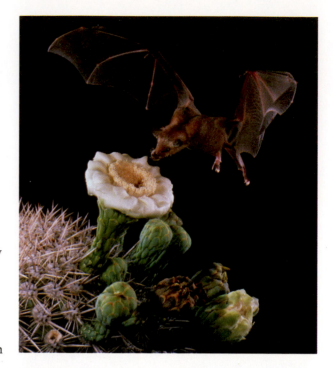

Figure 17.13 Pollination by bats. Flowers that are pollinated by bats are typically large and contain large nectar rewards to entice the bats to visit. Pictured here is a long-nosed bat (*Leptonycteris curasoae*) pollinating a flower of the saguaro cactus (*Carnegiea gigantea*). Photo by Dr. Merlin D. Tuttle/Science Source.

reproduce. For many years researchers wondered what prevented the female moths from laying so many eggs that all the seeds would be consumed, leaving the flower with no fitness. After careful observation, they discovered that the plant is tolerant of up to six moth eggs per flower. Under these conditions, the plant develops 62 percent of its flowers to maturity. However, if there are more than six eggs per flower, some species of yuccas selectively abort most of their flowers and retain only 17 percent of their flowers to

Figure 17.14 A yucca moth. A female yucca moth transfers pollen to the style of a yucca flower after she lays her eggs in the ovary of the flower. Photo by Robert and Linda Mitchell.

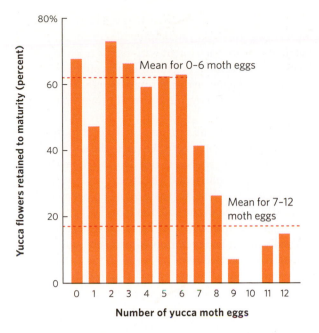

Figure 17.15 Controlling consumption of yucca seeds by yucca moth caterpillars. When female yucca moths lay 0 to 6 eggs in a yucca flower, the average plant develops 62 percent of its flowers to maturity. If a moth lays 9 to 12 eggs, the average plant only develops 17 percent of its flowers to maturity; the rest are selectively aborted. This response favours yucca moths that lay fewer eggs. Data from O. Pellmyr and C. J. Huth, Evolutionary stability of mutualism between yuccas and yucca moths, *Nature* 372 (1994): 257–260.

maturity, as shown in Figure 17.15. When a flower is aborted, all moth eggs and larvae within the aborted flower dies. By aborting flowers whenever the moths lay more than six eggs, the plant favours those moths that do not lay too many eggs in each flower.

SEED DISPERSAL

The seeds and fruits produced by plants span a wide range of sizes, from tiny dandelion seeds that float on the wind to massive fruits such as coconuts. Many of the smallest seeds are easily dispersed by wind, but larger seeds are typically dispersed by animal mutualists. As we have seen, animals act in this role if they receive a benefit in return. The most common benefit is food in the form of seeds or nutritious fruit that surrounds the seeds. When animals eat the fruits of plants, the seeds commonly pass through the digestive system unharmed and are viable after being excreted. However, when animals eat the seeds of plants, the seeds are digested and are no longer viable.

It may seem like a contradiction that plants would depend on seed-eating animals to help disperse their seeds. After all, if the seeds are eaten, there are no seeds left to disperse. However, in most cases, plants produce large numbers of seeds and not all seeds that are carried away get eaten. Many of them are stored in the ground, which means that the seeds are planted and able to germinate if they are not consumed. For example, squirrels store acorns in numerous locations and some of these acorns are never retrieved. Similarly, a single Clark's nutcracker (*Nucifraga columbiana*)—a bird that specializes on the seeds of the whitebark pine (*Pinus albicaulis*)—collects approximately 32 000 pine seeds in a season and stores them in thousands of different locations. Because this number of seeds is three to five times what the bird requires to meet its energy needs, many of the stored seeds are never eaten but rather germinate into new pine trees in locations far from the parent tree.

Some species of plants entice animals to disperse their seeds by surrounding them with a fruit that contains substantial nutrients. For example, many forest herbs including the painted trillium (*Trillium undulatum*) produce seeds that have a lipid- and protein-rich package—known as an *elaiosome*—attached to the seed (Figure 17.16). Ants collect the

Figure 17.16 Ants eating elaiosomes. Some forest herbs produce seeds with a packet of lipid- and protein-rich tissue attached that is known as an elaiosome. Shown here are seeds of the bloodroot (*Sanguinaria canadensis*). The ants carry these seeds to their nest, consume the elaiosome, and discard the seeds outside of the nest where they can subsequently germinate. Photo by Alexander Wild.

seeds and take them back to their nests. After consuming the elaiosomes, they discard the seeds from the nest. This process disperses the seeds away from the parent plant. The ants receive nutrition and the plants receive dispersal of their seeds.

When plants surround their seeds with a large fruit, the fruit can serve as a substantial reward to animal dispersers. If the seeds inside the consumed fruit have a hard coat that resists digestion, they pass through the animal's digestive system and are still able to germinate. One of the most striking examples is the African tree known as *Omphalocarpum procerum*. The tree produces fruits as large as a person's head and only the African elephant can break it open. Moreover, the seeds cannot germinate in the soil unless they have passed through an elephant's digestive system. The tree is highly dependent on elephants to consume and disperse its seeds, but, as elephant populations decline, the tree is losing its seed disperser.

A key element to this strategy for seed dispersal is to remain inedible or hidden until the seeds are fully developed. As a result, many fruits are green and relatively camouflaged while the seeds are developing. At this stage, the fruit is typically quite unpalatable. However, once the seeds have fully developed, the fruit becomes ripe, the tissues of the fruit become palatable, and it commonly changes colour to be highly visible to animal dispersers.

Mutualisms can change when conditions change

Although individuals of two or more species might interact in a way that allows both to receive fitness benefits, we need to recognize that each individual participates in the mutualism to improve its own fitness, not that of its partner. Therefore, when conditions alter the costs and benefits for each species, the interaction can change to something that is no longer a mutualism. Environmental conditions also influence the mutualistic relationship between grasses and their arbuscular mycorrhizal fungi in Grassland National Park, Saskatchewan. In moist conditions that are typical for early summer, the fungi provide grasses with nutrients. In the drier conditions of late summer, fungi no longer provide any nutrients to the grasses, and the mutualistic relationship breaks down.

FROM POSITIVE TO NEGATIVE INTERACTIONS

When one species in a relationship provides a benefit to another species at some cost but no longer receives a benefit in return, the interaction can shift from a positive, mutualistic interaction to a negative interaction such as herbivory, predation, or

parasitism. For example, we discussed the role that mycorrhizal fungi play in helping plants obtain nutrients. In an experiment with citrus trees in highly fertile soils, researchers examined the effect of eliminating the mycorrhizal fungi by treating the soil with a fungicide. They found that eliminating the fungus made the trees grow up to 17 percent faster. This suggests that the fungus posed a cost to the tree without providing a benefit. Because the soil was fertile, the citrus trees were not dependent on the fungi to provide nutrients. The trees could grow well by collecting nutrients on their own but, because the fungi still existed in the soil, the trees still provided the fungi with the products of photosynthesis. The normally mutualistic relationship had changed into a parasitic interaction.

A similar situation exists for cleaner fish. You may recall that cleaner fish remove ectoparasites from larger fish. Both species benefit and the interaction is a mutualism. However, it turns out that cleaner fish also like to consume mucus and scales of larger fish, which is harmful to the larger fish because mucus and scales are costly to produce and may offer protection against infection. Researchers working on coral reefs in the Caribbean examined whether the feeding decisions of a cleaner fish, the Caribbean cleaning goby (*Elacatinus evelynae*), changed when there were differences in the number of ectoparasites carried by the longfin damselfish (*Stegastes diencaeus*). They sampled the number of ectoparasites on the damselfish off the coasts of six different islands, and then observed the cleaner fish to determine the percentage of mucus and scales in their diet. You can see their data in Figure 17.17. When damselfish populations had a high number of parasites, the cleaner fish ingested a small percentage of mucus and scales. However, when the damselfish had a low number of parasites, the cleaner fish ingested a much higher percentage of mucus and scales. Therefore, when the parasites are rare on the damselfish, the cleaner fish switch from being mutualistic to predatory and consume a higher percentage of mucus and scales.

DEALING WITH CHEATERS

A mutualistic relationship can evolve over time into a relationship in which one species receives a benefit but does not provide one in return—a behaviour known as cheating. Natural selection would favour cheating as long as it provides a net increase in the cheater's fitness, although we would also expect natural selection to favour mechanisms that enable organisms to defend themselves against cheaters. We saw an example of this with yuccas and yucca moths. When a moth lays so many eggs that the hatching larvae will eat all of the developing seeds, some

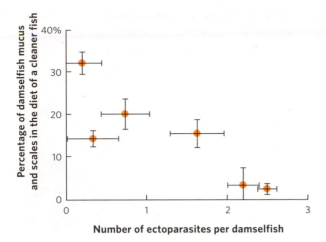

Figure 17.17 Switching from mutualism to predation. In populations of damselfish where ectoparasites are abundant, the cleaner fish primarily consume parasites of the damselfish and very little of the damselfish's mucus and scales. This is a mutualistic interaction. In populations of damselfish where the ectoparasites are not abundant, the diet of the cleaner fish includes a substantial percentage of the mucus and scales from the damselfish. This becomes more of a predatory interaction. Error bars are standard errors. Data from K. L. Cheney and I. M. Côté, Mutualism or parasitism? The variable outcome of cleaning symbioses, *Biology Letters* 1 (2005): 162–165.

species of yucca can respond by aborting the flower, thereby killing the moth larvae. In this way, the yucca punishes yucca moths that act as cheaters.

A similar situation exists in relationships between plants and mycorrhizal fungi. When they operate as mutualists, both species provide a benefit at some cost. If a fungus reduces the benefit it provides to a

plant, the plant should respond by providing a smaller benefit to the fungus. In 2009, researchers conducted a study to see if a plant could discriminate between different fungi and send products of photosynthesis to the most beneficial fungal mutualist. To test this question, the researchers planted individual wild onion (*Allium vineale*) plants in two pots; half of the plant's roots went into a pot of soil containing a fungal species that provides large benefits to the plant and the other half of the roots went into a second pot containing a fungal species that provides no benefit to the plant. This experiment is illustrated in Figure 17.18a. After the onion roots grew for 9 weeks, researchers measured how much photosynthetic product the plant was sending to each fungus. They did this by placing a bag around each plant and pumping in a special form of CO_2 that contained a rare isotope of carbon known as ^{14}C. Using this radioactive form of carbon, they could track the movement of carbon from the plant to each fungal species. They found that onion plants sent more products of photosynthesis to the beneficial fungi, as you can see in Figure 17.18b. This means that plants can discriminate among different fungi and preferentially provide greater benefits to the most beneficial fungi.

Mutualisms can affect communities

When we think about mutualisms, we often focus on how the relationship helps each of the interacting species; it often improves the fitness and abundance of each participant. For example, corals cannot

Figure 17.18 Favouring the most beneficial partner. **(a)** In an experiment, the wild onion plant was grown with its roots into two separate pots: one contained a beneficial mycorrhizal fungus and the other contained a nonbeneficial mycorrhizal fungus. After 9 weeks of growth, researchers were able to determine whether the plant allocates more of its photosynthetic products to the beneficial or nonbeneficial fungus. **(b)** Because the plant can distinguish between the two fungal species, it is able to allocate more of its photosynthetic products to the more beneficial fungus. Error bars are standard errors. Data from J. D. Bever et al., Preferential allocation to beneficial symbiont with spatial structure maintains mycorrhizal mutualism, *Ecology Letters* 12 (2009): 13–21.

survive without zooxanthellae and many plants cannot produce offspring without pollinators. In addition to affecting abundance, mutualisms can also affect other ecological levels. In this section, we will examine how mutualisms, and the interruption of mutualisms, can alter the distributions of species, the diversity of communities, and the functioning of ecosystems.

EFFECTS ON SPECIES DISTRIBUTIONS

Given the benefits of mutualisms, we might expect natural selection to favour mutualistic relationships among species and to expand the distribution of species engaged in mutualisms. In contrast, we would expect the disruption of a mutualism to cause a decline of the species involved and a reduction in their distribution. We can see an example of this in the plant known as garlic mustard (*Alliaria petiolata*), a member of the mustard family with leaves that smell like garlic when crushed. Native to Europe and Asia, it was introduced to North America more than a century ago for human consumption. When garlic mustard grows in North American forests, it causes young trees to grow poorly and makes them less likely to reach adult size. For a number of years, researchers sought to understand the mechanism underlying garlic mustard's harmful effects.

In 2006, researchers discovered that garlic mustard was interfering with the mutualism between the forest trees and arbuscular mycorrhizal fungi in the soil. To demonstrate this link, they examined how well three species of trees grew when raised in soil where garlic mustard had invaded and in soil where it had not invaded. As you can see in Figure 17.19a, the increases in biomass of sugar maple, red maple, and white ash trees were many times greater when grown in soil that had not been invaded by garlic mustard. Moreover, when the researchers examined the percentage of tree roots colonized by mycorrhizal fungi, they found that soil collected from forests with garlic mustard had little or no colonization by mycorrhizae, as illustrated in Figure 17.19b.

Additional experiments revealed that the roots of garlic mustard produce chemicals that inhibit the growth of mycorrhizal fungi. Given that tree species depend on fungal mutualism to different degrees, garlic mustard has the largest negative effect on those tree species that have the greatest dependence on the fungi. Because garlic mustard disrupts vital mutualisms, it has the potential to alter the distribution of a large number of other species as it spreads across North America.

EFFECTS ON COMMUNITIES

Mutualisms that alter the abundance and distribution of one or more species can have widespread effects on

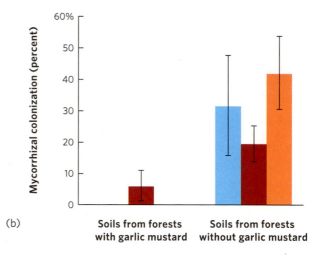

Figure 17.19 Disrupting a mutualism with garlic mustard. When researchers grew three species of trees in soils from forests with and without garlic mustard, they found that soils from forests with garlic mustard caused **(a)** much smaller increases in biomass and **(b)** little or no colonization by mycorrhizal fungi. Error bars are standard errors. Data from K. A. Stinson et al., Invasive plant suppresses the growth of native tree seedlings by disrupting belowground mutualisms, *PLOS Biology* 4 (2006): 727–731.

the rest of the community. A community can be affected in several ways; the mutualism might cause a change in species diversity or it might alter the abundance of individuals within species in the community.

Altering Species Diversity

A generalist mutualist species interacts with many other species and offers widespread benefits. In the case of the cleaner fish, for example, a particular species of cleaner fish may remove parasites from many different species of larger fish. If this behaviour

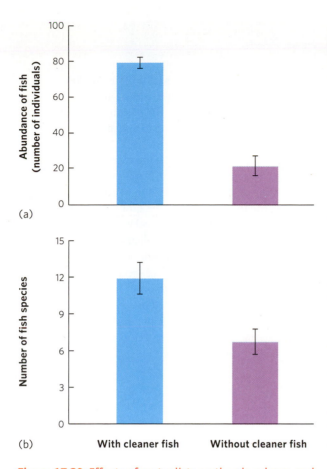

(a)

(b)

With cleaner fish   **Without cleaner fish**

Figure 17.20 Effects of mutualists on the abundance and diversity of a coral reef fish community. After 18 months of removing cleaner fish, **(a)** the abundance of individual fish declined by about three-quarters and **(b)** the number of species declined by about half. Error bars are standard errors. Data from A. S. Grutter et al., Cleaner fish drives local fish diversity on coral reefs, *Current Biology* 13 (2003): 64–67.

helps species to persist on a coral reef, then the removal of cleaner fish should cause a decline in the total number of large fish and the number of species. Researchers working on coral reefs in Australia tested this question by removing a species of cleaner fish (*Labroides dimidiatus*) from nine small reefs. They designated another nine reefs as controls. After 18 months, they counted both the number of individual fish and the number of fish species on each reef. As you can see in Figure 17.20, removing the cleaner fish caused the number of individual fish to decline by about three-quarters and the number of species to decline by half, which suggests that cleaner fish play a critical role in maintaining populations of reef fish.

Initiating a Chain of Interactions

In some cases, the community does not lose species when a mutualism is disturbed, but the abundance of many species changes through a chain of interactions. Earlier in the chapter we discussed the mutualistic relationship between arbuscular mycorrhizal fungi

and plants. Fungi provide water and minerals to plants, and plants provide fungi with sugars. We also considered the mutualistic relationship between flowering plants and insects. Insects consume a plant's pollen and nectar, and in turn fertilize plants by transporting pollen between individual plants. To determine the links between these mutualistic relationships, researchers investigated how a change in fungal abundance might affect plant and insect communities.

The researchers began by setting up 20 blocks in an experimental grassland in Alberta. They divided each block into two 2-by-2-m plots, separated by a 0.5-m strip. One plot in each block was treated with a fungicide to reduce growth of arbuscular mycorrhizal fungi. The second plot in each block served as a control. After 3 years, the researchers examined how the fungicide treatment affected the plants and the pollinator community.

As you can see in Figure 17.21, bumblebees were the major pollinators of plants in the plots without fungicide. In plots treated with fungicide, bumblebees were replaced as major pollinators by small bees, hoverflies, and beeflies. What caused this change in the pollinator community? The fungicide treatment affected field chickweed (*Cerastium arvense*), which grew more flowering stems in fungicide-treated plots compared to control plots.

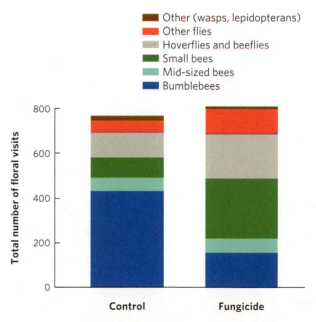

Figure 17.21 Community effects of a mutualism. Reducing the mutualistic interaction between arbuscular mycorrhizal fungi and plants also impacted the pollinator community, which depends on the plants. Smaller insects were more common pollinators on treated plots. Control plots had more visits from large bumblebees. Data from J.F. Cahill, Jr., E. Elle, G.R. Smith, and B.H. Shore, Disruption of a belowground mutualism alters interactions between plants and their floral visitors, *Ecology* 89 (2008): 1791–1801.

Comparing Two Groups That Do Not Have Normal Distributions

As we saw in the case of the cleaner fish, researchers who study mutualisms often need to test whether the mutualistic interaction really provides a benefit to each species in the interaction. To do so, they need to conduct statistical tests that compare how each species performs both with and without the presence of the other species. In Chapter 15 we discussed the use of *t*-tests to compare the means of two groups. As you may recall, *t*-tests require that the data collected follow a normal distribution, which are the bell-shaped curves that we discussed in Chapter 2. In some cases, however, the data from two groups do not have normal distributions, so we cannot use the *t*-test. In these cases, we need to use the Mann-Whitney rank sum test, which is named for the statisticians who developed it.

The Mann-Whitney rank sum test begins by ranking the data—from low to high—and then taking the sum of those ranks. For example, consider the following set of data for the number of species occurring in eight coral reefs where researchers have removed the cleaner fish and eight coral reefs where researchers have allowed the cleaner fish to remain. If we were to take all 16 numbers and place them in order from the lowest value to the highest value, we could give each value a rank from 1 to 16. Whenever we have more than one occurrence of a value, we assign both values the average rank.

NUMBER OF SPECIES OBSERVED WITHOUT CLEANER FISH	RANKS	NUMBER OF SPECIES OBSERVED WITH CLEANER FISH	RANKS
(GROUP 1)	(GROUP 1)	(GROUP 2)	(GROUP 2)
3	1	7	6.5
4	2	8	8.5
5	3	9	10.5
6	4.5	10	12.5
6	4.5	10	12.5
7	6.5	11	14
8	8.5	12	15
9	10.5	14	16

Therefore, chickweed flowers made up a larger proportion of the available flowers in the fungicide-treated plots. The increased availability of chickweed made the plant community in fungicide-treated plots more attractive to smaller insects, since the large bumblebees prefer flower species that were not affected by the fungicide.

This experiment demonstrated that reducing the mutualistic interaction between arbuscular mycorrhizal fungi and flowering plants affects not only the plant community. The effects reach further and also change the insect community that depends on the flowers. In turn, the altered insect community will affect the rate at which different plant species are being pollinated, which will influence seed production and, in the long run, population growth and community structure. As this example demonstrates, mutualisms affect more than the species involved in the interactions; they can have far-reaching effects throughout the entire community.

EFFECTS ON ECOSYSTEM FUNCTION

Mutualisms can also have ecological effects at the level of the ecosystem. As we discussed in Chapter 1, researchers working on ecosystems examine the movement of energy and matter among many sources including both the biotic and abiotic realms. For

The next step is to sum the ranks for group 1 (denoted as R_1) and group 2 (denoted as R_2).

$$R_1 = 1 + 2 + 3 + 4.5 + 4.5 + 6.5 + 8.5 + 10.5$$
$$= 40.5$$

$$R_2 = 6.5 + 8.5 + 10.5 + 12.5 + 12.5 + 14 + 15 + 16$$
$$= 95.5$$

We can then use either value of R to calculate a test statistic known as U. We will use R_2:

$$U = R_2 - (n_2 \times (n_2 + 1) \div 2)$$

where n_1 is the number of observations in group 1 and n_2 is the number of observations in group 2. Using this formula,

$$U = 95.5 - (8 \times (8 + 1) \div 2) = 59.5$$

Now that we know the value of U, we need to calculate the mean and standard deviation of U for the entire collection of data from both groups. The mean value of U is:

$$m_U = (n_1 n_2) \div 2$$

$$m_U = (8 \times 8) \div 2 = 32$$

The standard deviation of U is:

$$\sigma_U = \sqrt{\frac{n_1 n_2 \times (n_1 + n_2 + 1)}{12}}$$

$$\sigma_U = \sqrt{\frac{8 \times 8 \times (8 + 8 + 1)}{12}} = 9.52$$

Using our calculated values of U, m_U, and σ_U we can calculate z, which is defined as:

$$z = \frac{U - m_U}{\sigma_U}$$

$$z = \frac{59.5 - 32}{9.52} = 2.89$$

We can look up this value in a table of z values in the Statistical Tables appendix, where we find that the probability value is 0.002. Since this probability is less than 0.05, we can conclude that the groups are significantly different from each other. This means that removing the cleaner fish causes a significant decrease in the number of fish species that live on a coral reef.

YOUR TURN When we calculate the value of z, we actually use the absolute value of z. To convince yourself that it does not matter whether we use R_1 or R_2, redo the above calculations based on the formula for U when using R_1:

$$U = R_1 - (n_1 \times (n_1 + 1) \div 2)$$

example, researchers working with arbuscular mycorrhizal fungi investigated how a grassland ecosystem of 15 species would respond to different numbers of fungal species in the soil. When they examined the total amount of phosphorus that the plants took up from the soil, they found that plants living in soils that contained more fungal species took up a greater amount of phosphorus, as shown in Figure 17.22a. They also quantified the total biomass of roots in the soil and shoots above the soil. They found that a higher number of fungal species in the soil caused substantial increases in the biomass of the roots and shoots in the ecosystem. You can see these data in Figures 17.22b and 17.22c. Communities are

also influenced by the responses of different plant species to arbuscular mycorrhizal fungi. In an experiment carried out at Guelph University, 64 different plant species were grown either with or without mycorrhizal fungi. The results showed that arbuscular mycorrhizal fungi formed mutualistic interactions with some plant species and parasitic interactions with other plant species. The researchers suspected that this variation in plant responses to the presence of arbuscular mycorrhizal fungi may have strong effects on plant diversity and ecosystems. As illustrated in these examples, mutualisms not only affect individual species, but they can also have large effects on the functioning of ecosystems.

Figure 17.22 Ecosystem effects of fungal diversity. When researchers manipulated increases in the species diversity of mycorrhizal fungi, they observed an increase in **(a)** the total amount of phosphorus in the plants, **(b)** the total biomass of roots, and **(c)** the total biomass of shoots. Error bars are standard errors. Data from A. G. A. van der Heijden et al., Mycorrhizal fungal diversity determines plant biodiversity, ecosystem variability, and productivity, *Nature* 396 (1998): 69–72.

Throughout this chapter, we have seen how species can interact in a variety of mutualisms. These interactions fulfill a wide range of needs including resource acquisition, places to live, defence, pollination, and dispersal. These interactions can have important effects on communities and ecosystems and they can even have major conservation implications, as you will see in the "Ecology Today" feature that follows.

ECOLOGY TODAY CONNECTING THE CONCEPTS

DEALING WITH THE DEATH OF DISPERSERS

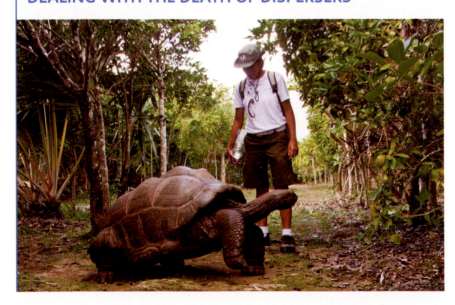

A substitute mutualist. Aldabra tortoises were introduced onto a small island in the Republic of Mauritius to act as a seed disperser for a population of critically endangered ebony trees. Photo by Olivier Naud.

The Republic of Mauritius provides an example of how important mutualisms can be to the persistence of species in nature. Mauritius, a group of islands located in the Indian Ocean southeast of the African continent, is the only place on Earth where the dodo bird (*Raphus cucullatus*) once lived. A large, flightless bird, the dodo was an easy source of food for the sailors who visited the islands, and was also harmed by the introduction of many non-native species. By the late 1600s, about 200 years after humans first set foot on Mauritius, the dodo was extinct. Numerous other species unique to Mauritius have also been driven extinct, including two species of giant tortoises: the high-backed tortoise (*Cylindraspis triserrata*) and the domed tortoise (*C. inepta*). Not only are the islands missing these species, they are also missing the services that these species provided.

Over subsequent centuries, as humans settled the islands, they cut down large swaths of tropical forest for lumber, firewood, and space for sugar cane fields. Today, only about 2 percent of the original forest remains and some of the tree species continue to decline. For example, *Syzygium mamillatum* is a critically endangered tree species that consists of just over a hundred individuals. Scientists studying this problem realized that many declining tree species had probably once relied on the dodos and tortoises for seed dispersal. These species consumed the fruit, and the seeds were released after they passed through the animals' digestive systems. When the animals defecated, typically far away from the trees that produced the fruits, the seeds were ready to germinate. With these mutualists extinct, fewer fruits were consumed and any released seeds remained close to the parent tree. Seeds that germinate near the parent tree compete with the parent and there are high rates of disease among conspecific trees concentrated in one place. Without dispersers, the seeds of the rare trees cannot disperse and recolonize open areas that have been logged. Researchers wondered if it was possible to bring back the process of seed dispersal even without the animals that originally did the dispersing.

Researchers came up with a radical idea. They hypothesized that perhaps non-native species could be introduced to fill the role of the lost mutualists and help save the endangered trees. To play the role of the dodo, they brought in domestic turkeys (*Meleagris gallopavo*), which—similar to dodos—have a large gizzard that grinds the fruits they swallow. The researchers hypothesized that when the dodo consumed the fruits, the grinding of the gizzard broke down the fruits and released the seeds, but maintained the viability of the seeds as they passed through the bird's digestive system. To play the role of the two species of extinct tortoises, they brought in giant Aldabra tortoises (*Aldabrachelys gigantea*) from the nearby Seychelles Islands.

To test the potential for these species to act as substitute seed dispersers, the researchers fed fruits of the *S. mamillatum* tree to the turkeys and Aldabra tortoises. When fruits were fed to the turkeys, the large gizzards not only broke apart the fruit but also ground the seeds into small pieces, which meant that turkeys were not suitable replacements for the dodos. However, when fruits of the tree were fed to the Aldabra tortoises, the researchers found that 16 percent of the seeds inside the fruits remained whole in tortoise feces. Although these seeds had lower germination success than seeds that did not pass through the tortoises' gut, they produced seedlings that grew taller and had more leaves. This meant that the Aldabra tortoises might serve as an effective substitute disperser.

In a follow-up study, the researchers tested whether the tortoises could serve as a new mutualist for critically endangered trees if they were released into the wild. To test this idea, they decided to introduce Aldabra tortoises to a small, 25-ha island that is part of the Republic of Mauritius. This island contains another rare tree species, the ebony tree (*Diospyros egrettarum*), which was once abundant and produces large, 16-g fruits. Logging for firewood had made this tree rare on the island and large areas contained no new ebony tree seedlings. Without dispersers, the fruits landed near the parent tree and stayed there. After introducing the tortoises into pens in 2000, the researchers confirmed that the tortoises had no negative effects on the plant community. In 2005, 11 tortoises were allowed to roam freely throughout the island. In 2011, the investigators reported some remarkable results. Few fruits were now found near the parent trees because the tortoises were

Seed dispersing tortoises. By consuming the large ebony fruits and dispersing the seeds through their feces, the tortoises spread the seeds throughout the island. They may represent the best hope for bringing the ebony tree back from the brink of extinction. Photo by Dennis Hansen.

consuming most of them and, because the seeds passed through the tortoises when they defecated, the seeds were relocated to many places around the island. New seedlings showed up everywhere, including open areas where the tree had not existed for many decades. Moreover, ebony seeds that passed through the tortoise gut germinated better than unconsumed seeds.

After more than a decade of research, these results suggest that, although we cannot bring back extinct species, we may be able to substitute some species to resurrect mutualisms so that the original extinctions do not cause subsequent extinctions of their mutualistic partners.

SOURCES: Hansen, D. M., et al. 2008. Seed dispersal and establishment of endangered plants on oceanic islands: The Janzen-Connell model, and the use of ecological analogues. *PLOS One* 3: 1–13.

Griffiths, C. J., et al. 2011. Resurrecting extinct interactions with extant substitutes. *Current Biology* 21: 1–4.

SUMMARY OF CHAPTER CONCEPTS

■ **Mutualisms can improve the acquisition of water, nutrients, and places to live.** Mutualisms can be categorized as generalists, which interact with many species, or specialists, which interact with few other species. When both species require each other to persist, they are obligate mutualists. When the interaction is beneficial but not critical to the persistence of either species, they are facultative mutualists. Mutualisms for resources include the algae and fungi that compose lichens and the corals and zooxanthellae that build coral reefs. Plants also participate in this type of mutualism by interacting with endomycorrhizal fungi, ectomycorrhizal fungi, and *Rhizobium* bacteria. In most animals, protists can play an important role in digesting food. Other animals construct habitats that they share with other species in exchange for other benefits.

■ **Mutualisms can aid in defence against enemies.** Plants make use of defensive mutualisms in a number of ways, including mutualisms with aggressive insects such as ants, and with endophytic fungi that produce chemicals harmful to herbivores. Animals that interact as mutualists to defend against enemies include cleaner fish that remove parasites from large fish and oxpecker birds that remove ticks from mammals.

■ **Mutualisms can facilitate pollination and seed dispersal.** Pollinators allow many species of plants to be fertilized and some plants have evolved traits that favour a particular type of pollinator. When this happens, the plants and the pollinators can coevolve. Numerous plants also depend on mutualisms to disperse their seeds. In some cases, seeds are dispersed as the result of animals storing them far from the parent plant. In other cases, animals consume the fruit of plants and the seeds are dispersed after passing through their digestive systems.

■ **Mutualisms can change when conditions change.** Although mutualisms benefit all species in the interaction, a positive mutualism can switch to a neutral or negative interaction when conditions change. In some cases, species can respond to cheaters in a mutualism by only rewarding individuals that provide benefits in return.

■ **Mutualisms can affect communities.** Mutualisms can increase or decrease the abundance of participating species. An absent mutualist can cause another species to be completely eliminated, thereby affecting the distribution of a species. Mutualists can also affect communities either by directly altering the number of species or by initiating a chain of interactions through a community. At the ecosystem level, mutualists can also have effects such as moving nutrients into producers and increasing the total biomass of producers.

REVIEW QUESTIONS

1. Compare and contrast obligate mutualists and facultative mutualists.

2. What is the difference between endomycorrhizal and ectomycorrhizal fungi?

3. What is the function of protozoan mutualists in the digestive systems of termites?

4. If one species provides a habitat as part of a mutualistic relationship, what is the probable effect on the abundance and distribution of the other species?

5. Explain how ants and acacia trees fulfill the definition of a mutualistic relationship.

6. If cleaner fish consume both parasites and scales from larger fish, what would determine whether the interaction is best categorized as mutualism or parasitism?

7. How might natural selection favour the evolution of flower sizes, shapes, and colours that attract a specific group of pollinators?

8. Explain why a plant would not make its fruit very colourful until the seeds are fully developed.

9. What might prevent a species in a mutualism from reaping the benefits from a partner without providing a benefit in return?

10. In addition to affecting the abundance of species, what other levels of ecology can be affected by a mutualism?

PRACTICING ECOLOGY: THE BENEFITS OF A BACKUP PLAN

Megan Frederickson at the University of Toronto studies the factors that maintain mutualistic relationships between plants and animals. Photo by Gabriel Miller.

Relationships do not always work out. When it comes to facultative mutualisms, sometimes a mutualistic partner does not arrive. For example, predators or competitors may limit the abundance of one partner. In an extreme case, as we saw in the case of the dodo in "Ecology Today," one mutualistic partner can go extinct and leave the other partner to fend for itself. In such instances, natural selection should favour adaptations that allow a species to exist by itself, but such adaptations may be costly in terms of resources. One solution may be phenotypic plasticity: An individual exhibits a backup adaptation only when its mutualistic partner is absent. Do such plastic adaptations exist in nature?

Cordia nodosa, a species of acacia tree found in Peru, has mutualistic relationships with several ant species, although only one ant species colonizes a particular tree at any one time. Like most acacia trees, *C. nodosa* forms little pockets in its stem to house ants that defend the plant from herbivores and it provides nectar that the ants can consume. However, the ants are not always present to defend the tree and not all ant species that colonize *C. nodosa* offer the same level of protection since some species are better at repelling herbivores than others. Aside from the protection that ants offer, the trees can also produce antiherbivore chemical compounds in their tissues, known as phenolics. They can also grow small hairs on their leaves, known as trichomes, which make it difficult for herbivorous insects to chew the leaf. Since such defences are costly to produce, we might hypothesize that *C. nodosa* only produces these defences when ants are not available to defend the tree.

Megan Frederickson from the University of Toronto and her colleagues tested this hypothesis by exposing *C. nodosa* trees to herbivores in the presence and absence of ants. Working in the rainforest of Peru, the researchers planted 22 trees in the understorey. They collected ant colonies from local acacia trees and placed the colonies on half of their planted trees. They excluded ant colonies from the other half of their planted trees by spraying the surrounding soil with an insecticide. After letting trees grow for 47 weeks, they measured phenolics and trichomes on three randomly chosen leaves from each tree. Using a microscope, they measured the number of trichomes on 1 cm^2 of each leaf's surface. They then dried and ground 0.25 g of each leaf and measured the concentration of phenolics in the leaf tissue through a chemical analysis. You can view their data in the following table.

TRICHOME DENSITY (NUMBER/cm^2) AND PHENOLIC CONCENTRATION (mg/g LEAF) OF ACACIA LEAVES WITH AND WITHOUT ANTS

TRICHOME DENSITY		PHENOLIC CONCENTRATION	
WITH ANTS	WITHOUT ANTS	WITH ANTS	WITHOUT ANTS
4.52	8.38	3.42	3.80
3.71	3.66	3.66	3.71
3.36	6.21	2.85	2.81
3.36	6.69	3.65	2.64
3.91	5.75	3.99	2.26
4.02	1.47	3.57	2.63
4.18	4.64	4.97	2.08
5.78	11.61	6.70	4.43
6.84	3.21	3.35	3.80
7.31	28.30	2.61	6.80
10.25	3.45	7.53	12.96

GRAPH THE DATA

1. Calculate the means and standard errors of the means for each of the four columns.

2. Create a bar graph showing the mean and standard error of trichome density with and without ants.

3. Create a second bar graph showing the mean and standard error of phenolic concentrations with and without ants.

INTERPRETING THE DATA

1. Conduct a Mann-Whitney rank sum test to determine if trichome density differs between acacia leaves with and without ants (see "Analyzing Ecology" on page 406).

2. Conduct a second Mann-Whitney rank sum test for phenolic concentrations.

3. Explain how the results of the statistical tests support or refute the hypothesis that *C. nodosa* only produces trichomes and phenolics when ants are not available to defend the plant from herbivores.

4. In the same study, the researchers found that trees always suffered more herbivore damage in the absence of ants. Given the results from the statistical tests, does this suggest that there is selection pressure for *C. nodosa* to maintain the ant mutualism over time?

SOURCE: Frederickson, M. E., A. Ravenscraft, L. M. A. Hernández, G. Booth, V. Astudillo, and G. A. Miller. 2013. What happens when ants fail at plant defence? *Cordia nodosa* dynamically adjusts its investment in both direct and indirect resistance traits in response to herbivore damage. *Journal of Ecology*, 101:400–409.

CHAPTER 18

Community Structure

A Web of Interactions in Social Spiders

If you have ever walked through a forest or a field, you have undoubtedly seen a spider perched on a web, waiting for an unsuspecting insect to come by. Most spiders live solitary lives except when they mate. However, some species of social spiders live in large groups composed of thousands of spiders. These spiders build webs that are typically 1 to 4 m wide, and they occasionally make massive webs up to 100 m wide. These webs are so large that a single web can serve as a habitat for more than 100 species of invertebrates that include many other species of spiders.

The interactions that occur in such a web are diverse. Some of the species that join the community of invertebrates are spiders that catch prey. The scraps they leave behind can be consumed by the spider species that builds the web. Other species that join the community are predators of the web-building spider species. In addition, the individual spiders that build and maintain the web have specific behavioural traits. For example, the webs of the tangle web spider (*Anelosimus studiosus*), which lives in riparian areas from the northeastern United States to southern Argentina, are inhabited almost entirely by females. These female spiders have different personalities; some are quite aggressive whereas others are docile. Aggressive individuals are better at capturing prey and defending against predators whereas docile individuals spend more of their time building the massive web.

> "Some species of social spiders live in large groups that make massive webs up to 100 m wide."

The composition of the spider community and the specific behavioural traits of the spiders that build the web both influence how long the community can persist. Although docile individuals grow and reproduce better in the presence of other species of spiders, colonies containing only docile individuals ultimately go extinct sooner. In 2012, researchers discovered that the higher extinction rate in webs containing docile spiders is related to the diversity of other spider genera that live in the web. When there are more spider genera, it is more likely the web will contain spiders that are predatory on the docile females; these predatory genera will ultimately drive the docile females to extinction.

In contrast, webs that contain a higher proportion of aggressive females successfully maintain a lower diversity of spider genera because they attack new spiders that attempt to intrude. As a result, these webs are less likely to contain predatory spider genera. Ultimately, a mixture of docile spiders and aggressive spiders strikes the best balance between the need to catch and subdue prey and the need to avoid too many aggressive interactions.

The tangle web spiders. As social spiders, docile and aggressive individuals work together to build massive webs that catch a wide variety of prey including grasshoppers. Photo by Alexander Wild.

The web of social spiders is both a physical web that catches prey and a network of species interactions. The outcome of these interactions depends not only on the specific species present but also on the traits of individuals. In this chapter, we will see how the interaction of species and environmental conditions influence both the number of species and the abundance of each species. We will also examine how some species have disproportionate effects on the diversity of a community.

SOURCES: J. N. Pruitt and M. C. O. Ferrari, Intraspecific trait variants determine the nature of interspecific interactions in a habitat-forming species, *Ecology* 92 (2011): 1902–1908.

J. N. Pruitt et al.,Behavioural trait variants in a habitat-forming species dictate the nature of its interactions with and among heterospecifics, *Functional Ecology* 26 (2012): 29–36.

The massive web of social spiders. Tangle web spiders live in large social groups that build massive spider webs. This web discovered in Lake Tawakoni State Park in Texas was built by the tangle web spider and more than 50 other species of spiders. These webs serve as habitats for a high diversity of invertebrates. Photo by TOM PENNINGTON/Fort Worth Star-Telegram/MCT/Landov.

CHAPTER CONCEPTS

- Communities can have distinct or gradual boundaries.

- The diversity of a community incorporates both the number and relative abundance of species.

- Species diversity is affected by resources, habitat diversity, keystone species, and disturbance.

- Communities are organized into food webs.

- Communities respond to disturbances with resistance, resilience, or switching among alternative stable states.

In Chapter 1 we defined a community as an assemblage of species living together in a particular area. We also discussed how ecologists working at the community level focus on a multitude of species interactions, and how these interactions affect the number of species and the relative population size of each species. In Chapters 14 through 17 we discussed the major types of species interactions. We are now ready to expand our understanding of communities in which these interactions take place. In this first of two chapters on ecological communities, we will examine community structure, which includes the composition of species in a community, the relative abundance of each species, and the relationships among the species. We will start by looking at a classic debate over whether or not communities are distinct entities. We will then examine how ecologists quantify patterns of species diversity in communities based on the number of species and the relative abundance of each species in the community. All of these species function as producers and consumers in a *food web,* which helps to determine the number of species, the relative

abundance of each species, and the stability of the community.

Communities can have distinct or gradual boundaries

If we are to understand ecological communities, we first need to understand how communities change across the landscape and how they are categorized. As we will see, because many species move between communities, drawing boundaries around a community can be difficult. For example, many birds migrate every spring and autumn, and amphibians spend their larval life in aquatic communities and their adult life in terrestrial communities. In this section, we will discuss community boundaries and investigate how communities are affected when community boundaries are either distinct or gradual.

COMMUNITY ZONATION
One of the most noticeable features of communities is that the species composition of the community

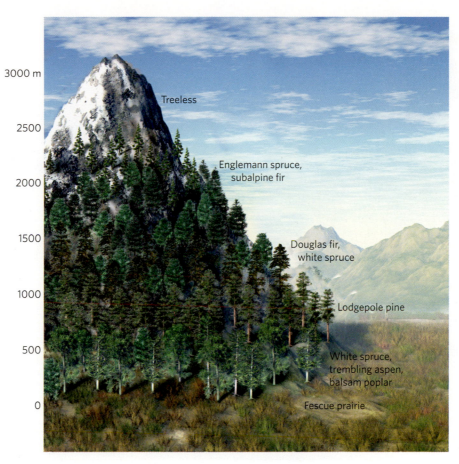

Figure 18.1 Zones in terrestrial communities. As one climbs the eastern slopes of the Rocky Mountains, the vegetation changes from prairie grasses to large trees. At higher altitudes, trees are smaller before they disappear at the tree line.

changes as one moves across the landscape. With the change in environmental conditions, some species become better able to survive and compete. For example, if you were to walk from the base to the top of a mountain on the eastern side of the Rocky Mountains in Alberta, you could observe striking changes in the vegetation, as illustrated in Figure 18.1. You would walk out of a prairie environment dominated by grasses, such as fescue prairie (see Figure 16.11). Walking up the low foothills, you would first encounter white spruce (*Picea glauca*), trembling aspens, and balsam poplar (*Populus balsamifera*). In the higher foothills, the forest is dominated by lodgepole pine (*Pinus contorta*). As you approach the top of the mountain, you would pass through a forest of Douglas fir and white spruce (*Picea glauca*), followed by a forest of Engelmann spruce (*Picea engelmanni*) and subalpine fir (*Abies lasiocarpa*). The zones in which each species flourishes reflect different tolerance ranges for temperature and moisture availability, as well as different abilities to compete with other species. There are similar

changes in the animal species that live at different elevations on the mountains. The changes in plants and animals at different elevations create continuous changes in community composition from the base of the mountains to the highest peaks.

Zonation also occurs in aquatic communities, as depicted in Figure 18.2. We saw an example of this in Chapter 16 when we discussed competition between species of barnacles on rocky coasts of Great Britain. Due to a combination of competition and resistance to desiccation, stellate barnacles live in the upper intertidal zone whereas the rock barnacles live in the lower intertidal zone. Occupation of different zones is a pattern that is repeated among a wide variety of species in intertidal communities. For example, we find kelp forests below the intertidal zone, and a variety of seaweeds, mussels, anemones, rock barnacles, and hermit crabs occupy the lower and middle intertidal zones. In the upper intertidal zone we find limpets, which are a type of gastropod, and stellate barnacles. Higher up on the shore an area known as the splash zone is home to more limpets and

Figure 18.2 Zonation of communities along ocean shorelines. As one moves from deep to shallow water in the rocky shoreline of northeastern England, the species composition of the communities changes dramatically.

a type of snail known as the periwinkle (*Littorina littorea*). As we saw in the case of the two barnacle species, the distribution of species in the different zones of the coastline reflects a combination of tolerance to changing abiotic conditions and the outcome of species interactions that include competition, predation, and herbivory.

CATEGORIZING COMMUNITIES

Ecologists generally categorize communities as they do biomes—either by their dominant organisms or by physical conditions that affect the distribution of species. In North America, for example, we might examine a beech-maple forest community in Ontario, a pine savannah community in Virginia, or a sagebrush community in Wyoming. Each of these communities is named for the dominant plants that are present. In aquatic systems, we might focus on physical characteristics—such as a stream community, lake community, or wetland community— or on the dominant group of organisms in that system, such as a coral reef community.

When ecologists study communities, they rarely study every species in the community, which could number in the hundreds or thousands, but rather focus on a subset of species that live in an area. For example, researchers studying a wetland community might study only bacteria, algae, snails, crustaceans, insects, amphibians, or fish. They might also examine several of these groups simultaneously.

ECOTONES

When we try to categorize communities, it is clear that some communities have distinct boundaries that are either natural or constructed by humans. For example, the natural boundary between a lake community and a forest community can be clearly drawn at the water's edge. This boundary exists because there is a sharp change in the environmental conditions as we leave one community and enter the adjacent community. Similar sharp changes in environmental conditions occur, for example, when there is an abrupt change in the type of soil due to the underlying geology of an area or when an individual

Figure 18.3 Ecotones. Ecotones are regions where two communities come together with a relatively sharp boundary, indicated by a rapid replacement of species. This ecotone, in Greater Sudbury, Ontario, occurs where a forest and a field come together. Photo by Don Johnston/age fotostock.

moves from north-facing to south-facing slopes of mountains that have large differences in temperatures and moisture. Humans can also create distinct community boundaries. For example, an area of forest that has been cleared for agriculture creates a clear boundary between the field community and the forest community. Sharp changes in environmental conditions over a relatively short distance, accompanied by a major change in the composition of species, creates a boundary known as an **ecotone** (Figure 18.3). Although some species move between the adjacent communities that come together to form the ecotone, most species live in one of the communities and spread into the ecotone. As a result, ecotones typically support a large number of species, including those from the adjoining habitats and species adapted to the ecotone's special conditions.

One way to document the existence of an ecotone is to use a line-transect survey to determine the abundances of different species along an environmental gradient. When an ecotone is present, we expect to observe sharp changes in the distributions of species as we leave one community and enter the adjacent community. An example can be found in plant communities on serpentine soils. These soils are derived from underlying rock that contains heavy metals, such as nickel, chromium, and magnesium, that are toxic to many plants. Serpentine rock outcrops exist in small patches across the landscape in many different parts of the world. Serpentine soils are typically low in nutrients such as

nitrogen and phosphorus. Because of the harsh conditions, most plant species cannot live in serpentine soils, although some species of plants have evolved an ability to tolerate them (Figure 18.4).

In one study in southern Oregon, researchers conducted a line-transect survey that stretched from nonserpentine soils to serpentine soils. They measured the concentrations of several metals and quantified the presence of numerous plant species. Concentrations of chromium, nickel, and magnesium all increased in the shift from nonserpentine soils to serpentine soils, as you can see in the bottom half of

Figure 18.4 Serpentine soils. In areas where serpentine rocks emerge to the soil surface, the soils contain low concentrations of nutrients and high concentrations of metals such as nickel, chromium, and magnesium. At this site in southern British Columbia, the forest abruptly ends where the serpentine soils begin because the trees are not able to survive in the serpentine soils. Photo by Gary Lewis.

Ecotone A boundary created by sharp changes in environmental conditions over a relatively short distance, accompanied by a major change in the composition of species.

Figure 18.5. When they examined the distribution of the plants, as shown in the top half of the figure, they found that species such as black oak (*Quercus kelloggii*) and poison oak (*Rhus diversiloba*) did not grow on the serpentine soils, whereas canyon live oak (*Quercus chrysolepis*) and ragwort (*Senecio integerrimus*) were found almost entirely in the ecotone where the two soils come together, and species such as fireweed (*Epilobium minutum*) and knotweed (*Polygonum douglasii*) were found only in serpentine soils. These species show distinct boundaries as one moves across the gradient of nonserpentine to serpentine soils. As you can also see in the transect data, a few species such as hawkweed (*Hieracium albiflorum*) and fescue (*Festuca californica*) were found across the entire gradient. From these data we can conclude that while some species are largely restricted to either serpentine or nonserpentine soils, the highest number of species occurs within the ecotone.

COMMUNITIES WITH INTERDEPENDENT VERSUS INDEPENDENT SPECIES DISTRIBUTIONS

As we have noted, a community is often described by the dominant species that live within it. Although we know that species interact with each other, for many years ecologists questioned whether species in a community were found together because they depended on each other or because they had similar habitat needs.

Interdependent communities are those in which species depend on each other to exist. In the case of the social spider, for example, the various species of spiders determine the growth and reproduction of all spiders and determine the life span of the community. In the early twentieth century, plant ecologist Frederic Clements proposed that most communities function as interdependent communities and that they act as *superorganisms*. He compared individual species to the different parts of an organism's body, all of which require each other for survival.

Independent communities are those in which the species do not depend on each other to exist. Independent communities are composed of species that live in the same place because they have similar adaptations and nutrient requirements. Although each species has a somewhat different range of conditions under which it can live, the community that we see in a given location reflects the overlapping ranges of the species that exist within it. In other words, similar habitat requirements happen to put various species in the same place at the same time. Plant ecologist Henry Gleason rejected Clements's superorganism metaphor and proposed that most communities consist of species with independent distributions.

Differentiating Between Interdependent and Independent Species Distributions

How can we determine whether a community is comprised of an interdependent or an independent group of species? One approach has been to study which species recolonize areas after a disturbance, such as fire. If species are interdependent, we would

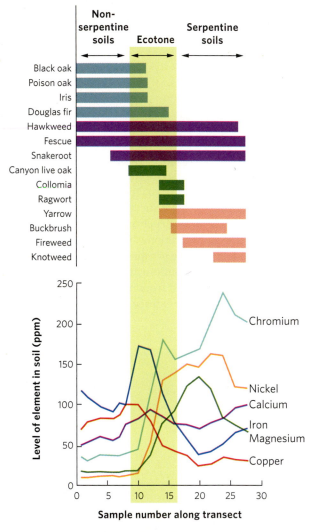

Figure 18.5 A line transect of serpentine and nonserpentine soils. As one moves along the transect, the concentration of metals in the soil exhibits a sharp increase midway along the transect. The presence of plant species along the gradient, as indicated by the bars, demonstrates that some species primarily live in nonserpentine soils (shown in blue), a few species primarily live in the ecotone (shown in green), and several species primarily live in the serpentine soils (shown in pink). Three of the species are able to live across the entire transect (shown in purple). Data from C. D. White, "Vegetation–Soil Chemistry Correlations in Serpentine Ecosystems" (Ph.D. dissertation, University of Oregon, 1971).

Interdependent communities Communities in which species depend on each other to exist.

Independent communities Communities in which species do not depend on each other to exist.

expect the same group of species to recolonize the disturbed area at the same time. If species are independent, we should see varied patterns of recolonization, with some species occurring early after the disturbance in some instances and appearing later in others. This is what researchers observed when they studied areas in the boreal mixed-wood forest in Alberta that had been disturbed by fire between 38 to 59 years ago. White spruce recolonized some areas less than 20 years after the fire, while it did not begin to grow for almost 60 years in other areas. These findings indicate that boreal mixed-wood forests in Alberta consist of independent species.

Another approach to differentiate between interdependent and independent species distributions has been the use of line-transect studies. If species distributions are independent, we should be able to observe gradual changes in species composition as we move along a line transect. Each species will appear and disappear at different points along the line because of the unique habitat requirements of each species.

A classic study using this approach was conducted during the 1950s by Robert Whittaker who surveyed the distributions of plant species in the Great Smoky Mountains on the border between Tennessee and North Carolina (Figure 18.6a). At different elevations that varied in temperature and moisture, he calculated the number of stems present for each plant species. As you can see in Figure 18.6b, different species appeared and disappeared at different points along the moisture gradient, and each species achieved its peak abundance at different points. This study and other studies by Whittaker provided strong evidence that when environmental conditions change gradually along a gradient, species in the community commonly show gradual changes in abundance that are independent of each other.

Observing species distributions along environmental gradients is one way to test the independence of species distributions. However, even when we see abrupt changes in species distributions we cannot be sure whether species in a community

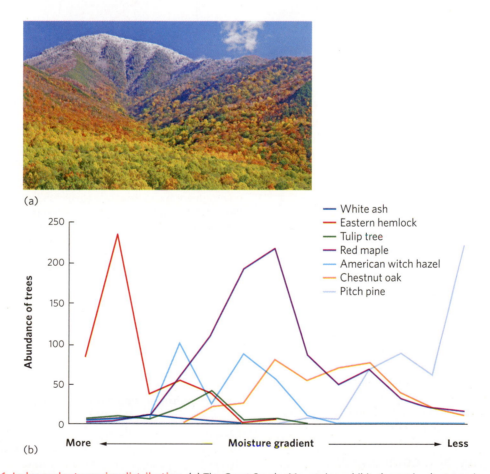

(a)

(b)

Figure 18.6 Independent species distribution. (a) The Great Smoky Mountains exhibit change in plant species composition with changes in elevation and soil moisture. (b) When a tree survey was conducted across a moisture gradient within an elevation of 1067 to 1372 m, researchers found that each species has its greatest abundance at different points along the moisture gradient. Moreover, each species appears and disappears at different points along the moisture gradient. This led researchers to reject the hypothesis that plant communities are distinct entities in which the existence of one species depends on the presence of one or more other species. Data from R. H. Whittaker, Vegetation of the Great Smoky Mountains, *Ecological Monographs* 26 (1956): 1–80. Photo by Chuck Summers/Contemplative Images.

are independent or interdependent. An abrupt transition may reflect an abrupt change in abiotic conditions where many species simply cannot exist under the changed conditions. For example, fish and other aquatic organisms are not restricted to live in a lake because they are necessarily interdependent; they are restricted to the lake because they lack the adaptations to live on land. To determine whether species are interdependent and therefore rely on each other for their existence, we need to examine how well species perform when other species are removed. If species rely on each other to persist, then removing a species from a community should cause other species in the community to decline. If they don't need each other to persist, then removing a species should leave other species in the community unaffected or even improve their fitness if they happen to compete with the species that is removed.

Recent studies have increasingly found that some species do rely on each other to persist, particularly in communities that experience harsh environmental conditions, such as extreme high or low temperatures or low moisture. For example, in a large experiment conducted in 11 locations around the world, researchers examined how 115 plant species growing at either low or high elevation in alpine tundra biomes responded when a neighbouring species was removed. The researchers then measured the percentage of plants that survived as well as the percentage of plants that produced flowers. Removing neighbouring plants at low elevation sites caused an increase in the survival of the remaining plants, as shown in Figure 18.7a. This also caused an increase in the percentage of plants that produced flowers or fruit, which you can see in Figure 18.7b. In contrast, removing neighbouring plants at high elevation sites caused a decrease both in the survival of the remaining plants and in the percentage of plants that produced flowers. Plants living under the harsher conditions of high elevations were helped by the presence of neighbouring species because those species reduced the harsh winds, provided shade, or offered protection from herbivores. Such experiments have taught ecologists that while most communities appear to be composed of species with independent distributions, species living under harsh environmental conditions frequently depend on other species in the community.

The diversity of a community incorporates both the number and relative abundance of species

To understand the processes that influence the structure and functioning of communities, we need to quantify how communities differ from place to place. In this section, we will examine patterns in

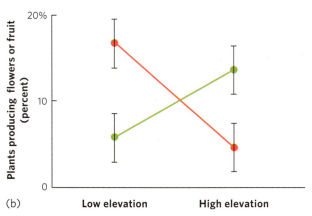

Figure 18.7 Interdependence of species under extreme environmental conditions. In alpine tundra biomes around the world, researchers measured how plants at low and high elevations responded to having neighbouring plants removed in terms of **(a)** survival and **(b)** the percentage of plants that produced flowers. At low elevations, the removal of neighbours caused plants to survive better and produce more flowers or fruit, which suggests that the plants are competing. At high elevations, removal of neighbours caused the plants to experience lower survival and produce fewer flowers, which suggests the plants in the community facilitate each other. Error bars are standard errors. Data from R. M. Callaway et al., Positive interactions among alpine plants increase with stress, *Nature* 417 (2002): 844–848.

species richness, which refers to the number of species in a community. We will also examine patterns in the abundance of individuals for each species and investigate how ecologists measure the diversity of species in a community in terms of richness and abundance. This is an important issue because ecologists often want to compare the diversity of species among communities or assess the effects of human activities on individual abundance and the species diversity of a community.

Species richness The number of species in a community.

PATTERNS OF ABUNDANCE AMONG SPECIES

Abundance can be examined either in absolute terms or in relation to other species. **Relative abundance** is the proportion of individuals in a community represented by each species. When ecologists count the number of individuals of each species in a community, they frequently find that only a few species have low or high abundance whereas most species have intermediate abundance. Frank Preston developed one way to visualize this pattern in a series of classic papers in which he examined the variation in abundance of different species. For example, Preston examined the abundance of different bird species near Westerville, Ohio, and plotted their abundances, as shown in Figure 18.8a. He found that only a few species of birds in the community had fewer than 2 individuals or more than 100 individuals; most had 4 to 64 individuals.

To create these graphs, Preston used the *y axis* to represent the number of species and the *x* axis to represent the number of individuals that comprise each species. The key to visualizing the patterns of abundance in communities was to use categories of abundance, for example, < 2, 2 to < 4, 4 to < 8, 8 to < 16, and so on. When these categories are plotted on a $\log_2$ scale, the first value in each category translates into 0, 1, 2, 3, etc. Data plotted in this way produces a normal, or bell-shaped, distribution such that a few species have high abundance, many species have moderate abundance, and a few species have low abundance. A normal, or bell-shaped, distribution that uses a logarithmic scale of the *x* axis is a **log-normal distribution**.

Log-normal distributions of species abundances can be found across a wide variety of communities and taxonomic groups. For instance, Canada's Department of Fisheries and Oceans carries out annual fish surveys on the Scotian Shelf using trawl nets. When researchers analyzed the catches from the 2002 surveys and plotted individual abundance against the number of fish species, shown in Figure 18.8b, they also found a log-normal distribution of abundance. Many studies since Preston's classical work have also observed log-normal distributions of species abundances.

RANK-ABUNDANCE CURVES

Another way to envision the relationship between the number of species and the relative abundance of each species is to use *rank-abundance curves*. **Rank-abundance curves** plot the relative abundance of each species in a community in rank order from the most abundant species to the least abundant species. Rank abundance curves are particularly helpful when we want to illustrate how communities differ in species richness and *species evenness*. **Species evenness** is a comparison of the relative abundance of

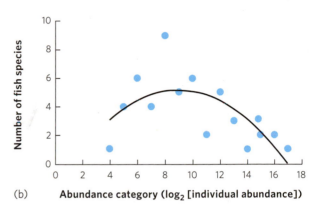

(a)

(b)

Figure 18.8 Log-normal distributions of species abundance. Log-normal distributions are found by plotting the abundance categories of each species on a $\log_2$ scale and then plotting the number of species that contain each abundance category. **(a)** A 10-year survey of birds in Westerville, Ohio, found that few species had extremely low or high abundances; most bird species had intermediate abundances. **(b)** A 2002 survey of fish species on the Scotian Shelf found a similar pattern, with few species having low or high abundance. Most fish species showed intermediate abundance. Data from (a) F. W. Preston, The canonical distribution of commonness and rarity: Part I, *Ecology* 43 (1962): 185–215; (b) J. S. Gray et al. Are there differences in structure between marine and terrestrial assemblages?, *Journal of Experimental Marine Biology and Ecology* 330 (2006): 19–26.

each species in a community. The greatest evenness occurs when all species in a community have equal abundances, and the lowest evenness occurs when one species is abundant and the remaining species are rare. To plot a rank-abundance curve, we rank each species in terms of its abundance; the most abundant species receives a rank of 1, the next most abundant species receives a rank of 2, and so on.

Relative abundance The proportion of individuals in a community represented by each species.

Log-normal distribution A normal, or bell-shaped, distribution that uses a logarithmic scale on the *x* axis.

Rank-abundance curve A curve that plots the relative abundance of each species in a community in rank order from the most abundant species to the least abundant species.

Species evenness A comparison of the relative abundance of each species in a community.

Consider two hypothetical communities. Community A has five species with relative abundances of 0.5, 0.3, 0.1, 0.06, and 0.04, and Community B has five species with relative abundances of 0.2, 0.2, 0.2, 0.2, and 0.2. From these numbers, we can see that both communities have the same species richness, but Community A has lower species evenness than Community B. If we plot these data as rank-abundance curves, we create the graph shown in Figure 18.9a. Both curves extend equally far to the right, which confirms that they have the same species richness. However, the slope of Community A is considerably steeper than the slope of Community B. This steeper slope occurs because the five species in Community A range from a very high relative abundance to a very low relative abundance. In contrast, all five species in Community B have equal relative abundances, which, when plotted, produces a flat line. Rank-abundance curves allow us to determine quickly which communities have greater species richness and evenness.

Ecologists use rank abundance curves to determine how communities differ in richness and evenness. For example, researchers in Ontario examined rank-abundance curves for ground beetles from tilled and untilled agricultural fields. They collected beetles over 3 years in traps placed in the fields and determined the abundance of every ground beetle species caught in the traps. They then ranked each species based on abundance and plotted the ranks against the abundance of each species, as shown in Figure 18.9b. As you can see in this figure, the rank-abundance curve for untilled fields extends further to the right than the curve for beetles in tilled fields, which indicates that untilled fields contain more species of ground beetles than tilled fields. If we consider the slopes of the curves across the ranks that the two types of fields have in common—from ranks 1 to 42—we see that the slopes of the curves differ only slightly. This indicates that the beetle community in tilled fields has approximately the same evenness as the community in the untilled fields.

Species diversity is affected by resources, habitat diversity, keystone species, and disturbance

We have seen that communities differ in the number of species they contain, but we have not yet addressed

---

ANALYZING ECOLOGY

Calculating Species Diversity

Ecologists agree that communities with more species and greater evenness have higher species diversity. Although rank abundance curves are a helpful way to visualize differences in richness and evenness among communities, they do not provide a specific value for the species diversity within a community. To quantify the species diversity of a community, we need a method of calculating a single value for species diversity. Over the years, numerous indices of species diversity have been

created. We will consider the two that are most common and equally valid: *Simpson's index* and *Shannon's index*. Both indices incorporate species richness, which we abbreviate as S, and evenness. However, they do so in different ways.

To see how we calculate species diversity with either index, we can begin with data from three communities for which we have the absolute abundance for each of the species. From these data we can then calculate relative abundance for each of the five species in the community, which is denoted as p_i. With these relative abundance data, we can calculate both Simpson's index and Shannon's index.

THE ABUNDANCE OF DIFFERENT MAMMAL SPECIES IN THREE COMMUNITIES

SPECIES	COMMUNITY A		COMMUNITY B		COMMUNITY C	
	ABSOLUTE ABUNDANCE	RELATIVE ABUNDANCE (p_i)	ABSOLUTE ABUNDANCE	RELATIVE ABUNDANCE (p_i)	ABSOLUTE ABUNDANCE	RELATIVE ABUNDANCE (p_i)
MOUSE	24	0.24	20	0.20	24	0.25
CHIPMUNK	16	0.16	20	0.20	24	0.25
SQUIRREL	8	0.08	20	0.20	24	0.25
SHREW	34	0.34	20	0.20	24	0.25
VOLE	18	0.18	20	0.20	0	0.00
TOTAL ABUNDANCE	100	1.0	100	1.0	100	1.0
S	5	5	5	5	4	4

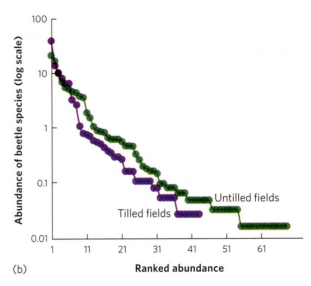

Figure 18.9 Rank-abundance curves. (a) These rank-abundance curves represent two hypothetical communities with identical species richness of five species. However, the two communities differ in species evenness. Community A has one species with a high relative abundance and other species that have moderate to low abundance, which means this community has low species evenness. In contrast, Community B contains five species that are all equally abundant so there is high evenness. **(b)** These rank-abundance curves for beetles living in Ontario fields show that tilled fields have a lower species richness than fields that were not tilled. Species evenness was comparable in tilled and untilled fields. Data from S. Belaoussoff et al., Assessing tillage disturbance on assemblages of ground beetles (Coleoptera: Carabidae) by using a range of ecological indices, *Biodiversity and Conservation* 12 (2003): 851–882.

Simpson's index, a measurement of species diversity, is given by the following formula:

$$\frac{1}{\sum_{i=1}^{s}(p_i)^2}$$

In words, this formula means that we square each of the relative abundance values, sum these squared values, and then take the inverse of this sum. For example, for Community A, Simpson's index of species diversity is

$$\frac{1}{(0.24)^2 + (0.16)^2 + (0.08)^2 + (0.34)^2 + (0.18)^2} = \frac{1}{.24} = 4.21$$

Simpson's index can range from a minimum value of 1, which occurs when a community only contains one species, to a maximum value equal to the number of species in the community. This maximum value only occurs when all the species in the community have equal abundances.

Shannon's index (H'), also known as the **Shannon-Wiener index**, is another measurement of species diversity given by the following formula:

$$H' = -\sum_{i=1}^{s}(p_i)(\ln p_i)$$

In words, this formula means that we multiply each of the relative abundance values by the natural log of the relative abundance values, sum these products, and then take the negative of this sum. For example, in Community A, Shannon's index is

$$-[(0.24)(\ln 0.24) + (0.16)(\ln 0.16) + (0.08)(\ln 0.08) + (0.34)(\ln 0.34) + (0.18)(\ln 0.18)]$$

$$-[(-0.34) + (-0.29) + (-0.20) + (-0.37) + (-0.31)]$$

$$= 1.51$$

Shannon's index can range from a minimum value of 0, which represents a community that contains only one species, to a maximum value that is the natural log of the number of species in the community. As we saw with Simpson's index, the maximum species diversity value occurs when all species in the community have the same relative abundances.

YOUR TURN Calculate Simpson's index and Shannon's index for communities B and C. Based on your calculations, how does species evenness and species richness affect the values of each index?

Simpson's index A measurement of species diversity, given by the following formula: $\dfrac{1}{\sum_{i=1}^{s}(p_i)^2}$

Shannon's index (H') A measurement of species diversity, given by the following formula:

$$H' = -\sum_{i=1}^{s}(p_i)(\ln p_i)$$

Also known as **Shannon-Wiener index.**

the more fundamental question of why they differ. On a global scale, there is great variation in the number of species found in different locations. For example, naturalists have known for centuries that more species live in tropical regions than in temperate or boreal zones. We will discuss global patterns of species richness in later chapters and consider the effects of long-term processes such as continental drift, the time that has passed since glaciation, the emergence of new islands, the dispersal of species, and the evolution of new species. In this section, we will focus on how species richness of communities within any given region of the world is affected by the amount of resources available, the diversity of the habitat, the presence of *keystone species,* and the frequency and magnitude of disturbances.

RESOURCES

The number of species in a community can be affected by the amount of available resources.

Researchers have examined the effects of resources on species diversity by determining correlations between productivity and species richness in nature. They have also looked at how species richness changes when the productivity of a community is experimentally manipulated.

Natural Patterns of Productivity and Species Richness

For decades ecologists have examined how the amount of biological productivity correlates with the number of species in communities. To do this they commonly measure productivity in terms of the biomass of producers or consumers that is generated over time. Among hundreds of studies that have been conducted on animals and plants in both aquatic and terrestrial environments, researchers have found a wide range of patterns, as illustrated in Figure 18.10a. In rare cases, researchers have observed a U-shaped curve in which increased productivity is associated

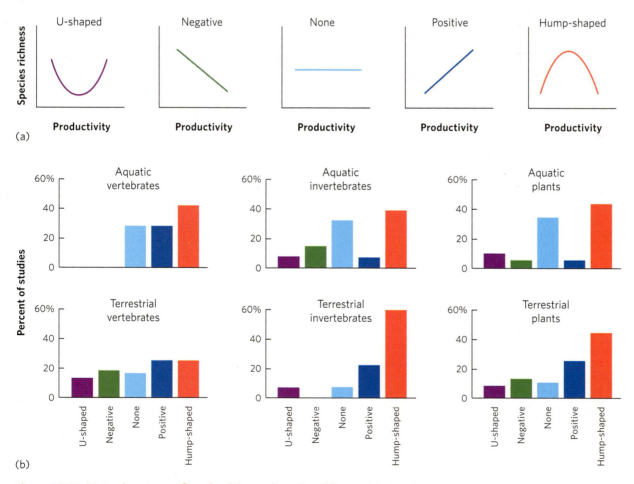

Figure 18.10 Natural patterns of productivity and species richness. (a) The relationship between productivity and species richness can be described by one of five types of curves. **(b)** Among six categories of organisms, a hump-shaped curve is the most commonly observed relationship between productivity and species richness for aquatic vertebrates, aquatic and terrestrial invertebrates, and aquatic and terrestrial plants. Data from G. G. Mittelbach et al., What is the observed relationship between species richness and productivity?, *Ecology* 82 (2001): 2381–2396.

with an initial decrease in species richness that is followed by an increase in species richness. In some studies, conducted in different biomes and in different parts of the world, diversity decreases with increasing productivity. In other cases, the correlation is positive; increases in productivity are associated with increases in species richness. In still other studies, there is no relationship between productivity and species richness. Finally, in some studies, the relationship is best described by a hump-shaped curve; initial increases in productivity are associated with an increase in species richness but further increases in productivity are associated with a decrease in species richness.

To gain a better sense of the relationship between productivity and species richness, a group of researchers compiled data from the hundreds of studies that have been conducted to determine the most common type of correlations. You can view their results in Figure 18.10b. Among studies of vertebrates, a hump-shaped curve was the most commonly observed relationship in aquatic systems, although hump-shaped and positive relationships were observed with similar frequency in terrestrial systems. For aquatic and terrestrial invertebrates and plants, the most commonly observed relationship was a hump-shaped curve. In short, although a variety of relationships have been observed in individual studies, the most common relationship across all

studies is a hump-shaped curve. This means that a site with medium productivity has a higher species richness than sites with either low or high productivity.

Manipulations of Productivity and Species Richness

Since most communities exhibit a hump-shaped relationship between productivity and species richness, we ought to be able to predict how species richness will be affected if communities experience increased productivity. Numerous experiments have been conducted in communities of plants in which productivity has been manipulated by adding soil nutrients such as nitrogen and phosphorus. A common observation in such experiments is that species diversity declines over time. In a classic study that took place in England, known as the Park Grass experiment, researchers in 1856 began adding different types of fertilizers to several plots of grass each year while leaving other plots as unfertilized controls (Figure 18.11a). Within 2 years of starting the experiment, species richness in the fertilized plots began to decline. This decline continued for several decades and researchers determined that the amount of decline was related to the number of different nutrients that were added, as shown in Figure 18.11b. The annual addition of nutrients continues today, over 150 years later, which

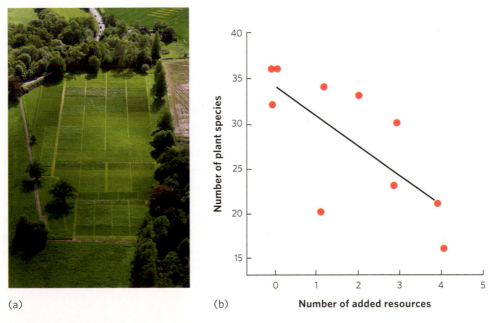

(a)   (b)

Figure 18.11 Fertility effects on species richness. (a) Researchers of the Park Grass experiment have manipulated the addition of different types of nutrients in grassland plots over time and observed changes in species richness. (b) Plots that received more types of resources—such as nitrogen, phosphorus, and micronutrients—experienced a greater decline in species richness. Data from W. S. Harpole and D. Tilman, Grassland species loss resulting from reduced niche dimension, *Nature* 446 (2007): 791–793. Photo by Centre for Bioimaging/ Rothamsted Research/Science Source.

makes the Park Grass experiment one of the longest running experiments in the world. Although the original experiment contained treatments that were not replicated, subsequent experiments with replicated treatments have supported the conclusions of the Park Grass experiment.

The Park Grass experiment and others like it have demonstrated that added fertility commonly causes a decline in the species richness of producers such as plants and algae. Typically, the total biomass of producers increases when a fertilizer is added, but this added fertilizer causes a few species to dominate the community while rare species—which are often competitively inferior—begin to decline until they eventually disappear from the community. This observation has important global ramifications because humans continue to add nutrients inadvertently to aquatic and terrestrial communities through fertilizer runoff and air pollution.

Although we know that species richness declines with increased habitat fertility, the reasons have been unclear. In the case of plant communities, ecologists have hypothesized that the increase in nutrients caused competitively dominant plants to cast more shade on the competitively inferior plants. In 2009, researchers reported the results of an experiment with four manipulations, as shown in Figure 18.12: a control community, a community that received

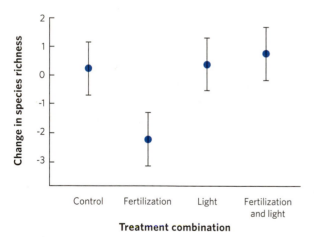

Figure 18.12 Reversing the effects of added nutrients on species richness. When fertilizer alone is added to a grassland community, species richness declines. When light is added under the canopy of the grasses, species richness is not affected. However, when fertilizer is added in combination with light under the canopy, there is no decline in richness. These results suggest that adding fertilizer causes richness to decline because increased fertility favours the tall and competitively dominant plants, which in turn shade the competitively inferior plants. Error bars are 95 percent confidence intervals. Data from Y. Hautier et al., Competition for light causes plant biodiversity loss after eutrophication, *Science* 324 (2009): 636–638.

added nutrients, a community that received added light in the plant understorey, and a community that received added nutrients and added light in the plant understorey. Researchers found that while adding nutrients alone caused a decline in species richness, as expected from past experiments, adding light alone had no effect. However, adding both nutrients to the soils and light to the understorey reversed the decline in species richness that was seen with the addition of nutrients alone. These results confirm that soil fertility drives down species richness because it promotes growth of taller, competitively superior plants that cast shade over less-competitive plants.

HABITAT DIVERSITY

The number of species in a community can also be affected by the diversity of the habitat. Because different habitats provide places to feed and breed for different species, it seems reasonable that communities with a higher diversity of habitats—which should offer more potential niches—will also have a higher diversity of species.

In a classic study examining the role of habitat diversity, Robert and John MacArthur investigated whether there was a relationship between the diversity of habitats among different regions of the United States and Panama and the diversity of birds in each location. To accomplish this, they measured the density of foliage at different heights above the ground in these different regions, from 0.2 to 18.3 m, and calculated the diversity of foliage heights using the Shannon index. Then they surveyed the number of breeding bird species in each area and calculated the diversity of birds, again using the Shannon index. When they graphed the relationship between the two sets of data, as shown in Figure 18.13, they found that habitats with greater foliage height diversity supported higher diversity of bird species.

KEYSTONE SPECIES

All species play a role in a community, but some species, known as **keystone species**, substantially affect the structure of communities despite the fact that individuals of that species may not be particularly numerous. The concept of the keystone species is a metaphor that comes from the field of architecture. In arches that are built of stone, the

Keystone species A species that substantially affects the structure of communities despite the fact that individuals of that species might not be particularly numerous.

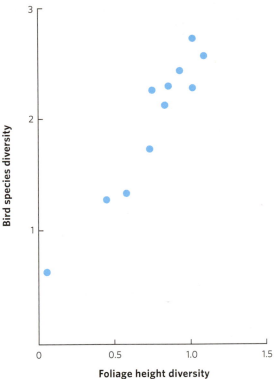

Figure 18.13 Habitat diversity and species diversity.
Using the Shannon index, researchers quantified the diversity of foliage height and the diversity of bird species in different locations around the United States and Panama. They found that more diverse habitats contained a greater diversity of birds. Data from R. H. MacArthur and J. W. MacArthur, On bird species diversity, *Ecology* 42 (1961): 594–598.

centre stone is known as the keystone (Figure 18.14). The keystone does not carry much of the arch's mass, but without the keystone the arch would collapse. Similarly, removing a keystone species can cause a community to collapse.

Keystone species affect communities in a wide variety of ways, some in their roles as predators, parasites, herbivores, or competitors. We saw an example of herbivores acting as a keystone species in Chapter 16 when we discussed an experiment in which researchers sprayed several field plots with an insecticide and left other plots unsprayed (see Figure 16.13). In the unsprayed plots, beetles consumed the competitively superior goldenrod and the competitively inferior plants survived and grew better. In the sprayed plots, the beetles remained rare and this allowed the goldenrod plants to dominate the community. In this community, the beetle was a keystone species because its presence completely altered the structure of the field community.

A similar scenario can be seen in the intertidal communities along the coast of Washington State. In a classic experiment, Robert Paine built cages in the intertidal zone to exclude predatory sea stars (*Pisaster ochraceus*) from feeding on a variety of herbivores, including mussels, barnacles, limpets, and snails. Adjacent plots of similar size were left as uncaged controls. As you can see from Paine's data in Figure 18.15a, the control plots experienced little change in species composition from 1963 to 1973. In the caged plots, however, nearly 20 species declined to the point that they were eventually eliminated from the area. In their place, a single species of mussel (*Mytilus californianus*) came to dominate the rock surface. Because the mussels are the superior competitors for space on the rocks, mussels dominate the community when sea stars are absent (Figure 18.15b). When the sea stars are present, they consume large numbers of mussels, and this creates open areas of rock that many species of inferior competitors can colonize (Figure 18.15c). In short, the sea star is a keystone species in the intertidal community.

Keystone species can also affect communities by influencing the structure of a habitat. In such cases, keystone species are sometimes called *ecosystem engineers*. We saw an example of this with the social spiders discussed at the beginning of this chapter.

Figure 18.14 Keystone. A keystone in an arch is the final stone that is inserted to prevent the arch from collapsing. Similarly, a keystone species is one whose removal causes a dramatic change in the community, despite not being particularly abundant. Photo by Nick Hawkes/Alamy.

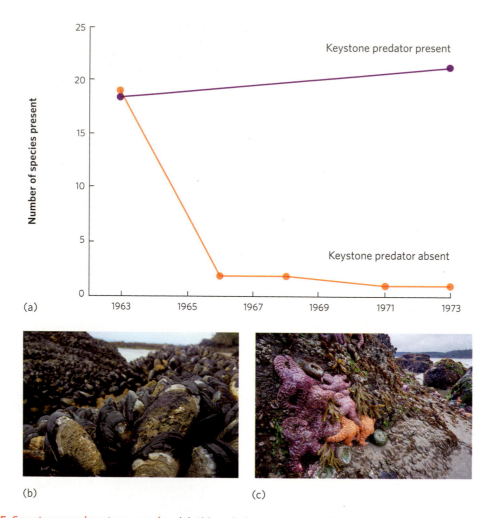

Figure 18.15 Sea stars as a keystone species. (a) Although they are not very abundant, sea stars act as keystone species in intertidal communities along the coast of Washington State because they remove competitively dominant mussels. **(b)** In areas where sea stars were removed, the competitively superior mussels came to dominate the community and caused a decline in the diversity of the competitively inferior species. **(c)** In areas where the sea stars remained, the community retained a high diversity of intertidal species. Data from R. T. Paine, Intertidal community structure: Experimental studies on the relationship between a dominant competitor and its principal predator, *Oecologia* 15 (1974): 93–120. Photos by (b) Jonathan Hucke; (c) Gary Luhm/DanitaDelimont.com/Newscom.

One of the best-known ecosystem engineers is the beaver. Beavers build dams in streams that block the flow of water and cause large ponds to develop (Figure 18.16). Because the flowing stream is converted into a nonflowing pond, a different community of plants and animals colonize and persist in beaver ponds than in streams. Similarly, alligators create numerous large depressions in wetlands, known as *gator holes,* that many other species use, including fish, insects, crustaceans, birds, and mammals. Although alligators and beavers are not particularly numerous, each of these species has a major effect on its community, which makes it a keystone species.

DISTURBANCES

As we discussed in previous chapters, some species are well adapted to environments that experience frequent and large disturbances, such as hurricanes,

fires, or intense herbivory. When environments are rarely disturbed or the disturbances have a low intensity, populations can continue to grow, resources become less abundant, and the ability to compete becomes more important for the persistence of a species. Conversely, frequently disturbed habitats typically support species that are adapted to disturbances. However, when habitats experience disturbances at some intermediate frequency, both types of species can persist and the total number of species can be higher than it would be at either extreme. The **intermediate disturbance hypothesis** tells us that more species are present in a community that

Intermediate disturbance hypothesis The hypothesis that more species are present in a community that occasionally experiences disturbances than in a community that experiences frequent or rare disturbances.

Figure 18.16 Ecosystem engineers. While beavers are not very abundant, the dams they build, like this one in Wood Buffalo National Park, Alberta—the largest beaver dam in the world—have a major effect on community structure. Beaver dams flood terrestrial communities, which kills most terrestrial plants. In addition, the aquatic community changes from species that live in flowing streams to those that live in nonflowing ponds. Photos by (left) Tom & Pat Leeson; (right) Parks Canada/Wood Buffalo National Park/S. Anderson.

occasionally experiences disturbances than in a community that experiences frequent or rare disturbances. As depicted in Figure 18.17a, when disturbances in a community are of low frequency or intensity, species richness is relatively low. However, when disturbances are moderate in frequency or intensity, species richness is relatively high. When disturbances are high in frequency or intensity, species richness declines.

The intermediate disturbance hypothesis has been observed in many different communities. One

classic study was conducted by Jane Lubchenco, who manipulated the density of herbivorous periwinkle snails in tidal pools to determine how an

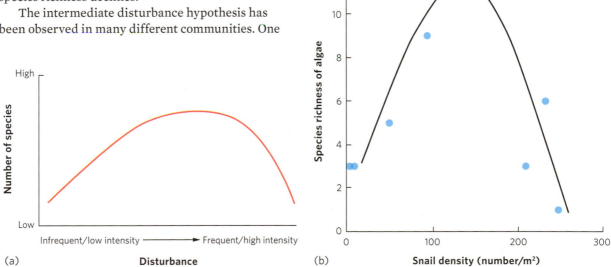

Figure 18.17 The intermediate disturbance hypothesis. (a) When disturbances are frequent or of high intensity, only those species that are adapted to these conditions can persist. When disturbances are rare or of low intensity, competition becomes more intense and only those species that are well adapted to compete can persist. At intermediate levels of disturbance, species from both extremes can persist, which results in a higher species richness. **(b)** In a study that manipulated snail density to create different intensities of disturbance from herbivory, there was strong support for the intermediate disturbance hypothesis. Data from J. Lubchenco. Plant species diversity in a marine intertidal community: Importance of herbivore food preference and algal competitive abilities, *American Naturalist* 112 (1978): 23–39.

increase in the magnitude of an herbivory disturbance would affect the species richness of algae in the pools. Her data, shown in Figure 18.17b, revealed a low richness of algal species at low snail densities, a high richness of algal species at moderate snail densities, and a low richness of algal species at high snail densities. Many similar types of studies have been conducted over several decades and a 2012 analysis of these studies found strong support for the intermediate disturbance hypothesis with respect to species richness.

Communities are organized into food webs

Even the simplest communities are composed of a large number of species. To understand the relationships among the species, ecologists have found it helpful to categorize species in a community into *food chains* and food webs. **Food chains** are linear representations of how species in a community consume each other and therefore how they transfer energy and nutrients from one group to another in an ecosystem. Food chains greatly simplify species interactions in a community. In contrast, **food webs** are complex and realistic representations of how species feed on each other in a community and include links among many species of producers, consumers, detritivores, scavengers, and decomposers. In constructing food webs, ecologists draw arrows that indicate consumption—and therefore the movement of energy and nutrients—from one group to another. You can see an example of a food web in Figure 18.18.

Food webs help us understand the feeding relationships of ecological communities, which is important because feeding relationships help determine whether a species can exist in a community and whether it will be rare or abundant. The species richness in a food web can often be quite high, which makes it challenging to comprehend how the abundance of one species is affected by other species in the community. To simplify this challenge, ecologists categorize species into *trophic levels*.

TROPHIC LEVELS

You may recall the concepts of *producer* and *consumer* from our overview of basic ecological concepts in Chapter 1. These represent broad categories of **trophic levels**, which are the levels in a food chain or food web of an ecosystem. All organisms at a particular trophic level obtain their energy in a similar way. Producers are the autotrophs, including algae such as phytoplankton, and plants that convert light energy and CO_2 into carbohydrates through photosynthesis. Producers form the first trophic level of a food web. Consumers can be subdivided into *primary consumers, secondary consumers,* and *tertiary consumers.* **Primary consumers** are those species that eat producers. In the aquatic food web portrayed in Figure 18.18, the primary consumers include zooplankton that eat algae and snails that eat plants. **Secondary consumers** are those species that eat primary consumers. In our lake food web these secondary consumers include small fish that eat the zooplankton and ducks that eat the snails. Some communities support **tertiary consumers**, which eat secondary consumers; for example, large fish in a lake are tertiary consumers because they consume small fish. In addition to these producers and consumers, food webs also include consumers of dead organic matter such as scavengers, detritivores, and decomposers. One of the challenges in thinking about placing species into specific trophic levels is that many species are **omnivores**, meaning that they can feed at several trophic levels. For example, crayfish consume algae, which would make them primary consumers, but they also feed on insects and detritus, which would make them secondary consumers and detritivores.

Within a given trophic level, we can group species that feed on similar items into **guilds**. For example, a grassland might contain a wide variety of primary consumers that feed on plants in different ways; these consumers can be categorized into guilds of leaf eaters, stem borers, root chewers, nectar sippers, or bud nippers. Members of a guild feed on similar items but they need not be closely related. In Chapter 16, for example, we discussed how ants and rodents in the deserts of the southwestern United States compete for seeds of the same size. Though ants and rodents are not closely related, their competition for the same seeds would place them in the same guild.

Food chain A linear representation of how different species in a community feed on each other.

Food web A complex and realistic representation of how species feed on each other in a community.

Trophic level A level in a food chain or food web of an ecosystem.

Primary consumer A species that eats producers.

Secondary consumer A species that eats primary consumers.

Tertiary consumer A species that eats secondary consumers.

Omnivore A species that feeds at several trophic levels.

Guild Within a given trophic level, a group of species that feeds on similar items.

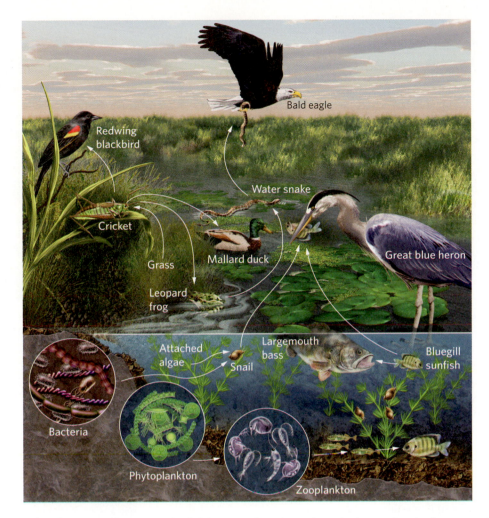

Figure 18.18 Food webs. Food webs capture the full diversity of species and species interactions in the community. In this North American lake, the major producers are the plants along the shoreline and the phytoplankton in the water. These producers are eaten by a variety of primary consumers, as indicated by the arrows that illustrate the flow of energy from an organism to the species consuming it. This food web also includes secondary consumers, such as bluegill sunfish, tertiary consumers, such as the great blue heron, and decomposers, such as bacteria.

DIRECT VERSUS INDIRECT EFFECTS

As we have seen, because communities can include a large number of species and species interactions, a change in the abundance of any one species can affect the abundance of the other species. The simplest way that one species can affect the abundance of another is through a **direct effect**, which occurs when two species interact without involving other species. We have discussed a variety of direct effects in our chapters on predation, parasitism, competition, and mutualism.

Because so many species are interconnected in a food web, the direct effects of one species on another often set off a chain of events that continue to affect still other species in the community. When two species interact in a way that involves one or more intermediate species, we call it an **indirect effect**.

When indirect effects are initiated by a predator, we call it a **trophic cascade**. Indirect effects are widespread in ecological communities. For example, in Chapter 14 we discussed how scientists introduced the *Myxoma* virus onto Macquarie Island to help reduce a population of European rabbits that had been introduced by sailors in the 1800s. As the virus

---

Direct effect An interaction between two species that does not involve other species.

Indirect effect An interaction between two species that involves one or more intermediate species.

Trophic cascade Indirect effects in a community that are initiated by a predator.

killed the rabbits, there were fewer rabbits remaining to consume the vegetation, which resulted in an increase in the island's vegetation. In short, the virus caused an indirect effect on the vegetation.

Exploitative competition between two animals might at first glance seem like a direct effect because each species has a negative effect on the other. However, it is actually an indirect effect because the two competitors are interacting with each other by feeding on a common resource. In the case of the ants and rodents in the southwestern United States, for instance, the presence of ants causes a reduction in available seeds and this reduction causes the rodent population to decline. In short, the negative effect of one competitor on another is mediated through the abundance of a third species, which is a shared resource.

Indirect effects are common within communities, but sometimes they can occur between communities. For example, fish prey on aquatic insects, including the aquatic larvae of dragonflies. As a result, ponds containing fish typically have fewer dragonfly larvae. Dragonfly larvae eventually metamorphose into dragonfly adults that consume bees and flies, which are common plant pollinators. You can see this food web in Figure 18.19a. When researchers examined the dragonfly populations of ponds with fish and ponds without fish, they confirmed that the ponds without fish had more larval and adult dragonflies. Because of this higher abundance of dragonflies, the ponds without fish also had fewer bees and flies visiting flowering plants along the shoreline, as shown in Figure 18.19b. The scarcity of bees and flies reduced potential seed production in at least one plant species. Thus, a direct effect between fish and dragonflies caused a cascade of indirect effects on the surrounding terrestrial community.

Ecologists previously assumed that indirect effects occurred only when changes in one species altered the density of another species, which initiated a trophic cascade. However, more recent research has shown that trophic cascades can also be initiated when one species causes changes in the traits of another species. We now look at a variety of indirect effects in more detail.

Density-Mediated Indirect Effects

When indirect effects are caused by changes in the density of an intermediate species, we call them **density-mediated indirect effects**. For example, increased densities of sea stars in intertidal communities cause a decline in mussels, which allows other species such as snails to occupy the limited open space on the rocks. The positive indirect effect of sea stars on snails occurs because the sea stars

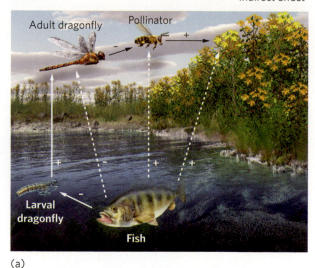

(a)

(b)

Figure 18.19 Indirect effects between communities. (a) A food web of species from the aquatic and terrestrial communities includes fish that consume larval dragonflies, adult dragonflies that consume pollinators, and pollinators that pollinate plants. Solid lines represent direct effects. Dashed lines with a minus sign represent a negative indirect effect and dashed lines with a plus sign represent a positive indirect effect. (b) Researchers found that ponds without fish had higher dragonfly populations, which reduced the numbers of pollinators available to visit flowering plants on the surrounding shoreline. Error bars are standard errors. Data from T. M. Knight et al., Trophic cascades across ecosystems, *Nature* 237 (2005): 880–883.

reduce the density of mussels. Similarly, the introduction of *Myxoma* virus reduced the density of rabbits on Macquarie Island, which allowed an increase in the growth of the plants that rabbits eat.

Density-mediated indirect effect An indirect effect caused by changes in the density of an intermediate species.

Trait-Mediated Indirect Effects

More recently ecologists have come to appreciate that communities can also experience **trait-mediated indirect effects**, which are indirect effects that are caused by changes in the traits of an intermediate species. This commonly happens when a predator causes its prey to change its feeding behaviour, which in turn alters the amount of food consumed by the prey. For example, the reintroduction of wolves into Yellowstone National Park has caused elk to feed in more protected areas. This behaviour alters the amount of plant consumption that occurs in these areas, not because the density of elk has changed but because the behaviour of the elk has changed.

A similar example can be found in a community of spiders, grasshoppers, and grasses. Grasshoppers are insects that readily consume grasses and serve as prey for some species of spiders. There is more grass present when spiders are present in the community. The increase in grass could occur because spiders eat the grasshoppers and therefore lower the grasshoppers' density, or because the spiders scare the grasshoppers so that they hide more and spend less time feeding. To determine which process is operating, researchers conducted an experiment in which they placed cages around small areas of grass in a field and then conducted one of four manipulations by adding: (1) no animals (the control), (2) grasshoppers, (3) grasshoppers and lethal spiders, and (4) grasshoppers and nonlethal spiders. Although the nonlethal spiders could not kill the grasshoppers because their mouthparts had been glued shut, the grasshoppers would still recognize them as potential predators, which should cause grasshoppers to reduce their feeding time.

When researchers added the lethal spiders, the grass experienced less herbivory, as you can see in Figure 18.20a. At first glance, this appears to be a density-mediated indirect effect because spiders kill grasshoppers. However, the researchers found that the density of grasshoppers was not affected by the presence of lethal spiders, probably because grasshopper reproduction offset spider predation. When the researchers added nonlethal spiders with glued mouthparts, the number of grasshoppers was identical to when lethal spiders were added and the grass still experienced reduced herbivory. In short, nonlethal spiders, which can alter the behavioural traits of the grasshoppers but not grasshopper density, had the same positive indirect effect on the grass as lethal spiders, which alter both the density and the traits of grasshoppers. These results illustrated that the mere presence of spiders can alter the behavioural traits of grasshoppers and initiate a trait-mediated indirect effect.

(a)

(b)

Figure 18.20 Trait-mediated indirect effects. (a) When researchers created simplified communities with only grass and grasshoppers, the amount of grass damage was relatively high. When lethal spiders were added, the grasshoppers caused less damage to the grass, which indicates that spiders had an indirect positive effect on the grass. When spiders with glued mouthparts were added, there was a similar reduction in herbivory because the grasshoppers spent less time feeding. These data confirm that the indirect effect is not mediated by reductions in grasshopper density but by changes in grasshopper traits. **(b)** After the grasshoppers died, the bodies of grasshoppers raised with predators were lower in nitrogen, which caused the decomposition of the surrounding dead grass to occur at a slower rate. Error bars are standard errors. Data from A. P. Beckerman, M. Uriarte, and O. J. Schmitz, Experimental evidence for a behavior-mediated trophic cascade in a terrestrial food chain, *Proceedings of National Academy of Science* 94 (1997): 10735–10738; D. Hawlena et al., Fear of predation slows plant-litter decomposition, *Science* 336 (2012): 1434–1438.

Ecologists are beginning to discover that trait-mediated effects can be far-reaching. For example, in 2012 researchers reported that grasshoppers that are exposed to the threat of spider predation not only feed less, but their bodies also contain less nitrogen.

Trait-mediated indirect effect An indirect effect caused by changes in the traits of an intermediate species.

Because nitrogen is typically a limited resource in soils, as we discussed in Chapter 3, researchers suspected that the amount of nitrogen in grasshopper bodies would affect the rate of soil decomposition when grasshoppers die and become part of the detritus. To test this hypothesis, researchers again raised grasshoppers with no spiders and spiders with glued mouthparts. After being exposed to the two treatments, the grasshoppers were killed and their carcasses were mixed with dead grass. When researchers measured the rate of grass decomposition, they found that it was approximately three times faster with grasshoppers that had been reared without spiders than grasshoppers that had been reared with spiders, as shown in Figure 18.20b. These results demonstrate that trait-mediated effects can be far-reaching throughout communities.

TOP-DOWN AND BOTTOM-UP EFFECTS

The amount of available resources as well as the amount of predation and parasitism a species experiences can affect the abundance of species. The same can be said for entire trophic groups. When the abundances of trophic groups in nature are determined by the amount of energy available from the producers in a community, we say there is

bottom-up control of the community. When the abundance of trophic groups is determined by the existence of predators at the top of the food web, we say there is **top-down control** of the community. You can see both of these concepts illustrated in Figure 18.21. If we return to our lake example, we can start with four trophic groups that consist of large fish, small fish, zooplankton, and phytoplankton. If an increase in phytoplankton causes an increase in the zooplankton, small fish, and large fish, the abundance of the trophic groups experiences bottom-up control. If an increase in the abundance of large fish causes a decrease in the small fish, an increase in the zooplankton upon which the small fish feed, and a decrease in phytoplankton, the abundance of the trophic groups experiences top-down control.

Bottom-up control When the abundances of trophic groups in nature are determined by the amount of energy available from the producers in a community.

Top-down control When the abundance of trophic groups is determined by the existence of predators at the top of the food web.

Figure 18.21 Top-down and bottom-up control of communities. When communities are controlled from the bottom up, an increase in the abundance of producers results in an increase in the abundance of higher trophic levels. When communities are controlled from the top down, a trophic cascade occurs from the top down: The trophic level just below the top predator decreases in abundance and the next lower trophic level increases in abundance.

For many years, ecologists debated whether communities are more commonly under top-down or bottom-up control. If we think of food webs as having three broad trophic levels, top-down control by predators would reduce the abundance of herbivores and this would result in an abundance of vegetation. In a classic paper published in 1960, Nelson Hairston, Frederick Smith, and Lawrence Slobodkin suggested that because most communities contain an abundance of vegetation, trophic groups must be controlled from the top of the food web.

Their hypothesis caused a great deal of debate among ecologists and inspired researchers to survey patterns of abundance in nature and to conduct manipulative experiments. For example, when researchers surveyed the abundance of zooplankton and phytoplankton, they found that ponds with more phytoplankton also had more zooplankton, which suggested that zooplankton abundance is controlled from the bottom up. However, when they conducted manipulative experiments in which they added small fish that eat zooplankton, the abundance of zooplankton decreased, which caused a trophic cascade that allowed the phytoplankton to increase. This suggested that the abundance of species in the community was controlled from the top down. Over the past 2 decades, it has become clear that many communities are simultaneously controlled both from the top down by predators and from the bottom up by resources. For example, the number of zooplankton in a lake can be influenced both by the amount of phytoplankton available for consumption and by the number of small fish available to consume the zooplankton.

Communities respond to disturbances with resistance, with resilience, or by switching among alternative stable states

Communities commonly experience disturbances from both natural and human causes. Whatever the reason behind the disturbance, the key question is whether the community will be affected and, if it is, whether it will bounce back or become a substantially different community. To address this question, we need to consider the concepts of *community stability* and *alternative stable states*.

COMMUNITY STABILITY

As we have seen throughout this chapter, communities can experience large changes when the abundances of particular species change. Given enough time, most altered communities can rebound and resemble their original structure of species richness, composition, and relative abundance. The ability of a community to maintain a particular structure is known as **community stability**. There are two aspects of community stability, *community resistance* and *community resilience*. **Community resistance** is a measure of how much a community changes when acted upon by some disturbance, such as the addition or removal of a species. For example, we might measure community resistance after a predator is removed by looking at how much the abundance of herbivores increases. The smaller the change, the greater the resistance of the community to the disturbance. **Community resilience** is the time it takes after a disturbance for a community to return to its original state.

Does having more species help communities bounce back faster from disturbances? This is an interesting ecological question and a question of practical importance as human activities increasingly alter the species diversity of biological communities. One of the most extensive experiments on this question was conducted in grasslands where researchers manipulated the diversity of plant species in a Minnesota prairie ecosystem by 1, 2, 4, 8, or 16 species. For 11 years they monitored the abundances of more than 700 species of invertebrate herbivores and invertebrate predators and parasitoids. During the experiment, when the environment varied considerably, researchers asked which level of plant diversity provided the greatest community stability. In this study, community stability was measured as the amount of year-to-year variation in the abundance and species richness of the herbivores and predators.

In 2011, the researchers reported that increasing the number of plant species provided increased stability of the communities, as you can see in Figure 18.22. When they examined the herbivores, shown in Figure 18.22a, they found that an increase in plant species richness caused an increase in the stability of herbivore richness and herbivore abundance. When they examined the predators and parasitoids, shown in Figure 18.22b, they found that an increase in plant species richness caused an increase in the stability of richness, but there was no effect on the stability of abundance. Plant diversity

Community stability The ability of a community to maintain a particular structure.

Community resistance The amount that a community changes when acted upon by some disturbance, such as the addition or removal of a species.

Community resilience The time it takes after a disturbance for a community to return to its original state.

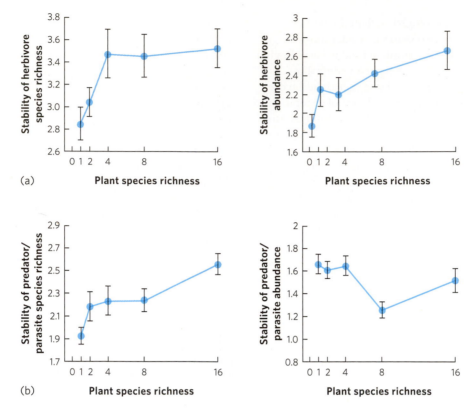

Figure 18.22 Species richness and community stability. Researchers manipulated the species richness of plants and then monitored the stability of the abundance and species richness of invertebrate herbivores, predators, and parasitoids. **(a)** An increase in the diversity of plants caused greater stability of richness and abundance of herbivores. **(b)** An increase in the diversity of plants also caused greater stability of richness of predators and parasitoids, but had no effect on the stability of abundance. In all cases, stability is defined as the inverse of the coefficient of variation, which is the standard deviation divided by the mean. Error bars are standard errors. Data from N. M. Haddad et al., Plant diversity and the stability of foodwebs, *Ecology Letters* 14 (2011): 42–46.

led to increased stability of higher trophic levels because communities with high plant diversity provided a more consistent availability of food and habitat for the herbivores, predators, and parasitoids.

ALTERNATIVE STABLE STATES

Sometimes when a community is perturbed, it does not bounce back to its original state. Rather, it settles into an **alternative stable state**, which occurs when a community is disturbed so much that the species composition and relative abundance of populations in the community change, and the new community structure is resistant to further change. For a community to move from one stable state to an alternative stable state typically requires a large disturbance, such as the removal of a keystone species or a dramatic change in the environment.

The lichen woodlands in Quebec's Parc des Grands-Jardins provide a good example of how a community can move to an alternative stable state. Eight thousand years ago, dense spruce-moss forest covered the area of today's park. This changed when fires became more frequent about 2500 years ago and

prevented species associated with the spruce-moss forest from regenerating. An infestation by the spruce budworm (*Choristoneura fumiferana*) added to the disturbance caused by the frequent fires. Both of these impacts prevented regrowth of the spruce-moss forest. Instead, an open, thinly wooded forest with a dense ground cover of lichen developed between 580 to 1440 years ago. These lichen woodlands have existed in this state since then and form an alternative stable state to the spruce-moss forest. The switch between the two states was caused by fire and insect infestation.

Alternative stable states also occur in aquatic environments. For example, intertidal communities on the coast of Maine are typically dominated by the brown rockweed algae *Ascophyllum nodosum*. During

---

Alternative stable state When a community is disturbed so much that the species composition and relative abundance of populations in the community change, and the new community structure is resistant to further change.

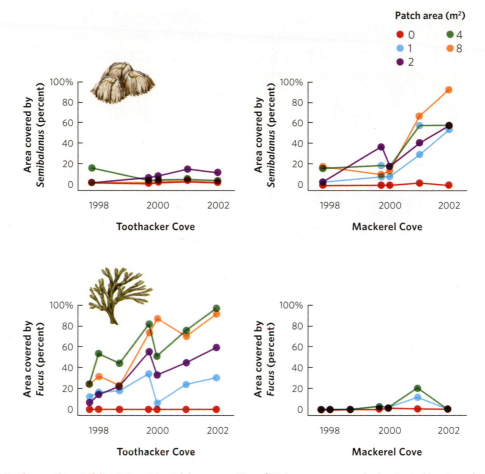

Figure 18.23 **Alternative stable states.** Intertidal communities of Maine are commonly dominated by *A. nodosum* algae. When researchers scoured areas of different sizes in 1996 and then monitored the colonization of the scoured areas, they found that the largest scoured areas switched to alternative stable states. In Toothacker Cove, a site with a southern exposure, the communities became dominated by *F. vesiculosus* algae. In Mackerel Cove, a site with a northern exposure, the communities became dominated by barnacles. Data from P. S. Petraitis and S. R. Dudgeon, Divergent succession and implications for alternative states on rocky intertidal shores, *Journal of Experimental Marine Biology and Ecology* 326 (2005): 14–26.

the winter, however, the scouring action of ice can scrape areas free of the rockweed algae and these areas become dominated either by another brown algae, *Fucus vesiculosus,* or by the barnacles *Semibalanus balanoides.* To determine whether these alternative communities are dependent on large disturbances, researchers simulated ice scouring on the coast by clearing areas of 1, 2, 4, and 8 m². As shown in Figure 18.23, small scoured areas did not change to alternative communities, but the largest scoured areas did; the north-facing site, Mackerel Cove, became dominated by barnacles because it was more protected from large waves. The south-facing site, Toothacker Cove, became dominated by *F. vesiculosus* because the site was more exposed to waves. The switch to an alternative state occurred because *A. nodosum* is a poor disperser so it is slow to

reach newly scoured sites compared to *F. vesiculosus* and barnacles. However, *A. nodosum* is a superior competitor for space, so over time *A. nodosum* may eventually be able to take over the open areas. The researchers found that the brown algae and barnacles were able to continue dominating sites for at least 20 years.

In this chapter we have seen that communities are composed of large numbers of interacting species that operate within food webs. The interconnected nature of food webs means that changes in the abundance of one species can potentially affect many other species in the community. In "Ecology Today: Connecting the Concepts," we will see that understanding the connectedness of species was a key to discovering the effects of pollution on amphibians.

ECOLOGY TODAY CONNECTING THE CONCEPTS

LETHAL EFFECTS OF PESTICIDES AT NONLETHAL CONCENTRATIONS

Leopard frog. Low concentrations of an insecticide that cannot kill the tadpoles of the leopard frog can kill other species in the community. This sets off a chain of events that alters the food web and indirectly prevents the leopard frogs from achieving metamorphosis. Photo by Lee Wilcox.

Appreciating the connectedness of species in a food web can help us understand some of the effects our activities have on ecological communities. Pesticides, for example, provide important benefits in protecting crops and improving human health, and their widespread use has made them common in ecological communities. Maintaining pesticide concentrations below levels lethal to nonpest organisms is the key to ensuring that pesticides harm only pests and not other species. Pesticides are tested on a range of species in the laboratory to determine concentrations that cause death. However, laboratory tests do not consider the fact that species are part of a food web. Therefore, researchers have asked if a pesticide that is not directly lethal to a species can nonetheless have indirect impacts that cause mortality.

Some of the best-studied pesticides are those used to control terrestrial insects that damage crops as well as insects such as mosquitoes that carry infectious diseases. The most commonly applied insecticide is malathion, which impairs the nervous system of animals and is highly lethal to insects and other invertebrates. Based on single-species laboratory tests, scientists thought that concentrations found in nature were not lethal to vertebrate animals.

In a series of experiments, researchers examined the effects of malathion in large outdoor tanks that contained many of the components of natural wetlands including algae, zooplankton, and tadpoles (see Figure 1.18b). It had been long known that insecticides such as malathion were highly toxic to zooplankton, which are tiny crustaceans, but little was known about how large reductions in zooplankton might affect other species in the food web.

In one such experiment, researchers added tadpoles of wood frogs and leopard frogs (*R. pipiens*) to tanks and then manipulated malathion in one of four ways: no malathion to serve as a control, 50 parts per billion (ppb) at the beginning of the experiment, 250 ppb at the beginning of the experiment, or 10 ppb once per week. These manipulations allowed the researchers to ask whether single, large applications and multiple, small applications could alter the food web in ways that could affect the amphibians.

Soon after malathion was applied, the zooplankton populations declined to very low numbers in all of the tanks treated with the insecticide. This decline was not surprising since zooplankton were known to be quite sensitive to insecticides, but it set off a chain of events that ultimately affected the amphibians. Zooplankton feed on phytoplankton. When malathion was added and the zooplankton were killed, the phytoplankton populations rapidly grew over the next few weeks. This rapid growth, known as an algal bloom, caused the water to turn pea green.

Effects of an insecticide on an aquatic community. **(a)** In the absence of an insecticide, zooplankton are abundant and they consume phytoplankton. This consumption keeps the water relatively clear and allows sunlight to reach the bottom of a wetland where periphyton grow and serve as food for tadpoles. **(b)** When an insecticide is added, the zooplankton populations are dramatically reduced, which allows an increase in phytoplankton. Abundant phytoplankton cause the water to turn green and prevent sunlight from reaching the periphyton at the bottom of the wetland. With less light available, the periphyton grows poorly, which limits tadpole growth. The number of individuals pictured in each group indicates the relative change in biomass.

When the water turned green with phytoplankton, sunlight could no longer reach the bottom of the water column, where another group of algae, known as *periphyton,* grows attached to objects such as soil and leaves. The periphyton could not grow very well with less available sunlight, so within a few weeks after the start of phytoplankton bloom, the periphyton declined in abundance, especially in those communities that received 10 ppb of malathion every week. In this treatment, the decimated zooplankton populations could never exhibit resilience because the weekly additions of malathion continually reduced the rebounding populations.

Periphyton is the primary source of food for amphibians; without it, they cannot grow and metamorphose. Fortunately, the wood frogs develop rapidly. In the experiment, wood frogs metamorphosed after only 5 weeks, which meant that they were able to metamorphose before the bloom of phytoplankton caused a decline in their periphyton food source. As a result, wood frog growth and survival were not affected by the insecticide.

The leopard frogs, however, were not so fortunate. Because leopard frogs need 7 to 10 weeks of growth as tadpoles before they can metamorphose, they experienced the full effect of the declining periphyton when the insecticide was present. In fact, the leopard frogs had such stunted growth in the tanks with the weekly addition of 10 ppb of malathion that almost half of them failed to metamorphose before their aquatic environment dried in late summer.

This research demonstrated for the first time that a low concentration of an insecticide that is not directly lethal to amphibians can be indirectly lethal. The insecticide initiated a chain of events that moved through the food web and caused nearly half of the leopard frogs to die. Moreover, the smallest application of the insecticide, which was applied each week, had a much larger effect on the leopard frogs than single applications containing 25 times more insecticide.

These results underscore the importance of understanding ecology at the community level. Quantifying the insecticide's direct effects on the amphibians suggested that the animals would not be harmed, but tests that incorporated a food web approach with indirect effects found that the insecticide could kill nearly half of the animals.

SOURCES: Relyea, R. A. 2005. The impact of insecticides and herbicides on the biodiversity and productivity of aquatic communities. *Ecological Applications* 15: 618–627.

Relyea, R. A., and N. Diecks. 2008. An unexpected chain of events: Lethal effects of pesticides on frogs at sublethal concentrations. *Ecological Applications* 18: 1728–1742.

SUMMARY OF CHAPTER CONCEPTS

■ **Communities can have distinct or gradual boundaries.** Communities frequently exist in zones that reflect changes in biotic and abiotic conditions. Based on the aggregations of species, ecologists categorize communities in terms of the dominant biological or physical conditions. Although some communities have distinct boundaries as a result of sharp changes in environmental conditions, most communities contain species with geographic ranges that are independent of each other, which produces gradual changes in the composition of communities as one moves across the landscape.

■ **The diversity of a community incorporates both the number and relative abundance of species.** Species richness is the number of species in a community whereas species evenness is the similarity of relative abundance among species in a community. Rank-abundance curves are graphical representations of richness and evenness. When we plot categories of individual abundance against the number of

species that fall within each category, communities commonly exhibit a log-normal distribution, which means that few species have either few or many individuals and that most species have an intermediate number of individuals.

■ **Species diversity is affected by resources, habitat diversity, keystone species, and disturbance.** Increases in the abundance of resources and in the number of different resources can alter species diversity. Increases in habitat diversity offer a greater diversity of niches, which favours higher species diversity. Keystone species can alter the composition of species in a community because they have large effects even though they are not particularly numerous. Disturbances also affect the species diversity of communities, with the greatest diversity occurring in communities experiencing disturbances that are intermediate in frequency or intensity.

■ **Communities are organized into food webs.** Food chains are linear representations of how different species in a community feed on each other,

whereas food webs are more complex and realistic representations. Species in a food web can be categorized in trophic levels and guilds according to how they obtain their energy: producers, primary consumers, secondary consumers, and tertiary consumers. Because species exist within food webs, the abundance of each species is affected not only by direct effects but also by density- and trait-mediated indirect effects. Because of these indirect effects, communities can experience trophic cascades from the top of the food chain down, and from the bottom up.

■ **Communities respond to disturbances with resistance, resilience, or switching among alternative stable states.** Resistant communities show little or no response to a disturbance. Resilient communities can be affected by a disturbance but they bounce back to their original states relatively quickly. Some communities respond to large disturbances by moving to alternative stable states in which the community persists with a different composition of species for a relatively long time.

REVIEW QUESTIONS

1. What are the differences between the interdependent community and independent community perspectives of Frederic Clements and Henry Gleason?

2. Given that the distributions of many animals are determined by the species composition of the plant community, what might you predict about the diversity of animals in an ecotone?

3. How did Robert Whittaker's data on plant species composition along gradients of temperature and moisture in mountains provide evidence against the idea of interdependent communities?

4. Explain why ecologists consider both species richness and species evenness when they quantify species diversity.

5. Compare and contrast log-normal abundance distributions and rank-abundance curves.

6. Why does an increase in total resources or number of added resources commonly lead to declines in species diversity?

7. Explain why we typically find more species under conditions of intermediate disturbance compared to either rare or frequent disturbance.

8. Why might you expect communities with fewer species to contain fewer trophic levels?

9. Explain the difference between density- and trait-mediated indirect effects.

10. What does it mean when a community is said to exhibit top-down control versus bottom-up control?

PRACTICING ECOLOGY: INVASIVE CRAYFISH AND TROPHIC CASCADES

Jonathan Moore at Simon Fraser University in British Columbia studies how the invasion or extinction of species alters the functioning of aquatic ecosystems. Photo by Susan Johnson.

As we learned in "Ecology Today" in this chapter, altering a community can cause trophic cascades with substantial effects on the biomass, abundance, and life history of multiple species. Introducing an invasive predator is one example of such an alteration. An omnivorous invasive predator can have an indirect positive effect on primary production by reducing the abundance of herbivores as well as a direct negative effect on primary production by consuming producers. Given these two potential effects, what net effect will an invasive omnivore have on primary production?

The signal crayfish (*Pacifastacus leniusculus*), native to northwest streams of North America, is currently invading streams of central and southern California. The crayfish prefer to feed on invertebrates such as snails, caddisflies, and stoneflies. Given that these invertebrates eat algae and leaf litter, we might hypothesize that crayfish feeding on invertebrates should generate a trophic cascade that leads to an increase in algae. However, after the preferred invertebrate prey become scarce, the crayfish should switch to feeding on algae. Hence, we might also hypothesize that after their initial invasion, crayfish will exert direct negative effects on algal biomass.

Jonathan Moore from Simon Fraser University in British Columbia and his colleagues tested these hypotheses with two studies conducted in the Scott Creek watershed of central California. The first study tested the short-term effects of crayfish during an initial invasion by manipulating the number of crayfish in enclosures placed in streams without crayfish. The researchers added different amounts of crayfish to each enclosure. To measure the short-term effect of invasive crayfish on algal biomass, they placed several small ceramic tiles in each enclosure, and measured the accumulated biomass of algae after 75 days. The second study measured the long-term effect of invasive crayfish on algal biomass by comparing algal growth on tiles placed in streams within the watershed, some of which had been invaded by crayfish for as long as 90 years and others that had not been invaded. You can view their data in the table.

EFFECTS OF CRAYFISH BIOMASS (g/m^2) ON ALGAL BIOMASS (µg/m^2) IN SHORT- AND LONG-TERM EXPERIMENTS

SHORT-TERM EFFECTS OF CRAYFISH (FIRST STUDY)		LONG-TERM EFFECTS OF CRAYFISH (SECOND STUDY)	
CRAYFISH BIOMASS	ALGAL BIOMASS	CRAYFISH BIOMASS	ALGAL BIOMASS
2.6	0.99	3.7	0.52
6.3	3.12	1.2	4.63
1.5	1.47	0.8	0.46
5.2	1.29	0.0	0.62
1.1	0.85	0.7	1.50
11.5	1.79	0.0	1.11
9.2	2.45	2.3	0.14
6.9	1.81	5.1	0.61
1.4	1.87	0.6	1.69
4.8	2.09	0.0	0.21

GRAPH THE DATA

1. Create a scatterplot graph for the short-term effects of crayfish, with crayfish biomass on the *x* axis and algal biomass on the *y* axis. Visually estimate and draw a regression line.

2. Create a second scatterplot graph for the long-term effects of crayfish, with crayfish biomass on the *y* axis and algal biomass on the *x* axis. Visually estimate and draw a regression line.

INTERPRET THE DATA

1. Explain how the data support or refute the hypothesis that in the short term the crayfish invasion causes an indirect positive effect on algae.

2. In the same study, the researchers measured herbivorous invertebrate prey biomass in the enclosures and found that herbivore biomass decreased as crayfish biomass increased. Does this provide evidence for the existence of a trophic cascade?

3. Explain how the data support or refute the hypothesis that following their initial invasion, crayfish will cause a direct negative effect on the algae.

SOURCE: Moore, J. W., S. M. Carlson, L. A. Twardochleb, J. L. Hwan, J. M. Fox, and S. A. Hayes. 2012. Trophic tangles through time? Opposing direct and indirect effects of an invasive omnivore on stream ecosystem processes. *PLoS One* 7:1–10.

Community
Succession

Retreating Glaciers in Alaska

In the panhandle of Alaska near Juneau, a stretch of land has undergone incredible changes during the past 200 years. In 1794, Captain George Vancouver found an inlet that headed toward modern-day Juneau. To the north of the inlet was a massive glacier that measured more than 1 km thick and 32 km wide. When the naturalist John Muir visited the glacier site in 1879, 85 years after Captain Vancouver's expedition, he was shocked to find that the glacier had receded nearly 80 km and left behind a large bay. Moreover, while most bays in Alaska were heavily forested, the shores of this new bay were relatively barren. As Muir explored, he found several locations with large stumps that were the remnants of hemlock, black cottonwood (*Populus trichocarpa*), and Sitka spruce trees that had been sheared off by the glacier as it advanced centuries earlier. Muir had no way of knowing that he was looking at one of the fastest melting glaciers in the world.

> "When the naturalist John Muir visited the glacier site in 1879, 85 years after Captain Vancouver's expedition, he was shocked to find that the glacier had receded nearly 80 km and left behind a large bay."

John Muir's writings about the natural world were widely read. In 1916, William Cooper, an ecologist who was particularly intrigued by Muir's stories of Alaska, made the first of several expeditions to the site that came to be known as Glacier Bay. On these expeditions, he surveyed the plants growing at various sites along the shore of the bay, including sites where the glacier existed during Vancouver's visit, sites where the glacier existed during Muir's visit, and sites where the glacier had receded only recently. He also set up quadrats around the bay to record the changes in vegetation at various locations that he planned to check when he returned on subsequent expeditions.

Cooper reasoned that sites exposed for the longest duration would have the most time to grow back into the kind of forest that had existed before the glaciers had advanced centuries earlier. In contrast, he expected recently exposed sites to be bare rock and gravel, representing the first stages of a developing forest. On these recently exposed sites, Cooper found mosses, lichens, herbs, and low shrubs. Sites that had been exposed for 35 to 45 years had tall species of willow and alder shrubs (*Alnus sinuata*) and black cottonwood trees. Sites older than 100 years contained Sitka spruce trees, and sites older than 160 years had hemlock in the understorey. By examining changes in the plant community at sites that had been exposed for different lengths of time, he was able to hypothesize how the forests of Alaska

Glacier Bay, Alaska. In 1794, a glacier that was more than 1200 m thick covered all but the inlet of the bay. Since then, the glaciers have melted and retreated, leaving behind a large body of water. Photo by Ron Niebrugge/ wildnatureimages.com.

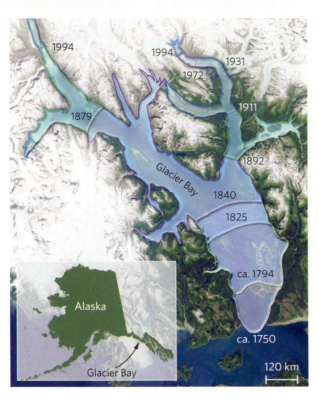

1994
1994
1931
1972
1879
1911
Glacier Bay
1892
1840
1825
ca. 1794
ca. 1750

Alaska

Glacier Bay

120 km

The rate of glacier retreat. Historic observations by Captain George Vancouver, John Muir, William Cooper, and current researchers indicate that the glacier at Glacier Bay has retreated rapidly during the past 200 years.

responded to the massive disturbance of an advancing and retreating glacier.

The observations of Vancouver, Muir, and Cooper paved the way for more than a century of subsequent ecological research at Glacier Bay. In fact, the original terrestrial quadrats set up by Cooper are still monitored today, as are local streams that have been created by the melting glacier. As we will see in this chapter, long-term changes in ecological communities follow predictable patterns, which are important for us to understand the ways in which terrestrial and aquatic communities change over time and the processes that underlie these changes.

SOURCES: W. S. Cooper, The recent ecological history of Glacier Bay, Alaska II: The present vegetation cycle, *Ecology* 4 (1923): 223–246.

C. L. Fastie, Causes and ecosystem consequences of multiple pathways of primary succession at Glacier Bay, Alaska, *Ecology* 76 (1995): 1899–1916.

CHAPTER CONCEPTS

- Succession occurs in a community when species replace each other over time.

- Succession can occur through different mechanisms.

- Succession does not always produce a single climax community.

Because ecological communities typically experience little change over weeks or months, it is tempting to think of a community as a static collection of species. However, communities commonly experience changes in species composition and relative abundance over longer periods. Such changes in communities over time are especially evident when a community experiences a major disturbance, like the receding glacier at Glacier Bay. Other examples include a field that has been plowed, a forest that has experienced an intense fire, or a pond that has dried during a drought but subsequently refills. In these cases, the community slowly rebuilds and, given sufficient time, it often resembles the original community that existed prior to the disturbance. In this chapter, we will explore how the composition of species in communities changes over time. We will also discuss the evidence that scientists have used to quantify these changes in both terrestrial and aquatic communities and describe how modern studies have altered original hypotheses. Understanding such changes is critical for scientists trying to predict how communities might respond to future environmental disturbances, such as

hurricanes and fires, or the results of anthropogenic activity such as logging and mining. Such an understanding is also critical for scientists who try to predict the impacts of larger disturbances such as global climate change.

Succession occurs in a community when species replace each other over time

The process of **succession** in a community is the change in species composition over time. For example, when a field is plowed but not planted, it soon becomes colonized by grasses and wildflowers. In climates with sufficient precipitation, the grasses and wildflowers will eventually be replaced by shrubs and then by large trees. Each stage of community change during the process of succession is known as a **seral stage**. The earliest species to arrive at a site are

Succession The process by which the species composition of a community changes over time.

Seral stage Each stage of community change during the process of succession.

known as **pioneer species**. These species typically have the ability to disperse long distances and arrive quickly at a disturbed site. The final seral stage in this process of succession is known as the **climax community**. A climax community is generally composed of the group of organisms that dominate in a given biome. As we will see, a climax community can be achieved through a variety of different sequences over time. Also, the climax community may continue to experience change.

We begin this section by considering two methods of observation that ecologists employ to study succession: direct and indirect. We will then examine how these techniques help scientists examine successional patterns in a variety of terrestrial and aquatic environments. While most studies of succession focus on the changes in plant communities, there are also associated changes in the species of animals that depend on plants for food and habitat. Much less studied is the succession that occurs on dead organic material such as rotting logs and animal carcasses.

OBSERVING SUCCESSION

Observing succession in a community can take different amounts of time depending on the life histories of the species involved. For example, the succession of decomposers on a dead animal can happen in a matter of weeks or months. In contrast, the succession of a forest can take hundreds of years. When succession occurs over long time horizons, modern scientists can sometimes return to sites that have been studied by other scientists decades or even centuries earlier and observe how the species composition of an area has changed over time. In other cases, when scientists have no historical data for an area, they must use more indirect methods to estimate the pattern of succession. As we will see, taking multiple direct and indirect approaches to examine succession provides the most complete picture of how communities change over time.

Direct Observations

The clearest way to document succession in a community is by direct observation of the changes over time. We have already described the research at Glacier Bay, a well-known example of directly observed succession. Another example comes from the small island of Krakatau, Indonesia.

In 1883 a massive volcanic eruption on Krakatau blew away nearly three-quarters of the island, as depicted in Figure 19.1a. The remaining part of the island was covered with a layer of volcanic ash that obliterated all life. Researchers began to visit the island after the eruption to document how and when various species would return and whether the developing community would eventually be similar to the community that had existed on the island before the eruption. By 1886 the researchers observed that 24 species of plants had colonized Krakatau, as you can see in Figure 19.1b. Ten of these species were sea-dispersed plants that are common on tropical shores throughout the region. Many of the other plants were wind-dispersed grasses and ferns whose seeds and spores had been blown to the island from surrounding islands. With more time, the seeds of trees arrived. By 1920 most of the island had developed into a forest community. The presence of forest habitats made the island more hospitable to many species of birds and bats. Bird and bat species that were fruit eaters brought a variety of seeds with them in their digestive systems. Many of the seeds that were excreted while the birds and bats were on the island subsequently germinated. After the 1920s most of the new plant species on the island were animal-dispersed species brought by visiting birds and bats. Today, scientists continue direct observations to monitor how succession on the island is affected by additional disturbances such as volcanic eruptions, erosion of ash deposits by the ocean waves, and strong storms that pass through the region.

Indirect Observations

Because it is a challenge to directly observe succession in many types of communities, researchers have sought ways to determine the patterns of succession indirectly. The two most common methods attempt to look back in time from the present day. One approach examines regional communities that began succession at different times. Hudson Bay is an ideal place to make indirect observations of succession. At the end of the last ice age approximately 20 000 years ago, a massive ice sheet covered all of Hudson Bay and, indeed, most of Canada. As this immense ice sheet melted, the land below, no longer pushed down by the heavy ice, began to rise. Although most of the ice sheet has now melted, the land continues to rise at a rate of about 1 m per century. As a consequence of this process, the shore of Hudson Bay contains a community that has been established relatively recently. As one moves uphill away from the shore, the communities become progressively older. A sequence of communities that exists over time in a given area is known as a **chronosequence**. You can view the chronosequence for the eastern coast of Hudson Bay in (Figure 19.2). Sea sandwort (*Honckenya peploides*) is the first plant to colonize the new shore line. It is quickly replaced by American

---

Pioneer species The earliest species to arrive at a site.

Climax community The final seral stage in the process of succession.

Chronosequence A sequence of communities that exist over time at a given location.

(a)

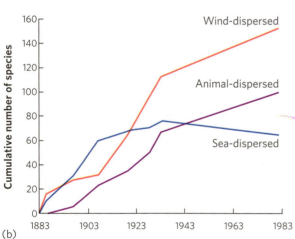

(b)

Figure 19.1 Direct observations of succession. (a) The island of Krakatau in Indonesia experienced a massive volcanic eruption in 1883 that blew away much of the original island and obliterated all life. **(b)** Since then researchers have documented the colonization of new species on the island over time. Initially, most plants colonizing the island were sea- and wind-dispersed. As the forests developed, birds and bats that came to the island carried the seeds of many other species of plants. Data from R. J. Whittaker et al., Plant recolonization and vegetation succession on the Krakatau Islands, Indonesia, *Ecological Monographs* 59 (1989): 59–123.

dune grass (*Leymus mollis*) and sea pea (*Lathyrus japonicas*). Other species, such as lichens and white spruce, only occur in old communities high above the shoreline.

It is also possible to look back in time by examining pollen and other plant parts preserved in distinct layers of lake and pond sediments. Flowering plants produce pollen grains with distinct sizes and shapes. When these pollen grains travel through the air and land on the surface of a lake, they sink and, over time, become preserved in layers of sediment at the bottom of the lake. To determine the age of each of these layers, researchers take a sample that

penetrates through many layers of mud on the lake bottom and use a technique known as *carbon dating* that identifies the age of the pollen in each layer. Dating the pollen helps them determine changes in the species of plants around the lake over hundreds or even thousands of years.

SUCCESSION IN TERRESTRIAL ENVIRONMENTS

Researchers investigating terrestrial environments have focused primarily on the succession of plant communities. In terrestrial environments, we can categorize succession into two types that are based on their starting conditions: *primary succession* and

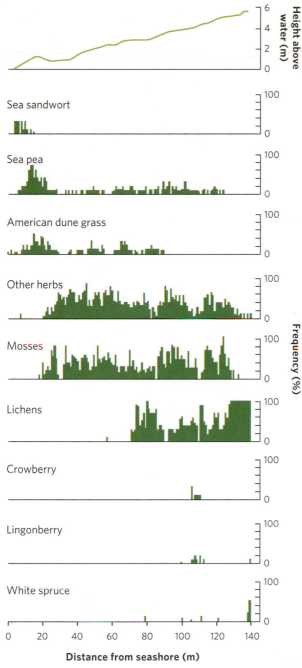

Distance from seashore (m)

Figure 19.2 Chronosequences. Along the coast of Hudson Bay, communities increase in age the farther one moves away from the water. Based on this relationship, researchers can examine the plant life on land of different ages as a way of estimating how succession proceeds over time. Data from A.-C. Laliberté and S. Payette, Primary succession of subarctic vegetation and soil on the fast-rising coast of eastern Hudson Bay, Canada, *Journal of Biogeography* 35 (2008): 1989–1999.

secondary succession. In both cases, we will begin by discussing a simplified version of terrestrial succession that is represented by an ordered progression over time. Later in this section we will see that succession in terrestrial communities can be much more complex.

Primary Succession

Primary succession is the development of communities in habitats that are initially devoid of plants and organic soil, such as sand dunes, lava flows, and bare rock. These inhospitable environments are colonized by species that require no soil, such as lichens and mosses that can live on the surfaces of rocks, and drought-tolerant grasses that are able to colonize dry sand dunes (Figure 19.3). The species that first colonize these places produce bits of organic matter that combine with the processes of rock weathering and microbial activity to create soils that make the site more hospitable for other species.

Secondary Succession

Secondary succession is the development of communities in habitats that have been disturbed and

Primary succession The development of communities in habitats that are initially devoid of plants and organic soil, such as sand dunes, lava flows, and bare rock.

Secondary succession The development of communities in habitats that have been disturbed and contain no plants but still contain an organic soil.

(a)

(b)

Figure 19.3 Primary succession. Primary succession occurs when organisms colonize sites that are initially devoid of life. **(a)** Bare rocks, such as these rocks in Banff National Park in Banff, Alberta, are initially colonized by lichens and mosses. **(b)** Sand dunes, like these dunes along the shores of Whaler Island in Clayoquot Sound, Vancouver Island, British Columbia, are initially colonized by grasses able to tolerate dry, sandy soils with little organic matter. Photos by (a) Kevin Maskell/Alamy, and (b) Chris Cheadle/Alamy.

contain no plants but still have an organic soil. For example, secondary succession occurs in fields that have been plowed or forests that have been uprooted by a hurricane. Such habitats typically contain well-developed soils, plant roots, and seeds, all of which contribute to rapid growth of new plants after the disturbance.

Secondary succession can be observed in a chronosequence of abandoned agricultural fields. An excellent example of this is Duke Forest, an outdoor laboratory in the Piedmont region of North Carolina composed of 1900 ha of abandoned agricultural fields and forests that Duke University purchased in 1931. To understand the process of succession in this area, ecologist Henry Oosting visited multiple farm sites, each abandoned for different lengths of time. He found a clear pattern of succession that starts with annual plants and ends with large deciduous trees, as illustrated in Figure 19.4. Even before abandonment, crabgrass (*Digitaria sanguinalis*) is common in the fields. In the first summer after abandonment, crabgrass and horseweed dominate the fields. By the second summer, ragweed (*Ambrosia artemisiifolia*) and heath asters (*Symphyotrichum ericoides*) dominate the fields, and by the third summer broomsedge (*Andropogon virginicus*) dominates. Next come shrubs, followed by pines. Within about 25 years, pines eventually crowd out earlier successional species. As the decades pass, deciduous tree species arrive and start displacing the pines. After about 100 years, deciduous trees dominate and constitute the last seral stage in the successional sequence.

We distinguish between primary and secondary succession by the starting point of the community. For example, in the midwestern United States, an area containing bare rock will pass through seral stages of lichens and mosses, annual weeds, perennial weeds, shrubs, several species of pioneer tree species, and then a beech-maple forest. In contrast, an area containing bare soil will start with annual weeds and then pass through many of the same seral stages on its way to form a beech-maple forest. As we will see later in the chapter, this beech-maple forest can vary somewhat in the composition of dominant tree species.

Sometimes the distinction between primary and secondary succession is unclear because the intensity of a disturbance can vary. For instance, because a tornado that levels a large area of forest usually does not harm soil nutrients, seeds, and living roots, succession follows quickly. In contrast, a severe fire that burns through organic layers of the forest soil requires the community to start over from scratch in a way that resembles primary succession; there is soil, but it contains few seeds or roots that can immediately sprout after the fire.

The Complexity of Terrestrial Succession

A given biome has a characteristic climax community, but the sequence of seral stages through which a single site passes on its way to this climax community can differ depending on the initial conditions. For example, consider the succession of a field, a sand dune, and a wetland near Georgian Bay in Ontario. The abandoned field is typically colonized by asters that include horseweed and goldenrod. The sand dune is typically colonized by reedgrass and bluestem grasses, which are perennial grasses that can stabilize the sand dunes and add organic matter to the soil. The wetland supports plants such as cattails, which produce organic matter; over many years this organic matter can fill in the wetland and create conditions that allow terrestrial plants to colonize the site. In all three cases, the progression of seral stages begins with a different community but ends with the same climax community of a forest dominated by oak, pine, and maple trees.

Field:	Year 1	Year 2	Years 3–25	Years 25–100	Years 100–200	Years 200+
Crabgrass	Crabgrass, horseweed	Ragweed, heath aster	Broomsedge, perennial flowers, shrubs, pines	Pine forest, hardwood understorey	Remnant pines with young oak and hickory trees	Oak-hickory climax forest

Figure 19.4 Secondary succession. Secondary succession begins with a well-developed soil. In a temperate seasonal forest, such as this one from the Piedmont region of North Carolina, the climax community is a forest dominated by deciduous trees.

Although it is tempting to think that terrestrial succession is a simple linear process, modern studies have demonstrated that this is often not the case. The sequence of seral stages can be quite variable. For example, the concept of chronosequences helps ecologists observe how a community changes in space and over time. However, this approach relies on the assumption that older sites pass through the same stages as the younger sites. It also assumes that sites of different ages do not differ in other aspects, such as historic abiotic conditions, soil fertility, and human or natural disturbances. When chronosequences go back hundreds of years, it can be difficult or impossible to confirm that the sites did not differ in ways that affect succession. For example, researchers have found that sites of similar age in a given area sometimes contain important differences in species composition because of local disturbances such as a tornado. Modern researchers will observe the changes in species composition, but they may not know that a tornado had passed through the area.

The research at Glacier Bay provides a good example of the complexity of succession. For decades, ecologists presented a simplified scenario of how the communities in Glacier Bay have changed over time. Based on Cooper's original observations, they identified primary succession as a simplified linear process that began with lichens, mosses, and herbs. Next is a seral stage containing low shrubs, followed by a stage containing tall shrubs that include willows, cottonwoods, and alders. The next stage is dominated by spruce trees and the final seral stage is dominated by hemlock trees.

More recent research has used tree rings to date the colonization of successional sequences and has reached a different conclusion regarding the path of succession. The number of growth rings tells us the age of the tree. From the tree's age, we can determine when the tree first appeared at a site relative to when the retreating glacier opened the site to succession. For example, we now know that the older sites probably never had a seral stage that included cottonwood trees, whereas the younger sites currently pass through a stage of abundant cottonwoods. Furthermore, spruce and hemlock trees rapidly colonized sites that were exposed in the early 1800s, but not sites that were exposed in the late 1800s and early 1900s. You can view the data for Sitka spruce trees in Figure 19.5. This suggests that the soils in the older and younger sites may not have experienced the same changes as they underwent succession.

Because the assumptions underlying chronosequences may not always be supported, the best approach is to use a combination of research methods including chronosequences, pollen records, and long-term studies of single locations undergoing

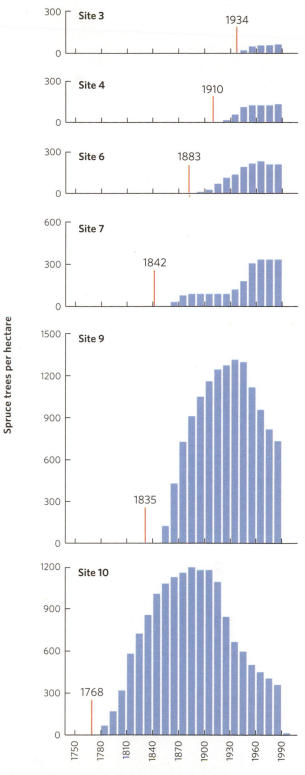

Figure 19.5 Complex succession. Using tree rings to date the Sitka spruce trees from different sites in Glacier Bay, we can see that older sites experienced rapid colonization by spruce trees after the glaciers retreated, whereas the younger sites experienced slow colonization by spruce trees after the glaciers retreated. Vertical red lines indicate the approximate year of glacial retreat at each site. Data from C. L. Fastie, Causes and ecosystem consequences of multiple pathways of primary succession at Glacier Bay, Alaska, *Ecology* 76 (1995): 1899–1916.

succession. Collectively, these studies provide the most accurate description of terrestrial succession.

Animal Succession

Ecologists have traditionally focused on changes in plant species when describing succession in terrestrial environments. However, the changes in the plant community cause substantial changes in the habitats that are available to animals, which in turn cause changes in the animal community. For example, a classic study by David Johnston and Eugene Odum examined the distribution of bird species in the Piedmont region of Georgia, along the same successional seral stages that Oosting surveyed in the Piedmont region of North Carolina. As illustrated in Figure 19.6, grasshopper sparrows (*Ammodramus savannarum*) and eastern meadowlarks (*Sturnella magna*) dominate the early-succession stages that contain annual plants. With the colonization of shrubs into the abandoned fields comes the arrival of many different bird species, including field sparrows (*Spizella pusilla*) and yellowthroats (*Geothlypis trichas*). As we move from the pine forest to the oak-hickory climax forest, other species appear that include the red-eyed vireo (*Vireo olivaceus*) and the wood thrush (*Hylocichla mustelina*). Although some species of birds specialize on a narrow range of the plant successional stages, many species inhabit multiple seral stages.

In the black spruce forests in western Quebec, researchers have also observed changes in the animal community that coincide with plant succession. They collected beetles in different forests that burned at various times in the past, ranging from 0 to 340 years ago. The researchers found that each successional stage was characterized by different groups of beetles. Early-succession forests that had been burned in the past 2 years were dominated by harpaline ground beetles whereas older forests that had burned 20 to 60 years ago contained mainly flat ground beetles. Mature forests that had burned 70 to 170 years ago were dominated by snail-eating beetles, but old-growth forests that had burned more than 170 years ago contained mostly grease bugs.

SUCCESSION IN AQUATIC ENVIRONMENTS

Succession also occurs in aquatic environments. In this section, we will examine how succession proceeds in three types of aquatic environments: intertidal zones, streams, and ponds.

Intertidal Communities

In contrast to most terrestrial communities, succession in intertidal communities can occur much faster after a disturbance, in part because the generation time of the dominant species is much shorter. In intertidal communities, powerful waves that occur during storms commonly remove organisms that are attached to boulders. For example, ecologist Wayne Sousa in a classic experiment examined the succession of different species of algae on boulders of the intertidal zone in southern California. He examined some boulders that were not overturned by storms and others that had been overturned and had areas of bare rock exposed that could be colonized by algae. As shown in Figure 19.7, Sousa observed that the first species to arrive was a green alga known as sea lettuce (*Ulva lactuca*). Over the next year, sea lettuce came to dominate the rocky

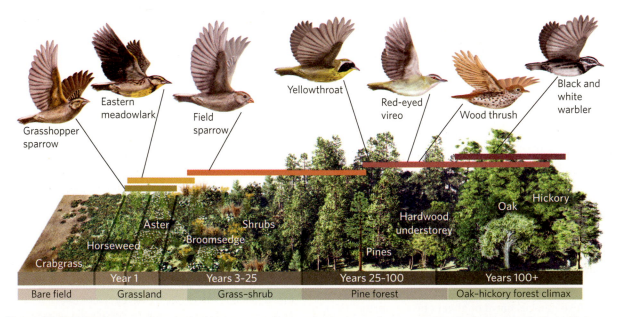

Figure 19.6 Animal succession. As the plant community experiences secondary succession, the habitat available to birds changes. In response to changes in the habitat, the species of birds that live in the community change. Data from D. W. Johnston and E. P. Odum, Breeding bird populations in relation to plant succession on the Piedmont of Georgia, *Ecology* 37 (1956): 50–62.

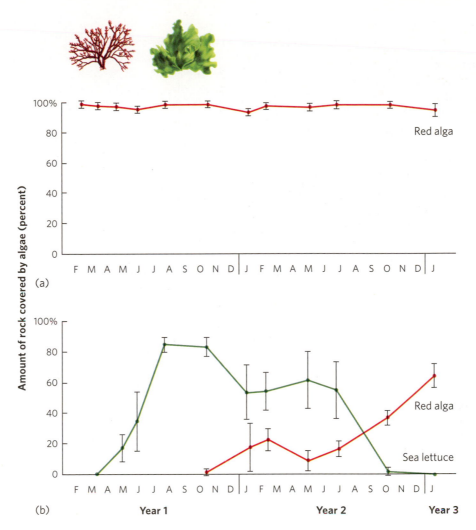

(a)

(b)   Year 1   Year 2   Year 3

Figure 19.7 Intertidal succession. (a) On undisturbed boulders in the intertidal zone, communities are dominated by a species of red alga, shown in the drawing on the left. **(b)** When strong waves overturn boulders, bare areas of rock are exposed and quickly colonized by sea lettuce, shown in the drawing on the right. Sea lettuce initially dominates the bare rock surface, as measured by the percent of the rock covered by the algae. As time passes, crabs begin to consume the sea lettuce, which creates space that allows the red algae to colonize the rock and ultimately dominate the boulder's surface. Error bars are standard errors. Data from W. P. Sousa, Experimental investigations of disturbance and ecological succession in a rocky intertidal algal community, *Ecological Monographs* 49 (1979): 227–254.

habitat and largely prevented a competing species of red alga, *Gigartina canaliculata,* from colonizing. However, as the sea lettuce became more dominant, it attracted crabs that eat it, which cleared areas on the boulders for the less edible red algae to colonize. Over time, red algae came to dominate the community.

In the Saint Lawrence estuary, algae and animals can also be removed by ice. Within a year, crevices in the rocks are the first places that are resettled by wrack and barnacles. During the second year, mussels colonize the deepest part of the crevice. After 4 to 5 years, the mussels increasingly replace barnacles.

Stream Succession

Like intertidal habitats, streams also experience rapid succession largely because organisms can move downstream from sites that are less disturbed.

Streams can experience major disturbances during heavy rainfall that increases both the volume of water the streams carry and the speed at which the water moves. With the greater speed of the rushing water, sand and rocks can tumble downstream and wipe out most plants, animals, and algae. In one study, researchers looked at the effects of a flood event in Sycamore Creek in Arizona. The floodwater scoured the creek and eliminated nearly all of the algae and 98 percent of the invertebrates, leaving behind rocks and bare sand. The researchers then monitored how the community changed over the subsequent 2 months; you can view their data in Figure 19.8. Within just a few days after the flood, the stream became colonized with several species of algae known as diatoms. Within 5 days, the diatoms covered nearly 50 percent of the creek bottom and after 13 days the

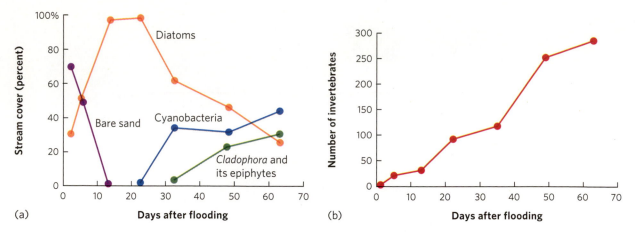

(a) **Days after flooding**

(b) **Days after flooding**

Figure 19.8 Stream succession. After a major flood event, Sycamore Creek in Arizona experienced a nearly complete elimination of algae, leaving behind only rocks and bare sand. **(a)** Within just a few days, several species of diatoms came to dominate the bare sand and rocks in the stream. Later in the summer, other groups of algae colonized the stream and became more abundant, including cyanobacteria and a species of green filamentous alga (*Cladophora*) and its epiphytes. **(b)** As the different types of algae returned, adult flying insects from the terrestrial environment began to lay their eggs in the stream. As a result, the stream experienced a rapid increase in the number of invertebrates. Data from S. G. Fisher et al., Temporal succession in a desert stream ecosystem following flash flooding, *Ecological Monographs* 52 (1979): 93–110.

diatoms covered nearly 100 percent. After 3 weeks, cyanobacteria started to colonize the stream, followed by a species of filamentous green alga (*Cladophora glomerata*) along with associated diatoms that live as epiphytes on this green alga. As the three types of

algae rebounded, adult insects from the surrounding terrestrial environment began to lay eggs in the stream. This returned a diverse group of larval insect species to the stream, which remained there until they metamorphosed into terrestrial adults.

(a)

The lake edge is colonized by plants.

The plants expand laterally across open water and begin accumulating dead plant matter known as peat.

The plants continue to expand laterally and the peat layer becomes thicker.

Lateral expansion of plants eventually covers the lake surface and peat sediments fill in the basin.

Gradual change over millennia

(b)

An open lake experiences a drought of one or more decades and the water level falls.

Plants colonize the newly exposed lake sediments.

As the drought ends and the water level rises, living plants detach from loose lake sediments and float, thereby covering the lake's surface.

Peat sediments eventually fill the basin.

Intermittent periods of large change over millennia

Figure 19.9 Succession in shallow lakes and ponds. (a) The classical explanation for succession in these habitats describes a gradual and steady accumulation of organic matter that eventually fills in the basin and converts it to a terrestrial habitat. **(b)** More recent studies have demonstrated that the process can occur in occasional large bursts when droughts allow vegetation to extend along the dry part of the basin. When water becomes abundant again, the extended vegetation floats on the water's surface and grows in thickness. Multiple drought events allow the vegetation to expand; eventually it covers the water's surface and fills in the basin. Based on A. W. Ireland, Drought as a trigger for rapid state shifts in kettle ecosystems: Implications for ecosystem responses to climate change, *Wetlands* 22 (2012): 989–1000.

Quantifying Community Similarity

When ecologists examine the species living in different communities, they often quantify composition and richness. Although such data tell us about the species living in each community, they do not provide a measure of comparison. To address this need, ecologists have developed several indices of community similarity that can range from zero to one; a value of zero indicates that two communities have no species in common, whereas a value of one indicates that two communities have an identical composition of species.

One of the most common ways to quantify similarity is Jaccard's index of similarity, developed by the Swiss botanist Paul Jaccard in 1901. Jaccard's index is calculated using the following equation:

$$J = \frac{X}{A + B + X}$$

where A represents the number of species that are *only* present in Community A, B represents the number of species that are *only* present in

Community B, and X represents the number of species present in *both* communities. For example, consider the table below that lists the species of fish found in each of three stream communities that are at different stages of succession.

We can now use Jaccard's index to calculate the similarity between Community A and Community B.

$$J = \frac{X}{A + B + X}$$

$$J = \frac{3}{1 + 4 + 3}$$

$$J = 0.33$$

This value indicates that there is relatively low similarity in the species composition of Communities A and B.

YOUR TURN Use Jaccard's index to calculate the similarity between Communities A and C and between Communities B and C. Based on these calculations, which communities are the most similar to each other?

FISH SPECIES	STREAM COMMUNITY A	STREAM COMMUNITY B	STREAM COMMUNITY C
RAINBOW TROUT	X	X	
BROOK TROUT		X	X
BROWN TROUT	X	X	
MUDMINNOW			X
COMMON SHINER	X	X	
CREEK CHUB	X		
WHITE SUCKER		X	X
JOHNNY DARTER		X	X
SMALLMOUTH BASS			X
MOTTLED SCULPIN		X	X

Lake Succession

For decades ecologists have explained pond and lake succession using a paradigm of slow transformation, as depicted in Figure 19.9a. If we imagine a pond or shallow lake created by a receding glacier thousands of years ago or a beaver dam converting a stream into a pond, we start with a basin full of water. Over time, the erosion of soil and the growth and death of organisms in the lake form sediments that gradually fill the basin. In addition, plants living along the shoreline slowly extend themselves out into the water to form a floating mat of vegetation. Underneath this live mat of vegetation is an accumulating layer of partially decomposed vegetation known as *peat*. Eventually, the entire basin has a floating mat of vegetation that continually contributes detritus to the peat layer in the basin. Microbial decomposition of the dead vegetation is slow because the water underneath the floating mat has little oxygen. As a result, detritus accumulates on the bottom of the basin, and over hundreds or thousands of years the pond or lake becomes a bog. In North American bogs, sphagnum moss, sedges, and shrubs—such as leatherleaf and cranberry—become established along the edges and add to the development of a soil with progressively more terrestrial qualities. At the edges of the bog, shrubs may be followed by black spruce and tamarack trees, which eventually give way to birch, maple, and fir trees, depending on the locality. Ecologists once believed that this succession reflects the way that many aquatic habitats are very gradually transformed—organic detritus accumulates until the soil rises above the water table and a terrestrial habitat emerges. Not all lakes experience succession into terrestrial habitats, but it is common for shallow lakes and ponds to do so over long periods of time.

In 2012, researchers proposed a new model of pond and lake succession. This model is based on studies that took core samples of a bog. By using carbon dating, researchers were able to identify when different species appeared. Contrary to the classic model of slow and continuous succession, they discovered that ponds and lakes can experience long periods of several hundred years in which little succession occurs, followed by brief episodes of rapid change. During times of prolonged drought that last for a decade or more, water levels drop and vegetation grows down onto the newly exposed shoreline. As you can see in Figure 19.9b, when the drought ends and the water level rises again, the mat of vegetation releases its hold from the lake bottom and floats up to the water's surface. With continued growth of the vegetation, this floating mat becomes thicker and deposits dead organic matter into the water below it. For example, after studying a 16-ha bog in Pennsylvania, the researchers estimated that 50 percent of the former lake became covered by bog vegetation in just a few decades during a severe drought in the late sixteenth century. In short, lake succession does not always have to be slow and steady; it can happen in brief bursts during rare periods of drought.

CHANGE IN SPECIES DIVERSITY

Across both terrestrial and aquatic habitats, the process of succession exhibits consistent effects on species richness. When succession follows a major disturbance that eliminates most or all of the species that had been present in a community, species richness necessarily starts at or near zero and then must increase over time with succession. In most cases of succession, we see species richness increase rapidly at first, followed by a plateau and a small decline, as shown in Figure 19.10. Oosting's survey of the Duke Forest, for example, found a rapid increase in the species richness of woody plants during the first 25 years and then a gradual decline in the rate of increase over the next 125 years. Similar patterns of

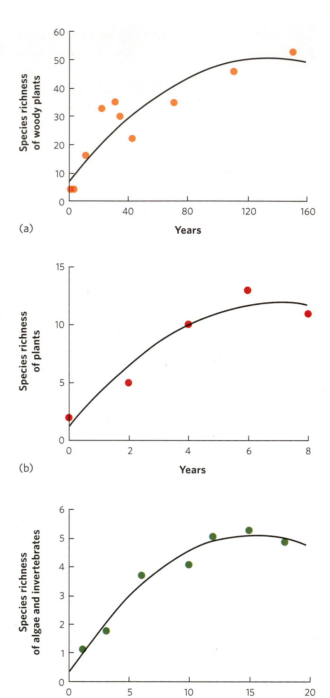

Figure 19.10 Succession effects on species richness. Across a variety of communities, succession is associated with a rapid increase in species richness that slows over time and eventually plateaus. Examples shown here include **(a)** woody plants in old fields in North Carolina, **(b)** plants in restored peatland in eastern Quebec, and **(c)** algae and invertebrates from the intertidal boulders of southern California. Data from (a) H. J. Oosting, An ecological analysis of the plant communities of Piedmont, North Carolina, *American Midland Naturalist* 28 (1942): 1–126; (b) M. Poulin et al., A new approach for tracking vegetation change after restoration: A case study with peatlands, *Restoration Ecology* 21 (2013): 363–371; (c) W. P. Sousa, Disturbance in marine intertidal boulder fields: The nonequilibrium maintenance of species diversity, *Ecology* 60 (1979): 1225–1239.

TABLE 19.1
GENERAL CHARACTERISTICS OF EARLY-SUCCESSION AND LATE-SUCCESSION PLANTS

TRAIT	EARLY SUCCESSION	LATE SUCCESSION
Number of seeds	Many	Few
Seed size	Small	Large
Mode of dispersal	Wind or stuck to animals	Gravity or eaten by animals
Seed viability	Long	Short
Root:shoot ratio	Low	High
Growth rate	Fast	Slow
Size at maturity	Small	Large
Shade tolerance	Low	High

richness over time can be seen in the number of plant species in a restored peatland in Quebec and the number of algae and invertebrates observed by Sousa in an intertidal community.

Succession can occur through different mechanisms

Now that we have an idea of what succession looks like across a variety of communities, we can explore how succession actually occurs. In this section we consider the traits of early- and late-succession species, compare the different mechanisms of succession, and then examine succession studies to determine which mechanisms are common in ecological communities.

TRAITS OF EARLY- VERSUS LATE-SUCCESSION SPECIES
Early-succession and late-succession species possess different traits important to their performance. For example, pioneer species of terrestrial plants are typically better at dispersing seeds to newly created or disturbed sites. They produce many small seeds that are easily dispersed by wind or that stick to passing animals. These seeds can also persist in the soil for years and then germinate when a disturbance occurs. When they do germinate, these early-succession plants invest more into their shoots than their roots and grow rapidly and reproduce quickly. They are also typically quite tolerant of the harsh abiotic conditions that can exist in newly disturbed sites, including full sun and widely fluctuating temperatures and water availability. However, they are not tolerant of the high-shade conditions of late-succession plant communities.

In contrast, climax species produce a relatively small number of large seeds that disperse poorly; some simply drop to the ground whereas others are consumed by animals. The seeds have a relatively short viability and once they germinate they grow slowly, but their shade tolerance as seedlings and their large size as mature plants give them a competitive edge over early-succession species.

As succession progresses, we see a shift in the balance between adaptations that promote dispersal,

rapid growth, and early reproduction, and adaptations that enhance competitive ability. Table 19.1 summarizes traits of early-succession and late-succession plants.

The traits of early-succession and late-succession plants are different because they face inherent trade-offs, similar to the life history trade-offs that we discussed in Chapter 8. To test for trade-offs between early-succession and late-succession species, researchers examined the mass of a single seed from each of nine species of trees. They raised the seeds under low-light conditions to simulate the high amount of shade in mature forests. After 12 weeks, they examined the mortality of the germinated seedlings, which you can view in Figure 19.11. The large-seeded species, which are common in old

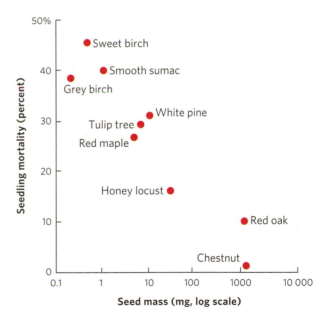

Figure 19.11 Trade-offs between seed mass and seedling mortality under high-shade conditions. Species with large seeds, which are common in late-succession forests, produce seedlings that survive well in high shade. Species with small seeds, which are common in early-succession forests, produce seedlings that have a poor survival rate in high shade. Data from J. P. Grime and D. W. Jeffrey, Seedling establishment in vertical gradients of sunlight, *Journal of Ecology* 53 (1965): 621–642.

forests, experienced low rates of mortality under low-light conditions since large seeds provide their seedlings with ample nutrients to get started in the light-limited environment of the forest floor. In contrast, small-seeded species, which are common pioneer species that disperse well in large numbers, experienced high rates of mortality under low-light conditions. As a result, pioneer species cannot establish in mature forests.

FACILITATION, INHIBITION, AND TOLERANCE

The ability to disperse and the ability to persist under existing abiotic and biotic conditions determine which species will appear in different seral stages during succession. Organisms that disperse well and grow rapidly have an initial advantage and therefore dominate the early stages of succession. Species that disperse slowly or grow slowly once they colonize an area typically become established later in succession. Early-succession species can also modify the environment in ways that affect whether late-succession species can become established. We must therefore consider whether a species has a positive, negative, or neutral effect on the probability of a second species becoming established. The mechanisms can be categorized as *facilitation, inhibition,* and *tolerance.*

Facilitation is a mechanism of succession in which the presence of one species increases the probability that a second species can become established. Early-succession species do this by altering the environmental conditions of the site in a manner that makes it more suitable for other species to establish and less suitable for themselves. For example, alder shrubs, which are legumes, live in a mutualistic relationship with nitrogen-fixing bacteria in their roots, as we discussed in Chapter 17. This relationship produces additional nitrogen in the soil, which facilitates the establishment of nitrogen-limited plants such as spruce trees. Over time, the spruce trees eventually grow tall and cast a deep shade that is not a favourable environment for the alder shrubs.

Inhibition is a mechanism of succession in which one species decreases the probability that a second species will become established. Common causes of inhibition include competition, predation, and parasitism. That is, individuals of one species can inhibit those of other species by outcompeting them for resources, eating them, or attacking them with noxious chemicals or antagonistic behaviour. Early in succession, inhibition can prevent movement toward a climax community, while late in succession, inhibition can prevent the pioneer species from colonizing and surviving. For example, in a mature forest in southeastern Canada, adult maple and beech trees cast a deep shade that prevents pioneer tree species from surviving.

When inhibition occurs in a seral stage, the outcome of an interaction between two species depends on which species becomes established first. The arrival of one species at a site that affects the subsequent colonization of other species is known as a **priority effect**. We can see an example of the priority effect in the subtidal habitats of South Australia with bryozoans, a group of tiny invertebrate animals that live in colonies attached to rocks and that feed by filtering the water. If bryozoans become established first, they can prevent the establishment of tunicates and sponges—two other groups of filter-feeding animals that attach themselves to rocks. Sometimes the priority effect occurs because the first species to arrive has grown to a competitively superior adult stage whereas the second species to arrive is in the competitively inferior immature stage. For example, if a beech tree gets established in a forest and grows to an adult stage, it will cast deep shade on the ground below that prevents the seedlings of most other tree species from conducting sufficient photosynthesis to survive. In short, the mechanism of inhibition can make the path of succession dependent on which species arrives at the site first.

Tolerance is a mechanism of succession in which the probability that a species can become established depends on its dispersal ability and its ability to persist under the physical conditions of the environment. For example, species that can tolerate the stressful environmental conditions of early succession—such as low moisture or more extreme temperature fluctuations—can become established quickly and dominate early stages of succession. Similarly, plants that can tolerate high-shade environments can become established in forests that cast deep shade. These species do not alter the environment in ways that either facilitate or inhibit other species, but once a stress-tolerant species becomes established, it can be affected by interactions with other species. For example, superior competitors

---

Facilitation A mechanism of succession in which the presence of one species increases the probability that a second species can become established.

Inhibition A mechanism of succession in which one species decreases the probability that a second species will become established.

Priority effect When the arrival of one species at a site affects the subsequent colonization of other species.

Tolerance A mechanism of succession in which the probability that a species can become established depends on its dispersal ability and its ability to persist under the physical conditions of the environment.

that arrive later will eventually replace the stress-tolerant species.

TESTS FOR THE MECHANISMS OF SUCCESSION

For many years ecologists debated which of the three succession mechanisms were the most important determinants of the pattern of species replacement over time. Knowing the dominant mechanisms would allow scientists to predict how changes in communities proceed over time, especially since succession is often not a simple linear progression toward a climax community. To address this question, a great deal of research had to be done across a variety of biomes. Here we discuss two such studies: one conducted in an intertidal community and one conducted in a forest community.

Succession in Intertidal Communities

Sousa's research in southern California showed that green algae prevented the colonization of red algae

(see Figure 19.7.) This suggested that the succession of intertidal communities is determined by inhibition. However, the intertidal communities off the Oregon coast consist not only of large species of algae—known as macroalgae—but also of several common species of invertebrates that include the little brown barnacle (*Chthamalus dalli*), the common acorn barnacle (*Balanus glandula*), and several species of limpets, which are gastropods that consume algae. Observation of these communities indicated that they were dominated by the common acorn barnacle and a brown macroalgae, *Pelvetiopsis limitata*.

To determine how these communities came to be dominated by the two species, researchers scraped areas of rock clean and watched the succession of species over the subsequent 2.5 years. As illustrated in Figure 19.12a, the bare rocks were first colonized by the little brown barnacle. However, as more time passed, the larger acorn barnacle became established

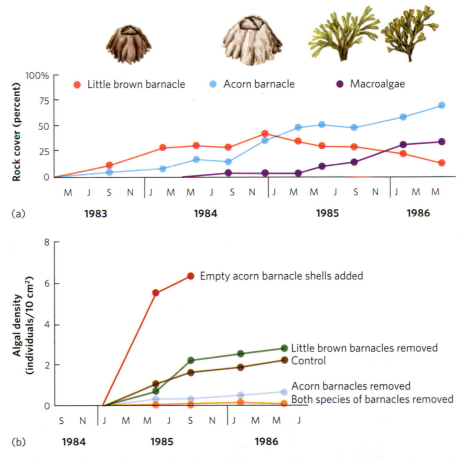

Figure 19.12 Tests of succession mechanisms in an intertidal community. (a) When rocks were scraped bare of all organisms, the first species to dominate was the little brown barnacle. Over time, the acorn barnacle became more abundant, followed by an increase in several species of macroalgae. **(b)** To determine which mechanisms were responsible for these successional changes, the presence of each barnacle species was manipulated. The acorn barnacle was critical to facilitate the macroalgae, whereas the little brown barnacle was not. Data from T. M. Farrell, Models and mechanisms of succession: An example from a rocky intertidal community, *Ecological Monographs* 61 (1991): 95–113.

on the sites. As it became established, it slowly crushed the smaller little brown barnacle. Over time, the little brown barnacle became rare. Thus, the colonization of the acorn barnacle fit the tolerance model, but the decline of the brown barnacle fit the inhibition model. As the acorn barnacle became abundant, numerous species of macroalgae colonized the site and also became abundant. In fact, after 3 years, communities in the cleared sites closely resembled the control sites that had not been cleared.

Which mechanisms of succession caused the increase of acorn barnacles and macroalgae? To answer this question, researchers conducted another experiment in which they removed different species from the community. They hypothesized that although the little brown barnacle would be the first to arrive, the acorn barnacle would eventually arrive and be a superior competitor for space. They also hypothesized that the presence of the acorn barnacles facilitated the colonization and survival of the macroalgae. The researchers began the experiment by once again scraping multiple sites clear of all organisms. Each site then received one of five manipulations: (1) no subsequent removal of any organisms, (2) removal of any little brown barnacles that colonized, (3) removal of any acorn barnacles that colonized, (4) removal of both species of barnacles that colonized, or (5) attachment of empty acorn barnacle shells. The final manipulation was used to test whether the mere physical presence of the acorn barnacle facilitated the colonization and growth of macroalgae by providing the macroalgae with protected crevices to which they could attach. The density of the macroalgae was then monitored for 2 years.

The results of this experiment are shown in Figure 19.12b. Starting near the bottom of the figure, you can see that the removal of the acorn barnacle, or both species of barnacles, caused a lower abundance of macroalgae than the control. In other words, the acorn barnacle facilitates the macroalgae. In contrast, the removal of the little brown barnacle had little effect on the density of the macroalgae. This confirmed that the little brown barnacle neither facilitates nor inhibits the macroalgae. However, the addition by researchers of empty acorn barnacle shells to an intact community that already included live acorn barnacles facilitated a large increase in macroalgae. Additional experiments revealed that the acorn barnacle facilitates the macroalgae by providing crevices where young macroalgae can attach to the rocks without being consumed by herbivorous limpets.

Collectively, these experiments in the intertidal zone indicate that the little brown barnacle persists because it is a good disperser that can rapidly colonize the frequently disturbed intertidal rocks and tolerate the conditions present on a bare rock. This is an example of tolerance. However, once the acorn barnacle arrives, it outcompetes the little brown barnacle, so inhibition also occurs later in succession. Finally, the succession of the macroalgae depends on the acorn barnacle, which is an example of facilitation.

Succession in Forest Communities

Like intertidal communities, forest communities can also exhibit a mixture of successional mechanisms. At Glacier Bay, for example, it was long hypothesized that each seral stage facilitates species in the subsequent stage. This made sense because later seral stages had soils containing greater organic matter, more nitrogen, and more moisture. If facilitation were the most important mechanism of succession, we would expect that a given species would grow and survive well in a seral stage where it dominates but have a difficult time growing and surviving in earlier stages.

To test this hypothesis, a group of researchers planted spruce seeds and seedlings in four seral stages of Glacier Bay: an early-succession pioneer stage containing lichens, mosses, and herbs; a low-shrub stage dominated by mountain avens (*Dryas drummondii*); an alder stage dominated by dense thickets of tall alder shrubs; and a spruce stage dominated by spruce trees. You can see the results in Figure 19.13a. When spruce seeds were planted in bare plots within each seral stage, germination of these seeds was high in the pioneer stage, low in the low-shrub and alder stages, and high in the spruce stage. When spruce seedlings were planted in plots within each seral stage, seedling growth was high in the pioneer and low-shrub stages, none survived in the alder stage, and growth was low in the spruce stage, as shown in Figure 19.13b. In short, the spruce seeds and seedlings were capable of germinating and growing quite well in the earliest seral stages at Glacier Bay, yet spruce was a rare plant in these early stages. Moreover, whereas spruce seeds and seedlings grew well in the low-shrub and alder stages, seedling growth was inhibited in the spruce stage.

These observations refuted the hypothesis that each seral stage facilitates the next stage, but it led researchers to question why spruce trees were not common in the earlier stages. They decided to measure how many spruce seeds were dispersed by wind to each site. They discovered that very few spruce seeds arrived to colonize the pioneer and low-shrub stages. There were many more spruce seeds dispersed into the alder stage and even more dispersed into the spruce stage, as shown in Figure 19.13c. The researchers concluded that the dominance of spruce in later seral stages had little to

(a)

(b)

(c)

Figure 19.13 Testing the mechanism of spruce succession. Across four successional seral stages in Glacier Bay, Alaska, researchers examined **(a)** germination success of planted spruce seeds and **(b)** biomass of 2-year-old spruce seedlings. They also measured **(c)** the number of spruce seeds arriving naturally at each seral stage. These data demonstrate that while spruce trees can germinate and grow well in earlier seral stages, they are absent from the early seral stages because very few seeds get dispersed to these sites. Error bars are standard errors. Data from F. S. Chapin III et al., Mechanisms of primary succession following deglaciation at Glacier Bay, Alaska, *Ecological Monographs* 64 (1994): 149–175.

occurs. A site can follow different pathways of succession and end up at the same climax community. However, the species composition of a climax community can still exhibit variation over space and time within a given biome. The climax community can also be an ephemeral condition because disturbances can wipe it out. In this section, we will examine how climax communities can change over time and how their composition varies along environmental gradients. We will also investigate how alternative climax communities can be favoured in the presence of different soils and by disturbances of different sizes and intensities.

CHANGES IN CLIMAX COMMUNITIES OVER TIME

When succession occurs in a community, we typically observe changing environmental conditions and a progression from small life forms to large life forms. For example, primary succession on land begins with lichens and mosses and progresses to grasses and herbs. When sufficient moisture is available, as is the case in eastern North America, succession can continue to a stage that includes large trees. As succession occurs, the abiotic conditions are rapidly altered; areas with trees have less light at ground level, lower ground temperatures during hot summer days, and higher soil moisture. However, once a point is reached where the community contains the largest plants it can support, changes in environmental conditions occur more slowly. As a result, the changes in the community become less dramatic once the climax community develops.

When environmental conditions become relatively stable, the composition of plant species that dominate the community also becomes relatively stable. However, the species found in a climax community can continue to change. For example, northern deciduous forests have a climax community dominated by large trees, but the composition of large tree species can slowly change over time. Initially these large trees are mostly oak, hickory, and tulip poplar (*Liriodendron tulipifera*). However, over time

do with facilitation or inhibition, but was due to differences in how many dispersing spruce seeds arrived in each location.

Succession does not always produce a single climax community

Ecologists have traditionally viewed succession as proceeding through a series of stages that end with a climax community, which remains constant over space and time unless another major disturbance

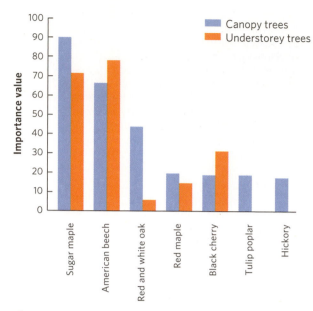

Figure 19.14 Changing species composition in a climax forest. There are many species of trees in the canopy of a forest in Pennsylvania, as indicated by the importance value of each species. However, in the understorey there are few oak trees and no tulip poplar or hickory trees because when they are young these species are not tolerant of deep shade and they are susceptible to herbivory by deer. This suggests that the future forest canopy will experience a substantial change in the composition of dominant trees. Data from Z. T. Long et al., The impact of deer on relationships between tree growth and mortality in an old-growth beech-maple forest, *Forest Ecology and Management* 252 (2007): 230–238.

the dominant species can change to sugar maple and beech. In an old-growth forest in Pennsylvania, researchers surveyed the adult trees that lived in the canopy and the sapling trees that lived in the understorey. After surveying the trees, they calculated the *importance value* of each tree, which incorporates both the abundance and total area of the trunks for each species. As you can see in Figure 19.14, the researchers discovered that the canopy contained a variety of species with high importance values that included sugar maple, American beech, tulip poplar, several species of oaks, and hickories.

In contrast to this adult distribution, there were few oaks and no tulip poplars or hickories in the understorey. The oaks, tulip poplars, or hickories were not in the understorey because these species are not very tolerant of the deep shade that occurs in a mature forest and they are also susceptible to browsing by deer. In comparison, maple and beech trees are very tolerant of shade and less susceptible to deer browsing. As a result, maple and beech trees can survive and grow in the understorey of large trees while other species cannot. Over time, the climax community of large trees will experience a gradual

shift in the composition of the dominant species; as the current canopy species gradually die, there will be no younger oaks, tulip poplars, or hickories to replace them. In short, the composition of a climax community of large trees in a northern deciduous forest can continue to change over time.

CHANGES IN CLIMAX COMMUNITIES OVER SPACE

We have seen that the composition of a climax community can vary over time. Climax communities can also vary in composition as one moves along environmental gradients. For instance, in the 1930s plant ecologists described the climax vegetation of much of Wisconsin as a sugar maple–basswood (*Tilia americana*) forest. However, they later determined that the climax forest exhibited differences in various locations around the state. In the southern part of the state, beech trees were more common while in the north, birch, spruce, and hemlock trees were more common. In the drier regions bordering prairies to the west, oaks became prominent. On drier upland sites, quaking aspen, black oak, and shagbark hickory—long recognized as successional species on moist, well-drained soils—came to be accepted as climax species.

TRANSIENT CLIMAXES

Sometimes a climax community is not persistent, a phenomenon known as a **transient climax community**. A transient climax occurs when a site is frequently disturbed so that the climax community cannot continue to perpetuate itself. A common example of a transient climax occurs in small wetlands, sometimes known as *vernal pools,* which fill with water each spring and then either dry up in summer or freeze solid in winter (Figure 19.15). Although drying and freezing events eliminate most species that comprise a pond community, some species have resting stages and persist in the soil until the wetland refills again in the spring. For example, many species of zooplankton produce resting eggs that can persist in the dry bottom of a wetland and then hatch when the wetland fills with water again. Similarly, some species of snails can aestivate, as discussed in Chapter 4, which allows them to live under the soil surface of the dry pond with their metabolic processes largely shut down. When the water returns, plants, animals, and microbes come back to life from their resting stages. Many other species that live on land as adults, such as frogs, salamanders, aquatic beetles, and dragonflies, lay their eggs in the wetland. In this way, the community once again starts the process of succession, only to be destroyed by summer drying or winter freezing.

Transient climax community A climax community that is not persistent.

(a)

(b)

Figure 19.15 Transient climaxes. Vernal pool communities undergo rapid succession of the aquatic community, but **(a)** the climax community that is achieved throughout the spring and summer is frequently destroyed by **(b)** drying in the late summer or freezing in the winter. This vernal pool is in the French Creek Wildlife Area of Wisconsin. Photos by Lee Wilcox.

CREATING GAPS IN A CLIMAX COMMUNITY

Sometimes climax communities contain species that are not considered climax species. These species fill relatively large gaps created by small-scale disturbances in an area. In mature forests, for example, adult trees eventually die and leave a gap in the canopy that lets in sunlight (Figure 19.16). If the gap is not large, the surrounding branches of neighbouring trees will likely grow in and close the gap. However, if the gap is large, the area of intense sunlight provides local conditions favouring species from earlier seral stages that have widely dispersed seeds and an ability to grow rapidly under high sunlight conditions. As a result, a mature forest that contains mostly climax species of trees can also contain a few early-succession trees. Gaps can occur in a variety of terrestrial and aquatic biomes. For example, in Chapter 18 we saw how gaps of increasing size in the intertidal zone favoured the formation of alternative stable states (see Figure 18.23).

Figure 19.16 Gaps in a climax community. When a large tree dies and falls to the ground, it opens a gap in the forest that allows large amounts of light to reach the forest floor and favours the growth of species from earlier seral stages. This photo is from the Corcovado National Park in Costa Rica. Photo by Chris Gallagher/Science Source.

CLIMAX COMMUNITIES UNDER EXTREME ENVIRONMENTAL CONDITIONS

As we have seen throughout this chapter, the composition of a climax community is determined by the environmental conditions that develop over time, including temperature, light, nutrients, and moisture. In some areas, however, additional environmental conditions also play a role, such as in communities affected by fire or grazing.

Biomes in which fires occur at regular intervals favour the persistence of fire-tolerant species. For example, we saw previously that boreal forests in the Yukon Territory experience periodic fires that kill all trees. Some pine species thrive after fires because their cones release seeds only when triggered by the heat of a fire. Under such conditions, pine seedlings grow rapidly because there is little or no competition from other understorey species. As a result, forest succession reaches a climax that is dominated by pine trees. When a successional stage persists as the final seral stage due to periodic fires, we call it a **fire-maintained climax community**.

The chaparral vegetation found in California is another community that has a fire-maintained climax. As we discussed in Chapter 6, the California chaparral is an example of the woodland/shrubland biome, which has cool, wet conditions in the winter and hot, dry conditions in the summer. As a result, plants can produce thick layers of detritus that become very susceptible to fires during the dry summers.

When a successional stage persists as the final seral stage due to intense grazing, we call it a **grazer-maintained climax community**. Grazers can create a different climax community because they preferentially consume the most palatable plants and do not consume the less palatable or better-defended plants. In the Canadian Arctic, for example, intense grazing by greater snow geese (*Chen caerulescens atlantic*) can reduce the cumulative biomass of the dominant grasses by between 50 to 92 percent by the end of the summer (Figure 19.17).

In western North America, cattle grazing allows invasion by cheatgrass (*Bromus tectorum*), a grass native to Europe, Asia, and North Africa that was introduced to North America in the late 1800s. Cheatgrass is able to colonize when grazing removes much of the competing grasses. Once it establishes,

Figure 19.17 Grazer-maintained climax. In the Canadian Arctic, intense grazing by greater snow geese can reduce dominant grass species. Photo by Sorin Papuc/Alamy.

the detritus it produces is quite susceptible to fire. The grazing promotes the invasion of cheatgrass and the repeated fires that occur promote its long-term persistence. In this case, the change in climax community is maintained by a combination of grazing and fire.

Throughout this chapter, we have discussed how communities change over time after a major disturbance. We have emphasized that communities are always changing, both when we move from early to late seral stages and even when a community appears to have attained a climax state. By observing both what happens after a disturbance and the underlying mechanisms that cause the changes, we obtain new insights into the processes that regulate the structure of communities. It is also important to remember that the climate that influences succession is also changing, which means that succession toward a climax stage is actually heading toward a perpetually moving target.

Fire-maintained climax community A successional stage that persists as the final seral stage due to periodic fires.

Grazer-maintained climax community When a successional stage persists as the final seral stage due to intense grazing.

PROMOTING SUCCESSION ON A STRIP MINE

Strip mine succession. When strip mines are no longer used for the extraction of coal, grasses are commonly planted to rapidly hold the soil and prevent erosion, such as at this site in Muskingum County, Ohio. Over time, other plants will slowly colonize the area and succession will occur. Photos by Michael Hiscar, Office of Surface Mining Reclamation and Enforcement, U.S. Department of the Interior.

Coal has long been a major source of energy in North America and Europe. For several centuries humans have mined coal near the surface of the ground by removing the upper layers of soil and rock and then digging out the coal underneath. This process, known as strip mining, surface mining, or mountaintop-removal mining, is an efficient way of mining coal, but these mines eventually run out of coal and leave an immense barren landscape. Many areas of the world require the coal mining industry to return the barren land to a more natural condition.

Since the ground may have little or no organic soil remaining after strip mining, it is a challenge to manipulate the succession of communities to reclaim the stripped land. Prevention of soil erosion is critical and can be accomplished by rapidly colonizing the area with plants that will hold the soil and ultimately undergo succession in a way that promotes species richness and improves the functioning of the ecosystem. To promote succession on former mined areas, we need to understand how it works by observing the succession changes in plant communities.

In Spain, researchers examined a chronosequence of 26 strip mines in the northern region of the country that had been abandoned for 1 to 32 years. The mines all had similar soils and climate. When the researchers examined changes in species richness, they found that mines abandoned for 1 year had 8 plant species and that species richness peaked at 28 species after 10 years of being abandoned. This peak was caused by the persistence of some of the pioneer species combined with the colonization by later succession

species. After 10 years, species richness exhibited a gradual decline; mines abandoned for 32 years had only 7 to 8 species. As with other communities discussed in this chapter, most of the species in the young abandoned sites were annual wind-dispersed plants. These species were able to tolerate the harsh conditions of the newly abandoned mine. After 10 years there was an increase in the number of perennial herbs, and woody plants started to colonize the area. These included nitrogen-fixing plants that contributed large amounts of organic matter and nitrogen to the soil, which facilitated subsequent woody plants, including trees that began to dominate after 20 years. These changes correlated with increased nitrogen in the soil, which suggested that the plants that formed the earliest communities facilitated the colonization and growth of later species.

Understanding how succession proceeds helps scientists develop recommendations for speeding up succession. For example, researchers in Germany examined how succession would be affected under three different manipulations: let the site naturally undergo succession; sow the area with a mixture of herb and grass seeds; or cover the area with newly cut hay, which helps to reduce erosion and provides the seeds of dozens of herbs and grasses contained in the soil. After 4 years, the plots experiencing natural succession had only 35 percent plant cover while the plots treated with either sown seeds or hay had more than 80 percent plant cover. These substantial differences in plant cover affected soil erosion; plots with sown seeds or green hay had channels that had eroded less than 5 cm deep while plots undergoing natural succession had channels that had eroded up to 1.5 m deep. The plots experiencing natural succession also had lower species richness during the first year and a lower index of similarity, although the richness of the three plots converged by the end of the 9-year experiment. Collectively, these data confirm that while primary succession will naturally occur on abandoned strip mines, we can use our knowledge of succession to speed up the process and rapidly move a barren landscape toward a much more

natural community. However, as we have learned throughout this chapter, terrestrial communities may require centuries of succession after a disturbance such as strip mining before they can approximate the communities that were originally present.

SOURCES: Alday, J. G., et al. 2011. Functional groups and dispersal strategies as guides for predicting vegetation dynamics on reclaimed mines. *Plant Ecology* 212: 1759–1775.

Baasch, A., et al. 2012. Nine years of vegetation development in a postmining site: Effects of spontaneous and assisted site recovery. *Journal of Applied Ecology* 49: 251–260.

SUMMARY OF CHAPTER CONCEPTS

■ **Succession occurs in a community when species replace each other over time.** The process of succession can be observed through either direct observations over time or indirect observations that use chronosequences or parts of organisms such as pollen that have been naturally preserved over time. Succession occurs on land, where we can distinguish between primary and secondary succession, and in the water. Succession often does not follow a simple linear path of species replacements, but there is a common pattern of rapidly increasing species richness over time that plateaus and can subsequently exhibit a small decline.

■ **Succession can occur through different mechanisms.** The mechanisms of succession can be categorized as facilitation, inhibition, and tolerance. More than one mechanism can operate in a community experiencing succession and the traits of species help determine the mechanisms that occur and where each species occurs along the successional stages.

■ **Succession does not always produce a single climax community.** As succession proceeds, the environment continues to change until conditions reach a point of relative stability and the dominant species appear to be persistent. However,

the climax community can continue to experience slow changes over time. The climax community can also differ within a region along environmental gradients such as temperature and moisture. Some climax communities are transient because they experience regular disturbances that reset succession, such as vernal pools that dry each summer. Extreme conditions, including fires and intense grazing, can also alter climax communities to produce a different composition of dominant organisms.

REVIEW QUESTIONS

1. How do ecologists distinguish between primary and secondary succession?

2. When using a chronosequence to document the pathway of succession, what are the assumptions?

3. How would a trade-off between dispersal ability and competitive ability affect which types of species could colonize small as opposed to large gaps in a community?

4. Why should we not expect a single climax community within a biome?

5. Compare and contrast the concepts of facilitation, inhibition, and tolerance in the context of ecological succession.

6. If two plant species have similar dispersal and competitive abilities, what factor might help determine which species occupies an early seral stage?

7. Compare the classic and modern explanations for the succession of ponds and lakes.

8. When a community experiences succession, why does the species

richness of a community initially increase, then plateau, and finally decline?

9. Why do early- and late-succession species tend to possess different adaptations?

10. Why are transient climaxes not stable?

PRACTICING ECOLOGY: DETERMINING CONSERVATION BASELINES

Mark Vellend and his lab at the Université de Sherbrooke study the historical and contemporary processes that alter plant communities. Photo by Jenny L. McCune.

To reverse human effects on Earth, we often try to restore ecosystems to a state that existed before human development of the land occurred. In North America, baselines for restoration goals usually focus on conditions that existed before European settlement. However, this assumes that First Nations—who settled the land long before Europeans—had little or no influence on the landscape. However, if First Nations did in fact alter and manage the landscape before the arrival of Europeans, then similar management strategies will need to be employed in the present day to sustain a pre-European baseline. Given that there are often few records of ecosystem species composition in the period before European settlement, how can we determine what influence the First Nations had on the landscape?

Southeastern Vancouver Island and adjacent islands offer an opportunity to address this question. The islands were settled by Europeans in the mid-1800s but had been occupied by First Nations for thousands of years. Current management efforts in the region aim to restore tree diversity to pre-European settlement status, yet there are few data on the diversity of tree species that existed during that time. Anecdotal evidence suggests that First Nations managed the landscape by frequently setting fires that would have reduced the abundance of fire-intolerant tree species, including grand fir (*Abies grandis*), bigleaf maple (*Acer macrphyllum*), and Pacific red cedar (*Thuja plicata*). These managed fires, in contrast, would have promoted the abundance of fire-resistant species such as Douglas fir (*Pseudotsuga menziesii*). The Douglas fir trees also would have gained dominance over slower-growing, fire-resistant trees such as Garry oak (*Quercus garryana*). Since the European arrival, settlers have largely prevented forest fires, which pose severe risks to property. Hence, we might hypothesize that managed forest fires played an important role in determining the tree diversity prior to, but not after, European settlement. If this hypothesis is correct, we might predict that fire-resistant species will exhibit a contemporary decrease and fire-intolerant species will exhibit a contemporary increase, relative to pre-European settlement tree diversity.

Mark Vellend and colleagues from the University of British Columbia tested this hypothesis by comparing modern surveys of tree diversity with historical land survey records. When Europeans settled the land, they defined the boundaries of landowner's property by recording the location of large trees. When Europeans arrived on the land in 1859, surveyors documented 225 tree landmarks throughout 176 km² of the Cowichan Valley in southeast Vancouver Island. In 2007, the researchers determined the location of each tree landmark and identified the 10 closest tree species. They then compared the relative abundance of the historical and contemporary trees to determine if the composition of the forest had changed over time. You can view their data in the table. Note that rare species are listed as "Other."

PERCENT TREE ABUNDANCE IN HISTORICAL (1859) AND CONTEMPORARY (2007) TREE SURVEYS OF THE COWICHAN VALLEY, VANCOUVER ISLAND

TREE SPECIES	1859	2007
BIGLEAF MAPLE	6	14
OAK	3	3
CEDAR	8	14
GRAND FIR	3	10
DOUGLAS-FIR	80	51
OTHER	0	8

GRAPH THE DATA

1. Create a pie chart showing the percent abundance of each tree species in the Cowichan valley in 1859.

2. Create a second pie chart showing the percent abundance of each tree species in 2007.

INTERPRET THE DATA

1. Which tree species increased in abundance between 1859 and 2007? Which tree species decreased in abundance?

2. Given what you know about the fire tolerance of each tree species, explain how the data support or refute the hypothesis that managed forest fires played an important role in determining the tree diversity prior to European settlement.

3. Current restoration efforts are aimed at conserving Garry oaks and other rare species on Vancouver Island. After examining the change in Garry oak abundance between 1859 and 2007, explain how this approach is beneficial or detrimental in achieving a pre-European settlement restoration goal.

SOURCE: Bjorkman, A. D., and M. Vellend. 2010. Defining historical baselines for conservation: Ecological changes since European settlement on Vancouver Island, Canada. *Conservation Biology* 24: 1559–1568.

CHAPTER 20

Movement of Energy in Ecosystems

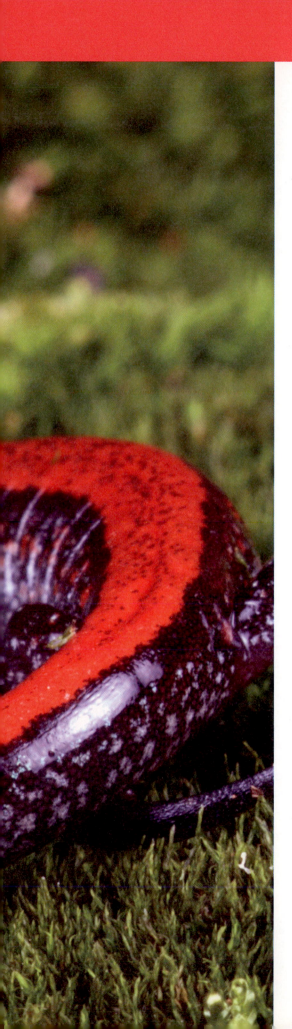

Worming Your Way into an Ecosystem

Earthworms play an important role in terrestrial ecosystems. As they burrow in the ground to consume detritus, they aerate compact soils, which allows water to better percolate into the ground. Earthworms seem to be everywhere in North America; you may have seen them on sidewalks and roads during rainy days or in the soil if you have ever worked in a garden. Therefore it is surprising that many common species of earthworms are not native to North America but were introduced from Europe and Asia in the eighteenth century.

> "The northern temperate and boreal forests of North America did not previously have any earthworms, but the introduced worms are now having a profound effect on the movement of energy in these northern ecosystems."

The northern temperate and boreal forests of North America did not previously have any earthworms, but the introduced worms are now having a profound effect on the movement of energy in these northern ecosystems.

Scientists hypothesize that northern forests completely lack native species of earthworms because the glaciers that advanced over the region during past ice ages eliminated all life. Since the ice receded 10 000 years ago, many other animal species have since returned to the area. However, the native earthworms that survived in the southern United States have a very slow rate of dispersal and have not yet arrived in many habitats in the northern United States. When there are no worms present, the leaf litter in these forests is primarily decomposed by soil fungi and microbes. However, this changed when European colonists inadvertently introduced European and Asian worms to the northern regions of North America. These worms have a high rate of dispersal and they tolerate a wide range of ecological conditions.

The introduced earthworms are still in the process of spreading into northern forests, assisted by the construction of logging roads and anglers who use the worms as bait. As consumers of dead leaves, the invasive worms can consume much of the energy available in the detritus and move many of the nutrients deeper into the soil where young plants cannot access them. The worms also leave much less energy for other organisms, such as the soil fungi that specialize on decomposing dead leaves. This is a major shift in the movement of energy. In addition to moving nutrients away from the forest floor, earthworm activity leaves a thinner layer of leaves and other organic matter which leads to drier soil conditions. These changes in abiotic conditions dramatically alter the soil food web. For example, in 2007 researchers reported that the recent arrival of introduced worms into the forests of the Kananaskis Valley in Alberta caused a decrease in the thickness of the litter layer from 3 cm to almost zero. This, in turn, resulted in a 75 percent reduction in the density of soil arthropods, such as springtails and mites. Such a decline in arthropods also means that there is less energy available for predators of these arthropods, such

The red-backed salamander. The introduction of earthworms from Europe and Asia to northern forests that historically lacked earthworms has caused a decline in leaf litter and soil arthropods. This can lead to a decline in the abundance of red-backed salamanders. Photo by George Grall/National Geographic Stock.

Introduced earthworms. Northern temperate and boreal forests have lacked native earthworms since glaciers eliminated them during past ice ages. Today these forests are being invaded by species of earthworms that have been inadvertently introduced from Europe and Asia. Photo by Oxford Scientific/Getty Images.

as the red-backed salamander (*Plethodon cinereus*). In areas with high densities of introduced earthworms, the decline in leaf litter and soil insects resulted in an 80 percent reduction in the number of salamanders. While adult salamanders can consume earthworms, young salamanders consume the much smaller soil insects.

Salamanders aren't the only consumer affected by the earthworm invasion. In 2012 researchers reported that a forest-dwelling ovenbird (*Seiurus aurocapilla*) was declining in areas of Wisconsin and Minnesota where the introduced earthworms had invaded. While the exact mechanisms were not clear in this case, the researchers suspected that the decline in the ovenbirds was in part because earthworms reduced the availability of insects on the forest floor, something similar to what caused the decline of the red-backed salamander.

The story of the invasive earthworms demonstrates that species depend on energy flowing between producers, detritivores, and consumers. Changes in energy flow between trophic groups can have major impacts on the species that inhabit an ecosystem. In this chapter, we will explore the flow of energy through food webs and the dynamics of energy movement through the ecosystem.

SOURCES: S. R. Loss et al., Invasions of non-native earthworms related to population declines of ground-nesting songbirds across a regional extent in northern hardwood forests of North America, *Landscape Ecology* 27 (2012): 683–696.

N. Eisenhauer et al., Invasion of a deciduous forest by earthworms: Changes in soil chemistry, microflora, microarthropods and vegetation, *Soil Biochemistry & Biochemistry* 39 (2007): 1099–1110.

CHAPTER CONCEPTS

- Primary productivity provides energy to the ecosystem.
- Net primary productivity differs among ecosystems.
- The movement of energy depends on the efficiency of energy flow.

In Chapter 1 we noted that the ecosystem approach to ecology focuses on the transfer of energy and matter among living and nonliving components within and between ecosystems. The amount of energy that fuels ecosystems and the efficiency with which it is transferred through trophic levels determine the number of trophic levels in communities and ecosystems. The amount of energy available and the efficiency of its transfer determine the biomass of organisms that exist at each trophic level and the amount of energy that is left behind for scavengers, detritivores, and decomposers. In this chapter, we will discuss the movement of energy in ecosystems including the importance of primary producers and the flow of energy throughout the food web. In the next chapter, we will concentrate on how matter, in the form of key chemical elements, cycles around ecosystems.

Primary productivity provides energy to the ecosystem

The vast majority of all energy that moves through ecosystems originates as solar energy that powers the photosynthesis of producers. As we saw in Chapter 1, some communities—such as those that form around thermal vents in the deep ocean—rely on chemosynthesis as their source of energy. Regardless of origin, producers harness energy and form the basis of food webs. Producers use this energy for respiration, growth, and reproduction; the amount used for growth and reproduction represents the energy that is available to consumers. In this section, we will examine how ecologists quantify the amount of energy that producers use for different functions, how researchers make these measurements in different types of ecosystems, and how the amount of energy available in producers affects the growth and reproduction of consumers.

PRIMARY PRODUCTIVITY
Primary productivity is the rate at which solar or chemical energy is captured and converted into

---

Primary productivity The rate at which solar or chemical energy is captured and converted into chemical bonds by photosynthesis or chemosynthesis.

chemical bonds by photosynthesis or chemosynthesis. Primary productivity tells us how much energy is available in an ecosystem. A related concept is the **standing crop** of an ecosystem, which is the biomass of producers present in the ecosystem at a given area at a particular moment in time. For example, the standing crop of a forest is the total mass of trees, shrubs, herbs, and grasses that is present in an area of the forest on a particular day. Ecosystems with high primary productivity may or may not have a high standing mass. In lakes where algae experience high productivity, consumers may eat it nearly as quickly as it grows, and so the standing crop of algae will remain low.

Ecologists identify two types of primary productivity: *gross primary productivity* and *net primary productivity*. **Gross primary productivity (GPP)** is the rate at which energy is captured and assimilated by producers in an area. We often express productivity in units of Joules (J) or kilo Joules (kJ) per square metre per year. From this total, producers use some of the assimilated energy

for their own metabolism, which is measured in terms of the amount of respiration by the producers. The rest of the assimilated energy is converted into producer biomass that includes growth and reproduction. The rate of energy that is assimilated by producers and converted into producer biomass in an area is **net primary productivity (NPP)**. We can also show this in the form of an equation:

$$NPP = GPP - \text{Respiration}$$

Considering the rate at which the Sun provides energy, photosynthesis is not a very efficient process. As illustrated in Figure 20.1, approximately

---

Standing crop The biomass of producers present in a given area of an ecosystem at a particular moment in time.

Gross primary productivity (GPP) The rate at which energy is captured and assimilated by producers in an area.

Net primary productivity (NPP) The rate of energy that is assimilated by producers and converted into producer biomass in an area.

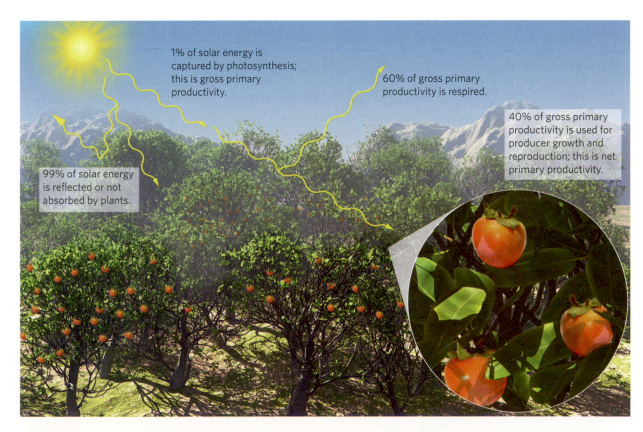

1% of solar energy is captured by photosynthesis; this is gross primary productivity.

60% of gross primary productivity is respired.

40% of gross primary productivity is used for producer growth and reproduction; this is net primary productivity.

99% of solar energy is reflected or not absorbed by plants.

Figure 20.1 Gross primary productivity and net primary productivity. About 99 percent of solar energy is reflected by producers or passes through their tissues without being absorbed. Only about 1 percent is captured by photosynthesis for gross primary productivity (GPP). Of the total GPP, about 60 percent is used by producers for respiration. The remaining 40 percent is available to producers for growth and reproduction, which is known as net primary productivity (NPP).

99 percent of all solar energy that is available to producers either reflects off them or passes through their tissues without being absorbed. Of the 1 percent of the solar energy that producers absorb and use for photosynthesis—that is, gross primary productivity—approximately 60 percent is used for respiration. This means that only 40 percent of absorbed solar energy is used for net primary productivity—the growth and reproduction of producers. For example, if we measure the productivity of a forest in North America in units of kilograms of carbon per square metre per year, such a forest might have a GPP of 2.5 kg C/m²/year. Of this total, the forest will use about 1.5 kg C/m²/year for respiration, leaving 1.0 kg C/m²/year for growth and reproduction, which represents the forest's NPP.

MEASURING PRIMARY PRODUCTIVITY

Primary productivity is the foundation for the flow of energy through food webs and ecosystems. Being able to measure primary productivity allows us to track how it changes in an ecosystem over time and how it varies in different ecosystems around the globe. Since primary productivity is a rate, we can measure it by quantifying the change in the biomass of producers over time, the movement of carbon dioxide over time, or the movement of oxygen over time. In terrestrial ecosystems, we commonly measure the productivity of plants, but in aquatic systems we might measure the productivity of plants, large kelps, or tiny species of algae. As we will see, the choice of how to measure

primary productivity depends on the particular ecosystem being studied.

When measuring primary productivity, it is critical to determine whether our measurements represent gross or net primary productivity. In some cases we can measure net primary productivity and respiration separately and then use these values to estimate gross primary productivity. We will now discuss a variety of common methods for measuring primary productivity in both terrestrial and aquatic ecosystems.

Measuring Changes in Producer Biomass

One of the simplest ways to measure NPP is by measuring producer biomass in an area at the beginning and at the end of a growing season (Figure 20.2). For example, in prairie ecosystems researchers commonly measure the amount of new plant growth that has accumulated by the end of the summer growing season, while in aquatic ecosystems, ecologists measure the biomass of large aquatic producers such as kelps. Researchers can harvest a season's growth, let it dry, and then weigh it to determine how much growth occurred during the season. When scientists measure a harvest to determine biomass, they either assume that no substantial herbivory or tissue mortality has occurred during the period of producer growth or they estimate the amount of biomass lost to these two processes and include this amount in the estimate of net primary productivity.

When we harvest, dry, and weigh plants from aquatic or terrestrial ecosystems, researchers may only

(a)

(b)

Figure 20.2 Harvesting primary productivity. Net primary productivity is commonly measured by harvesting a defined area of an ecosystem at the end of the growing season and determining the accumulated biomass of producers. This technique can be used on **(a)** small scales, such as by these students, who are measuring the aboveground biomass of plants, or on **(b)** large scales, such as the harvesting of crops in a field. Photos by (a) Jamie Shady and (b) Leonid Shcheglov/Shutterstock.

(a) Light

(b) Dark

Figure 20.3 Measuring CO_2 uptake and release. **(a)** In a sealed container illuminated by a light, sensors can detect a decline in CO_2 in the air as a leaf conducts photosynthesis. This decline occurs because plant photosynthesis consumes CO_2 at a faster rate than plant respiration produces CO_2. **(b)** When the light is turned off, sensors detect an increase in CO_2 in the air because the leaf only undergoes respiration.

harvest the part of the plant that lives aboveground. However, the amount of biomass belowground for some plants can be substantial. For example, perennial grasses with extensive root systems have twice as much belowground biomass as aboveground biomass. In contrast, most trees have about five times more aboveground biomass than belowground biomass.

Measuring underground biomass is quite a challenge. Retrieving the rhizomes, tubers, and roots of plants from the terrestrial soils or the aquatic benthos is often not feasible because they are either very deep or the root systems are composed of many fine roots that break off when harvested. In many plants, these fine roots die and are replaced with new fine roots, which makes it difficult to estimate how much biomass a group of producers has accumulated. It is also difficult to account for plants in mutualistic relationships with mycorrhizal fungi. These plants may send carbohydrates to the fungi, but because these carbohydrates are no longer a part of the plant, it is hard to get an accurate estimate of the plant's productivity. Therefore, we have to be cautious when making conclusions about differences in productivity when we measure only changes in the aboveground biomass to estimate net primary productivity.

Measuring CO_2 Uptake and Release

Because producers take up CO_2 during photosynthesis and produce CO_2 during respiration, we can measure primary productivity in terrestrial ecosystems by quantifying the uptake and release of CO_2 by plant leaves. One way to measure these changes in CO_2 is by placing a small plant or a leaf into a sealed container with a highly sensitive CO_2 sensor, as illustrated in Figure 20.3. When the container is placed in front of a light that simulates sunlight, the plant consumes CO_2 as it conducts photosynthesis and produces carbohydrates. The plant simultaneously produces CO_2 as it metabolizes some of its carbohydrates through respiration. Since CO_2 uptake from photosynthesis exceeds CO_2 release from respiration, the net uptake of CO_2 represents net primary productivity.

We can also estimate GPP using this technique. Since GPP is the sum of net primary productivity and respiration, we can determine GPP by combining our estimate of NPP with an estimate of how much respiration is occurring in a plant. We can do this by placing leaves or whole plants in chambers without sunlight. If we measure the plant's rate of respiration in the dark and add it to the plant's NPP, we can estimate the gross primary productivity using the following rearranged equation:

$$GPP = NPP + Respiration$$

Because these experiments are conducted using some chambers placed in the light and other chambers placed in the dark, they are sometimes referred to as a light-dark bottle experiment.

Several other techniques can also be used to measure CO_2 uptake and release. For example, researchers can place a plant or a leaf into a sealed container and then add CO_2 into the container with a rare isotope of carbon, such as ^{14}C. They can track the net movement of ^{14}C from the air and into the plant tissues during photosynthesis and from the plant tissues back to the air during respiration. This net movement of ^{14}C provides a measure of net primary productivity.

On a larger scale, such as a grassland or forest, researchers have developed techniques to measure CO_2 uptake and release using towers that sample the concentration of CO_2 at different heights above the ground. By measuring the concentration of CO_2 moving up from the vegetation compared to the concentration of CO_2 in the atmosphere, researchers can estimate the rates of photosynthesis and respiration that are occurring in an area.

Measuring O_2 Uptake and Release

In aquatic ecosystems, the dominant producers are typically algal cells. Because most algae are tiny, it is not always feasible to harvest and measure their biomass as a way to measure net primary productivity. Quantifying changes in CO_2 concentration is also not a viable way to measure primary productivity because CO_2 dissolved in water rapidly converts into bicarbonate ions, as we discussed in Chapter 2. However, because producers release O_2 during photosynthesis and take up O_2 during respiration, we can estimate NPP and GPP by measuring the uptake and release of O_2.

To estimate primary productivity using O_2 concentration in the water, we can use a process that is similar to estimating primary production based on CO_2. That is, we can conduct a light-dark bottle experiment. We begin by immersing two bottles under the surface of the water to collect the algae. One bottle is clear, allowing sunlight to penetrate. In this bottle we measure the net increase in O_2 production that occurs with the combined effects of photosynthesis and respiration by the algae. The other bottle is opaque, so sunlight cannot penetrate. In this bottle, the algae cannot photosynthesize but can only respire and drive down the concentration of O_2. Because the clear bottle measures NPP and the opaque bottle measures only respiration, we can estimate GPP by adding the values obtained from each.

Remote Sensing

The techniques we have discussed measure gross and net primary productivity on relatively small spatial scales, ranging from a single leaf to a small area of land to a small volume of water. What if we want to assess productivity on very large spatial scales, including changes in productivity across continents or oceans? One solution to this challenge is *remote sensing*. **Remote sensing** is a technique that allows measurement of conditions on Earth from a distant location, typically using satellites or airplanes that take photographs of large areas of the globe. These images allow scientists to determine how different wavelengths of light are reflected or absorbed. As we discussed in Chapter 3, chlorophyll pigments absorb wavelengths in the red and blue range but reflect wavelengths in the green range. Therefore, satellite images of ecosystems that show a pattern of high absorption of blue and red light and high reflectance of green light indicate ecosystems with high standing crops of producers. Changes in producer biomass over time can then be used to estimate net primary productivity.

SECONDARY PRODUCTION

Primary productivity is the foundation of the food web because it represents the source of energy for herbivores. To understand how energy moves from producers to consumers, we need to consider a number of different pathways, shown in Figure 20.4. We begin with producers that herbivores consume. Herbivores consume only a small fraction of the total amount of producer biomass available and they can only digest a portion of the energy they consume. For example, many fruits contain hard seeds that herbivores cannot digest, so the seeds are excreted whole. We saw an example of this at the end of Chapter 17 when we discussed Aldabra tortoises that had been introduced to an island in the Republic of Mauritius. These tortoises consume fruits of the ebony tree and disperse the seeds when they defecate. The portion of consumed energy that is excreted or regurgitated is known as **egested energy**. In contrast, the portion of energy that a consumer digests and absorbs is known as **assimilated energy**.

Of the energy assimilated by a consumer, the portion used for respiration is known as **respired energy**. The rest can be used for growth and

---

Remote sensing A technique that allows measurement of conditions on Earth from a distant location, typically using satellites or airplanes that take photographs of large areas of the globe.

Egested energy The portion of consumed energy that is excreted or regurgitated.

Assimilated energy The portion of energy that a consumer digests and absorbs.

Respired energy The portion of assimilated energy a consumer uses for respiration.

Figure 20.4 The path from primary to secondary productivity. Herbivores gather energy by ingesting producers. Of the energy gained by ingestion, some is lost in the form of egested energy, which represents the indigestible tissues of the producers. The rest of the energy is assimilated. Of the assimilated energy, some is used for respiration. The rest is used for growth and reproduction, which is secondary productivity.

reproduction. For example, consumers such as fruit-eating birds need a great deal of energy for respiration to maintain a constant body temperature, whereas fruit-eating tortoises of the same mass need much less energy for respiration because they are ectotherms. As a result, birds devote more of their assimilated energy to respiration than tortoises do. Because the rate of assimilated energy of consumers is the amount of energy that is used for respiration, growth, or reproduction, it is analogous to the concept of GPP for producers.

If we consider the assimilated energy of consumers and subtract the energy used for respiration, we are left with the energy used for biomass accumulation; the rate of biomass accumulation of consumers in a given area is called **net secondary productivity**. Because net secondary productivity depends on primary productivity as its source of energy, we might expect that an increase in net primary productivity would cause an increase in net secondary productivity. Researchers have compiled net primary and net secondary productivity estimates from a wide range of terrestrial biomes. As shown in

Figure 20.5a, increases in net primary productivity are positively correlated with increases in net secondary productivity. In aquatic ecosystems, researchers have also found that increases in the productivity of phytoplankton—a dominant producer in many aquatic biomes—are positively correlated to the productivity of the herbivorous zooplankton that eat the algae, which you can see in Figure 20.5b.

Quantifying net secondary productivity has many of the same challenges as quantifying net primary productivity, including the need to account for herbivores that are removed from the ecosystem by predators or diseases. In addition, the positive correlation between net primary and secondary productivity suggests an important role of bottom-up control of communities, a concept we discussed in Chapter 18. However, we know that the top-down effects of predators also affect some communities.

Net secondary productivity The rate of consumer biomass accumulation in a given area.

Figure 20.5 Secondary productivity. (a) In terrestrial ecosystems, an increase in primary productivity is positively correlated to an increase in secondary productivity. **(b)** In aquatic ecosystems, a similar relationship is found between the primary productivity of phytoplankton and the secondary productivity of the herbivorous zooplankton that consume the phytoplankton. Data from S. McNaughton et al., Ecosystem-level patterns of primary productivity and herbivory in terrestrial habitats, *Nature* 341 (1989): 142–144; M. Brylinsky and K. H. Mann, An analysis of factors governing productivity in lakes and reservoirs, *Limnology and Oceanography* 18 (1973): 1–14.

Net primary productivity differs among ecosystems

Now that we understand net primary productivity and how it affects the herbivores that consume it, we can examine how net primary productivity varies among ecosystems around the world. Understanding these patterns is important because ecosystems with greater primary productivity should generally support greater secondary productivity, which means that the most productive places will likely have either a high abundance or a high diversity of consumers. In this section, we will investigate patterns of primary productivity in different ecosystems and discuss the abiotic factors that determine these patterns of net primary productivity.

PRIMARY PRODUCTIVITY AROUND THE WORLD

When we look at patterns of NPP around the world in Figure 20.6, we see that net primary productivity varies with latitude. The most productive terrestrial ecosystems occur in the tropics and productivity declines as we move to temperate and polar regions. The most productive ocean ecosystems are found along coasts, while primary productivity is low in the open ocean. We can also find these general patterns in the distribution of primary productivity across Canada, as you can see in Figure 20.6. The most productive ecosystem is the temperate rain forest on Vancouver Island and along British Columbia's southern coast, while the boreal forests show medium levels of primary productivity. We find the lowest primary productivity in the tundra region across Canada's North.

We can also consider the differences in NPP among different ecosystems that exist on the continents or in the oceans. These ecosystem categories largely coincide with the biomes described in Chapter 6. Figure 20.7 shows that tropical rainforests are the most productive terrestrial ecosystems. The least productive ecosystems include those that are very cold, such as tundra, and those that are very dry, such as deserts. Among the freshwater ecosystems, lakes and streams on average are on the low end of NPP, although NPP can vary widely in these systems. In contrast, marine ecosystems exhibit a wide range of productivity, ranging from high productivity in coral reefs and salt marshes to low productivity in the open ocean. Many abiotic factors are responsible for these patterns of productivity.

DRIVERS OF PRODUCTIVITY IN TERRESTRIAL ECOSYSTEMS

As you may recall from our discussion of terrestrial biomes in Chapter 6, dominant plant forms are determined by patterns of annual temperature and precipitation. Similarly, temperature and precipitation are major drivers of NPP. The most productive terrestrial ecosystems occur in tropical areas because those areas have the most intense sunlight, warm temperatures throughout the year, high amounts of precipitation, and rapidly recycled nutrients that support growth. At higher latitudes, such as in temperate and polar regions, productivity is much lower due to shorter periods of sunlight and lower temperatures during the winter. In the deserts that occur at 30° N and 30° S latitude, productivity is primarily constrained by the lack of precipitation.

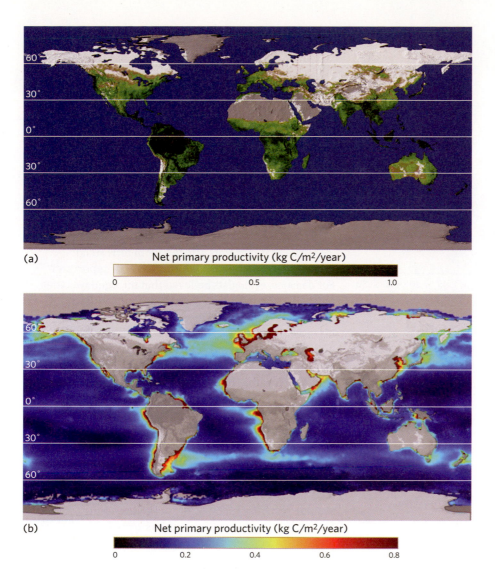

(a)

Net primary productivity (kg C/m²/year)

0   0.5   1.0

(b)

Net primary productivity (kg C/m²/year)

0   0.2   0.4   0.6   0.8

Figure 20.6 Global patterns of net primary productivity. (a) Among terrestrial ecosystems, the highest productivity occurs in the tropics, which experience warm temperatures throughout the year and have abundant rainfall. (b) The most productive aquatic ecosystems include the shallow water near the edges of continents and large islands. Data from http://eoimages.gsfc.nasa.gov/images/news/NasaNews/ReleaseImages/20030917/Image1m.jpg and http://geo.arc.nasa.gov/sge/ecocast/publications/press/globalgarden/earth_observatory_2.htm

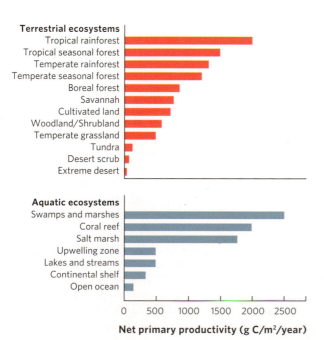

Figure 20.7 The net primary productivity of different ecosystems around the world. Among terrestrial ecosystems, tropical rainforests are the most productive whereas deserts are the least productive. Among aquatic ecosystems, swamps, marshes, and coral reefs are the most productive whereas the open ocean is the least productive. Data from R. H. Whittaker and G. E. Likens, Primary production: The biosphere and man, *Human Ecology* 1 (1973): 357–369.

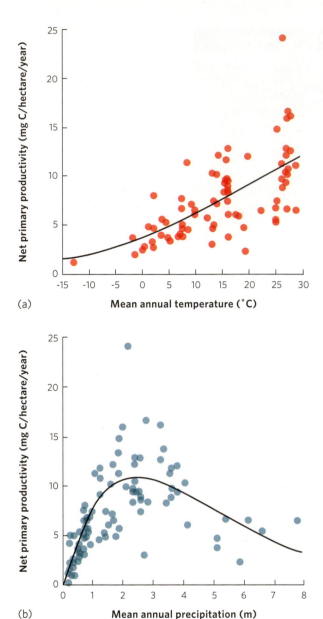

Figure 20.8 The effects of temperature and precipitation on the net primary productivity of terrestrial ecosystems. **(a)** Locations on Earth that have higher mean annual temperatures have higher NPP. **(b)** Locations that have higher mean annual precipitation have higher NPP but at locations with the highest amounts of precipitation, increased leaching and reduced decomposition of organic matter cause NPP to decline. Data from E. A. G. Schuur, Productivity and global climate revisited: The sensitivity of tropical forest growth to precipitation, *Ecology* 84 (2003): 1165–1170.

One way to examine the effects of temperature and precipitation on NPP is by looking at a large number of studies that have measured NPP in different parts of the world. For example, Figure 20.8 displays the results of studies from 96 locations. In Figure 20.8a, you can see that an increase in mean annual temperature is positively correlated with an increase in NPP. This reflects the fact that lower latitudes with warmer

temperatures favour plant growth and have a longer growing season. In Figure 20.8b, you can see that an increase in mean annual precipitation shows a positive correlation with NPP until 3 m of annual precipitation is reached. Ecosystems that receive 3 m or more of precipitation experience a decline in NPP because nutrients leach away from the soil, and the rates of decomposition are reduced because of waterlogged soils. When organic matter is broken down more slowly, fewer nutrients are available in the soil for primary productivity. In short, while temperature and precipitation are the primary drivers of NPP in terrestrial ecosystems, they also influence the availability of nutrients in ways that affect NPP.

Given that nutrients affect NPP in terrestrial ecosystems, which nutrients are the most important? For decades, ecologists believed that nitrogen was the most important element that constrained NPP in terrestrial ecosystems. However, ecologists began to discover that some terrestrial ecosystems are also limited by phosphorus or a combination of nitrogen and phosphorus. To obtain a more accurate overview of whether nitrogen or phosphorus limited NPP, researchers in 2007 compiled data from 141 separate terrestrial experiments that had manipulated nitrogen, phosphorus, or both. For each experiment, they determined the ratio of NPP in treatments with added nutrients compared to NPP in the control, which had no added nutrients. With a response ratio for each study, the researchers could then determine the average response ratio for all studies in three categories of terrestrial ecosystems: grasslands, forests and shrublands, and tundra. As you can see in Figure 20.9, all three categories of terrestrial ecosystems experienced increases in NPP with the addition of nitrogen or phosphorus. In grasslands and tundra, adding both nitrogen and phosphorus caused a greater increase in NPP than either alone. From these data, we can see that nitrogen and phosphorus are both important nutrients that constrain NPP.

DRIVERS OF PRODUCTIVITY IN AQUATIC ECOSYSTEMS

We have seen that terrestrial ecosystems are primarily constrained by temperature, precipitation, and nutrients. In addition to these factors, aquatic ecosystems are also constrained by light since the transmission of light down through the water is required for photosynthesis. Indeed, abundant light is one reason that coral reefs, which exist in shallow tropical waters, are such productive ecosystems. However, within aquatic ecosystems that have similar temperatures and similar light levels, the most important driver of NPP is the amount of nutrients.

The limiting role of nutrients can be seen in a variety of aquatic ecosystems. For example, in the open

Figure 20.9 Increases in net primary productivity when nutrients are added to terrestrial ecosystems. Based on 141 experiments conducted around the world, researchers examined how NPP responds to different nutrient additions. To quantify the change in NPP, they determined the ratio of NPP in treatments with added nutrients compared to NPP in the control, which had no added nutrients. The addition of nitrogen and phosphorus increased the NPP of grassland, forest, shrublands, and tundra ecosystems. When both nutrients were added, the response ratio of NPP was often larger than when either nutrient was added separately. Error bars are standard errors. Data from J. J. Elser et al., Global analysis of nitrogen and phosphorus limitation of primary producers in freshwater, marine, and terrestrial ecosystems, *Ecology Letters* 10 (2007): 1135–1142.

ocean, the remains of dead animals sink to the bottom where they decompose and release nutrients. Because this regeneration of nutrients is far below the ocean's surface, the surface of the open ocean experiences low NPP. Small streams also are typically low in nutrients and experience low NPP. Moreover, if the small streams are in a forest, they receive little sun because of the shade from trees, and this restricts their productivity. As we discussed in Chapter 6, a large fraction of the energy and nutrients that exist in small streams enters the stream in the form of allochthonous inputs such as dead leaves that drop in from the surrounding terrestrial environment. In contrast, estuaries and coral reefs receive abundant nutrients in the form of runoff from rivers and the adjacent land, which allows these ecosystems to have very high primary productivity. Among all aquatic ecosystems, NPP is most commonly constrained by the availability of phosphorus and nitrogen, although silicon and iron can be limiting in some areas of the open ocean.

Limitation by Phosphorus and Nitrogen

For many years the conventional wisdom was that phosphorus is the most important nutrient that limits the NPP of aquatic ecosystems. For example, in a classic experiment David Schindler and his colleagues selected an hourglass-shaped lake in Ontario and placed a plastic curtain at the constriction that divided the lake into two halves (Figure 20.10a). On one side

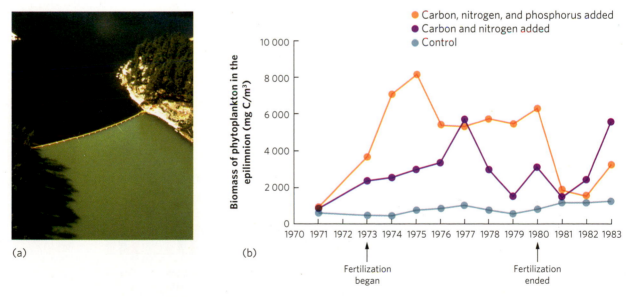

(a)　　　(b)

Figure 20.10 Adding phosphorus to half of a lake. **(a)** Researchers split a lake in northwestern Ontario in half and then added carbon and nitrogen to one side and carbon, nitrogen, and phosphorus to the other side. **(b)** Compared to a reference lake that received no nutrient additions, the side that received additions of carbon and nitrogen from 1973 to 1980 experienced a modest increase in NPP, as measured by the biomass of phytoplankton in the epilimnion. The side of the lake that received additions of carbon, nitrogen, and phosphorus experienced a large increase in NPP. When the addition of phosphorus stopped in 1980, NPP in the fertilized side of the lake declined to levels that were similar to the control. Data from D. L. Findlay and S. E. M. Kasian, Phytoplankton community responses to nutrient addition in Lake 226, Experimental Lakes Area, Northwestern Ontario, *Canadian Journal of Fisheries and Aquatic Sciences* 44 (Suppl. 1) (1987): 35–46. Photo courtesy of David W. Schindler, from *Science*, 184: 897–899 [1974].

the researchers added carbon and nitrogen, and on the other side they added carbon, nitrogen, and phosphorus. These additions were continued from 1973 to 1980 as the researchers monitored the two sides of the lake, as well as a second lake that had an initial fertility similar to the divided lake to serve as a control. In the half of the lake that received carbon and nitrogen, there was a modest increase in NPP compared to the control lake, measured by the growth of cyanobacteria. However, in the half of the lake that received carbon, nitrogen, and phosphorus, there was a large increase in NPP, as shown in Figure 20.10b. After the treatments were terminated in 1980, the NPP of the side of the lake with added phosphorus rapidly declined. This experiment confirmed that human inputs of excess phosphorus—in the form of fertilizers that wash off farms and various household detergents—can have a major effect on the productivity of aquatic ecosystems.

To gain a broader perspective on how nitrogen and phosphorus affect aquatic ecosystems, let's revisit the average NPP responses from experiments conducted around the world. As part of the same 2007 study discussed earlier, researchers compiled data from 928 experiments that manipulated the addition of nitrogen, phosphorus, or both in freshwater and marine ecosystems. Among freshwater ecosystems, as shown in Figure 20.11a, nitrogen and phosphorus both caused an increase in NPP, although phosphorus has a much larger effect on the benthic ecosystem of lakes than does nitrogen. Among marine ecosystems, shown in Figure 20.11b, nitrogen and phosphorus additions had similar effects on NPP in ocean ecosystems with soft bottoms such as estuaries comprised of seagrass and algae. However, the addition of nitrogen had a much larger effect on NPP than that of phosphorus in ecosystems with hard bottoms, such as coral reefs and rocky intertidal biomes, and in the open water of oceans. Across all freshwater and marine ecosystems, adding nitrogen and phosphorus together often caused an NPP response that was much larger than adding either nutrient separately. The results of this research suggest that the availability of nitrogen and phosphorus can constrain the NPP of aquatic ecosystems, as well as that of terrestrial ecosystems.

Limitation by Silicon and Iron in the Ocean

While the primary productivity of aquatic ecosystems is typically limited by the availability of nitrogen and phosphorus, primary productivity remains low in approximately 20 percent of the open ocean even though nitrogen and phosphorus are abundant there. This suggests that productivity in these areas is

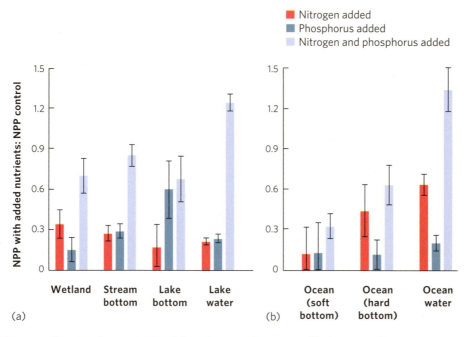

Figure 20.11 Increases in net primary productivity when nutrients are added to aquatic ecosystems. Using data from 928 experiments conducted around the world, researchers determined how NPP responds to different nutrient additions. NPP was examined as a ratio between productivity when nutrients were added and a control treatment in which nutrients were not added. In **(a)** freshwater and **(b)** marine ecosystems, adding nitrogen or phosphorus caused an increase in NPP, although nitrogen has a larger effect than phosphorus in the soft-bottom-ocean ecosystems and the ocean-water ecosystems. Error bars are standard errors. Data from J. J. Elser et al., Global analysis of nitrogen and phosphorus limitation of primary producers in freshwater, marine, and terrestrial ecosystems, *Ecology Letters* 10 (2007): 1135–1142.

limited by other nutrients that are unusually scarce, such as silicon and iron.

Silicon is the raw material for the silicate shells of diatoms (see Figure 2.5), which comprise most of the phytoplankton in some regions of the ocean. Silicon is lost from the surface waters when diatoms die and their dense shells sink to the ocean bottom. For example, the area west of southern South America, between 40° S and 50° S latitude, appears to have too little silicon, probably because silicon containing particles sink below the photic zone more rapidly than do nitrogen and phosphorus across this long stretch of the southern Pacific Ocean.

Iron is an important component in many metabolic pathways, but it is lost from the ocean's surface when it combines with phosphorus and precipitates. Rivers serve as a source of iron, which explains why low concentrations of iron appear to occur in regions of the ocean that are far from continents. In these areas, the only iron inputs occur in the form of windblown dust.

If iron limits the growth of phytoplankton in the ocean, then additions of iron should act as a fertilizer and cause large increases in primary productivity. In 1993, scientists conducted a large experiment in which they added 450 kg of iron—roughly the amount in an automobile—to over 64 km² of ocean off the west coast of South America. The treatment caused a 100-fold increase in the concentration of iron and within just a few days the concentration of phytoplankton tripled. Although this increase in phytoplankton was of a short duration, the experiment confirmed that a lack of iron limited the producers in this area of the ocean. In subsequent experiments conducted off the coast of Antarctica, researchers observed similar increases in productivity (Figure 20.12).

Fertilizing the ocean with iron has implications beyond confirming that iron is the limiting nutrient in these areas. If primary productivity in the ocean can be increased, it has the potential to draw down the amount of CO_2 in the water. Because CO_2 is exchanged between the water and the air, reducing CO_2 in the water would subsequently reduce the amount of CO_2 in the atmosphere. Since CO_2 is a greenhouse gas, some scientists have hypothesized that fertilizing the oceans with iron could be a way to counteract global warming.

Although adding iron increased primary production, in some cases the large increase in phytoplankton caused a subsequent large increase in the zooplankton populations that consume phytoplankton. When this happened, the zooplankton that was produced caused an increase in the amount of CO_2 due to increased respiration, thereby counteracting the beneficial effects of the

Figure 20.12 Fertilizing the oceans. In the Southern Ocean near Antarctica, researchers lower water-sampling equipment to measure the effects of adding iron to the surface waters. Photo © 2002 Kenneth Coale.

phytoplankton. In 2004, researchers conducted experiments that monitored the amount of carbon that precipitated out of the water column during iron additions and found that there was a small increase in the amount of exported carbon. Despite these somewhat promising results, the long-term effects of iron enrichment on CO_2 remain uncertain, as does the question of whether large-scale applications of iron in the ocean could have potential adverse effects on the ocean ecosystem.

The movement of energy depends on the efficiency of energy flow

As we have discussed, the organisms in each trophic level must obtain energy, assimilate some fraction of the energy they obtain, and then use some of the assimilated energy for respiration. Whatever remains goes to growth and reproduction. At each step, different organisms vary in the efficiency with which they obtain and retain energy. These efficiencies, combined with trophic interactions that affect the abundance of other trophic levels, affect the amount of energy and biomass that can be supported in a given trophic level. These factors also affect how many trophic levels can exist in an ecosystem. In this section, we will explore patterns in the amounts of energy and biomass that exist in different trophic levels. We will then examine the various efficiencies of energy transfer in trophic groups and how these differences affect the number of trophic levels found in an ecosystem.

TROPHIC PYRAMIDS
A useful way to think about the distribution of energy or biomass among the trophic groups in an ecosystem

Figure 20.13 **The energy pyramid of Cedar Bog Lake.** Most of the energy assimilated in this lake ecosystem is found in producers such as algae. Because energy is lost as it is converted from one trophic level to the next, considerably less energy is found in primary consumers such as zooplankton and secondary consumers such as fish. Data from R. L. Lindeman, The trophic-dynamic aspect of ecology, *Ecology* 23 (1942): 399–417.

is by drawing a **trophic pyramid**, which is a chart composed of stacked rectangles representing the relative amount of energy or biomass in each trophic group. The first person to consider the flow of energy between trophic levels was Raymond Lindeman, an ecologist who conducted his doctoral dissertation on Cedar Bog Lake in Minnesota during the 1930s. As Lindeman collected data on the energy and biomass produced by different trophic levels in the lake, he concluded that energy must be lost as it moves from one trophic level to the next. He demonstrated that this was the case when he created a trophic pyramid that displayed the percentage of the total energy existing at each trophic level, known as a **pyramid of energy**. As you can see from his data in Figure 20.13, each trophic level in the lake contained less energy than the trophic level below it.

Pyramids of energy are just one way of representing the distribution of organisms in an ecosystem. We can also create a trophic pyramid that represents the standing crop of organisms present in different trophic groups, which is known as a **pyramid of biomass**. In terrestrial ecosystems, the distribution of biomass among trophic levels looks quite similar to the pyramid of energy. As Figure 20.14a shows, the greatest amount of biomass occurs in producers, with less biomass in primary and secondary consumers. For example, consider the

biomass in a forest ecosystem. The greatest portion of the biomass occurs in producers, which include trees, shrubs, and wildflowers. There is considerably less biomass in primary consumers, which include herbivorous birds, mammals, and insects. There is even less biomass in secondary consumers, which include hawks, owls, and carnivorous mammals. A similar scenario exists in the grassland ecosystem of Africa. All the grass in Africa piled together would dwarf a mound of all the grasshoppers, gazelles, zebras, wildebeests, and other animals that eat grass. That mound of herbivores, in turn, would dwarf the relatively tiny mound of all the lions, hyenas, and other carnivores that feed on them.

In aquatic ecosystems, the pyramid of biomass has a very different shape. In these ecosystems, the major producers are the phytoplankton, the tiny algae that float or swim through the water. Unlike trees and shrubs, algae have short lives with rapid reproduction,

Trophic pyramid A chart composed of stacked rectangles representing the amount of energy or biomass in each trophic group.

Pyramid of energy A trophic pyramid that displays the total energy existing at each trophic level.

Pyramid of biomass A trophic pyramid that represents the standing crop of organisms present in different trophic groups.

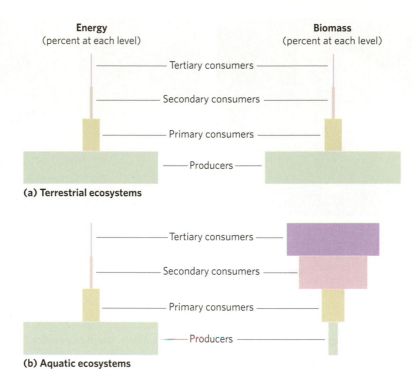

Figure 20.14 Pyramids of energy and biomass. (a) In terrestrial ecosystems, pyramids of energy and biomass have similar shapes because most of the energy and most of the standing biomass is found in the producers. **(b)** In aquatic ecosystems, most of the energy is still found in the producers, but these producers are primarily tiny algae that do not live very long because they are rapidly consumed by herbivores. This continual rapid consumption results in a large biomass of consumers in these systems.

and they are consumed in large numbers. As a result, although the productivity of algae is much higher than the consumers of algae, the standing biomass of algae is often much lower than the standing biomass of the consumers of algae. This creates an inverted pyramid, as illustrated in Figure 20.14b.

THE EFFICIENCIES OF ENERGY TRANSFERS
As we discussed, the amount of energy that moves from one trophic level to another determines how much energy or biomass can exist at each trophic level. The amount of energy that is transferred between adjacent trophic levels depends on several steps that occur within each trophic level and include consumption, assimilation, and production. To determine the amount of energy transfer, we need a way to quantify the movement of energy in each of these steps. When we consider the transfer of energy from one trophic level to another, we can examine either the absolute amount of energy that is transferred or the percentage of available energy that is transferred, which is a measure of efficiency.

Consumption Efficiency
Turning back to Figure 20.4, we see that the first step in the transfer of energy from one trophic level to another is the consumption of energy at the lower

trophic level. Some of the total amount of energy available in the lower trophic level is consumed and the rest is left to become dead organic matter. The percentage of energy or biomass in a trophic level that is consumed by the next higher trophic level is known as the **consumption efficiency**. Consumption efficiency is calculated using the following equation:

$$\text{Consumption efficiency} = \frac{\text{Consumed energy (J)}}{\text{Net production energy of the next lower trophic level (J)}}$$

For example, herbivores might only consume 1 J of energy for every 10 J of energy available in a field of wildflowers because many of the plant species possess defences against herbivores. In this case, the consumption efficiency of the herbivores would be 10 percent. In ecosystems containing producers with fewer defences against herbivores, the consumption efficiency would be much higher so that more energy would enter the consumer trophic level.

Consumption efficiency The percentage of energy or biomass in a trophic level that is consumed by the next higher trophic level.

(a)

(b)

Figure 20.15 Undigested energy. **(a)** Owls such as this barn owl (*Tyto alba*) consume small rodents but they do not digest the hair and bones. **(b)** Instead owls regurgitate a pellet that contains these indigestible materials. Consumers that eat large amounts of indigestible materials have low assimilation efficiencies. Photos by (a) Frederic Desmette/Biosphoto and (b) Philippe Clement/naturepl.com.

Assimilation Efficiency

As we have discussed, energy that is consumed is subsequently either assimilated or egested. In the case of plants, especially terrestrial plants, many components such as cellulose and lignin are not easily digested. Similarly, the feathers, bones, exoskeletons, and hair of animals are not easily digestible by the predators that consume them. Owls provide an excellent example. All species of owls are predators that commonly feed on small mammals. They swallow their prey whole, digest it, and then regurgitate the hair and small bones in the form of compact pellets (Figure 20.15).

The percentage of consumed energy that is assimilated is known as the **assimilation efficiency**. We can calculate assimilation efficiency as follows:

$$\text{Assimilation efficiency} = \frac{\text{Assimilated energy (J)}}{\text{Consumed energy (J)}}$$

Assimilation efficiency varies a great deal among trophic levels. For example, herbivores that feed on seeds, such as many species of birds and rodents, have assimilation efficiencies as high as 80 percent. In contrast, herbivores, such as horses that feed on grasses and woody vegetation, have assimilation efficiencies of 30 to 40 percent. Secondary consumers, which are predators on the herbivores, typically have high assimilation efficiencies, ranging from 60 to 90 percent. These high assimilation efficiencies occur because prey tissues are generally more digestible than plant tissues.

Net Production Efficiency

Ultimately, we want to know how much assimilated energy is converted into the growth and reproduction of the organisms in a trophic level. The percentage of assimilated energy that is used for growth and reproduction is the **net production efficiency**. Net production efficiency is calculated using the following equation:

$$\text{Net production efficiency} = \frac{\text{Net production energy (J)}}{\text{Assimilated energy (J)}}$$

Put another way, net production efficiency is the percentage of assimilated energy that remains after respiration. For active homeothermic animals that must spend a large fraction of their energy to maintain a constant body temperature, move about, circulate their blood, and balance their salts, net production efficiency can be as low as 1 percent. In contrast, sedentary poikilothermic animals, particularly aquatic species, channel as much as 75 percent of their assimilated energy into growth and reproduction.

Understanding net production efficiency also has practical applications. For example, livestock producers understand that if they keep their animals indoors at high densities during the winter, the animals will spend less of their assimilated energy on respiration to maintain a constant body temperature. If they are kept warm, the animals will devote more of their assimilated energy to growth, and faster growing livestock result in more income for the livestock producer.

Assimilation efficiency The percentage of consumed energy that is assimilated.

Net production efficiency The percentage of assimilated energy that is used for growth and reproduction.

Ecological Efficiency

Now that we understand the three efficiencies that occur within each trophic level, we can explore the efficiency of energy transfer between adjacent trophic levels. **Ecological efficiency**, also known as **food chain efficiency**, is the percentage of net production from one trophic level compared to the next lower trophic level.

$$\text{Ecological efficiency} = \frac{\text{Net production energy of a trophic level (J)}}{\text{Net production energy of the next lower trophic level (J)}}$$

When we consider all of the efficiencies in the chain of events that starts with consumption and ends with net production, we can see that each plays a role in determining the ecological efficiency between trophic levels. To confirm that this is true, we can multiply the efficiencies in each step of the process to produce the equation for ecological efficiency:

$$\text{Ecological efficiency} = \begin{array}{l} \text{Consumption energy (J)} \times \\ \text{Assimilation efficiency} \times \\ \text{Net production efficiency} \end{array}$$

$$\text{Ecological efficiency} = \frac{\text{Consumed energy (J)}}{\text{Net production energy of the next lower trophic level (J)}} \times$$

$$\frac{\text{Assimilated energy (J)}}{\text{Consumed energy (J)}} \times$$

$$\frac{\text{Net production energy (J)}}{\text{Assimilated energy (J)}}$$

$$\text{Ecological efficiency} = \frac{\text{Net production energy (J)}}{\text{Net production energy of the next lower trophic level (J)}}$$

Because energy is lost at each of these steps, ecological efficiencies are typically quite low and range from 5 to 20 percent. Given this range, ecologists often use 10 percent as a rule of thumb.

An ecological efficiency of 10 percent between adjacent trophic levels means that only 10 percent of the total energy present in the producer trophic level will be found in primary consumers, and only 1 percent will be found in secondary consumers. As you can see, an ecological efficiency of 10 percent makes it difficult to have long food chains in an ecosystem because there is not enough energy to support additional higher trophic levels. The only way to support additional trophic levels is by increasing the amount of absolute energy entering at the producer level or by increasing the ecological efficiencies moving between adjacent trophic levels.

A 10 percent ecological efficiency also has relevance to humans and where we feed along a food chain. For example, when we act as a primary consumer and eat plants, we can assimilate 10 percent of the energy available from plants. However, when we act as a secondary consumer and consume herbivores, we assimilate 1 percent of the original energy that was in the plants. As a result, we would expect a diet composed of more plants and less meat to dramatically increase the amount of food available to humans.

Ecological Efficiency and the Number of Trophic Levels

Aquatic ecosystems typically have more trophic levels than terrestrial ecosystems, in part because of differences in ecological efficiencies. In terrestrial ecosystems, producers are composed primarily of plants that range in size from wildflowers to trees. In many of these plants, a large proportion of tissue is dedicated to deterring herbivores from consuming them. Other plants contain an enormous proportion of biomass that herbivores cannot consume, such as the wood of trees. As a result, there is a low consumption efficiency in terrestrial ecosystems, so a large fraction of the producer biomass ultimately becomes detritus. In contrast, aquatic ecosystems are composed primarily of unicellular algae that contain fewer defences and are relatively easy for herbivores to digest. Therefore, algae provide higher consumption and assimilation efficiencies to their herbivores, which leads to higher ecological efficiencies. These higher ecological efficiencies mean that a higher fraction of the ecosystem's energy can move up the food chain and support additional trophic levels in aquatic ecosystems compared to terrestrial ecosystems.

In addition to having higher ecological efficiencies, aquatic ecosystems also contain herbivores that are generally quite small, such as the tiny zooplankton that consume algae. These zooplankton are consumed by secondary consumers such as small species of fish. These small fish are consumed by medium fish, which in turn are consumed by large fish. Because the producers are mostly single-cell algae, and because each successive consumer is only a bit larger than the diet it consumes, aquatic ecosystems can often contain five trophic levels.

Ecological efficiency The percentage of net production from one trophic level compared to the next lower trophic level. *Also known as* **Food chain efficiency.**

Quantifying Trophic Efficiencies

To help us understand how we calculate trophic efficiencies, we can use data collected from aquatic and terrestrial ecosystems to calculate consumption efficiency, assimilation efficiency, net production efficiency, and ecological efficiency.

ENERGY	TERRESTRIAL ECOSYSTEM	LAKE ECOSYSTEM	STREAM ECOSYSTEM
NET PRODUCTION AVAILABLE IN LOWER TROPHIC LEVEL (J)	1000	1000	400
CONSUMED ENERGY (J)	250	650	260
ASSIMILATED ENERGY (J)	120	450	100
NET PRODUCTION ENERGY (J)	50	190	40

We can calculate these efficiencies using the data from the terrestrial ecosystem.

$$\frac{\text{Consumption}}{\text{efficiency}} = \frac{\text{Consumed energy (J)}}{\text{Net production energy of the next lower trophic level (J)}}$$

$$\frac{\text{Consumption}}{\text{efficiency}} = \frac{250 \text{ J}}{1000 \text{ J}} = 25\%$$

$$\frac{\text{Assimilation}}{\text{efficiency}} = \frac{\text{Assimilated energy (J)}}{\text{Consumed energy (J)}}$$

$$\frac{\text{Assimilation}}{\text{efficiency}} = \frac{120 \text{ J}}{250 \text{ J}} = 48\%$$

$$\frac{\text{Net production}}{\text{efficiency}} = \frac{\text{Net production energy (J)}}{\text{Assimilated energy (J)}}$$

$$\frac{\text{Net production}}{\text{efficiency}} = \frac{50 \text{ J}}{120 \text{ J}} = 42\%$$

$$\frac{\text{Ecological}}{\text{efficiency}} = \frac{\text{Net production energy of a trophic level (J)}}{\text{Net production energy of the next lower trophic level (J)}}$$

$$\frac{\text{Ecological}}{\text{efficiency}} = \frac{50 \text{ J}}{1000 \text{ J}} = 5\%$$

YOUR TURN Use the data for the lake and stream ecosystems to calculate the four different efficiencies for each ecosystem.

Based on these calculations, why do the two aquatic ecosystems have higher ecological efficiencies than the terrestrial ecosystem? Which efficiencies cause the stream ecosystem to have a lower net production energy than the lake ecosystem?

We can contrast this with the situation that occurs in terrestrial ecosystems. In terrestrial ecosystems, the producers are relatively large, ranging in size from wildflowers to large trees. In addition, many of the major herbivores such as deer and antelope are quite large and can be consumed only by very large secondary consumers such as wolves and lions. Because terrestrial ecosystems contain many large producers and large herbivores, they are less likely to have a fourth and fifth trophic level. In short, the low ecological efficiency of terrestrial ecosystems combined with the large size of many producers and herbivores results in terrestrial ecosystems that commonly contain only three or four trophic levels.

RESIDENCE TIMES

Ecological efficiency tells us the proportion of energy that moves from producers into the higher trophic levels. We also want to examine the rate of energy movement between trophic levels, which tells us how long energy remains in a given trophic level and, therefore, how much energy can accumulate at that level. We can define the length of time that energy spends in a given trophic level as the **energy residence time**. Energy residence time is directly related to the amount of energy that exists in a trophic level; the longer the residence time, the greater the accumulation of energy in that trophic level. The average residence time of energy at a particular trophic level equals the energy present in the tissues of organisms divided by the rate at

Energy residence time The length of time that energy spends in a given trophic level.

which energy is converted into biomass, or net productivity:

$$\text{Energy residence time (years)} = \frac{\text{Energy present in a trophic level (J/m}^2)}{\text{Net productivity (J/m}^2/\text{year)}}$$

If we substitute biomass for energy in this equation, we can determine **biomass residence time**, which is the length of time that biomass spends in a given trophic level.

$$\text{Biomass residence time (years)} = \frac{\text{Biomass present in a trophic level (kg/m}^2)}{\text{Net productivity (kg/m}^2/\text{year)}}$$

For example, plants in humid tropical forests produce dry matter at an average rate of 1.8 kg/m^2/year and have an average living biomass of 42 kg/m^2. Inserting these values into the preceding equation gives a biomass residence time of 23 years. Average residence times for primary producers range from more than 20 years in forest ecosystems to less than 20 days in aquatic phytoplankton-based ecosystems. This much shorter biomass residence time in aquatic ecosystems is the reason these ecosystems frequently have inverted pyramids of biomass, like the one shown in Figure 20.14b; the biomass produced by algae is rapidly consumed by zooplankton.

These residence times track the movement of energy that is consumed from one trophic level to another by primary, secondary, and tertiary consumers, but these estimates do not take into account the residence time of the dead organic matter that is consumed by scavengers, detritivores, and decomposers. We calculate the residence time in dead organic matter using another variation of the equation for energy residence time:

$$\text{Dead organic matter residence time (years)} = \frac{\text{Dead organic matter present in a trophic level (kg/m}^2)}{\text{Dead organic matter productivity (kg/m}^2/\text{year)}}$$

For example, the residence time of dead leaf litter is 3 months in humid tropical ecosystems, 1 to 2 years in dry tropical ecosystems, 4 to 16 years in temperate forest ecosystems, and more than 100 years in temperate mountains and boreal ecosystems. As you can see from the preceding equation, these differences in the residence times of leaf litter are a function of how much litter falls each year and how rapidly decomposition can occur. In lowland tropical regions, warm temperatures and abundant moisture allow rapid decomposition of litter, while the colder and drier conditions of temperate and boreal ecosystems slow decomposition, which causes litter to accumulate.

STOICHIOMETRY

In addition to obtaining energy, organisms also must have the correct balance of nutrients to grow and reproduce. Ideally, the ratio of nutrients an organism needs must match the ratio of nutrients it consumes, but sometimes this is a challenge. The study of the balance of nutrients in ecological interactions, such as between an herbivore and a plant, is called **ecological stoichiometry**. Understanding ecological stoichiometry is useful in explaining variation in the ecological efficiencies that we have just discussed.

The balance of nutrients required by different species depends on their biology. For example, diatoms have high requirements for silicon because they produce glass shells for protection (see Figure 2.5). In contrast, vertebrates require large amounts of calcium and phosphorus for growing bones and scales. Fruit-eating birds and mammals often must supplement their diets with snail shells or bits of limestone to take in sufficient amounts of calcium and phosphorus. Growth rates and other life-history traits also can influence the nutrient composition of organisms. For instance, if we compare two types of zooplankton, we find that slowly growing marine copepods (Figure 20.16a) have nitrogen to phosphorus ratios as high as 50:1 whereas rapidly growing freshwater water fleas (Figure 20.16b) have ratios below 15:1. The faster growing water fleas have a lower ratio because they must maintain high concentrations of phosphorus in their tissues to synthesize the large amounts of proteins necessary for rapid growth.

To understand how stoichiometry affects the efficiencies of energy transfer between adjacent trophic groups, consider again the water flea that has tissues containing a 15:1 ratio of nitrogen and phosphorus. When the ratio of ingested nutrients does not match the ratio of needed nutrients, consumers must process larger amounts of food to obtain sufficient amounts of the most limiting nutrient. If the water flea consumed algae that contained a 30:1 ratio of nitrogen and phosphorus, it would have to consume twice as much algae to meet its phosphorus need. Moreover, when it consumed algae containing a 30:1 ratio, it would have to excrete

---

Biomass residence time The length of time that biomass spends in a given trophic level.

Ecological stoichiometry The study of the balance of nutrients in ecological interactions, such as between an herbivore and a plant.

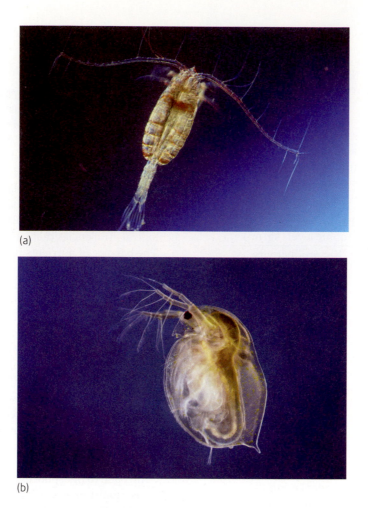

(a)

(b)

Figure 20.16 Effects of growth rate on stoichiometry. (a) Slow-growing organisms, such as this marine copepod, have tissues with a high ratio of nitrogen to phosphorus of approximately 50:1. **(b)** Fast-growing organisms, such as this water flea (*Daphnia magna*), require much more phosphorus so that they can produce nucleic acids. As a result, their tissues have a lower ratio of nitrogen to phosphorus that is approximately 15:1. Photos by (a) Image Quest Marine and (b) Laguna Design/Science Source.

the excess nitrogen. In this example, the assimilation efficiency of the water flea, in terms of nitrogen, declines and therefore the ecological efficiency of the trophic level declines. As you can see, tracking the ecological stoichiometry helps us understand why there can be a low ecological efficiency whenever a nutrient-poor producer is consumed by an herbivore that requires a nutrient-rich diet. In short, understanding ecological stoichiometry helps us explain why ecological efficiencies can vary between trophic levels.

In this chapter, we have explored how energy moves through ecosystems. We have seen that this process begins with producers capturing the energy of the Sun and converting it into producer biomass. The rate of primary production influences the rate of production for higher trophic levels. We have seen that the amount of primary productivity differs among ecosystems around the world because of variations in temperature, precipitation, light, and nutrients. The amount of energy that passes through the trophic levels of ecosystems depends on the efficiency of each step in the chain, with all of the steps producing an overall ecological efficiency between adjacent trophic levels and different residence times of energy, biomass, and dead organic matter.

ECOLOGY TODAY CONNECTING THE CONCEPTS

FEEDING AN OCEAN OF WHALES

Humpback whales. Researchers have determined the number of whales living off the West Coast of the United States, such as this humpback whale. When whale numbers were combined with data on the diet of each species and the ecological efficiencies between adjacent trophic levels, the researchers estimated that whales consume approximately 12 percent of the NPP of the ocean off the West Coast of the United States. Photo by Masa Ushioda/age fotostock.

The ocean contains a diversity of species feeding at different trophic levels, and whales sit at or near the top trophic level. Although whales are a top predator, it has been a long-standing challenge for us to understand how much NPP is required to provide the energy that whale populations need. Knowing the NPP requirement would tell us how much energy passes through this trophic level. It would also allow us to estimate how much NPP whales used prior to the commercial hunting that has caused many whale populations to decline during the past century. The reduction in consumed NPP could potentially be consumed by other species in the ecosystem.

Given that most whales live across large areas of the oceans, how does one begin to estimate the percentage of NPP that whales consume? A group of researchers decided to tackle this challenge by starting with information about the whales and working down to the producers. First, they settled on a particular study area: the region of ocean within 550 km from the western coast of the United States. Next, they needed to estimate how many whales lived in this

area. Fortunately, line transects had been conducted in this region of the ocean for 15 years and these transects provided abundance estimates for 21 species of whales that included dolphins, porpoises, sperm whales (*Physeter macrocephalus*), and humpback whales (*Megaptera novaeangliea*). By knowing the typical mass of an individual whale for each species and how much energy whales of different sizes consume, assuming an assimilation efficiency of 80 percent, the researchers could estimate the total energy consumed annually by each species of whale.

The next step was to determine the diet of each whale species. Each species has a unique diet of prey comprised of different proportions of primary consumers, such as krill, and secondary consumers, such as fish. Using existing data on whale diets, the researchers knew the proportion of primary and secondary consumers that each whale species eats. They then used the ecologist's rule of thumb that adjacent trophic levels passed along energy with an ecological efficiency of 10 percent. Based on this rule of thumb, they could determine how much NPP was required to produce the prey consumed by the whales.

The final step was to determine how much NPP was available in this region of the ocean. They measured NPP using satellites and the remote sensing techniques for detecting chlorophyll concentrations that we discussed earlier in the chapter. Once NPP was quantified, they determined that whales consumed about 12 percent of the ocean's NPP. For comparison, this is about half of the NPP required to support fish populations that are commercially harvested.

The percentage of NPP currently used by this group of mammals is likely to increase dramatically in the coming decades. For example, during the 1980s and early 1990s, many dolphins were accidentally killed by commercial fishing operations. Because this source of mortality has declined in recent years, we anticipate an increase in the number of some types of whales, such as dolphins. This may cause a higher proportion of the NPP to be consumed. Moreover, because many of the larger species of whales are now protected around the world, populations are rising. Scientists expect populations of some of the larger species to eventually triple or quadruple, which will result in an even greater consumption of the ecosystem's NPP. Although researchers do not yet know how this large increase in the biomass of top predators will affect the ocean's distribution of energy among the trophic levels, such an increase will undoubtedly be investigated in the decades to come.

SOURCE: Barlow, J., et al. 2008. Cetacean biomass, prey consumption, and primary production requirements in the California Current ecosystem. *Marine Biology Progress Series* 371: 285–295.

SUMMARY OF CHAPTER CONCEPTS

■ **Primary productivity provides energy to the ecosystem.** Primary productivity is the process of capturing solar or chemical energy and converting it into chemical bonds by photosynthesis or chemosynthesis over a given amount of time. We can distinguish between gross primary productivity, which is the total amount of energy assimilated, and net primary productivity, which is the assimilated energy that is converted into producer biomass. Net primary productivity can be measured in numerous ways including measuring plant biomass, measuring CO_2 uptake and release in terrestrial ecosystems, measuring O_2 uptake and release in aquatic ecosystems, and using remote sensing. Across all ecosystems, the amount of net primary productivity has a direct positive relationship with the amount of net secondary productivity.

■ **Net primary productivity differs among ecosystems.** Net primary productivity differs a great deal among ecosystems around the world. In terrestrial ecosystems, major drivers of this productivity include temperature, precipitation, nitrogen, and phosphorus. In aquatic ecosystems, major drivers include temperature, light, nitrogen, and phosphorus.

■ **The movement of energy depends on the efficiency of energy flow.** The energy of ecosystems exists in different trophic levels and moves between these trophic levels with different efficiencies. Pyramids of energy exhibit similar distributions among ecosystems, with producers having the most energy and each higher trophic group possessing less energy. Pyramids of biomass show a similar trend in terrestrial ecosystems, but they are often inverted in aquatic ecosystems. To understand how energy moves between trophic levels, we can calculate the consumption efficiency, assimilation efficiency, and net production efficiency, all of which can be multiplied to determine the overall ecological efficiency of transferring energy between adjacent trophic groups. The efficiencies can be affected by the stoichiometry of consumer tissues relative to the consumer's diet and can affect the number of links in an ecosystem's food chain. Efficiencies can also affect the residence time of energy and biomass in ecosystems.

REVIEW QUESTIONS

1. Why is the efficiency of energy transfer between two trophic levels generally quite low?

2. How do ecologists distinguish between gross primary productivity and net primary productivity?

3. Compare and contrast the measurement of primary productivity in terrestrial versus aquatic ecosystems.

4. What factors limit the net primary productivity of terrestrial ecosystems?

5. What factors limit the net primary productivity of aquatic ecosystems?

6. Why might assimilation efficiencies be much higher for herbivores eating seeds than for herbivores eating leaves?

7. Why do aquatic and terrestrial ecosystems differ in their pyramids of biomass?

8. How do changes in stoichiometry affect assimilation efficiencies?

9. Why are residence times much longer in forest ecosystems than in aquatic phytoplankton-based ecosystems?

10. How might the carrying capacity for humans on Earth change if the human population ate more plant products than animal products?

PRACTICING ECOLOGY: VIRUSES AND MARINE NUTRIENT CYCLING

Chris Suttle and his lab at the University of British Columbia study the ecological importance of viruses in the open ocean. Photo by Jerome Payet.

Viruses are microscopic organisms that occur in extremely high numbers. In marine environments viruses are present at abundances of tens of millions per millilitre and are integral members of the food web. At this abundance, viruses contain a significant portion of the nutrients found in marine ecosystems. They also increase the mortality rate of other species. Given these attributes, how do viruses alter the cycling of nutrients in the ocean?

While some viruses infect large organisms in marine environments, most infect bacteria in the photic zone. When a virus infects a bacterium, it replicates and eventually causes the cell wall to break open, which releases replicated viruses into the water. At the same time, nutrient-rich compounds such as proteins and nucleic acids are also released from the bacteria. These compounds are quickly absorbed by other bacteria and by organisms such as phytoplankton. Since bacteria account for 70 percent of all living carbon in the ocean, scientists think this process of infection, nutrient release, and reabsorption may be a significant factor in marine nutrient cycling. If so, we might hypothesize that the presence of viruses will increase the amount of nutrients released into the water relative to water without viruses.

Chris Suttle from the University of British Columbia and his colleagues tested this hypothesis by measuring nutrient release in water with and without viruses. They collected samples of surface water from False Creek in Vancouver, a moderately saline environment connected to English Bay. They filtered the water samples through a 0.2 μm filter, which is a mesh size that collects bacteria and phytoplankton but allows viruses to pass through. While the bacteria and phytoplankton were still alive, the researchers placed half the cells into water containing the viruses that passed through the filter. The remaining bacteria and phytoplankton were placed in sterile water containing no viruses. To measure the amount of nutrients released in the water, the researchers measured the concentration of nitrogen over 25 hours. You can view their data in the table.

NITROGEN CONCENTRATION (μmol/L) IN WATER WITH AND WITHOUT VIRUSES OVER 25 HOURS

TIME (HOURS)	WITH VIRUSES	WITHOUT VIRUSES
0	3.15	3.38
6	3.30	3.29
10	3.35	3.36
19	2.87	2.40
25	2.99	1.46

GRAPH THE DATA

1. Create a line graph showing time on the *x* axis and nitrogen concentration in the presence of viruses on the *y* axis.

2. Add a second line showing the changes in nitrogen concentration over time in the absence of viruses.

INTERPRET THE DATA

1. Explain how the data support or refute the hypothesis that the presence of viruses will increase the amount of nutrients released into the water relative to water without viruses.

2. Why might nitrogen concentrations stay constant in the water containing viruses, while nitrogen concentrations decrease in the water lacking viruses?

3. In the same study, the researchers found that the concentration of algal cells and chlorophyll *a* also increased in water containing viruses. Given that most bacteria in the photic zone need oxygen for respiration, how might repeated bacterial infection benefit the bacteria population of marine environments?

SOURCES: Shelford, E. J., M. Middleboe, E. F. Møller, and C. A. Suttle. 2012. Virus-driven nitrogen cycling enhances phytoplankton growth. *Aquatic Microbial Ecology* 66: 41–46.

Wilhelm, S. W., and C. A. Suttle 1999. Viruses and nutrient cycles in the sea. *BioScience* 49: 781–788.

CHAPTER 21

Movement of Elements in Ecosystems

Living in a Dead Zone

Each summer, as the Mississippi River flows into the Gulf of Mexico, an area develops where animals can't survive. While fish, crawfish, and crabs remain abundant in other parts of the Gulf, the summer algal bloom in this area makes it uninhabitable. In many cases, the rapidly increasing populations of algae contain green pigments that turn the water green. When the algae contain red pigments, the algal bloom is called a red tide.

Algal blooms can have both direct and indirect effects on aquatic organisms. A direct effect occurs when the species of algae or cyanobacteria that bloom produce toxins. At high algal densities, these toxins can accumulate to concentrations that impair the survival, growth, and reproduction of other species living in the area.

> "The resulting dead zone in the Gulf of Mexico can cover more than 22 000 km² during the summer— an area about four times the size of Prince Edward Island."

An indirect effect of algal blooms occurs after the algae, also known as phytoplankton, bloom and die. While live algae produce oxygen during photosynthesis, zooplankton and bacteria that consume the enormous biomass of dead algae use large amounts of oxygen. This leads to a dramatic reduction in oxygen in the water and causes many of the animals in the water to die from oxygen deprivation. Aquatic ecosystems that experience algal blooms and large animal die offs are called dead zones.

What causes the large algal blooms? Researchers have discovered that many rivers carry large amounts of nutrients such as nitrogen and phosphorus that come from fertilizers that run off lawns and agricultural fields when it rains. This water runoff enters streams and rivers that join before emptying into the ocean. Other sources of nutrients include wastewater with a range of components including detergents that contain phosphorus and sewage that is released from wastewater treatment systems when they are overwhelmed by large rain events. The nutrients that rivers dump into the ocean allow rapid algal growth, which causes an algal bloom.

The Mississippi River drains 41 percent of the contiguous United States, so it carries nutrients gathered from a very large area. The resulting dead zone in the Gulf of Mexico can cover more than 22 000 km² during the summer—an area about four times the size of Prince Edward Island. As autumn weather comes, fewer nutrients enter the Mississippi River and the temperatures in the Gulf of Mexico become cooler, conditions that make it more difficult for algal populations to increase. As a result, the dead zone disappears each winter.

Human activities cause most dead zones, although some do have natural causes. The abundance of dead zones around the world is growing rapidly. In the 1910s, only 4 dead zones were known to exist. This number increased to 49 dead zones in the 1960s and 87 in the 1980s. The number of dead zones

The effects of a dead zone on fish populations. Because algae that bloom will eventually die, the decomposition uses up nearly all of the oxygen in the water. These fish died from a dead zone that occurred in Lake Trafford, Florida. Photo by Steven David Miller/naturepl.com.

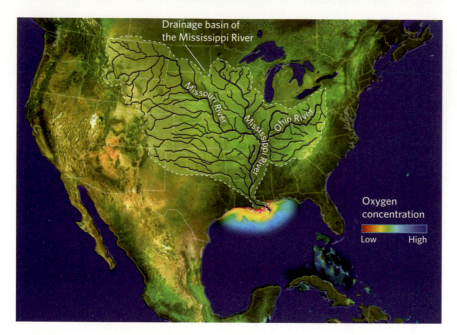

Drainage basin of the Mississippi River

Missouri River

Mississippi River

Ohio River

Oxygen concentration

Low   High

A dead zone at the mouth of the Mississippi River. The Mississippi River drains 41 percent of the contiguous United States and carries nutrients from fertilizers leached from people's yards and agricultural fields and from wastewater from communities located throughout the region. These nutrients facilitate the rapid growth of algae that ultimately cause a dead zone, which you can see at the mouth of the river. Dead zone after NOAA.

rose to 305 in 1995 and increased further to 405 by 2008, the most recent date for which there is an estimate. For example, a dead zone occurs in the lower St. Lawrence estuary in Quebec, where waters deeper than 275 metres can become *hypoxic*, meaning the water is low in oxygen. Similarly, the bottom of Lake Erie becomes hypoxic each summer. Around the world, dead zones cover a total area of more than 205 000 km^2.

The existence of dead zones illustrates why we need to understand how nutrients—including water, nitrogen,

and phosphorus—move within and between ecosystems and the important role that decomposition plays in recycling these nutrients. In this chapter, we will examine the movement of nutrients and how nutrients are regenerated in terrestrial and aquatic ecosystems.

SOURCES: N. N. Rabalais, R. E. Turner, and W. J. Wiseman, Jr., Gulf of Mexico hypoxia, a.k.a. The dead zone, *Annual Review of Ecology and Systematics* 33 (2002): 235–263.

R. J. Diaz and R. Rosenberg, Spreading dead zones and consequences for marine ecosystems, *Science* 321 (2008): 926–929.

CHAPTER CONCEPTS

- The hydrologic cycle moves many elements through ecosystems.

- The carbon cycle is closely tied to the movement of energy.

- Nitrogen cycles through ecosystems in many different forms.

- The phosphorus cycle moves between land and water.

- In terrestrial ecosystems, most nutrients regenerate in the soil.

- In aquatic ecosystems, most nutrients regenerate in the sediments.

Unlike energy—which moves through ecosystems—elements such as hydrogen, oxygen, carbon, nitrogen, and phosphorus cycle among the biotic and abiotic components of ecosystems. The movements of these elements within ecosystems are affected by chemical, physical, and biological processes. To understand the movement of these elements, which exist in a variety of chemical forms, it is helpful to think of different pools in which a given element resides, as well as the different processes that are responsible for moving an element from one pool to another. For example, two important pools for carbon are the CO_2 that exists in the atmosphere and the biomass of producers that use carbon to build their tissues. In this example, the

process that causes carbon to move from the atmosphere to producers is photosynthesis.

Organisms contain large amounts of hydrogen, oxygen, and carbon. However, as we noted in Chapter 2, organisms also need seven major nutrients: nitrogen, phosphorus, sulphur, potassium, calcium, magnesium, and iron. Some elements are required in much smaller amounts; these include silicon, manganese, and zinc. In this chapter, we will examine the biogeochemical cycles of some of the major elements on Earth by looking at interactions among biological, geological, and chemical processes. We will also explore how human activities are currently altering these cycles in ways that have far-reaching effects on ecosystems.

The hydrologic cycle moves many elements through ecosystems

Throughout the previous chapter, we saw that water plays a critical role at every level of ecological study. In Chapter 2 we discussed how water is a key compound involved in the many chemical transformations that happen in the living and nonliving components of ecosystems. Chapter 5 examined the processes that determine global climates, including patterns of precipitation, and in Chapter 6 we discussed the various types of aquatic biomes. We begin this chapter by tracking the movement of water through ecosystems. Once we understand how water moves through ecosystems, we will look at the role water plays as it moves elements through ecosystems.

THE HYDROLOGIC CYCLE

The movement of water through ecosystems and atmosphere, known as the **hydrologic cycle**, is driven largely by evaporation, transpiration, and precipitation. The largest pool of water, approximately 97 percent of all water on Earth, is found in the oceans. The remaining water exists in polar ice caps and glaciers, lakes, streams, rivers, wetlands, underground aquifers, and soil.

The hydrologic cycle is illustrated in Figure 21.1. Evaporation of water occurs from bodies of water, soil, and plants that experience evapotranspiration, which we discussed in Chapter 5. Solar energy provides the energy for the process of evaporation and evapotranspiration, which changes water from a liquid to a gas in the form of water vapour. There is a limit to the amount of water vapour that the atmosphere can contain. As additional water continues to evaporate, the water vapour in the atmosphere condenses into clouds that ultimately create precipitation in the form of rain, hail, sleet, or snow.

When precipitation drops from the atmosphere, it can take several paths. Some precipitation falls directly onto the surface of aquatic ecosystems and the rest falls onto terrestrial ecosystems. Water that falls onto terrestrial ecosystems can travel along the surface of the ground or it can infiltrate the ground where it is either absorbed by plants or moves deeper into the ground and becomes part of the underlying groundwater. The surface runoff and some of the groundwater will eventually find their way back into water bodies, thereby completing the cycle.

Hydrologic cycle The movement of water through ecosystems and atmosphere.

Solar energy heats Earth and causes evaporation.

Water evaporated from aquatic and terrestrial habitats combines with water from plant evapotranspiration to condense into clouds.

Water returns to Earth as precipitation.

Precipitation that falls on land either runs off along the surface or infiltrates the soil. In the soil, this water may evaporate, be taken up by plants, or enter groundwater. Excess water ultimately returns to the ocean.

Figure 21.1 The hydrologic cycle. The movement of water is driven by the energy of the Sun, which causes evaporation from soil and water bodies, and evapotranspiration from plants. Evaporated water condenses into clouds that eventually return the water to Earth as precipitation. Water from precipitation either runs off on the surface or infiltrates the soil. Runoff flows along the surface of the ground until it enters streams and rivers. Water in the soil is taken up by plants or enters the groundwater. Ultimately, excess water returns to the ocean.

The amount of evaporation must balance the amount of precipitation or all of the water would continually accumulate in one part of the cycle. When we consider the hydrologic cycle on a global scale, we find that precipitation exceeds evaporation in terrestrial ecosystems, whereas evaporation exceeds precipitation in aquatic ecosystems. To help maintain an overall balance, the excess water that is evaporated from aquatic ecosystems is transported by the atmosphere and falls onto terrestrial ecosystems. At the same time, the excess water that falls on terrestrial ecosystems is transported in the form of runoff and groundwater into aquatic ecosystems.

HUMAN IMPACTS ON THE HYDROLOGIC CYCLE

Water cycles through Earth's biosphere with no net loss or gain over the long term. Therefore any change in one part of the water cycle influences the other parts. For example, in large developed areas, construction materials such as roofing and paved parking lots are impervious to water infiltration. The amount of water that can percolate into the soil is significantly reduced, and we see an increase in surface runoff. Less water is able to infiltrate the soil for plants to use or to replenish the groundwater that many people rely on for drinking water. An increase in surface runoff also increases soil erosion. A similar effect occurs when we reduce the amount of plant biomass in a terrestrial ecosystem, as occurs during logging. Where there are fewer trees and other plants, much less precipitation is taken up by plant roots and subsequently released to the atmosphere through evapotranspiration. Consequently, the amount of surface runoff increases, which often causes severe soil erosion and flooding (Figure 21.2). Finally, when we pump out groundwater for irrigation or household use, we sometimes reduce the amount of groundwater at a rate that exceeds its replenishment. For example, while lakes and rivers supply most Canadian households with their water, about 30 percent of the Canadian population relies on groundwater for domestic use. Agriculture and industry also withdraw groundwater, which makes up between 27 percent and 75 percent of provincial groundwater consumption. In 2006, Natural Resources Canada initiated a study to assess the size and usage of 30 aquifers across the country. After having completed the assessment of 12 aquifers, the ministry indicated that increasing water extraction by growing populations and expanding agriculture and industry threatens to exceed the rate of replenishment in 9 of them.

Figure 21.2 Altering the hydrologic cycle. When forests are logged, as in this site in Haiti, fewer plant roots are available to hold soil and the soil absorbs less rainwater. These changes cause increased surface runoff, more severe floods, and large amounts of soil erosion. Photo by REUTERS/Daniel Morel/Landov.

Humans also alter the hydrologic cycle through activities that contribute to global warming. Scientists expect that as air and water temperatures rise, there will be an increase in the rate of water evaporation. An increase in the evaporation rate will cause water to move through the hydrologic cycle more quickly, potentially leading to an increased intensity of rain and snowstorms in various parts of the world.

The carbon cycle is closely tied to the movement of energy

Because all organisms are composed of carbon, the movement of carbon in ecosystems largely follows the same paths as the movement of energy. In this section, we consider the many pools and processes that are involved in the carbon cycle. We will then examine how human activities have altered the carbon cycle.

THE CARBON CYCLE

To understand how the carbon cycle works, we need to consider six types of transformations: photosynthesis, respiration, sedimentation and burial, exchange, extraction, and combustion. The carbon cycle is illustrated in Figure 21.3.

We begin our examination with the processes of photosynthesis and respiration. As we have discussed in previous chapters, producers use photosynthesis in terrestrial and aquatic ecosystems to take CO_2 from the air and water and to convert it to carbohydrates. These carbohydrates are used to make other compounds including proteins and fats. The carbon that is locked up in producers can then be transferred to consumers,

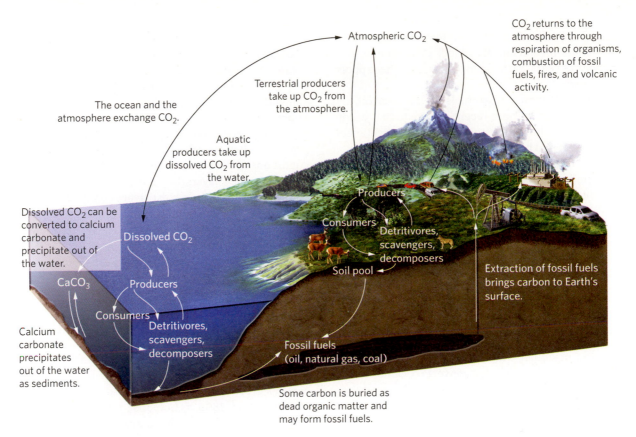

Figure 21.3 The carbon cycle. In the carbon cycle, producers take up CO_2 from the atmosphere and the water. They transfer assimilated carbon to consumers, detritivores, scavengers, and decomposers. These organisms return CO_2 to the atmosphere and oceans through respiration. Throughout an ecosystem, CO_2 is exchanged between the atmosphere and the ocean and between the ocean and sediments. Carbon that has been stored underground for long periods turns into fossil fuels, which can be extracted. CO_2 is returned to the atmosphere through combustion of fossil fuels, burning in terrestrial ecosystems, and volcanic activity.

scavengers, detritivores, and decomposers. All of these trophic groups experience respiration, which releases CO_2 back into the air or water.

In some habitats, such as the waterlogged sediments of swamps or marshes, oxygen is not available to serve as a terminal electron acceptor for respiration. Under such anaerobic conditions, some species of archaea use carbon compounds to accept the electrons. For example, some archaea use methanol (CH_3OH) during respiration to produce CO_2 in the following reaction:

$$4 CH_3OH \rightarrow CO_2 + 2 H_2O + 3 CH_4$$

As you can see from this reaction, the products are carbon dioxide, water, and methane. The methane that is released from swamps during anaerobic respiration is known as swamp gas. The production of methane through the process of anaerobic respiration is a growing concern because methane is a greenhouse gas and, on a per-molecule basis, it is 72 times more effective at absorbing and radiating infrared radiation back to Earth than CO_2.

Carbon dioxide is also exchanged between aquatic ecosystems and the atmosphere, as shown in Figure 21.3. The exchange occurs in both directions at a similar magnitude, which means there is little net transfer over time. As we discussed in Chapter 2, when CO_2 diffuses from the atmosphere into the ocean, some of it is used by plants and algae for photosynthesis and some is converted to carbonate (CO_3^{2-}) and bicarbonate ions (HCO_3^-). The carbonate ions can then combine with calcium in the water to form calcium carbonate ($CaCO_3$). Calcium carbonate has a low solubility in water, so it precipitates out of the water and becomes part of the sediments at the bottom of the ocean. Over millions of years, the calcium carbonate sediments that accumulate in the ocean bottoms, combined with the calcium carbonate skeletons from tiny marine organisms, can develop into massive sources of carbon in the form of rock known as dolomite and limestone. Humans mine dolomite and limestone for use in making concrete and fertilizer as well as for numerous industrial processes.

Carbon can also be buried as organic matter before it fully decomposes. Over millions of years, some of this organic matter is converted to fossil fuels such as oil, gas, and coal. The rate of carbon burial is slow and it is offset by the rate of carbon released into the atmosphere by the weathering of limestone rock and during volcanic eruptions. Because the process of sedimentation and burial can lock up carbon for millions of years, carbon moves through these pools very slowly.

The extraction of fossil fuels such as coal, oil, and natural gas represents a recent change to the carbon cycle. Most of human history saw little extraction of carbon that had been buried for millions of years. During the past 2 centuries, however, humans have extracted carbon at an increasing rate to meet growing energy demands. The extraction of fossil fuels moves fossil carbon from underground to Earth's surface, and the combustion of these fossil fuels alters the carbon cycle significantly.

Combustion of carbon sources produces CO_2 that goes into the atmosphere. Some combustion is natural, such as fires occurring in forests and grasslands. Other combustion is caused by humans, such as burning land to prepare it for agriculture and burning fossil fuels to generate electricity and heat or to power machinery. Like respiration and decomposition, combustion converts organic compounds to CO_2.

HUMAN IMPACTS ON THE CARBON CYCLE

Now that we understand the pools and processes in the carbon cycle, we can examine how human activities have altered it. In Chapter 4 we discussed the recent rise in atmospheric CO_2 seen in measurements made atop Mauna Loa on the island of Hawaii. These measurements have documented an increase in CO_2 from 316 ppm in 1958 to 394 ppm in 2012—a 25 percent increase in just 54 years.

Although these measurements at Mauna Loa began only in 1958, human activities have affected CO_2 concentrations in the atmosphere for much longer. To measure the concentrations of CO_2 that

were in the atmosphere hundreds of thousands of years ago, researchers have travelled to some of the coldest places on Earth. In locations such as Ellesmere Island in the Arctic and Antarctica, snowfall slowly compresses into ice that has tiny bubbles of air trapped inside. Because these bubbles represent tiny samples of the air from thousands of years ago, they can tell us about the climate conditions in the distant past. Each year, ice is formed by adding a new layer; the surface layers contain the youngest ice and the deepest layers contain the oldest ice. To sample the air that is trapped in the ice, researchers drill far down into the ice and extract long cylinders of ice known as ice cores (Figure 21.4).

Ice cores represent ice that has been formed as far back as 500 000 years ago. After determining the age of different ice core layers, researchers melt each layer, which allows the release of trapped air bubbles so the concentration of CO_2 can be measured. You can view some of the data from these ice cores in Figure 21.5. As you can see, CO_2 concentrations in the atmosphere during the past 400 000 years have varied a great deal, from about 180 to 300 ppm. However, since 1800, as humans increasingly burned fossil fuels, CO_2 concentrations have increased exponentially to the current value of 394 ppm. This means that the current concentration of CO_2 in our atmosphere is 31 percent higher than the highest concentrations that existed during the past 400 000 years.

The rise in atmospheric CO_2 is of great importance to humans because CO_2 is a greenhouse gas that absorbs infrared radiation and radiates some of it back to Earth. Having CO_2 in the atmosphere helps keep our planet warm, but an excessive amount of CO_2 and other greenhouse gases in the atmosphere could cause our planet to become much warmer than

(a)

(b)

Figure 21.4 Ice cores. (a) Researchers from the Catlin Arctic Survey 2010 drill far down into ancient ice to collect ice cores that have been created in layers over the past 500 000 years. **(b)** Ice cores contain air bubbles that were trapped at the time the ice was formed. By melting each layer, researchers can determine the concentration of CO_2 that was in the atmosphere over the past 500 000 years. Photos by (a) Martin Hartley/eyevine/Redux; and (b) George Steinmetz/Corbis.

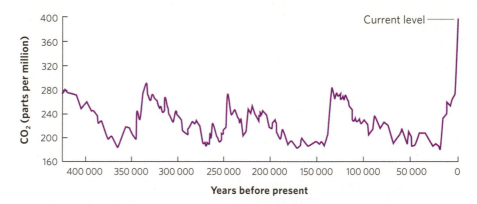

Figure 21.5 Atmospheric concentrations of CO$_2$ over time. Using measurements of air bubbles from ice cores and, in modern times, direct measurements, researchers have documented that for more than 400 000 years CO$_2$ concentrations never exceeded 300 ppm. During the past 200 years the concentration of CO$_2$ in the atmosphere has increased rapidly and is now approaching 400 ppm. Data from http://climate.nasa.gov/evidence.

it has been in a very long time. We know that the mean temperature on Earth is now 0.8°C warmer than it was when the first temperature measurements were taken in the 1880s. While a mean increase of 0.8°C may not seem like much, we find some dramatic changes in specific locations. Some regions, such as parts of Antarctica, have experienced cooler temperatures while other regions, such as the high latitudes of Alaska, Canada, and Russia, are now 4°C warmer than they were a century ago.

These high-latitude regions contain large deposits of frozen peat, which is a mixture of dead sphagnum moss and other plants. Peat thaws and decomposes more easily with warmer temperatures. Because peat decomposes under anaerobic conditions, the decomposition produces methane, which is a greenhouse gas. This means that the rise in temperatures due to increased atmospheric CO$_2$ causes the release of additional greenhouse gases from the decomposing peat that exacerbate the problem. Temperature increases can have numerous effects around the world, such as reducing the size of the polar ice sheets (see Chapter 5), altering the length of plant growing seasons, and changing the timing of plant and animal life histories. We will have much more to say about global warming in our discussion of global biodiversity conservation in Chapter 23.

Nitrogen cycles through ecosystems in many different forms

Nitrogen is an important component of amino acids, the building blocks of proteins, and nucleic acids, which are the building blocks of DNA. Nitrogen exists in many different forms and exhibits a complex set of pathways. In this section, we will examine the cycling of nitrogen and explore the ways that human activities have altered the nitrogen cycle.

THE NITROGEN CYCLE

A large pool of nitrogen gas (N$_2$) exists in the atmosphere, where it comprises 78 percent of all atmospheric gases. Nitrogen moves through five major transformations, shown in Figure 21.6: *nitrogen fixation, nitrification, assimilation, mineralization,* and *denitrification.*

Nitrogen Fixation

The process of converting atmospheric nitrogen into forms producers can use is known as **nitrogen fixation**. Nitrogen fixation converts nitrogen gas either into ammonia (NH$_3$), which is rapidly converted to ammonium (NH$_4^+$), or into nitrate (NO$_3^-$). The compound that is formed depends on whether nitrogen fixation occurs by organisms, lightning, or the industrial production of fertilizers.

As we discussed in previous chapters, some organisms are able to convert nitrogen gas into ammonia. Nitrogen fixation occurs in some species of cyanobacteria, in some free-living species of bacteria such as *Azotobacter,* and in mutualistic bacteria such as *Rhizobium* that live in the root nodules of some legumes and other plants (see Figure 17.4). Nitrogen fixation is an important source of required nitrogen, especially for early succession plants colonizing habitats that have little available nitrogen. The process of nitrogen fixation requires a relatively high amount of energy, which nitrogen-fixing organisms obtain either by metabolizing organic matter from the environment or by acquiring carbohydrates from a mutualistic partner.

As you can see in Figure 21.6, nitrogen fixation can also occur through abiotic processes. For example, lightning provides a high amount of energy that can convert nitrogen gas into nitrate in the atmosphere. Similarly, combustion that occurs during wildfires or when fuels are burned also produces nitrates. In both

Nitrogen fixation The process of converting atmospheric nitrogen into forms producers can use.

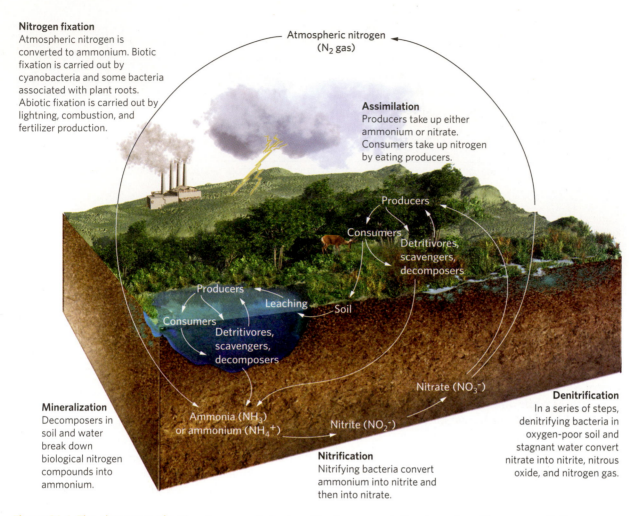

Nitrogen fixation
Atmospheric nitrogen is converted to ammonium. Biotic fixation is carried out by cyanobacteria and some bacteria associated with plant roots. Abiotic fixation is carried out by lightning, combustion, and fertilizer production.

Atmospheric nitrogen
(N_2 gas)

Assimilation
Producers take up either ammonium or nitrate. Consumers take up nitrogen by eating producers.

Producers

Consumers

Detritivores, scavengers, decomposers

Producers

Consumers

Leaching   Soil

Detritivores, scavengers, decomposers

Nitrate (NO_3^-)

Denitrification
In a series of steps, denitrifying bacteria in oxygen-poor soil and stagnant water convert nitrate into nitrite, nitrous oxide, and nitrogen gas.

Mineralization
Decomposers in soil and water break down biological nitrogen compounds into ammonium.

Ammonia (NH_3) or ammonium (NH_4^+)

Nitrite (NO_2^-)

Nitrification
Nitrifying bacteria convert ammonium into nitrite and then into nitrate.

Figure 21.6 The nitrogen cycle. The nitrogen cycle begins with nitrogen gas in the atmosphere. The process of nitrogen fixation converts it into a form that producers can use. The fixed nitrogen can then be assimilated into producers and consumers; it ultimately decomposes into ammonium through the process of mineralization. The ammonium can be converted into nitrite and then nitrate through the process of nitrification. Under anaerobic conditions, the nitrate can be converted into nitrogen gas through the process of denitrification.

cases, nitrates, which are suspended in the air after combustion, fall to the ground with precipitation.

The industrial production of fertilizers that improve crop productivity converts nitrogen gas to either ammonia or nitrates. Like all nitrogen fixation, this process requires a great deal of energy and is powered mostly by the combustion of fossil fuels. The manufacture of nitrogen fertilizer has developed into such a large commercial endeavour that fixation conducted by humans now exceeds the nitrogen fixation that occurs through all natural processes.

Nitrification

Another process in the nitrogen cycle is **nitrification**, which converts ammonium to nitrite (NO_2^-) and then from nitrite to nitrate (NO_3^-):

$$NH_4^+ \rightarrow NO_2^- \rightarrow NO_3^-$$

These conversions release much of the potential energy that is contained in ammonium. Each step is carried out by specialized bacteria and archaea in the presence of oxygen. The conversion of ammonium to nitrites in terrestrial and aquatic ecosystems is carried out by *Nitrosomonas* and *Nitrosococcus* bacteria whereas the conversion of nitrites to nitrates is carried out by *Nitrobacter* and *Nitrococcus* bacteria. Although nitrites are not an important nutrient for producers, they can take them up and use them.

Assimilation and Mineralization

Producers can take up nitrogen from the soil or water as either ammonium or nitrates. Once producers take up nitrogen, they incorporate it into their tissues, a process known as assimilation, which we described in Chapter 20. When primary consumers ingest producers, they can either assimilate nitrogen from the producers or excrete it as waste. The same process

Nitrification The final process in the nitrogen cycle, which converts ammonium to nitrite (NO_2) and then from nitrite to nitrate (NO_3^-).

occurs again with secondary consumers. Animal waste, as well as the biomass of producers and consumers that eventually die, is broken down by scavengers, detritivores, and decomposers. The fungal and bacterial decomposers break down biological nitrogen compounds into ammonia. The process of breaking down organic compounds into inorganic compounds is known as **mineralization**.

Denitrification

Because nitrates produced by nitrification are quite soluble in water, they readily leach out of soils and into waterways where they settle in the sediments of wetlands, rivers, lakes, and oceans. These sediments are typically anaerobic. Under anaerobic conditions, nitrates can be transformed back into nitrites, which are then transformed into nitric oxide (NO):

$$NO_3^- \rightarrow NO_2^- \rightarrow NO$$

This reaction is accomplished by bacteria such as *Pseudomonas denitrificans*. Additional chemical reactions under anaerobic conditions in soils and water subsequently convert nitric oxide to nitrogen gas, thereby completing the nitrogen cycle:

$$NO \rightarrow N_2O \rightarrow N_2$$

This process of converting nitrates into nitrogen gas is known as **denitrification**.

Denitrification is necessary for breaking down organic matter in oxygen-depleted soils and sediments. However, as you can see from the above reaction, it produces nitrogen gas (N_2). Since N_2 cannot be taken up by producers, the denitrification process causes nitrogen to leave the waterlogged soils and aquatic ecosystems in the form of a gas.

HUMAN IMPACTS ON THE NITROGEN CYCLE

Before human activities began dramatically altering the environment, the production of usable forms of nitrogen through the process of fixation was approximately offset, on a global scale, by the loss of usable nitrogen through denitrification. However, during the last 3 centuries, and especially during the past 50 years, human activities have nearly doubled the amount of nitrogen put into terrestrial ecosystems. These activities include combustion of fossil fuels that add nitric oxide to the air, production of nitrogen fertilizers, and planting nitrogen-fixing crops.

Burning fossil fuels puts nitric oxide into the air, where it is carried to ecosystems far away. As nitric oxide enters the atmosphere, it reacts with water in the air to form nitrates, which then fall to the ground during precipitation. Because nitrogen is often a limiting nutrient, we would expect the addition of nitrates to affect a variety of ecosystems.

Over the years, multiple teams of researchers have investigated whether adding nitrogen to terrestrial ecosystems in North America affects productivity and species richness. These study sites, which range north to south from Alaska to Arizona and west to east from California to Michigan, were recently compiled in an effort to determine whether their results showed a general pattern. When nitrogen was added in the form of nitrates and ammonium, all sites experienced an increase in primary productivity, shown in Figure 21.7a. This confirmed that nitrogen was a limiting resource at all sites. However, the sites differed in the proportion of species that were eliminated over time, as you can see in Figure 21.7b. The researchers explored a wide range of potential causes for this variation in species loss among sites. In addition to differences in temperature and soil among

---

Mineralization The process of breaking down organic compounds into inorganic compounds.

Denitrification The process of converting nitrates into nitrogen gas.

(a)

(b)

Figure 21.7 Effects of nitrogen addition on plant productivity and species richness. Twenty-three experiments at eight locations around the United States added nitrogen to plant communities. **(a)** All sites experienced increases in primary productivity. **(b)** All sites experienced a loss in species richness, but the magnitude of the loss varied greatly due to differences in existing nitrogen and differences in the cation exchange capacity of the soil. Error bars are standard errors. Data from C. M. Clark et al., Environmental and plant community determinants of species loss following nitrogen enrichment, *Ecology Letters* 10 (2007): 596–607.

the sites, they found that the sites with the largest increases in productivity experienced the largest reduction in species richness. Adding nitrogen to these communities commonly caused a few plant species to grow very large and to dominate the community. These large plants shaded the less-competitive smaller plants, which caused the smaller species to decline. These results demonstrate that the increases in nitrogen in the environment due to human activities can reduce the species diversity of ecosystems.

The phosphorus cycle moves between land and water

Phosphorus is a critical element for organisms because it is used in bones and scales, teeth, DNA, RNA, and ATP, a molecule involved in metabolism. As we discussed in Chapter 20, phosphorus is also a common limiting nutrient in aquatic and terrestrial ecosystems. For this reason, phosphorus is a component of most fertilizers manufactured to boost the growth of crops. In this section, we will examine the phosphorus cycle and explore how human activities have altered it in ways that affect ecosystems.

THE PHOSPHORUS CYCLE

The phosphorus cycle is considerably less complicated than the nitrogen cycle. As you can see in Figure 21.8, the atmosphere is not an important component of this cycle because phosphorus does not have a gas phase; phosphorus can only enter the atmosphere in the form of dust. Unlike nitrogen, phosphorus rarely changes its chemical form; it commonly moves as a phosphate ion (PO_4^{3-}). Plants take up phosphate ions from soil or water and incorporate them directly into various organic compounds. Animals eliminate excess phosphorus in their diets by excreting urine containing either phosphate ions or phosphorus compounds that are converted into phosphate ions by phosphatizing bacteria.

We begin our exploration of the phosphorus cycle by examining phosphate rocks, which are the major source of phosphate. As you can see in Figure 21.8, over time calcium phosphate ($Ca(H_2PO_4)_2$) precipitates out of ocean water and slowly forms sedimentary rock.

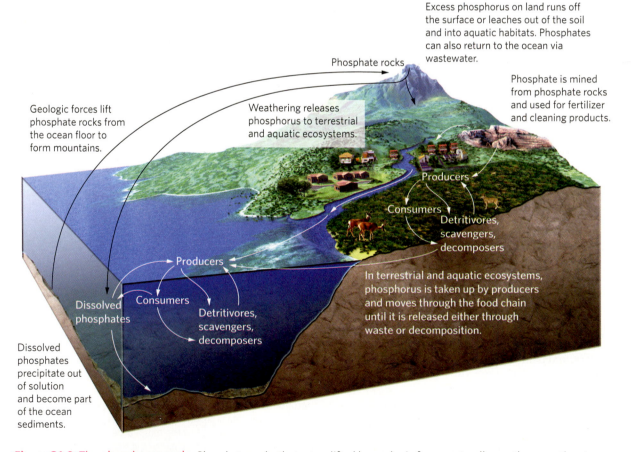

Excess phosphorus on land runs off the surface or leaches out of the soil and into aquatic habitats. Phosphates can also return to the ocean via wastewater.

Phosphate rocks

Phosphate is mined from phosphate rocks and used for fertilizer and cleaning products.

Geologic forces lift phosphate rocks from the ocean floor to form mountains.

Weathering releases phosphorus to terrestrial and aquatic ecosystems.

Producers

Consumers

Detritivores, scavengers, decomposers

In terrestrial and aquatic ecosystems, phosphorus is taken up by producers and moves through the food chain until it is released either through waste or decomposition.

Producers

Consumers

Detritivores, scavengers, decomposers

Dissolved phosphates

Dissolved phosphates precipitate out of solution and become part of the ocean sediments.

Figure 21.8 The phosphorus cycle. Phosphate rocks that are uplifted by geologic forces naturally weather over time to release phosphorus. These rocks are the source of phosphorus used in fertilizer and detergents. Phosphorus is taken up by producers and moves through the food chain until it is released either through waste or decomposition. Excess phosphorus on land runs off the surface or leaches out of the soil and into aquatic habitats. In the ocean, phosphorus combines with calcium or iron and precipitates out of the water column, ultimately to form phosphate rocks again.

Figure 21.9 Marine dead zones. In 2008, more than 400 marine dead zones were identified around the world. This represents an increase of 33 percent from 1995. Data from R. J. Diaz and R. Rosenberg, Spreading dead zones and consequences for marine ecosystems, *Science* 321 (2008): 926–929.

Later, some of this rock is uplifted by geologic forces. Exposed rocks experience weathering, which causes them to slowly release phosphate ions. Phosphate rocks are also mined for phosphate that is used in fertilizers and in a variety of detergents.

When phosphate ions enter terrestrial ecosystems, they can be either bound strongly to the soil or taken up by plants and passed through the terrestrial food web. Animal excretions and the decomposition of all terrestrial organisms release phosphorus back to the soil. Excess phosphorus that is not bound to the soil or taken up by plants either moves across the surface of the land during a rainstorm as runoff or leaches from the soil. When soil erosion occurs, the phosphorus that is bound to the soil is carried away with the eroding soil particles. In either case, the phosphorus can be carried to a variety of aquatic ecosystems.

When phosphate ions enter aquatic ecosystems, they are taken up by the producers and enter the food web in a manner that is similar to the terrestrial food web. In well-oxygenated waters, phosphorus readily binds with calcium and iron ions and precipitates out of the water column to become part of the sediments. Thus, marine and freshwater sediments act as a phosphorus sink by continually removing phosphorus from the water column. Under low-oxygen conditions, iron tends to combine with sulphur rather than phosphorus, so phosphorus remains more available in the water column. Over time, the phosphate that precipitates down to ocean sediments is converted into calcium phosphate rocks, and the phosphorus cycle begins again.

HUMAN IMPACTS ON THE PHOSPHORUS CYCLE
In Chapter 20 we discussed how phosphorus is commonly a limiting nutrient in terrestrial and aquatic ecosystems. Therefore, adding phosphorus to these ecosystems can have harmful effects. As we saw at the beginning of this chapter, phosphorus, sometimes in combination with excess nitrates, contributes to algal blooms that cause dead zones where rivers empty into oceans. This phenomenon happens in locations around the world, as shown in Figure 21.9. An increase in the productivity of aquatic ecosystems is called **eutrophication**. An increase in the productivity of aquatic ecosystems caused by human activities is called **cultural eutrophication**.

From the 1940s to the 1990s, household detergents contained phosphates to improve their cleaning effectiveness. These detergents became part of the wastewater that travelled through public sewage systems, ultimately emptying into rivers, lakes, and oceans. People began to realize that these detergents significantly increased phosphorus in waterways, which contributed to eutrophication and dead zones. Canada limited the amount of phosphorus allowed in laundry detergents in 1989, and further reduced the allowable amount in 2010, along with a ban on phosphorus in dishwasher detergents.

In terrestrial ecosystems, most nutrients regenerate in the soil

Now that we understand how the major nutrients cycle within and between ecosystems, we change our focus so we can examine how nutrients are regenerated in

Eutrophication An increase in the productivity of aquatic ecosystems.

Cultural eutrophication An increase in the productivity of aquatic ecosystems caused by human activities.

ecosystems. The regeneration of nutrients is a key factor in the productivity of ecosystems because the nutrients that are taken up by producers must be replaced. In this section, we will examine the regeneration of nutrients in terrestrial ecosystems, which experience nutrient regeneration by the weathering of bedrock or by the breakdown of organic matter.

THE IMPORTANCE OF WEATHERING

Terrestrial ecosystems experience a constant loss of nutrients because many are leached out of the soil and transported away in streams and rivers. To maintain a stable level of productivity, the loss of nutrients from an ecosystem must be balanced by an input of nutrients. For some nutrients, such as nitrogen, inputs come from the atmosphere. For most other nutrients, such as phosphorus, the inputs come from the weathering of bedrock beneath the soil. As we described in Chapter 5, weathering is the physical and chemical alteration of rock material near Earth's surface. Substances such as carbonic acid in rainwater and organic acids produced by the decomposition of plant litter react with minerals in the bedrock and release various elements that are essential to plant growth.

Determining the rate of weathering can be difficult because bedrock often exists far below the surface of the soil. One solution has been to measure the nutrients that enter a terrestrial ecosystem from precipitation and the nutrients that leave an ecosystem by leaching out of the soil and into a stream. You can see a diagram of nutrient inputs and outputs in Figure 21.10. If the system is in equilibrium—if nutrient inputs equal nutrient outputs—the difference between the nutrients entering the system by precipitation and particulate matter and the amount of nutrients leaving the system by leaching and runoff should equal the amount of nutrients made available by weathering.

Ecologists commonly determine the rate of nutrient regeneration through weathering by quantifying nutrient inputs and outputs from a *watershed*. A **watershed** is an area of land that drains into a single stream or river, as illustrated in Figure 21.11. In a watershed, scientists can estimate the rate of weathering by measuring the net movement of several highly soluble nutrients, such as calcium (Ca^{2+}), potassium (K^+), sodium (Na^+), and magnesium (Mg^{2+}). These nutrients easily leach out of the soil and move into streams where their concentrations can be measured as the stream leaves the watershed.

An example of this approach was reported in 2012 for 21 small watersheds in Quebec. Each watershed contained a small lake with a stream flowing out of the lake that drained the watershed. The watersheds were all forested and had little human activity, so the researchers assumed that nutrient movement was at equilibrium. Researchers monitored inputs of calcium, potassium, sodium, and magnesium entering each

Watershed An area of land that drains into a single stream or river.

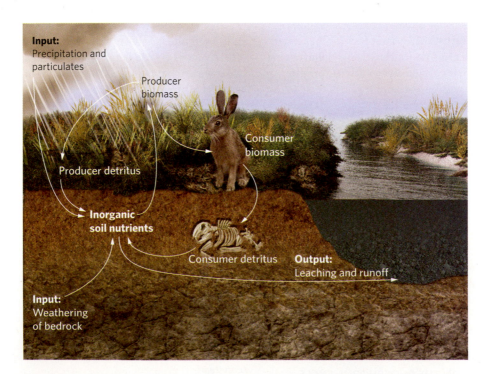

Figure 21.10 Quantifying nutrient weathering. Soil nutrients cycle through producers, consumers, and detritus. This cycle also has inputs from precipitation and weathering, and outputs of groundwater and surface water runoff.

Figure 21.11 A watershed.
A watershed is an area of land that drains down into a single stream or river.

watershed through precipitation. In addition, they determined the amount of each element present in the soil and bedrock. Finally, the researchers monitored the amount of each element coming out of the watershed in the stream that drained it. By knowing the inputs and outputs of a watershed, scientists were able to determine its weathering. When they placed their data on a map of the study area, shown in Figure 21.12, they

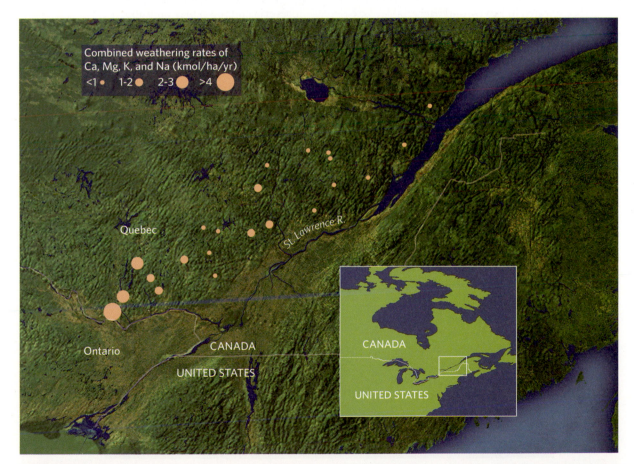

Figure 21.12 Weathering rates in 21 watersheds in Quebec. When researchers quantified the combined inputs and outputs of calcium, magnesium, potassium, and sodium, they found that the rates of weathering differed a great deal across the landscape. After D. Houle et al., Soil weathering rates in 21 catchments of the Canadian Shield, *Hydrology and Earth System Sciences* 16 (2012): 685–607.

also found that weathering rates varied geographically. The rate of weathering was highest in the southwest region of the province, probably because of regional differences in temperatures, precipitation, and soil conditions.

THE BREAKDOWN OF ORGANIC MATTER

Although weathering of inorganic nutrients provides nutrients to terrestrial ecosystems, weathering is a very slow process. Therefore, primary production largely depends on rapid regeneration of nutrients from detritus. To understand how nutrients are regenerated in these systems, we need to examine the process of decomposition. As we discussed in Chapter 1, scavengers consume dead animals and detritivores break up organic matter into smaller particles. However, decomposition is the process that breaks down organic matter into smaller and simpler chemical compounds and is conducted primarily by bacteria and fungi.

As illustrated in Figure 21.13, the breakdown of plant matter in a forest occurs in four ways: soluble minerals and small organic compounds leach out of organic matter, large detritivores consume organic matter, fungi break down the woody components and other carbohydrates in leaves, and bacteria decompose almost everything. Leaching removes 10 to 30 percent of soluble substances from organic matter, which includes most salts, sugars, and amino acids. What remains behind are the complex carbohydrates, such as cellulose, and other large organic compounds, such as proteins and lignin. Lignin determines the toughness of leaves and gives wood many of its structural qualities. The lignin content of plants is a particularly important determinant of decomposition rate because it resists decomposition more than cellulose. Some portion of the lignins, as well as other plant compounds that resist decomposition, may never be broken down in the surface soils but can form fossil fuels when buried for millions of years.

Large detritivores, including millipedes, earthworms, and wood lice, also play an important role in decomposition. These animals can consume 30 to 45 percent of the energy available in leaf litter, but they consume a much lower fraction of the energy available in wood. The importance of large detritivores is twofold; they decompose organic matter directly and they macerate organic matter into smaller pieces of detritus. By breaking the organic matter into smaller pieces, the detritivores cause it to have a greater surface area to volume ratio. This gives bacteria and fungi more surfaces on which to act and increases the rate of decomposition.

Bacteria and fungi play an important role in decomposition because they help convert organic matter into inorganic nutrients. Fungi play a special role because the hyphae of fungi can penetrate the

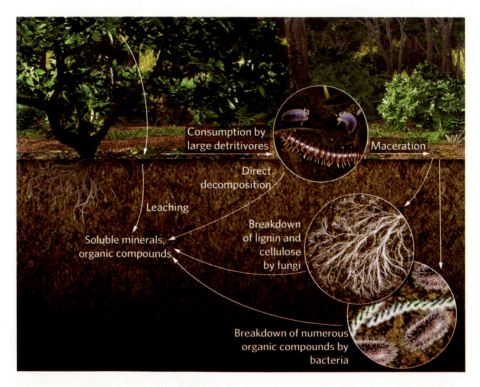

Figure 21.13 Decomposition of organic matter. Organic matter decomposes through the leaching action of water, consumption by invertebrates, mineralization by fungi, and mineralization by bacteria.

Figure 21.14 Fungi decomposing a dead log. Fungi play a key role in the decomposition of organic matter in terrestrial ecosystems by penetrating the dead tissues of plants. The fruiting bodies of this oyster bracket fungus (*Pleurotus ostreatus*), commonly referred to as mushrooms, are emerging from a dead log in Belgium. Photo by Philippe Clement/naturepl.com.

tissues of leaves and wood that large detritivores and bacteria cannot penetrate on their own. If you have ever walked through a forest, you may have seen the fruiting bodies of many different fungi, including the impressive shelf fungi that emerge from the sides of dead logs (Figure 21.14). Some fungi and bacteria can break down cellulose and lignin, which are two major components of organic matter that few other organisms can decompose. They secrete enzymes that break down the plant matter into simple sugars and amino acids that they then absorb.

In terrestrial ecosystems, 90 percent of all plant matter produced in a given year is not consumed directly by herbivores but is ultimately decomposed. Many plants resorb some of the nutrients from their leaves before the leaves are dropped. The aboveground dead plant biomass, combined with the organic matter of dead animals and animal waste, drops onto the soil surface where nutrients are leached. Detritivores break the organic matter into smaller pieces and decomposers break it down even further. Here decomposition is primarily aerobic, and plant roots and their associated mycorrhizal fungi have ready access to the nutrients that are released by the decomposers.

Because plant growth and decomposition are biochemical processes, nutrient cycling between producers and decomposers in terrestrial ecosystems is influenced by temperature, pH, and moisture. The rate of decomposition is also affected by the ratio of carbon and nitrogen in the organic matter. As we discussed in Chapter 20, differences in the stoichiometry of an organism's food can affect the consumption of the food and the number of consumers that can be supported by the food supply.

In the case of decomposition, if the decomposers require high amounts of nitrogen, then low nitrogen availability in the organic matter can cause slower rates of decomposition.

A study of leaf decomposition in Costa Rica provides insights into how several factors affect the rate of decomposition. In this study, researchers gathered leaves recently shed from 11 species of tropical trees and placed them into coarse-mesh bags. The bags were positioned on the ground in the forest where invertebrates had access to the leaves while the leaves remained in the bag. These invertebrates shredded the leaves into smaller pieces with an increased surface area, which facilitated more rapid decomposition. The researchers weighed the leaf bags at different times over a 230-day period to determine the rate of leaf decomposition. You can view the data for several of these species in Figure 21.15a. Leaf species with curves that decline faster over time, such as the Brazilian firetree (*Schizolobium parahyba*), have the highest rate of decomposition. Such species have relatively low amounts of lignin and cellulose and they have high leaf solubility, which allows more nutrients to be leached out of the leaf. The daily rate of mass loss, which we denote as k, can be calculated for each curve. Species that decompose faster have a higher value of k. In "Analyzing Ecology: Calculating Decomposition Rates of Leaves," we discuss these calculations in more detail.

To examine how leaf traits affected decomposition rate, measured again as the daily rate of mass loss (k), the researchers focused on distinct time periods: from day 0 to 50 when most decomposition would occur due to leaching, and from day 51 to day 230 when most decomposition would be due to invertebrates, fungi, and bacteria. During the first 50 days, there was a positive relationship between the decomposition rate and the fraction of soluble compounds in the different leaf species that can be leached, as shown in Figure 21.15b. During the latter portion of the experiment, there was a negative relationship between decomposition rate and the ratio of lignin to nitrogen, as you can see in Figure 21.15c. This means that leaf species with a relatively high amount of lignin experience a slower rate of decomposition. From this study we see that the decomposition of organic matter depends on both the environmental conditions and the chemical traits of the organic matter.

Once researchers calculated values of k for each species, they could examine how both precipitation and the traits of the leaves affect the leaf decomposition rate. For example, they manipulated the amount of precipitation that some of the mesh bags received. When they reduced precipitation to 50 or 25 percent of normal levels, it caused a 10 to 20 percent decline in the rate of decomposition.

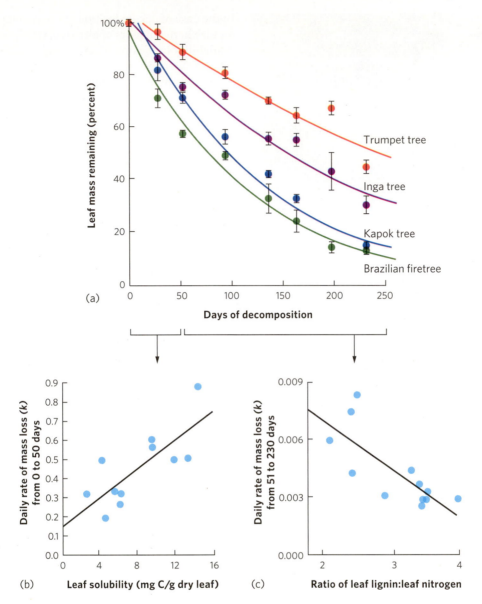

Figure 21.15 Decomposition rates of leaves in a tropical forest. Researchers collected recently fallen leaves and placed them into mesh bags to determine the rate of daily mass loss over time in the forest. **(a)** A sample of 4 of the 11 species examined reveals a range of different decomposition rates. Error bars are standard errors. **(b)** During the first 50 days of decomposition, the decomposition rates of the 11 leaf species were positively correlated with the solubility of compounds that could be leached from each species. **(c)** After 50 days, the decomposition rates of the 11 leaf species were considerably lower and negatively correlated to the amount of lignin in each leaf species relative to the amount of nitrogen. Data from W. R. Wieder et al., Controls over leaf litter decomposition in wet tropical forests, *Ecology* 90 (2009): 3333–3341.

DECOMPOSITION RATES AMONG TERRESTRIAL ECOSYSTEMS

Because environmental conditions are a key determinant for rates of decomposition, terrestrial ecosystems differ a great deal in their decomposition rates. As you saw in Figure 21.15, in a tropical forest it takes 250 days to decompose approximately 70 percent of the leaves from an Inga tree. In comparison, at boreal and subarctic sites in Saskatchewan and Manitoba, it takes 12 years to decompose the same proportion of aspen leafs and pine needles. This difference reflects the fact that the colder temperature, lower moisture, and lower pH typical of boreal and subarctic forests slow down the rate of decomposition. Comparative studies of temperate and tropical forests show that detritus in the tropics decomposes more rapidly because of warmer temperatures and higher amounts of precipitation. For example, we can compare the amount of dead plant matter on the forest floor—including leaves, branches, and logs—versus the total biomass of vegetation and detritus in a forest. The proportion of dead plant matter is about 20 percent in

Calculating Decomposition Rates of Leaves

Researchers often want to examine the rate of leaf litter decomposition to determine how fast nutrients can be regenerated so they are available to producers, such as algae that form the base of an aquatic food web, plants that form the basis of a terrestrial food web, or last year's crops that decompose and provide nutrients to the current year's crop. As we have seen, when researchers want to estimate decomposition rates of leaves to determine how rapidly nutrients are regenerated in an ecosystem, they commonly weigh samples and put them into mesh bags that are positioned on the ground where they will decompose. Over time, the researchers collect, dry, and weigh the bags to determine how much mass remains and how much has been decomposed. The rate of decomposition commonly follows a negative exponential curve; there is initially a rapid loss in mass that slows over time. This negative exponential curve can be described by the following equation:

$$m_t = m_o \, e^{-kt}$$

where m_t is the mass of leaf litter that remains at a particular time, m_o is the original mass of leaf litter, e is the base of the natural log, k is the daily rate of mass loss, and t is time, which is measured in days. The decay constant k is the key parameter in this equation because it determines the shape of the curve; leaves that decompose at a faster rate have a larger value of k.

Ecologists can estimate the value of k in this equation with statistical software that determines the line that best fits a set of data for decomposition over time. Once the statistical software determines the values for the daily decomposition rate, we can then estimate the amount of leaf litter at any point in time provided that the decomposition occurs under similar environmental conditions. For example, if we start with 100 g of leaves and $k = 0.01$, we can estimate the mass of leaves that has not decomposed after 10, 50, and 100 days:

After 10 days: $m_t = m_o \, e^{-kt} = 100 \, e^{-(0.01)(10)}$
$= 90 \text{ g}$

After 50 days: $m_t = m_o \, e^{-kt} = 100 \, e^{-(0.01)(50)}$
$= 61 \text{ g}$

After 100 days: $m_t = m_o \, e^{-kt} = 100 \, e^{-(0.01)(100)}$
$= 37 \text{ g}$

YOUR TURN Estimate the remaining mass of leaves that has not decomposed after 10, 50, and 100 days for two other leaf species, one with a daily decomposition rate of $k = 0.05$, and the other with a daily decomposition rate of $k = 0.10$.

---

temperate coniferous forests, 5 percent in temperate hardwood forests, and only 1 to 2 percent in tropical rainforests. Of the total organic carbon in terrestrial ecosystems, more than 50 percent occurs in soil and litter in northern forests, but less than 25 percent occurs in tropical rain forests where the majority of the organic matter exists in the living biomass.

These differences in litter decomposition rates mean that tropical forests have a much larger proportion of the total organic matter in living vegetation than in detritus. This has important implications for tropical agriculture and conservation. For example, when tropical forests are cleared and burned, a large fraction of the nutrients are mineralized by burning and by subsequent high rates of decomposition. Together, these processes create an abundance of nutrients during the first 2 to 3 years of crop growth, but any surplus nutrients not taken up by the crops quickly leach away.

Traditionally, tropical areas burned for agricultural fields would be farmed for 2 to 3 years and then left to undergo natural succession for 50 to 100 years to rebuild the fertility of the soil. In modern times, however, many regions have human populations that are too dense to allow rotation of agriculture into different areas over several decades. Without rotation, the soils cannot replenish their nutrients and the fertility of the land rapidly degrades.

In aquatic ecosystems, most nutrients regenerate in the sediments

Because most cycling of elements takes place in an aqueous medium, the chemical and biochemical processes involved are similar in terrestrial and aquatic ecosystems. However, the location of decomposition differs between terrestrial and aquatic ecosystems. In terrestrial ecosystems, nutrients regenerate close to the location where they are taken up by producers. In aquatic ecosystems, most nutrients regenerate in sediments, which are often far below the surface waters that contain dominant producers such as phytoplankton. In addition, terrestrial ecosystems commonly experience aerobic decomposition whereas the sediments and deep waters of aquatic ecosystems commonly experience anaerobic decomposition, which is considerably slower. In this section we will investigate how decomposition operates in streams and wetlands, which receive much of their energy from leaves that blow in from the surrounding terrestrial environment. We will examine the important role of sedimentation in rivers, lakes, and oceans and explore how stratification of water affects the movement of regenerated nutrients in lakes and oceans.

ALLOCHTHONOUS INPUTS TO STREAMS AND WETLANDS

As we have discussed in previous chapters, streams and small forested wetlands receive a major portion of their energy from the surrounding terrestrial environment in the form of dead leaves that fall into the water. The decomposition process of leaves in a stream is similar to the process on land. As the leaves settle onto the bottom of the stream, the first stage is the leaching of soluble compounds followed by the shredding of the leaves into smaller pieces by invertebrates such as amphipods, isopods, and larval caddisflies. At the same time that leaves are being shredded, fungi and bacteria are working to decompose the leaves much as they do on land.

Given the similar processes in both the terrestrial and stream ecosystems, it is perhaps not surprising that the rate of leaf decomposition depends on the temperature of the water and the traits of the leaves. To determine the rate at which leaves decompose in streams, aquatic ecologists follow a protocol that is similar to the protocol used by terrestrial ecologists. They collect newly fallen leaves and place a weighed amount of dry leaves into a mesh bag that is submerged in the stream. The mesh bag allows aquatic invertebrates to enter the bag without losing any of the leaves. Over time, the bags are removed from the stream, dried, and reweighed to determine the amount of leaf mass that remains.

To determine how leaf traits affect leaf decomposition rates, researchers placed coarse-mesh bags of nine leaf species into a stream located in the Black Forest of Germany and removed the bags over time. The nine leaf species differed a great deal in their nitrogen, phosphorus, and lignin content. The researchers discovered that leaf decomposition rate was not related to the amount of nitrogen or phosphorus in the leaves but it was strongly associated with the lignin content of the leaves, as shown in Figure 21.16a.

The researchers also were interested in knowing how important the invertebrates were in the decomposition process. To answer this question, they placed a second set of leaves into fine-mesh leaf bags that prevented invertebrates from entering. As you can see in Figure 21.16b, leaf decomposition was about 20 percent higher in bags where invertebrates had been allowed to enter.

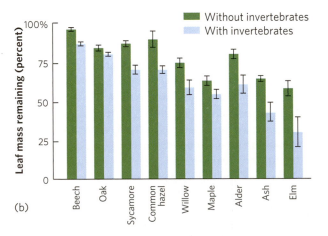

Figure 21.16 Leaf decomposition in a stream. (a) When nine species of leaves in mesh bags were added to a stream in Germany, the decomposition rate was slower in leaf species that contained more lignin in their tissues. **(b)** The researchers also assessed the contribution of aquatic invertebrates to leaf decomposition by using large-mesh and fine-mesh leaf bags. After 55 days, the leaves in the fine-mesh bags, which excluded invertebrates, experienced 20 percent less decomposition compared to the large-mesh bags, which allowed invertebrates to enter. Error bars are standard errors. Data from M. H. Schindler and M. O. Gessner, Functional leaf traits and biodiversity effects on litter decomposition in a stream, *Ecology* 90 (2009): 1641–1649.

The small wetlands that exist in forests also receive a large proportion of their energy from leaves. Similar to leaves in streams, the decomposition of leaves in forested wetlands is largely tied to the lignin content. Moreover, the breakdown rate of the leaves has widespread effects on the entire food web and functioning of the ecosystem. Nutrients can also travel in the reverse direction and be transferred from aquatic to terrestrial systems. For example, this happens regularly around streams in British Columbia that are home to migrating salmon. As we saw in Chapter 8, salmon swim from the oceans up streams and rivers to spawn. During this migration, terrestrial predators, such as bears, catch fish and bring them ashore to feed on them. Bears usually do not consume all of the fish, and leave the fish carcasses onshore to decompose. This decomposition adds nutrients to the terrestrial habitats, which leads to increased plant growth and higher plant diversity along the streams.

DECOMPOSITION AND SEDIMENTATION IN RIVERS, LAKES, AND OCEANS

In most rivers, lakes, and oceans, organic matter sinks to the bottom and accumulates in deep layers of sediments. While some nutrients are recycled in the surface waters when animals excrete waste or when microbes in the surface water decompose organic matter, most organic matter sinks to the sediments. As a result, most nutrients must come from the sediments, although they will return slowly to the productive surface waters.

The process of nutrient regeneration in sediments of aquatic ecosystems helps us understand many patterns of ecosystem productivity. For example, you might recall from Chapter 20 that the least productive aquatic ecosystems are the deep oceans where the benthos is far from the surface waters (see Figure 20.6). Shallow oceans are more productive in part because the sediments are much closer to the surface waters and can therefore regenerate nutrients more quickly through decomposition. The upwelling of water from the deep sediments to the surface brings nutrients from the site of regeneration to the site of algal productivity. In addition, some of the most productive regions of the ocean occur at the upwelling of water along the coasts of continents where currents draw deep, nutrient-rich water up to the surface (see Figure 5.11).

STRATIFICATION OF LAKES AND OCEANS

The stratification of water, a phenomenon that we discussed in Chapter 6, further affects the availability of nutrients in aquatic ecosystems. The vertical mixing of water from the sediments to the surface can be hindered whenever surface waters have a different temperature and therefore a different density than that found in deep waters. The vertical mixing can affect primary production in two opposing ways. On the one hand, mixing can bring the deep, nutrient-rich water to the surface where phytoplankton can use it. On the other hand, mixing can carry phytoplankton down to the deep water, where they cannot sustain themselves because of the low light conditions. Under such conditions, primary production may shut down altogether in the deep water and little primary production will occur in the nutrient-rich waters at the surface.

Stratification happens in temperate and tropical lakes when the surface waters are warmed by the summer sunlight while the deeper waters stay cold and dense. Such stratification does not happen in polar lakes because their surfaces never become warm enough. In estuaries and oceans, stratification of the water happens when an input of less dense freshwater from rivers or melting glaciers is positioned above a layer of the more dense ocean saltwater.

Occasionally, ecosystems that stratify experience periods of vertical mixing. For example, temperate lakes in the spring and fall experience changes in the temperature of surface waters that eventually match the temperature of the deeper waters. When the temperatures become equal, spring and fall winds that blow along the surface of the lakes cause the entire lake to mix. Mixing can also happen in oceans. In areas where the surface water is less salty than the deep water, sunlight can slowly cause the surface waters to evaporate and leave the salt behind. At some point, the surface water becomes saltier than the deeper water and the surface water sinks, thereby causing the ocean water to circulate.

In this chapter, we have learned that elements cycle within and between ecosystems and that this cycling determines the availability of elements to organisms. We have also seen that human activities commonly affect element cycles in ways that are harmful to ecosystem function. Finally, we have seen how elements regenerate through weathering and the decomposition of organic matter. In "Ecology Today" we will apply these concepts to understand how logging and global change alters the functioning of an entire forest.

ECOLOGY TODAY CONNECTING THE CONCEPTS

CYCLING NUTRIENTS IN NEW HAMPSHIRE

Monitoring nutrient flows from a watershed. At the Hubbard Brook Experimental Forest in New Hampshire, the soils lie on bedrock that prevents water from percolating down, so all nutrients leached from the soils find their way into the streams that drain the watershed. At the bottom of the watershed, researchers constructed a catchment device to monitor water volumes and nutrient flow from the watershed. Photo by U.S. Forest Service, Northern Research Station.

As we have seen throughout this chapter, producers play a key role in assimilating nutrients from the environment and holding the nutrients in their tissues to be passed on to consumers and decomposers. However, assessing the degree to which plants affect nutrient cycling in an ecosystem is a daunting task, in part because ecosystems are large and complex. Fifty years ago, researchers at the Hubbard Brook Experimental Forest in New Hampshire took on this challenge by selecting a number of forested watersheds and logging some of them. Then they monitored the changes in how nutrients moved in each watershed.

The Hubbard Brook Experimental Forest was an ideal location for this grand experiment. A layer of impenetrable rock under the soil prevents water from percolating into the deep groundwater. Because of this, all of the precipitation falling on the watershed is

either taken up by plants or passes over and through the soil, ultimately ending up in the stream that leaves the watershed. Therefore, all water and nutrients not held by the plants or the soil exit the watershed through the stream, which allows the researchers to monitor the stream to measure the amount of water and nutrients leaving the ecosystem.

In 1962, a massive experiment was initiated in which researchers removed all of the trees from an entire watershed and sprayed it with herbicides for several years to suppress plant growth. As a control, adjacent watersheds were not logged. Since that time, ecologists have tracked how the ecosystem has responded to this disturbance.

Without plants to take up water and nutrients, the movement of elements in the ecosystem with suppressed plant growth changed dramatically. For example, the amount of water leaving the watershed

Logging a watershed. To determine the role of plants in the cycling of nutrients, researchers at the Hubbard Brook Experimental Forest logged an entire watershed and then applied herbicides for several years to prevent plants from growing. Other watersheds were logged and allowed to regrow while still others were left unmanipulated as controls. Photo by U.S. Forest Service, Northern Research Station.

in the stream increased severalfold. In addition, because the nitrates available in the soil were no longer being used by plants, there was a large increase in the amount of nitrates that leached out of the soil and into the water of the stream. In forests that were not logged, there was a net gain of soil nitrogen over time at a rate of 1 to 3 kg per ha; this increase came from precipitation and nitrogen fixation. However, in the logged watershed, there was a net loss of nitrogen at a rate of 54 kg per ha. Because nitrates do not bind well to soil, they leached out of the soil and into the stream after logging.

Many other nutrients were also affected by the logging. For example, the researchers tracked the movement of calcium ions that come from precipitation and weathering of the bedrock. They found the vast majority of calcium ions came from the decomposition of detritus whereas less than 10 percent of the calcium came from weathering. When the region experienced years of acid precipitation, the inputs of calcium from precipitation and weathering could not keep pace with the calcium that leached from the soil. As a result, the forest experienced a net loss of calcium.

Today the Hubbard Brook Experimental Forest continues to provide insights into nutrient cycling. For example, in 2012 researchers reported that over a 5-decade period, the amount of nitrates leaving forested watersheds through the streams each year declined by more than 90 percent. They hypothesized that these declines could be due to changes in the species composition of the trees, climate change, or a recovery from past events that removed many of the trees. When they examined these alternatives, they discovered that changes in the composition of tree species only contributed to a small part of the decline in nitrates leaving the watershed. The forest today has fewer sugar maple trees, which have leaves that decompose and release nitrates rapidly, and more beech trees, which have leaves that decompose and release nitrates slowly. The researchers determined that climate change was responsible for about 40 percent of the reduction in nitrates in the stream; warmer temperatures in the late fall and early spring in recent years give plants a longer time to take up the nitrates, so fewer nitrates leach out of the soil and into the stream.

The researchers were surprised to find that 50 to 60 percent of the reduction in nitrates over the 5 decades was due to a recovery from historic disturbances, which they defined as events that removed more than 20 percent of the trees. At Hubbard Brook, these events include logging that took place in 1906 and 1917, a hurricane in 1938, and an ice storm in 1998. Just as we saw with the clear-cutting of a watershed in 1962, any removal of trees causes a short-term flush of nitrates out of the soils and into the stream. As the forests begin to come back, however, the soil recovers and holds on to more of the available nitrates. As the soil accumulates nitrates, the amount of nitrates leached into the stream declines.

The long-term experiment at Hubbard Brook shows us how natural events and human activities can dramatically alter the movement of elements within and between ecosystems in ways that can have major consequences for terrestrial and aquatic ecosystems.

SOURCES: Likens, G. E. 2004. Some perspectives on long-term biogeochemical research from the Hubbard Brook Ecosystem Study. *Ecology* 85: 2355–2362.

Bernal, S., et al. 2012. Complex response of the forest nitrogen cycle to climate change. *PNAS* 109: 3406–3411.

SUMMARY OF CHAPTER CONCEPTS

■ **The hydrologic cycle moves many elements through ecosystems.** Water evaporates from water bodies, soil, and plants and moves as water vapour into the atmosphere. This water vapour condenses into clouds and eventually falls back to Earth as precipitation. This precipitation can be taken up by plants, run across the surface of the ground, or infiltrate into the groundwater. This water moves to streams and lakes and ultimately makes its way back to the oceans. Humans can alter this cycle by reducing infiltration due to the construction of impermeable surfaces, by logging trees and thereby increasing runoff, and by causing global warming, which increases the rate of evaporation.

■ **The carbon cycle is closely tied to the movement of energy.** Carbon exists in the atmosphere in the form of CO_2. This CO_2 can be used by terrestrial producers and, after dissolving into the water, can be used by aquatic producers. The producers, consumers, scavengers, detritivores, and decomposers in terrestrial and aquatic ecosystems can produce CO_2 when they respire. Carbon can also leave the water by sedimentation and be buried both in the water and on land. Buried carbon can be extracted in the form of fossil fuels. The combustion of fossil fuels and the combustion of organic matter during fires release CO_2 into the atmosphere. Humans can alter this cycle primarily by affecting the extraction and combustion of carbon.

■ **Nitrogen cycles through ecosystems in many different forms.** Nitrogen gas in the atmosphere can be converted into ammonia and nitrates by lightning, nitrogen-fixing bacteria, and the manufacture of fertilizers. Producers can take up these forms of nitrogen and assimilate them. This nitrogen is then transferred through terrestrial and aquatic food webs. During decomposition, nitrogen in organisms and their wastes can be converted to ammonia by the process of mineralization. Ammonia can be converted into nitrites and nitrates through the process of nitrification and nitrates can be converted to nitrous oxide and nitrogen gas by the process of denitrification. Humans alter this cycle primarily by manufacturing and applying large amounts of fertilizer and combusting fossil fuels, which produces nitric oxide in the air that later mixes with precipitation and falls to the ground as nitrates. These activities alter the fertility of terrestrial and aquatic environments.

■ **The phosphorus cycle moves between land and water.** Most phosphorus is released by weathering from rocks. This phosphorus is taken up by terrestrial and aquatic producers, which pass it to consumers, scavengers, detritivores, and decomposers. The phosphorus from excretions and decomposed organisms dissolves in the water of soil or in the water of streams, rivers, lakes, and oceans. In the ocean, phosphorus precipitates into sediments that are slowly converted into rocks. Humans affect the phosphorus cycle primarily by mining rocks for fertilizer. This phosphorus-rich fertilizer can alter the fertility of terrestrial and aquatic habitats and lead to algal blooms.

■ **In terrestrial ecosystems, most nutrients regenerate in the soil.** The nutrients in terrestrial ecosystems are primarily regenerated in soils. Some nutrients such as phosphorus are regenerated by weathering of rocks and all nutrients are regenerated by the decomposition of dead organic matter. Because decomposition rates are faster under warm temperatures and high precipitation, tropical ecosystems have high decomposition rates and low amounts of dead organic matter. Boreal and other cold ecosystems have low rates of decomposition and large amounts of dead organic matter.

■ **In aquatic ecosystems, most nutrients regenerate in the sediments.** In many streams and some wetlands, allochthonous inputs of leaves from the surrounding terrestrial environment are the major source of nutrients. The rate of leaf decomposition is determined primarily by the temperatures of the water and the lignin content of the leaves. In rivers, lakes, and streams, much of the organic matter can settle out of the water and onto the bottom sediments where they decompose. When lakes and oceans stratify, the movement of nutrients that have been released by decomposition are hindered from moving to the more productive surface waters. Although nitrogen and phosphorus are the most common limiting nutrients in aquatic ecosystems, some regions of the ocean are limited by the availability of other nutrients, including silicon and iron.

REVIEW QUESTIONS

1. How does energy from the Sun drive the movement of water from the oceans to the continents and back to the oceans again?

2. How might the ocean reduce the effects of fossil fuel combustion on CO_2 concentrations in the atmosphere?

3. Why is methane gas commonly produced in swamps?

4. Given that the bottom of the ocean is anaerobic, what process in the nitrogen cycle is likely to be occurring in this location?

5. How might nitrogen-fixing bacteria living in symbiosis with a plant affect the types of environments in which the plant could live?

6. Why is the weathering of bedrock responsible for such a small fraction of the nutrients available to plants?

7. Why do tropical and temperate soils have different rates of nutrient regeneration?

8. Explain why agricultural soils in boreal Canada retain their nutrients for many more years than agricultural soils in tropical South America.

9. How might global warming cause the release of CO_2 from boreal forest soils?

10. What is the chain of events in which the dumping of raw sewage into the Mississippi River leads to fish kills in the dead zone in the Gulf of Mexico?

PRACTICING ECOLOGY: LEAF AFTER DEATH OF SALMON

John Richardson and his lab at the University of British Columbia study the factors that alter stream community composition and ecosystem function. Photo by Felix Fornoff.

As we learned in this chapter, forest streams receive most of their energy and nutrients from terrestrial inputs, such as leaf litter and soil runoff. However, inputs can vary widely in nutritional quality. For example, most leaf litter is low in nutrients relative to other organic matter such as dead animals. When aquatic decomposers consume nutrient-poor organic matter, they garner fewer resources that can be used for growth and reproduction than when they consume nutrient-rich organic matter. Given this difference in resource use, how might decomposition rates on nutrient-poor inputs change in a stream if nutrient-rich inputs were to arrive?

Streams along the northwest coast of North America receive seasonal nutrient inputs of leaf litter, but they can also receive massive inputs of dead salmon. Each year, several species of salmon in the Pacific Ocean, such as the chum salmon (*Oncorhynchus keta*), migrate to inland freshwater streams to breed. As you may recall from the introduction of Chapter 8, many species of salmon die after breeding. Once the salmon die in the streams, many species of insects decompose their carcasses. Since these nutrient-rich carcasses are likely to attract decomposers away from nutrient-poor leaf litter, we might hypothesize that litter decomposes slower when salmon carcasses are present.

John Richardson and his lab at the University of British Columbia tested this hypothesis by manipulating the presence of salmon carcasses in Mayfly Creek, which is a small forest stream east of Vancouver. They diverted water from the creek into artificial stream channels, where they placed three mesh bags each containing 10 g of red alder (*Alnus rubra*) leaf litter into each channel. In half of the channels, the researchers placed 1 kg of salmon carcass. In the other half of the channels, they placed a stone of similar size as a control. Insect decomposers naturally entered the stream channels from the main creek. After 62, 85, and 111 days, they removed a single litter bag from each channel. They collected and weighed all insect decomposers in each leaf pack, and then determined the percentage of mass lost from the original biomass of the litter. You can view their data in the table.

MASS OF INSECT DECOMPOSERS (mg/LEAF BAG) AND PERCENT LITTER MASS REMAINING IN MESH BAGS

DAY	MASS OF INSECT DECOMPOSERS		PERCENT LITTER REMAINING	
	SALMON CARCASS	NO SALMON CARCASS	SALMON CARCASS	NO SALMON CARCASS
62	1.61	2.98	69.66	69.78
85	3.90	23.64	65.64	57.93
111	5.74	19.74	61.96	51.23

GRAPH THE DATA

1. Create a line graph with days on the *x* axis and insect mass on the *y* axis. Draw a separate line for each treatment.

2. Create a second line graph with days on the *x* axis and percent leaf litter mass remaining on the *y* axis. Draw a separate line for each treatment.

INTERPRET THE DATA

1. Explain how the data support or refute the hypothesis that leaf litter will decompose more slowly in the presence of salmon carcasses.

2. During the study, a small rodent invaded one of the artificial stream channels and removed a salmon carcass. The researchers found that the biomass of insects decomposing the litter in the channel increased immediately. Explain how this suggests that the effects observed in this experiment were a result of salmon carcasses attracting insect decomposers away from leaf litter.

3. In the same study, the researchers found that the biomass of microbes increased on leaf litter decomposing in the presence of salmon carcasses. This probably occurred because the nutrients released from the carcass promoted growth of microbes throughout the stream channels and on the leaf litter. Since this microbial growth is a nutrient-rich resource that detritivores can consume, predict how the rate of leaf litter decomposition may change after the salmon carcasses disappear from the stream.

SOURCE: Zhang, Y., J. N. Negishi, J. S. Richardson, and R. Kolodziejczyck. 2003. Impacts of marine-derived nutrients on stream ecosystem functioning. *Proceedings of the Royal Society of London* B 270: 2117–2123.

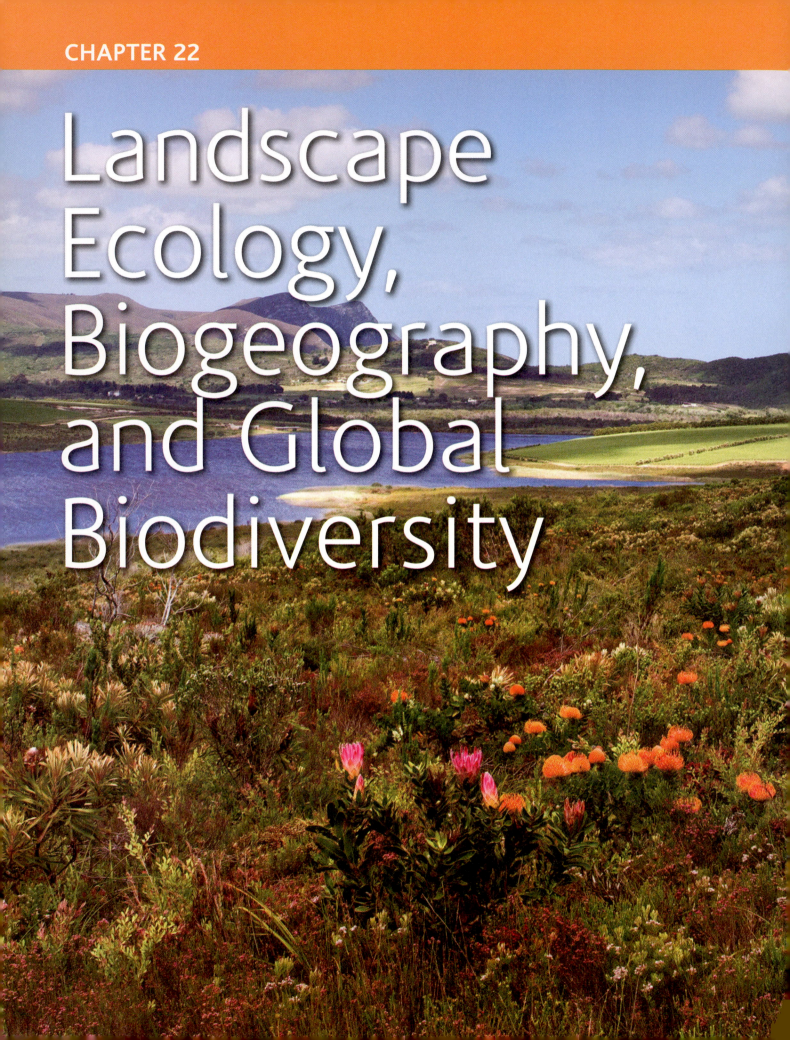

CHAPTER 22

Landscape Ecology, Biogeography, and Global Biodiversity

The Magnificent Biodiversity of the Cape Floristic Region

The Cape Floristic Region on the southwestern tip of Africa is home to more than 9000 species of plants, which makes it one of the most species-rich places on Earth. For comparison, Florida is about the same size and is positioned at the same latitude but has only about 4200 species of plants. Moreover, nearly 70 percent of plant species on the cape are found nowhere else on Earth.

Why is this region such a hotspot of biodiversity? Research on the phylogenies of the most species-rich plant genera and families revealed that their ancestors all originated in the cape region. Therefore, the evolution of so many species is somehow connected to factors existing in southwestern Africa. Some scientists hypothesized that the high diversity of the region was the outcome of historic changes in climate. Approximately 5 million years ago, as the positions of the continents of the world were shifting, the climate in the Cape Floristic Region became cooler

> "The heterogeneity of soils offers a diversity of habitat conditions that favoured the evolution of new species."

and drier. Perhaps these climate changes favoured a period of rapid speciation in plants. However, when researchers tested this hypothesis by examining when speciation events occurred in the plant phylogenies, they found that much of the speciation had predated the changes in climate. Thus, the climate hypothesis was refuted.

Other scientists proposed that the region's great diversity is the result of having plants with highly specialized pollinators and narrow flowering periods. Both factors would allow populations of existing plant species to become reproductively isolated from each other and to diverge genetically over time. This would favour an increased rate of speciation. Researchers tested this idea by determining whether closely related species differ more in flowering time or in their pollinators than we would expect to occur at random. If these differences occurred more than would be randomly expected, it would suggest that they were the driving factor in the evolution of new species from a common ancestor. In 2011, researchers refuted this hypothesis, too, when they found that closely related plants often have similar flowering times and use similar pollinators.

A final hypothesis proposed that the mountainous topography of the region created conditions that favoured the evolution of new plant species. This topography contains tremendous heterogeneity in soil types that would favour the evolution of new species adapted to each. In support of this hypothesis, researchers discovered that closely related species of plants prefer very different soil types. These preferences suggested that the heterogeneity of soils offered a diversity of habitat conditions that favoured the evolution of new species. Two other factors also contributed to the diversity of plants in this region. Because many mountainous areas are

Species diversity in South Africa. The Cape Floristic Region, on the southwestern tip of South Africa, has an unusually high richness of plant species. Photo by Hoberman/UIG/age fotostock.

isolated, evolving populations would also be reproductively isolated, further favouring evolutionary divergence from conspecifics living in other areas. In addition, for the past several million years the climate in the region has been stable; this has slowed the rate of species extinctions, which has contributed to the high number of species today.

The story of the Cape Floristic Region demonstrates the importance of exploring the historical causes of species richness. It illustrates how a diversity of habitats facilitates a diversity of species, and it highlights the fact that we sometimes need to think about ecology at large spatial scales to understand the distribution and composition of species on Earth. In this chapter, we will explore ecology at large spatial scales to examine the distribution of species both across landscapes and around the world.

SOURCE: J. Schnitzler et al., Causes of plant diversification in the Cape Biodiversity Hotspot of South Africa, *Systematic Biology* 60 (2011): 343–357.

CHAPTER CONCEPTS

- Landscape ecology examines ecological patterns and processes at large spatial scales.

- The number of species increases with area.

- The equilibrium theory of island biogeography incorporates both area and isolation.

- On a global scale, biodiversity is highest near the equator and declines toward the poles.

- The distribution of species around the world is affected by Earth's history.

Throughout this book we have considered the role of physical conditions and of species interactions at particular locations. We have seen how these factors can affect the ecology of individuals, populations, communities, and ecosystems. When we explored these topics, we often found it helpful to focus on relatively homogeneous areas of land or water. However, as we moved across landscapes and across continents, we observed that terrestrial and aquatic ecosystems vary from place to place. In this chapter, we will consider much larger areas, from landscapes that include a variety of habitats to entire continents that contain a range of climates. In taking this large-scale approach, our goal is to understand why we find different numbers of species and often very different types of species in different places around the globe and why we sometimes find very similar types of species on widely separated continents. Once we understand the patterns of biodiversity, we can understand the processes that affect diversity and develop plans to conserve it, which is the topic of the next chapter.

Landscape ecology examines ecological patterns and processes at large spatial scales

If you were to look across an expansive landscape from the window of a plane, whether in the mountains of the Cape Floristic Region or anywhere else, you would undoubtedly see a diversity of terrestrial and aquatic habitats comprising a wide range of sizes and shapes

(see Figure 4.2). The field of **landscape ecology** considers the spatial arrangement of habitats at different scales and examines how they influence individuals, populations, communities, and ecosystems. In this section, we will discuss the sources of habitat heterogeneity and how this heterogeneity generates local and regional biodiversity.

CAUSES OF HABITAT HETEROGENEITY

The current heterogeneity of habitats reflects recent and historical events caused by both natural forces and human activities. Natural causes of habitat heterogeneity include geologic events such as the eruption of volcanoes and the advance and retreat of glaciers across continents. These events leave their mark on the landscape by moving large amounts of rock and soil and by creating water bodies and later changing their locations.

Long-lasting influences of historical processes on the current ecology of an area are known as **legacy effects**. An interesting legacy effect of glaciers that you can observe today is the presence of eskers, which are the remnants of long, winding streams of water that once flowed inside or under the glaciers. Over time, these glacial streams deposited soil and rock in their paths. After glaciers melt away, these old stream

Landscape ecology The field of study that considers the spatial arrangement of habitats at different scales and examines how they influence individuals, populations, communities, and ecosystems.

Legacy effect A long-lasting influence of historical processes on the current ecology of an area.

Figure 22.1 A natural legacy effect of the landscape. This long, winding esker near Whitefish Lake, Northwest Territories, was formed by a stream that once ran through a glacier thousands of metres thick. Particles of soil settled into the bed of the stream, which today forms the long ridge of land that is the esker. Photo by Gramber Photography/age fotostock.

courses appear as long and winding hills (Figure 22.1). These hills harbour different microhabitats that favour unique plant and animal communities.

Natural forces continue to cause habitat heterogeneity in modern times. At both local and regional scales, catastrophes such as tornadoes, hurricanes, floods, mudslides, and fires can alter vegetation structure, which causes changes in populations and communities that depend on it. Although catastrophic events have always occurred naturally, human activity has influenced them. For example, the frequency and intensity of fires has changed over the years in the boreal forest of the Lake Nipigon region in northwestern Ontario. As you can see in Figure 22.2, different patches of the forest burned during different decades, creating a mosaic of unburned and burned patches of various ages across the landscape. Because of fire suppression, regions in the south experienced fewer and smaller fires. The patterns of burning across the landscape depended on the origin of the fires, landscape characteristics such as the amount of combustible plant litter and local wind patterns, and human activities such as fire suppression.

As we saw in Chapter 18, some animals—for example beavers and alligators—are ecosystem engineers and can alter the habitats in a landscape. Humans are the most extensive ecosystem engineers; they build homes, offices, and factories, construct dams and irrigation channels, channelize waterways for improved navigation, and clear forests for lumber, paper, and agriculture. Logging provides a particularly clear example of a human activity that produces a mosaic of habitat types across the landscape. In the western United States, it is common practice to log medium-sized swaths of forest scattered throughout the landscape. This practice helps minimize soil erosion and other damaging effects of large-scale clear-cutting. It also produces a mosaic of forest patches of different ages. Logging is just one of many human activities that causes major effects on the landscape that can remain for many years.

Human activities can also cause habitat heterogeneity. For instance, during the first century ce, the Romans built small villages and farms in France. These farms were abandoned by the fourth

Figure 22.2 Habitat heterogeneity after fire. In northwestern Ontario, fires burned different patches of the forest over time and left behind a heterogenous mix of unburned and burned areas of various ages. Data from D. Senici et al., Spatiotemporal variations of fire frequency in central boreal forest, *Ecosystems* 13 (2010): 1227–1238.

Fire by Decade
1920
1930
1940
1950
1960
1970
1980
1990
2000

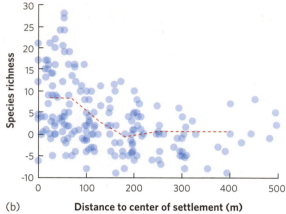

(a)

(b)

Figure 22.3 Legacy effects of humans on the landscape. (a) Ruins of Roman farming villages are present throughout northern France. These Gallo-Roman ruins are in the forest of Tronçais at the site of "Petits Jardins" in Isle-et-Bardais, Allier. As the stone building materials weathered, they likely contributed calcium and phosphorus to the soil and altered the abiotic conditions. In addition, the Romans introduced numerous plant species. **(b)** Collectively, these changes created a legacy of more plant species living close to the old settlements with fewer species living 500 m away. The number of plant species represents the difference from the mean number of plant species found far from the settlements. The dashed line represents the changing mean species richness as one moves away from the centre of the settlement. Photo by Laure Laüt. Data from E. Dambrine et al., Present forest biodiversity patterns in France related to former Roman agriculture, *Ecology* 88 (2007): 1430–1439.

century and the land reverted to forest (Figure 22.3a). Researchers studied the soil conditions and plant species located between 0 and 500 m from the former Roman settlements. They discovered that sites closer to the settlements had higher soil pH, more available phosphorus, and greater plant species richness, including many weedy species. The researchers attributed these legacy effects, which are shown in Figure 22.3b, to two causes. The slow breakdown of the ancient building materials contributed calcium and phosphorus to the soil. In addition, the Romans introduced numerous plant species to the area. In short, human habitation from 1600 years ago continues to have strong legacy effects on the modern forest.

RELATIONSHIPS BETWEEN HABITAT HETEROGENEITY AND SPECIES DIVERSITY

When we quantify the number of species in a given area, we include a greater variety of habitats at the landscape level than at the local scale and therefore generally observe a larger number of species. For example, ecologists in Alberta recently surveyed bird species in 206 sites in the boreal forest. At each site they measured the physical characteristics of the site itself, such as elevation and mean annual temperature and precipitation. They also measured habitat characteristics, such as the proportion of area affected by human activities. Each site supported an average of almost 29 bird species. However because different bird groups preferred different habitat characteristics, the entire landscape contained 134 bird species. For example, species diversity of insectivorous birds was

highest in areas dominated by woody plants. In contrast, raptor diversity was highest in forests with an open canopy. Most bird species showed higher diversity at sites with large areas changed by human activities. Since groups of birds favour different types of habitat, a landscape containing a larger number of habitat types will feature a higher regional diversity of bird species compared to a landscape with only a few habitat types. Because this heterogeneity of habitat types across the landscape supports a high richness of bird species, conserving a variety of habitats over a large area is critical to the conservation of bird species diversity.

LOCAL AND REGIONAL SPECIES DIVERSITY

The study of birds across the boreal forest in Alberta highlights the fact that species diversity can be measured at different spatial scales. For example, if we consider the number of species in a relatively small area of homogeneous habitat, such as a stream, we are looking at **local diversity** or **alpha diversity**. If we consider the number of species in all of the habitats that comprise a large geographic area, we would be looking at **regional diversity** or **gamma diversity**. In the Alberta study, the regional diversity

---

Local diversity The number of species in a relatively small area of homogeneous habitat, such as a stream. *Also known as* Alpha diversity.

Regional diversity The number of species in all of the habitats that comprise a large geographic area. *Also known as* Gamma diversity.

would be the entire list of 134 birds that the researchers identified across all 206 sites.

If each species occurred in all habitats within a region, then the species diversity at the local and regional scales would be identical. However, if species prefer particular habitats, then the number of species at a local scale would be less than the number of species at the regional scale. Moreover, the list of species in each local habitat would be different from each other. Ecologists refer to the number of species that differ in occurrence between two habitats as **beta diversity**. For example, imagine two sites in Alberta: Site A contains five species not found in Site B and Site B contains three species not found in Site A. Because the two sites differ by a total of eight species, the beta diversity is eight. The greater the difference in species between two habitats, the greater the beta diversity.

The collection of species that occurs within a region is the **regional species pool**, which serves as the source of species for all of the local sites within the region. The actual species that live in each local site depend on the species that exist in the regional pool and how well biotic and abiotic conditions at the local scale match the niche requirements of species in the regional pool. Therefore, species in the regional species pool are sorted among localities according to their adaptations and interactions, a process we call **species sorting**.

An example of species sorting can be seen in an experiment in which researchers set up artificial wetlands and manipulated a variety of conditions, including the fertility of the soil and the amount of flooding the soil experienced. The researchers then sowed the seeds of 20 wetland plant species into each wetland to see which plants would germinate and persist over the next 5 years. Of the original 20 species, 1 species failed to germinate in any of the wetlands and 5 others were unable to persist. Of the remaining 14 species, each wetland contained only 3 to 5 species by the end of the experiment. Moreover, there were particular combinations of plant species that survived under each wetland condition. These results confirmed that differences in local conditions cause the sorting of species from the regional species pool.

The number of species increases with area

When we consider large-scale ecology, we see that larger aquatic or terrestrial areas contain more species. In this section, we discuss examples of this phenomenon from a variety of systems and then explore the mathematical relationship between area and species richness. We will then consider the

(a)

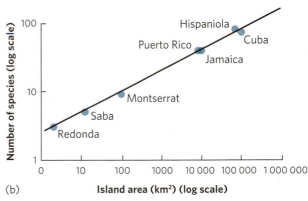

(b)

Figure 22.4 A species–area curve for amphibians and reptiles. On the islands of the West Indies, larger islands contain more species of amphibians and reptiles. **(a)** When plotted as untransformed data, the relationship is a plateauing curve. **(b)** When plotted on a log scale, the relationship is a straight line. Data from R. H. MacArthur and E. O. Wilson, *The Theory of Island Biogeography* (Princeton University Press, 1967).

underlying processes that cause the positive relationship between area and species richness.

SPECIES–AREA RELATIONSHIPS

In a classic study of species diversity, Robert MacArthur and E. O. Wilson examined the patterns of species richness in taxonomic groups living on islands of different sizes. When they plotted the species richness of different islands against the size of the islands, they found that larger islands had a greater species richness. For example, as you can see in Figure 22.4a, the number of amphibian and reptile

---

Beta diversity The number of species that differ in occurrence between two habitats.

Regional species pool The collection of species that occurs within a region.

Species sorting The process of sorting species in the regional pool among localities according to their adaptations and interactions.

species living on islands in the West Indies increases with island size.

The observation that larger areas tend to contain more species has led to the concept of the *species–area curve*. The **species–area curve** is a graphical relationship in which increases in area (A) are associated with increases in the number of species (S). This curve can be described by the following equation:

$$S = cA^z$$

where c and z are constants fitted to the data. To make it easier to work with this equation in the form of a graph, we can take the logarithm of both sides:

$$\log S = \log c + z \log A$$

This is the equation for a straight line with a y intercept of log c and a slope of z. For example, Figure 22.4b plots the same amphibian and reptile data using axes on a log scale. A similar linear relationship can be seen in another collection of data by MacArthur and Wilson in which they examined the number of bird species living on the Sunda Islands in Malaysia, the Philippines, and New Guinea. As you

Figure 22.5 A species–area curve for birds. Among the Sunda Islands in Malaysia, Philippines, and New Guinea, islands with a larger area contain more species of birds when both variables are plotted on a log scale. Data from R. H. MacArthur and E. O. Wilson, An equilibrium theory of insular zoogeography, *Evolution* 17 (1963): 373–387.

can see in Figure 22.5, when the data are plotted on a log scale, the number of bird species linearly increases with island size.

Species–area curve A graphical relationship in which increases in area (A) are associated with increases in the number of species (S).

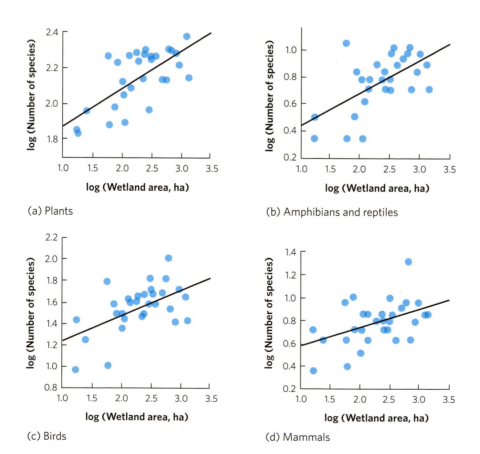

(a) Plants

(b) Amphibians and reptiles

(c) Birds

(d) Mammals

Figure 22.6 Species richness increases with habitat area. In a survey of 30 wetlands in Ontario, an increase in wetland area was found to be associated with an increase in species richness in **(a)** plants, **(b)** amphibians and reptiles, **(c)** birds, and **(d)** mammals. Data from C. S. Findlay and J. Houlahan, Anthropogenic correlates of species richness in southeastern Ontario wetlands, *Conservation Biology* 11 (1997): 1000–1009.

Across many different groups of organisms, the relationship between log A and log S typically has a slope within the range of $z = 0.20$ to 0.35 across scales of 1 m² to the area of an entire state or province or a country of a modest size. This relatively narrow range of slope values across different studies and different taxa suggests that the relationships between species richness and island area reflect similar processes.

Oceanic islands provide good examples of the species–area relationship, but the pattern can also be observed in a wide range of ecosystems. For example, researchers in Ontario, surveyed the species living in 30 wetlands of different sizes. As shown in Figure 22.6, they found that as wetland area increased, there was a corresponding increase in the number of species of plants, amphibians, reptiles, birds, and mammals. As you can see, the positive correlation between area and species richness is common in nature.

HABITAT FRAGMENTATION

While natural processes such as fires and hurricanes can cause habitats to become fragmented, human activities have caused widespread fragmentation of large habitats throughout the world. For example, in the Central American nation of Costa Rica, the forests in 1940 covered much of the country and existed as a large, continuous habitat. As human populations and their associated activities increased, the forests were cleared on a continuing basis, as shown in Figure 22.7. By 2005, much of the forested area had been cleared and the forests that remained existed in many small fragments.

Fragmentation of a large contiguous habitat creates several effects: the total amount of habitat decreases, the number of habitat patches increases, the average patch size decreases, the amount of edge habitat increases, and patch isolation increases. Conversely, in the habitat matrix between the fragments, such as cleared fields between forest fragments, the total area increases and the matrix becomes more continuous. Ecologists are particularly interested in how the creation of isolated habitat patches with different sizes and shapes influences biodiversity and how habitat corridors and the quality of the matrix between habitat fragments affect the rate of species turnover.

Figure 22.7 The fragmentation of old forests. In Costa Rica, the forest—shown in dark green—was once largely continuous. Over time, it was cleared by human activities, and today exists only in a number of smaller fragments. After United Nations Environment Program, Food and Agricultural Program of the United Nations, and the United Nations Forum on Forests Secretariat, 2009, Vital Forest Graphics, 69 pp.

Estimating the Number of Species in an Area

As we have seen, ecologists often need to estimate how many species exist in an area. Since counting every individual is rarely possible, we must take a sample of an area to estimate how many species are present. One way to estimate the number of species in a sample is to graph the number of species that we observe as we increase our sample size. One would expect that the more we sample, the closer we will get to knowing the actual number of species in an area. Eventually, the curve will reach a plateau so that any additional sampling will not discover additional species. A graph of the number of species observed in relation to the number of individuals sampled is known as a **species accumulation curve**.

A second way to estimate the number of species in an area is based on the number of rare species that are detected. When we sample many individuals in a community, we typically end up with a high number of individuals for each of the common species but we may detect only one or two individuals of rare species.

We can estimate the number of species actually living in an area (S) by sampling the community to determine how many species are observed (S_{obs}), how many species are represented by two

individuals (f_2), and how many are represented by one individual (f_1). In essence, species represented by two individuals are likely to be more common in the community than species represented by single individuals. We can therefore estimate the number of actual species that live in a community by starting with how many species we have observed and multiplying this by a ratio that incorporates the number of species represented by one versus two individuals:

$$S = S_{obs} + \frac{f_1(f_1 - 1)}{2(f_2 + 1)}$$

As you can see in this equation, the estimated number of species in a community increases with an increase in the number of species represented by a single individual and decreases with an increase in the number of species represented by two individuals. For example, consider the table on the following page that lists the number of individuals observed for 12 species sampled from three pond communities.

Using the data for Community A, we can estimate the actual number of species. In this community, we have observed 12 species. Two individuals were observed for each of two species

Species accumulation curve A graph of the number of species observed in relation to the number of individuals sampled.

Effects of Fragment Size

The reduction in habitat size that comes with fragmentation commonly causes a decline in species diversity. This happens because each fragment supports smaller populations of species than existed in the original larger habitat and, as discussed in Chapter 13, smaller populations experience higher rates of extinction (see Figure 13.13). For example, in eastern Venezuela a large river was dammed to create a 4300-km^2 lake known as Lago Guri. The region had been composed of a mixture of grazed fields and tropical forests, but after the dam caused the region to flood, hundreds of high points in the landscape became islands in the newly formed lake, as shown in Figure 22.8a. The smallest islands did not contain enough prey to sustain the large vertebrate predators, so predators isolated on the small islands eventually went extinct. These extinctions caused a chain of events that affected the species that remained on the islands. The extinction of the large predators allowed

the herbivores on the islands, such as leaf-cutter ants, iguana lizards, and howler monkeys (*Alouatta seniculus*), to increase. The population explosion of these herbivores affected the island plants. As you can see in Figure 22.8b, greater abundance of herbivores on smaller islands caused higher mortality in existing sapling trees and reduced the recruitment of new trees into the sapling stage.

Effects of Fragment Edges

As you may recall from Chapter 18, when habitats are fragmented they produce ecotones, which are regions with distinct species composition and environmental conditions over a relatively short distance. When a habitat is fragmented, an ecotone is formed along the edges of the habitat. Increases in fragmentation cause an increase in the amount of edge habitat compared to the amount of edge present in the original unfragmented habitat. For example, consider a single square habitat with an area of 1 ha compared to the

SPECIES	COMMUNITY A	COMMUNITY B	COMMUNITY C
WOOD FROGS	32	65	42
GREEN FROGS	45	24	45
NEWTS	12	14	5
SPOTTED SALAMANDERS	15	17	12
RAMSHORN SNAILS	2	23	67
POND SNAILS	25	14	35
AMPHIPODS	2	14	2
ISOPODS	1	1	2
FINGERNAIL CLAMS	9	1	0
DIVING BEETLES	1	1	0
WATER BOATMEN	5	1	0
DRAGONFLIES	1	1	0

whereas one individual was observed for each of three species. Using our equation above:

$$S = S_{obs} + \frac{f_1(f_1 - 1)}{2(f_2 + 1)}$$

$$S = 12 + \frac{3(3 - 1)}{2(2 + 1)}$$

$$S = 12 + \frac{6}{6}$$

$$S = 13$$

Based on these results, we conclude that although we observed 12 species, we estimate that the community actually contains 13 species.

YOUR TURN Using the equation, estimate the number of species that are present in Community B and Community C. Based on your results, how does the number of species represented by single individuals affect the estimated number of species in the community?

(a)

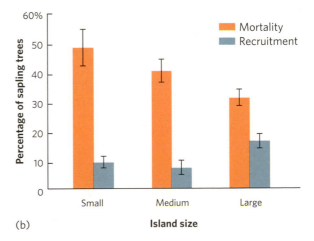

(b)

Figure 22.8 Fragmentation of terrestrial habitats from the flooding of Lago Guri. In eastern Venezuela, a dam was constructed to form Lago Guri in 1978 and the area behind the dam was flooded. **(a)** Flooding caused the hills of the region to become islands in the lake. **(b)** On the smallest islands, large predators were no longer present, so the populations of herbivores increased, with a major effect on the vegetation. Sapling trees on small islands experienced the highest mortality and the lowest recruitment. Error bars are standard errors. Photo by Peter Langer/DanitaDelimont.com. Data from J. Terborgh et al., Vegetation dynamics of predator-free land-bridge islands, *Journal of Ecology* 94 (2006): 253–263.

100 m

Total area = 1 ha
Total edge = 400 m

Total area = 1 ha
Total edge = 1600 m

Figure 22.9 The effect of habitat fragmentation on edge habitat. If we have two habitats of the same total area, the habitat that is fragmented has much more edge.

same total area divided into 16 smaller square habitats. As illustrated in Figure 22.9, the single large habitat would have 400 m of edge whereas the 16 smaller habitats would have 1600 m of edge. Round habitats have the lowest ratio of edge to area while long, skinny ovals or rectangles have much higher edge to area ratios.

An increase in edge habitat changes both the abiotic conditions and the species composition of a habitat. In a fragmented forest, for instance, the edge of a newly created fragment experiences greater sunlight, warmer temperatures in the summer, and higher rates of evaporation. These changes can make the edge of the forest less suitable to many forest species and more suitable to others.

Species that prefer edge habitat increase in abundance when fragmentation occurs, and this can affect other species living in the fragment. For example, the marbled murrelet (*Brachyramphus marmoratus*) is a threatened seabird that prefers nesting sites close to the forest edge, and nests on large trees in old growth forests on the British Columbia coast (Figure 22.10). Murrelets can reach nests close to forest edges more easily than nests inside the forest. However, researchers studying this bird have recently shown that chicks in nests close to forest edges are more likely to be preyed upon by Steller's jays (*Cyanocitta stelleri*), which prefer fragmented habitats and forest edges. The researchers recommended a plan that would reduce fragmentation and create larger protected areas to reduce nest predation on the murrelets.

Fragmentation can also have consequences for human health. You may recall from Chapter 15 that human exposure to Lyme disease depends on a complex food web that starts with mice and chipmunks serving as hosts for newly hatched ticks that carry the pathogenic bacteria. In forest fragments in the northeastern United States, the abundance of many vertebrate animals has declined, but the abundance of the white-footed mouse has increased, probably because most of the mouse's competitors and predators cannot live in the smallest

forest fragments. When researchers surveyed mouse populations from forest fragments of different sizes, they found that the smallest forest fragments had the highest mouse densities, and thus the highest tick densities. The researchers also found that mice living in the smallest fragments had the highest proportion of ticks infected with the Lyme disease bacterium. In short, the fragmentation of forests due to human activities has created a landscape that makes us more likely to be exposed to Lyme disease.

Corridors, Connectivity, and Conservation

In Chapters 11 and 12 we touched on the importance of considering geographic scales larger than the local area. In our discussion of metapopulations, we emphasized that many populations are divided into habitat fragments and that the regional populations persist because each patch is connected by the occasional dispersal of individuals between patches. We saw an example of this in Chapter 11 when we discussed the recovery of the collared lizard. This lizard depended on patches of open habitat, known as

Figure 22.10 Edge habitat and nest predation. Marbled murrelets nest in old-growth forests on the coast of British Columbia. While nests on the forest edges are easier to access for the murrelets, such nests are also more likely to be preyed upon by Steller's jays, which are more common in fragmented and edge habitat. Photo by AP Photo/Courtesy of David Forthoffer, HO.

glades, as well as corridors between the glades through which they could disperse. Corridors facilitate movement, which can save declining populations of many species from local extinction. They also increase gene flow and genetic diversity within populations, which counteract the negative effects of genetic bottlenecks and genetic drift. Corridors can simply be pieces of preserved habitat or they can be constructed, such as corridors built to let animals cross over a highway (Figure 22.11).

Although corridors can rescue declining populations by adding new colonists that bring genetic variation, they can also have unintended downsides. For example, corridors built to help conservation of a particular species can facilitate the movement of predators (including poachers), competitors, and pathogens that are harmful to conservation efforts. Therefore, resource managers must carefully consider the costs and benefits of developing corridors among habitat before spending time and money to implement this strategy.

The importance of corridors is greatest for those organisms that require a continuous connection to move between habitat fragments. However, organisms such as birds and flying insects can pass over stretches of inhospitable habitat matrix and therefore may not need a continuously connected corridor. Instead, these species can move between large patches of favourable habitat if they have access to small intervening patches where they can stop to rest or forage. These small intervening patches that dispersing organisms can use to move between large favourable habitats are known as **stepping stones**.

The role of habitat corridors and stepping stones that can facilitate the movement of organisms among

Figure 22.12 Wildlife corridors in India. The original habitat of Asian elephants is now fragmented into several protected areas. Conservation groups are currently trying to preserve strips of land between these protected areas to serve as corridors through which elephants and other species can move. Photo by Jagdeep Rajput/DanitaDelimont.com.

patches of fragmented habitat has spurred major efforts at preservation. India, for example, is home to nearly 60 percent of all Asian elephants (*Elephas maximus*), which live in several national parks and protected areas. These areas are the fragmented remains of a much larger and contiguous habitat. The World Land Trust and the Wildlife Trust of India are working together to protect important corridors between the protected habitat fragments to ensure the long-term persistence of the elephants (Figure 22.12). Although elephants are charismatic animals that can draw attention to conservation needs, these corridors are likely to assist in the conservation of other charismatic species including tigers (*Panthera tigris*) and Himalayan black bears (*Ursus thibetanus*), as well as many less-charismatic species.

The Matrix of Habitats Between Fragments
As you may recall from our discussion of metapopulations in Chapter 11, the quality of the matrix habitat that lies between favourable fragments of habitat is a key factor that helps determine whether organisms can move between these favourable fragments. Some habitats in the matrix may contain favourable conditions for the movement of organisms whereas others might be inhospitable. For example, researchers in Colorado mapped the locations of meadows, willow thickets, and coniferous forests in a valley and then studied the movements of more than 6000 individual butterflies from six taxonomic groups. Butterflies feed in meadows and when moving from one meadow fragment to another they must pass through a matrix containing willow thickets and coniferous forests. The researchers

Figure 22.11 Building corridors. Road construction through forests is one way that humans have fragmented large habitats. To help wildlife move between forest fragments, some roads, such as this one in Alberta, have overhead corridors with surfaces of soil and natural vegetation. Photo by Joel Sartore/National Geographic Stock.

Stepping stones Small intervening habitat patches that dispersing organisms can use to move between large favourable habitats.

captured butterflies in meadows and wrote a unique number on a wing of each butterfly before releasing it. Researchers then recaptured the butterflies to determine whether they exhibited a preference for dispersing through the willow thickets or through the coniferous forests. There was no preference in two of the six taxonomic groups. These two groups reflected the two extremes in flying ability; one group rarely moved out of a meadow while the other group was composed of excellent fliers that easily navigated both intervening habitats. In contrast, individuals from the other four taxonomic groups were 3 to 12 times more likely to move between meadows via willow thickets than through coniferous forests. This confirms that some species of butterflies have distinct preferences for dispersing through particular habitats and these preferences are related to their ease of dispersal. The Alpine Parnassius butterfly we discussed in Chapter 13 provides another example. As we saw, this butterfly lives in a patchwork of meadows on the eastern slopes of the Rocky Mountains in Alberta. When researchers marked individual butterflies and followed their movements, they found that the butterflies move easily over open meadows but are much less likely to fly through forested areas. From these examples we can see that the matrix habitats can affect the movements of organisms among fragments.

The equilibrium theory of island biogeography incorporates both area and isolation

When Robert MacArthur and E. O. Wilson examined the relationship between island size and species richness, they noticed that although the size of the island was strongly associated with the number of species, there was commonly a large amount of variation around the species–area curve that suggested other processes were also affecting the number of species living on the island. They considered the locations of the islands from which they collected data and observed that the distance from a source of colonizing species also affects the number of species living on an island. Islands closer to a source of colonizing species—for example a mainland—appeared to receive more colonizing species. In this section, we will examine the combined effects of island area and isolation on species richness and then explore a graphical model that helps us understand these patterns. Finally, we will explore how this information can be used to design nature reserves.

THE EVIDENCE

To test their hypothesis that species richness is determined by both island area and isolation, MacArthur and Wilson collected data on different types of organisms from groups of islands throughout

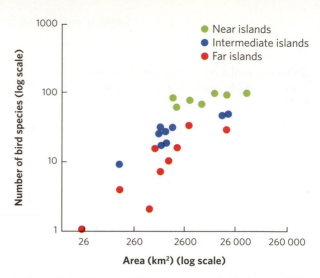

Figure 22.13 Effects of island area and isolation on bird species richness. In the South Pacific, researchers examined the number of birds living on islands that were various distances from New Guinea. On islands of a similar distance from New Guinea, large islands contained more bird species than small islands. On islands of a similar area, islands near to New Guinea contained more bird species than islands far from New Guinea. Data from R. H. MacArthur and E. O. Wilson, An equilibrium theory of insular zoogeography, *Evolution* 17 (1963): 373–387.

the world. For example, they examined the number of bird species that existed on 25 islands in the South Pacific. The nearest source of new species for these islands was the large island of New Guinea, and these islands could be categorized as being near, intermediate, or far from New Guinea. When the researchers plotted the number of bird species for each island as a function of island area, shown in Figure 22.13, they found that both island area and the distance from New Guinea affected the number of species living on the island. Within each of the distance categories, larger islands contained more species. For islands of a given area, near islands contained more bird species than far islands.

Although oceanic islands provide a good test of how colonization and extinction affect the number of species, we can also consider other types of distinct habitats including islands of habitat that exist on continents. For example, James Brown and his colleagues examined the number of mammal species living on mountaintops in the southwestern United States. These mountaintops include alpine tundra and conifer forest habitats that are surrounded by lower elevation habitats of woodlands, grasslands, and desert scrub. Twenty-six species of mammals in the region prefer to live in the mountaintop habitats. Researchers asked whether the number of mammal species on each mountaintop was affected by the area of the mountaintop or the isolation of the mountaintop from two sites containing sources of colonizing species: the southern Rocky Mountains and the Mogollon Rim, a mountain range that cuts across northern Arizona. As you can see in Figure 22.14, the researchers discovered more species of mammals on mountaintops with the largest areas and fewer species of mammals on mountaintops that were farthest away from the two sources of colonizers.

The observations on oceanic islands and mountaintop islands suggest that island area and isolation are important factors in determining species richness. However, a manipulative experiment was needed to demonstrate that these processes do, in fact, cause the observed pattern. In a classic experiment, Daniel Simberloff, a graduate student of E. O. Wilson, worked on a set of tiny islands in the Florida Keys. These islands typically contain a single mangrove tree and no animals other than insects, spiders, and other arthropods. Prior to the experiment, Simberloff and his team documented that islands located closer to sources of colonizing species contained more species than islands located far away from sources of colonizing species. The researchers built tents over selected islands, as you can see in Figure 22.15a, and then fumigated the islands with an insecticide that killed nearly every arthropod. After removing the tents, they returned to the islands every few weeks for a year to track how many species recolonized the islands from large islands nearby. As you can see in Figure 22.15b, the researchers found that each island was rapidly recolonized by arthropods. Moreover, the final number of species was close to the original number of species prior to the fumigation and the more isolated islands once again contained fewer species of arthropods. The actual list of species was different, but the richness was similar. This discovery was important because it experimentally

(a)

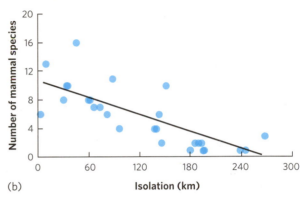

(b)

Figure 22.14 Effects of area and isolation on mountaintop mammals. For mammals that live on mountaintop habitats in the southwestern United States, researchers found **(a)** more mammal species living on mountaintops of greater area and **(b)** fewer mammal species living on mountaintops that are more distant from a source of colonists. Data from M. V. Lomolino et al., Island biogeography of montane forest mammals in the American Southwest, *Ecology* 70 (1989): 180–194.

(a)

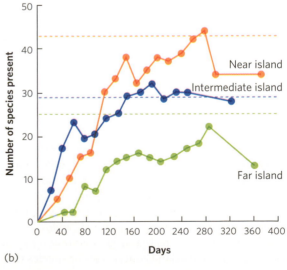

(b)

Figure 22.15 An experimental test of island biogeography theory. **(a)** Researchers in the Florida Keys constructed a scaffold frame around islands and covered the scaffold with tarps to act as a tent when fumigating the islands. Fumigating the islands removed most of the arthropods. **(b)** Over the course of a year, researchers returned to determine how many species of arthropods had recolonized. By the end of the experiment, the islands had nearly recovered their original number of species, as indicated by the dashed lines, with the near islands containing a higher number of species than the far islands. Photo by Daniel Simberloff. Data from D. S. Simberloff and E. O. Wilson, Experimental zoogeography of islands: The colonization of empty islands, *Ecology* 50 (1969): 278–296.

demonstrated that isolation was a key factor in determining how many species could live on an island.

THE THEORY

Based on repeated observations that both island size and island isolation affect the number of species living on an island, Robert MacArthur and E. O. Wilson developed the *equilibrium theory of island biogeography*. The **equilibrium theory of island biogeography** states that the number of species on an island reflects a balance between the colonization of new species and the extinction of existing species.

To understand the equilibrium theory of island biogeography, we need first to understand the factors that affect the colonization of species onto an island and the extinction of species from an island. As we discussed in Chapter 11, species differ in their ability to disperse depending on their overall size and mode of locomotion. So now imagine that we have an island that is completely uninhabited by any species. With no species on the island, a large number of species on the nearby mainland could potentially colonize it. As a growing number of species colonize the island, the pool of species that have not yet colonized the island shrinks and those that remain on the mainland are probably not very good at dispersing. Therefore, the rate of new species colonizing an island declines as a function of how many species have already colonized the island. If the island contained every species that is found on the nearby mainland, the rate of new colonizing species would fall to zero. In Figure 22.16, the relationship between species richness and rate of colonization by new species is shown as an orange line.

Next we need to consider the factors that affect the rate of extinction on an island. Since the extinction rate is expressed as the number of species that will go extinct over a period of time, the extinction rate must be affected by how many species are present. In the simplest case, when there are no species, there can be no extinctions. As the island begins to be colonized, it holds a few species that could potentially go extinct. As more species live on the island, more species are subject to possible extinction, and so the extinction rate increases. The extinction rate should also be affected by harmful species interactions on the island. For example, competition, predation, and parasitism are all more likely to increase as the total number of species increases. The blue line in Figure 22.16 shows that the extinction rate ranges from zero when no species are present on the island to a maximum rate of extinction when the island contains every species that it can possibly support.

Given that the island continues to experience colonization of new species and extinction of existing species, these two opposing forces will eventually reach an equilibrium point where the rate of

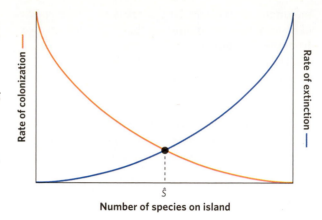

Figure 22.16 Achieving an equilibrium number of species on an island. As the number of species living on an island increases, the rate of colonization by new species from the regional pool declines. At the same time, the rate of extinctions of species living on the island increases. The equilibrium number of species, $\hat{S}$, occurs where the two curves cross and the opposing processes are balanced.

colonization is equal to the rate of extinction. On our graphical model, equilibrium occurs where the two curves cross. Following the dashed line down from the equilibrium point, we see that the number of species on the island at equilibrium is designated as $\hat{S}$ on the *x* axis. In considering this figure, it is important to remember that the model predicts the number of species present at equilibrium, but it does not predict a particular composition of species at equilibrium. Even at equilibrium new species continue to colonize the island while species already on the island continue to go extinct. At equilibrium, there is still a continuous turnover of species on the island. We saw an example of this in the islands in the Florida Keys that were fumigated by Simberloff and Wilson. Over time, the species richness of the islands came to equilibrium, with the number of species depending on the distance between the island and the mainland. Once an island reached equilibrium, it contained a relatively fixed number of species, though the composition of species changed as the islands continued to experience extinctions and colonizations of invertebrate species.

Using the graphical model in Figure 22.16 as a foundation, we can make predictions about the combined effects of island size and isolation on the number of species on an island when it is at equilibrium. Let's first consider the effects of island size. We would expect smaller islands to sustain smaller populations of each species. As we have discussed in several previous chapters, smaller populations typically have higher extinction rates.

Equilibrium theory of island biogeography A theory stating that the number of species on an island reflects a balance between the colonization of new species and the extinction of existing species.

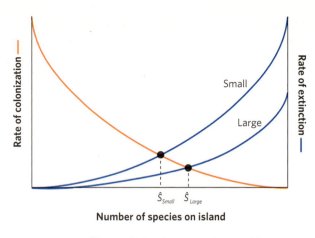

Figure 22.17 Effects of island size on the equilibrium number of species. Because smaller islands support smaller populations that are more prone to extinction, smaller islands have steeper extinction curves. As a result, smaller islands contain fewer species at equilibrium ($\hat{S}_{Small}$) than larger islands ($\hat{S}_{Large}$).

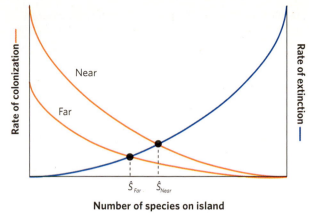

Figure 22.18 Effects of island isolation on the equilibrium number of species. Islands that are far from a source of new colonizing species experience lower rates of colonization than islands that are close. As a result, farther islands contain fewer species at equilibrium ($\hat{S}_{Far}$) than near islands ($\hat{S}_{Near}$).

Therefore, smaller islands should experience higher rates of extinction, as shown by the curve labelled "Small" in Figure 22.17. When an island is at equilibrium, extinction rates balance colonization rates. Therefore the number of species on a small island at equilibrium should be lower than the number of species on a large island at equilibrium. If we follow the two equilibrium points on the figure down to the *x* axis, we can see that the number of species at equilibrium is lower for small islands than for large islands.

Next we can consider the effects of island isolation. An island that is near a continent should experience higher rates of colonization by new species than an island that is far from a continent. Once again using Figure 22.16 as our starting point, we can make predictions about how island isolation should affect the number of species at equilibrium. In Figure 22.18, we can see that the far island has a lower colonization curve than the near island. This creates two equilibrium points between colonization and extinction rates. If we follow the dashed lines down from each equilibrium point, we see that the number of species at equilibrium is lower for far islands than for near islands.

When we combine the effects of island size and island isolation from a continent, we arrive at several predictions. Looking at Figure 22.19, we see that the combinations of large versus small islands and near versus far islands creates four different possible equilibrium points. Once again, we can follow the dashed line down from each equilibrium point to determine how many species are predicted to be on a particular type of island when colonization and extinction rates are at equilibrium. Small islands that are far from a major source of new species should contain the fewest species. In contrast, large islands that are near a major source of new species should

contain the most species. In short, the combination of island size and isolation determines the number of species that the island can hold.

APPLYING THE THEORY TO THE DESIGN OF NATURE RESERVES

Understanding the effects of island size, shape, and distance from sources of colonizing species has helped scientists to design nature reserves to protect biodiversity, as depicted in Figure 22.20. For example, we know that large areas typically contain more species than small areas because they contain a greater diversity of habitats. Large areas can support larger populations of each species, which lowers the rate of species extinction. Based on these observations, we can say that setting aside large nature reserves will

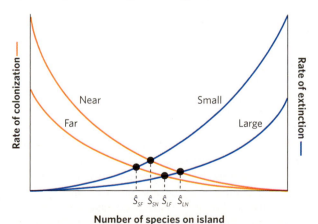

Figure 22.19 The combined effects of island size and isolation on species richness. When we simultaneously consider island size and isolation, we find that the small islands that are far from a continent should have the lowest number of species at equilibrium ($\hat{S}_{SF}$) whereas large islands that are close to a continent should have the highest number of species ($\hat{S}_{LN}$).

Figure 22.20 Nature reserve design. The number, shape, size, and proximity of reserves affect the probable success of preserving a region's biodiversity. As noted in this figure, some designs are likely to be better than others for preserving biodiversity. Adapted from J. M. Diamond, The island dilemma: Lessons of modern biogeographic studies for the design of nature reserves, *Biological Conservation* 7 (1975): 129–146 and J. C. Williams et al., Spatial attributes and reserve design models, *Environmental Modeling and Assessment* 10 (2005): 163–181.

better protect biodiversity than will small nature reserves. Similarly, when faced with the option of setting up a single large reserve or several small reserves of the same total area, the single large reserve is often the better option to preserve biodiversity. However, a single habitat poses several risks. When a species of conservation concern is located in a single large habitat, it is more likely that the species will be destroyed by natural disaster or disease.

When we are faced with creating multiple reserves, we need to consider the proximity of the reserves to each other. Reserves need to be close enough for organisms to disperse between them but far enough away to reduce the ability for predators and diseases to move between them. Setting this distance is a compromise between these positive and negative outcomes, and the best solution will differ among organisms with different dispersal abilities. Sometimes the ideal distance will be set for the species of greatest conservation interest. A similar challenge exists when reserves are joined by corridors that provide strips of hospitable habitat between adjacent reserves.

Because we know that increased edge can have detrimental effects on species that prefer to live in the interior of habitats, it is also important that we consider reserve shape. Round reserves have the lowest ratio of edge to area while long, skinny ovals or rectangles have much higher edge to area ratios that will favour edge species. Although reserve size, number, shape, and proximity are all important considerations, often the

ideal scenario is not possible; nature reserves typically represent a compromise between the areas that are desirable and the areas that are available.

On a global scale, biodiversity is highest near the equator and declines toward the poles

We have examined the local and regional processes that can affect the number of species in a given location on Earth. However, patterns of biodiversity also exist at the global scale. One of the most striking patterns is that species richness of all taxa combined is highest near the tropics and declines toward the poles. For example, a hectare of forest typically has fewer than 5 species of trees in boreal regions, 10 to 30 species of trees in temperate regions, and up to 300 species of trees in tropical regions. These latitudinal trends in diversity are pervasive and extend even to the oceans. In this section, we will explore the patterns of diversity within and across latitudes. We will then discuss the two general hypotheses for these patterns of biodiversity and explore three important processes.

PATTERNS OF DIVERSITY

In the Northern Hemisphere, the number of species in most groups of animals and plants increases from north to south. For example, when a researcher counted the number of mammal species that one could find in blocks of habitat that were 241 km on a side, he found that there were fewer than 20 mammal species per block in northern Canada but more than 50 species per block in the southern United States. You can view his data in Figure 22.21a.

The number of mammals also increases as we move from east to west in North America. For instance, there are typically 50 to 75 species per sample block in the east whereas there are 90 to 120 species per block in the west. Such a pattern is likely due to the greater amount of habitat heterogeneity in the extensive mountain ranges of western North America. This greater heterogeneity of environments in the west apparently provides suitable conditions for a greater number of species, similar to the pattern we saw for plants in the Cape Floristic Region at the beginning of this chapter.

The pattern of species richness for breeding land birds in North America resembles the pattern for mammals, but the patterns for reptiles, trees, and amphibians are strikingly different, as you can see in Figures 22.21b–d. The richness of reptile species declines fairly uniformly as temperature decreases toward the north. However, trees and amphibians are more diverse in the moister eastern half of North America than in the drier, more mountainous western regions.

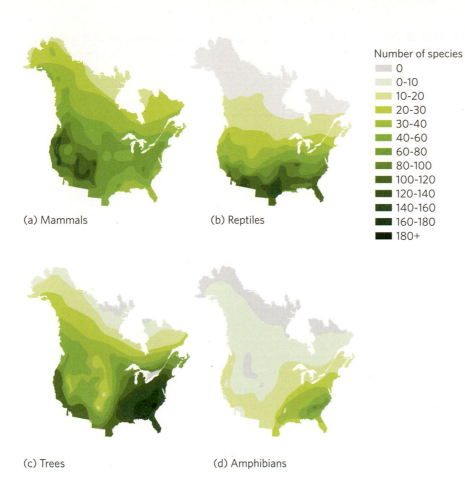

(a) Mammals

(b) Reptiles

(c) Trees

(d) Amphibians

Number of species
- 0
- 0-10
- 10-20
- 20-30
- 30-40
- 40-60
- 60-80
- 80-100
- 100-120
- 120-140
- 140-160
- 160-180
- 180+

Figure 22.21 Patterns in North American diversity. When a researcher examined the number of species living in Canada and the United States, he found strong patterns in species richness. The contour lines on the map indicate the numbers of species found in blocks of land that were 241 km on a side. (a) Mammals increase in species richness from north to south and from east to west. (b) Reptiles increase in species richness from north to south. (c, d) For both trees and amphibians, the highest number of species occurs in the southeast where there is a combination of high precipitation and warm temperatures. Data from D. J. Currie, Energy and large-scale spatial patterns of animal- and plant-species richness, *American Naturalist* 137 (1991): 27–49.

The general pattern of increased species richness at lower latitudes that we observed for trees and vertebrate animals also holds true for the oceans. For example, in 2010 researchers compiled millions of records of marine organisms around the world including whales, dolphins, fish, and corals. When they examined the global patterns of diversity using sample blocks of 880 km on a side, as shown in Figure 22.22, they found that the highest diversity occurred in the tropics including Central America and Southeast Asia whereas the lowest diversity occurred near the poles. Within a given latitude, they also found a higher diversity near the coasts and lower diversity in the open ocean.

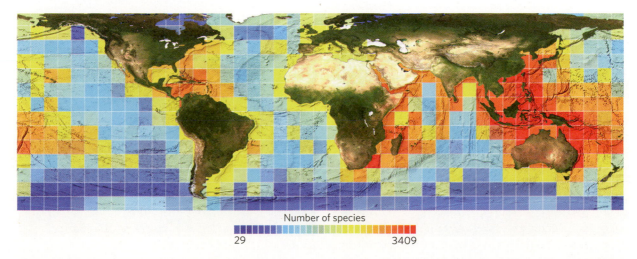

Number of species

29   3409

Figure 22.22 Patterns of marine biodiversity. In the ocean, the greatest number of species exists in the tropics and declines toward the poles. At a given latitude, more species exist near the coasts and fewer species exist in the open ocean. The number of species is calculated using sample blocks of 880 km per side. After D. P. Tittensor et al., Global patterns and predictors of marine biodiversity across taxa, *Nature* 466 (2010): 1098–1103.

PROCESSES THAT UNDERLIE PATTERNS OF DIVERSITY

For more than a century ecologists have investigated reasons for the decline in species richness as we move from the equator to the poles. Historically, ecologists have considered two general hypotheses. According to one hypothesis, species are continually created over time and without limit. Because the world's current temperate and polar regions experienced repeated advances and retreats of glaciers during the Ice Age, species in these regions have been eliminated or driven toward refuges closer to the equator. In contrast, because the world's tropical regions did not experience glaciation, habitats in these areas have remained stable much longer and so have had more time to accumulate species.

The second hypothesis proposes that the number of species reflects an equilibrium between the processes that create new species and the processes that drive species extinct, similar to the state of equilibrium described by the theory of island biogeography. According to this hypothesis, the higher number of species in the tropics is the result of higher rates of speciation or lower rates of extinction there compared to the temperate and polar regions. Similarly, variation in the number of species as one moves across a given latitude should also reflect an equilibrium between processes that create new species and destroy existing species. We will now consider three processes that play substantial roles in determining the number of species that occur in an area: ecological heterogeneity, solar energy and precipitation on land, and water temperatures in the oceans.

The Role of Ecological Heterogeneity

At any given latitude, we find more species in areas where there is greater ecological heterogeneity, such as heterogeneity in the soils and plant life. For example, grasslands contain vegetation that is less heterogeneous in growth form than shrublands or deciduous forests. Surveys of breeding birds in North America find an average of 6 species in grasslands, 14 species in shrublands, and 24 species in floodplain deciduous forests. You may recall that we discussed a similar phenomenon in Chapter 18 when we looked at the positive relationship between the diversity of vegetation height and the diversity of birds (see Figure 18.13). This same pattern has also been observed in lizards of the southwestern United States, where lizard diversity is associated with vegetation diversity, and in the plants of the Cape Floristic Region of Africa, where plant diversity is associated with the heterogeneity of the soils.

Although more productive terrestrial habitats tend to have more species, habitat heterogeneity can also play a role in determining species richness when two habitats have similar levels of productivity. For instance, habitats with less variation in vegetation growth form, such as grasslands, have fewer animal species than habitats with similar productivity but more variation in vegetation. This principle also applies to plants. Marshes are highly productive but have a relatively uniform landscape and thus contain relatively few species of plants. While desert vegetation is less productive than marsh vegetation, the greater heterogeneity of the desert landscape provides room for more species of plants (Figure 22.23).

The Role of Solar Energy and Precipitation on Land

The number of species found at any location as one moves from the poles to the tropics is positively correlated to the amount of solar energy and precipitation at that location. Ecologists have combined solar energy and precipitation into a

(a)

(b)

Figure 22.23 Habitat heterogeneity. Although marshes are more productive than deserts, the relative uniformity of the marsh environment has produced a low diversity of species whereas the heterogeneous landscape of deserts has produced a higher diversity of species. Pictured here are **(a)** Algonquin Provincial Park, Ontario and **(b)** the Sabino Canyon Recreation Area in Arizona. Photos by (a) Michael P. Gadomski/Science Source, and (b) Ron Niebrugge/Alamy.

measure of *potential evapotranspiration*. **Potential evapotranspiration (PET)** is the amount of water that could be evaporated from the soil and transpired by plants, given the average temperature and humidity. Because PET integrates temperature and solar radiation, it provides an index to the overall energy input into the environment. As you can see in Figure 22.24, PET correlates with the species richness in North American vertebrates relatively well. In each group of vertebrates, the initial rise in PET is associated with a rise in species richness, which reflects increasing diversity from north to south within the continent. However, species richness eventually plateaus at very high levels of PET. These high PET levels occur as we move from east to west at middle latitudes where the increasing temperature fails to improve the capacity of the environment to support additional vertebrate species. In the arid western parts of the continent, for instance, increasing temperature eventually becomes a stressor.

The correlations between PET and species diversity for terrestrial vertebrates has given rise to the idea that there is a causal relationship between the two variables. This hypothesized causal relationship is known as the **energy–diversity hypothesis**, which states that sites with higher amounts of energy are able to support more species. Higher amounts of energy would also support higher abundances of individuals from each species, which should reduce the rate of extinction. Moreover, higher energy input might accelerate the rate of evolutionary change and thereby increase the rate of speciation. While these ideas are attractive, none of these mechanisms has yet been verified experimentally.

The Role of Water Temperature in the Oceans

As we have seen, biodiversity in marine environments is greater in the tropics than it is at higher latitudes. However, this pattern does not appear to be driven by greater productivity in the tropics. Though marine productivity is greatest in temperate latitudes (see Figure 20.6), the high productivity in those regions is seasonal. As you may recall, temperature differences and stratification of the water in temperate zones make nutrients readily available during the seasonal mixing of the water column but scarce during periods when the water is stratified. In contrast, tropical marine environments experience relatively stable temperatures that lead to relatively small nutrient fluctuations and low but steady productivity. In 2010 researchers tested whether the patterns in species diversity across the oceans were better explained by productivity, mean water temperature, or variation in water temperature. They discovered that the only significant predictor of marine biodiversity across latitudes was the mean temperature of the sea surface. Because a higher mean temperature is a

Potential evapotranspiration (PET) The amount of water that could be evaporated from the soil and transpired by plants, given the average temperature and humidity.

Energy–diversity hypothesis A hypothesis that sites with higher amounts of energy are able to support more species.

(a) Birds

(b) Mammals

(c) Amphibians

(d) Reptiles

Figure 22.24 The relationship between species diversity and potential evapotranspiration. For all four groups of vertebrates, the potential evapotranspiration of a sample is strongly correlated to the number of species found in the sample block. Data from D. J. Currie, Energy and large-scale spatial patterns of animal- and plant-species richness, *American Naturalist* 137 (1991): 27–49.

250 Mya   100 Mya   60 Mya

Figure 22.25 Continental drift. About 250 Mya, all of Earth's landmasses were joined together into Pangaea. Pangaea later split into Gondwana and Laurasia, which in turn split into modern-day continents. Many of these landmasses subsequently joined together in new configurations.

measure of greater total energy, this pattern further supports the energy–diversity hypothesis.

The patterns of species richness across and within latitudes are potentially affected by all three processes of habitat heterogeneity, temperature and precipitation on land, and mean temperature in the oceans. In all three cases, the mechanisms involved suggest that the global distribution of species richness is the outcome of an equilibrium between the processes that create new species and the processes that cause the extinction of existing species.

The distribution of species around the world is affected by Earth's history

When we consider the patterns of species diversity on Earth, we need to remember that Earth was formed 4.5 billion years ago and life arose during the first billion years. This means that the species on Earth today are the result of billions of years of evolution and that current diversity arose in response to past environmental conditions. In this section, we will examine how the movement of continents and historic changes in climate affected the distribution of species.

CONTINENTAL DRIFT
During Earth's history, the continents have repeatedly come together and drifted apart. The movement of landmasses across the surface of Earth is called **continental drift**. Continental drift occurs because the continents are essentially giant islands of low-density rock that are moved by convection currents of semi-molten material. About 250 Mya, all of Earth's landmasses were joined together as a single landmass, which scientists have named **Pangaea**, illustrated in Figure 22.25. By 150 Mya, Pangaea had separated into a northern landmass, known as **Laurasia**, and a southern landmass, known as

Gondwana. Laurasia subsequently split into North America, Europe, and Asia while Gondwana split into South America, Africa, Antarctica, Australia, and India. Ultimately, India collided into Asia, which caused the land to rise and created the Himalayas. In the Northern Hemisphere, a widening Atlantic Ocean separated Europe from North America, but a land bridge had already formed on the other side of the world between North America and Asia. More recently, Europe and Africa joined about 17 Mya and a land bridge formed between North and South America at the Isthmus of Panama 3 to 6 Mya.

An important consequence of continental drift was the changing opportunities for dispersal among continents. Once separated, continents could independently evolve species in different regions of Earth. For instance, because Australia has experienced a long period of isolation from other continents, it has evolved many unique groups of species that include a wide variety of marsupial animals, such as koalas and kangaroos, and plants, such as eucalyptus trees. When continents later joined together, groups of organisms that were unique to one landmass were able to move into new areas. You may recall from Chapter 3 that camels originated in North America about 30 Mya. When North America became connected to Asia at the Bering Land Bridge in Alaska, camels dispersed into

Continental drift The movement of landmasses across the surface of Earth.

Pangaea The single landmass that existed on Earth about 250 Mya and subsequently split into Laurasia and Gondwana.

Laurasia The northern landmass that separated from Pangaea about 150 Mya and subsequently split into North America, Europe, and Asia.

Gondwana The southern landmass that separated from Pangaea about 150 Mya and subsequently split into South America, Africa, Antarctica, Australia, and India.

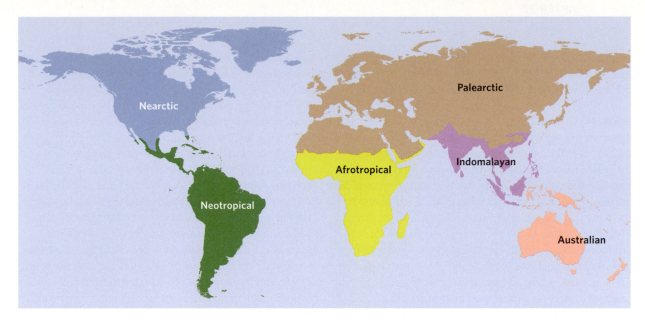

Figure 22.26 Biogeographic regions. Terrestrial regions of the world can be categorized in terms of distinctive groups of plants and animals that largely reflect the history of continental drift.

Asia and then moved into Africa where they diversified into the modern-day Bactrian camels and dromedary camels. At about the same time, other camels dispersed into South America where they further diversified into llamas, guanacos, vicuñas, and alpacas. In short, the history of connections between the continents endures in the distributions of present-day animals and plants.

BIOGEOGRAPHIC REGIONS

Because continental drift has allowed individual continents independently to evolve groups of organisms for long periods, we can see patterns of species distributions on each continent. Alfred Wallace, a contemporary of Charles Darwin and codiscoverer of evolution by natural selection, first described these patterns and delineated six major zoogeographic regions that are based on the distributions of animals. These regions, illustrated in Figure 22.26, are still recognized today. Botanists also recognize six major biogeographic regions, based on the distributions of plants, with boundaries that closely coincide with those of the zoogeographic regions.

The Northern Hemisphere is divided into the **Nearctic region**, which roughly corresponds to North America, and the **Palearctic region**, which corresponds to Eurasia. Through most of the past 100 million years the continents of these regions maintained connections across what is now Greenland between North America and Eurasia and across the Bering Strait between Alaska and Russia. Consequently, these two regions share many groups of animals and plants. European forests seem familiar to tourists from North America, and vice versa; although few species are the same, both regions have representatives of many of the same genera and families.

The Southern Hemisphere is divided into four biogeographic regions. The **Neotropical region** corresponds to South America and the **Afrotropical region**, also known as the **Ethiopian region**, corresponds to most of Africa. Farther to the east is the **Indomalayan region**, also known as the **Oriental region**, which includes India and Southeast Asia. The final biogeographic region is the **Australasian region**, which includes Australia, New Zealand, and New Guinea. In Wallace's delineation, the continent of Antarctica was not included.

Each of these regions contains unique groups of species that reflect their long histories of isolation from the rest of the terrestrial world and the subsequent exchange of species after they were joined together. For example, when the Isthmus of Panama joined the Nearctic and Neotropical regions, there was a large, though uneven, exchange of mammals. Many North American lineages moved into South America and caused the extinction of most of the marsupial animals that had come to South America during an earlier connection with Australia. In

Nearctic region The biogeographic region of the Northern Hemisphere that roughly corresponds to North America.

Palearctic region The biogeographic region of the Northern Hemisphere that corresponds to Eurasia.

Neotropical region The biogeographic region of the Southern Hemisphere that corresponds to South America.

Afrotropical region The biogeographic region of the Southern Hemisphere that corresponds to most of Africa. *Also known as Ethiopian region.*

Indomalayan region The biogeographic region of the Southern Hemisphere that corresponds to India and Southeast Asia. *Also known as Oriental region.*

Australasian region The biogeographic region of the Southern Hemisphere that corresponds to Australia, New Zealand, and New Guinea.

contrast, only a few mammals moved into North America from South America; the most prominent are the opossums, a group of marsupial mammals.

HISTORIC CLIMATE CHANGE

As we saw in Chapter 5, the position of the continents and the circulation of water around the continents influence the variation in climates around the world. Because the continents were drifting over the past 250 million years, it is not surprising that the climates of the world experienced dramatic changes. For example, the polar regions have not always been as cold as they are today. Polar regions have always received relatively little solar energy due to the angle of the sun. However, when oceans extended from the tropics to polar areas, the ocean currents distributed heat more evenly and temperate climates could be found closer to the poles.

We also know from fossil evidence that large portions of North America and Europe once had tropical climates. Tropical forests reached into Russia and Canada and warm temperate forests covered the Bering Land Bridge from Alaska to Asia. Moreover, the Antarctic land connection between South America and Australia supported temperate vegetation and abundant animal life. However, as Antarctica drifted over the South Pole and as North America and Eurasia gradually encircled the northern polar ocean, the Arctic Ocean became largely enclosed between North America and Eurasia. This created a circumpolar ocean current around Antarctica, which caused cooler temperatures at high latitudes. As a result, Earth's climates became more strongly differentiated from the

equator to the poles. Tropical environments contracted into a narrow zone near the equator, and temperate and boreal climate zones expanded. These climate changes had profound effects on the geographic distributions of plants and animals.

About 2 Mya, the gradual cooling of the planet gave way to a series of dramatic oscillations in climate known as the Ice Age. Climate changes during this time had dramatic effects on habitats and organisms in most parts of the world. Alternating periods of cooling and warming caused the advance and retreat of ice sheets at high latitudes over much of the Northern Hemisphere and cycles of cool, dry climates and warm, wet climates in the tropics. Ice sheets came as far south as Ohio and Pennsylvania in North America and covered much of northern Europe, driving vegetation zones southward, possibly restricting tropical forests to isolated refuges where conditions remained moist.

A striking example of this disruption is the migration of forest trees in eastern North America and Europe. At the peak of the most recent glacial period, many tree species were restricted to southern refuges, but about 18 000 years ago after the ice began to retreat and the forests started to spread north again. Pollen grains deposited in the lakes and bogs left by retreating glaciers record the coming and going of plant species. These records show that the composition of plant associations changed as species migrated back north over different routes across the landscape. The migrations of some representative tree species from their southern refuges are mapped in Figure 22.27. As you can see in this figure, spruce trees shifted

Thousands of years before present

18   14   12   10   8   6   0.5

Spruce

Oak

Hemlock

Figure 22.27 Changing tree distribution following the retreat of glaciers. As the glaciers began retreating in North America 18 000 years ago, the trees began moving north to colonize the newly exposed habitats. Data from G. L. Jacobson et al., in W. F. Ruddiman and H. E. Wright, Jr. (eds.), *North America During Deglaciation* (Geological Society of America, 1987), pp. 277–288.

northward just behind the retreating glaciers. Oaks expanded out of their southern refuges to cover most of the eastern part of temperate North America, from southern Canada to the Gulf Coast. Hemlocks expanded into the Great Lakes Region and into central Canada.

The forests of Europe suffered even more extinctions from the spread of glaciers than those in North America because populations were blocked from shifting southward by the Alps and the Mediterranean Sea. Several northern European tree species went extinct. Many species that survived were restricted to refuges in southern Europe, from which they expanded after the glaciers receded beginning about 18 000 years ago. Recent research suggests that many tree species in Europe have not yet expanded fully into their potential ranges, which suggests that the European flora has not returned to an equilibrium state.

Throughout this chapter, we have seen that to understand the distribution of species, we need to consider ecology at both the landscape scale and the global scale. In doing so, we see that biodiversity is maintained by large-scale factors including area, isolation, and historical events that continue to show their effects today. In "Ecology Today" we will see that this knowledge has helped scientists design a massive nature reserve to help protect the biodiversity of Florida.

ECOLOGY TODAY CONNECTING THE CONCEPTS

TAKING A LONG WALK FOR CONSERVATION

The Florida panther. Many species, such as this Florida panther—an endangered subspecies of mountain lion—are expected to benefit from the Florida Wildlife Corridor project that will link together large areas of protected habitats. Photo by Tom & Pat Leeson.

As we have seen throughout this chapter, the distribution and persistence of biodiversity depends on a variety of factors. Knowing the factors responsible for maintaining diversity helps ecologists design nature reserves to protect biodiversity. For example, we know that large habitats typically contain more species than small habitats, so setting aside large nature reserves will better protect biodiversity than setting aside small ones. Similarly, when faced with the option of setting up a single large reserve or several small reserves even of the same total area, the single large reserve is often the better option to preserve biodiversity. However, a single habitat poses more risk from a disease or natural disaster that could destroy the species we are trying to conserve. If we have the opportunity to create multiple reserves, we need to position reserves close enough for organisms to disperse between habitats but far enough away to reduce the ability of predators and diseases to move easily between them. Because increased edge habitat can have detrimental effects on species that prefer to live in the interior of habitats, we also need to consider reserve shape.

Although we have a good idea about the features of an ideal nature reserve, in reality the design of nature reserves and the corridors that connect them is largely determined by existing habitats, their configurations, and land ownership. An excellent example of this can be found in a conservation effort known as the Florida Wildlife Corridor, a visionary plan to create a continuous connection of protected habitat from the southern tip of Florida all the way up to Georgia. This would provide a large number of species, including endangered species such as the Florida panther, with a tremendous amount of land to inhabit for a long time to come. The ultimate goal is to connect a number of current federal nature reserves, including the Everglades National Park, the Big Cypress National Preserve, the Ocala National Forest, and the Okefenoke National Wildlife Refuge, with a variety of properties owned by the state of Florida, conservation organizations such as The Nature Conservancy, and private landowners including large cattle ranches.

With such a multitude of landowners, the Florida Wildlife Corridor is an excellent case study of the challenges presented by the design of nature reserves. It is as much a sociological problem as an ecological problem, so the promoters of the corridor had to think about a public relations campaign that was as big as the 1600-km swath of land down the middle of the state. In 2012, four leaders of the project decided to travel the entire length of the proposed corridor in what they referred to as "1000 miles in 100 days." Moving on foot and in kayaks, the conservationists traversed swamps, grasslands, and forests. They called attention to their effort by using daily photo and video updates on their websites. Along the way, they were joined by state and federal biologists, conservationists, politicians, and private ranch owners. At every stop, they emphasized that preserving this swath of land through the middle of Florida provided benefits not only to the organisms that lived in undeveloped lands but also to the residents of Florida who rely on tourism as well as on the water and agriculture that comes out of this region. On Earth Day 2012, the 100-day expedition was completed at the Stephen C. Foster State Park in southern Georgia.

The effort to protect the land in a north-south corridor through Florida is far from complete, but there is a growing momentum by many people to find a way to protect the lands that lie between the state and federal reserves while at the same time retaining the long-held traditions of agriculture and ranching on the private lands. If such protection can be achieved, then the state will have a total reserve that is both large and interconnected and this should result in the long-term persistence of species in the region.

SOURCES: Lindenmayer, D., et al. 2008. A checklist for ecological management of landscapes for conservation. *Ecology Letters* 11: 78–91.

The Florida Wildlife Corridor (http://www.floridawildlifecorridor.org/about)

The Florida Wildlife Corridor. Conservationists hope to link public and private land through the state of Florida in a way that promotes the movement of wild organisms between habitats and facilitates their long-term persistence.

SUMMARY OF CHAPTER CONCEPTS

■ **Landscape ecology examines ecological patterns and processes at large spatial scales.** Habitat heterogeneity exists across the landscape as a result of both natural processes and human activities, both in the past and in modern times. This heterogeneity allows more species to exist at the landscape scale than at local scales. Species diversity can change with spatial scale; we quantify alpha diversity at the local scale, gamma diversity at the regional scale, and beta diversity across habitats. The process of species sorting determines which species from the regional species pool are found in a local community.

■ **The number of species increases with area.** The number of species found in areas of different sizes can be described by the species–area curve. As the area sampled grows, we see an increase in species richness because we begin to sample new habitats that contain different species. Habitat fragmentation divides a large habitat into several small habitats and thereby increases the amount of habitat edge. These effects cause the decline in the abundance of many species, but favour those species that prefer habitat edges.

■ **The equilibrium theory of island biogeography incorporates both area and isolation.** Species diversity is highest on large islands that are close to a source of colonizing species because these islands have a low rate of species extinction and a high rate of colonization by new species. In contrast, species diversity is lowest on small islands that are far from a source of colonizing species because these islands have a high rate of species extinction and a low rate of colonization by new species.

■ **On a global scale, biodiversity is highest near the equator and declines toward the poles.** Species diversity can also vary within a given latitude belt as a result of habitat heterogeneity, temperature, and precipitation. In terrestrial ecosystems, the best predictor of species diversity is potential evapotranspiration. The energy–diversity hypothesis states that sites with higher amounts of energy are able to support more species. In marine ecosystems, the best predictor of species diversity is the temperature at the surface of the sea.

■ **The distribution of species around the world is affected by Earth's history.** Continental drift is a major factor in determining the distribution of species. As continents have come together and separated, they have experienced unique evolutionary lineages that today are recognized as distinct biogeographic regions. The drift of the continents has also affected historic climate changes including widespread changes in the distribution of tropical, temperate, and polar regions around the world.

REVIEW QUESTIONS

1. Explain how the fragmentation of a landscape can have both positive and negative effects on biodiversity.

2. Why is it important to consider the quality of the matrix habitat that exists between habitat fragments?

3. How do local sampling, regional habitat diversity, and the isolation of continents affect the slope of the species–area relationship?

4. Compare and contrast alpha, gamma, and beta diversity.

5. Why might we expect that regions with high species diversity will also exhibit high niche diversity?

6. Why do small islands that are distant from a mainland typically have fewer species than large islands that are close to a mainland?

7. How has continental drift affected global patterns of biodiversity?

8. In what way might the age and area of a region affect its species richness?

9. How does knowledge of historic climatic patterns affect our interpretation of present-day patterns of species diversity?

10. What is the relationship between the length of time the modern continents have been connected and the similarity among the species on each continent?

PRACTICING ECOLOGY: MAINTAINING NIGHTSHADE

Nash Turley, a graduate student at the University of Toronto Mississauga, studies the ecology and conservation of plants and herbivorous insects. Photo by Daniel M. Evans.

Large reserves often maintain high species richness and abundance of species. However, this is not necessarily beneficial for plants at the bottom of the food chain because, while large reserves attract seed dispersers that increase the spread of many plant species, they also attract herbivores. Herbivory in a reserve can quickly destabilize ecological interactions, particularly if herbivores consume plant populations faster than they can regenerate. However, the shape and structure of a reserve can alter the fate of plant populations by changing the amount of suitable habitat for both plants and herbivores. What design for a reserve can best maintain both plant and herbivore populations?

American black nightshade (*Solanum americanum*) is a common herbaceous flower found in savannahs throughout southern and western North America. Although its leaves and berries are severely toxic to many organisms, birds can consume its berries and grasshoppers readily consume its leaves. Since grasshoppers are more abundant and more active in warmer habitats, herbivory by grasshoppers increases with average ground temperature. Land managers charged with designing and conserving savannah habitats must take this relationship into consideration. This is particularly true in southwest North America, where managers are restoring savannahs by clear-cutting plots of forest within large forest landscapes. Due to the cooling effects of tree shade, ground temperature increases farther from the forest edge. Hence, we might hypothesize that habitats with more area away from a forest edge will have higher ground temperatures, more grasshoppers, and higher rates of herbivory by grasshoppers on nightshade.

As a graduate student at the University of Toronto Mississauga, Nash Turley and his colleagues tested this hypothesis at the Savannah River Site reservation in the state of South Carolina. In 2000, researchers removed trees from 14 plots on the reservation to cultivate savannah habitat. All plots were 14 km², half of the plots were rectangular while the other half—known as "winged plots"—were composed of a central square with two smaller rectangles of cleared forest jutting out on opposite sides. Because of their shape, winged plots contained a higher amount of area closer to the forest edge. As a result, the winged plots had lower average ground temperatures. In early spring, the researchers planted nightshade along multiple transects in each plot. At the end of the summer, the researchers counted the number of grasshoppers per plot and measured the average percentage of damaged nightshade leaves on each plant. You can view their data in the table.

MEAN GRASSHOPPER ABUNDANCE (NUMBER/PLOT) AND PERCENT LEAF DAMAGE IN RECTANGULAR AND WINGED PLOTS

GRASSHOPPER ABUNDANCE		PERCENT LEAF DAMAGE	
RECTANGULAR	WINGED	RECTANGULAR	WINGED
62.5	29.0	5.3	1.9
35.2	45.7	5.5	2.4
39.0	30.7	8.5	3.2
40.2	30.2	9.3	5.9
30.7	44.2	11.2	6.4
69.4	42.8	13.2	6.8
65.1	28.7	14.5	9.6

GRAPH THE DATA

1. Calculate the means and standard errors of grasshopper abundance and percent of leaf damage for rectangular and winged plots.

2. Create a bar graph showing the means and standard errors of grasshopper abundance in rectangular and winged plots.

3. Create a second bar graph showing the means and standard errors of percent leaf damage in rectangular and winged plots.

INTERPRET THE DATA

1. Explain how the data support or refute the hypothesis that habitats with more area away from a forest edge will have more grasshoppers and higher rates of nightshade herbivory.

2. In the same study, the researchers also found that the number of berries produced by individual nightshade plants was lower in rectangular plots than winged plots. What does this suggest about the trajectory of nightshade population abundance in the two plot types?

3. Grasshoppers can easily disperse among habitat patches by flying. Nightshade seeds are dispersed between habitat patches by birds that eat the plant's berries. Given each organism's dispersal mechanism, which is likely to be more important to the persistence of its population: the shape of its habitat or the presence of corridors linking suitable habitats?

SOURCE: Evans, D. M., N. E. Turley, D. J. Levey, and J. J. Tewksbury 2012. Habitat patch shape, not corridors, determines herbivory and fruit production of an annual plant. *Ecology* 93:1016–1025.

CHAPTER 23

Global Conservation of Biodiversity

Protecting Hotspots of Biodiversity

The biodiversity of the world faces a wide range of threats from a growing human population that has caused species to go extinct at a rapid rate. To reverse this downward spiral, conservationists seek ways to protect aquatic and terrestrial ecosystems so that threats from human activities can be reduced or eliminated. A common approach to protecting species is to protect their habitats. But there are limits to how much habitat can be protected: not all habitats are available for purchase, limited funds are available for purchasing habitats, and political and economic factors often play a role in determining whether or not particular habitats can be set aside. With all these limitations, how should we prioritize the protection of the world's biodiversity?

> **"How should we prioritize the protection of the world's biodiversity?"**

In 1988, Norman Myers noted that small isolated areas such as tropical islands contain a large proportion of endemic species, which are species that have a relatively restricted distribution and are not found in other parts of the world. Therefore, a large proportion of the world's terrestrial species is located in a relatively small portion of the world's land. Myers argued that we should concentrate our conservation efforts on these species-rich areas, which he defined as biodiversity hotspots, because saving these areas should save the most species.

Myers identified 10 locations on Earth as biodiversity hotspots. Shortly afterward, the group Conservation International adopted Myers' approach and decided that hotspots should be defined as areas that contained at least 1 500 endemic plant species and that had experienced at least 70 percent vegetation loss due to human activities. These criteria would identify areas of high species richness facing substantial threats. The group assumed that regions with high plant diversity also contained high animal diversity. Conservation International identified 34 biodiversity hotspots around the world, including the Caribbean Islands, the Cape Floristic Region of South Africa, and several sites in the islands of Southeast Asia. Collectively, these locations represent 2.3 percent of the Earth's land surface but contain 50 percent of the world's plants (more than 150 000 species) and 42 percent of the world's vertebrate animals (nearly 12 000 species). A similar effort is underway for aquatic hotspots, although conservation of declining fish and invertebrate diversity remains in the early stages.

While identifying hotspots based on high numbers of endemic species is certainly reasonable, some scientists have suggested other approaches. One such approach would be to focus on areas that simply have high species richness. For example, the Amazon rainforest has a very high number of species, but most are not endemic to small geographic areas. Another approach would be to prioritize species-rich locations facing the highest current or future threats of species extinctions, such as locations that have, or are projected to have, a rapidly growing human population. Each approach supports a different list of protection priorities. Focusing conservation priorities on places with high diversity, however, automatically excludes low-diversity locations with species that people care about, such as the bison,

Conserving biodiversity. Efforts are being made to protect the world's biodiversity, including this critically endangered morningside tree frog (*Polypedes fastigo*) from Sri Lanka. Photo by Fletcher & Baylis/Science Source.

Biodiversity hotspots. Thirty-four biodiversity hotspots have been identified for terrestrial sites around the world. These sites contain at least 1500 endemic plant species and have experienced at least a 70 percent decline in their vegetation. While the criteria used are based on plants, these areas also contain a high diversity of animals. Drawn from http://specieslist.com/images/external/ci-hotspots.jpg.

wolves, and grizzly bears that live in western North America. It also places an emphasis on species richness rather than on the important functions that many ecosystems provide. As an example, although wetlands typically have low plant diversity, they are incredibly important for flood control and water filtration.

It is clear that we can take a variety of approaches in how we prioritize the conservation of biodiversity. All of these approaches seek to help us invest our limited resources to save the most biodiversity that we possibly can. In this chapter, we will focus on the services that

we receive from biodiversity, the causes of the decline in biodiversity, and the efforts that are being made to save it.

SOURCES: G. Bacchetta et al., A new method to set conservation priorities in biodiversity hotspots, *Plant Biosystems* 146 (2012): 638–648.

P. Kareiva and M. Marvier, Conserving biodiversity coldspots, *American Scientist* 91 (2003): 344–351.

N. Myers et al., Biodiversity hotspots for conservation priorities, *Nature* 403 (2000): 853–858.

C. D. Orme et al., Global hotspots of species richness are not congruent with endemism or threat, *Nature* 436 (2005): 1016–1019.

CHAPTER CONCEPTS

■ The value of biodiversity arises from social, economic, and ecological considerations.

■ Although extinction is a natural process, its current rate is unprecedented.

■ Human activities are causing the loss of biodiversity.

■ Conservation efforts can slow or reverse declines in biodiversity.

Throughout this book, we have examined factors that affect the distribution of species around the world. We have seen that these distributions are a result of the abiotic conditions a species can tolerate,

the positive and negative interactions that occur among species, the ability to disperse to suitable habitats, and historical processes that include the movement of the world's continents. We have also examined how human activities have affected particular species and how conservation efforts have tried to minimize these impacts. In this final chapter, we will take a broad view of the decline in biodiversity around the globe. We begin by considering the many different ways in which people value biodiversity. We will then compare current rates of biodiversity decline with historic rates and examine the ways in which human activities contribute to these declines. Finally, we will discuss efforts underway to slow or even reverse the decline of biodiversity.

The value of biodiversity arises from social, economic, and ecological considerations

A case for conserving the world's biodiversity can reflect a range of different values. For example, the **instrumental value of biodiversity** focuses on economic values that species can provide, such as lumber for building or crops for eating. In contrast, the **intrinsic value of biodiversity** recognizes that species have inherent value, in and of themselves, that is not tied to any economic benefit. Of course, species and ecosystems can simultaneously have both instrumental and intrinsic values.

INSTRUMENTAL VALUES

The total benefit provided by biodiversity is difficult to estimate because much of the world's biodiversity remains undiscovered and the values of each species and ecosystem can be difficult to estimate. However, researchers have attempted to do so. For example, the total economic benefit from boreal forests in Canada has been estimated at $93 billion. To put this in perspective, this is about 9 percent of Canada's annual gross domestic product. At the global level, estimates of the total benefit of biodiversity range from $3 trillion to $54 trillion. We can group the instrumental values of biodiversity into four categories of services: *provisioning, regulating, cultural,* and *supporting.*

Provisioning Services
Provisioning services are benefits of biodiversity that provide products humans use, including lumber, fur, meat, crops, water, and fibre. In many cases, plants and animals from the wild have been cultivated or domesticated and then selectively bred to enhance their valuable qualities. Provisions also include pharmaceutical chemicals that come from plants and animals; nearly 70 percent of the top 150 pharmaceutical drugs originated in chemicals that are produced in nature. A prominent example of the economic benefits of such a drug is the chemical Taxol, which is used to fight cancer. Today Taxol is synthesized in the laboratory, but it originally

Figure 23.1 Provisioning services. The Pacific yew is one of many species that has served as a critical source of pharmaceutical chemicals that improve human health. Photo by Inga Spence/Alamy.

came from the Pacific yew tree (*Taxus brevifolia*) (Figure 23.1). This single chemical currently has more than $1.5 billion in annual sales around the world. Over the past 25 years, more than 800 natural chemicals have been identified in the search to provide treatments for everything from cancer to contraception and there is no indication that the pace of these discoveries is slowing down.

Regulating Services
Regulating services are benefits of biodiversity that include climate regulation, flood control, and water

Instrumental value of biodiversity A focus on the economic value a species can provide.

Intrinsic value of biodiversity A focus on the inherent value of a species, not tied to any economic benefit.

Provisioning services Benefits of biodiversity that humans use, including lumber, fur, meat, crops, water, and fibre.

Regulating services Benefits of biodiversity that include climate regulation, flood control, and water purification.

purification. For example, wetlands absorb large amounts of water and so prevent flooding from water runoff during rainy periods. The plants living in the wetlands also remove contaminants from the water and make it more suitable for drinking. The CO_2 that is taken out of the air by producers on land and in the ocean is another regulating service. Of the 8 gigatons of carbon that are put into the air each year by human activities, about 4 gigatons are taken out of the air by producers, which reduces the effect humans have on global temperatures due to global warming.

Cultural Services

Cultural services are benefits of biodiversity that provide aesthetic, spiritual, or recreational value. For example, cultural services include the benefits that people obtain when they go hiking, camping, boating, or birdwatching. People pay to visit beautiful natural areas, such as Banff National Park in Canada or the Florida Everglades in the United States. Sometimes areas are preserved because income from tourists can exceed what would be received from clearing a forest or from using the land for housing and industry. Many tropical countries have capitalized on this attraction by establishing parks and support services for tourists. In Palo Verde National Park in Costa Rica, for example, monkeys and beautiful tropical birds draw tourists to areas where the species are protected (Figure 23.2). Diversity itself is often the attraction in tropical rainforests and coral reefs because these ecosystems contain hundreds of different species of trees, birds, corals, or fish.

Supporting Services

Supporting services are benefits of biodiversity that allow ecosystems to exist, such as primary production, soil formation, and nutrient cycling. As we have seen in Chapters 20 and 21, these processes are essential for the existence of species and ecosystems. There would be no ecosystems without producers that capture the energy of the Sun and then transfer it to all other trophic levels. Similarly, the formation of soil and the cycling of nutrients both play key roles in the persistence of existing ecosystems.

INTRINSIC VALUES

In contrast to instrumental values, the intrinsic values of biodiversity do not provide any economic benefits to humans. Instead, people who place intrinsic value in biodiversity feel religious, moral, or ethical obligations to preserve the world's species. However, it becomes very difficult to prioritize conservation efforts only by arguing that all species are intrinsically valuable. In reality, we can consider both instrumental and intrinsic values when deciding how to focus our conservation efforts.

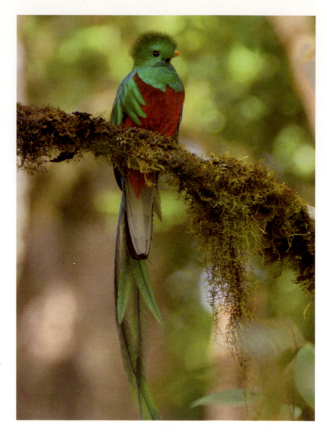

Figure 23.2 Attracting tourists to Palo Verde National Park. Striking examples of biodiversity, such as this resplendent quetzal (*Pharomachrus mocinno*), attract tourists from around the world. Photo by Rolf Nussbaumer Photography/Alamy.

Although extinction is a natural process, its current rate is unprecedented

The current number of species on Earth is difficult to estimate. We do know that 1.3 million species have received Latin names and about 15 000 new species are described each year. While estimates for the total number of species range from 3 to 100 million, depending on the assumptions used, most scientists agree that there are about 10 million species. Some species are declining in abundance and facing extinction as humans continue to alter terrestrial and aquatic ecosystems. However, as we have seen throughout this book, some extinctions are natural. Therefore, we need to understand the historic versus modern rates of extinctions. In this section, we will explore the past and present rates of extinction and then examine how specific groups of organisms are

---

Cultural services Benefits of biodiversity that provide aesthetic, spiritual, or recreational value.

Supporting services Benefits of biodiversity that allow ecosystems to exist, such as primary production, soil formation, and nutrient cycling.

faring. As part of this discussion, we will consider both declines in species diversity and declines in genetic diversity.

BACKGROUND EXTINCTION RATES

Over the past 500 million years, the world has experienced five **mass extinction events**, which are defined as events in which at least 75 percent of the existing species go extinct within a 2-million-year period. During these events, large numbers of species, genera, and families around the world went extinct, as illustrated in Figure 23.3. In each of these events, the extinctions were not instantaneous but rather occurred over periods of about 2 million years. During the first mass extinction, about 443 Mya, most species lived in the oceans. An ice age caused sea levels to drop and the ocean chemistry to change, which resulted in 86 percent of species going extinct. The second mass extinction happened 359 Mya when much of the ocean lacked oxygen—for reasons that are unclear—and 75 percent of all species went extinct. During the third mass extinction—248 Mya—an astounding 96 percent of all species then present on Earth went extinct. Although researchers have constructed multiple hypotheses to explain this third mass extinction, we are still uncertain about the cause. The fourth mass extinction, which occurred 200 Mya, caused 80 percent of the world's species to go extinct. Hypotheses for the causes of this fourth extinction include increased volcanic activity, asteroid collisions with Earth, and climate change.

The best-known mass extinction is the fifth, which happened 65 Mya and included the extinction of the dinosaurs. This event is attributed to several factors. First, volcanic eruptions and changes in climate caused long periods of cold weather. This was followed by a massive asteroid that struck the Yucatan Peninsula in Mexico. The asteroid is estimated to have been 10 km wide and struck with a force more than 1 billion times that of the atom bomb dropped on Hiroshima during World War II. The explosion created a massive, 180-km-wide crater. Scientists hypothesize that the explosion put so much dust into the atmosphere that it blocked the Sun's rays, making Earth much less hospitable to dinosaurs along with many other groups such as flowering plants. During this time, 76 percent of the species on Earth went extinct.

As you can see from this history of natural mass extinction events, only a small percentage of all species that ever lived on Earth are present today. In fact, over the past 3.5 billion years, it is estimated that 4 billion species have existed on Earth and 99 percent of these species are now extinct. However, as you can see in Figure 23.3, after each mass extinction event, new species evolved and, overall, the number of species has increased with time.

A POSSIBLE SIXTH MASS EXTINCTION

We have seen that a mass extinction is defined as the extinction of 75 percent of species within a 2-million-year period. Many scientists have hypothesized that the increase in the human population during the past 10 000 years may have initiated a sixth mass extinction event. To evaluate this hypothesis, we need to quantify the rates of extinction during the first five mass extinctions and then compare them with the current rate of extinction. Researchers have addressed this question by looking at groups of organisms, such as mammals, for which there are very good fossil data on extinctions. When researchers compare the extinction rate for mammals during the most recent 500 years to the rate of mammal extinctions over 500-year intervals in the past, they find that the current extinction rate exceeds the historic extinction rate. In fact, the United Nations Convention on Biological Diversity estimates that the extinction rate during the last 50 years has been as much as 1000 times higher than the historic rate. Should this rate continue for hundreds or thousands of years, it could qualify as a mass extinction event.

GLOBAL DECLINES IN SPECIES DIVERSITY

When we think about the decline in the world's biodiversity, we often focus on the last few centuries, from the Industrial Revolution to today—a time in which we have drastically altered our world. However,

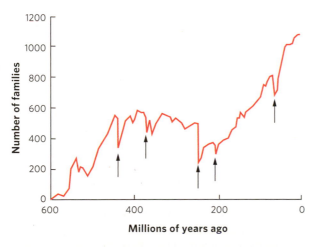

Figure 23.3 Mass extinctions. Over the past 500 million years, there have been five mass extinctions. During these periods, the world experienced a significant decline in the number of families, which also means there were declines in the number of genera and species. Continued speciation during subsequent years has helped to offset these extinctions. Data from J. J. Sepkoski, Jr., Ten years in the library: New data confirm paleontological data, *Paleobiology* 19 (1993): 43–51.

Mass extinction events Events in which at least 75 percent of the existing species go extinct within a 2-million-year period.

human impacts on biodiversity can be seen much farther back in time. For example, researchers in 2009 examined the diversity of North American mammals from 30 million years ago to 500 years ago. This span of time is interesting because humans arrived in North America from Asia approximately 10 000 years ago.

Based on a rich fossil history of mammals, the researchers defined different geographic regions of North America and determined the number of fossil species in each region. With these data they created species–area curves for different time periods in North America. Compared to the period before the arrival of humans—150 000 to 11 500 years ago—the species–area curve for the period after human arrival—11 500 to 500 years ago—was significantly lower, as you can see in Figure 23.4. Simply put, the arrival of humans coincided with a 15 to 42 percent decline in mammal diversity, depending on the geographic region examined. The losses include 56 species and 27 genera of large mammals, including horses, a giant ground sloth, camels, elephants, the sabre-toothed tiger, and a species of lion.

Explanations for these extinctions include rapid climate change following the retreat of the glaciers, hunting pressure from the human population, and epidemic diseases carried from Asia by domesticated animals. Many scientists hypothesize that most of the large mammals were driven extinct by humans who hunted them. We know that the diversity of mammal species had been substantially reduced during the

period of human occupation of North America prior to the Industrial Revolution although we can't be sure of the reason. Any current impacts are adding to previous extinctions.

Whether the world's current extinction rate of species will approach the magnitude of a mass extinction depends on how many of the species currently on Earth go extinct during the next few centuries. To assess how different groups of organisms are currently faring, the International Union for Conservation of Nature (IUCN) has defined categories that describe whether a species is persisting well, declining, or extinct. *Extinct* describes a situation in which a species was known to be in the wild in the year 1500 but no individuals remain alive today. *Extinct in the wild* is a category used when the only individuals remaining are in captivity, such as animals living in zoos. *Threatened* species are those whose populations face a high risk of extinction in the future. This category includes species that are considered "endangered." *Near-threatened* species are those that will likely become threatened in the future. In contrast, *least-concern* species are those that have abundant populations and are not likely to become threatened in the future. In some cases, the status of a species has not been determined or there are simply insufficient data to make a reliable determination.

Assessing the status of species from a large taxonomic group is not easy. Groups such as birds, mammals, amphibians, and reptiles contain thousands of species and often a large percentage of them have not been studied well enough to know if a species is abundant or if it is declining. Making such a determination requires substantial time and money for each species. Currently, our best data to assess the decline in biodiversity are for conifers, birds, mammals, amphibians, and fish. Assessments for these taxonomic groups were produced by the IUCN from 2009 to 2012, and you can see a summary of these data in Figure 23.5.

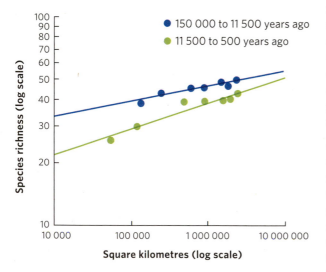

Figure 23.4 Declining North American mammals. When researchers created species–area curves of North American mammals from different geographic regions before and after human arrival, they discovered that from 11 500 to 500 years ago the number of mammals declined anywhere from 15 to 42 percent, depending on the geographic region examined. Data from M. A. Carrasco et al., Quantifying the extent of North American mammal extinction relative to the pre-anthropogenic baseline, *PLoS One* 4 (2009): e8331.

Conifers

Conifers include pines, spruces, firs, cedars, and redwoods, and the survival prospects of 95 percent of these species have been assessed. This high level of assessment is possible in part because the group has a relatively low number of species and large trees and shrubs are more easily assessed for population declines. Of the 607 species of conifers, none has gone extinct. Of those species that have sufficient data for assessment, 57 percent are categorized as of least concern, 13 percent are near-threatened, and 30 percent are threatened.

Birds

Some of the best data on species assessments come from birds because they are relatively easy to monitor and they have been studied for a long time. The 2012 assessment found sufficient data for assessment of

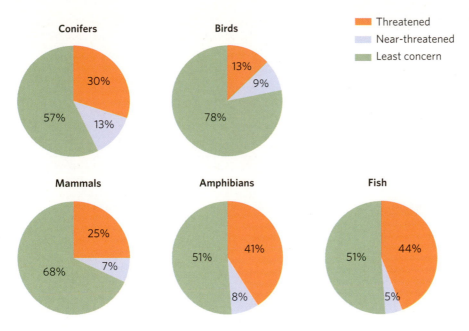

Figure 23.5 The global status of conifers, birds, mammals, amphibians, and fish. Species threatened by the risk of future extinction include 30 percent of conifers, 13 percent of birds, 25 percent of mammals, 41 percent of amphibians, and 44 percent of fish. Data from International Union for Conservation of Nature, 2012 http://www.iucnredlist.org/about/summary-statistics#Tables_3_4.

99 percent of the more than 10 000 species of birds on Earth. Since the year 1500, 130 of these (1.3 percent) have gone extinct, including three species that went extinct since the year 2000. Of the remaining species for which there are reliable data, 78 percent have abundant populations, 9 percent are near-threatened, and 13 percent are threatened with extinction.

Mammals

Of the 5490 species of mammals that have lived on Earth since the year 1500, 85 percent have sufficient data to assess their status. During the past 500 years, 78 mammal species (1.4 percent) have gone extinct. Of those remaining for which there are reliable data, researchers found that 68 percent of the species are categorized as of least concern, 7 percent are near-threatened, and 25 percent are threatened.

Amphibians

Amphibians have been particularly hard hit in recent decades from several causes that include habitat loss and the spread of the deadly chytrid fungal disease that we discussed in Chapter 15. Of the more than 6300 species of amphibians assessed, only 75 percent contained sufficient data to assess their conservation status. Of those with sufficient data, 51 percent of the species are categorized as of least concern, 8 percent are near-threatened, and a staggering 41 percent of amphibian species are threatened.

Because amphibians are not as visible as birds and mammals, scientists are still discovering new species at a rapid rate. For example, from 2004 to 2012, researchers discovered 1666 new species of amphibians, which represents nearly one-fourth of the world's described amphibian species. One challenge with so many new discoveries is that little is known about the population status of these species. Researchers expect hundreds more species to be discovered in the future and therefore our estimates of each category will continue to be updated.

Fish

Like amphibians, fish have also experienced a tremendous number of declines. The IUCN considered more than 3000 species of fish in their most recent assessment and 431 were found to be data deficient. Of those species with reliable data, 51 percent of the species are categorized as of least concern, 5 percent are near-threatened, and 44 percent are threatened.

The data from conifers, birds, mammals, amphibians, and fish suggest that if human impacts on these species continue we should expect a large number of species extinctions in future centuries. Although our best data come from these five groups, researchers hypothesize that the patterns of decline observed in these groups are representative of many other groups for which data on the status of species are relatively poor. This prediction is supported by preliminary efforts to assess the status of other major groups. Although less than 10 percent of all species of flowering plants, reptiles, and insects have been assessed, 46 to 74 percent of the currently assessed species are categorized as threatened.

As we have discussed in previous chapters, these declines in species richness are a concern not only because of the risk of losing species but also because of the effect these declines have on communities and ecosystems. As we discussed in Chapter 17, experimental increases in the number of mycorhizal fungi cause increases in the biomass of plants (see Figure 17.22). In Chapter 18 we also observed that a

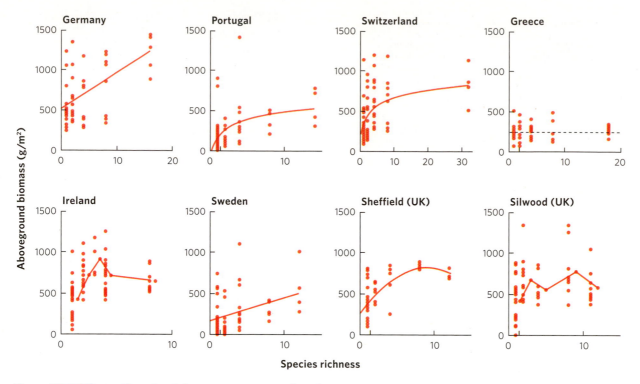

Figure 23.6 Effects of species richness on ecosystem function. Researchers have observed positive relationships between species richness and ecosystem function in seven of eight locations in Europe, although the shape of the relationship differs among locations. Data from D. U. Hooper et al., Effects of biodiversity of ecosystem functioning: A consensus of current knowledge, *Ecological Monographs* 75 (2005): 3–35.

decline in species richness can cause communities to be less stable over time by affecting community resistance or resilience (see Figure 18.22). Declines in species richness can also cause a decline in the functioning of ecosystems. For example, a review of all research examining patterns between experimental manipulations of plant species richness and the aboveground biomass of plants—just one measure of an ecosystem's function— has found that there is commonly a positive relationship, although there are exceptions. You can see examples of these relationships in Figure 23.6.

GLOBAL DECLINES IN GENETIC DIVERSITY

In addition to the decline in species diversity around the world, we have also seen a decline in the genetic diversity of many existing species. As we discussed in Chapter 7, causes of decline in genetic diversity include declining population sizes, inbreeding depression, and the bottleneck effect. Smaller populations do not possess the same amount of genetic diversity as large populations. These declines in genetic diversity reduce the probability that a population contains genotypes able to survive changing environmental conditions, including changes in climate and emerging infectious diseases.

The decline in the genetic diversity of livestock and crops has a direct and immediate effect on

humans. The livestock that we consume or use for labour and transportation include just seven species of mammals (cattle, pigs, sheep, goats, buffalo, horses, and donkeys) and four species of birds (chickens, turkeys, ducks, and geese). Humans have bred these species for a wide variety of different traits including size, strength, quality of meat, and ability to persist under challenging environmental conditions such as drought and diseases (Figure 23.7).

During the past century, many livestock varieties have not been maintained because larger, modern livestock operations favour a relatively few breeds. In Europe, for instance, about half of the livestock breeds present in 1900 are now extinct. Of those breeds remaining, 43 percent are at a serious risk of extinction. In North America, 80 percent of the livestock breeds that have been evaluated are either declining in abundance or facing extinction. Around the world, of the 7000 breeds of 35 domesticated species of birds and mammals, more than 10 percent are already extinct and another 21 percent are at risk of extinction. Such a rapid decline in genetic diversity means there is considerably less diversity to draw from should we need to breed domesticated animals that can live in new locations or in changing environments, or that can withstand new diseases. Simply put, reduced genetic variation reduces our future options.

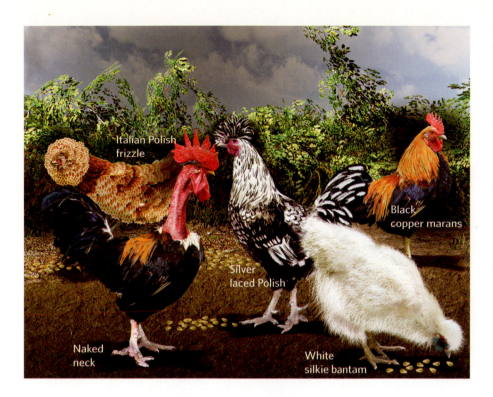

Figure 23.7 Genetic diversity of chickens. Over the past few centuries, people have bred a wide variety of livestock breeds such as chickens to suit their particular needs or local environmental conditions. Today most of these varieties are extinct because large livestock operations have focused on just a few breeds to maximize the production of livestock.

Declines in genetic diversity are also occurring in plant species that are important to humans. Humans historically consumed more than 7000 species of plants, but today consume only about 150 species. Moreover, just 12 species of plants make up the vast majority of people's diets; among these are wheat, rice, and corn. In some cases—for example, corn— modern varieties look very different from their ancestors (Figure 23.8). In earlier times, humans bred varieties that grew well under particular local environmental conditions. However, as agricultural practices changed, irrigation and fertilizer made it possible to reduce the harshness of the growing environment and small farms gave way to much larger operations that favoured only the top-producing varieties. As a result, many of the older,

local varieties of crops are no longer available. For example, Canadian farmers grew about 5000 varieties of apples a century ago but today, only 15 varieties are still grown commercially. Similarly, 80 percent of the corn varieties that existed in Mexico in 1930 and 90 percent of the wheat varieties that existed in China in 1949 are now gone.

Losing the genetic diversity of these crops reduces our options when we need to respond to challenges such as new pathogens that attack a crop. For example, in the 1970s a fungus attacked corn fields in the southern United States and was able to kill half the corn crop because the plants all came from a single variety. Fortunately, another variety of corn possessed a fungus-resistant gene and breeders were able to produce a new variety that is resistant to the fungus.

Figure 23.8 Genetic diversity of crops. **(a)** Corn originated from a wild plant in Mexico and looks very different from the modern cultivated corn plant. This photo shows an ear of teosinte on the left, an ear of maize on the right, and an ear of their F1 hybrid in the centre. **(b)** From the teosinte ancestor, humans have bred a wide range of genetic varieties to perform well under different conditions, including these varieties grown in Oaxaca State, Mexico. Today, the tremendous variety of genetic crops is at risk of being lost. Photos by (a) John Doebley and (b) Philippe Psaila/Science Source.

Each heat-sealed aluminum foil packet contains approximately 500 seeds.

Each storage box holds up to 500 seed packets.

Each vault holds hundreds of boxes.

Vaults

The concrete-lined cave is held at –18°C.

Control room

Entrance

North Pole

Greenland Svalbard

Iceland

Norway

Atlantic Ocean

Figure 23.9 The Svalbard Global Seed Vault. This facility was built on an island north of the Norwegian mainland to preserve the genetic diversity of crop plants so that humans will be able to use this genetic variation long into the future.

To protect the genetic diversity of plant varieties, many countries have been archiving seed varieties of different crop species in thousands of storage facilities around the world. Concern that such facilities could be destroyed by natural disasters or war led to the construction of the Svalbard Global Seed Vault. Located on an island in the Arctic region north of the Norwegian mainland, the facility is a 125-m tunnel built into a mountain with rooms for seeds on each side of the tunnel, as illustrated in Figure 23.9. This facility protects seeds from virtually all catastrophes. The vault has a total capacity of 1.5 million samples; as of 2013, the facility contained more than 700 000 samples of crop seeds from nearly every country. The Svalbard Global Seed Vault and many other seed storage facilities around the world preserve the ability to call upon the genetic diversity of plant species far into the future.

Human activities are causing the loss of biodiversity

The current rapid decline in biodiversity is caused by the rapid increase in human populations and our many activities. Throughout this book we have mentioned how humans have altered the environment and how this has affected populations, communities, and ecosystems. Virtually all areas within temperate latitudes that are suitable for agriculture have been plowed or fenced, 35 percent of the land area is used for crops or permanent pastures, and countless additional hectares are grazed by livestock. Tropical forests are being logged at a rate of 10 million ha each year. Semiarid subtropical regions, particularly in sub-Saharan Africa, have been turned into deserts by overgrazing and harvesting firewood. Rivers and lakes are badly contaminated in many parts of the world. Gases from chemical industries and the burning of fossil fuels pollute our atmosphere. In this section, we take a global view of human impacts including habitat loss, overharvesting, introduced species, pollution, and global climate change. While each of these factors is important, keep in mind that many of these factors occur simultaneously.

HABITAT LOSS

The destruction and degradation of habitat has been the largest cause of declining biodiversity. On the south

Figure 23.10 Deforestation of a tropical forest. The Atlantic coast forest in the Brazilian state of Bahia once covered the vast majority of the area. Over time, most of the forest was cleared, leaving only a small fraction of forest today. Maps by J. R. Mendonça, Projeto Mata Atlântica Nordeste, Convênio CEPLAC/New York Botanical Garden.

coast of British Columbia, for example, only 26 percent of the old-growth forest remains. Old-growth forests in other provinces were also logged intensively during the nineteenth century. Many of these logged areas are now forested again, although these younger forests do not provide habitat to all of the same species as the original old-growth forests did. Today, many areas of the tropics are experiencing a similar pattern of deforestation. For instance, the Atlantic forest along the coast of Brazil has been cleared so much in the past 70 years that only a small fraction of the original forest remains, as you can see in Figure 23.10. This deforestation has critically endangered many endemic birds and mammals, such as the golden lion tamarin (*Leontopithecus rosalia*), which lives in one tiny area of Brazil (Figure 23.11). Because such endemic species live nowhere else in the world, intense conservation efforts have focused on saving them from extinction by protecting what little habitat remains.

To gain insight into the global scale of habitat change, researchers have assessed how forests have been changing in more modern times. From 1980 to 2000, continued loss of forests has occurred in many regions including the Amazon, Russia, and Southeast Asia. However, there has been an increase in forests in the United States, Europe, and Northeast Asia. You can view a map of these changes in Figure 23.12. The species composition that currently exists in regions that have experienced an increase in forest cover, however, is often quite different from what existed originally, particularly in cases when a single species of tree is planted due to its high commercial value.

As we have discussed in previous chapters, habitat loss also leads to smaller habitat sizes and increased habitat fragmentation. In Chapter 22 we saw that a reduction in habitat size can lead to reduced population sizes and make it more likely a population will go extinct. This process is thought to be a significant reason that many national parks have lost species of mammals over the past 50 years, despite being protected from most harmful human activities. In addition, fragmented habitats have a high proportion of edge habitat that can alter the abiotic conditions of the interior habitat and favour edge species. We saw an example of this in Chapter 22, when we discussed predation by Steller's jays—which prefer habitat edges—on marbled murrelets.

Forests are not the only habitats that have been changing as a result of human activities. For example, in Manitoba, 6000 km² were once covered by tallgrass prairie. Today, only 3 km² of the historical range remain. The situation is similar in Ontario, where less than 16 km² remain of the historical extent of more than 800 km². Because the remaining areas are small fragments, many local populations of prairie plants and animals are threatened or are being displaced by non-native species. A similar story exists for wetland habitats. Scientists estimate that approximately 18 percent, or 1 430 000 km², of Canada's landmass was covered by wetlands. However, drainage for agriculture and other uses during the last 200 years has reduced this area by 200 000 km². As we have

Figure 23.11 Golden lion tamarin. Habitat destruction along the Atlantic coast forest, in the Brazilian state of Bahia, has caused the decline of many endemic species, including the golden lion tamarin. Photo by Tom and Pat Leeson.

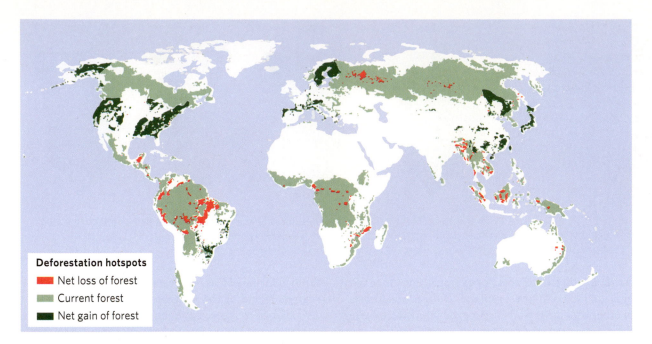

Figure 23.12 Changes in forest cover. While some regions of the world experienced a decline in forest cover from 1980 to 2000, other regions experienced an increase. Data from Millennium Ecosystem Assessment, *Ecosystems and Human Well-being: Synthesis* (Island Press, Washington, D.C., 2005).

Deforestation hotspots
- Net loss of forest
- Current forest
- Net gain of forest

seen, these habitat losses have a large negative effect on the world's biodiversity.

OVERHARVESTING

Human advances in techniques for logging trees, plowing grasslands, and capturing animals more efficiently has allowed us to harvest species at rapid rates and drive some species to extinction. For instance, during the past 3 centuries, commercial hunters in North America have hunted to extinction the Steller's sea cow (*Hydrodamalis gigas*), great auk (*Pinguinus impennis*), passenger pigeon (*Ectopistes migratorius*), and Labrador duck (*Camptorhynchus labradorius*). Each of these once-abundant species was valued for food or feathers and they were easily killed.

Extinction caused by overhunting and overfishing is not a recent phenomenon. Wherever humans have colonized new regions, some elements of the fauna have suffered. For example, researchers examined the skeletal remains of animals in archaeological sites around the Mediterranean region to see how human diets changed over thousands of years. At a site in current-day Italy, they found that early human populations initially ate large quantities of tortoises and shellfish, which were easy to catch. As you can see in Figure 23.13, as supplies of those foods were depleted over time, people switched to hunting hares, partridges, and other small mammals and birds.

Similar scenarios occurred when humans colonized other parts of the world. When Australia was colonized 50 000 years ago, several species of

large mammals, flightless birds, and a species of tortoise soon disappeared. On Madagascar, a large island off the southeastern coast of Africa, the arrival of humans about 1500 years ago caused the demise of 14 species of lemurs and between 6 to 12 species of elephant birds—giant flightless birds found only on that island. At about the same time, a small population of fewer than 1000 Polynesian colonists in New Zealand hunted 11 species of moas—another group of large flightless birds—so that these birds went extinct in less than a century. In each of these cases, humans encountered island species that were unaccustomed to predation pressure. Failure to recognize danger and a lack of defences made these species particularly vulnerable to human hunting.

Overharvesting of species continues in modern times. In some cases, the harvesting is part of an illegal trade in plants and animals, an endeavour that is valued at $5 billion to $20 billion annually. For example, many animal skins are sold for furs and some cultures believe that certain animal body parts have medicinal value. Some species of rare trees, such as big-leaf mahogany (*Swietenia macrophylla*), are sold for their lumber, while species of rare flowers, such as endangered species of orchids, are sold for their beauty.

Governments frequently regulate the harvest of plants and animals in the wild to ensure that the harvested species will persist for future generations to enjoy. However, harvesting regulations must balance not only the good of the harvested population but also

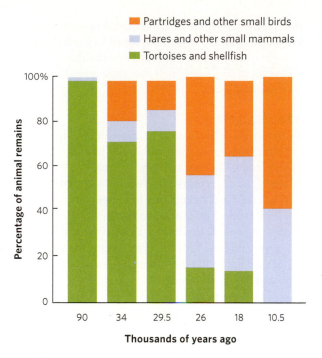

Figure 23.13 Historic overharvest. By examining the bones of consumed animals found in archaeological excavations, researchers have determined that early humans in Italy initially consumed animals that were easy to catch, such as tortoises and shellfish. Once these animals became rare due to overharvesting, people turned to eating hares, partridges, and other species of mammals and birds. Data from M. C. Stiner et al., Paleolithic population growth pulses evidenced by small animal exploitation, *Science* 283 (1999): 190–194.

the human employment that the harvest supports. As a result, some regulations set harvest levels that do not stop populations from declining. In marine environments, for example, modern fishing techniques have made it much easier to harvest enormous numbers of fish and shellfish. These techniques include fishing lines that are several kilometres long with thousands of baited hooks, nets that can surround schools of fish up to 2 km in diameter and 250 m deep, and huge trawlers that can scrape large areas of the ocean bottom. The ability to fish more efficiently and to cover larger areas has caused a decline of many fish populations around the world. When a fishery no longer has a population that can be fished, it is called a collapsed fishery. A recent analysis of commercially captured fish and other seafood found that since the 1950s there has been a steady increase in the percent of collapsed fisheries. Some of the most complete data exists for fish and shellfish in North America. As you can see in Figure 23.14, the percentage of collapsed fisheries increased to 14 percent by 2007. Some regions, such as the eastern Bering Sea off the shores of Alaska, have very few collapsed fisheries. In contrast, collapsed fisheries occur in more than 25 percent of species assessed off the coast of the northeastern United States

and more than 60 percent of species assessed off the coast of eastern Canada.

For example, consider the case of the Atlantic cod, a species of fish caught by commercial trawlers. On the Grand Banks fishery off the coast of Newfoundland,

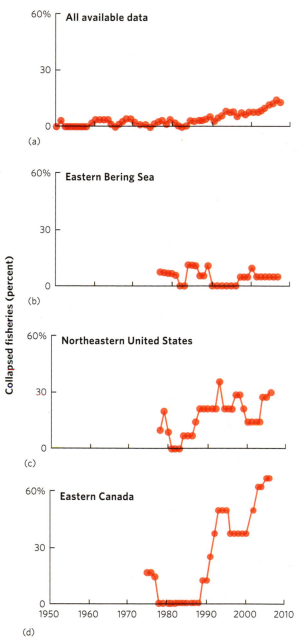

Figure 23.14 Collapsed fisheries. Over the past 60 years, there has been a steady increase in the percent of fish and other seafood populations that are categorized as collapsed. **(a)** Around the world, 14 percent of assessed species are considered collapsed. In different regions of North America, these percentages differ a great deal including **(b)** a low percentage of collapsed fisheries in the eastern Bering Sea, **(c)** a modest percentage in the water off the coast of the northeastern United States, and **(d)** a high percentage off the coast of eastern Canada. Data from B. Worm et al., Rebuilding global fisheries, *Science* 325 (2009): 578–585.

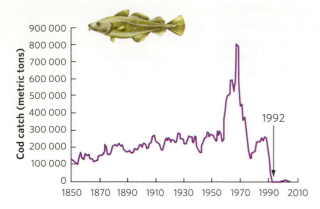

Figure 23.15 A collapse of the Atlantic cod fishery. From 1850 to 1960, there was a slow increase in the catch of Atlantic cod by commercial fishermen off the coast of Newfoundland. New technologies in the 1970s and 1980s allowed much larger catches of cod, but this led to an overharvest of the fish and a collapse of the population in 1992 that persists today. Figure from Millennium Ecosystem Assessment, *Ecosystems and Human Well-being: Synthesis* (Island Press, Washington, D.C., 2005).

the amount of cod caught from 1850 to 1960 slowly increased from 100 000 to 300 000 metric tons, as shown in Figure 23.15. During the 1970s and 1980s, new technologies—including advanced sonar, GPS, and larger trawlers—allowed a large increase in the number of cod caught, with a peak catch of 800 000 metric tons. However, the population crashed to very low levels in the early 1990s, leading the Canadian government to close the fishery in 1992. In 2012, 20 years after the ban was imposed, scientists reported that there were finally signs that the cod population was beginning to rebound.

The decline in cod also occurred in New England. Because it was not as severe as the decline off the Newfoundland coast, fishing continued although the quota of cod that could be caught was reduced. By 2010, the U.S. government greatly reduced the amount of cod that could be caught by commercial anglers, in the hope that the population would rebound. An assessment in 2011 found that the cod population had been very slow to respond, and so limits were reduced even further for 2013 through 2016. U.S. cod fishermen lobbied for higher cod quotas so that they could continue to make a living. However, the government biologists argued that if the cod limits were not lowered substantially, there would soon be no cod left to catch. This debate mirrored the Canadian experience of 2 decades earlier. At that time, despite low cod numbers, the Canadian government was under pressure from cod fishermen to allow continued fishing. The government agreed, but soon the number of fish fell so much that cod fishing had to be completely stopped and 35 000 Canadian fishermen lost their livelihood. Although

reducing the harvest of overharvested species has real economic impacts on those employed in the industry, failure to restrict the harvest hastens the decline of the species to levels at which there are no individuals left to harvest.

INTRODUCED SPECIES

Another cause of declines in biodiversity is the increasing number of species that are introduced from one region to another. Some of these introductions are intentional, such as when tropical plants are sold for houseplants in temperate parts of the world. Often, however, species are introduced by accident, as we saw in Chapter 15 with the many pathogens that have moved between continents and have caused emerging infectious diseases. Although only about 5 percent of introduced species become established in a new region, those that do can have a variety of effects. Some introduced species provide important benefits, such as the common honeybee that was introduced to North America from Europe in the 1600s. Other introduced species can have substantial negative effects on native species, like the brown tree snake introduced to the island of Guam, which we discussed in Chapter 14 (see Figure 14.2). In that case, the snake caused the decline or extinction of nine species of birds, three species of bats, and several species of lizards. In general, introduced species that compete with native species rarely cause extinction of native species whereas introduced species that act as predators or pathogens on native species can cause large population declines and extinctions of native species.

One of the highest-profile introduced species today is the silver carp (*Hypophthalmichthys molitrix*), a species of fish that has been introduced around the world because it consumes excess algae in water treatment plants and aquaculture operations. Brought to the United States in the 1970s, the carp escaped captivity in the 1980s when floodwaters washed them out of ponds and into the Mississippi River. The species rapidly spread through the Mississippi River and its tributaries, including the Illinois River. The primary concern is that the Illinois River connects the Mississippi River to Lake Michigan and, thus, the carp has the potential to invade the entire Great Lakes ecosystem. In 2010, the carp's DNA was detected in Lake Michigan, which suggests that the carp may be spreading throughout the Great Lakes. The silver carp is such a voracious consumer of algae that scientists worry it will compete with native consumers of algae, which serve as a key link in the food chain for many species of commercially important fish. The carp also has the unusual behaviour of jumping out of the water when a boat passes by. Since the silver carp can reach a mass of 18 kg and jump up to 3 m out of the water, it

Figure 23.16 Silver carp. Silver carp jump into the air and pose a danger to boaters like this biologist on the Illinois River near Starved Rock State Park. Photo courtesy of Chris Olds, USFWS.

poses a serious safety risk to boaters (Figure 23.16). It will take researchers several years before they can assess the full impact of the silver carp on North American waters.

Some of the most complete data on introduced species exist for the Nordic countries of Sweden, Finland, Norway, Denmark, and Iceland. As you can see in Figure 23.17, the number of introduced species in this area has rapidly increased since 1900. Across

terrestrial, freshwater, and marine ecosystems, there are currently more than 1600 introduced species in the Nordic region. Similarly, during the past 200 years in North America thousands of species have been introduced, many of which have spread rapidly and are considered invasive species. According to the Center for Invasive Species and Ecosystem Health, North America currently has a large number of invasive species that include nearly 200 pathogens, 300 vertebrates, 500 insects, and 1600 plants. One introduced species with largely unappreciated negative effects is the domestic house cat. In 2013, researchers examined predation by free-ranging domestic cats in the United States and found that the cats killed 1.4 to 3.7 billion birds and 7 to 21 billion mammals each year. Collectively, these data suggest that introduced species have the potential to cause widespread effects on native species and ecosystems. As the movement of people, cargo, and species becomes more common among the regions of the world, the unique species compositions originally found in different regions are slowly become more similar, a process known as **biotic homogenization**.

Biotic homogenization The process by which unique species compositions originally found in different regions slowly become more similar due to the movement of people, cargo, and species.

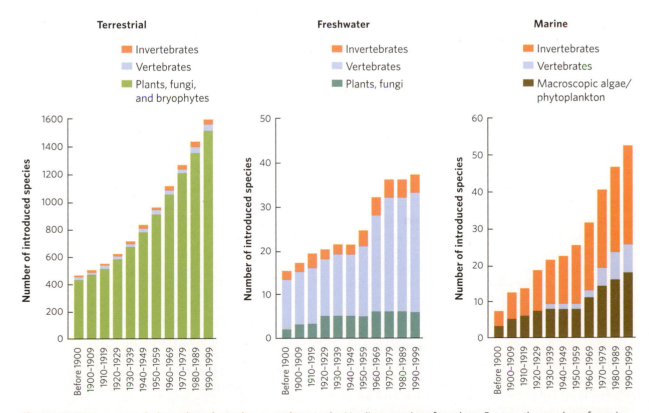

Figure 23.17 Increases in introduced species over time. In the Nordic countries of northern Europe, the number of species introduced to terrestrial, freshwater, and marine ecosystems has increased over time. Data from Secretariat of the Convention on Biological Diversity, *Global Biodiversity Outlook 2*, (Montréal, 2006, 81 + vii pages).

POLLUTION

We have talked about many types of pollution throughout this book. For example, in Chapter 2 we discussed acid precipitation and in Chapter 21 we looked at the effects of adding excess nutrients to bodies of water. Both examples demonstrated the harmful effects of pollutants on biodiversity.

The group of chemicals collectively called pesticides is a common type of pollutant. Pesticides include insecticides that kill insects, herbicides that kill plants, and fungicides that kill fungi. These chemicals are designed to target a particular type of pest; ideally they will not harm nonpests in the ecosystem. However, some pesticides do kill nonpests, either directly, by being toxic, or indirectly, by altering food webs, as we saw in the wetland study discussed at the end of Chapter 18. In that study, a very small amount of an insecticide, while not directly toxic to tadpoles, was toxic to zooplankton. The death of zooplankton set off a chain of events in the food web that prevented the tadpoles from obtaining enough food to metamorphose before the pond dried. By indirect effect, the insecticide caused the death of approximately half of the tadpoles.

A classic study on pesticides examined the role of the insecticide DDT on birds of prey. During the 1950s and 1960s in North America, populations of many predatory birds, particularly the peregrine falcon (*Falco peregrinus*), bald eagle, osprey (*Pandion haliaetus*), and brown pelican (*Pelecanus occidentalis*), declined drastically. Several of these species disappeared from large areas and the peregrine falcon disappeared east of the Great Plains. The causes of these population declines were traced to pollution of aquatic habitats by DDT, a pesticide that was widely used to control crop pests and mosquito vectors of malaria after World War II.

When this pesticide enters a water body, it binds to particles, including algae, to become about 10 times more concentrated than it was in the water. When the algae are consumed by zooplankton, DDT accumulates in fat tissues until the zooplankton have a concentration that is about 800 times higher than in the water, as shown in Figure 23.18. The process of increasing the concentration of a contaminant as it moves up the food chain is known as **biomagnification**. When small fish eat the zooplankton and large fish eat the small fish, DDT can be further concentrated about 30-fold. Finally, when a fish-eating bird such as an osprey consumes a large fish, DDT becomes concentrated another 10-fold. In short, at the top of the food chain the insecticide is 276 000 times more concentrated in the body of the fish-eating bird than it was in the water. Such a high concentration in predatory birds interferes with their physiology in a way that causes the eggs they lay to have very thin shells. When the parents sit on these

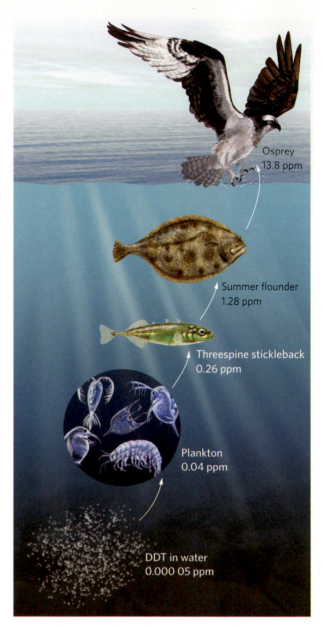

Figure 23.18 Biomagnification of DDT. When DDT was widely used to control insects, it could be found in very low concentrations in the water. However, its concentration increases in particles with which it binds in the water such as algae. At each higher trophic level, the insecticide becomes more concentrated. Data from G. M. Woodwell, C. F. Wurster, Jr., and Peter A. Isaacson, DDT residues in an East Coast estuary: A case of biological concentration of a persistent insecticide, *Science* 156 (1967): 821–824.

thin-shelled eggs, the egg shells break, and the embryos die. This eggshell thinning caused the populations of predatory birds to plummet during the 1960s.

Understanding the connection between DDT and declining bird populations led the United States

---

Biomagnification The process by which the concentration of a contaminant increases as it moves up the food chain.

Contaminant Half-Lives

As we have mentioned, an important factor in determining whether a contaminant such as a pesticide or radioactive compound affects the environment is the length of time the chemical persists in the environment. A helpful way to assess this persistence is to measure the time required for the chemical to break down to half of its original concentration, which we call the **half-life**. To calculate the half-life, we start by recognizing that most chemicals have a breakdown rate that follows a negative exponential curve, as shown in the accompanying figure for a hypothetical chemical. If we were to start with 8 mg of the chemical and want to ask how many days it takes until the chemical breaks down to only 4 mg, we see that it takes 10 days. To break down by half again, from 4 to 2 mg, it takes another 10 days.

It is relatively easy to determine the half-life when we have a line graph. However, it is more difficult when we only have two data points: the amount of a chemical we start with and the amount of chemical remaining at some later point in time. In such cases, we can make use of an equation that calculates half-lives:

$$t_{1/2} = t \ln(2) \div \ln(N_0 \div N_t)$$

where $t_{1/2}$ is the half-life of the chemical, N_0 is the initial amount of the chemical, and N_t is the amount of chemical after some amount of time (t) has elapsed. For example, using data from the figure, we begin with 8 mg of a chemical and after 30 days we have 1 mg remaining. Using these data, we can calculate a half-life:

$$t_{1/2} = 30 \ln(2) \div \ln(8 \div 1)$$

$$t_{1/2} = 20.8 \div 2.08$$

$$t_{1/2} = 10 \text{ days}$$

Note that this is the same answer we obtained by estimating the half-life directly from the curve.

YOUR TURN Modern pesticides are often designed to break down rapidly so their effects on nontarget species in the ecosystem are minimized. Imagine that you spray a wetland with an insecticide to kill mosquitoes that carry the West Nile virus and this causes a concentration in the water of 50 parts per billion (ppb). Twenty-four hours later you sample the water and you find that the water now contains 10 ppb of the chemical. Using these data, calculate the chemical's half-life.

Half-life The time required for a chemical to break down to half of its original concentration.

government to ban the use of DDT and related pesticides in 1972. In Canada, usage of DDT became restricted in 1970 and was forbidden in 1985. Alternative insecticides have since been developed. With the help of many biologists who hand-reared hundreds of predatory birds, populations of species such as the peregrine falcon and bald eagle have rebounded. DDT is still used in many tropical countries, although now in a limited manner, such as in houses to control the mosquitoes that carry malaria.

GLOBAL CLIMATE CHANGE
Changing global climates play both future and current roles in the biodiversity of species. In Chapter 5 we discussed the greenhouse effect and how gases such as water vapour, CO_2, methane, and nitrous oxide allow our planet to be warmed naturally by absorbing infrared radiation emitted by Earth and then reradiating a portion back toward Earth. During the past century, human activities have increased the concentration of these greenhouse gases in the atmosphere (see "Ecology Today" in Chapter 4), as well as the gases used as refrigerants known as chlorofluorocarbons (CFCs). One result of the increase in greenhouse gases has been an increase in the average temperature on Earth. From 1880, when the earliest measurements were taken, to 2013, the temperature on Earth has increased an average of

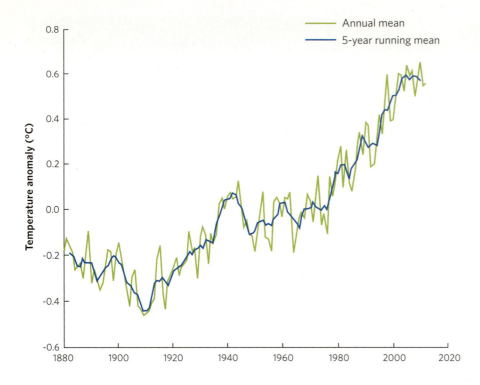

Figure 23.19 Global warming over time. Based on thousands of measurements around the world, scientists have observed a 0.8°C increase in the average temperature of our planet. Because of year-to-year variation in temperatures, the pattern of increasing temperatures is clearer when we examine the 5-year-running mean temperatures. Temperature anomaly is a comparison of each year's temperature compared to the mean temperature observed from 1951 to 1980. Data from http://data.giss.nasa.gov/gistemp/graphs_v3/.

0.8°C, as illustrated in Figure 23.19. While this is an average, some parts of the world such as northern Canada and Alaska have experienced temperature increases as high as 4°C.

Global warming has caused a number of effects that can be readily observed. For example, we might expect warmer temperatures to cause more of the world's ice to melt. As we noted at the end of Chapter 5, the massive polar ice cap in the Arctic has decreased in mass by 45 percent during the past 30 years. Similarly, the decline in ice from 2002 to 2010 has been more than 1300 metric gigatons for Antarctica and more than 1700 gigatons for Greenland. You can view these data in parts *a* and *b* of Figure 23.20. All of this melting ice, combined with an overall warming of the oceans that causes water to expand, should cause a rise in sea levels. Researchers have examined water gauges for ocean tides since 1870 and, as expected, have observed

(a)

(b)

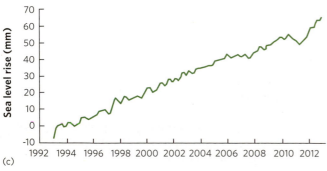

(c)

Figure 23.20 Global warming and melting ice. In both (a) Antarctica and (b) Greenland, the amount of ice has steadily declined from 2002 to 2010. Data are plotted relative to the mean ice mass throughout the entire time period. (c) Measurements of sea level from 1993 to 2013 demonstrate a continual increase in sea level compared to the sea level in 1993. Data from http://climate.nasa.gov/key_indicators#globalTemp.

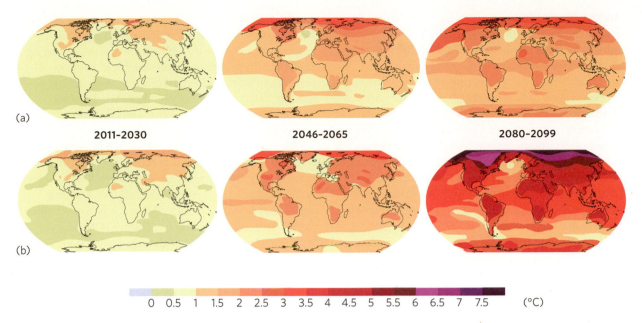

(a)

2011–2030   2046–2065   2080–2099

(b)

0 0.5 1 1.5 2 2.5 3 3.5 4 4.5 5 5.5 6 6.5 7 7.5   (°C)

Figure 23.21 **Predicted changes in global temperatures.** Depending on whether we assume **(a)** low or **(b)** high increases in CO_2 in future years, the temperature is predicted to change. The changes in temperature are relative to the mean temperature observed from 1961 to 1990. Data from G. A. Meehl et al., Global climate projections. In *Climate Change 2007: The Physical Science Basis. Contribution of Working Group I to the Fourth Assessment Report of the Intergovernmental Panel on Climate Change,* ed. S. Solomon, et al. (Cambridge University Press, 2007).

a steady increase in sea level; today the sea level is 0.2 m higher than it was in 1870. During the past 20 years, the sea level has been rising more than 3 mm per year, as shown in Figure 23.20c. At this rate, the sea will rise 0.3 m every 100 years, which is enough to dramatically affect habitats on islands and along coastlines.

As we have seen throughout this book, these changes in global temperatures are already affecting species. In Chapter 8 we discussed how many species of plants now flower earlier in the spring than they did in past decades and how some species of birds and amphibians now breed earlier. In Chapter 11 we saw that warming ocean temperatures have caused a major shift in the fish species that live in the North Sea. Several of the species that historically lived in the North Sea have moved farther north to cooler waters while a number of species that had lived in more southern waters moved up into the North Sea; this changed the species composition of the community.

Global climate change has not yet led to the widespread extinction of species. However, it is difficult for scientists to predict what will happen to global climates in the future and how these changes might affect species and ecosystems. To accomplish this, researchers have developed computer models that make predictions about how temperature and precipitation will change over future decades. One of the critical factors is how much we continue to increase the amount of CO_2 in the atmosphere. As you can see in Figure 23.21, researchers predict that a small increase in CO_2 will cause the northern latitudes

to experience an additional 4°C rise in average temperatures by the end of this century. A large increase in CO_2 will raise temperatures by an additional 7°C. Researchers predict that these changes will cause extreme weather events such as hurricanes and droughts to occur more frequently. In addition, some regions of the world will receive more annual precipitation than is currently typical while other regions will receive less. While specific predictions vary with different models, the distributions of organisms in nature will probably shift as the Earth's climates change over the next century.

Conservation efforts can slow or reverse declines in biodiversity

We have seen that human activities have caused a decline in the world's biodiversity. We now look at what can be done to slow or even reverse these declines. Over the long term, we need to stabilize the size of the human population because human activities have caused most species declines during the past several centuries. Over the short term, we need to reduce human-caused sources of mortality and low reproduction in species so that these species can persist long into the future. During the past several decades, scientists have been developing effective strategies to preserve biodiversity, although these approaches are often neither easy nor inexpensive. In this section, we will focus on habitat protection, habitat management, reduced harvest, and species reintroductions.

HABITAT PROTECTION

Because the largest contributor to the loss of biodiversity is the loss of habitat, one of the major factors in conserving biodiversity has been habitat protection. The goal in protecting a habitat is commonly the preservation of a large enough area to support a **minimum viable population (MVP)**, which is the smallest population size of a species that can persist in the face of environmental variation. The population must also be distributed widely enough to prevent local catastrophes, such as hurricanes and fires, from threatening the entire species. At the same time, some degree of population subdivision may help prevent the spread of disease from one subpopulation to another.

The task of protecting a suitable habitat becomes more complicated when a population's habitat requirements change with the seasons or when it undertakes large-scale seasonal migrations. In the Serengeti ecosystem of East Africa, for example, patterns of rainfall distribution and plant growth vary seasonally. Huge populations of wildebeests, zebras, and gazelles undertake long-distance seasonal migrations in search of suitable grazing areas. Because migratory populations use the entire area of the Serengeti ecosystem over the course of a year, the preservation of only one section would not meet their needs. For similar reasons, the massive herds of bison,

also known as buffalo, can never be fully restored to North American prairies because their seasonal migration routes are now blocked by miles of fencing and agricultural fields. Bison survive in a few small reserves in western North America—such as the Greater Yellowstone Ecosystem or Wood Buffalo National Park—but most of the lands they once occupied can never be recovered.

Wood Buffalo National Park, shown in Figure 23.22, is Canada's largest national park, covering 44 807 km^2 in Alberta and the Northwest Territories. The park was established in 1922 to protect the last free-roaming buffalo herds in the area. In addition, it is home of the world's largest beaver dam (see Figure 18.16) and the last natural nesting areas of the whooping crane (see Figure 7.7). The park also contains the Peace–Athabasca Delta, which is one of the largest freshwater deltas in the world. There are vast salt plains and large areas of undisturbed boreal forest. The management plan for the park seeks to maintain the area in a natural, self-sustaining condition. For instance, natural fires are allowed to burn as long as they do not endanger communities or park facilities. Because the park is listed as one of the UNESCO's World Heritage

Minimum viable population (MVP) The smallest population size of a species that can persist in the face of environmental variation.

Figure 23.22 The Wood Buffalo National Park. While initially set up to protect buffalo herds, the immense area of the park is now managed to protect a range of unique habitats and areas that are managed as a natural ecosystem. Photos by John David McKinnon.

Sites, another management goal is to help preserve the cultural heritage of the aboriginal groups that have used parklands for more than 8000 years. Aboriginal groups continue to use the park for hunting, fishing, and trapping.

The need to preserve habitat is recognized around the world, and many countries are setting aside such areas. The amount of terrestrial habitat that is being protected has continually increased over the past 40 years, as shown in Figure 23.23a. In fact, 57 percent of the world's countries have protected at least one-tenth of their land. Though the need to protect marine habitats has been recognized much more recently, as you can see in Figure 23.23b, it has also increased over time.

Even when lands are set aside, many countries cannot afford to protect them from squatters and poachers, and often governments allow mining and logging within protected lands. However, involving local people in the design and management of protected areas has been particularly effective since the benefits of conservation, which often include the income generated by ecotourism, become tangible and economically compelling to local people. The most successful efforts typically include large geographic areas, low human densities, a positive public attitude toward the preservation effort, and effective law enforcement to protect species from being poached.

REDUCED HARVESTING

When overharvesting is identified as a cause of decline in a species, reducing the harvest is an obvious approach to protection. However, this is complicated when it involves people's livelihoods. For example, we saw that the collapse of the Atlantic cod fishery had a major negative economic impact on the fishing industry that had depended on cod for generations and on the businesses supported by the cod fishing industry. Although some species can take a long time to recover their population size, reducing the harvest of a declining species often leads to a return to abundance. For instance, the northern elephant seal (*Mirounga angustirostris*) was hunted so much that by the late 1800s it was thought to be extinct (Figure 23.24). Small populations were subsequently discovered in the 1890s and the total population was estimated to be about 100 individuals. In the early 1900s, the seal was given protection by

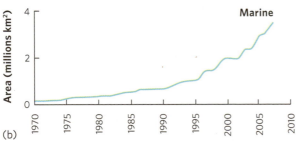

(a)

(b)

Figure 23.23 Protected terrestrial and marine areas. **(a)** Terrestrial ecosystems have a longer history of being protected and this level of protection has increased over the past 40 years. **(b)** Marine ecosystems began receiving protection much later, but the amount of protected marine habitats has also grown over time. A small amount of additional terrestrial and marine habitat has been protected, but the initial year of protection is unknown so it is not shown. Data from Secretariat of the Convention on Biological Diversity, 2010. *Global Biodiversity Outlook 3* (Montréal, 2010, 94 pages) http://www.cbd.int/doc/publications/gbo/gbo3-final-en.pdf.

Figure 23.24 The northern elephant seal. Once hunted nearly to extinction, legal protection by Mexico and the United States has allowed the population to increase to more than 150 000 today. These seals are located on the coast of central California. Photo by Wild Nature Photos/Animals Animals/Earth Scenes.

the United States and Mexico. Once protected, the elephant seal population started to rebound rapidly and today there are more than 150 000 individuals, distributed throughout much of the former range of this species in California and Mexico. Similar successes have occurred for the American crocodile (*Crocodylus acutus*), the whooping crane (*Grus americana*), and the bald eagle.

SPECIES REINTRODUCTIONS

Sometimes a species comes so close to extinction that it requires human intervention to bring it back from the brink. We saw an example of this when we discussed the recovery program for the black-footed ferret at the end of Chapter 13. Such efforts can require a great deal of effort and money, and the recovery may take decades.

A classic example of a species reintroduction is the case of the California condor, a large vulture that feeds on dead animals (Figure 23.25). During the latter half of the twentieth century, condor numbers dwindled due to a variety of causes including illegal shooting, illegal collection of eggs from condor nests, consumption of lead bullet fragments from animals the condor scavenged, and consumption by the condor of dead coyotes and rodents that had been poisoned. When the condor population in southern California sank to fewer than 30 individuals in the late 1970s, biologists made the difficult decision to bring the entire population into captivity. From 1982 to 1987, the 22 remaining wild birds were captured and brought to special breeding facilities located at zoos in Los Angeles and San Diego where they were protected from the threats that had taken such a toll in the wild.

Condors presented an extreme challenge for recovery. Individuals take 6 to 8 years to reach maturity and they typically only lay one egg per year in the wild. However, researchers found that if they removed the single egg, a female would lay another. In fact, a female condor could be tricked into laying up to three eggs per year. Biologists incubated the surplus eggs and reared the birds using puppets to feed the chicks to minimize human contact. As the chicks grew up, they were released back into the wild. At the same time, efforts were made to reduce the threats to the condors. These substantial efforts have paid off; in 2013, there were more than 400 condors alive both in captivity and in the wild following reintroductions into California, Arizona, and Mexico. In addition to increasing the condor population, this high-profile case heightened local residents' awareness of conservation issues and resulted in the preservation of large tracts of habitat in mountainous regions of southern California. People have also come to understand that restoring the condor population can

Figure 23.25 The California condor. This large vulture was on the brink of extinction in the 1980s when only 22 birds remained in the wild. A captive breeding program, followed by reintroductions into the wild, have brought the numbers up to 400 birds today living both in captivity and in the wild. Photo by Diana Kosaric/age fotostock.

be compatible with other land uses. Recreation does not have to be banned from condor habitats as long as human access to nesting sites is restricted. Legal hunting doesn't harm condors as long as steel rather than lead bullets are used. Finally, ranching does not threaten condor populations as long as coyote and rodent control programs are condor-safe.

Species reintroductions can cost millions of dollars and are usually directed toward species that appeal to the public such as black-footed ferrets, condors, and wolves. Some people may question the wisdom of spending millions of dollars to save one species. However, efforts made to save a single species, such as setting aside habitat and reducing the prevalence of poisons, often have positive effects on a large number of species. We will see an excellent example of this in "Ecology Today," which details the ecosystem effects of reintroducing the grey wolf to Yellowstone National Park.

Throughout this book, we have examined the factors that determine the distribution and abundance of species throughout the world. We have seen that there is an amazing diversity of species on Earth that impresses us with its beauty and plays essential roles in the ecosystems upon which we depend. The decline in the world's biodiversity is at a critical stage because human activities threaten populations of many species. However, by appreciating the importance of the world's species and understanding their ecology we can take the steps necessary to slow population declines and species extinctions, and to find ways to coexist.

ECOLOGY TODAY CONNECTING THE CONCEPTS

RETURNING WOLVES TO YELLOWSTONE

The wolves and elk of Yellowstone. Following the reintroduction of wolves, the elk population has declined and the ecosystem has changed. Photo by Larry Thorngren.

As we have seen in this chapter, restoring biodiversity can be a difficult challenge, and nowhere has this been more apparent than in the attempt to return the grey wolf to the Greater Yellowstone Ecosystem. The challenges in this case include ecological, economic, social, and political factors. After all, not everyone is excited about bringing back a top predator with a reputation for being a bloodthirsty killer.

The grey wolf once roamed throughout most of North America, but fear of wolves and concerns about their preying on livestock led to government programs that eliminated the wolf from nearly all of the United States and southwestern Canada. These programs caused the wolf to be eradicated from all of the lower 48 states except for northern Minnesota by 1925. In the 1940s, Aldo Leopold, a prominent professor of wildlife management, first proposed returning the wolves to Yellowstone National Park. By the 1960s, public attitudes toward wolves started to shift as many people began to see the intrinsic value of restoring ecosystems to a more natural state. In 1973, the Endangered Species Act was passed and it classified grey wolves as an endangered species, which required the development of a plan that would increase their abundance. Because Yellowstone was a large protected area, it was an obvious place for a possible reintroduction.

The plan to reintroduce wolves had supporters and detractors. Public opinion polls found that the general public and the tourists who visited the park favoured the idea. However, local hunters worried that the wolves would reduce the populations of large game such as elk. Moreover, the local livestock ranchers worried about their herds because wolves sometimes kill cattle and sheep. For nearly 2 decades, the reintroduction of wolves was debated, studied, and pushed in different directions by politicians representing various constituents. Throughout the early 1990s, the U.S. Fish and Wildlife Service held 130 town meetings in the area and received thousands of written comments on a proposed reintroduction plan. They also estimated the impacts of a wolf reintroduction to Yellowstone. Their best estimate was that an eventual population of 100 wolves would kill about 20 cattle, 70 sheep, and 1200 wild ungulates such as elk, deer, and antelope each year from a population of 95 000 wild ungulates. The economic benefit from increased tourism to the area because of the wolves was estimated at $23 million, which is a substantial instrumental value. After a series of court hearings and repeated judgments in favour of the reintroduction plan, 31 wolves were captured in Canada and released into Yellowstone in 1995 and 1996.

Nearly 2 decades later, the effects of the reintroduction have been felt throughout the ecosystem. When the wolves were eradicated in the early 1900s, the elk population rapidly increased and their feeding nearly eliminated aspen and cottonwood trees along rivers. The return of the wolves, and their rapid increase to a population of more than 220 individuals by 2001, reduced the elk population by half and caused the elk to alter their habitat use. Rather than spending much of their time along the rivers and streams during the summer, the elk moved to the higher elevations where the risk of wolf predation was lower. As a result, aspen and cottonwood trees are now thriving. Because these trees are a favourite food of beavers, the increase in aspens and cottonwoods has attracted more beavers into the region, which has led to an increase in beaver dams that form large ponds. In addition, the wolf reintroduction caused a severe reduction in the population of coyotes since they are prey for wolves. This likely had a cascading effect on the many prey species that coyotes consume. The abundant carcasses of elk and other prey of the wolves also benefited populations of scavengers such as ravens and golden eagles, which are once again common in Yellowstone. In short, the reintroduction of the wolf caused the ecosystem to move back to more closely resemble its earlier condition. In 2008, the Yellowstone wolf population was deemed to have recovered well enough to be removed from the federal list of endangered species. The wolves are now hunted in limited numbers and the wolf population has persisted. It will be interesting to see if hunting affects the size of the wolf population and what continued effects the wolves will have on the ecosystem.

The return of the wolf to the Greater Yellowstone Ecosystem demonstrates that restoring biodiversity is not a simple task. Restoration plans need to consider not only the ecology of the reintroduction but also the social, political, and economic factors. The plans must be evaluated by all interested parties and the process can often take a long time. With patience, however, species and ecosystems can eventually be restored.

(a)

(b)

(c)

SOURCES: Fritts, S. H., et al. 1997. Planning and implementing a reintroduction of wolves to Yellowstone National Park and Central Idaho. *Restoration Ecology* 5: 7–27.

Ripple, W. J., and R. L. Betscha. 2011. Trophic cascades in Yellowstone: The first 15 years after wolf reintroduction. *Biological Conservation* 145: 205–213.

Smith, D. W., et al. 2003. Yellowstone after wolves. *BioScience* 53: 330–340.

Effects of wolves on the Yellowstone ecosystem. With the reduction of elk populations due to wolf introductions, aspen and cottonwood trees along river banks have recovered. These photos were taken at Soda Butte Creek in **(a)** 1997, **(b)** 2001, and **(c)** 2010. Photos by (a) National Park Service; (b) and (c) William J. Ripple.

SUMMARY OF CHAPTER CONCEPTS

■ **The value of biodiversity arises from social, economic, and ecological considerations.** Instrumental values represent the material benefits that species and ecosystems provide to humans including food, medicines, water filtration, and pollination. Intrinsic values recognize that species and ecosystems are valuable regardless of any benefit to humans.

■ **Although extinction is a natural process, its current rate is unprecedented.** Historically there have been five mass extinctions, each followed by continued speciation. The current rate of species extinction is higher than background rates, which suggests that we may be in the early stages of a sixth mass extinction. Many species are also experiencing a rapid decline in genetic diversity, although efforts are being made to preserve the genetic diversity of livestock and crops.

■ **Human activities are causing the loss of biodiversity.** One of the most important factors contributing to species declines is the loss of habitat. Other impacts include overharvesting, introducing exotic species, and polluting the environment with contaminants. Global warming and global climate change present a rising threat. Some changes have already occurred and species may not be able to accommodate more substantial climate shifts in the future.

■ **Conservation efforts can slow or reverse declines in biodiversity.** Major efforts are being made to respond to the decline in biodiversity. Increasing amounts of terrestrial and marine habitat are being protected, harvest regulations are being adjusted to slow population declines, and species approaching extinction are being reintroduced where high-quality habitat exists and the threats can be reduced.

REVIEW QUESTIONS

1. What factors can be used to identify biodiversity hotspots?

2. Compare and contrast instrumental values versus intrinsic values of species and ecosystems.

3. In what ways can preserving biodiversity provide economic benefits?

4. What data are needed to determine if the current rate of extinction is similar to past mass extinctions?

5. Why should we be concerned with preserving both species diversity and genetic diversity?

6. What is the evidence that humans have historically contributed to extinctions through overharvesting?

7. In what ways might introduced species contribute to the extinction of native species?

8. Why do we need to consider the process of biomagnification when assessing the risk of a pesticide to wildlife?

9. What effects of global climate change are already being observed?

10. Explain the ecological, economic, and social challenges that can arise when considering a species reintroduction.

PRACTICING ECOLOGY: A CRITICAL THRESHOLD FOR COD

Boris Worm at Dalhousie University in Nova Scotia studies global trends in marine conservation. Photo by Danny Abriel, Dalhousie University.

When a species declines due to overexploitation, ecologists often seek ways to reverse the decline. However, the decline of a species to a very low abundance can fundamentally alter the interactions in a food web. In some circumstances, the population of a declining species may cross a critical threshold where food web interactions change so severely that the food web no longer favours the persistence of that species. In these cases, attempts to conserve or restore the species in decline may fail. To improve our conservation efforts, how can we identify the causes and consequences of critical thresholds?

As we have learned in this chapter, the Atlantic cod has severely declined in abundance due to overfishing, particularly on the eastern shores of Canada. Despite recent moratoriums on fishing, the population has failed to rebound. Ecologists think that cod populations may experience a critical threshold; below this threshold, populations cannot rise in abundance. This threshold may occur because of interactions between cod and herring. Adult cod are predators of many small fish, such as Atlantic herring (*Clupea harengus*). However, herring consume the eggs and fry of Atlantic cod. When cod populations are large, adult cod consume large quantities of herring, which should reduce herring predation on cod eggs and fry. When cod populations fall, herring populations should increase and cause higher predation on cod eggs and fry, and this higher predation reduces the number of cod that reach adulthood. We might therefore hypothesize that overfishing of cod has altered the food web by making herring populations large enough to continually suppress cod population growth.

Boris Worm and his colleagues at Dalhousie University tested this hypothesis by monitoring cod and herring abundance off the eastern coasts of Canada. They compiled 25 years of data on cod and herring biomass from a variety of sources, such as government reports and commercial fishing records. They examined four fishing regions: Georges Bank off the coast of Nova Scotia, the Gulf of Maine extending from Massachusetts to Cape Sable, the northern Gulf of St. Lawrence off the coast of Quebec, and the southern Gulf of St. Lawrence. You can view their data in the table.

BIOMASS (mg/km^2) OF COD AND HERRING IN FOUR FISHING REGIONS OFF THE CANADIAN EAST COAST

YEAR	GEORGES BANK		GULF OF MAINE		NORTHERN GULF OF ST. LAWRENCE		SOUTHERN GULF OF ST. LAWRENCE	
	COD	HERRING	COD	HERRING	COD	HERRING	COD	HERRING
1980	9.42	2.41	4.08	1.59	6.35	4.28	30.51	2.71
1985	8.18	1.83	5.14	2.10	3.18	7.77	14.87	8.10
1990	4.82	4.48	3.47	4.89	0.47	5.67	6.24	9.07
1995	1.58	8.47	1.90	9.49	0.40	3.96	4.26	6.35
2000	0.58	11.46	0.84	12.48	0.37	4.19	3.01	6.34
2005	1.49	9.13	11.37	10.46	0.53	3.89	1.40	7.30

GRAPH THE DATA

1. For each of the four fishing regions, create a line graph with year on the *x* axis and cod biomass on the *y* axis.

2. On each line graph, superimpose a second line graph that represents changes in herring biomass.

INTERPRET THE DATA

1. In which fishing regions are the population biomasses of cod and herring stable? In which fishing regions are the population abundances changing?

2. Explain how the data support or refute the hypothesis that the overfishing of cod altered the food web by making herring populations large enough to continually suppress cod population growth.

3. Given that cod is an important marine resource for humans, fishing industries are interested in increasing cod abundances. If cod populations are being suppressed by herring predation, what may be a potential management strategy to achieve the goal of the fishing industry?

SOURCE: Minto, C. and B. Worm. 2012. Interactions between small pelagic fish and young cod across the North Atlantic. *Ecology* 93: 2139–2154.

Reading Graphs

Ecologists often find it useful to display the data they collect in graphs. Unlike a table that contains rows and columns of data, graphs help us to visualize patterns and trends in the data. These patterns help us draw conclusions from our observations. Because graphs are such a fundamental part of ecology and most other sciences, you should become familiar with the major types of graphs that ecologists use. In this section we review different types of graphs. We also discuss how to create each type of graph and how to interpret the data that are presented in the graphs.

Scientists use graphs to present data and ideas

A graph is a tool that allows scientists to visualize data or ideas. Organizing information in the form of a graph can help us understand relationships more clearly. Throughout your study of ecology you will encounter many different types of graphs. In this section we will look at the most common types of graphs that ecologists use.

SCATTER PLOT GRAPHS

Although many of the graphs in this book may look different from each other, they all follow the same basic principles. Let's begin with an example in which researchers measured the mean temperature during the month of January in 56 cities throughout Canada. These cities span latitudes of 25° N to 50° N. The researchers wanted to determine if there was a relationship between the latitude of a city and the mean temperature in January.

We can examine this relationship by creating a *scatter plot graph,* as shown in Figure A.1. In the simplest form of a scatter plot graph, researchers look at two variables; they put the values of one variable on the

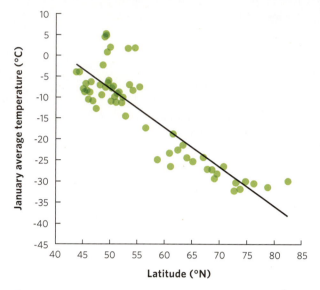

Figure A.2 (Text Figure from "Analyzing Ecology," Chapter 5) This scatter plot graph includes a line of best fit that helps us more easily see the relationship between latitude and mean January temperature for 56 cities in Canada.

x axis and the values of the other variable on the *y* axis. In our example, the latitude of the city is on the *x* axis and the mean January temperature of the city is on the *y* axis. By convention, the units of measurement tend to get larger as we move from left to right on the *x* axis and from bottom to top on the *y* axis. Using these two axes, the researchers could plot the latitude and temperature of each city on the scatter plot graph. As you can see in the figure, there is a pattern in the graph of mean January temperature declining as one moves north into the higher latitudes.

When two variables are graphed in this way, we can draw a line through the middle of the data points to describe the general trend of these points, as we have done in Figure A.2. Because such a line is drawn in a way that fits the general trend of the data, we call this the *line of best fit.* This line allows us to visualize a general trend. In our graph, the addition of a line of best fit makes it easier to identify a trend in the data; as we move from low to high latitudes we observe lower January temperatures. This is known as a negative relationship between the two variables because as one variable increases the other variable decreases.

When graphing data using a scatter plot graph, the line of best fit may be either straight or curved. For example, when researchers examined the conductivity of the water in ponds near roads, which is an indicator of the amount of road salt present, they found that as they moved away from the road there was an initial rapid decrease in conductivity that eventually levelled off, as illustrated in Figure A.3. In this case, the line of best fit is a curved line.

Sometimes scientists collect data that follow a curved line but they wish to view the data in terms of a straight line in order to compare the slopes and intercepts of lines produced by different sets of data. The slopes and

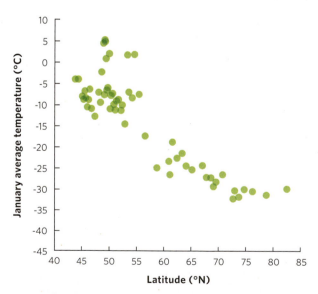

Figure A.1 (Text Figure from "Analyzing Ecology," Chapter 5) This scatter plot graph shows the relationship between latitude and mean January temperature for 56 cities in Canada.

Figure A.3 (Text Figure 2.12) This scatter plot graph shows the relationship between the distance from a pond to the nearest road and the amount of salt in the pond's water, which is measured in terms of conductivity. The line of best fit is curved.

intercepts can be calculated using the equation for a line: $y = mx + b$, where y is the dependent variable, x is the independent variable, m is the slope, and b is the intercept. In many cases, data that follow a curved line can be transformed to follow a straight line by using axes that contain the values arranged on a logarithmic scale. For example, when scientists examined the number of amphibian species living on islands of different sizes in the West Indies, the resulting scatter plot graph illustrated that the number of species initially increased rapidly as island area increased and then eventually levelled off, as shown in Figure A.4a. When the exact same data are plotted on a scatter plot graph using axes with a log scale, the line of best fit changes from a curved line to a straight line, as you can see in Figure A.4b.

LINE GRAPHS

A *line graph* is used to display data that occur as a sequence of measurements over time or space. For example, scientists have estimated the number of humans living on Earth from 8000 years ago through to the present time. Using all of the available data points, a line graph can be used to connect each data point over time, as shown in Figure A.5. In contrast to a line of best fit that fits a straight or curved line through the middle of all data points, a line graph connects one data point to another, so it can be straight or curved, or it can move up and down as it follows the movement of the data points.

Figure A.6 is an example of a line graph displaying data measurements over space. In this example, scientists tested soil samples for the concentration of chromium as they walked from one type of soil toward another type of soil. As they approached the serpentine soil, which is characterized by high concentrations of heavy metals, the concentration of chromium sharply increased.

Line graphs can also illustrate how several different variables change over time. When two variables contain different units or a different range of values, we can use two y axes. For example, Figure A.7 presents data on changes in the population sizes of two different animals on Isle Royale in the years 1955 to 2011. The left y axis represents the population changes in the wolf population whereas the right y axis represents the population changes in the moose population during the same time span.

(a)

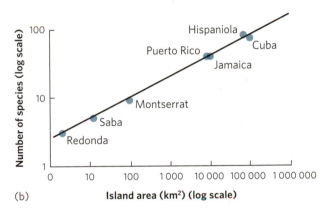

(b)

Figure A.4 (Text Figure 22.4) **(a)** This scatter plot shows the number of amphibian and reptile species on islands of different sizes in the West Indies. Note that the line of best fit is a curved line. **(b)** When the same data are plotted on a log scale, the line of best fit is a straight line.

BAR GRAPHS

A *bar graph* plots numerical values that come from different categories. For example, researchers conducted an experiment on a species of bird known as the European

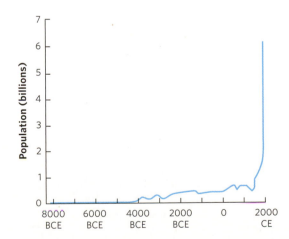

Figure A.5 (Text Figure 12.1) This line graph shows population growth over time by connecting a series of many data points.

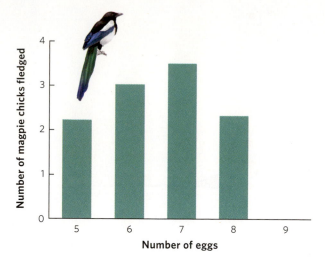

Figure A.6 (Text Figure 18.5) This line graph shows data measurement over space. It reflects changes in the concentration of chromium in the soil as researchers moved from nonserpentine soils to serpentine soils.

Figure A.8 (Text Figure 8.4) This bar graph shows the relationship between the number of eggs in nests of European magpies, plotted on the *x* axis, and the number of offspring that leave the nest as fledgling birds, plotted on the *y* axis.

magpie, which typically lays seven eggs. The researchers manipulated the number of eggs in a nest either by adding one or two eggs or by taking away one or two eggs. They then waited to see how many of the eggs hatched into birds that lived long enough to leave the nest as fledglings. The results of this experiment are illustrated in Figure A.8. As we can see in the figure, the researchers used the number of eggs as categories along the *x* axis and

then plotted the number of offspring that survived to fledgling, which is represented by the *y* axis.

Often a bar graph is used to display mean values based on multiple observations. In such cases, scientists often like to show the amount of variation in the data that comprise each of the mean values. We can do this by adding *error bars* to a bar graph. As an example, researchers who were interested in the effects of inbreeding measured the number of eggs that snails laid when they had mates available compared to when they did not have mates available. Because they measured the eggs laid by several different snails in each category,

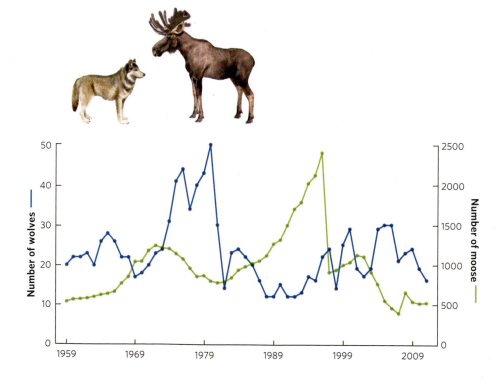

Figure A.7 (Text Figure 13.6) To illustrate changes in the population sizes of wolves and moose over time, researchers used a line graph containing two *y* axes. The *y* axis on the left represents the number of wolves and ranges from 0 to 50. In contrast, the *y* axis on the right represents the number of moose and ranges from 0 to 2500.

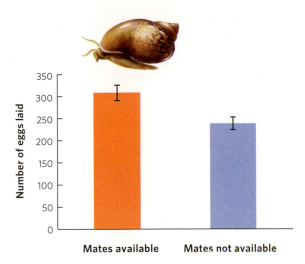

Figure A.9 (Text Figure 4.8) This bar graph shows the number of eggs produced by snails with mates available or not available. In this study, the researchers added error bars to convey the variation in egg laying among individuals. Error bars are standard errors.

they created a graph, shown in Figure A.9, that illustrates the mean number of eggs laid. However they also wanted to convey the amount of variation in eggs laid by these different individuals, so they also included error bars to each mean; in this case the error bars represent standard error of the mean.

Figure A.10 shows an example of a bar graph that presents two sets of data for each category. In this example, researchers measured the activity of six species of tadpoles. They made these measurements both in the presence and in the absence of predators. To include all this information, they drew two categories for each of the six species along the x axis: control or caged predator.

A bar graph is a very flexible tool and can be altered in several ways to accommodate data sets of different sizes or even several data sets that a researcher wishes to compare. When scientists measured the net primary productivity of different ecosystems, as shown in Figure A.11, they put the categories—various ecosystems—on the y axis and the plotted values—net primary productivity—on the x axis. This orientation makes it easier to accommodate the relatively large amount of text needed to name each ecosystem.

PIE CHARTS

A *pie chart* is a graph represented by a circle with slices of various sizes representing categories within the whole pie. The entire pie represents 100 percent of the data and each slice is sized according to the percentage of the pie that it represents. For example, Figure A.12 shows the percentage of mammals, amphibians, and fish from around the world that have been categorized as threatened, near-threatened, or of least concern from a conservation point of view. For each group of animals, each slice of the pie represents the percentage of species that fall within each conservation category.

TWO SPECIAL TYPES OF GRAPHS USED BY ECOLOGISTS

While scatter plot graphs, line graphs, bar graphs, and pie charts are used by many different types of scientists, ecologists also use two types of graphs that are not common in most other fields of science: *climate diagrams* and *age structure diagrams*. Although these two types of graphs are discussed within the text, we provide them here for review.

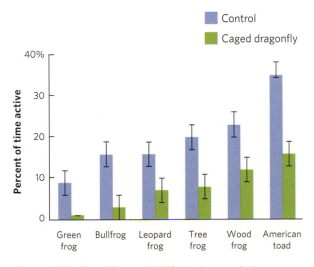

Figure A.10 (Text Figure 14.15) This bar graph shows activity of six tadpole species, each subdivided into two categories: tadpoles raised without predators or tadpoles raised in the presence of a caged predator. Error bars are standard errors.

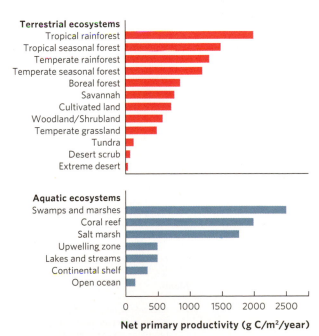

Figure A.11 (Text Figure 20.7) This bar graph shows net primary productivity of ecosystems around the world. In this case, the graph lists the categories on the y axis and plots the values on the x axis.

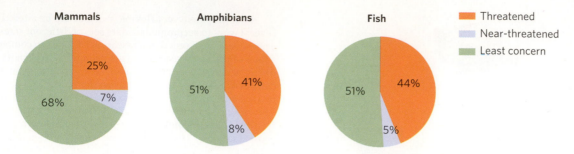

Figure A.12 (Text Figure 23.5) Pie charts illustrate the percentages that different groups compose. In this group of pie charts, we see the percentage of mammals, amphibians, and fish that fall into each of three conservation categories.

Climate Diagrams

Climate diagrams are used to illustrate the annual patterns of temperature and precipitation that help to determine the productivity of biomes on Earth. Figure A.13 shows two hypothetical biomes. By graphing the average monthly temperature and precipitation of a biome, we can see how conditions in a biome vary during a typical year. We can also observe the specific time period when the temperature is warm enough for plants to grow. In the biome illustrated in Figure A.13a, the growing season—indicated by the shaded region on the *x* axis—is mid-March through mid-October. In Figure A.13b, the growing season is mid-April through mid-September.

In addition to identifying the growing season, climate diagrams show the relationships among precipitation, temperature, and plant growth. In Figure A.13a, the precipitation line is above the temperature line in every month. This means that water supply exceeds demand, so plant growth is more constrained by temperature than by precipitation throughout the entire year. In Figure A.13b, the precipitation line intersects the temperature line. At this point, the amount of precipitation available to plants equals the amount of water lost by plants through evapotranspiration. When the precipitation line falls below the temperature line from May through September, water demand exceeds supply and plant growth will be constrained more by precipitation than by temperature.

Age Structure Diagrams

An age structure diagram is a visual representation of age distribution for both males and females in a country. Figure A.14 presents two examples. Each horizontal bar of the diagram represents a 5-year-age group and the length of a given bar represents the number of males or females in that age group.

While every nation has a unique age structure, we can group countries very broadly into three categories. Figure A.14a shows a country with many more young people than older people. The age structure diagram of a country with this population will be in the shape of a pyramid, with its widest part at the bottom, moving toward the smallest at the top. Age structure diagrams with this shape are typical

FIGURE A.13 (Text Figure 6.4) This figure shows two hypothetical climate diagrams. Climate diagrams illustrate patterns of temperature and precipitation throughout the year, the length of the growing season, and whether plant growth is limited by temperature or precipitation.

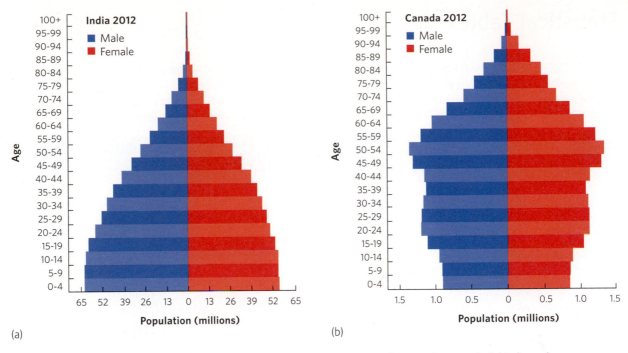

FIGURE A.14 (Text Figure 12.18) These diagrams portray the age structure of men and women in **(a)** India and **(b)** Canada.

of countries in the developing world, such as Venezuela and India.

A country with less of a difference between the number of individuals in the younger and older age groups has an age structure diagram that looks more like a column. With fewer individuals in the younger age groups, we can deduce that the country has little or no population growth. Figure A.14b shows the age structure of the Canadian population, which is similar to the age

structure of the populations of the United States, Australia, Sweden, and many other developed countries.

As you can see, we can use many different types of graphs to display data collected in ecological studies. This makes it easier to view patterns in our data and to reach the correct interpretation. With this knowledge of graph making, you should be prepared to interpret the graphs throughout the book and to do the graphing exercises at the end of each chapter.

Statistical Tables

STUDENT'S *T*-TEST DISTRIBUTION TABLE

You can identify the critical *t* value by selecting a column that contains your desired probability level and then selecting the row that has the appropriate degrees of freedom.

DEGREES OF FREEDOM	PROBABILITY LEVELS (α)		
	0.1	0.05	0.01
1	3.078	6.314	31.821
2	1.886	2.920	6.965
3	1.638	2.353	4.541
4	1.533	2.132	3.747
5	1.476	2.015	3.365
6	1.440	1.943	3.143
7	1.415	1.895	2.998
8	1.397	1.860	2.896
9	1.383	1.833	2.821
10	1.372	1.812	2.764
11	1.363	1.796	2.718
12	1.356	1.782	2.681
13	1.350	1.771	2.650
14	1.345	1.761	2.624
15	1.341	1.753	2.602
16	1.337	1.746	2.583
17	1.333	1.740	2.567
18	1.330	1.734	2.552
19	1.328	1.729	2.539
20	1.325	1.725	2.528
21	1.323	1.721	2.518
22	1.321	1.717	2.508
23	1.319	1.714	2.500
24	1.318	1.711	2.492
25	1.316	1.708	2.485
26	1.315	1.706	2.479
27	1.314	1.703	2.473
28	1.313	1.701	2.467
29	1.311	1.699	2.462
30	1.310	1.697	2.457
60	1.296	1.671	2.390
120	1.289	1.658	2.358
∞	1.282	1.645	2.326

CHI-SQUARE (χ^2) DISTRIBUTION TABLE

You can identify the critical χ^2 value by selecting a column that contains your desired probability level and then selecting the row that has the appropriate degrees of freedom.

DEGREES OF FREEDOM	PROBABILITY LEVELS (α)		
	.100	.050	.010
1	2.706	3.841	6.635
2	4.605	5.991	9.210
3	6.251	7.815	11.345
4	7.779	9.488	13.277
5	9.236	11.071	15.086
6	10.645	12.592	16.812
7	12.017	14.067	18.475
8	13.362	15.507	20.090
9	14.684	16.919	21.666
10	15.987	18.307	23.209
11	17.275	19.675	24.725
12	18.549	21.026	26.217
13	19.812	22.362	27.688
14	21.064	23.685	29.141
15	22.307	24.996	30.578
16	23.542	26.296	32.000
17	24.769	27.587	33.409
18	25.989	28.869	34.805
19	27.204	30.144	36.191
20	28.412	31.410	37.566
21	29.615	32.671	38.932
22	30.813	33.924	40.289
23	32.007	35.172	41.638
24	33.196	36.415	42.980
25	34.382	37.652	44.314
26	35.563	38.885	45.642
27	36.741	40.113	46.963
28	37.916	41.337	48.278
29	39.087	42.557	49.588
30	40.256	43.773	50.892

Z-DISTRIBUTION TABLE

To determine if a Z value is significant, begin by selecting the row that represents the first two digits of your calculated Z test statistic and then select the column that represents the third digit of your calculated Z test statistic. This gives you a probability value. By subtracting this probability value from 1, you can determine the α value.

Z	0.00	0.01	0.02	0.03	0.04	0.05	0.06	0.07	0.08	0.09
0.0	0.500	0.504	0.508	0.512	0.516	0.520	0.524	0.528	0.532	0.536
0.1	0.540	0.544	0.548	0.552	0.556	0.560	0.564	0.568	0.571	0.575
0.2	0.579	0.583	0.587	0.591	0.595	0.599	0.603	0.606	0.610	0.614
0.3	0.618	0.622	0.626	0.629	0.633	0.637	0.641	0.644	0.648	0.652
0.4	0.655	0.659	0.663	0.666	0.670	0.674	0.677	0.681	0.684	0.688
0.5	0.692	0.695	0.699	0.702	0.705	0.709	0.712	0.716	0.719	0.722
0.6	0.726	0.729	0.732	0.736	0.739	0.742	0.745	0.749	0.752	0.755
0.7	0.758	0.761	0.764	0.767	0.770	0.773	0.776	0.779	0.782	0.785
0.8	0.788	0.791	0.794	0.797	0.800	0.802	0.805	0.808	0.811	0.813
0.9	0.816	0.819	0.821	0.824	0.826	0.829	0.832	0.834	0.837	0.839
1.0	0.841	0.844	0.846	0.849	0.851	0.853	0.855	0.858	0.860	0.862
1.1	0.864	0.867	0.869	0.871	0.873	0.875	0.877	0.879	0.881	0.883
1.2	0.885	0.887	0.889	0.891	0.893	0.894	0.896	0.898	0.900	0.902
1.3	0.903	0.905	0.907	0.908	0.910	0.912	0.913	0.915	0.916	0.918
1.4	0.919	0.921	0.922	0.924	0.925	0.927	0.928	0.929	0.931	0.932
1.5	0.933	0.935	0.936	0.937	0.938	0.939	0.941	0.942	0.943	0.944
1.6	0.945	0.946	0.947	0.948	0.950	0.951	0.952	0.953	0.954	0.955
1.7	0.955	0.956	0.957	0.958	0.959	0.960	0.961	0.962	0.963	0.963
1.8	0.964	0.965	0.966	0.966	0.967	0.968	0.969	0.969	0.970	0.971
1.9	0.971	0.972	0.973	0.973	0.974	0.974	0.975	0.976	0.976	0.977
2.0	0.977	0.978	0.978	0.979	0.979	0.980	0.980	0.981	0.981	0.982
2.1	0.982	0.983	0.983	0.983	0.984	0.984	0.985	0.985	0.985	0.986
2.2	0.986	0.986	0.987	0.987	0.988	0.988	0.988	0.988	0.989	0.989
2.3	0.989	0.990	0.990	0.990	0.990	0.991	0.991	0.991	0.991	0.992
2.4	0.992	0.992	0.992	0.993	0.993	0.993	0.993	0.993	0.993	0.994
2.5	0.994	0.994	0.994	0.994	0.995	0.995	0.995	0.995	0.995	0.995
2.6	0.995	0.996	0.996	0.996	0.996	0.996	0.996	0.996	0.996	0.996
2.7	0.997	0.997	0.997	0.997	0.997	0.997	0.997	0.997	0.997	0.997
2.8	0.997	0.998	0.998	0.998	0.998	0.998	0.998	0.998	0.998	0.998
2.9	0.998	0.998	0.998	0.998	0.998	0.998	0.999	0.999	0.999	0.999
3.0	0.999	0.999	0.999	0.999	0.999	0.999	0.999	0.999	0.999	0.999
3.1	0.999	0.999	0.999	0.999	0.999	0.999	0.999	0.999	0.999	0.999
3.2	0.999	0.999	0.999	0.999	0.999	0.999	0.999	1.000	1.000	1.000
3.3	1.000	1.000	1.000	1.000	1.000	1.000	1.000	1.000	1.000	1.000

Answers to "Analyzing Ecology" and "Practicing Ecology"

Chapter 1

Analyzing Ecology

Insect abundance on uncaged trees:
    Mean = 3.0
    Variance = 1.0

Chapter 2

Analyzing Ecology

Standard deviation of the mean = 7.0
Standard error of the mean = 3.1

Practicing Ecology

Graph the Data

1. Mean percentage of SWS1 pigment in fry = 20.1; juveniles = 6.8; and adults = 6.2. Standard error of mean percentage of SWSI pigment in fry = 4.2; juveniles = 2.6; adults = 3.

2.

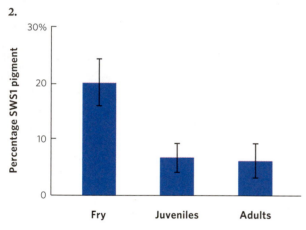

Interpret the Data

1. Based on the amount of SWS1 pigment in each age group, the fry of Nile tilapia is most sensitive to UV light.

2. This finding supports the hypothesis because the tilapia fry live in surface waters that contain ultraviolet light and have a higher percentage of SWS1 pigment than the tilapia juveniles and adults, which live in deeper waters where there is little ultraviolet light.

3. Given that zooplankton near the surface absorb ultraviolet light, their presence will be more noticeable to fish with visual sensitivity in the ultraviolet light spectrum. Therefore, having a higher sensitivity to ultraviolet would allow the tilapia fry to better locate and consume their preferred prey.

Chapter 3

Analyzing Ecology

Each of the three soil treatments would be considered a categorical variable because they fall into distinct

categories. In contrast, if we had considered three treatments such as 100 percent sand, 70 percent sand, and 40 percent sand, then we would have a continuous variable.

Practicing Ecology

Graph the Data

1.

2.

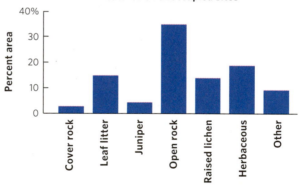

Interpret the Data

1. Habitat feature is a categorical variable because the values are distinct categories of habitat features. Percent area of each habitat feature is a continuous variable because the values can take on any numeric value.

2. Based on the differences between occupied and unoccupied sites, skinks prefer to occupy habitats that have cover rock.

3. Cover rock is likely to provide a cooler habitat during the day, which would lower body temperature to a more desirable level for the skinks. In contrast, habitat features such as open rock, juniper shrubs, and herbaceous material may warm to temperatures above the optimal level for skinks, so they avoid these habitat features.

4. Evidence that the preferred habitats of skinks are cooler than surrounding habitats supports the hypothesis that skinks select their habitat based on temperature.

Chapter 4

Analyzing Ecology

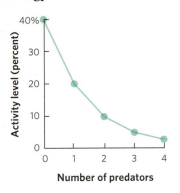

a. It is a negative correlation.

b. It is curvilinear.

Practicing Ecology

Graph the Data

1. Mean mouth sizes (± standard error of mean)

   Generalist: bloodworm diet = 3.35 (0.13), and shrimp diet = 2.83 (0.08)

   Benthic specialist: bloodworm diet = 3.22 (0.10), and shrimp diet = 2.91 (0.11)

   Pelagic specialist: bloodworm diet = 2.88 (0.10), and shrimp diet = 2.95 (0.08)

2.

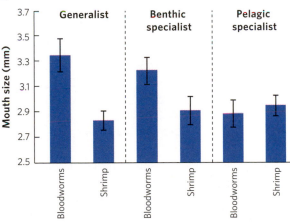

Interpret the Data

1. The generalist exhibits the greatest plasticity in mouth size.

2. This finding supports the hypothesis that the species with access to both benthic and pelagic prey should exhibit greater morphological plasticity than species restricted to only benthic or pelagic prey.

3. Because morphological plasticity of mouthparts offers a fitness advantage to fish in environments where the abundance of prey is variable, one might predict that the pelagic specialist would evolve morphological plasticity enabling it to consume either benthic or pelagic prey.

Chapter 5

Analyzing Ecology

Given the following equation:

$$\text{Temperature} = -1.2 \times \text{Latitude} + 43$$

At a latitude of 10 degrees, the mean January temperature is

$$\text{Temperature} = -1.2 \times 10 + 43 = 31°C$$

At a latitude of 20 degrees, the mean January temperature is

$$\text{Temperature} = -1.2 \times 20 + 43 = 19°C$$

At a latitude of 30 degrees, the mean January temperature is

$$\text{Temperature} = -1.2 \times 30 + 43 = 7°C$$

Practicing Ecology

Graph the Data

1.

2.

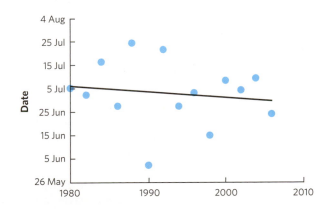

3. See regression lines in above graphs.

Interpret the Data

1. The slopes of both plots are negative, indicating that the mean mass of pregnant females has decreased with time and that the date of ice breakup has occurred slightly earlier with time.

2. As with all regressions, the data do not indicate causality. However, from what we know of polar bear feeding patterns, it seems likely that the date of ice

breakup is related to mean mass of pregnant females because earlier ice breakup would force polar bears to migrate away from their foraging grounds earlier in the year. This reduces the time that the polar bear can feed on seals, which lowers the energy reserves of pregnant females and the survival chances of mothers and their offspring.

3. The polar bears may be travelling to urban areas to forage because they have not built up enough energy reserves during the seal-hunting season to survive the ice-free season without additional food.

Chapter 6

Analyzing Ecology

Mean $= 15.6$
Median $= 17$
Mode $= 17$

The values differ because the mean measures the central tendency of the data by calculating the average value while the median ranks the data and uses the middle data point; the mode uses the commonly occurring value.

Practicing Ecology

Graph the Data

1.

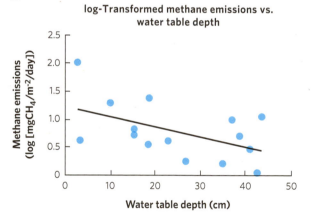

log-Transformed methane emissions vs. water table depth

2. See regression line of best fit in the graph above.

Interpret the Data

1. There is a negative regression between methane emissions and water table depth, which supports the hypothesis that reduced rainfall over peatlands will decrease the abundance of anaerobic bacteria, resulting in a decrease of CH_4 emission.

2. Based on the regression, one would predict that flooding above the surface of the peat would further reduce CH_4 emissions, leading to a further reduction in the rate of peat decomposition.

3. Beavers create dams that flood large expanses of land, which increases the saturation of soils. Hence, we can hypothesize that the presence of beavers increases the saturation of peatlands, reduces oxygen in the soil, increases anaerobic respiration, and subsequently increases methane emissions.

4. When rainfall decreases and peatlands become less saturated, we would expect the abundance of aerobic

bacteria to increase, leading to an increase in CO_2 gas emission.

Chapter 7

Analyzing Ecology

To determine the response to selection (R), we multiply the strength of selection on a trait (S) and the heritability of the trait (h^2).

S	h^2	R
0.5	0.7	0.35
1	0.7	0.7
1.5	0.7	1.05
2	0.9	1.8
2	0.6	1.2
2	0.3	0.6
2	0	0

Based on the above results, the responses to selection (R) are greater whenever the strength of selection increases or the heritability of the trait increases.

Practicing Ecology

Graph the Data

1.

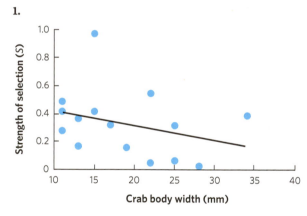

2. See regression line of best fit in the graph above.

Interpret the Data

1. The graph illustrates a negative relationship between strength of selection on snail shell thickness and crab body width. This suggests that older and larger crabs generate weaker selection on snail shell thickness than younger, smaller crabs, and supports the more general hypothesis that the strength of selection by crabs on snail shell thickness decreases with crab size.

2. The average thickness of snail shells should be thinner on the east side of the island. Crabs on the east side have larger bodies and are better at crushing snail shells of any thickness. This would decrease the strength of selection for snails with thicker shells that are harder for the crabs to crush.

3. Since the smaller crabs on the west side of the island would cause selection for snails with thicker shells, one would predict that *Littoria sitkana* would be more prevalent on the west side of the island. In contrast, since the larger crabs on the east side of the

island cause weak selection on shell thickness, one would predict that both snail species would be more prevalent on the east side.

Chapter 8

Analyzing Ecology

For Population A:

$$R^2 = 1 - \left(\frac{(21-20)^2 + (19-20)^2 + (13-12)^2 + (11-12)^2 + (5-4)^2 + (3-4)^2}{(21-12)^2 + (19-12)^2 + (13-12)^2 + (11-12)^2 + (5-12)^2 + (3-12)^2} \right) =$$

$$1 - \left(\frac{6}{262} \right) = 0.98$$

For Population B:

$$R^2 = 1 - \left(\frac{(22-20)^2 + (18-20)^2 + (14-12)^2 + (10-12)^2 + (6-4)^2 + (2-4)^2}{(22-12)^2 + (18-12)^2 + (14-12)^2 + (10-12)^2 + (6-12)^2 + (2-12)^2} \right) =$$

$$1 - \left(\frac{24}{280} \right) = 0.91$$

For Population C:

$$R^2 = 1 - \left(\frac{(24-20)^2 + (16-20)^2 + (16-12)^2 + (8-12)^2 + (7-4)^2 + (1-4)^2}{(24-12)^2 + (16-12)^2 + (16-12)^2 + (8-12)^2 + (7-12)^2 + (1-12)^2} \right) =$$

$$1 - \left(\frac{82}{338} \right) = 0.76$$

Because of the higher coefficient of determination, Population A provides the strongest confidence that there is a negative relationship between seed size and seed number.

Practicing Ecology

Graph the Data

1.

2.

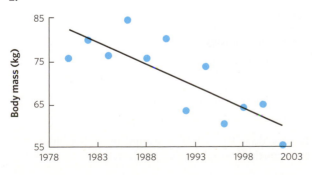

3. See regression lines of best fit in the graphs.

Interpret the Data

1. The data in the graphs indicate that mean horn length and body mass of individuals has decreased over time, which provides support for the hypothesis that sport hunting for rams with large horns has artificially selected for rams with shorter horns and small body mass.

2. Like all graphs, regressions do not indicate causality. The relationships of body mass and horn length over time are consistent with the hypothesis, but they do not definitively demonstrate that trophy hunting caused the changes in the sheep.

3. Harvesting of rams with phenotypes for large horns and rapid horn growth at an early age will probably artificially select for rams with slower-growing horns that are relatively small when the ram is fully grown.

Chapter 9

Analyzing Ecology

If there are four males and five females:

Average male fitness = 10 gene copies ÷ 4 males
= 2.5 gene copies/male

Average female fitness = 10 gene copies ÷ 5 females
= 2.0 gene copies/female

If there are six males and five females:

Average male fitness = 10 gene copies ÷ 6 males
= 1.7 gene copies/male

Average female fitness = 10 gene copies ÷ 5 females
= 2.0 gene copies/female

Over the long term, if either sex becomes rare, it will have higher fitness, which will subsequently cause it to become more common. Over the long term, this process will favour an equal sex ratio.

Practicing Ecology

Graph the Data

1.

2.

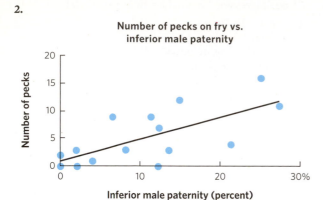

Number of pecks on fry vs. inferior male paternity

(y-axis: Number of pecks; x-axis: Inferior male paternity (percent))

Interpret the Data

1. The spread of data points around the regression line for inferior male paternity vs. pecks toward eggs is much smaller than the spread of data around the regression line for inferior male paternity vs. pecks toward fry. This indicates that cannibalism on fry is more strongly correlated with inferior male paternity than is cannibalism on eggs.

2. Although there is a positive regression in both graphs, the low R^2 value in the first graph indicates a weak correlation and only partial support for the hypothesis that cannibalism occurs more in nests with eggs that have been fertilized by inferior males.

3. This finding suggests that a superior male can only recognize that offspring are unrelated to him after the eggs hatch. This provides support for the current hypothesis, since it provides a potential explanation regarding why the amount of egg pecking did not strongly correlate with the percent of eggs that were fertilized by inferior males. It also provides evidence that the percentage of inferior male paternity caused superior males to cannibalize fry.

Chapter 10

Analyzing Ecology

If the fitness that primary helpers give to their parents declines to 1.0, their indirect fitness declines to 0.32 and their inclusive fitness declines to 0.73. Under this scenario, the secondary helper strategy would be the most favoured by natural selection.

MALE ROLE	YEAR 1			YEAR 2			DIRECT FITNESS	INCLUSIVE FITNESS
	B_1	r_1	INDIRECT FITNESS	B_2	r_2	P_{sm}		
PRIMARY HELPER	1.0 x 0.32 = 0.32			2.5 x 0.5 x 0.32 = 0.41			0.32 + 0.41 = 0.73	
SECONDARY HELPER	1.3 x 0.00 = 0.00			2.5 x 0.5 x 0.67 = 0.84			0.00 + 0.84 = 0.84	
DELAYER	0.0 x 0.00 = 0.00			2.5 x 0.5 x 0.23 = 0.29			0.00 + 0.29 = 0.29	

Practicing Ecology

Graph the Data

1. Siblings of genotype A: mean = 30.63, standard error of the mean = 0.72; siblings of genotype B: mean = 32.96, standard error of the mean = 1.79; siblings of genotype

C: mean = 34.07, standard error of the mean = 1.40; mixed non-siblings: mean = 26.92, standard error of the mean = 1.84.

2.

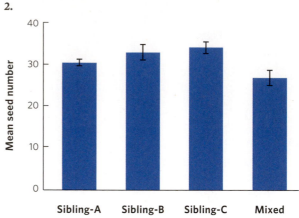

(y-axis: Mean seed number; x-axis categories: Sibling-A, Sibling-B, Sibling-C, Mixed)

Interpret the Data

1. On average, the mean seed number produced by the sibling genotype treatments was higher than the mean seed number produced by the mixed genotype treatment. This supports the hypothesis that morning glory siblings grown together will have greater seed production than nonsiblings grown together.

2. The mean seed number produced by the sibling-A treatment was less than the mean seed number produced by the sibling-B treatment, and both were less than the mean seed number produced by the sibling-C treatment. These differences suggest that the capacity of ivyleaf morning glories to recognize and respond to kin varies by genotype.

3. The data suggest there is genetic variation in the capacity to recognize and respond to kin. Since individuals that respond to kin consistently experience higher fitness, natural selection is likely to favour greater kin recognition.

4. Greater root mass in the mixed treatment indicates greater investment of energy in competition with neighbouring plants. Since seed number was also lower in the mixed treatment, this finding suggests that allocation of energy to root growth occurs at the expense of reproductive output, and that kin recognition likely explains Biernaskie's results.

Chapter 11

Analyzing Ecology

Given the equation:

$$N = M \times C \div R$$
$$N = 20 \times 48 \div 24$$
$$N = 40$$

Because these 40 crayfish were found in a 300 m² stretch of a stream, the density of the crayfish can be calculated as:

40 crayfish ÷ 300 m² = 0.13 crayfish/m²

Practicing Ecology

Graph the Data

1.

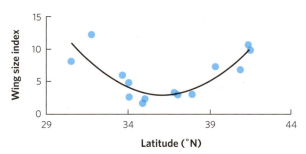

Fruit wing size index

2.

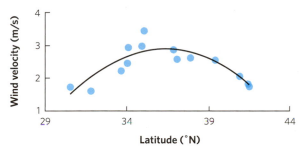

Wind velocity at fruit movement

3. See curvilinear regression lines in the graphs above.

Interpret the Data

1. The data indicate that mean fruit wing size increases near range edges, and the mean wind velocity needed to move fruits decreases near range edges. This supports the hypothesis that plants at range edges produce fruit that has larger wing sizes and greater dispersal distances.

2. Environmental conditions at the edges of geographic range are suboptimal for growth and fitness. In addition, the data suggest that the plants at the edge of their geographic range allocate a substantial amount of energy and resources towards growth of dispersal traits. Although this benefits the plant by dispersing seeds farther, it also might shift energy and resources away from reproductive efforts. Hence, we might hypothesize that seed production will decrease toward the edges of pink sand verbena's geographic range.

Chapter 12

Analyzing Ecology

AGE (x)	SURVIVAL RATE (n_x)	SURVIVOR- SHIP (l_x)	FECUNDITY (b_x)	$(l_x) \times (b_x)$	$(x) \times (l_x) \times (b_x)$
0	530	1.000	0.05	0.05	0.00
1	134	0.253	1.28	0.32	0.32
2	56	0.116	2.28	0.26	0.53
3	39	0.089	2.28	0.20	0.61
4	23	0.058	2.28	0.13	0.53
5	12	0.039	2.28	0.09	0.44
6	5	0.025	2.28	0.06	0.34
7	2	0.022	2.28	0.05	0.35

Net reproductive rate $(R_0) = \sum l_x b_x = 1.17$

Generation time $(T) = \dfrac{\sum x l_x b_x}{\sum l_x b_x} = \dfrac{3.13}{1.17} = 2.7$ years

Intrinsic rate of increase $(\lambda_a) = R_0^{\frac{1}{T}} = 1.17^{\frac{1}{2.7}} = 1.06$

Practicing Ecology

Graph the Data

1.

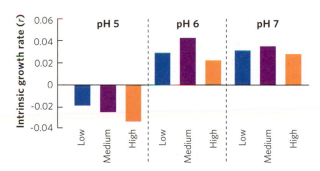

Intrinsic growth rates of *D. mendotae*

2.

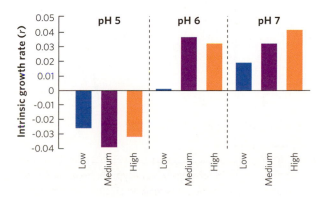

Intrinsic growth rates of *E. lacustris*

Interpret the Data

1. At pH 5, the intrinsic growth rates of *E. lacustris* are lower than *D. mendotae*, regardless of starting population density. This suggests that *E. lacustris* is more sensitive to lake acidification than *D. mendotae*.

2. As we move from low to high starting densities, the intrinsic growth rate increases for *E. lacustris* but not for *D. mendotae*, which provides support for the hypothesis that *E. lacustris* experiences positive density dependence.

3. Because *B. longimanus* preys on other zooplankton species, the presence of this species would decrease the population growth rates of *E. lacustris* and *D. mendotae*. However, as the pH of the lake rises, the population densities of *B. longimanus* would decrease. This would release the two species from the pressure of predation and allow them to increase their growth rate.

Chapter 13

Analyzing Ecology

The population sizes from year 1 through 15 are as follows:

YEAR	CHANGE IN POPULATION SIZE	TOTAL POPULATION SIZE
1		10.0
2	10.0	20.0
3	19.8	39.8
4	35.0	74.8
5	49.5	124.4
6	34.4	158.8
7	−42.6	116.2
8	−75.2	41.0
9	−7.3	33.7
10	21.9	55.6
11	40.5	96.1
12	47.0	143.1
13	6.2	149.2
14	−70.7	78.5
15	−42.5	36.0

Given that the product $rT = 1.1 \times 1 = 1.1$, we would expect the population will experience damped oscillations, which you can confirm from the graphed data.

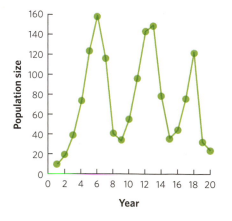

Practicing Ecology

Graph the Data

1.

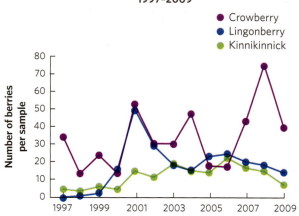

Mean number of berries per sample, 1997–2009

2.

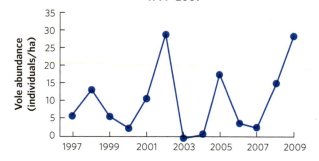

Number of voles per hectare, 1997–2009

Interpret the Data

1. Crowberry exhibits a year-to-year variation in abundance that is a similar to the vole's variation in abundance. This correlation suggests that crowberry regulates vole population size.

2. The abundance of crowberries peaks 1 year before the abundance of voles peaks. This suggests that voles exhibit a time lag in population growth. The lag probably occurs because survival over the winter depends on how much energy the voles can store during the previous fall. In years when berries are very abundant, the voles can store much more energy for the winter, and more individuals can survive to reproduce in the spring and summer.

3. Variation in the abundance of at least one berry species appears similar to the variation in abundance of voles; this correlation supports the hypothesis that food resources in the fall predict vole population size during the following summer.

Chapter 14

Analyzing Ecology

When we repeatedly sample the two distributions, we can conclude that they are significantly different if the two means do not overlap 95 percent of the time.

This would happen when the two distributions are approximately two standard deviations apart from each other.

Practicing Ecology

Graph the Data

1. Means and standard errors (SE) of understorey bird species: low deer density, mean = 2.03, SE = 0.49; high deer density, mean = 0.49, SE = 0.13. For canopy bird species, low deer density, mean = 1.00, SE = 0.25; high deer density, mean = 0.68; SE = 0.24.

2.

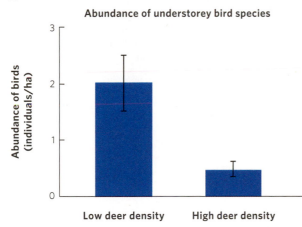

Abundance of understorey bird species

3.

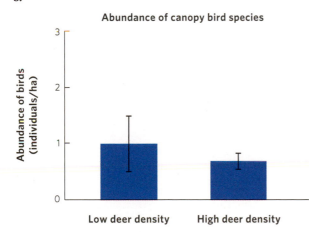

Abundance of canopy bird species

Interpret the Data

1. The data indicate that the abundance of understorey bird species is lower on islands with higher deer densities. In contrast, the abundance of canopy bird species is not affected by deer densities. This provides support for the hypothesis that deer abundance is negatively related with the abundance of understorey bird species due to effects of deer on understorey vegetation.

2. Islands with a high density of deer would likely act as sink populations for understorey bird species. If deer were removed from an island, populations of understorey bird species from nearby islands with low deer densities would likely colonize the island and increase population sizes.

3. Nest predators limit population sizes of understorey bird species. Hence, we might predict that removal of nest predators would likely allow population sizes of understorey bird species to increase. Eventually, the bird species would reach their carrying capacity, based on available food resources.

Chapter 15

Analyzing Ecology

To calculate the t-value:

$$t = \frac{\overline{X}_1 - \overline{X}_2}{\sqrt{\dfrac{s_1^2}{n_1} + \dfrac{s_2^2}{n_2}}} = \frac{10 - 5}{\sqrt{\dfrac{4}{8} + \dfrac{4}{8}}} = \frac{5}{\sqrt{1}} = 5$$

To calculate the degrees of freedom:

degrees of freedom = $8 + 8 - 2 = 14$

Based on the statistical table for a t-test (see the Statistical Tables appendix), the critical value of t for 14 degrees of freedom and an alpha value of 0.05 is 2.1. Since our t-value exceeds the critical value from the table, we conclude that the two means are significantly different.

Practicing Ecology

Graph the Data

1.

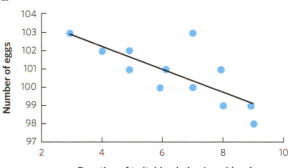

2. See regression line in the graph above.

Interpret the Data

1. There is a negative correlation between the duration of twitching behaviour exhibited by parasitized beetles and the number of eggs in emergent female wasps. This suggests that there is a cost of manipulating beetle behaviour, and supports the hypothesis that this energy expense comes at a cost of future fitness for the developing wasp.

2. If the wasps can continually produce eggs, wasps that emerge with a reduced number of eggs may compensate later in life by consuming resources and producing more eggs.

3. This suggests that wasps allocate energy consumed as pupae to egg production but not survival. In contrast, adults allocate energy consumed after emergence to survival.

Chapter 16

Analyzing Ecology

Use the following equation for a Chi-square test:

$$\chi^2 = \sum_{i=1}^{n} \frac{(O_i - E_i)^2}{(E_i)}$$

$$\chi^2 = \frac{(12 - 10)^2}{10} + \frac{(8 - 10)^2}{10}$$

$$\chi^2 = 0.8$$

Degrees of freedom =
(number of observed categories − 1) = (2 − 1) = 1

Using the Chi-square table in the Statistical Tables appendix, we can compare our calculated value (0.8) to the critical Chi-square value when we have 1 degree of freedom and an alpha value of 0.05. The critical Chi-square value is 3.841, which exceeds our Chi-square value. Thus, we conclude that our observed distribution of mice in the bare zone and in the grass is not significantly different from an equal distribution.

Practicing Ecology

Graph the Data

1. Mean and standard error (SE) for plots with similar traits: mean = 382, SE = 58; for plots with dissimilar traits: mean = 1062, SE = 114.

2.

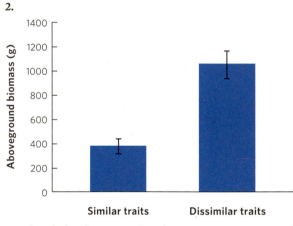

3. The *t*-value for comparing the two treatment means is −5.30.

Interpret the Data

1. From the *t*-table in the Statistical Tables appendix, we find that the critical *t*-value for a sample size of 10 and significance at $\alpha = 0.05$ is 2.23. Since our calculated *t*-value is lower than 2.23, we conclude that the two treatment means are statistically different from each other.

2. The data show that plots with species with similar traits had lower biomass than plots with high dissimilar traits. Since trait dissimilarity among plant species is assumed to positively correlate with

difference in resource-acquisition traits among species, this supports the hypothesis that total plant production in a grassland will increase with the diversity of resource-acquisition traits.

3. Instead of obtaining nitrogen from the soil, nitrogen-fixing plants obtain their nitrogen from the atmosphere, where the resource is abundant. The root nodules that fix nitrogen for the plant leak some of this nitrogen into the soil. In addition, nitrogen-fixing plants produce plant tissue that is rich in nitrogen. When this tissue senesces and decomposes, the nitrogen enters the soil and provides a source of nutrients for surrounding plants. In these ways, plants growing near nitrogen-fixing plants benefit from the added nutrients and tend to grow larger than they would if they were grown alone.

Chapter 17

Analyzing Ecology

$$U = R_1 - (n_1(n_1 + 1) \div 2)$$
$$U = 40.5 - (8(8 + 1) \div 2)$$
$$U = 40.5 - (72 \div 2)$$
$$U = 4.5$$

Using this value for R_1, we can now calculate z:

$$z = \frac{4.5 - 32}{9.52} = 2.89$$

When we look up the critical value of z in the Statistical Tables appendix, we use the absolute value of z. As a result, both R_1 and R_2 provide us with values of z that have the same absolute value.

Practicing Ecology

Graph the Data

1. Mean and standard errors (SE) of trichome density: with ants, mean = 5.20, SE = 0.65; without ants, mean = 7.58, SE = 2.24. Mean and SE of phenolic concentration: with ants, mean = 4.21, SE = 0.47; without ants, mean = 4.36, SE = 0.95.

2.

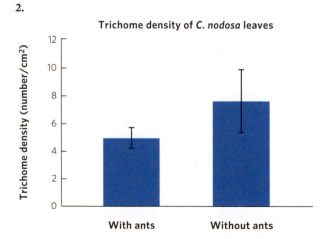

3.

Phenolic concentration of *C. nodosa* leaves

(Bar chart: Phenolic concentration (mg/g leaf) on y-axis from 0 to 6. "With ants" bar at approximately 4.2 with error bars; "Without ants" bar at approximately 4.4 with error bars.)

Interpret the Data

1. After ranking data, $U = 76.0$; $m_U = 60.5$; $\sigma_U = 15.2$; $z = 1.0$; $P = 0.15$

2. After ranking data, $U = 53.0$; $m_U = 60.5$; $\sigma_U = 15.2$; $z = -0.5$; $P = 0.29$

3. The Mann-Whitney rank-sum tests on trichome densities and phenolic concentration revealed P-values of 0.15 and 0.29, which are greater than the accepted cutoff of 0.05. This suggests that the expression of plant defence with and without ants does not differ. Hence, the data do not provide support for the hypothesis that *C. nodosa* produces more trichomes and phenolics when ants are not available to defend the plant.

4. Since plants suffered more damage in the absence of ants, this indicates that the back-up defences of trichome and phenolic production are less effective at removing herbivores than maintaining a mutualism with ants. Hence, we would predict that there is selection for the ant-tree mutualism to be maintained.

Chapter 18

Analyzing Ecology

Simpson's index
Community A:

$$\frac{1}{(0.24)^2 + (0.16)^2 + (0.08)^2 + (0.34)^2 + (0.18)^2} = \frac{1}{.24} = 4.21$$

Community B:

$$\frac{1}{(0.20)^2 + (0.20)^2 + (0.20)^2 + (0.20)^2 + (0.20)^2} = \frac{1}{.20} = 5.00$$

Community C:

$$\frac{1}{(0.25)^2 + (0.25)^2 + (0.25)^2 + (0.25)^2} = \frac{1}{.25} = 4.00$$

Shannon's index
Community A:

$-[(0.24)(\ln 0.24) + (0.16)(\ln 0.16) + (0.8)(\ln 0.08) + (0.34)(\ln 0.34) + (0.18)(\ln 0.18)] = 1.51$

Community B:

$-[(0.20)(\ln 0.20) + (0.20)(\ln 0.20) + (0.20)(\ln 0.20) + (0.20)(\ln 0.20) + (0.20)(\ln 0.20)] = 1.61$

Community C:

$-[(0.25)(\ln 0.25) + (0.25)(\ln 0.25) + (0.25)(\ln 0.25) + (0.25)(\ln 0.25)] = 1.39$

For both indexes, greater evenness with the same richness provides a higher index value. When evenness is identical, the community with the higher richness has a higher index value.

Practicing Ecology

Graph the Data

1.

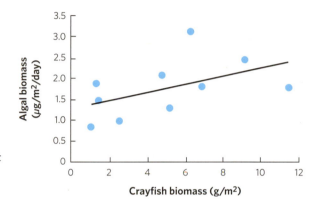

Short-term effects of crayfish on algal biomass

2.

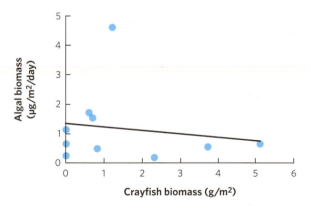

Long-term effects of crayfish on algal biomass

Interpret the Data

1. There is a positive regression between crayfish and algal biomass in the stream enclosures. This suggests that crayfish reduced the abundance of algal consumers, and provides support for the hypothesis that the short-term effects of crayfish cause a positive indirect effect on the algae.

2. The inverse relationship between herbivores and crayfish biomass further supports the hypothesis that the short-term positive effect of crayfish on algae is caused by a trophic cascade through the herbivores.

3. There is only a weak, negative regression between crayfish and algal biomass among natural sites varying in crayfish density. This does not provide strong support for the hypothesis that the long-term effect of the crayfish is to cause a direct negative effect on the algae.

Chapter 19

Analyzing Ecology

Comparing Communities A and B:

$$J = \frac{3}{1 + 4 + 3} = 0.33$$

Comparing Communities A and C:

$$J = \frac{0}{4 + 6 + 0} = 0.00$$

Comparing Communities B and C:

$$J = \frac{4}{3 + 2 + 4} = 0.44$$

Based on these calculations, Communities A and B are the most similar to each other.

Practicing Ecology

Graph the Data

1.

Percent abundance of tree species, 1859

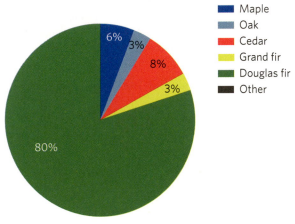

- ■ Maple
- ■ Oak
- ■ Cedar
- ■ Grand fir
- ■ Douglas fir
- ■ Other

2.

Percent abundance of tree species, 2007

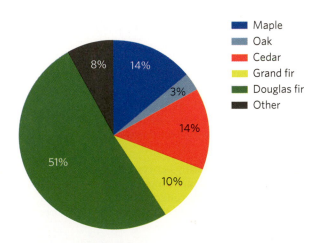

- ■ Maple
- ■ Oak
- ■ Cedar
- ■ Grand fir
- ■ Douglas fir
- ■ Other

Interpret the Data

1. Red maple, Pacific cedar, and grand fir trees all increased in abundance between 1859 and 2007, whereas Douglas fir decreased in abundance.

2. The tree species that increased in abundance are fire-intolerant, whereas the tree species that decreased in abundance are fire-tolerant. This indicates that fire suppression led to the change in tree composition over the years, and supports the hypothesis that managed forest fires played an important role in determining the tree diversity prior to European settlement.

3. The abundance of Garry oak was similar in 1859 and 2007. In addition, the abundance of rare species—indicated as "other"—increased during those years. If management efforts are aimed at restoring the landscape to a state that existed prior to European settlement, conserving Garry oak trees is not likely to achieve restoration goals.

Chapter 20

Analyzing Ecology

Lake ecosystem
   Consumption efficiency = 65%
   Assimilation efficiency = 70%
   Net production efficiency = 42%
   Ecological efficiency = 19%

Stream ecosystem
   Consumption efficiency = 65%
   Assimilation efficiency = 38%
   Net production efficiency = 40%
   Ecological efficiency = 10%

The two aquatic ecosystems have higher ecological efficiencies than the terrestrial ecosystem primarily due to higher consumption efficiencies.

   In this example, the lower ecological efficiency in the stream ecosystem compared to the lake ecosystem is caused by a lower assimilation efficiency in the stream ecosystem.

Practicing Ecology

Graph the Data

1.

2. See the graph.

Interpret the Data

1. The hypothesis was supported because the concentration of nitrogen decreased over time in water lacking viruses whereas nitrogen concentration stayed approximately constant over time in water containing viruses.

2. The bacteria and phytoplankton in the water are constantly using the nitrogen released from infected bacterial cells. This would keep the concentration of nitrogen relatively constant. In the absence of viruses, organisms would use the residual nitrogen in the water. This nitrogen would not be replenished, which would lead to a decrease in nitrogen over time.

3. Algal cells release oxygen that the bacteria use for their metabolism. Hence, infection of a small portion of bacterial cells would ultimately produce oxygen for other cells to live and generally benefit the bacteria population.

Chapter 21

Analyzing Ecology

We can calculate these answers using the following equation: $m_t = m_o\,e^{-kt}$

For $k = 0.05$

After 10 days: $m_t = m_o\,e^{-kt} = 100\,e^{-(0.05)(10)} = 61$ grams
After 50 days: $m_t = m_o\,e^{-kt} = 100\,e^{-(0.05)(50)} = 8.2$ grams
After 100 days: $m_t = m_o\,e^{-kt} = 100\,e^{-(0.05)(100)} = 0.7$ grams

For $k = 0.10$

After 10 days: $m_t = m_o\,e^{-kt} = 100\,e^{-(0.10)(10)} = 37$ grams
After 50 days: $m_t = m_o\,e^{-kt} = 100\,e^{-(0.10)(50)} = 0.7$ grams
After 100 days: $m_t = m_o\,e^{-kt} = 100\,e^{-(0.10)(100)} = 0.0$ grams

Practicing Ecology

Graph the Data

1.

Mass of insects in mesh bags

2.

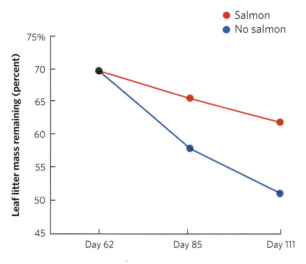

Mass of litter remaining in mesh bags

Interpret the Data

1. By the last sample date, the treatment with salmon carcasses contained less insect mass and a higher percentage of litter mass remained. This suggests that insect decomposers migrated to the salmon carcasses and away from the leaf litter, which supports the hypothesis that litter decomposes more slowly in the presence of salmon carcasses.

2. Since insect mass increased immediately after the carcass was removed, this suggests that the insects that had been decomposing the salmon carcass moved to the less nutrient-rich litter resources.

3. After the carcasses disappear from the stream, the increase in nutrient-rich microbial biomass would provide resources for insects. Hence, the mass of insects in the stream would not immediately decrease after the carcasses disappear. It is also likely that the rate of litter decomposition will accelerate as insects begin consuming the litter.

Chapter 22

Analyzing Ecology

For this problem, we will use the following equation:

$$S = S_{obs} + \frac{f_1(f_1 - 1)}{2(f_2 + 1)}$$

For Community B:

$$S = 12 + \frac{5 \times (5 - 1)}{2 \times (0+1)}$$

$$S = 12 + \frac{20}{2}$$

$$S = 22$$

For Community C:

$$S = 8 + \frac{0 \times (0 - 1)}{2 \times (2 + 1)}$$

$$S = 8 + \frac{0}{6}$$

$$S = 8$$

The greater the number of species represented by a single individual, the greater the estimated number of species in the community.

Practicing Ecology

Graph the Data

1. Mean and standard error (SE) of grasshopper abundance in rectangular plots: mean = 48.9, SE = 5.7; in winged plots: mean = 35.9, SE = 2.8. Mean and SE of percent leaf damage in rectangular plots: mean = 9.6, SE = 1.35; in winged plots: mean = 5.16, SE = 1.05.

2.

3.

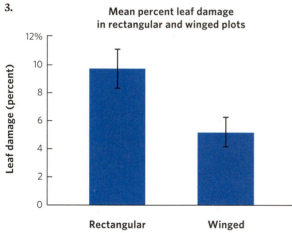

Interpret the Data

1. Both grasshopper abundance and percentage of leaves damaged were higher in rectangular plots. Because the shape of rectangular plots contained more area that was farther from the forest edge, the mean temperatures of these plots were higher. This suggests that the rectangular shape promoted higher grasshopper abundance. This supports the hypothesis that habitats with more area away from a forest edge will have more grasshoppers and higher rates of nightshade herbivory.

2. Since production of fewer berries reduces the chances for nightshades to reproduce, this suggests that nightshade populations in rectangular plots will decrease in abundance over multiple growing seasons relative to populations in winged plots.

3. Corridors linking habitats are important when the dispersal of a species is limited by its ability to move across a landscape. Since both grasshoppers and nightshades are not limited in this way, the shape of a habitat is likely to be more important for the persistence of both populations.

Chapter 23

Analyzing Ecology

We can make use of the equation that calculates half-lives:

$$t_{1/2} = t \ln(2) \div \ln(N_0 \div N_t)$$

$$t_{1/2} = (1) \ln(2) \div \ln(50 \text{ ppb} \div 10 \text{ ppb})$$

$$t_{1/2} = 0.69 \div 1.61$$

$$t_{1/2} = 0.43 \text{ days}$$

Practicing Ecology

Graph the Data

1.

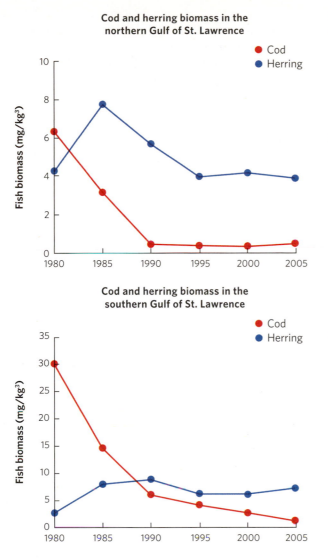

Cod and herring biomass in the northern Gulf of St. Lawrence

● Cod
● Herring

Fish biomass (mg/kg³)

Cod and herring biomass in the southern Gulf of St. Lawrence

● Cod
● Herring

Fish biomass (mg/kg³)

2. See herring data in plots above.

Interpret the Data

1. In the northern and southern Gulf of St. Lawrence, the cod and herring populations appear stable, with herring having a greater biomass than cod populations. In Georges Bank and the Gulf of Maine, populations of cod appear to be growing in biomass while populations of herring are declining in biomass.

2. In the four fishing regions, the biomass of cod and herring appear to be inversely related during part or all of the 25 years examined. In addition, cod populations in the northern and southern Gulf of St. Lawrence have remained lower than historical averages while herring populations have increased. This suggests—at least in some populations—that herring can suppress cod population growth when cod are low in abundance, and provides support for the hypothesis that overfishing of cod altered the food web by making herring populations large enough to continually suppress cod population growth.

3. To increase the abundance of cod in fishing regions, fishery managers might introduce adult cod. With sufficiently high numbers, adult cod can reduce herring populations to a level that will allow the cod population to grow. A second option is to increase fishing pressure on herring.

Suggested Readings

Chapter 1

Day, J. W., Jr., et al. 2009. Ecology in times of scarcity. *BioScience* 59: 321–331.

Dubilier, N., C. Bergin, and C. Lott. 2008. Symbiotic diversity in marine animals: The art of harnessing chemosynthesis. *Nature Reviews* 6: 725–740.

Johnson, C. K., et al. 2009. Prey choice and habitat use drive sea otter pathogen exposure in a resource-limited coastal system. *Proceedings of the National Academy of Sciences* 106: 2242–2247.

Marquis, R. J., and C. J. Whelan. Insectivorous birds increase growth of white oak through consumption of leaf-chewing insects. *Ecology* 75: 2007–2014.

Wilson, E. O. 1992. *The Diversity of Life*. W. W. Norton, New York.

Chapter 2

Cheng, C. H. C., and H. W. Detrich. 2007. Molecular physiology of Antarctic notothenioid fishes. *Philosophical Transactions of the Royal Society B*. 362: 2215–2232.

Hughes, T. P., et al. 2003. Climate change, human impacts, and the resilience of coral reefs. *Science* 301: 929–933.

Takei, Y. 2008. Exploring novel hormones essential for seawater adaptation in teleost fish. *General and Comparative Endocrinology* 157: 3–13.

Watanabe, M. E. 2005. Generating heat: New twists in the evolution of endothermy. *BioScience* 55: 470–475.

Chapter 3

Munns, R. 2002. Comparative physiology of salt and water stress. *Plant, Cell & Environment* 25: 239–250.

Osborne, C. P., and D. J. Beerling. 2006. Nature's green revolution: The remarkable evolutionary rise of C4 plants. *Philosophical Transactions of the Royal Society B* 361: 173–194.

Watanabe, M. E. 2005. Generating heat: New twists in the evolution of endothermy. *BioScience* 55: 470–475.

Chapter 4

Miner, B. G., et al. 2005. Ecological consequences of phenotypic plasticity. *Trends in Ecology and Evolution* 20: 685–692.

Pfennig, D. W., et al. 2010. Phenotypic plasticity's impacts on diversification and speciation. *Trends in Ecology and Evolution* 25: 459–467.

Chapter 5

Ducklow, H., et al. 2008. Ecological responses to climate change on the Antarctic Peninsula. *American Scientist* 96: 302–310.

Suplee, C. 1999. El Niño, La Niña. *National Geographic* 195 (3): 73–95.

Chapter 6

Ellis, E. C., and N. Ramankutty. 2008. Putting people in the map: Anthropogenic biomes of the world. *Frontiers in Ecology and the Environment* 6: 439–447.

Wolanski, E., et al. 2003. Mud, marine snow, and coral reefs. *American Scientist* 91: 44.

Chapter 7

Losos, J. B. 2007. Detective work in the West Indies: Integrating historical and experimental approaches to study island lizard evolution. *BioScience* 57: 585–597.

Moczek, A. P. 2011. The origins of novelty. *Nature* 473: 34–35.

Prud'homme, B., et al. 2011. Body plan innovation in treehoppers through the evolution of an extra wing-like appendage. *Nature* 473: 83–86.

Chapter 8

Janzen, D. H. 1976. Why bamboos wait so long to flower. *Annual Review of Ecology and Systematics* 7: 347–391.

Karban, R., C. A. Black, and S. A. Weinbaum. 2000. How 17-year cicadas keep track of time. *Ecology Letters* 3: 253–256.

Reznick, D. N., H. Bryga, and J. A. Endler. 1990. Experimentally induced life-history evolution in a natural population. *Nature* 346: 357–359.

Ricklefs, R. E., and M. Wikelski. 2002. The physiology–life history nexus. *Trends in Ecology and Evolution* 17(10): 462–468.

Stearns, S. C. 1992. *The Evolution of Life Histories*. Oxford University Press, Oxford.

Chapter 9

Busch, J. W., M. Neiman, and J. M. Koslow. 2004. Evidence for maintenance of sex by pathogens in plants. *Evolution* 58: 2584–2590.

Chapman, T. 2006. Evolutionary conflicts of interest between males and females. *Current Biology* 16: R744–R754.

Lively, C. M. 1996. Host–parasite coevolution and sex. *BioScience* 46: 107–114.

Munday, P. L., P. M. Buston, and R. R. Warner. 2006. Diversity and flexibility of sex-change strategies in animals. *Trends in Ecology and Evolution* 21: 89–95.

Reynolds, J. D. 1996. Animal breeding systems. *Trends in Ecology and Evolution* 11: 68–72.

Chapter 10

Clutton-Brock, T. H., et al. 1999. Selfish sentinels in cooperative mammals. *Science* 284: 1640–1644.

Foster, K. R., T. Wenseleers, and F. L. W. Ratnieks. 2006. Kin selection is the key to altruism. *Trends in Ecology and Evolution* 21: 57–60.

Griffin, A. S., and S. A. West. 2003. Kin discrimination and the benefit of helping in cooperatively breeding vertebrates. *Science* 302: 634–636.

Sherman, P. W., J. U. M. Jarvis, and S. H. Braude. 1992. Naked mole-rats. *Scientific American* 267: 72–78.

West, S. A., A. S. Griffin, and A. Gardner. 2007. Evolutionary explanations for cooperation. *Current Biology* 17: R661–R672.

Chapter 11

Brown, J. H., et al. 2004. Toward a metabolic theory of ecology. *Ecology* 85: 1771–1789.

Marquet, P. A., S. A. Naverrete, and J. C. Castilla. 1995. Body size, population density, and the Energetic Equivalency Rule. *Journal of Animal Ecology* 64: 325–332.

Neal, D. 2004. *Introduction to Population Biology.* Cambridge University Press, Cambridge.

Pearson, R. G., and T. P. Dawson. 2003. Predicting the impacts of climate change on the distribution of species: Are bioclimatic envelope models useful? *Global Ecology and Biogeography* 12: 361–371.

Pulliam, H. R. 2000. On the relationship between niche and distribution. *Ecology Letters* 3: 349–361.

Chapter 12

Kramer, A. M., et al. 2009. The evidence for Allee effects. *Population Ecology* 51: 341–354.

Murdoch, W. W. 1994. Population regulation in theory and practice. *Ecology* 75: 271–287.

Turchin, P. 1995. Population regulation: Old arguments and new synthesis. In N. Cappuccino and P. W. Price (eds.), *Population Dynamics: New Approaches and Synthesis,* pp. 19–40. Academic Press, New York.

Wolff, J. O. 1997. Population regulation in mammals. *Journal of Animal Ecology* 66: 1–13.

Chapter 13

Beckerman, A., et al. 2002. Population dynamic consequences of delayed life-history effects. *Trends in Ecology and Evolution* 17: 263–269.

Ehrlich, P. R., and I. Hanski (eds.). 2004. *On the Wings of Checkerspots: A Model System for Population Biology.* Oxford University Press, Oxford.

Hanski, I., and M. E. Gilpin (eds.). 1997. *Metapopulation Biology: Ecology, Genetics, and Evolution.* Academic Press, San Diego.

Honnay, O., et al. 2005. Forest fragmentation effects on patch occupancy and population viability of herbaceous plant species. *New Phytologist* 166: 723–736.

Lundberg, P., et al. 2000. Population variability in space and time. *Trends in Ecology and Evolution* 15: 460–464.

Chapter 14

Hanski, I., et al. 2001. Small rodent dynamics and predation. *Ecology* 82: 1505–1520.

Korpimäki, E., et al. 2004. The puzzles of population cycles and outbreaks of small mammals solved? *BioScience* 54: 1071–1079.

Myers, J. H. 1993. Population outbreaks in forest lepidoptera. *American Scientist* 81: 240–281.

Relyea, R. A. 2003. How prey respond to combined predators: A review and an empirical test. *Ecology* 84: 1827–1839.

Van Donk, E., et al. 2011. Induced defenses in marine and freshwater phytoplankton. *Hydrobiologia* 668: 3–19.

Chapter 15

Brown, J. K. M., and A. Tellier. 2011. Plant-parasite coevolution: Bridging the gap between genetics and ecology. *Annual Review of Phytopathology* 49: 345–367.

Hawley, D. M., and S. M. Altizer. 2011. Disease ecology meets ecological immunology: Understanding the links between organismal immunity and infection dynamics in natural populations. *Functional Ecology* 25: 48–60.

Lefevre, T., et al. 2009. The ecological significance of manipulative parasites. *Trends in Ecology and Evolution* 24: 41–48.

Chapter 16

Connell, J. H. 1961. The influence of interspecific competition and other factors on the distribution of the barnacle *Chthamalus stellatus. Ecology* 42: 710–723.

Grace, J. B., and D. Tilman. 2003. *Perspectives on Plant Competition.* Academic Press, San Diego.

Hardin, G. 1960. The competitive exclusion principle. *Science* 131: 1292–1297.

Schoener, T. W. 1983. Field experiments on interspecific competition. *American Naturalist* 122: 240–285.

Tilman, D. 1982. *Resource Competition and Community Structure.* Princeton University Press, Princeton.

Chapter 17

Bonfante, P., and I.-A. Anca. 2009. Plants, mycorrhizal fungi, and bacteria: A network of interactions. *Annual Review of Microbiology* 63: 363–383.

Bronstein, J. L. 2009. The evolution of facilitation and mutualism. *Journal of Ecology* 97: 1160–1170.

Egger, K. N., and D. S. Hibbett. 2004. The evolutionary implications of exploitation in mycorrhizas. *Canadian Journal of Botany* 82: 1110–1121.

Fenster, C. B., et al. 2004. Pollination syndromes and floral specialization. *Annual Review of Ecology, Evolution and Systematics* 35: 375–403.

Stachowicz, J. J. 2001. Mutualism, facilitation, and the structure of ecological communities. *BioScience* 51: 235–246.

Chapter 18

Halpern, B. S., et al. 2005. Predator effects on herbivore and plant stability. *Ecology Letters* 8: 189–194.

Lavergne, S., et al. 2010. Biodiversity and climate change: Integrating evolutionary and ecological responses of species and communities. *Annual Review of Ecology, Evolution, and Systematics* 41: 321–350.

Pace, M. L., et al. 1999. Trophic cascades revealed in diverse systems. *Trends in Ecology and Evolution* 15: 483–488.

Shurin, J. B., et al. 2002. A cross-ecosystem comparison of the strength of trophic cascades. *Ecology Letters* 5: 785–791.

Srivastava, D., et al. 2009. Diversity has stronger top-down than bottom-up effects on decomposition. *Ecology* 90: 1073–1083.

Chapter 19

Connell, J. H., and R. O. Slatyer. 1977. Mechanisms of succession in natural communities and their role in community stability and organization. *American Naturalist* 111: 1119–1144.

Ireland, A.W., et al. 2012. Drought as a trigger for rapid state shifts in kettle ecosystems: Implications for ecosystem responses to climate change. *Wetlands* (published online July 18, 2012).

Walker, L. R., et al. 2010. The use of chronosequences in studies of ecological succession and soil development. *Journal of Ecology* 98: 725–736.

Whittaker, R. J., M. B. Bush, and K. Richards. 1989. Plant recolonization and vegetation succession on the Krakatau Islands, Indonesia. *Ecological Monographs* 59: 59–123.

Zobel, D. B., and J. A. Antos. 1997. A decade of recovery of understory vegetation buried by volcanic tephra from Mount St. Helens. *Ecological Monographs* 67: 317–344.

Chapter 20

Elser, J. J., et al. 2007. Global analysis of nitrogen and phosphorus limitation of primary producers in freshwater, marine and terrestrial ecosystems. *Ecology Letters* 10: 1135–1142.

Hessen, D. O., et al. 2004. Carbon sequestration in ecosystems: The role of stoichiometry. *Ecology* 85: 1179–1192.

Schade, J. D., et al. 2005. A conceptual framework for ecosystem stoichiometry: Balancing resource supply and demand. *Oikos* 109: 40–51.

Schuur, E. A. G. 2003. Productivity and global climate revisited: The sensitivity of tropical forest growth to precipitation. *Ecology* 84: 1165–1170.

Chapter 21

Arrigo, K. R. 2005. Marine microorganisms and global nutrient cycles. *Nature* 437: 349–355.

Hättenschwiler, S., et al. 2005. Biodiversity and litter decomposition in terrestrial ecosystems. *Annual Review of Ecology, Evolution, and Systematics* 36: 191–218.

Kintisch, E. 2007. Should oceanographers pump iron? *Science* 318: 1368–1370.

Morgan, J. A., et al. 2004. CO_2 enhances productivity, alters species composition, and reduces digestibility of shortgrass steppe vegetation. *Ecological Applications* 14: 208–219.

Vitousek, P. M., et al. 1997. Human alteration of the global nitrogen cycle: Causes and consequences. *Issues in Ecology* 1: 1–17.

Chapter 22

Fahrig, L. 2003. Effects of habitat fragmentation on biodiversity. *Annual Review of Ecology and Systematics* 34: 487–515.

MacArthur, R. H. 1972. *Geographical Ecology: Patterns in the Distribution of Species*. Harper & Row, New York.

Magurran, A. E. 2004. *Measuring Biological Diversity*. Blackwell Publishing, Oxford.

Mittelbach, G. G., et al. 2007. Evolution and the latitudinal diversity gradient: Speciation, extinction and biogeography. *Ecology Letters* 10: 315–331.

Turner, M. G. 2005. Landscape ecology: What is the state of the science? *Annual Review of Ecology, Evolution, and Systematics* 36: 319–344.

Chapter 23

Cardillo, M., et al. 2005. Multiple causes of high extinction risk in large mammal species. *Science* 309: 1239–1241.

Firestone, R. B., et al. 2007. Evidence for an extraterrestrial impact 12,900 years ago that contributed to the megafaunal extinctions and the Younger Dryas cooling. *Proceedings of the National Academy of Sciences USA* 104: 16016–16021.

Secretariat of the Convention on Biological Diversity (2010). *Global Biodiversity Outlook 3*. Montreal, 94 pages. http://www.cbd.int/doc/publications/gbo/gbo3-final-en.pdf

Worm, B., et al. 2006. Impacts of biodiversity loss on ocean ecosystem services. *Science* 314: 787–790.

Year of Planet Earth. 2008. Feature on global climate change. *Nature* 451: 279–300.

Glossary

Abundance The total number of individuals in a population that exist within a defined area.

Acclimation An environmentally induced change in an individual's physiology.

Acid deposition Acids deposited as rain and snow or as gases and particles that attach to the surfaces of plants, soil, and water. *Also known as* Acid rain.

Acid rain *See* Acid deposition.

Acidity The concentration of hydrogen ions in a solution.

Active transport Movement of molecules or ions through a membrane against a concentration gradient.

Adaptation A characteristic of an organism that makes it well-suited to its environment.

Adiabatic cooling The cooling effect of reduced pressure on air as it rises higher in the atmosphere and expands.

Adiabatic heating The heating effect of increased pressure on air as it sinks toward the surface of Earth and decreases in volume.

Aestivation The shutting down of metabolic processes during the summer in response to hot or dry conditions.

Afrotropical region The biogeographic of the Southern Hemisphere that corresponds to most of Africa. *Also known as* Ethiopian region.

Age structure In a population, the proportion of individuals that occurs in different age classes.

Albedo The fraction of solar energy reflected by an object.

Algal bloom A rapid increase in the growth of algae in aquatic habitats, typically due to an influx of nutrients.

Allee effect *See* Positive density dependence.

Alleles Different forms of a particular gene.

Allelopathy A type of interference that occurs when organisms use chemicals to harm their competitors.

Allochthonous Inputs of organic matter, such as leaves, that come from outside of an ecosystem.

Alpha diversity *See* Local diversity.

Alternative stable state When a community is disturbed so much that the species composition and relative abundance of populations in the community change, and the new community structure is resistant to further change.

Altruism A social interaction that increases the fitness of the recipient and decreases the fitness of the donor.

Anaerobic Without oxygen. *Also known as* Anoxic.

Annual An organism that has a life span of one year.

Anoxic *See* Anaerobic.

Aphotic zone The area of the neritic and oceanic zones where the water is so deep that sunlight cannot penetrate.

Aposematism *See* Warning coloration.

Apparent competition When two species have a negative effect on each other through an enemy—including a predator, parasite, or herbivore.

Arbuscular mycorrhizal fungi A type of endomycorrhizal fungi that infects a tremendous number of plants including apple trees, peach trees, coffee trees, and grasses.

Area- and volume-based surveys Surveys that define the boundaries of an area or volume and then count all of the individuals in the space.

Artificial selection Selection in which humans decide which individuals will breed and the breeding is done with a preconceived goal for the traits of the population.

Asexual reproduction A reproduction mechanism in which progeny inherit DNA from a single parent.

Assimilated energy The portion of energy that a consumer digests and absorbs.

Assimilation efficiency The percentage of consumed energy that is assimilated.

Atmospheric convection currents The circulations of air between the surface of Earth and the atmosphere.

Australasian region The biogeographic of the Southern Hemisphere that corresponds to Australia, New Zealand, and New Guinea.

Autochthonous Inputs of organic matter that are produced by algae and aquatic plants inside an ecosystem.

Autotroph *See* Producer.

Basic metapopulation model A model that describes a scenario in which there are patches of suitable habitat embedded within a matrix of unsuitable habitat.

Batesian mimicry When palatable species evolve warning colouration that resembles unpalatable species.

Benthic zone The area consisting of the sediments at the bottoms of lakes, ponds, and oceans.

Beta diversity The number of species that differ in occurrence between two habitats.

Bicarbonate ion (HCO_3^-) An anion formed by the dissociation of carbonic acid.

Binary fission Reproduction through duplication of genes followed by division of the cell into two identical cells.

Biomagnification The process by which the concentration of a contaminant increases as it moves up the food chain.

Biomass residence time The length of time that biomass spends in a given trophic level.

Biome A geographic region that contains communities composed of organisms with similar adaptations.

Biosphere All of the ecosystems on Earth.

Biosphere approach An approach to ecology concerned with the largest scale in the hierarchy of ecological systems, including movements of air and water—and the energy and chemical elements they contain—over Earth's surface.

Biotic homogenization The process by which unique species compositions originally found in different regions slowly become more similar due to the movement of people, cargo, and species.

Blood shunting An adaptation that allows specific blood vessels to shut off so less of an animal's warm blood flows to the cold extremities.

Boreal forest A biome densely populated by evergreen needle-leaved trees, with a short growing season and severe winters. *Also known as* Taiga.

Bottleneck effect A reduction of genetic diversity in a population due to a large reduction in population size.

Bottom-up control When the abundances of trophic groups in nature are determined by the amount of energy available from the producers in a community.

Boundary layer A region of unstirred air or water that surrounds the surface of an object.

C_3 photosynthesis The most common photosynthetic pathway, in which CO_2 is initially assimilated into a three-carbon compound, glyceraldehyde 3-phosphate (G3P).

C_4 photosynthesis A photosynthetic pathway in which CO_2 is initially assimilated into a four-carbon compound, oxaloacetic acid (OAA).

Carbonate ion (CO_3^{2-}) An anion formed by the dissociation of carbonic acid.

Carrying capacity (K) The maximum population size that can be supported by the environment.

Caste Individuals within a social group sharing a specialized form of behaviour.

Categorical variable A variable that falls into a distinct category or grouping. *Also known as* Nominal variable.

Cation exchange capacity The ability of a soil to retain cations.

Census Counting every individual in a population.

Central place foraging Foraging behaviour in which acquired food is brought to a central place, such as a nest with young.

Chi-square test A statistical test that determines whether the number of observed events in different categories differs from an expected number of events, which is based on a particular hypothesis.

Chloroplasts Specialized cell organelles found in photosynthetic organisms.

Chromosomes Compact structures consisting of long strands of DNA that are wound around proteins.

Chronosequence A sequence of communities that exist over time at a given location.

Climate The typical atmospheric conditions that occur throughout the year, measured over many years.

Climate diagram A graph that plots the average monthly temperature and precipitation of a specific location on Earth.

Climax community The final seral stage in the process of succession.

Clones Individuals that descend asexually from the same parent and bear the same genotype.

Clustered dispersion A pattern of population dispersion in which individuals are aggregated in discrete groups.

Codominant When two alleles both contribute to the phenotype.

Coefficient of determination (R^2) An index that tells us how well data fit to a line.

Coefficient of relatedness The numerical probability of an individual and its relatives carrying copies of the same genes from a recent common ancestor.

Coevolution When two or more species affect each other's evolution.

Cohesion The mutual attraction among water molecules.

Cohesion–tension theory The mechanism of water movement from roots to leaves due to water cohesion and water tension.

Cohort life table A life table that follows a group of individuals born at the same time from birth to the death of the last individual.

Commensalism An interaction in which two species live in close association and one species receives a benefit while the other experiences neither a benefit nor a cost.

Community All populations of species living together in a particular area.

Community approach An approach to ecology that emphasizes the diversity and relative abundances of different kinds of organisms living together in the same place.

Community resilience The time it takes after a disturbance for a community to return to its original state.

Community resistance The amount that a community changes when acted upon by some disturbance, such as the addition or removal of a species.

Community stability The ability of a community to maintain a structure.

Competition An interaction with negative effects between two species that depend on the same limiting resource to survive, grow, and reproduce.

Competition coefficients Variables that convert between the number of individuals of one species and the number of individuals of the other species.

Competitive exclusion principle The principle that two species cannot coexist indefinitely when they are both limited by the same resource.

Concurrent circulation Movement of two fluids in the same direction on either side of a barrier through which heat or dissolved substances are exchanged.

Conduction The transfer of the kinetic energy of heat between substances that are in contact with one another.

Consumer An organism that obtains its energy from other organisms. *Also known as* a Heterotroph.

Consumption efficiency The percentage of energy or biomass in a trophic level that is consumed by the next higher trophic level.

Continental drift The movement of landmasses across the surface of Earth.

Continuous variable A variable that can take on any numeric value, including values that are not whole numbers.

Control A manipulation that includes all aspects of an experiment except the factor of interest.

Convection The transfer of heat by the movement of liquids and gases.

Convergent evolution A phenomenon in which two species descended from unrelated ancestors look similar because they have evolved under similar selective forces.

Co-operation When the donor and the recipient of a social behaviour both experience increased fitness from an interaction.

Coral bleaching Loss of colour in corals as a result of the corals expelling their symbiotic algae.

Coral reef A marine biome found in warm, shallow waters that remain 20°C year-round.

Coriolis effect The deflection of an object's path due to the rotation of Earth.

Correlation A statistical description of how one variable changes in relation to another variable.

Cosmopolitan Species with very large geographic ranges that can span several continents.

Cost of meiosis The 50 percent reduction in the number of a parent's genes passed on to the next generation via sexual reproduction versus asexual reproduction.

Countercurrent circulation Movement of two fluids in opposite directions on either side of a barrier through which heat or dissolved substances are exchanged.

Crassulacean acid metabolism (CAM) A photosynthetic pathway in which the initial assimilation of carbon into a four-carbon compound occurs at night.

Creek *See* Stream.

Crypsis Camouflage that either allows an individual to match its environment or breaks up the outline of an individual to blend in better with the background environment.

Cultural eutrophication An increase in the productivity of aquatic ecosystems caused by human activities.

Cultural services Benefits of biodiversity that provide aesthetic, spiritual, or recreational value.

Damped oscillations A pattern of population growth in which the population initially oscillates but the magnitude of the oscillations declines over time.

Data *See* Observations.

Decomposer Organism that breaks down dead organic material into simpler elements and compounds that can be recycled through the ecosystem.

Delayed density dependence When density dependence occurs based on a population density at some time in the past.

Demographic stochasticity Variation in birth rates and death rates due to random differences among individuals.

Demography The study of populations.

Denitrification The process of converting nitrates into nitrogen gas.

Density In a population, the number of individuals per unit area or volume.

Density dependent Factors that affect population size in relation to the population's density.

Density independent Factors that limit population size regardless of the population's density.

Density-mediated indirect effect An indirect effect caused by changes in the density of an intermediate species.

Deoxyribonucleic acid (DNA) A molecule composed of two strands of nucleotides that are wound together into a shape known as a double helix.

Dependent variable A factor that is being changed.

Determinate growth A growth pattern in which an individual does not grow any more once it initiates reproduction.

Deterministic model A model that is designed to predict a result without accounting for random variation in population growth rate.

Detritivore An organism that feeds on dead organic matter and waste products that are collectively known as detritus.

Diapause A type of dormancy in insects that is associated with a period of unfavourable environmental conditions.

Die-off A substantial decline in density that typically goes well below the carrying capacity.

Dilution effect The reduced, or diluted, probability of predation to a single animal when it is in a group.

Dioecious Plants that contain either only male flowers or only female flowers on a single individual.

Direct effect An interaction between two species that does not involve other species.

Direct fitness The fitness that an individual gains by passing on copies of its genes to its offspring.

Direct selection Selection that favours direct fitness.

Directional selection When individuals with an extreme phenotype experience higher fitness than the average phenotype of the population.

Dispersal The movement of individuals from one area to another.

Dispersal limitation The absence of a population from suitable habitat because of barriers to dispersal.

Dispersion The spacing of individuals with respect to one another within the geographic range of a population.

Disruptive selection When individuals with either extreme phenotype experience higher fitness than individuals with an intermediate phenotype.

Dominance hierarchy A social ranking among individuals in a group, typically determined through contests such as fighting or other contests of strength or skill.

Dominant An allele that masks the expression of the other allele.

Donor The individual who directs a behaviour toward another individual as part of a social interaction.

Dormancy A condition in which organisms dramatically reduce their metabolic processes.

Doubling time The time required for a population to double in size.

Dry climate A climate characterized by low precipitation and a wide range of temperatures, commonly found at approximately 30° N and 30° S latitudes.

Dynamic steady state When the gains and losses of ecological systems are in balance.

Ecological efficiency The percentage of net production from one trophic level compared to the next lower trophic level. *Also known as* Food chain efficiency.

Ecological envelope The range of ecological conditions that are predicted to be suitable for a species.

Ecological niche modelling The process of determining the suitable habitat conditions for a species.

Ecological stoichiometry The study of the balance of nutrients in ecological interactions, such as between an herbivore and a plant.

Ecological systems Biological entities that have their own internal processes and interact with their external surroundings.

Ecology The scientific study of the abundance and distribution of organisms in relation to other organisms and environmental conditions.

Ecosystem One or more communities of living organisms interacting with their nonliving physical and chemical environments.

Ecosystem approach An approach to ecology that emphasizes the storage and transfer of energy and matter, including the various chemical elements essential to life.

Ecotone A boundary created by sharp changes in environmental conditions over a relatively short distance, accompanied by a major change in the composition of species.

Ectoparasite A parasite that lives on the outside of an organism.

Ectomycorrhizal fungi Fungi characterized by hyphae that surround the roots of plants and enter between root cells but rarely enter the cells.

Ectotherm An organism with a body temperature that is largely determined by its external environment.

Egested energy The portion of consumed energy that is excreted or regurgitated.

El Niño–Southern Oscillation (ENSO) The periodic changes in winds and ocean currents in the South Pacific, causing weather changes throughout much of the world.

Electromagnetic radiation Energy from the Sun, packaged in small particle-like units called photons.

Emerging infectious disease A disease that is newly discovered, or has been rare and then suddenly increases in occurrence.

Endemic Species that live in a single, often isolated, location.

Endomycorrhizal fungi Fungi characterized by hyphal threads that extend far out into the soil and penetrate root cells between the cell wall and the cell membrane.

Endoparasite A parasite that lives inside an organism.

Endophytic fungi Fungi that live inside a plant's tissues.

Endotherm An organism that can generate sufficient metabolic heat to raise body temperature higher than the external environment.

Energy residence time The length of time that energy spends in a given trophic level.

Energy–diversity hypothesis A hypothesis that sites with higher amounts of energy are able to support more species.

Environmental sex determination A process in which sex is determined largely by the environment.

Environmental stochasticity Variation in birth rates and death rates due to random changes in the environmental conditions.

Epilimnion The surface layer of the water in a lake or pond.

Epistasis When the expression of one gene is controlled by another gene.

Equilibrium isocline The population size of one species that causes the population of another species to be stable. *Also known as* Zero growth isocline.

Equilibrium theory of island biogeography A theory stating that the number of species on an island reflects a balance between the colonization of new species and the extinction of existing species.

Ethiopian region *See* Afrotropical region.

Eusocial A type of social animal in which individuals live in large groups with overlapping generations, co-operation in nest building and brood care, and reproductive dominance by one or a few individuals.

Eutrophication An increase in the productivity of aquatic ecosystems.

Evaporation The transformation of water from a liquid to a gaseous state with the input of heat energy.

Evenly spaced dispersion A pattern of dispersion of a population in which each individual maintains a uniform distance between itself and its neighbours.

Evolution Change in the genetic composition of a population over time.

Experimental unit The object to which we apply an experimental manipulation.

Exploitative competition Competition in which individuals consume and drive down the abundance of a resource to a point that other individuals cannot persist.

Exponential growth model A model of population growth in which the population increases continuously at an exponential rate.

Extra-pair copulations When an individual that has a social bond with a mate also breeds with other individuals.

Facilitation A mechanism of succession in which the presence of one species increases the probability that a second species can become established.

Facultative mutualists Two species that provide fitness benefits to each other, but the interaction is not critical to the persistence of either species.

Fall bloom A rapid increase in the population of phytoplankton in fall that occurs with the infusion of nutrients from fall overturn.

Fall overturn The vertical mixing of lake water that occurs in fall, assisted by winds that drive the surface currents.

Fecundity The number of offspring produced by an organism per reproductive episode.

Field capacity The maximum amount of water held by soil particles against the force of gravity.

Fire-maintained climax community A successional stage that persists as the final seral stage due to periodic fires.

First law of thermodynamics Energy cannot be created or destroyed; it can only change form. *Also known as* the Law of conservation of energy.

Fitness The survival and reproduction of an individual.

Food chain A linear representation of how different species in a community feed on each other.

Food chain efficiency *See* Ecological efficiency.

Food web A complex and realistic representation of how species feed on each other in a community.

Founder effect When a small number of individuals leave a large population to colonize a new area and bring with them only a small amount of genetic variation.

Frequency-dependent selection When the rarer phenotype in a population is favoured by natural selection.

Freshwater wetland An aquatic biome that contains standing fresh water, or soils saturated with fresh water for at least part of the year, and which is shallow enough to have emergent vegetation throughout all depths.

Functional response The relationship between the density of prey and an individual predator's rate of food consumption.

Fundamental niche The range of abiotic conditions under which species can persist.

Gamma diversity *See* Regional diversity.

Gene pool The collection of alleles from all individuals in a population.

Generalist A species that interacts with many other species.

Generation time (T) The average time between the birth of an individual and the birth of its offspring.

Genetic drift A process that occurs when genetic variation is lost because of random variation in mating, mortality, fecundity, and inheritance.

Genotype The set of genes an organism carries.

Geographic range A measure of the total area covered by a population.

Geometric growth model A model of population growth that compares population sizes at regular time intervals.

Global climate change A phenomenon that refers to changes in Earth's climates including global warming, changes in the global distribution of precipitation and temperature, changes in the intensity of storms, and altered ocean circulation.

Glycerol A chemical that prevents the hydrogen bonds of water from coming together to form ice unless the temperatures are well below freezing.

Glycoproteins A group of compounds that can be used to lower the freezing temperature of water.

Gonads The primary sexual organs in animals.

Gondwana The southern landmass that separated from Pangaea about 150 Mya and subsequently split into South America, Africa, Antarctica, Australia, and India.

Good genes hypothesis The hypothesis that an individual chooses a mate that possesses a superior genotype.

Good health hypothesis The hypothesis that an individual chooses the healthiest mates.

Grazer-maintained climax community When a successional stage persists as the final seral stage due to intense grazing.

Greenhouse effect The process of solar radiation striking Earth, being converted to infrared radiation and being absorbed and re-emitted by atmospheric gases.

Greenhouse gases Compounds in the atmosphere that absorb the infrared heat energy emitted by Earth and then emit some of the energy back toward Earth.

Gross primary productivity (GPP) The rate at which energy is captured and assimilated by producers in an area.

Growing season The months in a location that are warm enough to allow plant growth.

Growth rate In a population, the number of new individuals that are produced in a given amount of time minus the number of individuals that die.

Guild Within a given trophic level, a group of species that feeds on similar items.

Gyre A large-scale water circulation pattern between continents.

Habitat The place, or physical setting, in which an organism lives.

Habitat corridor A strip of favourable habitat located between two large patches of habitat that facilitates dispersal.

Habitat fragmentation The process of breaking up large habitats into a number of smaller habitats.

Hadley cells The two circulation cells of air between the equator and 30° N and 30° S latitudes.

Half-life The time required for a chemical to break down to half of its original concentration.

Handling time The amount of time that a predator takes to consume a captured prey.

Haplodiploid A sex-determination system in which one sex is haploid and the other sex is diploid.

Herbivore An organism that consumes producers such as plants and algae.

Heritability The proportion of the total phenotypic variation that is caused by genetic variation.

Hermaphrodite An individual that produces both male and female gametes.

Heterotroph *See* Consumer.

Heterozygous When an individual has two different alleles of a particular gene.

Hibernation A type of dormancy that occurs in mammals in which individuals reduce the energetic costs of being active by lowering their heart rate and decreasing their body temperature.

Homeostasis An organism's ability to maintain constant internal conditions in the face of a varying external environment.

Homeotherm An organism that maintains constant temperature conditions within its cells.

Homozygous When an individual has two identical alleles of a particular gene.

Horizon A distinct layer of soil.

Horizontal transmission When a parasite moves between individuals other than parents and their offspring.

Hydrologic cycle The movement of water through ecosystems and atmosphere.

Hyperosmotic When an organism has a higher solute concentration in its tissues than the surrounding water.

Hypolimnion The deeper layer of water in a lake or pond.

Hyposmotic When an organism has a lower solute concentration in its tissues than the surrounding water.

Hypothesis An idea that potentially explains a repeated observation.

Ideal free distribution When individuals distribute themselves among different habitats in a way that allows them to have the same per capita benefit.

Inbreeding depression The decrease in fitness caused by matings between close relatives due to offspring inheriting deleterious alleles from both the egg and the sperm.

Inclusive fitness The sum of direct fitness and indirect fitness.

Independent communities Communities in which species do not depend on each other to exist.

Independent variable A factor that causes other variables to change.

Indeterminate growth A growth pattern in which an individual continues to grow after it initiates reproduction.

Indirect effect An interaction between two species that involves one or more intermediate species.

Indirect fitness The fitness that an individual gains by helping relatives pass on copies of their genes.

Indirect selection Selection that favours indirect fitness. *Also known as* Kin selection.

Individual A living being; the most fundamental unit of ecology.

Individual approach An approach to ecology that emphasizes the way in which an individual's morphology, physiology, and behaviour enable it to survive in its environment.

Indomalayan region The biogeographic of the Southern Hemisphere that corresponds to India and Southeast Asia. *Also known as* Oriental region.

Industrial melanism A phenomenon in which industrial activities cause habitats to become darker due to pollution and, as a result, individuals possessing darker phenotypes are favoured by selection.

Infection resistance The ability of a host to prevent an infection from occurring.

Infection tolerance The ability of a host to minimize the harm that an infection can cause.

Inflection point The point on a sigmoidal growth curve at which the population achieves its highest growth rate.

Inhibition A mechanism of succession in which one species decreases the probability that a second species will become established.

Instrumental value of biodiversity A focus on the economic value a species can provide.

Interdependent communities Communities in which species depend on each other to exist.

Interference competition When competitors do not immediately consume resources but defend them.

Intermediate disturbance hypothesis The hypothesis that more species are present in a community that occasionally experiences disturbances than in a community that experiences frequent or rare disturbances.

Interspecific competition Competition among individuals of different species.

Intertidal zone A biome consisting of the narrow band of coastline between the levels of high tide and low tide.

Intertropical convergence zone (ITCZ) The area where the two Hadley cells converge and cause large amounts of precipitation.

Intraspecific competition Competition among individuals of the same species.

Intrinsic growth rate (r) The highest possible per capita growth rate for a population.

Intrinsic value of biodiversity A focus on the inherent value of a species, not tied to any economic benefit.

Inverse density dependence *See* Positive density dependence.

Ions Atoms or groups of atoms that are electrically charged.

Isozymes Different forms of an enzyme that catalyze a given reaction.

Iteroparity When organisms reproduce multiple times during their life.

Joint equilibrium point The point at which the equilibrium isoclines for predator and prey populations cross.

Joint population trajectory The simultaneous trajectory of predator and prey populations.

J-shaped curve The shape of exponential growth when graphed.

Keystone species A species that substantially affects the structure of communities despite the fact that individuals of that species might not be particularly numerous.

Kin selection *See* Indirect selection.

Lake An aquatic biome that is larger than a pond and is characterized by nonflowing fresh water with some area of water that is too deep for plants to rise above the water's surface.

Landscape ecology The field of study that considers the spatial arrangement of habitats at different scales and examines how they influence individuals, populations, communities, and ecosystems.

Landscape metapopulation model A population model that considers both differences in the quality of the suitable patches and the quality of the surrounding matrix.

Latent heat release When water vapour is converted back to liquid, water releases energy in the form of heat.

Laterization The breakdown of clay particles, which results in the leaching of silicon from the soil, leaving oxides of iron and aluminum to predominate throughout the soil profile.

Laurasia The northern landmass that separated from Pangaea about 150 Mya and subsequently split into North America, Europe, and Asia.

Law of conservation of energy *See* First law of thermodynamics.

Law of conservation of matter Matter cannot be created or destroyed; it can only change form.

Leaching A process in which groundwater removes some substances by dissolving them and moving them down through the soil to lower layers.

Legacy effect A long-lasting influence of historical processes on the current ecology of an area.

Lek The location of an animal aggregation to put on a display to attract the opposite sex.

Liebig's law of the minimum Law stating that a population increases until the supply of the most limiting resource prevents it from increasing further.

Life expectancy *See* Longevity.

Life history The schedule of an organism's growth, development, reproduction, and survival.

Life tables Tables that contain class-specific survival and fecundity data.

Lifetime dispersal distance The average distance an individual moves from where it was hatched or born to where it reproduces.

Limnetic zone The open water beyond the littoral zone, where the dominant photosynthetic organisms are floating algae. *Also known as* the Pelagic zone.

Line-transect surveys Surveys that count the number of individuals observed as one moves along a line.

Littoral zone The shallow area around the edge of a lake or pond containing rooted vegetation.

Local diversity The number of species in a relatively small area of homogeneous habitat, such as a stream. *Also known as* Alpha diversity.

Local mate competition When competition for mates occurs in a very limited area and only a few males are required to fertilize all of the females.

Logistic growth model A growth model that describes slowing growth of populations at high densities.

Log-normal distribution A normal, or bell-shaped, distribution that uses a logarithmic scale on the x axis.

Longevity The life span of an organism. *Also known as* Life expectancy.

Lotic Characterized by flowing fresh water.

Lotka-Volterra model A model of predator–prey interactions that incorporates oscillations in the abundances of predator and prey populations and shows predator numbers lagging behind those of their prey.

Macroevolution Evolution at higher levels of organization including species, genera, families, orders, and phyla.

Mangrove swamp A biome that occurs along tropical and subtropical coasts, and contains salt-tolerant trees with roots submerged in water.

Manipulation The factor that we want to vary in an experiment. *Also known as* a Treatment.

Manipulative experiment A process by which a hypothesis is tested by altering a factor that is hypothesized to be an underlying cause of the phenomenon.

Mark-recapture survey A method of population estimation in which researchers capture and mark a subset of a population from an area, return it to the area, and then capture a second sample of the population after some time has passed.

Mass extinction events Events in which at least 75% of the existing species go extinct within a 2 million-year period.

Mate guarding A behaviour in which one partner prevents the other partner from participating in extra-pair copulations.

Mathematical model A representation of a system with a set of equations that correspond to hypothesized relationships among the system's components.

Mating system The number of mates each individual has and the permanence of the relationship with those mates.

Matric potential The potential energy generated by the attractive forces between water molecules and soil particles. *Also known as* Matrix potential.

Matrix potential *See* Matric potential.

Mesopredators Relatively small carnivores that consume herbivores.

Microcosm A simplified ecological system that attempts to replicate the essential features of an ecological system in a laboratory or field setting.

Microevolution The evolution of populations.

Microhabitat A specific location within a habitat that typically differs in environmental conditions from other parts of the habitat.

Mineralization The process of breaking down organic compounds into inorganic compounds.

Minimum viable population (MVP) The smallest population size of a species that can persist in the face of environmental variation.

Mixotroph An organism that obtains its energy from more than one source.

Moist continental mid-latitude climate A climate that exists at the interior of continents and is typically characterized by warm summers, cold winters, and moderate amounts of precipitation.

Moist subtropical mid-latitude climate A climate characterized by warm, dry summers and cold, wet winters.

Monoecious Plants that have separate male and female flowers on the same individual.

Monogamy A mating system in which a social bond between one male and one female persists through the period that is required for them to rear their offspring.

Müllerian mimicry When several unpalatable species evolve a similar pattern of warning colouration.

Mutation A random change in the sequence of nucleotides in regions of DNA that either comprise a gene or control the expression of a gene.

Mutualism An interaction between two species in which each species receives benefits from the other.

Mycorrhizal fungi Fungi that surround plant roots and help plants obtain water and minerals.

Natural experiment An approach to hypothesis testing that relies on natural variation in the environment.

Natural selection Change in the frequency of genes in a population through differential survival and reproduction of individuals that possess certain phenotypes.

Nearctic region The biogeographic of the Northern Hemisphere that roughly corresponds to North America.

Negative density dependence When the rate of population growth decreases as population density increases.

Negative feedbacks The action of internal response mechanisms that restores a system to a desired state, or set point, when the system deviates from that state.

Neotropical region The biogeographic of the Southern Hemisphere that corresponds to South America.

Neritic zone The ocean zone beyond the range of the lowest tidal level, and which extends to depths of about 200 m.

Net primary productivity (NPP) The rate of energy that is assimilated by producers and converted into producer biomass in an area.

Net production efficiency The percentage of assimilated energy that is used for growth and reproduction.

Net reproductive rate The total number of female offspring that we expect an average female to produce over the course of her life.

Net secondary productivity The rate of consumer biomass accumulation in a given area.

Niche The range of abiotic and biotic conditions that an organism can tolerate.

Nitrification The final process in the nitrogen cycle, which converts ammonium to nitrite (NO_2) and then from nitrite to nitrate (NO_3^-).

Nitrogen fixation The process of converting atmospheric nitrogen into forms producers can use.

Nominal variable *See* Categorical variable.

Nonrenewable resources Resources that are not regenerated.

Numerical response A change in the number of predators through population growth or population movement due to immigration or emigration.

Obligate mutualists Two species that provide fitness benefits to each other and require each other to persist.

Observations Information, including measurements, that is collected from organisms or the environment. *Also known as* Data.

Oceanic zone The ocean zone beyond the neritic zone.

Omnivore A species that feeds at several trophic levels.

Optimal foraging theory A model describing foraging behaviour that provides the best balance between the costs and benefits of different foraging strategies.

Optimum The narrow range of environmental conditions to which an organism is best suited.

Oriental region *See* Indomalayan region.

Osmoregulation The mechanisms that organisms use to maintain proper solute balance.

Osmosis The movement of water across a semipermeable membrane.

Osmotic potential The force with which an aqueous solution attracts water by osmosis.

Overshoot When a population grows beyond its carrying capacity.

Palearctic region The biogeographic of the Northern Hemisphere that corresponds to Eurasia.

Pangaea The single landmass that existed on Earth about 250 Mya and subsequently split into Laurasia and Gondwana.

Parasite An organism that lives in or on another organism, but rarely kills it.

Parasite load The number of parasites of a given species that an individual host can harbour.

Parasitoid An organism that lives within and consumes the tissues of a living host, eventually killing the host.

Parent material The layer of bedrock that underlies soil and plays a major role in determining the type of soil that will form above it.

Parental investment The amount of time and energy given to an offspring by its parents.

Parity The number of reproductive episodes an organism experiences.

Parthenogenesis A form of asexual reproduction in which an embryo is produced without fertilization.

Pathogen A parasite that causes disease in its host.

Pelagic zone *See* Limnetic zone.

Perennial An organism that has a life span of more than one year.

Perfect flowers Flowers that contain both male and female parts.

Permafrost A phenomenon where layers of soil are permanently frozen.

pH A measure of acidity or alkalinity; defined as pH = –log (H+ concentration).

Phenotype An attribute of an organism, such as its behaviour, morphology, or physiology.

Phenotypic plasticity The ability of a single genotype to produce multiple phenotypes.

Phenotypic trade-off A situation in which a given phenotype experiences higher fitness in one environment whereas other phenotypes experience higher fitness in other environments.

Photic zone The area of the neritic and oceanic zones that contains sufficient light for photosynthesis by algae.

Photoperiod The amount of light that occurs each day.

Photorespiration The oxidation of carbohydrates to CO_2 and H_2O by Rubisco, which reverses the light reactions of photosynthesis.

Photosynthetically active region Wavelengths of light that are suitable for photosynthesis.

Phylogenetic trees Hypothesized patterns of relatedness among different groups such as populations, species, or genera.

Pioneer species The earliest species to arrive at a site.

Pleiotropy When a single gene affects multiple traits.

Podsolization A process occurring in acidic soils typical of cool, moist regions, where clay particles break down in the E or Ae horizon, and their soluble ions are transported down to the lower B horizon.

Poikilotherm An organism that does not have constant body temperatures.

Polar cells The atmospheric convection currents that move air between 60° and 90° latitudes in the Northern and Southern Hemispheres.

Polar climate A climate that experiences very cold temperatures and relatively little precipitation.

Polyandry A mating system in which a female mates with more than one male.

Polygamy A mating system in which a single individual of one sex forms long-term social bonds with more than one individual of the opposite sex.

Polygenic When a single trait is affected by several genes.

Polygyny A mating system in which a male mates with more than one female.

Polyploid A species that contains three or more sets of chromosomes.

Pond An aquatic biome that is smaller than a lake and is characterized by nonflowing fresh water with some area of water that is too deep for plants to rise above the water's surface.

Population The individuals of the same species living in a particular area.

Population approach An approach to ecology that emphasizes variation over time and space in the number of individuals, the density of individuals, and the composition of individuals.

Population cycles Regular oscillation of a population over a longer period of time.

Positive density dependence When the rate of population growth increases as population density increases. *Also known as* Inverse density dependence or the Allee effect.

Potential evapotranspiration (PET) The amount of water that could be evaporated from the soil and transpired by plants, given the average temperature and humidity.

Predator An organism that kills and partially or entirely consumes another individual.

Prediction A logical consequence of a hypothesis.

Primary consumer A species that eats producers.

Primary productivity The rate at which solar or chemical energy is captured and converted into chemical bonds by photosynthesis or chemosynthesis.

Primary sexual characteristics Traits related to fertilization.

Primary succession The development of communities in habitats that are initially devoid of plants and organic soil, such as sand dunes, lava flows, and bare rock.

Principle of allocation The observation that when resources are devoted to one body structure, physiological function, or behaviour, they cannot be allotted to another.

Priority effect When the arrival of one species at a site affects the subsequent colonization of other species.

Producer An organism that uses photosynthesis to convert solar energy into organic compounds or uses chemosynthesis to convert chemical energy into organic compounds. *Also known as* an Autotroph.

Profundal zone The area in a lake that is too deep to receive sunlight.

Promiscuity A mating system in which males mate with multiple females and females mate with multiple males and do not create a lasting social bond.

Provisioning services Benefits of biodiversity that humans use, including lumber, fur, meat, crops, water, and fibre.

Proximate hypothesis A hypothesis that addresses the immediate changes in an organism's hormones, physiology, nervous system, or muscular system.

Pyramid of biomass A trophic pyramid that represents the standing crop of organisms present in different trophic groups.

Pyramid of energy A trophic pyramid that displays the total energy existing at each trophic level.

Q_{10} The ratio of the rate of a physiological process at one temperature to its rate at a temperature 10°C cooler.

Queen The dominant, egg-laying female in eusocial insect societies.

Radiation The emission of electromagnetic energy by a surface.

Rain shadow A region with dry conditions found on the leeward side of a mountain range as a result of humid winds from the ocean causing precipitation on the windward side.

Random assortment The process of making haploid gametes in which the combination of alleles that are placed into a given gamete could be any combination of those possessed by the diploid parent.

Random dispersion A pattern of dispersion of a population in which the position of each individual is independent of the position of other individuals in the population.

Randomization An aspect of experiment design in which every experimental unit has an equal chance of being assigned to a particular manipulation.

Rank-abundance curve A curve that plots the relative abundance of each species in a community in rank order from the most abundant species to the least abundant species.

Realized niche The range of abiotic and biotic conditions under which a species persists.

Recessive An allele whose expression is masked by the presence of another allele.

Recipient The individual who receives the behaviour of a donor in a social interaction.

Recombination The reshuffling of genes that can occur as DNA is copied during meiosis and chromosomes exchange genetic material.

Red Queen hypothesis The hypothesis that sexual selection allows hosts to evolve at a rate that can counter the rapid evolution of parasites.

Regional diversity The number of species in all of the habitats that comprise a large geographic area. *Also known as* Gamma diversity.

Regional species pool The collection of species that occurs within a region.

Regression A statistical tool that determines whether there is a relationship between two variables and that also describes the nature of that relationship.

Regulating services Benefits of biodiversity that include climate regulation, flood control, and water purification.

Relative abundance The proportion of individuals in a community represented by each species.

Remote sensing A technique that allows measurement of conditions on Earth from a distant location, typically using satellites or airplanes that take photographs of large areas of the globe.

Renewable resources Resources that are constantly regenerated.

Replication Being able to produce a similar outcome multiple times.

Rescue effect When dispersers supplement a declining subpopulation and thereby prevent the subpopulation from going extinct.

Reservoir species Species that can carry a parasite but do not succumb to the disease that the parasite causes in other species.

Resource Anything an organism consumes or uses that causes an increase in the growth rate of a population when it becomes more available.

Respired energy The portion of assimilated energy a consumer uses for respiration.

Riparian zone A band of terrestrial vegetation alongside rivers and streams that is influenced by seasonal flooding and elevated water tables.

Risk-sensitive foraging Foraging behaviour that is influenced by the presence of predators.

River A wide channel of slow flowing fresh water.

Root pressure When osmotic potential in the roots of a plant draws in water from the soil and forces it into the xylem elements.

Rubisco *See* RuBP carboxylase-oxidase.

RuBP carboxylase-oxidase An enzyme involved in photosynthesis that catalyzes the reaction of RuBP and CO_2 to form two molecules of glyceraldehyde 3-phosphate (G3P). *Also known as* Rubisco.

Runaway sexual selection When selection for preference of a sexual trait and selection for that trait continue to reinforce each other.

Salinization The process of repeated irrigation, which causes increased soil salinity.

Salt marsh A saltwater biome that contains nonwoody emergent vegetation.

Sample standard deviation A statistic that provides a standardized way of measuring how widely data are spread from the mean.

Sample variance A measurement that indicates the spread of the data around the mean of a population when only a sample of the population has been measured.

Saturation The upper limit of solubility in water.

Saturation point The limit of the amount of water vapour the air can contain.

Scavenger An organism that consumes dead animals.

Sclerophyllous Vegetation that has small, durable leaves.

Search image A learned mental image that helps the predator locate and capture food.

Secondary consumer A species that eats primary consumers.

Secondary sexual characteristics Traits related to differences between the sexes in terms of body size, ornaments, colour, and courtship.

Secondary succession The development of communities in habitats that have been disturbed and contain no plants but still contain an organic soil.

Selection The process by which certain phenotypes are favoured to survive and reproduce over other phenotypes.

Selfishness When the donor of a social behaviour experiences increased fitness and the recipient experiences decreased fitness.

Self-thinning curve A graphical relationship that shows how decreases in population density over time lead to increases in the size of each individual in the population.

Semelparity When organisms reproduce only once during their life.

Semipermeable membrane A membrane that allows only particular molecules to pass through.

Senescence A gradual decrease in fecundity and an increase in the probability of mortality.

Sequential hermaphrodites Individuals that possess male or female reproductive function and then switch to possess the other function.

Seral stage Each stage of community change during the process of succession.

Sexual dimorphism The difference in the phenotype between males and females of the same species.

Sexual reproduction A reproduction mechanism in which progeny inherit DNA from two parents.

Sexual selection Natural selection for sex-specific traits that are related to reproduction.

Shannon's index (H') A measurement of species diversity, given by the following formula:

$$H' = -\sum_{i=1}^{s} (p_i)(\ln p_i)$$

Also known as Shannon-Wiener index.

Shannon-Wiener index *See* Shannon's index.

Simpson's index A measurement of species diversity, given by the following formula:

$$\frac{1}{\sum_{i=1}^{s}(p_i)^2}$$

Simultaneous hermaphrodites Individuals that possess male and female reproductive functions at the same time.

Sink subpopulations In low-quality habitats, subpopulations that rely on outside dispersers to maintain the subpopulation within a metapopulation.

Social behaviours Interactions with members of one's own species, including mates, offspring, other relatives, and unrelated individuals.

Soil The layer of chemically and biologically altered material that overlies bedrock or other unaltered material at Earth's surface.

Solar equator The latitude receiving the most direct rays of the Sun.

Solute A dissolved substance.

Source subpopulations In high-quality habitats, subpopulations that serve as a source of dispersers within a metapopulation.

Source–sink metapopulation model A population model that builds upon the basic metapopulation model and accounts for the fact that not all patches of suitable habitat are of equal quality.

Spatial structure The pattern of density and spacing of individuals in a population.

Specialist A species that interacts with one other species or a few closely related species.

Speciation The evolution of new species.

Species Historically defined as a group of organisms that naturally interbreed with each other and produce fertile offspring. Current research demonstrates that no single definition can be applied to all organisms.

Species accumulation curve A graph of the number of species observed in relation to the number of individuals sampled.

Species evenness A comparison of the relative abundance of each species in a community.

Species richness The number of species in a community.

Species sorting The process of sorting species in the regional pool among localities according to their adaptations and interactions.

Species–area curve A graphical relationship in which increases in area (A) are associated with increases in the number of species (S).

Spitefulness When a social interaction reduces the fitness of both donor and recipient.

Spring overturn The vertical mixing of lake water that occurs in early spring, assisted by winds that drive the surface currents.

S-shaped curve The shape of the curve when a population is graphed over time using the logistic growth model.

Stabilizing selection When individuals with intermediate phenotypes have higher survival and reproductive success than those with extreme phenotypes.

Stable age distribution When the age structure of a population does not change over time.

Stable limit cycle A pattern of population growth in which the population continues to exhibit large oscillations over time.

Standard error of the mean A measurement of variation in data that takes into account the number of replicates that were used to measure the standard deviation.

Standing crop The biomass of producers present in a given area of an ecosystem at a particular moment in time.

Static life table A life table that quantifies the survival and fecundity of all individuals in a population during a single time interval.

Stepping stones Small intervening habitat patches that dispersing organisms can use to move between large favourable habitats.

Stochastic model A model that incorporates random variation in population growth rate.

Stomata Small openings on the surface of leaves, which serve as the points of entry for CO_2 and exit points for water vapour.

Stratification The condition of a lake or pond when the warmer, less dense surface water floats on the cooler, denser water below.

Stream A narrow channel of fast flowing fresh water. *Also known as* a Creek.

Strength of selection The difference between the mean of the phenotypic distribution before selection and the mean after selection, measured in units of standard deviations.

Subpopulations When a larger population is broken up into smaller groups that live in isolated patches.

Subtropical desert A biome characterized by hot temperatures, scarce rainfall, long growing seasons, and sparse vegetation.

Succession The process by which the species composition of a community changes over time.

Supercooling A process in which glycoproteins in the blood impede ice formation by coating any ice crystals that begin to form.

Supporting services Functions that allow ecosystems to exist.

Survey Counting a subset of the population.

Susceptible-Infected-Resistant (S-I-R) model The simplest model of infectious disease transmission that incorporates immunity.

Symbiotic relationship When two different types of organisms live in a close physical relationship.

Taiga *See* Boreal forest.

Temperate grassland/cold desert A biome characterized by hot, dry summers and cold, harsh winters and dominated by grasses, nonwoody flowering plants, and drought-adapted shrubs.

Temperate rainforest A biome known for mild temperatures and abundant precipitation, and dominated by evergreen forests.

Temperate seasonal forest A biome with moderate temperature and precipitation conditions, dominated by deciduous trees.

Territory Any area defended by one or more individuals against the intrusion of others.

Tertiary consumer A species that eats secondary consumers.

The handicap principle The principle that the greater the handicap an individual carries, the greater its ability must be to offset that handicap.

Thermal inertia The resistance to a change in temperature due to a large body volume.

Thermal optimum The range of temperatures within which organisms perform best.

Thermal pollution Discharging water that is too hot to sustain aquatic species.

Thermocline A middle depth of water in a lake or pond that experiences a rapid change in temperature over a relatively short distance in depth.

Thermohaline circulation A global pattern of surface- and deepwater currents that flow as a result of variations in temperature and salinity that change the density of water.

Thermophilic Heat-loving.

Thermoregulation The ability of an organism to control the temperature of its body.

Tolerance A mechanism of succession in which the probability that a species can become established depends on its dispersal ability and its ability to persist under the physical conditions of the environment.

Top predators Predators that typically consume both herbivores and mesopredators.

Top-down control When the abundance of trophic groups is determined by the existence of predators at the top of the food web.

Torpor A brief period of dormancy that occurs in birds and mammals in which individuals reduce their activity and their body temperature.

Trait-mediated indirect effect An indirect effect caused by changes in the traits of an intermediate species.

Transient climax community A climax community that is not persistent.

Transpiration The process by which leaves can generate water potential as water evaporates from the surfaces of leaf cells into the air spaces within the leaves.

Treatment *See* Manipulation.

Trophic cascade Indirect effects in a community that are initiated by a predator.

Trophic level A level in a food chain or food web of an ecosystem.

Trophic pyramid A chart composed of stacked rectangles representing the amount of energy or biomass in each trophic group.

Tropical climate A climate characterized by warm temperatures and high precipitation, occurring in regions near the equator.

Tropical rainforest A warm and rainy biome, characterized by multiple layers of lush vegetation.

Tropical seasonal forest A biome with warm temperatures and pronounced wet and dry seasons, dominated by deciduous trees that shed their leaves during the dry season.

Tundra The coldest biome, characterized by a treeless expanse above permanently frozen soil.

Type I functional response A functional response in which a predator's rate of prey consumption increases in a linear fashion with an increase in prey density until satiation occurs.

Type II functional response A functional response in which a predator's rate of prey consumption begins to slow down as prey density increases and then plateaus when satiation occurs.

Type III functional response A functional response in which a predator exhibits low prey consumption under low prey densities, rapid consumption under moderate prey densities, and slowing prey consumption under high prey densities.

Ultimate hypothesis A hypothesis that addresses why an organism responds in a certain way to its environment in terms of the fitness costs and benefits of the response.

Upwelling An upward movement of ocean water.

Variance of the mean A measurement that indicates the spread of the data around the mean of a population when every member of the population has been measured.

Vector An organism that a parasite uses to disperse from one host to another.

Vegetative reproduction A form of asexual reproduction in which an individual is produced from the nonsexual tissues of a parent.

Vertical transmission When a parasite is transmitted from a parent to its offspring.

Visible light Wavelengths in between infrared and ultraviolet radiation that are visible to the human eye.

Warning colouration A strategy in which distastefulness evolves in association with very conspicuous colours and patterns. *Also known as* Aposematism.

Water potential A measure of the water's potential energy.

Watershed An area of land that drains into a single stream or river.

Weather The variation in temperature and precipitation over periods of hours or days.

Weathering The physical and chemical alteration of rock material near Earth's surface.

Wilting point The water potential at which most plants can no longer retrieve water from the soil, which is about −1.5 MPa.

Woodland/shrubland A biome characterized by hot, dry summers and mild, wet winters, a combination that favours the growth of drought-tolerant grasses and shrubs.

Zero growth isocline *See* Equilibrium isocline.

Zero population growth isocline Population sizes at which a population experiences zero growth.

Index

Note: Page numbers followed by f and t indicate figures and tables, respectively. **Boldfaced** page numbers indicate key terms.